Computer Oriented Numerical and Statistical Techniques

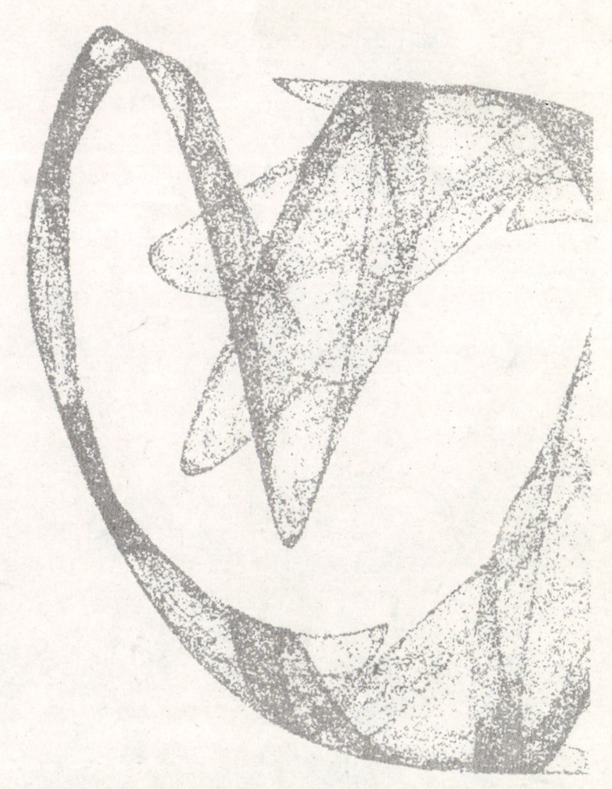

Computer Oriented Numerical and Statistical Techniques

FOR

(All Indian Universities, Engineering and Technical Colleges and Institutions, IIT's and IT Organizations)

R Singh • I Singh

Khanna Book Publishing Co. (P) Ltd.

Delhi

ISBN: 978-93-80016-78-8

First Edition 2010

Published by:
Khanna Book Publishing Co. (P) Ltd.
4C/4344, Ansari Road, Darya Ganj, New Delhi-110 002
Phone: 011-23244447 Telefax: 011-23244448
E-mail: contact@khannabooks.com
Website: www.khannabooks.com

Regd. Off: 1694-95, 1st Floor, Nai Sarak, Delhi-110006 (India)
Phone: 011-23283211 Telefax: 011-23256658

Cover Design & Layout by
Limited Colors, Delhi-92

Printed in India by:
R-Tech Offset Printers, Delhi

Dedicated to God
And
Our Parents

PREFACE

With the advent of modern digital computers, Numerical analysis is very different today. This book provides a fundamental introduction to numerical analysis suitable for undergraduate and postgraduate students in mathematics, computer science and engineering. This book presents a wide variety of Numerical methods, detailed description of the Statistical techniques, and several examples of the application to engineering, medicine, agriculture etc.

Beginning with the introduction to computer basics, numerical and error analysis and solution of transcendental and polynomial equations and Interpolation formulae, and Solution of linear equations has been described. Then different other techniques Numerical differentiation, Numerical integration, Solution of ordinary and partial differential equations have also been studied. Thereafter, various curve fitting techniques are solved graphically.

Due to importance of statistical techniques in design of engineering, and scientific problems, last chapters has been devoted to various statistical techniques like introduction to statistics, various Moment generating functions, Skewness, Kurtosis have been explained. We are also given the various measures of central tendency. Detailed analysis of techniques like Correlation, Linear, non-linear and multiple regression analysis.

Each and every technique described with the help of Algorithm, its Flowchart and Program using C and C++ and large number of solved and unsolved problems are given to have clear ideas.

Suggestions for improvement of the book will be gratefully accepted for future editions.

R Singh
I Singh

ACKNOWLEDGEMENT

We express our deep gratitude to our colleagues Prof. Harbhajan Singh, Prof. Sukhwinder Kaur, Prof. Mani arora, Mr. Navjeet Singh, Mr. Jagbir Singh, Mr. Avtar Singh Sandhu, Mr. Lovedeep Singh, Mr. Sukhwant Singh, Mr. Ikvinderpal Singh, Mrs. Mandeep Kaur and Mrs. Jyoti Maini for their valuable co-operation and guidance. We are grateful to the entire team of M/s Khanna Book Publishing Co. Pvt.Ltd. for then encouragement and motivation during this project.

R Singh
I Singh

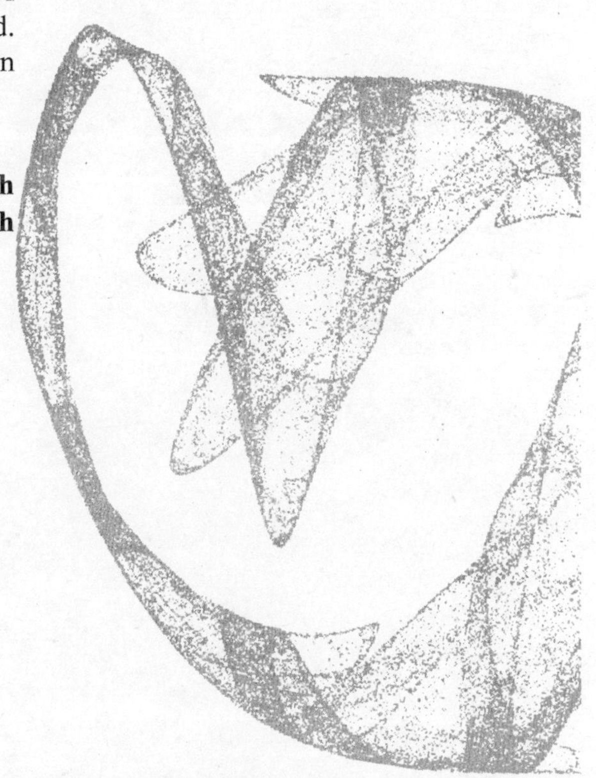

Syllabus

M.TECH/MCA/B.TECH / BCA / B.Sc (IT/CS/CM/ECO.) / B.A

Introduction

1. Numerical Methods, Numerical methods versus numerical analysis, Errors and Measures of Errors.
2. Non-linear Equations, Iterative Solutions, Multiple roots and other difficulties, Interpolation methods, Methods of bi-section, False position method, Newton Raphson - method.
3. Simultaneous Solution of Equations, Gauss Elimination Method Gauss Jordan Method, Gauss Seidel Method.
4. Interpolation and Curve Fitting, Lagrangian Polynomials, Newton's Methods: Forward Difference Method, Backward Difference Method Divided Difference Method.
5. Numerical Integration and different Trapezoidal Rule, Simpson's 3/8 Rule.
6. Numerical differentiation by Polynomial Fit.

Statistical Techniques

1. Measure of Central Tendency, Preparing frequency distribution tabs, Mean Arithmetic, mean geometric, Mean harmonic, Mean Medial Mode.
2. Measures of dispersion, Skewness and Kurtosis Range, Mean deviation. Standard deviation, co-efficient of variation, Moments Skewness Kurtosis.
3. Correlation.
4. Regression Linear Regression.
5. Least square fit linear trend, Non-linear trend.

   Y = axb

   Y = abx

   Y = aex

   Polynomial fit: Y = a + bx + cn2

CONTENTS

CHAPTER 1

Computer Basics for Numerical Problems

1.1 INTRODUCTION TO COMPUTERS

The computer is an information-processing system. It accepts information or data from the outside world and processes it to produce new information.

Technically, a computer is a programmable machine. This means it can execute a programmed list of instructions and respond to new instructions that it is given.

Hence, "The computer is an electronic device that accepts information (in the form of digitalized data) and manipulates it for some result based on a program or sequence of instructions on how the data is to be processed."

It consists of 2 parts-hardware and software . The computer processes input through input devices like mouse and keyboard. The computer displays output through output devices like color monitor and printer. The size of a computer varies considerably from very small to very big. The speed of computers also has a very large range. Computers have become indispensable in today's world. Millions of people use computers all over the world.

1.2 DEFINITIONS

A computer can be used to solve a problem by following set of stored instructions called a program. The problem solving process needs initial data, the operations that are to be performed and results in the form of output. The following steps are required for problem solving:

(i) Define the Problem: The first step is to give a clear, concise problem statement. The problem definition should clearly specify the desired input and output. This step demands that user should have full knowledge of the background of the problem. A stated goal will help in the organization of the remaining steps.

For Example

- To Compute the average of a set of data values.
- Given a quadratic equation, to compute its real and complex roots.

- To Print a table of weight and height values for a normal human being.

(ii) Formulate a Mathematical Model: Most technical problems can be modelled mathematically. An existing model must be chosen or a new one if no standard mathematical model exists. This step also requires full knowledge of the problem and its underlying mathematical concepts.

For Example

- Average = (Value1+ Value2 +Value3) / 3
- Root = (-b + sqrt(b*b – 4*a*c))/ (2*a)

(iii) Develop An Algorithm: The next step is to devise and describe a precise plan of what you want the computer to do. This plan, expressed as a sequence of operations, is called an **algorithm**.

Write the code based on the algorithm using a programming language: The above algorithm cannot be as such understandable by the computer. It first must be written in a programming language as a program. A compiler or translator will then convert program into machine code that the computer can understand and act upon. Learning a computer language is required first to be a programmer so as to translate the algorithm derived in step 3 into a program.

(iv) Testing the Program: If the program compiles correctly, use a simple set of test values to verify that the result is what you expected. If the results seem valid, test the program with a variety of "real" data sets. The mistakes if encountered in a program are called **bugs** and "debugging" code can take longer than writing the code.

To solve a problem some tools are used to simplify it and make it in the readable form. Following are the commonly used tools to solve the problem:

1.2.1 Algorithm

An algorithm is a finite sequence of instructions, logic, an explicit, step-by-step procedure for solving a problem. An algorithm may be expressed as a pseudo code that is shorthand notation of English sentences.

Problem: Add Two Numbers

Algorithm:

1. Start
2. Enter the value of A and B.
3. A=10,B=5.
4. Perform operation C = A + B.
5. Store the result in C.
6. Stop

1.2.2 Flow-chart

A diagram that uses graphic symbols to depict the nature and flow of the steps in a process. To solve a problem in a sequence or stepwise diagrammatically or pictorially is called Flow Chart.

"A flowchart is a common type of chart, which represents an algorithm or process, showing the steps as boxes of various kinds, and their order by connecting these with arrows. Flowcharts are used in analyzing, designing, documenting or managing a process or program in various fields."

Flowcharts are tools used by programmers because they facilitate breaking down a problem into its simplest steps, each symbol in a flowchart can represent only one step in the algorithm. Flowcharts use easy-to-understand symbols to represent actions on data and the flow of data. The following are the basic flowchart symbols:

(i) Terminator Or Start/Stop Box: The Terminator symbol is used to begin and end of each flowchart. It also has egg like shape called oval. It is used to indicates start/stop of a program.

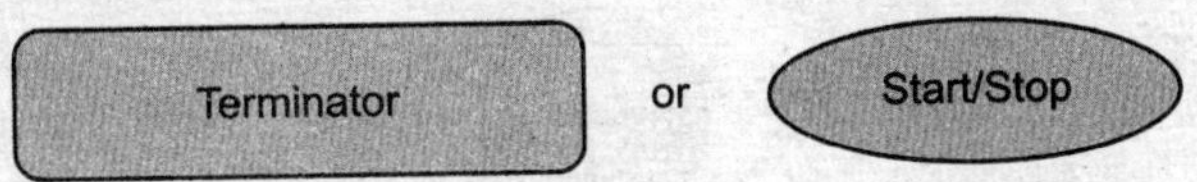

(ii) Input/Output Box: Input boxes are used for input value to program and Output boxes are used for output value to outside world. This is also called read/write box and it has parallelogram shape.

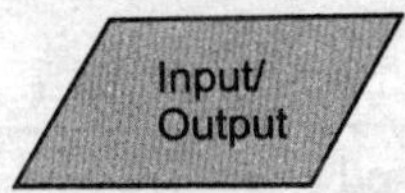

(iii) Process or Computation or Assignment Box: It has square shape or rectangular shape. It indicates information processing e.g., addition, subtraction, or movement of data to computer memory.

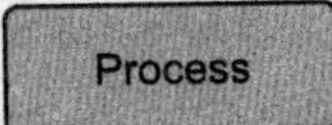

(iv) Decision or Condition Box: Decision diamonds create two data flows from one. A Decision diamond always has True/False or Yes/No written on the left and right sides. It is also called decision making box.

(v) Flow Line or Data Flow: Data Flows connect the different symbols, showing the order that is followed in the algorithm. Data Flows can only go in one direction, hence the arrow. When two or more data flows meet, they always meet at a connector. It has arrow shape.

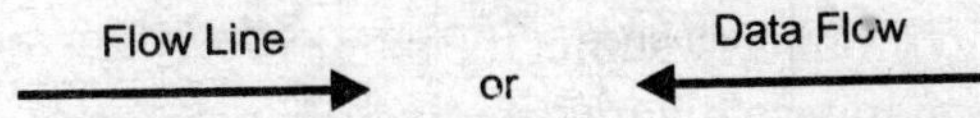

(vi) **Loop or Iterative Box:** This box has hexagon shape. It is used when a problem has repetitive steps or iterative steps i.e. a problem has solution with steps repeating again and again till a condition ends or definite number of times.

(vii) **Connector or Flow Linker:** This has circular shape. A small circle with any number or letter in it is used as a connector symbol. It connects various parts of a flow-chart which are far apart or spread over pages This makes the flow chart simple and in short. It means it removes complexity of the flowchart.

(viii) **Function or Procedure or Sub-program Box:** This symbol has a rectangle shape with two cuts or lines inside. It is used to denote a subprocess which is given elsewhere as indicated by connector symbol.

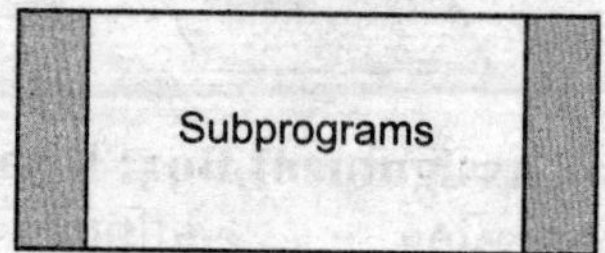

There are mainly three types of construct used in the programming which are useful in drawing the flow chart. The three programming constructs are known as:

Sequence construct that is used to carry out one step at a time, then next step and so on.

Selection construct performs an action based on the condition if it is true.

Iteration construct or looping construct performs an action as long as condition is true by carrying out the looping. When a problem has repetitive steps or iterative steps i.e. a problem has solution with steps repeating again and again till a condition ends or definite number of times.

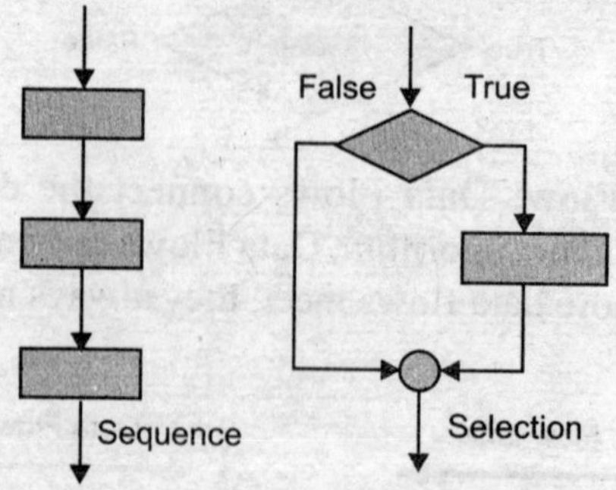

The third construct known as Iteration is derived from the selection construct by transferring the control to the start position as along as the condition remains true. Once the condition becomes false, the control will be transferred to the next level after decision making.

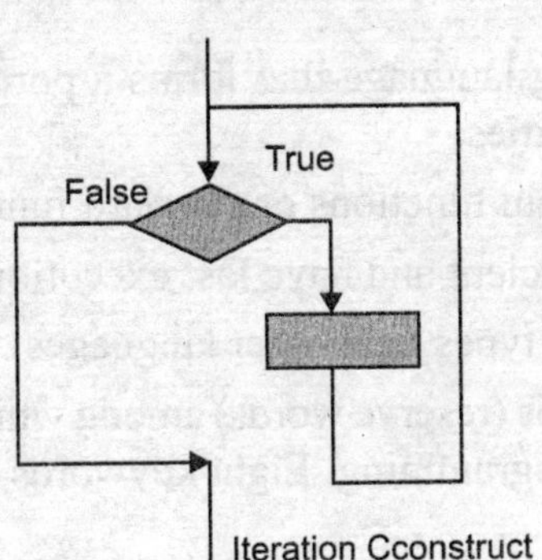

C LANGUAGE

1.3 INTRODUCTION TO C LANGUAGE

C is a general purpose structured powerful modern language. It is a High level scientific language as well as business oriented language. This language also has low level programming features. This language was designed for developing fast, reliable, efficient, economical and portable system software and application software (programs). Till the development of c, no language was simple and fully featured to fulfil all types of requirements, some languages was powerful but very complex and some was simple but not very powerful. But C fulfilled both the things i.e. simplicity and power to do versatile jobs.

In 1972 (some researcher say 1970) at BELL-laboratories (which is the part of AT & T) in U.S.A. by the scientist Dennis Ritchie developed a powerful software tool which has a big facility (compilation and linkage facility in combined form) called C-language.

After many years research, C language was developed and become popular. C-language has different versions from time to time.

Afterwards around 1983 an object oriented programming (OOP) software was developed from the C-language that is called C++. Between 1983 and 1986, Bjarne Stroustrup of Bell Labs began working on an extension to C, called C++. From the C++, Internet Programming Language Java was developed which is a completely platform independent. So we can say C is the basic beginner's language for the development of any system and application software. Latest version of the C language is C-99.

1.4 FEATURES OF C-LANGUAGE

C-Language has number of characteristics or features. Following are some advantages of 'C' language:

1. C is a system programming language that forms a portion of the operating system of the computer or its support utilities.
2. C-Language is rich in built-in functions or standard functions (library function)
3. C-programmes are very efficient and have fast execution speed.
4. C- Language has more data types than other languages .
5. C-Language has 40 keywords (reserve words) among which 32 are standard keywords which are used in daily routine programming. Eight keywords are optional which are used in low level programming.
6. C-Language has more operators than other language.
7. C-Language programs are highly portable. Portability means a C program written in one environment can be executed in another environment.
8. In C-Language you can take 0 to 15000 variables in a single program depending on the size of main memory and physical memory available
9. C-Language is a Structured Programming language. Structure programming means it has different modules and blocks like, if block, do block, for block, switch module etc.
10. C-Language is very simple to learn and use.
11. C-compiler gives reliable and accurate results. It has a facility of warning which guides for better and efficient programming.
12. C-Language is also called middle level language because it has both types of features .i.e. High level as well as Low level.
13. C is a Code Block language. A code block is a logically connected group of program statements that is treated as a unit. In C a code block is created by placing a sequence of statements between opening and closing curly braces.

1.5 CHARACTER SET

There are mainly four categories of the character set as shown in the figure below:

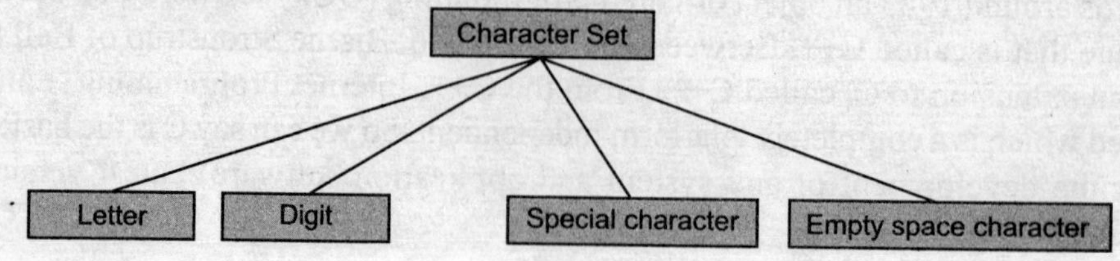

1. Letter or Alphabet

In the character set, character or alphabet are represented by A-Z or a-z. C- Language is case sensitive so it takes different meaning for small and upper case letters. There are total 26 letters used in C-programming.

2. Digit

In the character set digit are represented by 0-9 or by combination of these digits. There are total 10 digits used in the C-programming.

3. Special Characters

All the keyboard keys except alphabet, digits and white spaces are the special character. These are some punctuation marks and some special symbols used for special purpose.

Special characters are used for C-statements like to create an arithmetic statement +, -, * etc. , to create relational statement <, >, <=, >=, == etc. , to create assignment statement =, to create logical statement &&, || etc. are required.

4. Empty space character Or White Spaces

White spaces has blank space, new line return, Horizontal tab space, carriage ctrl, Form feed etc. are all used for special purpose.

1.6 KEYWORDS AND IDENTIFIERS

C-language has some reserve words which can not be used as variables or identifiers. These reserve words are keywords of C-language. These are the part of the C-Tokens.

There are mainly 40 keywords among which 32 are used by many C-compilers (also called Standard Keywords) for high level programming , whereas remaining 8 reserve words (also called Optional Keywords) are used by the programmer for the low level programming .

Identifiers are some words or names which identify whether it is a constant or variable. These are the data names used by the programmer. Identifier gives us the unique identification having unique sequence of character (from the character set) used for special purpose.

1.7 CONSTANT

Constant uses the secondary storage area. Constants are those quantities whose value does not vary during the execution of the program i.e. value is fixed. Constants in every language are the same. For example, in the C-language some valid constant are: 22.2 , -67.9, +23.32, "Raman", 'A', "1999" , "\n" etc

In C-language constants are mainly of two types:

I. Numeric constant

II. Non-numeric or Character Constant

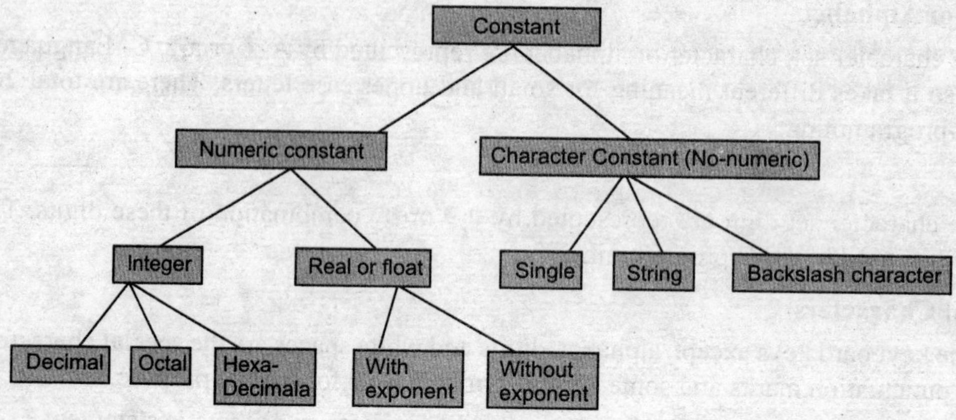

1.8 VARIABLES

Variables are those quantities whose value can vary during the execution of the program. Variables are basically memory locations which are given names and these locations are refereed in the program by variable names to read and write data in it. For example:

```
int a, b, c;
float a,b;
char c;
```

1.9 BASIC STRUCTURE OF A COMPLETE C-PROGRAM

A C-program can be developed from a general structure. The general structure of C-program is as shown below (which is also called overview of a C-program):

1. Documentation Section
2. Pre-processor directives or compiler directives Section or Macro
   - **Link Section**
   - **Definition Section**
3. Global Declaration Section
4. Main C-program function called main ( )
5. Beginning of the program by using left brace

   {

   - **Local declaration or local variable declaration;**
   - **Executable Part or C-executable statements;**

   }
6. End of the main program using right brace

7. Sub-program Section having user-defined functions.

```
Function 1( )
{ user defined program area }
Function 2( )
{ Sub Program or subroutine }
```

1.10 DATA TYPES

C-language becomes powerful with the help of number of data types linked with it.. Data type is the description of nature of data either in numeric forms (integer or real) or in character form (single character or string).

These are divided into four categories:

(a) Integer data type
(b) Float data type or Single precision real data type
(c) Double precision real or Double or Long float data type
(d) Character data type

(a) Integer Data Type: These are of integral type. These are without decimal point. These are signed or unsigned. These data types are small int (short int), int, long int

(b) Real (float) Data Type: These are also called precision real or single percussion value with a very small real number. These data type have at least one digit with a decimal point. These are also signed or unsigned.

(c) Double Data Type: These data types have ver large floating data, so are also called large real number.

(d) Character Data Type: These have either single character or combination of characters (called String). Characters are either signed or unsigned but mostly characters are used as unsigned type.

Type	Size(bytes)	Range
char or Signed char	1	-128 t0 127
Unsigned char	1	0 to 255
int or signed int	2	-32768 to 32767
Unsigned int	2	0 to 65535
short int or Signed short int	1	-128 t0 127
Unsigned short int	1	0 to 255
long int Or Signed long int	4	-2,147,483,648 to 2,147,483,647
Unsigned long int	4	0 to 4,294,967,295
float	4	3.4E-38 to 3.4E+38
Double	8	1.7E-308 to 3.4E+308
long double	10	3.4E-4932to1.1E+4932

1.11 OPERATORS

An operator is a symbol that tells the compiler to perform specific mathematical or logical manipulations.

Mainly operators are used to operate two or more than two operands depending upon their processing and given conditions.

The relationship of operator with the operands is shown as in the below common statement:

Operand1 Operator Operand2

Or

Operand1 Operator Operand2 Operator Operand3

Here Operand1, Operand2, Operand3 may be a variable, a constant or an expression etc. and Operator must be either arithmetic type or relational type or logical type etc. (as it will be discussed later on in detail). There are mainly eight operators used in C-language. The relationship is as shown below:

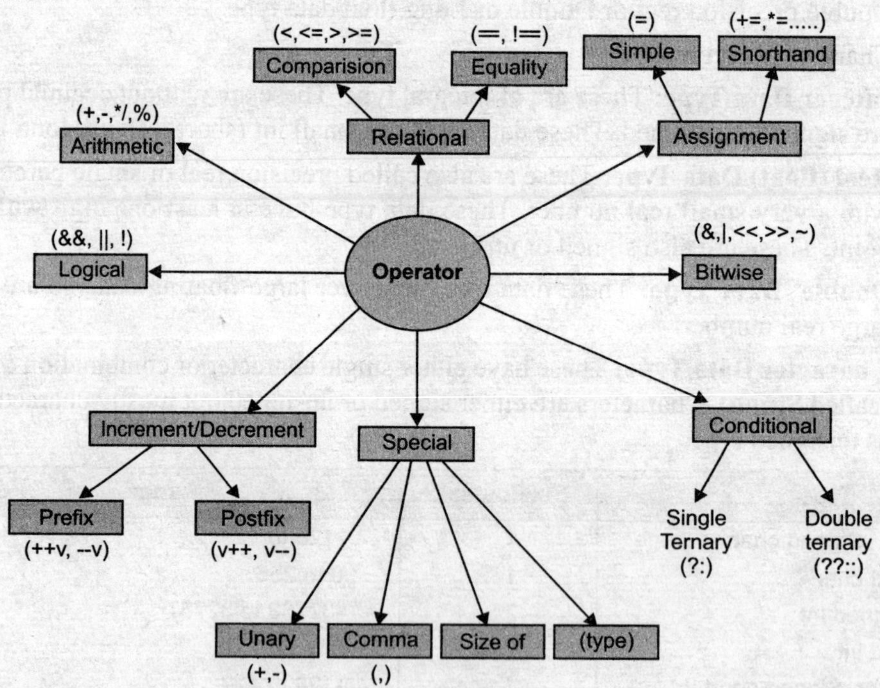

1.12 DECISION-MAKING STATEMENT

C supports many branching statements depending upon their flow of control and according their decision making policy. So it is also called decision-making statement. The various branching statements used are as:

1. **if statements**
2. **switch statement**
3. **conditional operator statement**

1. if statements

The if statement is a powerful decision making statement which can handle a single condition or group of statements. These have either true or false action.

There are four types of if statements:

(i) Simple if statement
(ii) if-else statement
(iii) nested if statement
(iv) else-if or ladder if or multi-condition if statement

(i) Simple if statement

Syntax:

```
if (condition)
{
true-statement block;
}
statement – x;
```

(ii) if-else statement

Syntax:

```
if (condition)
{
true statement block;
}
else
{
false statement block;
}
statement-x;
```

(iii) Nested if Statement

When an if statement occurs within another if statement, then such type of is called nested if statement.

Syntax :

```
if (condition-1)
{
if (condition-2)
{
st-1;
}
else
{
sf-1;
}
}
else
{
if (condition-3)
{
st-2;
}
else
{
sf-2;
}
}
statement-x;
```

(iv) Ladder if or else if statement

When in a complex problem number of conditions arise in a sequence, then we can use Ladder-if or else if statement to solve the problem in a simple manner.

Syntax :

```
if (condition-1)
{
st-1;
}
else if (condition-2)
{
```

```
st-2;
}
else if (condition-3)
{
st-3;
}
…
…
…
else if (condition-n)
{
st-n;
}
else
{
default-statement;
}
statement-x;
```

2. Switch Statement

It is also called Case statement because it has different cases and different blocks. It is also called multi-decision statement having multiple blocks.

Syntax:

```
switch (e or v)
{
case value1 :
block1;
break;
case value2 :
block2;
break;
case value3 :
block3;
break;
…
```

```
...
case value n :
block n;
break;
default :
block n+1;
}
statement-x;
```

3. Conditional Control Statement

This statement is based on conditional operator. This statement solves the problem's condition in a single line and is a fast executable operation. For this purpose, we can take combination of ? and :.

Syntax :

```
exp1 ? exp2 : exp3;
```

1.13 LOOPING STATEMENTS

When a single statement or a group of statements will be executed again and again in a program (in an iterative way), then such type processing is called loop.

There are four types of looping statements:

1. while statement
2. do-while statement
3. for statement
4. nested for statement

1. while statement

In this, first of all condition is checked and if it is true, then group of statements or body of loop is executed. It will execute again and again till condition becomes false.

Syntax :

```
while (test condition)
{
block of statements;
}
statement-x;
```

2. do-while statement

In this statement, first body of the loop is executed and then the condition is checked. If condition is true, then the body of the loop is executed. When condition becomes false, then it will exit from the loop.

Syntax :

```
do
{
block of statements;
}
while (condition);
statement-x;
```

3. for statement

It is one step loop, which initialize, check the condition and increment / decrement the step in the loop in a single statement.

Syntax :

```
for (initial value; test condition; increment/ decrement)
{
body of the loop;
}
statement-x;
```

4. Nested for statement

When a for statement is executed within another for statement, then it is called nested for statement. We can apply number of nested for statements in C-language.

Syntax :

```
for (initial value1; test condition1; increment1 / decrement1)
{
for (initial value2; test condition2; increment2 / decrement2)
{
inner- body of the loop;
}
outer-body of the loop;
}
statement-x;
```

1.14 ARRAY

Array is similar type of data collection name i.e. it is a collection of same data type elements. An array is a group of related data items, which share common name.

"It is the set of homogeneous data (same type)".

Suppose salary of the 50 employees named by salary variables name and we write it sal1,sal2,sal3......sal50. We repeat the variable name of employees by using the subscript. Because of use of single subscript or more than one subscript, it is also called subscripted variable. So Array's representation for the 10th employee is:

Salary [50];

Syntax :

```
Common name variable [size];
            or
array-name [subscript1] [subscript2];
```

Array representation can be done as:

```
storage-type data-type   array-name[size];
            or
storage-type   data-type   array-name[row size][column size];
```

1.15 STRINGS

Combination of characters (group of character) is called string. "I am a good boy" is a string. We can print or display the string by using the printf() function as:

printf ("\n I am a good boy \n");

Syntax :

```
char city [20];
printf ("%s", city);      or
char city[10][20];
for (i=0; i <10; i++)
    {
    printf ("%s", city[i]);
    }
```

1.15.1 Initializing String Variable

Syntax :

static data-type array-name[size] ={list of values};

For example, suppose you want to enter the string "BOMBAY" to the city variable , then the procedure is as:

static char city[10] = {'B', 'O', 'M', 'B', 'A', 'Y'};

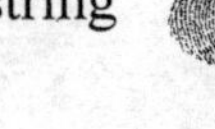

For example, suppose if you want to assign three different cities data values to the city string variable, then this can be done as:

static char city[2][20] = { {'B', 'A', 'T', 'A', 'L', 'A'}
{'A', 'M', 'R', 'I', ' T', 'S', 'A', 'R'}}

character data-type must be used in strings.

1.16 POINTERS

Pointers are memory addresses variable. A Pointer contains a memory address. Further we can say pointers are directly linked to the memory address. Actually pointer is a variable having address of another variable. If one variable contains the address of another variable, the first variable is said to point to the second. For example, suppose q is an integer variable and this variable have value 412 stored in the memory.

represented as:

Variable name	Memory value	Memory address		Memory variable
q	412	4000	←	p (pointer)

Here the value 412 is stored at address 4000. We give name to memory address as p. The p is a memory variable, which gives the address of the q variable having value 412. Then p is called pointer.

Syntax:

```
data-type *pointer-variable ;
```

1.17 STRUCTURE

A structure is heterogeneous collection of related fields. Here fields are called structure member or structure element.

For example, suppose if you want to store a book in library using C-program, then you give information about the title of the book, name of the author, number of pages in the book and the price of the book. Then before coding the problem you will examine it according to following table

Title of the Book	Name of Author	Pages of the Book	Price of the Book
solved programming in C	Rachhpal Singh	450	300
Fundamental of C	Rachhpal Singh	500	200
Let Us C	Yashvant K.	700	300

```
struct library
{
char    title [40] ;
```

```
char    author [40] ;
int     pages ;
float   price ;
} ;
void main()
{ }
```

1.18 UNION

A union is a variable that declares a set of multiple variables (called members or elements having different data-types) sharing the same memory area.

The compiler allocates sufficient memory to hold the largest variable of the union. All the members of union occupy the same memory locations called addresses. The old value of the already declared member will be destroyed and new value to the new member will be assigned in the memory by using the union concept.

Syntax:

```
union union-tag
{
        datatype-1  member-element-1;
        ...................................;
        dartatype-n   member-element-n;
}v1,v2,...vn;
```

SOLVED PROBLEMS

Problem 1: Program to find the area and circumference of the circle.

```
#include<stdio.h>
#include<conio.h>
#define PI 3.1427
void main()
{
int r;
float cir, area;
clrscr();
printf("\n ENTER THE RADIUS OF THE CIRCLE: ");
scanf("%d", &r);
```

```
cir = 2*PI*r;
area=PI*r*r;
printf("\n THE CIRCUMFERENCE OF CIRCLE IS= %f", cir);
printf("\n THE AREA OF CIRCLE IS= %f", area);
getche();
}
```

Output:

```
ENTER THE RADIUS OF THE CIRCLE: 5
THE CIRCUMFERENCE OF CIRCLE IS= 31.427
THE AREA OF CIRCLE IS= 78.5675
```

Problem 2: Program to compute the circumference and volume of a sphere.

```
#include<stdio.h>
#include<conio.h>
#define PI 3.1427
void main()
{
int r;
float cir, vol;
clrscr();
printf("\n ENTER THE RADIUS OF THE SPHERE: ");
scanf("%d", &r);
cir=4*PI*r*r;
vol=4/3*PI*r*r*r;
printf("\n CIRCUMFERENCE OF SPHERE IS=%f", cir);
printf("\n THE VOLUME OF SPHERE IS=%f", vol);
getche();
}
```

Output:

```
ENTER THE RADIUS OF THE SPHERE: 5
CIRCUMFERENCE OF SPHERE IS=314.27
THE VOLUME OF SPHERE IS= 523.78331
```

Problem 3: Program to compute simple interest and compound interest.

```
#include<stdio.h>
```

```
#include<math.h>
#include<dos.h>
void main()
{
float p, r, t, si, ci;
printf("\n Enter the value of Principle, Rate, Time: ");
scanf("%f%f%f",&p,&r,&t);
si=(p*r*t)/100;
ci=p*pow((1+r/100),t)-p;
printf("\n Simple interest is =%f",si);
printf("\n Compound interest is =%f",ci);
sleep(10);
}
```

Output:

Enter the value of Principle, Rate, Time: 1200 10 2

Simple interest is 240

Compound interest is 252

Problem 4: Program to find the greatest number between two numbers.

```
#include<stdio.h>
#include<conio.h>
void main ( )
{
int a,b,g;
clrscr();
printf ("\n Enter the value of two numbers: ");
scanf ("%d %d", &a, &b);
if (a > b)
{
g = a;
}
else
{
g = b;
}
```

```
printf ("\n Greatest Number is: %d", g);
getche ( );
}
```

Output:

Enter the value of two numbers: 40 36

Greatest Number is: 40

Problem 5: Program to find greatest number among three integer numbers.

```
#include<stdio.h>
#include<conio.h>
#include<dos.h>
main()
{
int a,b,c,g;
printf("\n Enter the three numbers:");
scanf("%d%d%d",&a,&b,&c);
if(a>b)
{
if(a>c)
{
g=a;
}
else
{
g=c;
}
}
else
{
if(b>c)
{
g=b;
}
else
{
```

```
g=c;
}
}
printf("\n Greatest number is: %d",g);
getch();
}
```

Output:

Enter the three numbers: 20 30 10

Greatest number is:30

Problem 6: Program to print the color according to the code.

```
#include<stdio.h>
#include<conio.h>
main()
{
int code;
printf(" \n  MAIN MENU  ");
printf(" \n1. For color Red ");
printf(" \n2. For color Green ");
printf(" \n3. For color White ");
printf(" \n4. For color Yellow ");
printf("\n Enter the color code: ");
scanf("%d", &code);
switch(code)
{
case 1:
printf("\n color is  RED ");
break;
case 2:
printf("\n color is GREEN");
break;
case 3:
printf("\n color is WHITE ");
break;
case  4:
```

```
printf("\n color is YELLOW ");
break;
default :
printf("\n Color does not found");
}
getch();
}
```

Output:

MAIN MENU
1. For color Red
2. For color Green
3. For color White
4. For color Yellow
Enter the color code: 2
color is GREEN

Problem 7: Program to find factorial of a number.

```
#include<stdio.h>
#include<dos.h>
#include<conio.h>
main()
{
int n,i,m;
clrscr();
printf("\n Enter the number : ");
scanf("%d",&n);
m=1;
for(i=1;i<=n;i++)
{
m=m*i;
}
printf(" \n Factorial of %d is =%d",n,m);
sleep(5);
}
```

Output:

Enter the number: 5
Factorial of 5 is = 120

Problem 8: Program to print

```
*
**
***
****
```

```
#include<conio.h>
#include<stdio.h>
main()
{
int a;
clrscr( );
printf("\n Enter any number :");
scanf("%d",&a);
for(int i=1;i<=a;i++)
{
for(int j=1;j<=i;j++)
printf("*");
printf("\n");
}
getch();
}
```

Program 9: Program to print

```
  *
 * *
* * *
```

```
#include<conio.h>
#include<stdio.h>
main()
{
int a,j,i,space;
clrscr( );
printf("\nEnter any number :");
scanf("%d",&a);
for(space=a-1,i=1;i<=a;i++,space—)
```

```
{
for(j=1;j<=space;j++)
cout<<' ';
for(j=1;j<=i;j++)
printf("* ");
printf("\n");
}
getch();
}
```

Problem 10: Program to find average of N numbers.

```
#include<stdio.h>
#include<conio.h>
main ( )
{
int n, a[50], i, s;
float av;
clrscr();
printf("\n Enter the value of n:");
scanf("%d",&n);
printf("\n Enter the different %d values:",n);
for (i = 0; i < n; i++)
{scanf( ("%d", &a[i]);
}
s=0;
for(i=0; i<n; i++)
{
s=s+a[i];
}
printf("\n Sum is %d",s);
av=(float)s/n;
printf("\n Average is %f",av);
getch();
}
```

Output:

```
Enter the value of n: 5
Enter the different 5 values:
20
30
40
10
50
Sum is 150
Average is 30.0
```

Problem 11: Program to search a number by using LINEAR SEARCH of Numeric data (NUMBERS).

```
# include <stdio.h>
# include <conio.h>
# include <dos.h>
main()
{
int c,i,n,b,s[30],s1;
clrscr();
printf("\n ENTER HOW MANY NUMBERS YOU WANT TO ENTER: ");
scanf("%d",&n);
printf("\n ENTER THE NUMBERS: ");
for(i=1;i<=n;i++)
{
scanf("%d",&s[i]);
}
clrscr();
printf("\n ENTER THE NUMBER YOU WANT  TO SEARCH: ");
scanf("%d",&s1);
i=1;
while(i<=n)
{
b=s[i]-s1;
if(b==0)
```

```
{
c=i;
break;}
else
{
c=0;
}
i++;
}
if(c==0)
{
printf("\n SEARCH UNSUCCESSFUL");
}
else
{
printf("\n SEARCH SUCCESSFUL AND LOCATION IS %d ",c);
}
sleep(3);
}
```

Output:

```
ENTER HOW MANY NUMBERS YOU WANT TO ENTER: 6
ENTER THE NUMBERS
88
66
44
22
55
11
ENTER THE NUMBER YOU WANT  TO SEARCH:
44
SEARCH SUCCESSFUL AND LOCATION IS 3
```

Problem 12: Program to illustrate the concept of displaying the address of a Pointer variable.

```
#include<conio.h>
#include<stdio.h>
```

```
main ( )
{
int x, *q ;
x = 220 ;
q = &x ;
printf ("/n % d is stored at address % u", x, &x) ;
printf ("/n % d is stored at address % u", x, q) ;
printf ("/n % d is stored at address % u", *q, q) ;
printf ("/n % d is stored at address % u", *& x, &x) ;
getch( ) ;
}
```

Output:

```
220 is stored at address  5010
220 is stored at address  5010
220 is stored at address  5010
220 is stored at address  5010
```

Problem 13: Program to reverse the string using pointer.

```
#include <stdio.h>
#include <conio.h>
#include<string.h>
main()
{
char a[20];
void f(char [20]);
void (*p)(char [20]);
clrscr();
printf("\n Enter the string  :");
scanf("%s", a);
p=&f;
(*p)(a);
getch();
}
void f(char *s)
{
```

```
int i,n;
char t;
n=strlen(s);
for(i=0;i<n/2;i++)
{
t=*(s+i);
*(s+i)=*(s+n-i-1);
*(s+n-i-1)=t;
}
printf("\n Reversed string is  :");
for(i=0;*(s+i)!=0;i++)
printf("\n %s",*(s+i));
}
```

Output:

Enter the string : rama
Reversed string is : amar

Problem 14: Program to read and write the student data by using structure in C language (illustrate the concept of accessing the value of structure variable).

```
#include<stdio.h>
#include<conio.h>
main ( )
{
struct    student
{
int     rn ;
char    name [20] ;
char    add [30] ;
float   fees ;
} ;
struct   student   st1;
printf (" \ n Enter the data of any student : ") ;
scanf (" % d % s % s % f ", &st1.rn,  st1.name,  st1.add,  &st1.fees) ;
printf (" \ n Roll No. of the student is % d",  st1.rn) ;
printf (" \ n Name of  the student is  % s",  st1.name) ;
```

```
printf (" \ n Address of the  student  is % s",  st1.add) ;
printf (" \ n Fees submitted by the student is %f",  st1.fees) ;
getch( ) ;
}
```

Output:

```
Enter the data of any student :
122
Geeta
Abc nagar, Asr
1500.20
Roll No. of the student is 122
Name of  the student is Geeta
Address of the  student  is Abc nagar, Asr
Fees submitted by the student is 1500.20
```

Problem 15: Program to read and display student data using Union.

```
#include<stdio.h>
#include<conio.h>
main()
{
union student
{
char grade;
int rollno;
float marks;
double fees;
}
union student s;
printf("\n Enter the garde, rollno, marks and fees of the student :");
scanf("\n %c%d%f%g", &s.grade, &s.rollno, &s.marks, &s.fees);
printf("\n Grade is: %c", s.garde);
printf("\n Rollno is: %d", s.rollno);
printf("\n Marks are : %f", s.marks);
printf("\n Fees paid  is: %g",s.fees);
getch ();
```

```
}
```

Output:

Enter the garde, rollno, marks and fees of the student :

A

122

90.50

12346.8888787

Grade is: A

Rollno is: 122

Marks are : 90.50

Fees paid is: 12346.8888787

UNSOLVED PROBLEMS

1. Program to find average of three numbers.
2. Program to find volume of a box.
3. Program to convert time in seconds to time hours, minutes and seconds.
4. Write a program to print the following outputs using for loops.

(a)
```
1
2   2
3   3   3
4   4   4   4
```

(b)
```
*   *   *   *
*   *   *
*   *
*
```

5. Write a C program to draw Reverse Pascal Triangle.
6. Write a C Program to find the leap year.
7. Write a program to print a Fibonacci series 1 2 3 5 8 .....upto 100.
8. Program to find the biggest number among N numbers.
9. Program to Sort an array (list) having N elements using Bubble Sort method.
10. Program to Calculate Matrix Multiplication, Matrix Addition and Matrix Transpose by using SWITCH statements (A menu driven program).
11. Program to count the negative elements from the list.
12. Program to transpose a matrix.

13. Program to find the sum of Diagonal Elements of a matrix (Trace a Matrix).
14. Write a C Program to reverse numbers in an array.
15. Write a C Program to remove duplicate numbers from an array.
16. Write a C Program to convert a square matrix into a unit matrix.
17. Program to find whether a sub string is present in the string or not, if it is present then find location of sub string in the string.
18. Program to search a string using Binary Search.
19. Program to find whether the string is PALINDROMEM or not.
20. Write a C Program to find all combination of characters of a word.
21. Write a C Program to find sum of row and columns in a matrix-using pointer.
22. Write a C Program to find the roots of quadratic equation using pointer.
23. Write a C Program to find max. & min. value from array using pointer.
24. Write a C Program to determine the area and perimeter of triangle using pointer.
25. Write a C Program to construct a pay bill structure of an employee in an organization.
26. Write a C Program to access one-dimensional array structure.
27. Write a C Program to compute the size of union.
28. Write a C program to design and print the DMC (Detail Marks card) of every student in a class of n students in a college.

C++ LANGUAGE

1.19 INTRODUCTION TO C++ LANGUAGE

C++ is an extension of C. It is an Object Oriented programming and is also flexible, but C has some big disadvantage i.e. when a project exceeds a certain size, C reaches its limits. So depending on the project (a program of 25,000 to 10,000 lines long) programmer manage it, but it is very difficult to grasp as a totality. In 1980, while working at Bell laboratories at Murray Hill, New Jersey, Bjarne Stroustrup addressed this problem by adding several extensions to the C language. Initially called "C with classes" the name was changed to C++ in 1983. The C++ developed by Stroustrup was a strong admirer of Simula-67, which was the first object oriented language and also was a strong supporter of C.

1.20 FEATURES OF OBJECT ORIENTED LANGUAGE

Object oriented language does not work around a single component or unit. It uses a vocabulary more than procedural language. It is group of all the elements or components linked with object approach.

Major features are:

1. Object
2. Classes
3. Inheritance and class hierarchy
4. Polymorphism
5. Encapsulation (Data Encapsulation)
6. Abstraction
7. Dynamic Binding
8. Message Passing
9. Reusability

1. Object

An object is an instance of a class. It can be uniquely identified by its name and it defines a state which is represented by the values of its attributes at a particular time.

The state of the object changes according to the methods which are applied to it. We refer to these possible sequence of state changes as the behaviour of the object. The behaviour of an object is defined by the set of methods which can be applied on it. Objects are further subdivided into physical objects and logical objects. For example a person, a place, a bank account, a table, a chair etc. are physical objects. An inventory, dictionary sale purchase files, library software are logical objects.

2. Class

A class is a data type whose variables are objects. Object as a variable has member functions as well as the ability to hold data values. A class is specified just like a structure with a name. The class is specified by keyword "class".

Syntax:

```
class    class-name or user-defined- name or tag-name
{
private :
data type1       data member1 ;
- - - - - - - - - - - - - - - - - - - -
data type n      data member n ;
member functions ( ) ;
implementation operations ;
list of friend functions ;
public :
```

```
data types data member ;
member function ( ) ;
implementation operations ;
protected :
data types data-members ;
implementation operations ;
list of friend functions ;
} ;
class class-name v1, v2, v3, ....... vn ;
```

3. Inheritance

Inheritance is process by which objects of one class acquire the properties of object of another class. It is in hierarchical order. For example crow is a flying bird which is further part of a bird. In other words classes has sub-classes and so on. For example, class of animals is divided into living organisim (human beings, animals, insects, birds etc.).

There are mainly five types of inheritance:

(i) Single inheritance
(ii) Multiple inheritance
(iii) Multilevel inheritance
(iv) Hierarchical inheritance
(v) Hybrid inheritance.

(i) Single Inheritance

A derived class inheriting from a single base class is called single inheritance.

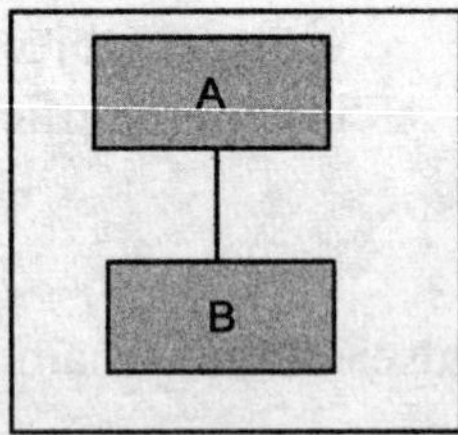

(ii) Multiple Inheritance

A derived class inheriting from multiple base classes is called multiple inheritance .i.e. suppose class B is derived from class A and then class B is further inherited by class C.

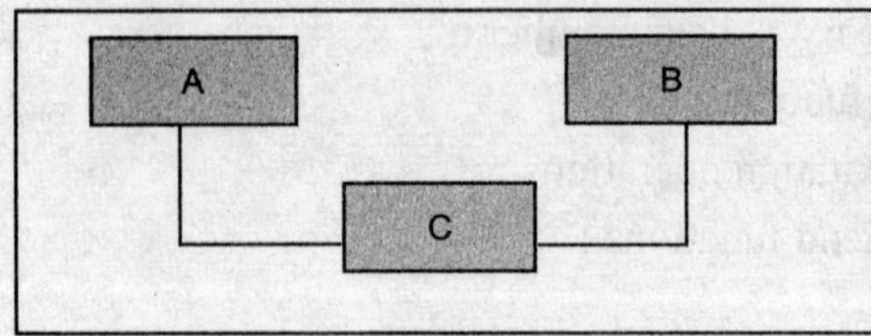

(iii) Multilevel Inheritance

A derived class inheriting from a class that inherits from another class is called multilevel inheritance.

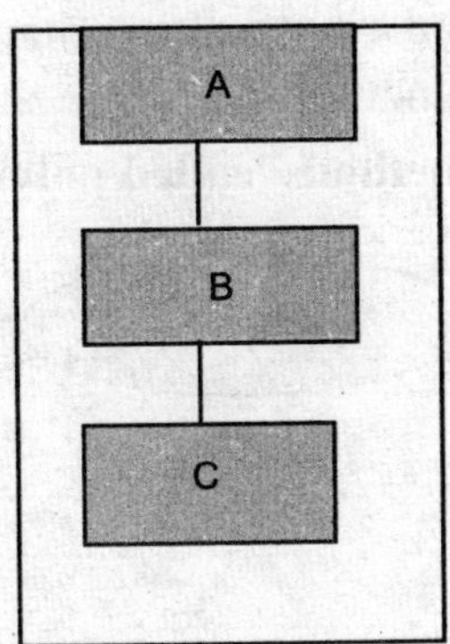

(iv) Hierarchical Inheritance

Number of derived classes inheriting from a single base class is called hierarchical inheritance.

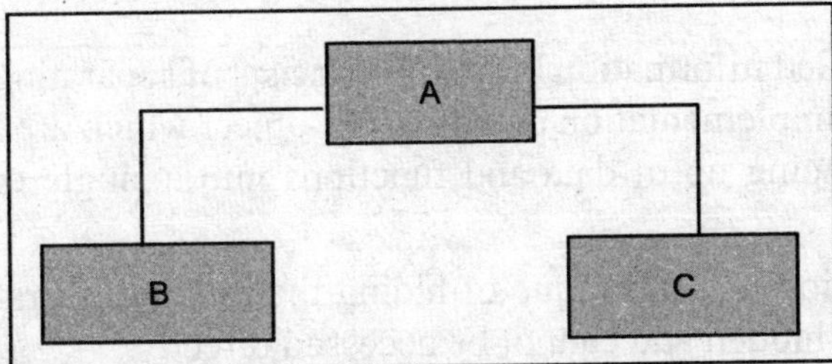

(v) Hybrid Inheritance

A derived class inheriting from multiple base classes and multiple base classes inheriting from a single base class is called hybrid inheritance.

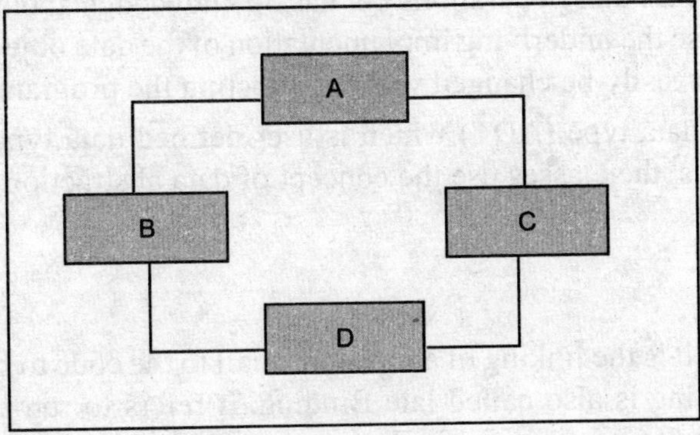

4. Polymorphism

Poly means "many" and morphism means "forms" is "many forms". So Polymorphism means the ability to take more than one form i.e. an operation may exhibit different behavior in different instances (one thing with several distinct forms).

DEF: "one interface multiple methods"called polymorphism.

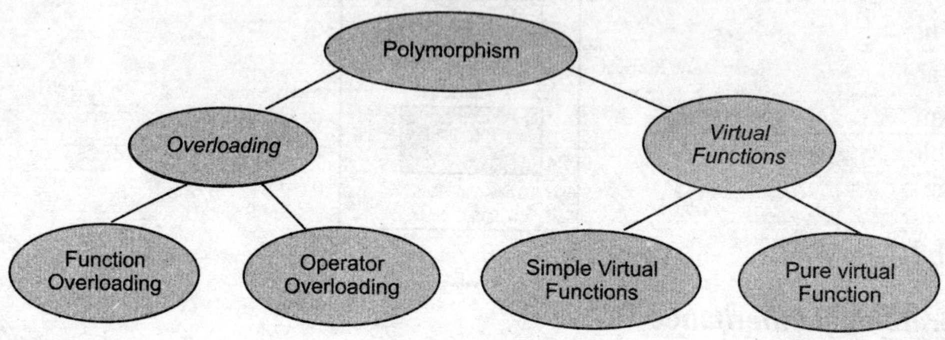

5. Encapsulation

Encapsulation is also called information hiding. It consists of separating the external aspects of an object from the internal implementation details of the object which are hidden from other objects. In other words the wrapping up of data and functions into a single unit (called class) is called Encapsulation.

Actually Encapsulation is a technique of hiding the internal operations of a class to support abstraction. Here data is hidden and cannot be accessed directly.

6. Abstraction

Abstraction refers to the act of representing essential features without including the background details or explanations.Data abstraction is concerned with separating the behavior of a data object from its representation or implementation. For example, a stack has four operation Push, pop, full and empty. As the user uses these operations but has no knowledge about how the stack is actually implemented. So because the underlying implementation of the data object is hidden from its users, the implementation can easily be changed without affecting the program that uses it.

The term Abstract data type (ADT) which is user defined data type is a major input in data structure. In other words, the classes use the concept of data abstraction called as ADT (Abstract data types).

7. Dynamic Binding

Binding means linking. It is the linking of a procedure call to the code to be executed in response to the call. Dynamic Binding is also called late Binding. It refers to run-time determination of the addresses of the function to be executed in the program. In Dynamic Binding, we can use a non-

constant pointer to postpone the allocation of memory until the program is running. Dynamic binding means the code associated with a given procedure call is known until the time of execution.

8. Message Passing

A message is a request to an object to invoke one of its methods. A message therefore contains

- the name of the method and
- the arguments of the method.

Consequently, invocation of a method is just a reaction caused by receipt of a message. This is only possible, if the method is actually known to an object. Method is the approach which passes the message from one location to another to call the required procedure.

9. Reusability

Once a class has been written, created and debugged, it can be distributed to other programmers for use in their own programs. This is called reusability.

1.21 KEYWORDS

C++ language has some reserve words which can not be used as variables. These reserve words are keywords of C++ language. There are mainly 54 keywords from which 47 are used by many C++ compilers.

1.22 INPUTTING USING CIN>> AND OUTPUTTING USING COUT<< STATEMENTS

As an extension to the standard stream (FILE) approach well known from the C programming language, C++ offers an I/O library based on class concepts.

Apart from defining the insertion (<<) and extraction(>>) operators, the use of the C++ I/O library offers the additional advantage of type safety in all kinds of standard situations. Objects (or plain values) are inserted into the iostreams. Compare this to the situation commonly encountered in C where the fprintf() function is used to indicate by a format string what kind of value to expect where.

NOTE: *Now in C++ using cout instead of printf and cin instead of scanf.*

Example:

In C

```
printf("Enter the number:");
scanf("%d",&n);
```

Compared to this latter situation **C++'s iostream** approach uses the objects where their values should appear, as in

```
cout << " Enter the number :";
cin>>n;
```

All concepts of C is same as in C++ like control statements, array and strings, pointer, structure and union .some new features are described below:

SOLVED PROBLEMS

Problem 1: Program to find the area and circumference of the circle.

```
#include<iostream.h>
#include<conio.h>
#define PI 3.1427
void main()
{
int r;
float cir, area;
clrscr();
cout<<"\n ENTER THE RADIUS OF THE CIRCLE: ";
cin>>r;
cir = 2*PI*r;
area=PI*r*r;
cout<<"\n THE CIRCUMFERENCE OF CIRCLE IS= "<< cir;
cout<<"\n THE AREA OF CIRCLE IS= "<< area;
getch();
}
```

Output:

```
ENTER THE RADIUS OF THE CIRCLE: 5
THE CIRCUMFERENCE OF CIRCLE IS= 31.42
THE AREA OF CIRCLE IS= 78.57
```

Problem 2: Program to compute the circumference and volume of a sphere.

```
#include<iostream.h>
#include<conio.h>
#define PI 3.1427
void main()
{
int r;
```

```
float cir, vol;
clrscr();
cout<<"\n ENTER THE RADIUS OF THE SPHERE: ";
cin>>r;
cir=4*PI*r*r;
vol=4/3*PI*r*r*r;
cout<<"\n CIRCUMFERENCE OF SPHERE IS="<< cir;
cout<<"\n THE VOLUME OF SPHERE IS="<< vol;
getch();
}
```

Output:

```
ENTER THE RADIUS OF THE SPHERE: 5
CIRCUMFERENCE OF SPHERE IS=314.27
THE VOLUME OF SPHERE IS= 392.83
```

Problem 3: Program to compute simple interest and compound interest.

```
#include<iostream.h>
#include<math.h>
#include<conio.h>
#include<dos.h>
void main()
{
float p, r, t, si, ci;
cout<<"\n Enter the value of Principle, Rate, Time: ";
cin>>p>>r>>t;
si=(p*r*t)/100;
ci=p*pow((1+r/100),t)-p;
cout<<"\n Simple interest is ="<<si;
cout<<"\n Compound interest is ="<<ci;
sleep(10);
}
```

Output:

```
Enter the value of Principle, Rate, Time: 1200  10  2
Simple interest is 240
Compound interest is 252
```

Problem 4: Program to find whether a number is even or odd.

```
#include<iostream.h>
#include<conio.h>
#include<dos.h>
void main( )
{
int n;
cout<<"\n Enter the number:";
cin>>n;
if(n%2==0)
cout<<"\n The number is Even";
else
cout<<"\n The number is Odd";
getch();
}
```

Output:

Enter the number: 66

The number is Even

Problem 5: Program to find the average of N numbers using for loop.

```
#include<iostream.h>
#include<conio.h>
void main()
{
int n, i, sum=0;
float av;
clrscr();
cout<<"\n Enter the value of N: ";
cin>>n;
for( i =1; i <=n; i++)
{
sum= sum+ i;
}
cout<<"\n Sum is:"<< sum;
av= (float) sum/n;
```

```
cout<<"\n Average is:"<< av;
getche(();
}
```

Output:

Enter the value of N: 10

Sum is: 55

Average is: 5.5

Problem 6: Program to print

```
*
**
***
****
```

```
#include<conio.h>
#include<iostream.h>
void main()
{
int a;
clrscr( );
cout<<"Enter any number :";
cin>>a;
for(int i=1;i<=a;i++)
{
for(int j=1;j<=i;j++)
cout<<'*';
cout<<endl;
}
getch();
}
```

Problem 7: Print the particular table.

```
#include<iostream.h>
#include<conio.h>
void main()
{
```

```
clrscr();
int m,n,j,k,t,i;
for(i=1;i<=20;i++)
{
cout<<"enter the required table";
cin>>m;
}
for(j=1;j<=10;j++)
{
n=m*j;
cout<<"\t"<<m<<"*"<<j<<"="<<n;
}
cout<<"\n";
getch();
}
```

Output:

```
enter the required table 5
5*1=5
5*2=10
5*3=15
5*4=20
5*5=25
5*6=30
5*7=35
5*8=40
5*9=45
5*10=50
```

Problem 8: To print a pattern (Downwards stars).

```
#include<iostream.h>
#include<conio.h>
void main()
{
int i,j,n;
clrscr();
```

```
cout<<"Enter number of lines=>";
cin>>n;
for(i=1;i<=n;i++)
{
for(j=n;j>=i;j—)
cout<<"* ";
cout<<endl;
}
getch();
}
```

Output:

```
* * *
 * *
  *
```

Problem 9: To print a pattern (Multiple stars).

```
#include<iostream.h>
#include<conio.h>
void main()
{
int i,j,n;
clrscr();
cout<<"Enter number of lines=>";
cin>>n;
for(i=1;i<=n;i++)
{
for(j=1;j<=n-i;j++)
cout<<" ";
for(j=1;j<=2*i-1;j++)
cout<<"*";
cout<<endl;
}
getch();
}
```

UNSOLVED PROBLEMS

1. Program to find average of three numbers.
2. Program to find volume of a box.
3. Program to convert time in seconds to time hours, minutes and seconds.
4. Write a program to print the following outputs using for loops.

(a)
```
1
2  2
3  3  3
4  4  4  4
```
(b)
```
*  *  *  *
*  *  *
*  *
*
```

5. Write a C program to draw Reverse Pascal Triangle.
6. Write a C Program to find the leap year.
7. Write a program to print a Fibonacci series 1 2 3 5 8 …..upto 100.
8. Program to find the biggest number among N numbers.
9. Program to Sort an array (list) having N elements using Bubble Sort method.
10 Program to Calculate Matrix Multiplication, Matrix Addition and Matrix Transpose by using SWITCH statements (A menu driven program).
11. Program to count the negative elements from the list.
12. Program to transpose a matrix.
13. Program to find the sum of Diagonal Elements of a matrix (Trace a Matrix).
14. Write a C Program to reverse numbers in an array.
15. Write a C Program to remove duplicate numbers from an array.
16. Program to search a string using Binary Search.
17. Program to find whether the string is PALINDROMEM or not.
18. Write a C Program to find all combination of characters of a word.
19. Write a C Program to determine the area and perimeter of triangle using pointer.
20. Write a C Program to construct a pay bill structure of an employee in an organization.
21. Write a C Program to access one-dimensional array structure.
22. Write a C Program to compute the size of union.
23. Write a C++ program with the employee class that makes the two employees. Set their age, year of service and salary. Also print their values.

24. Program to check whether number is Prime or not by using Class.
25. Program to enter and display student data by using Class.
26. Create a fully functional stack class and use the class for reversing string.
27. Program to compute marks to explain the concept of Multiple Inheritance.
28. Program to process student data to describe level of inheritance.

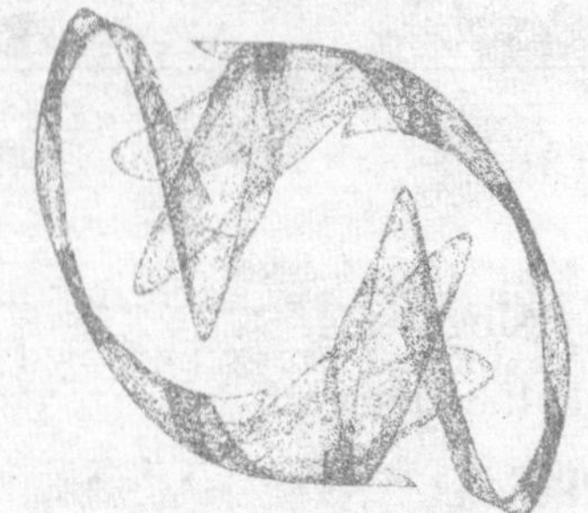

CHAPTER 2

Numerical Analysis

2.1 INTRODUCTION

Numerical analysis is the area of mathematics and computer science that creates, analyzes, and implements algorithms for solving numerically the problems of continuous mathematics. Such problems originate generally from real-world applications of algebra, geometry, and calculus, and they involve variables which vary continuously. These problems occur throughout the natural sciences, social sciences, medicine, engineering, and business. Here our aim to give a broad understanding of the theoretical basis for numerical analysis, including an awareness of the nature and types of numerical errors that occur in computer calculations, and extensive "hands-on" experience in the application of numerical methods. In this context, a wide variety of elementary but important problems will be considered, such as collocation and interpolation, differentiation, integration, initial value problems for differential equations, non-linear algebraic equations, and systems of linear algebraic equations.

Definition: *"Numerical analysis is the study of algorithms for the problems of continuous mathematics".*

Note that it different from the discrete mathematics. So this means it deals mainly with real variable or complex variable questions, numerical linear algebra over the real or complex fields, the solution of differential equations, and other related problems arising in the physical sciences and engineering.

Some problems in continuous mathematics can be solved exactly by an algorithm. These algorithms are called *direct methods*. Examples are Gaussian elimination for solving systems of linear equations and the simplex method in linear programming. However, no direct methods exist for most problems. We might then try to replace the continuous problem by a discrete problem; this process is called *discretization*. Another possibility is to use an iterative method. Such a method starts from a guess and finds successive approximations that hopefully converge to the solution.

Even when a direct method does exist, an iterative method may be preferable because it is more efficient.

Actually the overall goal of the field of numerical analysis is the design and analysis of techniques to give approximate but accurate solutions to hard problems, the variety of which is suggested by the following.

- Some of the advanced numerical methods are essential in making numerical weather prediction feasible.
- It has a big facility to compute the target at any line. Computing the trajectory of a spacecraft requires the accurate numerical solution of a system of ordinary differential equations.
- Any vehicles companies can improve the crash safety of their vehicles by using computer simulations of car crashes. Such simulations essentially consist of solving partial differential equations numerically.
- Any Hedge funds (private investment funds) use tools from all fields of numerical analysis to calculate the value of stocks and derivatives more precisely than other market participants.
- Number of airlines use sophisticated optimization algorithms to decide ticket prices, airplane and crew assignments and fuel needs. This field is also called operations research.
- Mostly all the Insurance companies use numerical programs for actuarial analysis.

Numerical analysis involves the study of methods of computing numerical data. In many problems this implies producing a sequence of approximations; thus the questions involve the rate of convergence, the accuracy (or even validity) of the answer, and the completeness of the response. (With many problems it is difficult to decide from a program's termination whether other solutions exist.) Since many problems across mathematics can be reduced to linear algebra, this too is studied numerically; here there are significant problems with the amount of time necessary to process the initial data. Numerical solutions to differential equations require the determination not of a few numbers but of an entire function; in particular, convergence must be judged by some global criterion. Other topics include numerical simulation, optimization, and graphical analysis, and the development of robust working code.

This chapter gives us role of Numerical analysis in computer science field. The goal of this chapter is to aware the readers about the scope and benefits of numerical analysis techniques i.e. how numerical analysis techniques are helpful for us to solve mathematical, scientific and engineering problems by means of basic arithmetic operations such as addition, subtraction, multiplication and division. Numerical analysis techniques are evolved rapidly after the evolution of fast computers. In the coming chapters we will discuss some numerical techniques that are useful for us.

2.2 ORIGINS OF NUMERICAL METHODS

Numerical algorithms are almost as old as human civilization. The Rhind Papyrus (~1650 BC) of ancient Egypt describes a rootfinding method for solving a simple equation. Archimedes of Syracuse (287-212 BC) created much new mathematics, including the "method of exhaustion" for calculating lengths, areas, and volumes of geometric figures. When used as a method to find approximations, it is in much the spirit of modern numerical integration; and it was an important precursor to the development of the calculus by Isaac Newton and Gottfried Leibnitz. A major impetus to developing numerical procedures was the invention of the calculus by Newton and Leibnitz, as this led to accurate mathematical models for physical reality, first in the physical sciences and eventually in the other sciences, engineering, medicine, and business.

These mathematical models cannot usually be solved explicitly, and numerical methods to obtain approximate solutions are needed. Another important aspect of the development of numerical methods was the creation of *logarithms* by Napier (1614) and others, giving a much simpler manner of carrying out the arithmetic operations of multiplication, division, and exponentiation. Newton created a number of numerical methods for solving a variety of problems, and his name is attached today to generalizations of his original ideas. Of special note is his work on rootfinding and polynomial interpolation. Following Newton, many of the giants of mathematics of the 18*th* and 19*th* centuries made major contributions to the numerical solution of mathematical problems. Foremost among these are Leonhard Euler (1707-1783), Joseph-Louis Lagrange (1736-1813), and Karl Friedrich Gauss (1777-1855). Up to the late 1800's, it appears that most mathematicians were quite broad in their interests, and many of them were interested in and contributed to numerical analysis.

An Early Mathematical Model: One of the most important and influential of the early mathematical models in science was that given by Newton to describe the effect of gravity. According to this model, the force of gravity on a body of mass m due to the Earth has magnitude

$$F = \frac{Gmm_e}{r^2}$$

where m_e is the mass of the Earth, r is the distance between the centers of the two bodies, and G is the universal gravitational constant. The force on m is directed towards the center of gravity of the Earth. Newton's model of gravitation has led to many problems that require solution by approximate means, usually involving the numerical solution of ordinary differential equations.

Following the development by Newton of his basic laws of physics, these were applied by many mathematicians and physicists to give mathematical models for solid and fluid mechanics. Civil and Mechanical Engineering use these models as the basis for most modern work on solid structures and the motion of fluids, and numerical analysis has become a basic part of the work of researchers in these areas of engineering. For example, building modern structures makes major use of finite element methods for solving the partial differential equations associated with models of stress; and computational fluid mechanics is now a fundamental tool in designing new airplanes. In the 19th century, phenomena involving heat, electricity, and magnetism were successfully modelled; and in the 20th century, relativistic mechanics, quantum mechanics, and other theoretical constructs were created to extend and improve the applicability of earlier ideas.

The field of numerical analysis predates the invention of modern computers by many centuries. In fact, many great mathematicians of the past were preoccupied by numerical analysis, as is obvious from the names of important algorithms like Newton's method, Lagrange interpolation polynomial, Gaussian elimination, or Euler's method. Linear interpolation was already in use more than 2000 years ago.

The father of Numerical analysis is a babylonian scientist. He designed a clay for numerical analysis caled Babylonian clay tablet YBC 7289 in the era 1800–1600 BCE..

To facilitate computations by hand, large books were produced with formulas and tables of data such as interpolation points and function coefficients. Using these tables, often calculated out to 16 decimal places or more for some functions, one could look up values to plug into the formulas given and achieve very good numerical estimates of some functions. The canonical work in the field is the NIST publication edited by Abramowitz and Stegun, a 1000-plus page book of a very large number of commonly used formulas and functions and their values at many points. The function values are no longer very useful when a computer is available, but the large listing of formulas can still be very handy.

The mechanical calculator was also developed as a tool for hand computation. These calculators evolved into electronic computers in the 1940s, and it was then found that these computers were also useful for administrative purposes. But the invention of the computer also influenced the field of numerical analysis, since now longer and more complicated calculations could be done. With the increasing availability of computers, the new discipline of scientific computing, or computational science, emerged during the 1980s and 1990s. The discipline combines numerical analysis, symbolic mathematical computations, computer graphics, and other areas of computer science to make it easier to set up, solve, and interpret complicated mathematical models of the real world.

2.3 ABOUT NUMERICAL ANALYSIS

The area of mathematics and computer science that creates, analyzes, and implements algorithms for obtaining numerical solutions to problems involving continuous variables. Such problems arise throughout the natural sciences, social sciences, engineering, medicine, and business. Since the mid 20th century, the growth in power and availability of digital computers has led to an increasing use of realistic mathematical models in science and engineering, and numerical analysis of increasing sophistication is needed to solve these more detailed models of the world. The formal academic area of numerical analysis ranges from quite theoretical mathematical studies to computer science issues. Numerical analysis is the area of mathematics and computer science that creates, analyzes, and implements algorithms for solving numerically the problems of continuous mathematics. Such problems originate generally from real-world applications of algebra, geometry and calculus, and they involve variables which vary continuously; these problems occur throughout the natural sciences, social sciences, engineering, medicine, and business. During the past half-century, the growth in power and availability of digital computers has led to an increasing use of realistic mathematical models in science and engineering, and numerical analysis of increasing sophistication has been needed to solve these more detailed mathematical models of the world. The formal academic area of numerical analysis varies from quite theoretical mathematical studies to computer science issues. With the growth in importance of using computers to carry out numerical procedures in solving mathematical models of the world, an area known as *scientific computing* or *computational science* has taken shape during the 1980s and 1990s. This area looks at the use of numerical analysis from a computer science perspective. It is concerned with using the most powerful tools of numerical analysis, computer graphics, symbolic mathematical computations, and graphical user interfaces to make it easier for a user to set up, solve, and interpret complicated mathematical models of the real world.

Numerical Analysis is one of the important branch of applied mathematics which devises and evaluates techniques for employing computers to solve numerical problems, and to study their convergence and errors. Numerical analysis is an applied mathematics mechanism to solve the entire mathematical problem with the well defined procedure. It was so generally used until the Institute of Numerical Analysis was founded at the University of California in Los Angles in 1947. Numerical analysis involves the study of methods of computing numerical data. By numerical analysis, we mean the theory of constructive methods in mathematics analysis. By constructive method we mean a procedure that permits us to obtain the solution of a mathematical problem with an arbitrary precision in a finite number of steps that can be prepared rationally (the number of steps depend on the desired accuracy). Numerical analysis is both a science and art. As a science, it is concerned with the process by which mathematical problem can be solved by the operations of arithmetic. As an art, numerical analysis is concerned with choosing that procedure, which is best, suited to the solution of a particular problem. One who is good in applied mathematics can definitely understand the numerical concept very easily. Scientists and Engineers approach higher mathematics in two ways –By using Numerical analysis and Analytical analysis. Numerical analysis is totally computer-based, while Analytical analysis is done by hand. Analytical methods are pure mathematical methods to solve a particular problem. As we integrate a given function by using an analytical technique as we use in pure mathematics, same function can also be integrated by using numerical integration techniques as by using Trapezoidal or Simpson's 1/3 or Simpson's 3/8 rule etc. As these methods may exists in analytical mathematics too but in case of analytical mathematics if we are interested to find definite integral +"f(x)dx having limits 'a' and 'b' when f(x) is not known explicitly but is known at certain points then analytical technique fails. The advantage of numerical analysis is that it can give a numerical answer to a problem, which has no analytical solution. It substitutes standard calculations for the parts of a problem that do not really exist. Thus, the results are only approximations or estimations.

Analytical analysis is more helpful when examining the exact behavior or properties of a mathematical function. But due to its slow speed of calculation we choose Numerical analysis. Analytical analysis gives birth to analytical methods that are used in pure mathematics and solve the problem by hand. On the other hand Numerical analysis gives us some powerful methods known as numerical methods to solve those problems also that are not solved easily and quickly by analytical methods. So in this book, we will only discuss numerical methods not the analytical methods. The advantage of numerical analysis is that it can give a numerical answer to a problem, which has no analytical solution. It substitutes standard calculations for the parts of a problem that do not really exist. Thus, the results are only approximations or estimations. We will now outline several important themes of numerical analysis. The overall goal is the design and analysis of techniques to give approximate solutions to hard problems. To fix ideas, the reader might consider the following problems and methods:

- If a company wants to put a toothpaste commercial on television, it might produce five commercials and then choose the best one by testing each one on a focus group. This would be an example of a Monte Carlo optimization method.
- To send a rocket to the moon, rocket scientists will need a rocket simulator. This simulator will essentially be an integrator for an ordinary differential equation.
- Every machine can works safely by using computer simulations. These simulations are essentially solving partial differential equations numerically.
- Some special techniques and tools are used to handle finicial matters by using their derivatives.
- Indian Railways reservation system use sophisticated optimization algorithms to decide ticket prices, route of the train, time managementn etc. This field is powerfull tool of the Computer optimization techniques known as operations research.
- Share broker use numerical programs for actuarial analysis.

Numerical analysis is concerned with all aspects of the numerical solution of a problem, from the theoretical development and understanding of numerical methods to their practical implementation as reliable and efficient computer programs. Most numerical analysts specialize in small subfields, but they share some common concerns, perspectives, and mathematical methods of analysis. These include the following:

1. When presented with a problem that cannot be solved directly, they try to replace it with a "nearby problem" that can be solved more easily. Examples are the use of interpolation in developing numerical integration methods and root-finding methods.
2. There is widespread use of the language and results of linear algebra, real analysis, and functional analysis (with its simplifying notation of norms, vector spaces, and operators).
3. There is a fundamental concern with error, its size, and its analytic form. When approximating a problem, it is prudent to understand the nature of the error in the computed solution. Moreover, understanding the form of the error allows creation of extrapolation processes to improve the convergence behaviour of the numerical method.
4. Numerical analysts are concerned with stability, a concept referring to the sensitivity of the solution of a problem to small changes in the data or the parameters of the problem. Consider the following example. The polynomial $p(x) = (x - 1)(x - 2)(x - 3)(x - 4)(x - 5)(x - 6)(x - 7)$, or expanded, $p(x) = x^7 - 28x^6 + 322x^5 - 1, 960x^4 - 6, 769x^3 - 13, 132x^2 + 13, 068x - 5, 040$has roots that are very sensitive to small changes in the coefficients. If the coefficient of x^6 is changed to –28.002, then the original roots 5 and 6 are perturbed to the complex numbers 5.459 0.540i—a very significant change in values. Such a polynomial $p(x)$ is called unstable or ill-conditioned with respect to the root-finding problem. Numerical methods for solving problems should be no more sensitive to changes in the data than the original problem to be solved. Moreover, the formulation of the original problem should be stable or well-conditioned.

5. Numerical analysts are very interested in the effects of using finite precision computer arithmetic. This is especially important in numerical linear algebra, as large problems contain many rounding errors.
6. Numerical analysts are generally interested in measuring the efficiency (or "cost") of an algorithm. For example, the use of Gaussian elimination to solve a linear system $Ax = b$ containing n equations will require approximately $2n^3/3$ arithmetic operations. Numerical analysts would want to know how this method compares with other methods for solving the problem.

2.4 IMPORTANCE OF NUMERICAL ANALYSIS

The area of numerical analysis is also known as numerical computation, numerical methods, numerical mathematics, computational mathematics, scientific computing or mathematical software, depending on which aspect of this broad subject is being emphasized. One can argue that the reason computers were invented was to perform numerical computations, and scientific computing continues to be one of the main driving forces in the development of advanced computing architectures. Several of the "Grand Challenges" in science are closely associated with numerical problems.

The broad long-term goal of numerical analysts is to develop good software for the solution of problems of a mathematical nature that arise in science, engineering and business. Some researchers focus on standard problems (such as solving differential equations), developing routines that can be used as tools by other scientists in the solution of practical problems, while others work closely with researchers from other areas on specific applications. With computers becoming both cheaper and increasingly more powerful, numerical simulation is being used more and more as a substitute for experiment. New designs of aircraft, jet engines, ships, dams, bridges, chemical plants, etc., are now being tested more numerically, as abstract computer models. Along with recognition that numerical "experiments" can be cheaper than real ones, this huge investment in numerical methods and algorithms reflects a growing faith in "*the computer*," especially amongst industrial design managers.

However, what comes out of a numerical simulation can be no better than what goes into it, possibly much worse. Not only do mathematical models of the physics need to be good, the numerical solutions of these models also need to be accurate and reliable. It is essential, therefore, to have good, reliable, computer software for solving many types of mathematical models accurately and efficiently. Without them, any faith in computers has no proper foundation. Because of the limitations of computer arithmetic, among other reasons, only approximate solutions can be obtained. Clearly, users of computer software want to have confidence in their numerical results. They want the software to be reliable, efficient and robust, and they want the answers to be accurate. The role of the numerical analyst is to derive efficient numerical methods for obtaining approximate solutions of a wide variety of mathematical equations, analyze the accuracy and stability properties of the methods and develop the associated computer software to implement the methods. Ideally, the computer software should be able to estimate and control the size of the error in the computed solution, so that the computed solution can be guaranteed to have the accuracy required by the user of the code. Thus, numerical analysis plays a vitally important role in all branches of science and engineering, as well as in industry.

2.5 ADVANTAGES OF NUMERICAL ANALYSIS

Following are the some advantages of Numerical Analysis:

1. Numerical analysis is most useful with complicated equations. Once the code is written for a type of equation, you can plug in different equations and get answers fast.
2. The other advantage is that as Numerical analysis is totally computer based and computers do not make mistakes, unlike humans.
3. The major advantage of numerical analysis is that we can solve any mathematical problem whether that problem has analytical solution or not.
4. We can reduce the errors in calculation/solutions by making use of suitable numerical techniques.

2.6 DISADVANTAGES OF NUMERICAL ANALYSIS

Following are the some advantages of Numerical Analysis:

1. The numerical solution is always an approximation to the solution i.e. does not give the exact value due to which result may vary slightly.
2. The results obtained by using numerical analysis are purely numerical in nature.
3. Numerical data used to solve a given problem are usually not exact and is correct up to certain number of decimal places, methods and procedures that are applied to solve a problem are also approximate.

2.7 USES OF NUMERICAL ANALYSIS

There are some problems that numerical analysis can solve easily and efficiently.

1. To solve a system of linear and non-linear equations.
2. Interpolate the approximate value within the set of table.
3. To fit a curve for a given table.
4. To find the derivatives when the values satisfying the function are given.
5. To integrate a function when the values satisfying the function are given.
6. To solve ordinary differential equation when initial values of the variables are given.
7. To find Eigen values and Eigen vectors of a matrix.
8. To establish a relationship between derivatives.

2.8 NUMERICAL ANALYSIS VS NUMERICAL METHODS

"Numerical analysis is the study of analyzing the numerical methods". In the study of a numerical method what is mainly involved the convergence of the method, and errors (both mainly rounding and truncation) produced by the method etc. The study of numerical methods has been done by Numerical analysts who are expert in their fields and study the numerical methods by using mathematical methods of analysis. Numerical analysis is the mathematical computation of numerical problem to get a required result in a required format up to given number of steps and up to a given error as tolerance. It is the study of numerical problem to get a required result. The main aim of numerical analysis is to produce such numerical methods that converge more rapidly and produce lesser number of errors. Numerical methods are algorithms in ***numerical analysis***.

Numerical analysis is the study of problems in which numbers and approximations play an especially significant role. Computers are particularly well suited to problems in numerical analysis because many such problems, while essentially involving common mathematical operations, require a lot of them. In the early days of computing, scientists monopolized computers with problems like this, which were far too intensive to be carried out by hand. Even today, problems in numerical analysis still occupy a good part of the cycles of some of the largest computers in the world. Hence, numerical analysis is a vast subject, and many numerical methods are as complicated and specific as the mathematical problems they solve. Numerical analysis is concerned with all the aspects of the numerical solution of a problem, from the theoretical development and understanding the numerical methods to their practical implementation as reliable and efficient computer programs. Numerical methods are at the heart of the numerical process. Numerical analysis provides the theoretical foundation for the numerical algorithms we rely on to solve a multitude of computational problems in science.

2.9 APPLICATIONS OF NUMERICAL ANALYSIS

The algorithms of numerical analysis are routinely applied to solve many problems in science and engineering. Examples are the design of structures like bridges and airplanes (as used in the computational physics and computational fluid dynamics), weather forecasting, climate models, the analysis and design of molecules (computational chemistry), and finding oil reservoirs. In fact, almost all supercomputers are continually running numerical analysis algorithms. As a consequence, efficiency plays an important role and a heuristic method may be preferred above a method with a solid theoretic foundation because it is more efficient. Generally, numerical analysis uses empirical results of computation runs to probe new methods and analyze problems, though it of course also employs mathematical axioms, theorems and proofs. *Computer aided design* (CAD) and *computer aided manufacturing* (CAM) are important areas within engineering, and some quite sophisticated techniques have been developed for CAD/CAM. A wide variety of numerical analysis is involved in the mathematical models that must be solved. The models are based on the basic Newtonian laws of mechanics; there are a variety of possible models, and research continues on designing such models. An important CAD topic is that of modeling the dynamics of moving mechanical systems. The mathematical model involves systems of both ordinary differential equations and algebraic equations (generally nonlinear). The numerical analysis of these mixed systems, called *differential-algebraic systems*, is quite difficult but important to being able to model moving mechanical systems. Building simulators for cars, planes, and other vehicles requires solving differential-algebraic systems in real-time.

Atmospheric modeling is important for simulating the behavior of the Earth's atmosphere, to understand the possible effect of human activities on our atmosphere. A large number of variables need to be introduced. These include the velocity $\mathbf{v}(x, y, z, t)$ in the atmosphere at position (x, y, z) and time t, the pressure $p(x, y, z, t)$, and the temperature $T(x, y, z, t)$. In addition, we must study various chemicals existing in the atmosphere and their interactions, including ozone, various chemical pollutants, carbon dioxide, and others. The underlying equations for studying $\mathbf{v}(x, y, z, t)$, $p(x, y, z, t)$, and $T(x, y, z, t)$ are partial differential equations; and the chemical kinetic interactions of the various chemicals are described using some quite difficulty ordinary differential equations. Many types of numerical analysis procedures are used in atmospheric modeling, including computational fluid mechanics and the numerical solution of differential equations.

Modern business makes much use of optimization methods in deciding how to allocate resources most efficiently. These include problems such as inventory control, scheduling, how best to locate manufacturing and storage facilities, investment strategies, and others.

Some applications of numerical methods are:

Linear Regression Models

Statistical models in which there is a linear-form relationship between an independent variable x and a variable y that depends on it. Least-squares estimators help to predict values of for values of x we have not observed experimentally.

Curve Fitting

The process of fitting a curve to a number of points is curve fitting. If the points for which we have values are located at meaningful places on the curve we are trying to fit, and we know values at enough points, interpolation helps us draw a smooth curve.

Scatter Plots

Statistical tools that help ascertain the relationship between an independent variable x and a variable y that depends on it. Using least-squares estimators to draw a best-fit line through a linear-form scatter plot helps with this.

Approximating Functions

The process of determining the value of a function at points for which exact values are not known. This can be done by constructing an interpolating polynomial of the appropriate degree.

Function Tables

Tables containing values of computationally expensive functions or models of complicated physical phenomena. Often it is too costly to compute and store values of a function with the granularity required at some later time. Thus, we store a limited number of points and interpolate between them.

Scientific Computing

An area in which solving equations is one of the most fundamental problems routinely performed.

2.10 COMMON PERSPECTIVES IN NUMERICAL ANALYSIS

Numerical analysis is concerned with all aspects of the numerical solution of a problem, from the theoretical development and understanding of numerical methods to their practical implementation as reliable and efficient computer programs.

Most Numerical analysts specialize in small sub areas, but they share common concerns, perspectives and mathematical methods for analysis. These include the following:

1. When presented with a problem that cannot be solved directly, then replace it with a "nearby problem" which can be solved more easily. Examples are the use of the interpolation in developing numerical integration methods and root finding methods.
2. There is a widespread use of the language and results of linear algebra, real analysis, and functional analysis (with its simplifying notations of norms, vector spaces, and operators).
3. There is a fundamental concern with the error, its size and its analytic form. When approximating a problem, it is prudent to understand the nature of the error in the computed solution. Moreover, understanding the forms of the error allows creation of extrapolation processes to improve the convergence behavior of the numerical method.
4. Stability is the concept referring to the sensitivity of the solution of a problem to small changes in the data or parameters of the problem. In developing numerical methods for solving problems, they should be no more sensitive to changes in the data than the original problem to be solved. Moreover, one tries to formulate the original problem to be stable or well conditioned.
5. Numerical analysts are very interested in the effects of using finite precision computer arithmetic. This is especially important in numerical linear algebra, as large problems contain many rounding errors.
6. Numerical analysts are generally interested in measuring the efficiency of the algorithms. What is the cost of particular algorithm? For example, the use of Gaussian elimination to solve a linear system $AX = B$ containing n equations will require approximately $(2/3)n^3$ arithmetic operations. How does this compare with other numerical methods for solving this problem?

2.11 MORE ABOUT NUMERICAL ANALYSIS

"Numerical Analysis is the study of algorithms for the problems of continuous mathematics."

- ***Algorithms.*** The center of numerical analysis – devising and analyzing algorithms to solve certain classes of problems.
- ***Continuous Mathematics.*** The variables involved are real or complex, as opposed to discrete. This work is *applied,* applied daily and successfully to thousands of applications on millions of computers around the world. "The problems of continuous mathematics" are the problems that science and engineering are built upon; without numerical methods, science and engineering as practiced today would come quickly to a halt. They are also the problems that preoccupied most mathematicians from the time of Newton to the twentieth century.

	Continuous Problems	VS	Discrete Problems
-Studied mainly by:	numerical analysts		computer scientists
-Most problems:	cannot be solved by finite algorithms		can be solved by finite algorithms.
-Most problems:	could not be solved exactly even if we could work in exact arithmetic		can be solved exactly.

Thus, the nature of numerical analysis implies that it must therefore be concerned with various types of computing errors:

- **Rounding errors:** Since real and complex numbers cannot be represented exactly on computers, part of the business of numerical analysis must be to approximate them. This is where rounding errors come in.
- **Deeper kinds of errors:** Propagation of rounding errors, and other errors related to convergence of approximations, such as errors due to truncation, discretization, iteration, etc. Machine architecture may also be important too.

2.12 HEART AND SOUL OF NUMERICAL ANALYSIS

To compute quantities that are typically uncomputable from an analytic point of view, and to do it with lightning speed!

- ***Even if rounding errors vanished, numerical analysis would remain.*** Approximating mere numbers, the task of floating-point arithmetic, is a rather small topic. The deeper business of numerical analysis is approximating *unknowns,* not known. *Rapid convergence of approximations is the aim*, and the pride of the field of numerical analysis is that, for many problems, algorithms have been invented that converge exceedingly fast.
- ***Numerical methods and applications.*** The progress and development of efficient algorithms, or numerical methods, has been a result of influence and motivation by applications and development of newer and more advanced computer systems, and vice versa. Some applications which often motivated discovery of new numerical methods include fluid dynamics, aeronautical engineering, stresses in mechanical structures, neutron diffusion in nuclear reactors, oil reservoir modeling, weather forecasting, chemical and biological systems, etc.

2.13 SOFTWARE FOR NUMERICAL ANALYSIS

Software to implement common numerical analysis procedures is very important. If it is to be shared by many users, it needs to be reliable, accurate, and efficient. Moreover, it needs to be written so as to be easily portable between different computers. Beginning around 1970, there has been a number of government sponsored research efforts to produce high quality numerical analysis software in particular problem areas. A more recent example of such a project is the LAPACK project, which contains state-of-the-art programs for basic problems in numerical linear algebra. Two important online numerical analysis libraries that contain many of these large scale numerical analysis programming projects can be found at the following internet sites:

Name	URL	Location
Netlib	www.netlib.org	Oak Ridge National Laboratory
GAMS	gams.nist.gov	National Institute of Standards and Technology

The most popular programming language for implementing numerical analysis methods continues to be *Fortran*, and it too continues to be updated to meet changing needs, with *Fortran 95* being the most recent standard. Other languages are also important, with the most important ones being *C, C++,* and *Java*. Another approach to supplying numerical analysis programs and programming tools has been to create higher level problem solving environments that contain numerical, programming, and graphical tools, including some quite sophisticated numerical analysis tools to handle many basic problems. Best known of these is *MATLAB* (The Mathworks, Inc.), a commercial package that has arguably become the most popular way to do numerical computing. For analytical mathematics computing, there are two popular commercial packages, *Maple* (Waterloo Maple, Inc.) and *Mathematica* (Wolfram Research, Inc.).

There are a number of computer programs used for performing numerical calculations:

- **MATLAB** is a widely-used program for performing numerical calculations it is a commercial package that has arguably become the most popular way to do numerical computing. It comes with its own programming language, in which numerical algorithms can be implemented.
- **GNU Octave** is a near-clone of MATLAB, providing most of the functionality of version 3 or 4 and can run scripts written for MATLAB with some effort.
- **R programming language** is a widely used system with a focus on data manipulation and statistics. Several hundred freely downloadable specialized packages are available.
- **Scilab**, an **open-source** (Scilab license) program for performing numerical calculations.
- **DADiSP** is popular program focused on **DSP** that combines the numerical capability of Matlab with a spreadsheet like interface.
- **IDL programming language.**
- **Perl Data Language,** a dialect of **Perl** used for data manipulation, statistics, and numerical simulation.

- **IGOR Pro**, a software package with emphasis on time series and image analysis, and curve fitting. It comes with its own programming language but can also be used interactively.
- **Mathcad** is a lower-cost alternative to MATLAB that has extensive support libraries available for specialized analysis applications. Mathcad differs in that it has a **WYSIWYG** interface and the ability to generate publication-quality mathematical equations.
- **XNUMBERS 4.4** — Multi Precision Floating Point Computing and Numerical Methods for EXCEL. Xnumbers is an **open source** Excel addin (xla) that performs multi-precision floating point arithmetic from 15 up to 200 significant digits. It is compatible with Excel 2000/XP and consists of a set of more than 300 functions for arithmetic, complex, trigonometric, logarithm, and exponential calculus.

QUERIES

1. What do you mean by Numerical analysis?
2. Why we need Numerical Analysis?
3. What are the benefits of using numerical methods over analytical methods?
4. Explain the importance of Numerical Analysis?
5. Discuss the uses of numerical analysis and methods.
6. Write down the Advantages and Disadvantages of Numerical Analysis?
7. Give a brief description between numerical analysis and numerical methods i.e. how both are related with each other?
8. Discuss the different applications of numerical methods.
9. How can you say that numerical analysis is both art and science?
9. Discuss Common perspectives in Numerical Analysis?
10. Differentiate between continuous and discrete problems?
11. Discuss the role of numerical analyst.
12. What are the various software used for numerical analysis?
13. How can you say Numerical Methods are stable?
14. Why MATLAB is best software for solving Numerical Problems?
15. What impact Numerical Analysis has in the Computer field?
16. Why Numerical Methods are best for Engineering?
17. What is the role of Numerical methods in Numerical analysis?

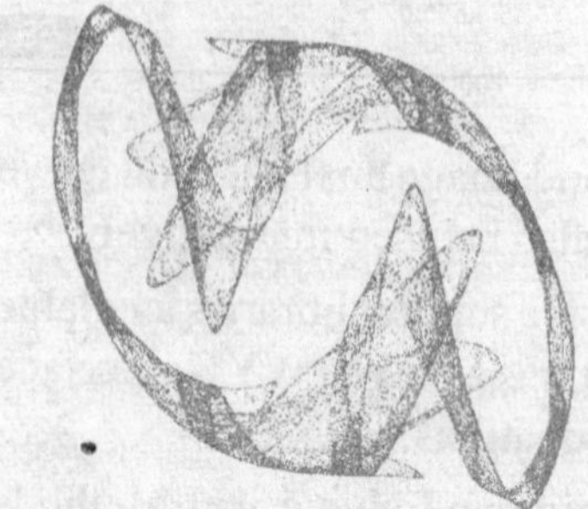

CHAPTER 3

Errors and their Analysis

3.1 INTRODUCTION

The object of the present chapter is to set forth some basic ideas and some methods relating to the approximate calculations and to give methods for estimating the accuracy of the results obtained. The study of errors forms an important part of numerical analysis. There are several ways in which error can be introduced in the solution of the problem. Error analysis is the study of kind and quantity of error that occurs, particularly in the fields of applied mathematics (particularly numerical analysis), applied linguistics and statistics. This chapter also covers how numbers are stored in computer memory, how arithmetic operations are performed on floating point numbers, propagation of errors etc.

Many computer users, who are not aware of how numbers are represented and processed within computer systems, tend to assume that the results of programs are accurate. For example, if a decimal number is printed as 123.45678 all the figures are assumed to be an accurate representation of the result of whatever calculation was performed.

When computer programs are being implemented and tested the errors which can occur include:

1. Errors in the syntax of the program language, i.e. keywords misspelt, operators missing, etc. The compiler will find such errors and present a more or less meaningful error message indicating the problem.
2. Errors in coding the program specification or algorithms into the programming language, i.e. the program does not do what the user had specified. This type of error is mainly due to inexperience and a lack of understanding of the language.
3. Errors in the program specification or algorithms. The set of rules and/or formulas and/or algebraic steps that the user has specified to solve the problem are incorrect, e.g. a mathematician has made a mistake in the that the user has specified to solve the problem are incorrect, e.g. a mathematician has made a mistake in the development of the equations. It is up to the user to ensure that algorithms being used are correct to the extent required.

4. Errors due to the storage and processing of numerical data, i.e. the accuracy of the hardware used to perform numerical calculations is finite. For example, real number data (with fractional components) is stored in floating point format with a typical accuracy between 8 and 20 significant decimal figures. Thus the results of programs using such floating point representation cannot be exact. In addition many mathematical functions are infinite series which have to be truncated This set of notes considers the type of error listed in 4 above.

3.2 NUMBERS AND THEIR ACCURACY

1. Approximate numbers

There are two types of numbers: Exact and Approximate.

Exact numbers are those that can be represented in finite number of digits. Examples of exact numbers are 1, 2, 3,........., 1/2,3/2,0.25 ....,etc.

Approximate numbers are those that cannot be represented in finite number of digits. Approximate numbers have lesser degree of accuracy as compared to exact numbers. Thus, an approximate value of **p** is 3.1416, or desire a better approximation, it is 3.14159265. But we can't write the exact value of **p**. Similarly, an approximate value of $\sqrt{2}$ is 1.41, or desire a better approximation, it is 1.414. But we can't write the exact value of $\sqrt{2}$.

2. Significant digits

The digits used to express a number are called significant digits. The digits 1, 2, 3, 4, 5, 6, 7, 8, 9 are significant digits. '0' is also a significant digit except when it is used to fix the decimal point or to fill the places of unknown or discarded digits. For example, each of the numbers 7845, 3.589, and 0.4758 contains 4 significant figures while the numbers 0.00386, 0.000587, 0.0000296 contain only three significant figures (since zeros only help to fix the position of the decimal point).

Similarly, in the number 0.0003090, the first four '0' s' are not significant digits since they serve only to fix the position of the decimal point and indicate the place values of the other digits. The other two '0' s' are significant. To be more clear, the number 3.0686 contains five significant digits.

A. The significant figure in a number **in positional notation** consists of

(i) All non-zero digits

(ii) Zero digits which

(a) lie between significant digits;

(b) lie to the right of decimal point and at the same time to the right of a non-zero digit;

(c) are specifically indicated to be significant.

B. The significant figure in a number written **in scientific notation** (e.g., M × 10k) consists of all the digits explicitly in M. A list is provided to help students understand how to calculate significant digits in a given number:

Number	Significant digits	Number of significant digits
3969	3, 9, 6, 9	04
3060	3, 0, 6	03
3900	3, 9	02
39.69	3, 9, 6, 9	04
0.3969	3, 9, 6, 9	04
39.00	3, 9, 0, 0	04
0.00039	3, 9	02
0.00390	3, 9, 0	03
3.0069	3, 0, 0, 6, 9	05
3.9×10^6	3, 9	02
3.909×10^5	3, 9, 0, 9	04
6×10^{-2}	6	01

3. Rounding-off

There are numbers with many digits, e.g.,22/7 =3.142857143 In practice, it is desirable to limit such numbers to a manageable number of digits, such as 3.14 or 3.143. This process of dropping unwanted digits is called rounding-off. Numbers are rounded-off according to the following rule: To round-off a number to n significant digits, discard all digits to the right of the nth digit and if this discarded number is

(i) less than 5 in $(n + 1)^{th}$ place, leave the nth digit unaltered. e.g., 7.893 to 7.89.

(ii) greater than 5 in $(n + 1)^{th}$ place, increase the nth digit by unity, e.g., 6.3456 to 6.346.

(iii) exactly 5 in $(n + 1)^{th}$ place, increase the nth digit by unity if it is odd, otherwise leave it unchanged.

e.g., 12.675 $\simeq$ 12.68 12.68512.68

The number thus rounded-off is said to be correct to n significant figures.

A list is provided for explanatory proposes:

Number	Rounded-off to		
	Three digits	Four digits	Five digits
00.543241	00.543	00.5432	00.54324
39.5255	39.5	39.52	39.526
69.4155	69.4	69.42	69.416
00.667676	00.668	00.6677	00.66768

3.3 REPRESENTATION OF FIXED AND FLOATING POINT NUMBERS

There are two types of arithmetic operations available in a computer. They are:

(i) Integer arithmetic

(ii) Real or floating point arithmetic.

Integer arithmetic deals with integer operands and is used mainly in counting and as subscripts. Real arithmetic uses numbers with fractional parts as operands and is used in most computations. Computers are usually designed such that each location, called **word,** in memory stores only a finite number of digits. Consequently, all operands in arithmetic operations have only a finite number of digits. Let us assume a hypothetical computer having memory in which each location can store 6 digits and having provision to store one or more signs. One method of representing real numbers in that computer would be to assume a fixed position for the decimal point and store all numbers after appropriate shifting if necessary with an assumed decimal point.

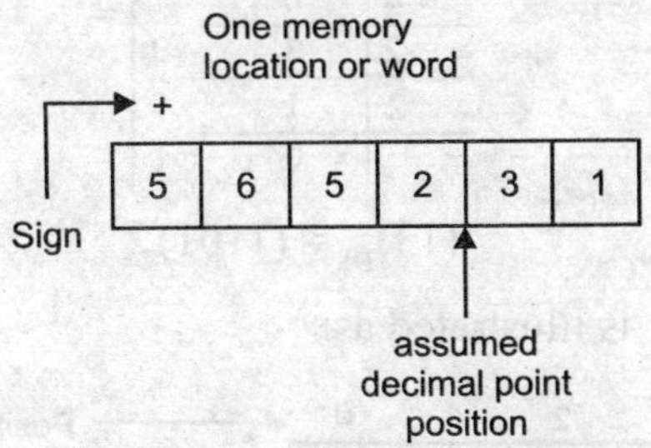

A memory location storing number 5652.31

In such a convention, the maximum and minimum possible numbers to be stored are 9999.99 and 0000.01, respectively, in magnitude. This range is quite inadequate in practice.

For this, a new convention is adopted that aims to preserve the maximum number of significant digits in a real number and also increase the range of values of real numbers stored. This representation is called the **normalized floating point** mode of representing and storing real numbers.

In this mode, a real number is expressed as a combination of a **mantissa** and an **exponent.** The mantissa is made less than 1 or ≥ 0.1 and the exponent is the power of 10 which multiplies the mantissa. For example, the number 43.76×10^6 is represented in this notation as .4376 E 8, where E 8 is used to represent 10^8. The mantissa is .4376 and the exponent is 8.

The number is stored in memory location as:

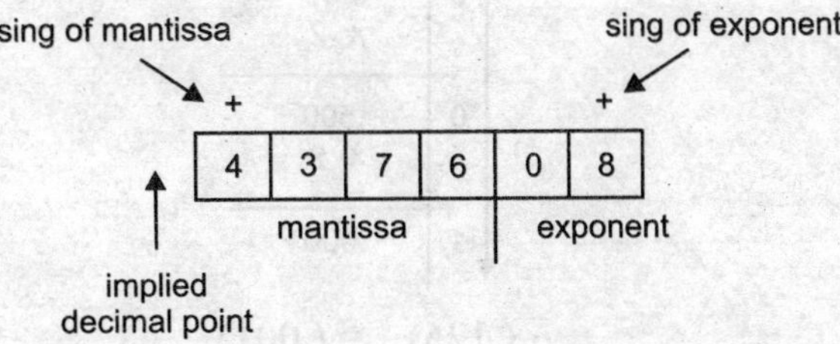

Moreover, the shifting of the mantissa to the left until its most significant digit is non-zero is called **normalization.** For example, the number .006831 may be stored as .6831 E–2 because the leading zeros serve only to locate the decimal point.

The range of numbers that may be stored is $.9999 \times 10^{99}$ to $.1000 \times 10^{-99}$ in magnitude, which is obviously much larger than that used earlier in fixed decimal point notation. This increment in range has been obtained by reducing the number of significant digits in a number by 2.

As our computer memory is organized as sequence of bytes so the way of storing a integer number is different from that of real number.

Before the storage of a decimal number (that has base value 10) into computer memory firstly this decimal number is converted into its equivalent binary form. As we have discussed that a decimal number exists into two forms: integer and real. So first we will discuss integer with the help of a simple example. Suppose we have a decimal number $(13)_{10}$ that have its equivalent binary form as:

2	13	
2	6	1
2	3	0
2	1	1
	0	1

$$(13)_{10} = (1101)_2$$

Now the binary number 1101 is illustrated as:

3	2	1	0	← Position
1	1	0	1	← Bit

$1\times2^3 + 1\times2^2 + 0\times2^1 + 1\times2^0 = 8+4+0+1 = (13)_{10}$

In this case **positional power** is taken to be positive.

Now we take case of a real number with an integer and fractional part. Both these parts are converted separately into its equivalent binary forms and then combined together. As we have solved the integer part before, now we just convert the fractional part into its equivalent binary form. Suppose we have a real number 13.125 that has fractional part .125 which will be converted into its equivalent binary form as:

	.125 x 2
0	.250 x 2
0	.500 x 2
1	.000

$$(.125)_{10} = (.001)_2$$

Now the question arises when to stop multiplication process the answer to this question is that when the fractional part in the multiplication comes out to be zero (shown in bolded form) we have to stop.

Now the binary number .001 is illustrated as:

	-1	-2	-3	← Position
·	0	0	1	← Bit

$$0\times2^{-1} + 0\times2^{-2} + 1\times2^{-3} = 0 + 0 + \frac{1}{8} = (.125)_{10}$$

In this case **positional power** is taken to be Negative.

Now after combining these binary conversions the decimal number 13.125 gives its equivalent binary form as 1101.001

3.3.1 Storage of Integer And Real Numbers

In this section we will discuss how integer and real numbers are stored in computer memory.

3.3.1.1 Storage of Integer Numbers

Before discussing storage of integer and real numbers we will discuss the concept of 1's and 2's complement.

1's Complement Representation

1's complement of a given binary number can be calculated by simply inverting 0's into 1's and 1's into 0's.

The 1's complement of eight bit binary number 10010101 will be 01101010. That is in 1's complement we simply have to change the 1's into 0's and 0's into 1's.

2's Complement Representation

2's complement of a binary number is calculated in two-step procedure as:

Step 1: Calculate 1's complement of given binary number.

Step 2: Add '1' to the right most bit (LSB) to the computed 1's complement number.

The 2's complement of binary number 10010101 will be computed as:

Step1: 1's complement of 10010101 is 01101010.

Step2: Add 1 to the right most bit (LSB) of 01101010, we get

```
01101010 (LSB)
       1
--------
01101011 (2's complement form)
```

Note: *If we have to convert a negative integer (for example -17) into its equivalent binary form then its better to take two's complement instead of one's complement of that number.*

As different computers have different formats for storing the integer numbers having different bit lengths (as 1-byte, 2-byte and 4-byte). But here we will discuss the format for storing 8 bit (1-byte) binary integer number which is as:

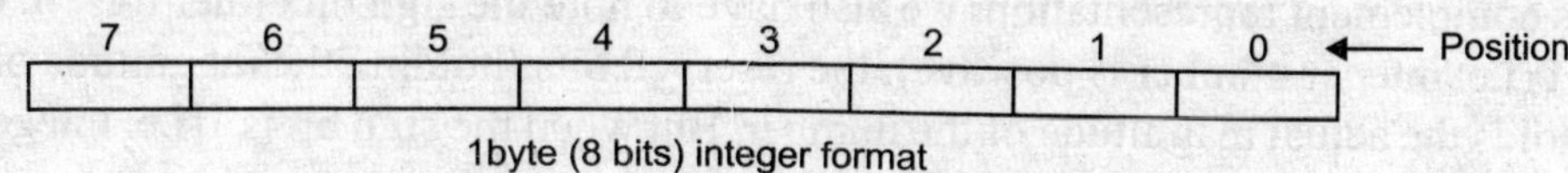

1byte (8 bits) integer format

Towards the left part (at position 7) the left most bit is stored that is used as sign bit and remaining bits are used to represent the magnitude of integer number to be stored. If the sign bit is 0 the number is assumed to be positive and if the sign bit is 1 the number is assumed to be negative.

Mainly we have three standards for representing integer numbers:

(i) Signed representation of integer

(ii) 1's complement representation of integer

(iii) 2's complement representation of integer

(i) Signed Representation

In signed representation of an integer number for one byte integer format (shown above) the 8th bit (MSB at position 7) is used for holding sign value (+ve or –ve) and remaining 7 bits are used for storing the binary value of integer number. For +ve number 8th bit takes binary value as 0 and for –ve number 8th bit takes binary value as 1. e.g. if we have to store two integer numbers 25 and –25 into this 8 bit integer format what we have to do, first we have to convert this number into its equivalent binary form.

In binary form the number 25 is written as 11001, which is a five bit number, for making it eight bit number we can attach three 0's to its left, which gives us 00011001. Now we have to place this number into memory as:

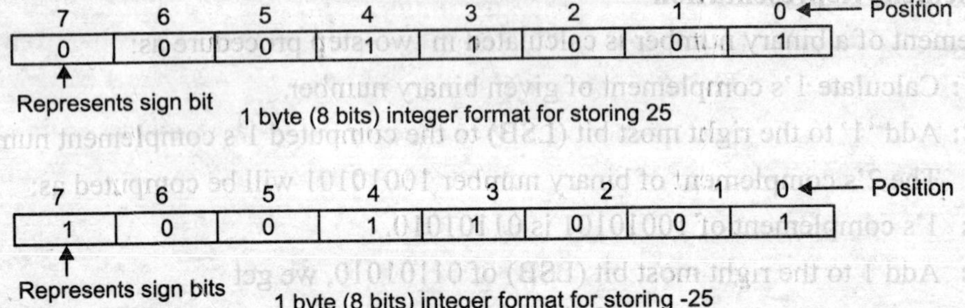

(ii) 1's Complement Representation of Integer

During 1's complement representation, we have to note the sign bit either it is '0' or '1'. If the sign bit is 0(i.e. integer number is positive), the reserved bits (holding the magnitude of the integer number) holds the actual magnitude of the number. But when the sign bit is 1(i.e. integer number is negative), the reserved bits (holding the magnitude of the integer number) holds the 1's complement of the magnitude of the number. E.g. the numbers 25 and -25 in 1-byte format will be represented as 00011001 and 11100110 respectively.

(iii) 2's Complement Representation of Integer

During 2's complement representation, we also have to note the sign bit either it is '0' or '1'. If the sign bit is 0(i.e. integer number is positive), the reserved bits (holding the magnitude of the integer number) holds the actual magnitude of the number. But when the sign bit is 1(i.e. integer number is negative), the reserved bits (holding the magnitude of the integer number) holds the 2's complement of the magnitude of the number. E.g. the numbers 25 and -25 in 1-byte format will be represented as 00011001 and 11100111.

Normally, we employ 2's complement representation in most of the computers to represent the integers.

3.3.1.2 Storage of Real Numbers

As we discussed real numbers are considered in two formats: fixed point and floating point. Fixed-point real numbers contains no exponent power whereas floating point numbers contains exponent powers. Due to the limitation of computer's memory to be small, very small and large numbers can't be represented in computer's memory. So to overcome this problem we represent the number, which is either too small or too large into exponential form by using a floating decimal point. This type of representation is known as **floating point representation**. Floating point representation can be easily understandable in the C/C++ language.

The process of shifting of the decimal (.) to the left of MSB (most significant bit) is called **normalization** or we can also say that a floating point number is said to be normalized if the most significant digit of the number is nonzero.

Now the question arises why we need normalization? The answer to this question is very simple and as.

The normalization is done to preserve the maximum number of useful (information carrying) digits. The numbers formed after normalization are known as normalized floating point numbers.

The normalized floating point number also has two parts: mantissa and exponent. The part before the decimal or fraction is known as mantissa and the part after the fraction is known as exponent.

A number is considered to be in normalized form if it satisfies the following conditions:

Conditions For Mantissa

For **+ve** numbers **0.1** $\leq$ mantissa < **1.0**

For **-ve** numbers **-1.0** < mantissa $\leq$ **–0.1**

Conditions For Exponent

If exponent increases +99 then it is the case of **overflow**.

If exponent decreases –99 then it is the case of **underflow**.

So we have to maintain the mantissa and exponent part within the above prescribed range.

If we have a number 0.000006364 which in normalized floating point form can be represented as 0.6364×10^{-5}. If our computer is of word length up to 6 digits and it also has the provision to store one or more signs, then we use first four digits four mantissa part and last two digits for exponent part. E.g. the number 0.6364e-5 is represented in memory as:

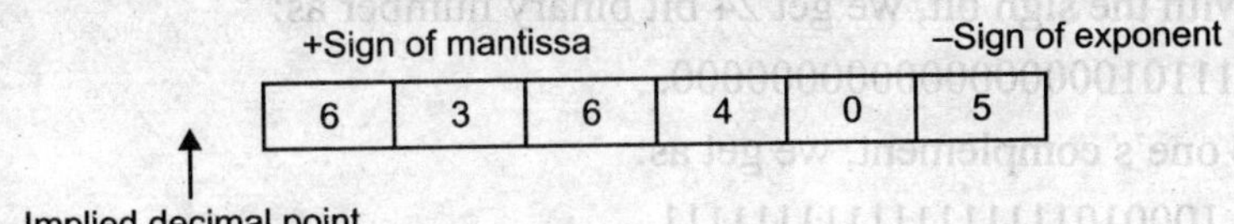

Similarly, a normalized floating point number having +ve exponent 0.5455×10^{15} written as 0.5455e15 or 0.5455e+15. The + sign is optional in case of +ve exponent power.

The real numbers are represented in computer memory as normalized binary floating point numbers, using four consecutive bytes(32 bits). The mantissa part is expressed in lower order three bytes out of these four bytes and exponent is represented in higher order fourth byte (also named as Most significant byte).

The standard format for storing floating point numbers is as:

→ Exponent part 8 bits	→ Mantissa part 24 bits ←	
Modified (or biased exponent)=8 bits	Mantissa sign=1 bit	Bits representing the mantissa=23 bits

3.3.1.3 IEEE 32 Bit-standard Format for Storing Floating Point Numbers

The sign of any floating-point number depends only on the sign of the mantissa not the exponent.

To illustrate the above discussion consider a number $(14.5)_{10}$ which can be represented in the above format as

$$(14.5)_{10} = (1110.1)_2$$

Now convert this binary number into normalized form as

$1110.1=.11101 \times 2^4$

Here mantissa = 11101

Which is five bit but our format requirement for mantissa is 23 bits to make it 23 bits add 23 – 5 = 18 zeroes to the right as it is not a fractional part.

So after adding zero's mantissa becomes 11101000000000000000000

Here true exponent = 4

Now modify this exponent by adding a constant to the true exponent which is determined in such a way that the minimum possible integer in 8 bits is raised to zero.

Thus in the above format the constant is 128. With the addition of 128, the true exponent (Which may vary between –128 to +127) is now modified to vary between 0 to 255. All these are positive values and therefore **this approach eliminates the use of sign bit.**

So modified exponent = 4 + 128 =132 = 10000100(in binary form)

As the number is +ve sign bit is taken to be 0 so finally we get the representation as:

10000100	0	11101000000000000000000

If the normalized floating point number is –ve real number we will follow the following procedure:

Combining with the sign bit, we get 24 bit binary number as:

011101000000000000000000.

By taking its one's complement, we get as:

100010111111111111111111

However the exponent remains the same and the representation of –14.5(taken in two's complement form) takes the form as:

10000100	1	11101000000000000000000

3.3.2 Possible Range of Values for Real Numbers

If we use the above format, the largest number that can be represented will have 0 as sign bit followed by all 1's for mantissa and 11111111 as the biased exponent, i.e. , 127 as the actual exponent. To determine the order of magnitude of largest numbers, in decimal number system, we have to solve following equation for x.

$$2^{127} = 10^x$$

By taking log on both sides, we have

$$127 \times \log 2 = x \times \log 10$$

$$x = 127 \times 0.301029 = 38.2 \cong 38$$

Therefore, the maximum range for representing real number (either positive or negative) is of the order 10^{+38}.

Similarly, the minimum positive number occurs when mantissa is positive, and the biased exponent is 00000000 i.e. -128 as actual exponent. So

$$2^{-128} = 10^x$$

By taking log on both sides, we have

$$-128 \times \log 2 = x \times \log 10$$

$$x = -128 \times 0.301029 = -38.49 \cong -38$$

Therefore, the minimum range for representing real number (either positive or negative) is of the order 10^{-38}.

Thus, real numbers will be of the order $10^{\pm 38}$.

3.3.3 Accuracy of Value

As we know lower 23 bits (in case if 4 byte standard format) are used for representing the magnitude of the mantissa. Now, for a given decimal real number with certain number of digits, we have to check the accuracy of the value during its representation in binary form. So, for this, we use the following equation

$$2^{23} = 10^x$$

By taking log on both sides, we have

$$23 \times \log 2 = x \times \log 10$$

$$x = 6.9 \cong 7$$

Thus, with 23 bits for mantissa, we can be certain of maximum accuracy of about 7 digits in the binary representation of a real number. It may happen that certain real numbers of six or even fewer digits such as 0.4 may not be accurately represented due to the problem of conversion.

SOLVED PROBLEMS

Problem 1: Represents IEEE 32 bit format of the number 13.125 in normalized floating point.

Solution: We will do this representation step by step as:

Step 1: Convert the given number into its equivalent binary form i.e. the binary equivalent of 13.125 is 1101.001

Step 2: Calculate the normalized floating point number of calculated binary number i.e. the normalized floating point number of 1101.001 is .1101001 × 2^4 (as before decimal there are 4 digits)

Step 3: Extend the mantissa to 23 bits by adding 0's to the right i.e. mantissa becomes 11010010000000000000000

Step 4: Add 128 to the biased exponent (4 in our case). So modified exponent becomes 132 & its binary equivalent 10000100.

Step 5: Check sign bit of given number. If it is positive take 0 as the sign bit otherwise take 1. As here given number (13.125) is positive so sign bit will be 0.

Step 6: Combine the result calculated in above steps to get a 32 bit format. So final representation of 13.125 will be 10000100011010010000000000000000. Underlined 0 represents sign bit.

Problem 2: Represents IEEE 32 bit format of the number -13.125 in normalized floating point.

Solution: We will do this representation step by step as:

Step 1: Convert the given number into its equivalent binary form i.e. the binary equivalent of 13.125 is 1101.001

Step 2: Calculate the normalized floating point number of calculated binary number i.e. the normalized floating point number of 1101.001 is .1101001 × 2^4 (as before decimal there are 4 digits)

Step 3: Extend the mantissa to 23 bits by adding 0's to the right i.e. mantissa becomes 11010010000000000000000

Step 4: Take 2's complement of above 23-bit pattern, we get 00101110000000000000000. Add 128 to the biased exponent (4 in our case). So modified exponent becomes 132 & its binary equivalent 10000100.

Step 5: Check sign bit of given number. If it is positive take 0 as the sign bit otherwise take 1. As here given number(-13.125) is negative so sign bit will be 1.

Step 6: Combine the result calculated in above steps to get a 32 bit format. So final representation of 13.125 becomes 10000100100101110000000000000000. Underlined 1 represents sign bit.

3.4 ARITHMETIC OF NORMALIZED FLOATING POINT REPRESENTATION

There are basically four types of arithmetic operations that are performed on normalized floating-point numbers. These are discussed in detail as below:

1. Addition
2. Subtraction
3. Multiplication
4. Division

Note: ***During addition, multiplication, subtraction and division resultant's mantissa part should be truncated(discussed in coming text) i.e. if we add two normalized numbers and both numbers have mantissa part having length k then the resultant mantissa should also be of length k. When decimal point (.) shifts towards left, exponent power increases, and vice-versa. We will clarify the concept by using examples.***

3.4.1 Addition (+)

If two numbers represented in normalized floating point notation are to be added, the exponents of the two numbers must be made equal and the Mantissa shifted appropriately. Before adding two normalized floating point numbers we have to follow some of the rules are as follows:

(i) Exponent powers of both normalized floating point numbers (addend and augend) should be the same. In case, these differs make the smallest exponent power equals to the largest by shifting the decimal positions to the left as much required.

(ii) After adding both the numbers, the resultant number's mantissa and exponent parts have to be checked, if mantissa part goes out of its range(as range discuss earlier) change it and set it within the prescribed range. During this setup, exponent values may also change in parallel and may exceed +99 then it becomes the case of overflow because computer stores only two digits for exponent. So the maximum two digits are accommodated for exponent.

3.4.1.1 Algorithm for Adding Two Normalized Floating Point Numbers

Let m_1, m_2 be the mantissas & e_1, e_2 be the corresponding exponents of two normalized floating point numbers. Let m, e be the mantissa and exponent of the resultant floating point number. Let d be any variable, denotes the difference between the exponent powers. Then the algorithm for addition is as follows:

1. Read m_1, e_1, m_2, e_2
2. Set $d = |e_1 - e_2|$
3. if $(e_1 > e_2)$ then
4. set $m_2 = m_2 / 10^d$

5. and set $e = e_1$
6. else
7. set $m_1 = m_1 / 10^d$
8. and set $e = e_2$
9. end if
10. set $m = m_1 + m_2$
11. if ($m \geq 1.0$) then
12. set m = m/10
13. set e = e + 1
14. end if
15. if (e > 99) then
16. Write : "Addition Overflows".
17. else
18. Write : m, e as "Resultant mantissa and resultant exponent"
19. end if
20. End

3.4.1.2 Program For Adding Two Normalized Floating Point Numbers in C

```
#include<stdio.h>
#include<math.h>
#include<conio.h>
void main()
{
int e1,e2,d,e;
float m1,m2,m;
clrscr();
printf("Enter the first(addend) floating point number:\n");
scanf("%f%d",&m1,&e1);
printf("Enter the second(addend) floating point number:\n");
scanf("%f%d",&m2,&e2);
d=fabs(e1-e2);
if(e1>e2)
{
m2=m2/pow10(d); // pow10(d) will Compute 10**d //
e=e1;
```

```
}
else
m1=m1/pow10(d);
e=e2;
m=m1+m2;
if(m>=1.0)
{
m=m/(float)10.0;
e=e+1;
}
if(e>99)
{
printf("Overflow occurs\n");
}
else
printf("Resultant m=%.4f  Resultant e =%d ",m ,e );
getch();
}
```

Output:

Enter the first (addend) floating point number:

0.6219 9

Enter the second (addend) floating point number:

0.7123 9

Resultant m=0.1334 Resultant e=10

3.4.1.3 Program for Adding Two Normalized Floating Point Numbers in C++

```
#include<stdio.h>
#include<math.h>
#include<iostream.h>
#include<conio.h>
#include<float.h>
#include<iomanip.h>
void main()
{
int e1,e2,d,e;
```

```
float m1,m2,m;
clrscr();
cout<<"Enter the first(addend) floating point number:\n";
cin>>m1>>e1;
cout<<"Enter the second(addend) floating point number:\n";
cin>>m2>>e2;;
d=fabs(e1-e2);
if(e1>e2)
{
m2=m2/pow10(d); // pow10(d) will Compute 10**d //
e=e1;
}
else
m1=m1/pow10(d);
e=e2;
m=m1+m2;
if(m>=1.0)
{
m=m/(float)10.0;
e=e+1;
}
if(e>99)
{
cout<<"Overflow occurs\n";
}
else
cout<<"Resultant m="<<setprecision(4)<<m<<"Resultant e ="<<e<<"\n" ;
getch();
}
```

Output:

Enter the first (addend) floating point number:

0.6219 9

Enter the second (addend) floating point number:

0.7123 9

Resultant m=0.1334 Resultant e=10

SOLVED PROBLEMS

Problem 1: Add 0.3459e4 and 0.7212e6.

Solution: In this example, exponent powers are different, so make e4 equals to e6 by shifting decimal two positions left. After shifting two positions, number becomes 0.0034e6. Since exponent powers of both the numbers are now equal so add these as

	0.0034 e6
	0.7212 e6
Sum	0.7246 e6

Here mantissa of sum is already within the range so no need to change here.

Problem 2: Add 0.6219e9 and 0.7123e9.

Solution: In this example, exponent powers are same so they can be added directly as:

	0.6219 e9
	0.7123 e9
Sum	1.3342 e9

Here mantissa of the sum is greater than 1 to set it within the prescribed range shift the decimal point to the left of 1 which increases the exponent by 1. So the result becomes 0.1334e10, which is within the range.

Problem 3: Add 0.9215e99 and 0.8134e99.

Solution: In this example, exponent powers are same so they can be added directly as:

	0.9215 e99
	0.8134 e99
Sum	1.7349 e99

Here mantissa of the sum is greater than 1 to set it within the prescribed range (**0.1 £ mantissa <1.0**) shift the decimal point to the left of 1 which increases the exponent power by 1. So the result becomes 0.1734e100, which also exceeds exponent power from +99. So, it becomes the case of **overflow.**

Problem 4: Add the following floating point numbers:

(i) .4546 E 5 and .5433 E 5

(ii) .4546 E 5 and .5433 E 7

(iii) .4546 E 3 and .5433 E 7

(iv) .6434 E 3 and .4845 E 3

(v) .6434 E 99 and .4845 E 99.

Solution:

(i) Here the exponents are equal. Mantissas are added Sum = .9979 E 5

(ii) Here exponents are not equal. The operand with the larger exponent is kept as it is

```
  .5433 E 7
+ .0045 E 7        | .4546 E 5 = .0045 E 7 |
-----------
  .5478 E 7
```

(iii) The addition will be as follows:

```
  .5433 E 7
+ .0000 E 7        | .4546 E 3 = .0000 E 7 |
-----------
  .5433 E 7
```

(iv) The exponents are equal but when the mantissas are added, the sum is 1.1279 E 3. As the mantissa has 5 digits and is > 1, it is shifted right one place before it is stored. Hence Sum = .1127 E 4

(v) Here, again the sum of the mantissas exceeds 1. The mantissa is shifted right and the exponent increased by 1, resulting in a value of 100 for the exponent. The exponent part cannot store more than two digits. This condition is called an **overflow** condition and the arithmetic unit will intimate an error condition.

3.4.2 Subtraction (-)

The operation of subtraction is nothing but the addition of a negative number. Thus the principles are the same. Before subtracting two normalized floating point numbers we also have to follow some rules:

(i) Exponent powers of both normalized floating-point numbers (Minuend and subtrahend) should be same. In case, these differs make the smallest exponent power equals to the largest by shifting the decimal positions to the left as much required.

(ii) After subtracting both the numbers, the resultant number's mantissa and exponent parts are to be checked, if mantissa part goes out of its range(as range discussed earlier) change it, and set it within the prescribed range. During this setup, exponent values may also change in parallel, if exponent reduces –99, it becomes the case of **underflow**.

3.4.2.1 Algorithm for Subtracting Two Normalized Floating Point Numbers

Let m_1, m_2 be the mantissas & e_1, e_2 be the corresponding exponents of two normalized floating point numbers. Let m, e be the mantissa and exponent of the resulting floating point number. Let d be any variable denotes the difference between the exponent powers. Then the algorithm for subtraction is as follows:

1. Read m_1, e_1, m_2, e_2
2. set $d = | e_1 - e_2 |$
3. if $(e_1 > e_2)$ then
4. set $m_2 = m_2 / 10^d$
5. and set $e = e_1$

6. else
7. set $m_1 = m_1 / 10^d$
8. and set $e = e_2$
9. end if
10. set $m = m_1 - m_2$
11. while (| m |< 0.1 and | m| > 1.0) do
12. set m = m*10
13. set e = e – 1
14. end while
15. if (e < -99) then
16. Write : " Underflow".
17. else
18. Write : m, e as "Resultant mantissa and resultant exponent"
19. end if
20. End

3.4.2.2 Program for Subtracting Two Normalized Floating Point Numbers in C

```
#include<stdio.h>
#include<math.h>
#include<conio.h>
void main()
{
int e1,e2,d,e;
float m1,m2,m;
clrscr();
printf("Enter the first(Minuend) floating point number:\n");
scanf("%f%d",&m1,&e1);
printf("Enter the second(Subtrahend) floating point number:\n");
scanf("%f%d",&m2,&e2);
d=fabs(e1-e2);
if(e1>e2)
{
m2=m2/pow10(d); // Compute 10**p //
e=e1;
}
```

```
else
m1=m1/pow10(d);
e=e2;
m=m1-m2;
do
{
m=m*10;
e=e-1;
}
while(abs(m<(0.1))&&abs(m>(0.0)));
if(e<-99)
{
printf("Underflow occurs\n");
}
else
printf("Resultant m=%.4f \n Resultant e = %d",m,e);
getch();
}
```

Output:

Enter the first (Minuend) floating point number:

0.9235 -99

Enter the second (addend) floating point number:

0.9213 -99

Underflow occurs

3.4.2.3 Program for Subtracting Two Normalized Floating Point Numbers in C++

```
#include<stdio.h>
#include<math.h>
#include<iostream.h>
#include<conio.h>
void main()
{
int e1,e2,d,e;
float m1,m2,m;
```

```
clrscr();
cout<<"Enter the first(Minuend) floating point number:\n";
cin>>m1>>e1;
cout<<"Enter the second(Subtrahend) floating point number:\n";
cin>>m2>>e2;;
d=fabs(e1-e2);
if(e1>e2)
{
m2=m2/pow10(d);  // Compute 10**p //
e=e1;
}
else
m1=m1/pow10(d);
e=e2;
m=m1-m2;
do
{
m=m*10;
e=e-1;
}
while(abs(m<(0.1))&&abs(m>(0.0)));
if(e<-99)
{
cout<<"Underflow occurs\n";
}
else
cout<<"Resultant m="<<setprecision(4)<<m<<"Resultant e = "<<e<<"\n" ;
getch();
}
```

Output:

Enter the first (Minuend) floating point number:

0.9235 -99

Enter the second (addend) floating point number:

0.9213 -99

Underflow occurs

SOLVED PROBLEMS

Problem 1: Subtract 0.3459e4 from 0.7212e6.

Solution: In this example, exponent powers differ so make e4 equals to e6 by shifting decimal two positions left after shifting number becomes .0034e6. Since exponent powers of both numbers are now equal so subtract as:

```
0.7212 e6
0.0034 e6
---------
0.7178 e6
```

Difference is within the range.

Problem 2: Subtract 0.6219e9 from 0.7123e9

Solution: In this example, exponent powers are same so they can be subtracted as:

```
0.7123 e9
0.6219 e9
---------
0.0904 e9
```

Here mantissa of the difference is less than 0.1 to set it within the prescribed range shift the decimal point to the right, which decreases the exponent by 1. So the result becomes 0.9040e8, which is within the range.

Problem 3: Subtract 0.9235e-99 from 0.9213e-99

Solution: In this example, exponent powers are same so they can be subtracted as:

```
0.9235 e-99
0.9213 e-99
-----------
0.0022 e-99
```

Here the mantissa of the difference is less than 0.1 to set it within the prescribed range shift the decimal point to two right positions right, which decreases the exponent by 2. So the result becomes 0.2200e-101, which shows the case of **underflow** as the exponent decreases from –99.

Problem 4: Subtract the following floating point numbers:

(i) 9432 E – 4 from .5452 E – 3

(ii) 5424 E 3 from .5452 E 3

(iii) 5424E – 99 from .5452 E – 99.

Solution:

(i)

```
  .5452 E – 3
– .0943 E – 3
-------------
  .4509 E – 3
```

(ii)

$$\begin{array}{r} .5452\ E\ 3 \\ -\ .5424\ E\ 3 \\ \hline .0028\ E\ 3 \end{array}$$

In a normalized floating point, the mantissa is ≥ 0.1

Hence, the result is .28 E 1

(iii)

$$\begin{array}{r} .5452\ E-99 \\ -\ .5424\ E-99 \\ \hline .0028\ E-99 \end{array}$$

For normalization, the mantissa is shifted left and the exponent is reduced by 1. The exponent would thus become – 100 with the first left shift, which can not be accommodated in the exponent part of the number. This condition is called an **underflow** condition and the arithmetic unit will signal an error condition

Problem 5: In normalized floating point mode, carry out the following mathematical operations:

(i) (.4546 E 3) + (.5454 E 8)

(ii) (.9432 E – 4) – (.6353 E – 5).

Solution:

(i)

$$\begin{array}{r} .5454\ E\ 8 \\ +\ .0000\ E\ 8 \\ \hline .5454\ E\ 8 \\ \hline \end{array}$$

| .4546 E 3 = .0000 E 8 |

(ii)

$$\begin{array}{r} .9432\ E-4 \\ -\ .0635\ E-4 \\ \hline .8797\ E-4 \\ \hline \end{array}$$

| .6353 E – 5 = .0635 E – 4 |

3.4.3 Multiplication (*)

Two numbers are multiplied in the normalized floating point mode by multiplying the mantissas and adding the exponents. After the multiplication of the mantissas, the resulting mantissa is normalized as in an addition or subtraction operation, and the exponent is appropriately adjusted.

Multiplication of two normalized floating point numbers is simple because during multiplication there is no restriction on the exponent powers i.e. they are not need to be same. In multiplication, mantissa parts are multiplied whereas exponent parts are added.

3.4.3.1 Algorithm for Multiplying Two Floating Point Numbers

Let m_1, m_2 be the mantissas & e_1, e_2 be the corresponding exponents of two normalized floating point numbers. Let m, e be the mantissa and exponent of the resulting floating point number. Then the algorithm for multiplication is as follows:

1. Read m_1, e_1, m_2, e_2
2. set m = m_1 * m_2
3. and set e = e_1 + e_2
4. while ((|m|< 0.1) or (|m|> 1.0))
5. if (|m|< 0.1) then
6. set m = m * 10.0
7. set e = e –1
8. else
9. set m = m/10.0
10. set e = e + 1
11. end if
12. end while
13. if (e > 99) then
14. write : "Overflow occurs"
15. Elseif (e < -99) then
16. write : "Underflow occurs"
17. End if
18. write : m, e as "Resultant mantissa and resultant exponent"
19. End

3.4.3.2 *Program for Multiplying Two Floating Point Numbers in C*

```
#include<stdio.h>
#include<math.h>
#include<conio.h>
#include<iomanip.h>
void main()
{
int e1,e2,d,e;
float m1,m2,m;
clrscr();
printf("Enter the first(Multiplier) floating point number\n");
scanf("%f%d",&m1,&e1);
printf("Enter the second(Multiplicand) floating point number\n");
scanf("%f%d",&m2,&e2);
```

```
m=m1*m2;
e=e1+e2;
if(m>=1.0)
{
m=m/10;
e=e+1;
}
if(m<0.1)
{
m=m*10;
e=e-1;
}
if(e>99)
{
printf("Overflow occurs\n");
}
else
printf("Resultant m=%.4f \n Resultant e = %d",m,e);
getch();
}
```

Output:

Enter the first (Multiplier) floating point number:

0.1985 20

Enter the second (Multiplicand) floating point number:

0.1150 23

Resultant m=0.2283 Resultant e=42

3.4.3.3 Program for Multiplying Two Floating Point Numbers in C++

```
#include<stdio.h>
#include<math.h>
#include<iostream.h>
#include<conio.h>
#include<iomanip.h>
void main()
{
```

```
int e1,e2,d,e;
float m1,m2,m;
clrscr();
cout<<"Enter the first(Multiplier) floating point number\n";
cin>>m1>>e1;
cout<<"Enter the second(Multiplicand) floating point number\n";
cin>>m2>>e2;
m=m1*m2;
e=e1+e2;
if(m>=1.0)
{
m=m/10;
e=e+1;
}
if(m<0.1)
{
m=m*10;
e=e-1;
}
if(e>99)
{
cout<<"Overflow occurs\n";
}
else
cout<<"Resultant m="<<setprecision(4)<<m<<"Resultant e="<<e<<"\n" ;
getch();
}
```

Output:

Enter the first (Multiplier) floating point number:

0.1985 20

Enter the second (Multiplicand) floating point number:

0.1150 23

Resultant m=0.2283 Resultant e=42

SOLVED PROBLEMS

Problem 1: Multiply 0.3492e7 by 0.6493e8.

Solution: 0.3492e7 × 0.6493e8 = (0.3492 × 0.6493) e(7+8)
= 0.22673556e15
= 0.2267e15 (truncated value)

Problem 2: Multiply 0.1985e20 by 0.1150e23.

Solution: 0.1985e20 × 0.1150e23 = (0.1985 × 0.1150) e(20+23)
= 0.0228275e43
= 0.0228e43 (truncated value)

Here the resultant mantissa is less than 0.1 so to set it within the prescribed range shift the decimal to one position right and decrease the exponent by 1. So final result is 0.2280e42.

Problem 3: Multiply 0.3496e70 by 0.6423e33.

Solution: 0.3496e70 × 0.6423e33 = (0.3496 × 0.6423) e(70+33)
= 0.22454808e103

Here the exponent part increases the limit +99 so in this case **overflow** occurs.

Problem 4: Multiply the following floating point numbers:

(i) .5543 E 12 and .4111 E – 15
(ii) .1111 E 10 and .1234 E 15
(iii) .1111 E 51 and .4444 E 50
(iv) .1234 E – 49 and .1111 E – 54.

Solution:

(i) .5543 E 12 × .4111 E – 15 = .2278 E – 3
(ii) .1111 E 10 × .1234 E 15 = .1370 E 24
(iii) .1111 E 51 × .4444 E 50 = .4937 E 100
The result overflows.
(iv) .1234 E – 49 × .1111 E – 54 = .1370 E – 104
The result underflows.

Problem 5: Apply the procedure for the following multiplications:

(i) $(.5334 \times 10^{9}) * (.1132 \times 10^{-25})$
(ii) $(.1111 \times 10^{74}) * (.2000 \times 10^{80})$

Indicate if the result is overflow or underflow.

Solution:

(i) .5334 E 9 × .1132 E – 25 = .6038 E – 17
(ii) .1111 E 74 × .2000 E 80 = .2222 E 153

Hence the above result overflows.

3.4.4 Division (÷)

Division of two normalized floating point numbers is also simple like multiplication. In division, the mantissa of the numerator is divided by that of the denominator. The denominator exponent is subtracted from the numerator exponent. The quotient mantissa is normalized to make the most significant digit non-zero and the exponent is appropriately adjusted. The mantissa of the result is chopped down to 4 digits.

During division, there is also no such restriction on the exponent powers i.e. they are not need to be same. In division, mantissa parts are divided whereas exponent parts are subtracted.

3.4.4.1 Algorithm for Division of Two Floating Point Numbers

Let m_1, m_2 be the mantissas & e_1, e_2 be the corresponding exponents of two normalized floating point numbers. Let m, e be the mantissa and exponent of the resulting floating point number. Then the algorithm for division is as follows:

1. Read m_1, e_1, m_2, e_2
2. $m = m_1 / m_2$
3. $e = e_1 - e_2$
4. while ((|m|< 0.1) or (|m|> 1.0))
5. if (|m|< 0.1) then
6. m = m * 10.0
7. e = e –1
8. else
9. m = m/10.0
10. e = e + 1
11. end if
12. end while
13. if (e > 99) then
14. write : "Overflow occurs"
15. Elseif (e < -99) then
16. write : "Underflow occurs"
17. end if
18. write : m, e as "Resultant mantissa and resultant exponent"
19. End

3.4.4.2 Program for Dividing Two Normalized Floating Point Numbers in C

```
#include<stdio.h>
#include<math.h>
```

```
#include<conio.h>
#include<iomanip.h>
void main()
{
int e1,e2,d,e;
float m1,m2,m;
clrscr();
printf("Enter the first(Dividend) floating point number:\n");
scanf("%f%d",&m1,&e1);
printf("Enter the second(Divisor) floating point number:\n");
scanf("%f%d",&m2,&e2);
m=m1/m2;
e=e1-e2;
if(m>=1.0)
{
m=m/10;
e=e+1;
}
if(m<0.1)
{
m=m*10;
e=e-1;
}
if(e>99)
{
printf("Overflow occurs\n");
}
else if(e<-99)
printf("Underflow occurs\n");
else
printf("Resultant m=%.4f \n Resultant e = %d",m,e);
getch();
}
```

Output:

Enter the first (Dividend) floating point number:

0.6434 10

Enter the second (Divisor) floating point number:

0.3212 6

Resultant m=0.2003 Resultant e=5

3.4.4.3 Program for Dividing Two Normalized Floating Point Numbers in C++

```
#include<math.h>
#include<iostream.h>
#include<conio.h>
#include<iomanip.h>
void main()
{
int e1,e2,d,e;
float m1,m2,m;
clrscr();
cout<<"Enter the first(Dividend) floating point number:\n";
cin>>m1>>e1;
cout<<"Enter the second(Divisor) floating point number:\n";
cin>>m2>>e2;
m=m1/m2;
e=e1-e2;
if(m>=1.0)
{
m=m/10;
e=e+1;
}
if(m<0.1)
{
m=m*10;
e=e-1;
}
if(e>99)
```

```
{
cout<<"Overflow occurs\n";
}
else if(e<-99)
cout<<"Underflow occurs\n";
else
cout<<"Resultant m="<<setprecision(4)<<m<<"Resultant e="<<e<<"\n" ;
getch();
}
```

Output:

Enter the first (Dividend) floating point number:
0.6434 10
Enter the second (Divisor) floating point number:
0.3212 6
Resultant m=0.2003 Resultant e=5

SOLVED PROBLEMS

Problem 1: Divide 0.6434e10 by 0.3212e6.

Solution: 0.6434e10 ÷ 0.3212e6 = (0.6434 ÷ 0.3212) e(10-6)
= 2.0031133e4

Here mantissa part exceeds the limit so after setting it within the range result becomes 0.2003e5 i.e. decimal shifts one position left and so exponent increases by 1.

Problem 2: Divide 0.7598e5 by 0.2139e-97.

Solution: 0.7598e5 ÷ 0.2139e-97 = (0.7598 ÷ 0.2139) e(5-(-97))
= 3.552127e102

Here mantissa part exceeds the limit so after setting it within the range result becomes 0.3552e103 i.e. decimal shifts one position left and so exponent increases by 1. The result also shows the case of **overflow.**

Problem 3: Divide 0.7598e-5 by 0.2139e97.

Solution: 0.7598e-5 ÷ 0.2139e97 = (0.7598 ÷ 0.2139) e(–5 – 97)
= 3.552127e – 102

Here mantissa part exceeds the limit so after setting it within the range result becomes 0.3552e-101 i.e. decimal shifts one position left and so exponent increases by 1. The result also shows the case of **underflow.**

Problem 4: Perform the following operations:

(i) .9998 E 1 ÷ .1000 E – 99

(ii) .9998 E – 5 ÷ .1000 E 98

(iii) .1000 E 5 ÷ .9999 E 3.

Solution:

(i) .9998 E 1 ÷ .1000 E – 99 = .9998 E 101

Hence the result overflows.

(ii) .9998 E – 5 ÷ .1000 E 98 = .9998 E – 104

Hence the result underflows.

(iii) .1000 E 5 ÷ .9999 E 3 = .1000 E 2.

Problem 5: Evaluate, applying normalized floating point arithmetic, for the following:

1 – cos x at x = .1396 radian

Assume cos (.1396) = .9903

Compare it when evaluated 2 sin^2 x/2

Assume sin .0698 = .6974 E – 1.

Solution: 1 – cos (.1396) = .1000 E 1 – .9903 E 0

= .1000 E 1 – .0990 E 1 = .1000 E – 1

Now, sin x/2 = sin (.0698) = .6974 E – 1

2 sin^2 x/2 = (.2000 E 1) × (.6974 E – 1) × (.6974 E – 1)

= .9727 E – 2

The value obtained by the alternate formula is closer to the true value .9728 E – 2.

Problem 6: For x = .4845 and y = .4800, calculate the value of $\frac{x^2 - y^2}{x + y}$ using normalized floating point arithmetic. Compare with the value of (x – y). Indicate the error in the former.

Solution: x + y = .4845 E 0 + .4800 E 0 = .9645 E 0

x^2 = (.4845 E 0) × (.4845 E 0) = .2347 E 0

y^2 = (.4800 E 0) × (.4800 E 0) = .2304 E 0

$x^2 - y^2$ = .2347 E 0 – .2304 E 0 = .0043 E 0

Now, $\frac{x^2 - y^2}{x + y}$ = .0043 E 0 ÷ .9645 E 0 = .4458 E – 2

Also, x – y = .4845 E 0 – .4800 E 0 = .0045 E 0 = .4500 E – 2

$$\text{Relative error} = \frac{.4500 - 0.4458}{.4500} = .93\%.$$

Problem 7: For e = 2.7183, calculate the value of e^x when x = .5250 E 1. The expression for e^x is

$$e^x = 1 + x + \frac{x^2}{2!} + \frac{x^3}{3!}$$

Solution: $e^{.5250}$ E 1 = $e^5 * e^{.25}$

e^5 = (.2718 E 1) × (.2718 E 1) × (.2718 E 1) × (.2718 E 1) × (.2718 E 1)

= .1484 E 3

Also, $$e^{.25} = 1 + (.25) + \frac{(.25)^2}{2!} + \frac{(.25)^3}{3!}$$

= 1.25 + .03125 + .002604 = .1284 E 1

Now, $e^{.5250}$ E 1 = (.1484 E 3) × (.1284 E 1) = .1905 E 3.

Problem 8: Find the solution of the following equation using floating point arithmetic with a 4 digit mantissa

$x^2 - 1000x + 25 = 0$

Give comments or the result so obtained.

Solution: $x^2 - 1000\,x + 25 = 0$

$$x = \frac{1000 + \sqrt{10^6 - 10^2}}{2}$$

Now, 10^6 = .1000 E 7 and 10^2 = .1000 E 3

$10^6 - 10^2$ = .1000 E 7

$\sqrt{10^6 - 10^2}$ = .1000 E 4

Roots are $\frac{(.1000 \text{ E } 4 + .1000 \text{ E } 4)}{2}$ and $\frac{(.1000 \text{ E } 4 - .1000 \text{ E } 4)}{2}$

which are .1000 E 4 and .0000 E 4 respectively. One of the roots becomes zero due to the limited precision allowed in calculation. Let us reformulate the problem and remember that in a quadratic equation $ax^2 + bx + c = 0$, the product of roots is given by c/a, so the smaller root may be obtained by dividing (c/a) by the larger root.

So, First root = .1000 E 4

and $$\text{Second root} = \frac{25}{.1000 \text{ E } 4}$$

$$= \frac{.2500 \text{ E2}}{.1000 \text{ E4}} = .250 \text{ E} - 1$$

Such a situation may be recognized in an algorithm by checking to see if $b^2 >> |\, 4\, ac\, |$.

3.5 CONSEQUENCES DUE TO FIXED OR NORMALIZED FLOATING POINT REPRESENTATION

3.5.1 Non-associativity Of Arithmetic

When the arithmetic operation is performed with the floating point numbers 0.6667 added 6 times gives 0.3997e1 whereas 0.6667 × 6 gives 0.4000e1. This happens because of truncation in the intermediate and final results in order to fit into the normalized floating point format. In other words, the equation 6x = x+x+x+x+x+x is not true!

0.6000e1 0.6667 = 0.6667 + 0.6667 + 0.6667 + 0.6667+ 0.6667 + 0.6667

So, we conclude that when we use floating point format the value pitfalls.

Another consequence of the floating point representation is that the associative and the distributive laws of arithmetic may not hold true. In other words,

$$(a+b)-c \neq (a-c)+b$$

$$a(b-c) \neq (ab-ac)$$

These are illustrated with examples below.

SOLVED PROBLEMS

Problem 1:

Let a = 0.5665e1, b = 0.5556e-1, c = 0.5644e1

(a+b) = 0.5665e1 + 0.5556e-1

= 0.5665e1 + 0.0055e1

= 0.5720e1

(a+b)–c = 0.5720e1 – 0.5644e1

= 0.0076e1

= 0.7600e-1

(a–c) = 0.5665e1 – 0.5644e1

= 0.0021e1

= .2100e-1

(a–c)+b = 0.2100e-1 + 0.5556e-1

= 0.7656e-1

So, we conclude that (a+b) –c ≠ (a–c) +b

Problem 2: Let a = 0.5555e1, b = 0.4545e1, c = 0.4535e1

(b–c) = 0.0010e1 = .1000e-1

$$a(b-c) = 0.5555e1 \times 0.1000e\text{-}1$$
$$= 0.0555e0 = 0.5550e\text{-}1$$
$$ab = 0.5555e1 \times 0.4545e1 = 0.2524e2$$
$$ac = 0.5555e1 \times 0.4535e1 = 0.2519e2$$
$$ab\text{-}ac = 0.0005e2 = 0.5000e\text{-}1$$

Thus a(b-c) ≠ ab-ac

3.5.2 The Concept of Zero in Floating Point Arithmetic

The number zero has a precise meaning in arithmetic. To fix our idea we will consider an example.

SOLVED PROBLEMS

Problem 1: The roots of quadratic equation $x^2 + 2x - 2 = 0$ are $-1 \pm \sqrt{3}$. The roots in floating point with 4-digit mantissa are 0.7320e0 and –0.2732e1. If we substitute 0.7320e0 for x in the expression $(x^2 + 2x)-2$ we should ideally obtain zero.

By using floating-point arithmetic, we get,

$$= (0.7320e0 \times 0.7320e0 + 0.2000e1 \times 0.7320e0) - 0.2000e1$$
$$= (0.5358e0 + 0.1464e1) - 0.2000e1$$
$$= -0.1000e\text{-}2$$

If instead of computing $(x^2 + 2x)$ first and then subtracting 2 from it, we compute $(2x - 2)$ and add it to x^2(using floating point arithmetic) we get –0.2000e-3.Similarly, if -0.2732e1 is substituted for x in $(x^2 + 2x) -2$ we get -0.1000e-2. Thus even though the roots are substituted in the quadratic expression the result is not zero. The main part to be noted is that in any computational algorithm it is not advisable to give a branching or looping instruction based on testing whether a floating point quantity is zero or not zero.

UNSOLVED PROBLEMS

1. What are approximate numbers and significant digits?
2. Discuss the representation of fixed and floating-point numbers.
3. What do you mean by normalization? Why we do normalization?
4. Discuss the terms underflow and overflow.
5. Prove the following consequence of the normalized floating point representation of numbers by taking x = .6667

$$6x \neq x + x + x + x + x + x.$$

6. Define normalized floating point representation of numbers and round off errors in representation. Find the sum of 0.123 × 103 and 0.456 × 102 and write the result in three digit mantissa form.

7. Represent 44.85 × 106 in normalized floating point mode.
8. Subtract the following two floating point numbers as
   (i) .36143448 E 7 – .36132346 E 7
   (ii) (.9682 E – 7) – (.3862 E – 9).
9. Explain underflow and overflow conditions of error in floating point's addition and subtraction.
10. Find the solution of the following equation using floating point arithmetic with 4-digit mantissa.
$$x2 - 7x + 4 = 0$$
Give comments on the results so obtained.
11. Represents the following floating-point numbers in IEEE 32 bit format.
   (i) . 11.6346 (ii) . 12.6566
   (iii) .-11.5346 (iv) . –14.6461
12. Add the following floating point numbers, Also discuss whether underflow or Overflow condition occurs.
   (i) 0.4546e5 and 0.5433e5 (ii) . 0.4546e5 and 0.5433e7
   (iii) 0.4546e3 and 0.5433e7 (iv) 0.6434e3 and 0.4845e3
   (v) . 0.6434e99 and 0.4845e99
13. Subtract the following floating-point numbers; also discuss whether underflow or overflow condition occurs.
   (i) .0.9432e-4 from 0.5452e-3 (ii) .0.5424e3 from 0.5452e3
   (iii) . 0.5452e-99 from 0.5424e-99
14. Multiply the following floating-point numbers; Also discuss whether underflow or overflow condition occurs.
   (i) . 0.5543e12 by 0.4111e-15 (ii) . 0.1111e10 by 0.1234e15
   (iii) . 0.1111e51 by 0.4444e50 (iv) . 0.1234e-49 by 0.1111e-54
15. Divide the following floating-point numbers, Also discuss whether underflow or overflow condition occurs.
   (i) . 0.9998e1 by 0.1000e-99 (ii) . 0.9998e-5 by 0.1000e98
   (iii) . 0.1000e5 by 0.9999e3
16. What type of consequences can occur due to fixed format representation?

3.6 ERROR AND THEIR TYPES

There are a number of ways to describe errors in measurements and calculations. An Error is defined as the difference between the actual value and the approximate value. Suppose that x represents the actual value and x_a is the approximate value of x the error may be calculated as:

$$\textbf{Error} = (x - x_a)$$

Not only the data of practical problems usually approximate, but also sometimes the methods or the processes through which we calculate the desired results are also approximate. So, we can say that an approximation value is one that involves approximate data, approximate methods, or both. It is therefore evident that error in a computed result may be due to one or both of two sources; **errors in the data and errors of calculation.**

Errors in the data cannot be remedied (solve easily), but errors of calculation can usually be made as small as we please. Thus, for example if in an expression p(pronounced as pie) is evolved which is replaced by its approximate value, we can decrease the error due to the approximation by taking p to as many figures as desired, and similarly in other cases, i.e. if we choose value as 3.1416 as compared to 3.14 we get small error in our desired result. So, from this we conclude that calculations are always carried out in such a manner as to make the errors of calculation negligible.

3.6.1 Why Errors?

The data which is used in solving problems are usually not exact. Computer cannot make exact form of such data items i.e. it take only approximate value by taking some significant digits. This is the main reason for errors. It is found that only data items are not taken as approximate but algorithms are also approximate i.e. in bisection method -we terminate the iterative cycle, if the results are approximately equal to the known results within the given error range.

So it means, if it is going to run for n steps (manually), but on computer it may be executed for steps which are <n. It has been discovered that the errors in the result may be due to data error or calculations errors. These errors can be reduced to some extent by using the normalized technique described above.

3.6.2 Sources of Error

The main purpose of numerical analysis and scientific computing is to develop ef-ficient and accurate methods to compute approximations to quantities that are difficult or impossible to obtain by analytic means.

Numerical results are affected by many types of errors. Some sources of error are difficult to influence; others can be reduced or even eliminated by, for example, rewriting formulas or making other changes in the computational sequence.

Errors are propagated from their sources to quantities computed later, some-times with a considerable amplification or damping. It is important to distinguish between the new error produced at the computation of a quantity (a source error), and the error inherited (propagated) from the data that the quantity depends on.

Errors in Given Input Data: Input data can be the result of measurements which have been influenced by systematic errors or by temporary disturbances. A rounding error occurs, for example, whenever an irrational number is shortened ("rounded off") to a fixed number of decimals. It can also occur when a decimal fraction is converted to the form used in the computer.

1. **Rounding Errors During the Computations:** The limitation of floating point numbers in a computer leads at times to a loss of information that, depending on the context, may or may not be important.

   Two typical cases are:

   (i) If the computer cannot handle numbers which have more than, say, s digits, then the exact product of two s-digit numbers (which contains 2s or 2s — 1 digits) cannot be used in subsequent calculations; the product must be rounded off.

   (ii) If, in a floating point computation, a relatively small term b is added to a, then some digits of b are "shifted out" (see Example 2.2.4), and they will not have any effect on future quantities that depend on the value of a + b.

2. **Truncation Errors:** These are errors committed when a limiting process is trun-cated (broken off) before one has come to the limiting value. A truncation error occurs, for example, when an infinite series is broken off after a finite number of terms, or when a derivative is approximated with a difference quotient (although in this case the term discretization error is better). Another example is when a nonlinear function is approximated with a linear function as in Newton's method. Observe the distinction between truncation error and rounding error.

3. **Simplifications in the Mathematical Model:** In most of the applications of math-ematics, one makes idealizations. In a mechanical problem, for example, one might assume that a string in a pendulum has zero mass. In many other types of problems it is advantageous to consider a given body to be homogeneously filled with matter, instead of being built up of atoms. For a calculation in economics, one might assume that the rate of interest is constant over a given period of time. The effects of such sources of error are usually more difficult to estimate than the types named in A, B, and C.

4. **"Human" Errors and Machine Errors:** In all numerical work, one must expect that clerical errors, errors in hand calculation, and misunderstandings will occur. One should even be aware that textbooks (!), tables and formulas may contain errors. When one uses computers, one can expect errors in the program itself, typing errors in entering the data, operator errors, and (more seldom).

5. **Algorithmic Errors:** If direct algorithms based on a finite sequence of operations are used, errors due to limited steps don't amplify the existing errors, but if infinite algorithms are used, exact results are expected only after an infinite number of steps. As this cannot be done in practice, the algorithm has to be stopped after a finite number of steps and the results are not exact.

Errors which are purely machine errors are responsible for only a very small part of the strange results which (occasionally with great publicity) are produced by computers. Most of the errors depend on the so-called human factor. As a rule, the effect of this type of error source cannot be analyzed with the help of the theoretical considerations of this chapter! We take up these sources of error in order to emphasize that both the person who carries out a calculation and the person who guides the work of others can plan so that such sources of error are not damaging. One can reduce the risk for such errors by suitable adjustments in working conditions and routines. Stress and tiredness are common causes of such errors. Our effort will be to minimize these errors so as to get the best possible results.

We begin by explaining the various kinds of errors and approximations that may occur in a problem and derive some results on error propagation in numerical calculations.

3.6.3 Various Types of Errors

We have mainly following types of errors occurred in numerical computations.

1. Data Errors
2. Truncation Errors
3. Round-off Errors
4. Absolute Errors
5. Relative Errors
6. Percentage Errors
7. Inherent Errors
8. Accumulated Errors

1. Data Errors

Data corruption refers to errors in computer data that occur during transmission or retrieval, introducing unintended changes to the original data. Computer storage and transmission systems use a number of measures to provide data integrity, the lack of errors. Data errors occur due to inaccurate measurements or observations by some measuring devices. Measuring devices involve vernier caliper, calculators, thermometer etc. Data errors mainly occur due to some limitations in such measuring devices which measures the quantity accurate to certain smallest value.

2. Truncation Errors

These types of errors are also one of the types of errors in the data. Truncation errors are caused by using approximate results or by replacing an infinite process with a finite one.

If we are using a decimal computer having a fixed word length of 4 digits, rounding-off of 24.659 gives 24.66, whereas truncation gives 24.66.

A truncation error is a type of algorithm error. Also,

Problem: Find the truncation error for

$$x = \frac{1}{5} \text{ if}$$

(i) The first three terms are retained in expansion.

(ii) The first four terms are retained in expansion.

Solution:

(i) Error = True value – Approximate value

$$= \left(1 + x + \frac{x^2}{2!} + \frac{x^3}{3!} + \ldots\ldots\right) - \left(1 + x + \frac{x^2}{2!}\right) = \frac{x^3}{3!} + \frac{x^4}{4!} + \frac{x^5}{5!} + \ldots\ldots$$

$$\text{Put } x = \frac{1}{5}$$

$$\text{error} = \frac{.008}{6} + \frac{.0016}{24} + \frac{.00032}{120} + \ldots\ldots$$

$$= .0013333 + .0000666 + .0000026 + \ldots = .0014025$$

(ii) Similarly the error for case II may be found.

Note:

(i) Before truncating a number to some digits first of all we have to convert that number into its normalized floating-point form and after that we have to truncate that number.

(ii) When we convert a number from one number system to another. For example, consider a number 9.1 whose equivalent decimal number comes out to be 1001.0001100110011 ....this decimal number shows repeating fraction and therefore the conversion must be terminated after some digits. Both the above cases introduce truncation error.

3. Round-off Errors

Rounding off the numbers is also truncation but with some adjustment of the last digit. Numbers are rounded off so as to cause the least possible error. Rounding according to the following rule attains this:

To round off a number to n significant figures, discard all digits to the right of the nth place. If the discarded number is less than half a unit in the nth place, leave the nth digit unchanged; if the discarded number is greater than half a unit in the nth place, add 1 to the nth digit. If the discarded number is exactly half a unit in the nth place, leave the nth digit unaltered if it is an even number, but increase it by 1 if it is an odd number; in other words, round off so as to leave the nth digit an even number in such cases. For example, following are some numbers that are rounded off correctly to four significant figures:

Actual number	Round-off number
15.63469	15.63
21.96621	21.97
02.49998	02.500
12.73371	12.73
16.275	16.27
50.495	50.50

These errors can be reduced, however, by

(i) changing the calculation procedure so as to avoid subtraction of nearly equal numbers or division by a small number.

(ii) retaining at least one more significant digit at each step and rounding-off at the last step. Rounding-off may be executed in two ways:

(a) Chopping: In chopping, extra digits are dropped by truncation of number. Suppose we are using a computer with a fixed word length of four digits, then a number like 12.92364 will be stored as 12.92.

We can express the number 12.92364 in the floating print form as

$$
\begin{aligned}
\text{True x} &= 12.92364 \\
&= 0.1292364 \times 10^2 = (0.1292 + 0.0000364) \times 10^2 \\
&= 0.1292 \times 10^2 + 0.364 \times 10^{-4+2} \\
&= f_x \,.\, 10^E + g_x \,.\, 10^{E-d} \\
&= \text{Approximate x + Error} \\
\text{Error} &= g_x \cdot 10^{E-d},\ 0 \le g_x \le d
\end{aligned}
$$

Here, g_x is the mantissa, d is the length of mantissa and E is exponent

Since $\quad 0 \le g_x < 1$

$\therefore$ Absolute error $\le 10^{E-d}$

Case I: If $g_x < 0.5$ then approximate $x = f_x \cdot 10^E$

Case II: If $g_x \ge .5$ then approximate $x = f_x \cdot 10^E + 10^{E-d}$

$$
\begin{aligned}
\text{Error} &= \text{True Value} - \text{Approximate Value} \\
&= f_x \cdot 10^E + g_x \cdot 10^{E-d} - f_x \cdot 10^E - 10^{E-d} \\
&= (g_x - 1) \cdot 10^{E-d}
\end{aligned}
$$

absolute error $\le 0.5. (10)^{E-d}$

(b) Symmetric round-off- In symmetric round-off, the last retained significant digit is rounded up by unity if the first discarded digit is ≥ 5, otherwise the last retained digit is unchanged.

4. Absolute Errors

Absolute error is the magnitude difference between the actual value and approximate value of a variable. Suppose if x is the actual value of the variable and x_a is the approximate value of the same variable. Then absolute error can be calculated as:

$$E_A = | x - x_a |$$

Where E_A represents the absolute error.

5. Relative Errors

The relative error is the absolute error divided by the actual value. So relative error can be calculated as:

$$E_R = \left| \frac{x - x_a}{x} \right| \quad \text{or} \quad E_R = \frac{E_A}{|x|}$$

6. Percentage Errors

The percentage error is 100 times the relative error. So percentage error can be calculated as:

$$E_{per} = 100\ E_R$$

Remark: It is to be noted that relative and percentage errors are independent of the unit of measurement, whereas absolute errors are expressed in terms of the unit used.

7. Inherent Errors

Errors which are already present in the statement of a problem before its solution are called inherent errors. Such errors arise either due to the fact that the given data is approximate or due to limitations of mathematical tables, calculators, or the digital computer. Inherent errors can be minimized by taking better data or by using high precision* computing aids. These errors are further classified as Data errors and Conversion errors.

Data errors arise due to inaccuracy of input data supplied for solving given problem, whereas the Conversion errors arise due to inherent limitation of the computing devices. Data errors can be reduced by taking better data.

Accuracy refers to the number of significant digits in a value, for example, 53.965 is accurate to 5 significant digits. Precision refers to the number of decimal positions or order of magnitude of the last digit in the value. For example, in 53.965, precision is 10^{-3}.

Problem: Which of the following numbers has the greatest precision?

4.1041, 4.12, 4.542012.

Solution: In 4.1041, precision is 10^{-4}

In 4.12, precision is 10^{-2}

In 4.320106, precision is 10^{-6}.

Hence, the number 4.542012 has the greatest precision.

8. Accumulated Errors

Accumulated errors are produced due to the accumulation of errors from one step to next steps i.e. the effect of an error increases with the succession. We explain the concept of accumulated error by taking an example.

Consider that we have a procedure:

$$A_{n+1} = 10A_n \text{ where } n = 0, 1, 2, 3\ldots\ldots\ldots$$

$$A_1 = 10A_0$$

$$A_2 = 10A_1$$

$$A_3 = 10A_2$$

....

....

Here if the actual value of A_0=1.99 and by approximation we use A_0=2.0 in place of 1.99 then the values of A_n variables comes out to be

Variable Name	Actual value	Approximate Value	Accumulated Error
A_0	1.99	2	2 – 1.99 = 0.01
A_1	19.9	20	20 – 19.9 = 0.1
A_2	199	200	200 – 199 = 1.0

Here from table we depicts that error gets accumulated at every next step. We see that an error of 0.01 leads to an error of 1.0 just after the two steps. This is how the error gets accumulated from one step to next. Such errors are not avoidable after later stages.

3.6.4 Difference Between Chopping and Rounding

The simplest way of reducing the number of significant digits in the representation of a number is merely to ignore the unwanted digits. This procedure, known as chopping was used by many early computers. A more common and better procedure is rounding which involves adding 5 to the first unwanted digit and then chopping. E.g. p chopped to four decimal places is 3.1415, but it is 3.1416 when rounded the representation 3.1416 is correct to five significant digits. The error involved in the reduction of number of digits is called round- off error, since p is 3.14159...... note that chopping produced more round- off error than rounding.

SOLVED PROBLEMS

Problem 1: Suppose 1.414 is used as an approximation to $\sqrt{2}$. Find the absolute and relative errors.

Solution:

$$\text{True value} = \sqrt{2} = 1.41421356$$

$$\text{Approximate value} = 1.414$$

$$\text{Error} = \text{True value} - \text{Approximate value}$$

$$= \sqrt{2} - 1.414 = 1.41421356 - 1.414$$

$$= 0.00021356$$

$$\text{Absolute error } e_a = |\text{ Error }|$$

$$= |\, 0.00021356 \,| = 0.21356 \times 10^{-3}$$

$$\text{Relative error } e_r = e_a / \text{True value} = 0.21356 \text{ X}10^{-3} \;/\; \sqrt{2}$$

$$= 0.151 \times 10^{-3}.$$

Problem 2: If 0.333 is the approximate value of 1/3 , Compute the absolute, relative, and percentage errors.

Solution:

$$\text{True value (X)} = 1/3$$

$$\text{Approximate value (X}') = 0.333$$

Absolute error $e_a = | X - X' |$

$= 1/3 - 0.333$

$= |0.333333 - 0.333| = .000333$

Relative error $e_r = e_a / X$

$= .000333/ .333333 = .000999$

Percentage error $e_p = e_r \times 100$

$= .000999 \times 100 = .099\%.$

Problem 3: An approximate value of ð is given by 3.1428571 and its true value is 3.1415926. Compute the absolute and relative errors.

Solution:

True value = 3.1415926

Approximate value = 3.1428571

Error = True value – Approximate value

= 3.1415926 – 3.1428571

= – 0.0012645

Absolute error e_a = | Error | = 0.0012645

Relative error $e_r = e_a$ /True value = 0.0012645 / 3.1415926

= 0.000402502.

Problem 4: There are three approximate values of the number 1/3 are given as 0.30, 0.33, and 0.34. Which one of these three is the best approximation?

Solution: The best approximation will be the one which has the least absolute error.

True value = 1/3 = 0.33333.

Case I:

Approximate value = 0.30

Absolute error = | True value – Approximate value |

= | 0.33333 – 0.30 |

= 0.03333

Case II:

Approximate value = 0.33

Absolute error = | True value – Approximate value |

= | 0.33333 – 0.33 |

= 0.00333.

Case III.

Approximate value = 0.34

Absolute error = | True value – Approximate value |

= | 0.33333 – 0.34 |

= | – 0.00667 | = 0.00667

Since the absolute error is **least** in case II, 0.33 is the best approximation.

Problem 5: Compute the relative error of the number 8.6 if both of its digits are correct.

Solution: Here, $e_a = .05$

therefore $e_a = 1/2 \times 10^{-1}$

$e_r = .05 / 8.6 = .0058.$

Problem 6: Compute the relative error if 2/3 is approximated as 0.667.

Solution: True value = 2/3

= 0.666666

Approximate value = 0.667

Absolute error $e_a = |\text{True value} - \text{approximate value}|$

$= |.666666 - .667| = .000334$

Relative error $e_r = .000334/.666666 = .0005.$

Problem 7: Compute the percentage error if 625.483 is approximated to three significant figures.

Solution:

$e_a = |625.483 - 625| = 0.483$

$e_r = e_a / 625.483 = .483 / 625.483 = .000772$

$e_p = e_r \times 100 = .077\%.$

Problem 8: Round-off the numbers 865250 and 37.46235 to four significant figures and compute the values e_a, e_r, e_p in each case.

Solution:

(i) Number rounded-off to four significant digits = 865200

$X = 865250$

$X' = 865200$

Error $= X - X' = 865250 - 865200 = 50$

Absolute error $e_a = |\text{error}| = 50$

Relative error $e_r = e_a / X = 50/ 865250 = 5.77 \times 10^{-5}$

Percentage error $e_p = e_r \times 100 = 5.77 \times 10^{-3}$

(ii) Number rounded-off to four significant digits = 37.46

$X = 37.46235$

$X' = 37.46$

Error $= X - X' = 0.00235$

Absolute error $e_a = |\text{error}| = 0.00235$

Relative error $e_r = e_a / X = 0.00235 / 37.46235 = 6.2729 \times 10^{-5}$

Percentage error $e_p = e_r \times 100 = 6.2729 \times 10^{-3}.$

Problem 9: Round-off the number 75462 to four significant digits and also compute the absolute error and percentage error.

Solution: Number rounded-off to four significant digits = 75460

$$\text{Absolute error } e_a = |\,75462 - 75460\,| = 2$$

$$\text{Relative error } e_r = e_a/75462 = 2 / 75462 = .0000265$$

$$\text{Percentage error } e_p = e_r \times 100 = .00265.$$

Problem 10: Calculate the absolute, relative, and percentage errors if x is rounded off to three decimal digits. Given x = 0.005998.

Solution: Number rounded-off to three decimal digits =.006

$$\text{Error} = .005998 - .006 = -\,.000002$$

$$\text{Absolute error } e_a = |\text{ error }| = .000002$$

$$\text{Relative error } e_r = e_a / .005998 = .000002 / .005998 = .0033344$$

$$\text{Percentage error } e_p = e_r \times 100 = .33344.$$

3.7 A GENERAL ERROR FORMULA OR GENERAL METHOD FOR OBTAINING AN ERROR OF A FUNCTION

In this section we derive a general formula for the error committed in using a certain formula or a functional relation. Let

$$U = f(x_1, x_2, \ldots\ldots\ldots, x_n)$$

be a function of several variables $x_p (p = 1,2,\ldots,n)$ and let the error in each x_p be Δx_p. Then the error ΔU in U is given by

$$U + \Delta U = f(x_1+\Delta x_1, x_2+\Delta x_2, \ldots\ldots\ldots\ldots, x_n+\Delta x_n).$$

By expanding the R.H.S by Taylor's series, we get

$$U + \Delta U = f(x_1, x_2, \ldots\ldots\ldots, x_n) + \sum_{p=1}^{n}\left(\frac{\partial f}{\partial x_p}\right).\Delta x_p + \text{terms involving } (\Delta x_p)^2$$

Assuming that the errors in x_p are small and that $\frac{\Delta x_p}{x_p} << 1$ so that the squares and higher powers of Δx_p can be neglected, the above relation yields

$$\Delta U \approx \sum_{p-1}^{n}\left(\frac{\partial f}{\partial x_p}\right).\Delta x_p$$

$$= \left(\frac{\partial f}{\partial x_p}\right).\Delta x_1 + \left(\frac{\partial f}{\partial x_p}\right).\Delta x_2 + \ldots\ldots + \left(\frac{\partial f}{\partial x_p}\right).\Delta x_n$$

We observe that this formula has the same form as that for the total differential of u. The formula for the relative error follows immediately:

$$E_R = \frac{\Delta U}{U} = \frac{\left(\frac{\partial f}{\partial x_1}\right)}{U} \cdot \Delta x_1 + \frac{\left(\frac{\partial f}{\partial x_2}\right)}{U} \cdot \Delta x_2 + \dots\dots + \frac{\left(\frac{\partial f}{\partial x_n}\right)}{U} \cdot \Delta x_n$$

Or percentage relative error, $E_{per} = \frac{\Delta U}{U} \times 100$

$$= \left[\frac{\left(\frac{\partial f}{\partial x_1}\right)}{U} \cdot \Delta x_1 + \frac{\left(\frac{\partial f}{\partial x_2}\right)}{U} \cdot \Delta x_2 + \dots\dots + \frac{\left(\frac{\partial f}{\partial x_n}\right)}{U} \cdot \Delta x_n \right] \times 100$$

The following example illustrates the use of this formula.

SOLVED PROBLEMS

Problem 1: If $U = f(x,y,z) = \frac{5xy^2}{z^3}$, find the relative maximum error when $\Delta x = \Delta y = \Delta z = 0.001$ and $x = y = z = 1$

Solution: Here $$U = \frac{5xy^2}{z^3}$$

Then $$\frac{\partial U}{\partial x} = \frac{5y^2}{z^3}$$

$$\frac{\partial U}{\partial y} = \frac{10xy}{z^3}$$

$$\frac{\partial U}{\partial z} = \frac{-15xy^2}{z^4}$$

$$\Delta U = \left(\frac{5y^2}{z^3}\right) \cdot \Delta x + \left(\frac{10xy}{z^3}\right) \cdot \Delta y - \left(\frac{15xy^2}{z^4}\right) \cdot \Delta z$$

In general, the errors Δx, Δy and Δz may be positive or negative, and hence we take the absolute values of the terms on the right side. This gives

$$(\Delta U)_{\max} \approx \left|\left(\frac{5y^2}{z^3}\right)\cdot \Delta x\right| + \left|\left(\frac{10xy}{z^3}\right)\cdot \Delta y\right| + \left|\left(\frac{15xy^2}{z^4}\right)\cdot \Delta z\right|$$

Now, let $\Delta x = \Delta y = \Delta z = .001$ and $x = y = z = 1$. Then, the relative maximum error$(E_R)_{\max}$ is given by

$$(E_R)_{\max} = \frac{(\Delta U)_{\max}}{U} = \frac{0.03}{5} = 0.006$$

Problem 2: If $U = f(x,y,z) = \dfrac{3x^2 y}{z}$, find the relative maximum error when $\Delta x = \Delta y = \Delta z = .001$ and $x = y = z = 1$.

Solution: Here
$$U = \frac{3x^2 y}{z}$$

Then $\dfrac{\partial U}{\partial x} = \dfrac{6xy}{z}$, $\dfrac{\partial U}{\partial y} = \dfrac{3x^2}{z}$, $\dfrac{\partial U}{\partial z} = \dfrac{-3x^2 y}{z^2}$

$$\therefore \qquad \Delta U = \left(\frac{6xy}{z}\right)\cdot \Delta x + \left(\frac{3x^2}{z}\right)\cdot \Delta y - \left(\frac{3x^2 y}{z^2}\right)\cdot \Delta z$$

Taking the absolute value of both sides since the errors Δx, Δy, Δz may be positive or negative, we have

$$|\Delta U| \leq \left|\left(\frac{6xy}{z}\right)\cdot \Delta x\right| + \left|\left(\frac{3x^2}{z}\right)\cdot \Delta y\right| - \left|\left(\frac{3x^2 y}{z^2}\right)\cdot \Delta z\right|$$

$$\therefore \qquad |\Delta U| \approx \left|\left(\frac{6xy}{z}\right)\cdot \Delta x\right| + \left|\left(\frac{3x^2}{z}\right)\cdot \Delta y\right| - \left|\left(\frac{3x^2 y}{z^2}\right)\cdot \Delta z\right|$$

$$\approx (6+3+3) \times 0.001 = 0.012, \text{ Also } U = \frac{3x^2 y}{z} = 3$$

Hence
$$(E_R)_{\max} = \frac{(\Delta U)_{\max}}{U} = \frac{0.012}{3} = 0.004$$

3.8 ERROR IN A SERIES APPROXIMATION

We know that the Taylor's series for $f(x)$ at $x = a$, with a remainder after n terms is:

$$f(x) = f[a + (x - a)]$$

$$= f(a) + (x-a)f'(a) + \frac{(x-a)^2}{2!} + ...+ \frac{(x-a)^{n-1}}{(n-1)!} f^{n-1}(a) + R_n(x)$$

Where $$R_n(x) = \frac{(x-a)^n}{n!} \cdot f^n(\xi), \quad a < \xi < x$$

$R_n(x) \longrightarrow$ Remainder term

If the series is convergent then $R_n \longrightarrow 0$ as $n \longrightarrow \infty$, Hence if $f(x)$ is approximated by the first n terms of the series then maximum error in $f(x) = R_n(x)$. On the other hand, if the accuracy required in the series is given, then it is possible to find n (total number of terms required in the series). The method is explained by using the following example.

SOLVED PROBLEMS

Problem 1: Find the number of terms required for the series $\tan^{-1}$

$$x = x - \frac{x^3}{3} + \frac{x^5}{5} - \ldots\ldots + \frac{(-1)^{n-1}x^{2n-1}}{(2n-1)} + \ldots\ldots$$

such that their sum gives the value of $\tan^{-1} x$ correct to 8 decimal places at $x = 1$.

Solution: We have given

$$\tan^{-1} x = x - \frac{x^3}{3} + \frac{x^5}{5} - \ldots\ldots + \frac{(-1)^{n-1}x^{2n-1}}{(2n-1)} + R_n(x)$$

where $$R_n(x) = \frac{(-1)^{n-1}x^{2n-1}}{2n-1}\tan^{-1}\xi, \; 0 < \xi < x$$

$\therefore$ Maximum absolute error (at $\xi = x$) $= \dfrac{x^{2n-1}}{2n-1}\tan^{-1} x$

$\therefore$ Maximum relative error $E_R = \dfrac{\text{Maximum absolute error}}{\tan^{-1} x}$

$$= \frac{x^{2n-1}}{2n-1}$$

Maximum relative error E_R at $x = 1$ is $\dfrac{1}{2n-1}$

If the result is true to 8 decimal places, then the relative error is less than 0.5×10^{-8}

$\therefore$ We have $\dfrac{1}{2n-1} < 0.5 \times 10^{-8}$

or $$2n - 1 > 2 \times 10^8$$

$$n > 0.5 + 10^8$$

$\therefore$ $$n = 10^8 + 1$$

Hence we have required $10^8 + 1$ terms in order to correction of $\tan^{-1} x$ at $x = 1$ up to 8 significant digits.

3.9 MEASURES OF ACCURACY

In this text, we will discuss some rules that tells us about the accuracy of results when we use truncated and rounding off values. These rules will measure the accuracy of two major types of errors namely absolute and relative errors. These rules will be proved numerically in the solved examples.

Let x is any number having exponent power n, then

Rule 1: Absolute error due to truncation

If x_a is the approximation of x then absolute error after truncating to m digits is

$$|x - x_a| \leq 10^{n-m}$$

Rule 2: Absolute error due to rounding-off

If x_a is the approximation of x then absolute error after rounding-off to m digits is

$$|x - x_a| \leq (0.5 \times 10^{n-m})$$

Rule 3: Relative error due to truncation

If x_a is the approximation of x then relative error after truncating error to m digits is

$$\left|\frac{x - x_a}{x}\right| \leq 10^{-m+1}$$

Rule 4: Relative error due to rounding-off

If x_a is the approximation of x then relative error after rounding-off error to m digits is

$$\left|\frac{x - x_a}{x}\right| \leq 0.5 \times 10^{-m+1}$$

Rule 5: If the number x is rounded to m decimal places then

$$\Delta x = 0.5 \times 10^{-m}$$

Rule 6: If x_a is the approximation of x correct to m significant digits then

$$\left|\frac{x - x_a}{x}\right| \leq 10^{-m}$$

Rule 7: If a number is correct to n significant figures and the first significant digit of the number is d_1 then relative error is given by

$$\left|\frac{x - x_a}{x}\right| \leq \left(\left(\frac{1}{d_1}\right) \times 10^{m-1}\right)$$

3.10 ERROR PROPAGATION

As we know to solve a numerical problem we have to solve various computational steps for achieving the result. During the computations the error from one step to next step propagates due to which we get the result in the form of approximation. In this text, we will discuss how error propagates during the application of arithmetic operations. We will consider all the basic arithmetic operations one by one.

Error propagation in Addition (+) & Subtraction (—)

Consider addition of two numbers, say, x and y.

As we know that $x - x_a = E_{Ax}$ and $y - y_a = E_{Ay}$

Where E_{Ax} and E_{Ay} are absolute errors in x and y and x_a, y_a are absolute values of x and y.

So
$$x + y = E_{Ax} + x_a + E_{Ay} + y_a$$
$$= (x_a + y_a) + (E_{Ax} + E_{Ay})$$

$\therefore$
$$\boxed{\text{Total error} = E_{Ax+Ay} = E_{Ax} + E_{Ay}}$$

Similarly, for subtraction

$$\boxed{\text{Total error} = E_{Ax-Ay} = E_{Ax} - E_{Ay}}$$

Note that the addition $E_{Ax} + E_{Ay}$ does not mean that error will increase in all cases. It depends upon the sign of individual errors. Same can happen in case of subtraction.

Since we don't know the sign of errors, we can only estimate error bounds. That is, we can say that

$$|E_{Ax \pm Ay}| \leq |E_{Ax}| + |E_{Ay}|$$

Therefore, the rule for addition and subtraction is: the magnitude of the absolute error of a sum (or difference) is equal to or less than the sum of the absolute errors of the operands.

This inequality is called the triangle inequality. The equality applies when the operands have the same signs, and the inequality applies if the signs are different.

Error propagation in Multiplication (×)

In multiplication, we have
$$x \times y = (x_a + E_{Ax}) \times (y_a + E_{Ay})$$
$$= x_a y_a + y_a E_{Ax} + x_a E_{Ay} + EA_x E_{Ay}$$

Errors are normally small and their products will be much smaller. Therefore, if we neglect the product of the errors, we get

$$x \times y = x_a y_a + y_a E_{Ax} + x_a E_{Ay}$$
$$= x_a y_a + x_a y_a \left(\frac{E_A}{x_a} + \frac{E_{Ay}}{y_a} \right)$$

Then,
$$\boxed{\text{Total error} = E_{AxAy} = x_a y_a \left(\frac{E_A}{x_a} + \frac{E_{Ay}}{y_a} \right)}$$

By applying triangle inequality, we have

$$E_{AxAy} \leq |\, x_a y_a| \left(\left|\frac{E_A}{x_a}\right| + \left|\frac{E_{Ay}}{y_a}\right| \right)$$

Error propagation in Division (÷)

In division, we have

$$\frac{x}{y} = \frac{(x_a + E_{Ax})}{(y_a + E_{Ay})}$$

Multiplying both numerator and denominator by $y_a - E_{Ay}$ and rearranging the terms, we get

$$\frac{x}{y} = \frac{(x_a y_a + y_a E_{Ax} - x_a E_{Ay} - E_{Ax} E_{Ay})}{(y^2{}_a - E^2{}_{Ay})}$$

By neglecting all terms that involve only product of errors, we have

$$\frac{x}{y} = \frac{(x_a y_a + y_a E_{Ax} - x_a E_{Ay})}{y^2{}_a}$$

$$= \left(\frac{x_a}{y_a}\right) + \frac{x_a}{y_a}\left(\frac{E_{Ax}}{x_a} - \frac{E_{Ay}}{y_a}\right)$$

Then $\qquad$ total error $= E_{\frac{Ax}{Ay}} = \left(\frac{x_a}{y_a}\right)\left(\frac{E_{Ax}}{x_a} - \frac{E_{Ay}}{y_a}\right)$

By applying triangle inequality, we have

$$E_{\frac{Ax}{Ay}} \leq \left|\frac{x_a}{y_a}\right| \left(\left|\frac{E_{Ax}}{x_a}\right| - \left|\frac{E_{Ay}}{y_a}\right| \right)$$

The errors E_{Ax} and E_{Ay} may be of any type it may be round off errors, truncation errors, empirical errors or any combination of these.

3.11 INVERSE PROBLEMS

Now we have to find errors in $x_1, x_2, ..., x_n$, where $X = f(x_1, x_2, ..., x_n)$, to have a desired accuracy.

Now, We have

$$\Delta X = \frac{\partial X}{\partial x_1}\Delta x_1 + \frac{\partial X}{\partial x_2}\Delta x_2 + ... + \frac{\partial X}{\partial x_n}\Delta x_n$$

$$\frac{\partial X}{\partial x_1}\Delta x_1 = \frac{\partial X}{\partial x_2}\Delta x_2 = ... = \frac{\partial X}{\partial x_n}\Delta x_n$$

$$\therefore \qquad \Delta X = n\frac{\partial X}{\partial x_1}\Delta x_1$$

$$\therefore \qquad \Delta x_1 = \frac{\Delta X}{n\left(\frac{\partial X}{\partial x_1}\right)}$$

Similarly, $\Delta x_2 = \dfrac{\Delta X}{n\dfrac{\partial X}{\partial x_2}}$ and so on.

SOLVED PROBLEMS

Problem 1: Choose the best approximation of the number 1/3 out of the three given approximations as 0.30, 0.33 and 0.34.

Solution: Best approximation is judged by considering the error.

As we know error is given by $|1/3 - 0.30| = 1/30$

Similarly, $|1/3 - 0.33| = 1/300$

Similarly, $|1/3 - 0.34| = 1/150$

Since the error term 1/300 is smallest out of other error terms, so 0.33 is the best approximation.

Problem 2: If actual value of a variable $x = 0.00550$ and approximate value to this number is 0.00541. Find the error.

Solution: Here $x = 0.00550$ and $x_a = 0.00541$

In floating point form, $x = 0.550 \times 10^{-2}$ and $x_a = 0.541 \times 10^{-2}$

Then error $= x - x_a$

$= (0.550 \text{ “ } 0.541) \times 10^{-2}$

$= 0.009 \times 10^{-2}$

$= 0.9 \times 10^{-4}$ (in normalized floating point form)

Problem 3: Find the relative error of the number 6.9 if both of its digits are correct.

Solution: Here $x = 6.9$, $n = 1$

According to rule 5, we have

$$\Delta x = 0.5 \times 10^{-m} = 0.5 \times 10^{-1} = 0.05$$

$\therefore$ Absolute error $E_A = 0.05$

Relative error, $E_R = \dfrac{E_A}{x} = \dfrac{0.05}{6.9} = 0.00724$

Problem 4: If x = 0.0004925 then what will be the absolute error if x is truncated to 3 decimal digits.

Solution: As $x = 0.0004925$

In floating point form, $x = 0.4925 \times 10^{-3}$

Then $x_a = 0.492 \times 10^{-3}$

(after truncating x to three decimal digits)

$$\text{Error} = x - x_a$$
$$= (0.4925 - 0.492) \times 10^{-3}$$
$$= 0.0005 \times 10^{-3}$$
$$\text{Absolute error} = |x - x_a|$$
$$= 0.0005 \times 10^{-3}$$
$$= 0.500 \times 10^{-6}$$ (in normalized floating point form)

According to rule 1 above result is less than or equals to$10^{n-m} = 10^{(-3-3)} = 10^{-6}$

Problem 5: If x = 0.0004925 then what will be the absolute error if x is rounded off to 3 decimal digits.

Solution: As $x = 0.0004925$

In floating point form, $x = 0.4925 \times 10^{-3}$

Then $x_a = 0.493 \times 10^{-3}$

(after rounding-off x to three decimal digits)

$$\text{Absolute error} = |x - x_a|$$
$$= |0.4925 \times 10^{-3}\text{"}\ 0.493 \times 10^{-3}|$$
$$= 0.0005 \times 10^{-3}$$
$$= 0.500 \times 10^{-6}$$ (in normalized floating point form)

According to rule 2 above result is less than or equals to $0.5 \times 10^{n-m} = 0.5 \times 10^{(-3-3)} = 0.5 \times 10^{-6}$

Problem 6: If x = 0.006299 then what will be the relative error if x is truncated to 3 decimal digits.

Solution: As $x = 0.006299$

In floating point form, $x = 0.6299 \times 10^{-2}$

Then $x_a = 0.629 \times 10^{-2}$

(after truncating x to three decimal digits)

$$\left|\frac{x-x_a}{x}\right| = \left|\frac{0.6299\times10^{-2} - 0.629\times10^{-2}}{0.6299\times10^{-2}}\right|$$

$= 0.00142 \qquad = 0.142\times10^{-2}$ (In floating point form)

According to rule 3 above result is less than or equals to $10^{-m+1} = 10^{(-3+1)} = 10^{-2}$

3

Problem 7: If x = 0.006299 then what will be the relative error if x is rounded off to 3 decimal digits.

Solution: As $x = 0.006299$

In floating point form, $x = 0.6299x10^{-2}$

Then $x_a = 0.630x10^{-2}$

(after rounding off x to three decimal digits)

$$\left|\frac{x-x_a}{x}\right| = \left|\frac{0.6299\times10^{-2} - 0.630\times10^{-2}}{0.6299\times10^{-2}}\right|$$

$= 0.0001$

$= 0.100 \times 10^{-5}$ (In floating point form)

According to rule 4 above result is less than or equals to $0.5 \times 10^{-m+1} = 0.5 \times 10^{(-3+1)} = 0.5 \times 10^{-2}$

Problem 8: Round off the number 865230 to four significant figures and compute the following types of errors.

(i) Absolute error

(ii) Relative error

(iii) Percentage error

Solution: Here $x = 865230$

Approximate value of x after rounded to four significant figures is

$x_a = 865200$

(i) Absolute error is given by

$$E_A = |x - x_a|$$

$$E_A = |865230 - 865200|$$

$$= 30$$

(ii) Relative error is given by

$$E_R = E_A/|x|$$

$$E_R = 30/865230 = 3.467 \times 10^{-5}$$

(iii) Percentage error is given by

$$\mathbf{E_{per} = 100\ E_R}$$

$$E_{per} = 100 \times 3.467 \times 10^{-5}$$

$$= 3.467 \times 10^{-3}$$

Problem 9: How many digits are required in computing $\sqrt{15}$ so that error does not exceed 0.1%?

Solution: The approximate value of $\sqrt{15}$ is 3.87(appx.)

$$R_n(x) = \frac{x^n}{n!} E_R \leq \frac{0.1}{100} = 0.001$$

The first digit of $\sqrt{15}$ is 3, ∴ According to rule 7, we have

$$d_1 = 3 \text{ and } E_R \leq \left(\left(\frac{1}{d_1}\right) \times 10^{m-1}\right)$$

$$Er \leq \left(\left(\frac{1}{3}\right) \times 10^{m-1}\right) \leq 0.001$$

$$10^{m-1} \geq \frac{1}{(3 \times 0.001)}$$

$\Rightarrow$ $$10^{m-1} \geq 333.33$$

$\Rightarrow$ $$10^{m-1} \geq 10^3$$

∴ $$m-1 \geq 3$$

∴ $$m \geq 4$$

Hence required numbers of digits are 4.

Problem 10: The solution of a problem as $x = 15.75$ and the absolute error of the solution is less than 1%. Find the range of values within which the exact value of the solution lies.

Solution: The given absolute error $= 1\% = \frac{1}{100} = 0.01$

Hence as $x_a = 15.75$

According to the problem as absolute error is less than 1% so we can write

$$|x - x_a| < 1\% \quad \text{i.e } |x - x_a| < 0.01$$

$$|x - 15.75| < 0.01$$

$$-0.01 < (x - 15.75) < 0.01$$

$$-0.01 + 15.75 < x < 0.01 + 15.75$$

$$15.74 < x < 15.76$$

Hence the exact value of x lies between the interval (15.74, 15.76)

Problem 11: The solution of a problem as $x = 65.85$ and the relative error of the solution is at the most than 1.5%. Find to five decimal digits the range of values within which the exact value of the solution lie.

Solution: The maximum relative error $= 1.5\% = \frac{1.5}{100} = 0.015$

Hence as $\quad x_a = 65.85$

According to the problem as relative error is at the most 2% so we can write

$$\left|\frac{x - x_a}{x}\right| < 1.5\%$$

$$\left|\frac{x - 65.85}{x}\right| < 0.015$$

$$-0.015 < \left(\frac{x - 65.85}{x}\right) < 0.015$$

if $\quad \left(\frac{x - 65.85}{x}\right) < 0.015$

then $\quad x - 65.85 < 0.015x \quad$ (since x is positive)

or $\quad x(1\text{-}0.015) < 65.85$

or $\quad x < \frac{65.85}{(1 - 1.015)}$

or $\quad x < \frac{65.85}{0.985}$

or $\quad \mathrm{x} < 66.852791 \quad$ ...(i)

if $\quad -0.015 < \left(\frac{x - 65.85}{x}\right)$

then $\quad x - 65.85 > -0.015x \quad$ (since x is positive)

or $\quad x(1 + 0.015) < 65.85$

or $\quad x > \frac{65.85}{1.015}$

or $\quad x > 64.876847 \quad$ ...(ii)

By using (i) and (ii), we get

$$64.87685 < x < 65.85279$$

Hence x lies between the interval (64.87685, 65.85279) after rounding off to five decimal digits.

Problem 12: Given $a = 7.00 \pm 0.05$, $b = 0.0435 \pm 0.0003$, $c = 1234 \pm 12$. Find the maximum value of absolute error in $a + b + c$.

Solution: Let S denotes the sum variable as

$$\begin{aligned} S &= a + b + c \\ &= 7.00 + 0.0435 + 1234 \\ &= 1241.0435 \\ \Delta S &= \pm(\Delta a + \Delta b + \Delta c) \\ &= \pm(0.05 + 0.003 + 12) \\ &= \pm 12.053 \end{aligned}$$

Hence the maximum value of absolute error in $a + b + c$ is 12.053

Problem 13: If $u = 3v^7 - 6v$, find the percentage error in u at $v = 1$, if the error in v is 0.05.

Solution: As given $u = 3v^7 - 6v$ and error in v is $\Delta v = 0.05$

So at $v = 1$, $\quad u = -3$

Also $$\frac{du}{dv} = 21v^6 - 6$$

$$\therefore \quad \Delta u = \left(\frac{du}{dv}\right)\cdot \Delta v = (21v^6 - 6)\,\Delta v$$

At $v = 1$, $\Delta u = (21\text{-}6)\,(0.005) = 0.75$

$$\therefore \quad E_R = \frac{\Delta u}{|u|} = \frac{0.75}{|-3|} = 0.25$$

$\therefore$ Relative error iv u at $v = 1$ is 0.25

Percentage error in u at $v = 1$ is given by $E_P = E_R \times 100 = 25\%$

Problem 14: Round off (i). 865230 (ii). 47.5689 to four significant digits. Also compute Absolute error, relative error and percentage error.

Solution:

(i) Let $x = 865230$

Then $\quad x_a = 865200$

(After rounded off to 4 significant digits)

$\therefore$ Absolute error $= |x - x_a| = |\,865230 - 865200\,| = 30$

$$\text{Relative error} = \left|\frac{x - x_a}{x}\right| = \frac{30}{865230} = 3.4673 \times 10^{-5}$$

Percentage error = relative error × 100 = 3.4673×10^{-3}

(ii) Let $x = 47.5689$

Then $$x_a = 47.57 \quad \text{(After rounded off to 4 significant digits)}$$

$$\therefore \quad \text{Absolute error} = |x - x_a| = |47.5689 - 47.57| = 0.0011$$

$$\text{Relative error} = \left|\frac{x - x_a}{x}\right| = \frac{0.0011}{47.5689} = 2.3124 \times 10^{-5}$$

$$\text{Percentage error} = \text{relative error} \times 100 = 2.3124 \times 10^{-3}$$

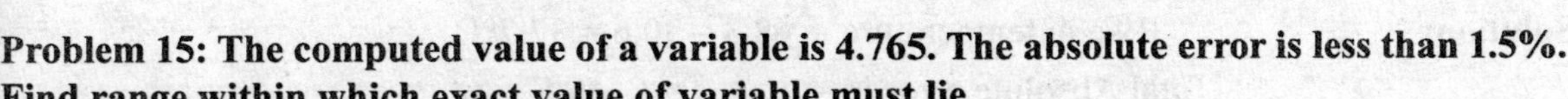

Problem 15: The computed value of a variable is 4.765. The absolute error is less than 1.5%. Find range within which exact value of variable must lie.

Solution: Let x = Exact value and x_a = Computed value

$$\text{Absolute error} = |x - x_a| = |x - 4.765| < 1.5\%$$

i.e. $$|x - 4.765| = 0.0015$$

$$\therefore \quad -0.015 < x - 4.765 < 0.015$$

or $$-0.015 + 4.765 < x < 0.015 + 4.765$$

$$4.750 < x < 4.780$$

So the required interval is (4.750, 4.780)

Problem 16: The computed value of a variable is 97.86. The relative error in the computed value is at most 2.5%. Find to four decimal places within which exact value may lie.

Solution: Let x = Exact value, x_a = Computed value

According to the problem

$$\left|\frac{x - x_a}{x}\right| < 2.5\% \Rightarrow \left|\frac{x - 97.86}{x}\right| < 0.025$$

$$\Rightarrow -0.025 < \frac{x - 97.86}{x} < 0.025$$

$$\Rightarrow -0.0245\,x < x - 97.86 < 0.025\,x$$

Case (i)

$$-0.025x < x - 97.86$$

$$\Rightarrow \quad -0.025x - x < x - 97.86$$

$$-1.025x < -97.86$$

$$\Rightarrow \quad 1.025x > 97.86$$

$$x > \frac{97.86}{1.025} = 95.47317 \quad \text{...(i)}$$

Case (ii)

$$x - 97.86 < 0.025x$$

$$x - 0.025x < 97.86$$

$$0.975x < 97.86$$

$$x < \frac{97.86}{0.975} = 100.36923 \qquad \text{... (ii)}$$

From (i) and (ii), exact value x must lie between 95.4732 and 100.3692.

Problem 17: The temperature rose from (30.6 ± 0.2)°C to (68.3 ± 0.3)°C. Find rise in temperature with proper error limits.

Solution: Rise in temperature = 68.3 – 30.6 = 37.7°C

$$\text{Total Absolute error in} = E_{Ax-Ay} = E_{Ax} + E_{Ay}$$

$$= 0.2 + 0.3 = 0.5$$

∴ Error limits are ± 0.5

Rise in temperature = (37.7 ± 0.5)°C.

Problem 18: Determine the value of $\frac{x^2 - y^2}{x + y}$ with $x = 0.4845$ and $y = 0.4800$ using normalized floating point arithmetic. Compare with value of $(x - y)$. Determine the relative error of the former.

Solution: Since $x = 0.4845e0$ and $y = 0.4800e0$

$$x^2 = x \times x = 0.4845e0 \times 0.4845e0$$

$$= (0.4845 \times 0.4845)e(0+0)$$

$$= 0.2347402e0$$

$$= 0.2347e0$$

$$y^2 = y \times y = 0.4800e0 \times 0.4800e0$$

$$= (0.4800 \times 0.4800)e(0+0)$$

$$= 0.2304e0$$

$$x^2 - y^2 = 0.2347e0 - 0.2304e0$$

$$= 0.0043e0$$

$$x + y = 0.4845e0 + 0.4800e0$$

$$= 0.9645e0$$

$$\frac{x^2 - y^2}{x + y} = \frac{0.0043e0}{0.9645e0} = \frac{0.4300e-2}{0.9645e0} = \frac{0.4300}{0.9646}e(-2-0)$$

$$= 0.4458268e\text{-}2$$

$$= 0.4458e\text{-}2$$

$$x - y = 0.4845e0 - 0.4800e0$$

$= 0.0045e0$

$= 0.4500e{-2}$

$$\text{Relative error} = \frac{0.4500e-2-0.4458e-2}{0.4500e-2}$$

$$= \frac{0.4500-0.4458}{0.4500}$$

$= 93\%$

Problem 19: What is the difference between 9.0 and 9.00?

Solution: 9.0 indicates that measurements is accurate upto first place of decimal, whereas 9.00 indicates that measurement to be correct upto second place of decimal.

Problem 20: The mass of milk in a polythene measured is 2.3 kg. Two quantities of milk masses 20.15 g and 20.17 g are added to the given milk. What is

(a) total mass of the whole milk

(b) the difference in masses of milk to correct significant figures.

Solution: Mass of milk, $m = 2.3$ kg

Mass of first quantity added, $m_1 = 20.15 \text{ g} = 0.02015$ kg and

Mass of second quantity added, $m_2 = 20.17 \text{ g} = 0.02017$ kg

(a) Total mass $= m + m_1 + m_2$

$= 2.3 + 0.02015 + 0.02017$

$= 2.34032$ kg.

As the result is correct only upto one place of decimal, therefore, on rounding off,

Total mass = 2.3 kg

(b) Difference in masses $= m_2 - m_1$

$= 20.17 - 20.15$

$= 0.02$ g (correct upto two places of decimal)

Problem 21: The length, breadth and thickness of a board are 4.234 m, 1.005 m, and 2.01 cm respectively. Compute the area and volume of the board to correct number of significant figures.

Solution: Length of board, $l = 4.234$ m; breadth, $b = 1.005$ m; thickness, $t = 2.01$ cm

$= 2.01 \times 10^{-2}$ m

$= 0.0201$ m

Area of board $= 2\,(l \times b + b \times t + t \times l\,)$

$= 2\,(4.234 \times 1.005 + 1.005 \times 0.0201 + 0.0201 \times 4.234)$

$$= 2(4.3604739)$$
$$= 8.7209478 \text{ m}^2$$

As area can contain a maximum of three significant digits, therefore, rounding off, we have,

$$\text{Area} = 8.72 \text{ m}^2$$

Also $$\text{volume} = l \times b \times t$$
$$V = 4.234 \times 1.005 \times 0.0201$$
$$= 0.0855289 \text{ m}^3$$
$$= 0.086 \text{ m}^3 \quad \text{(containing three significant digits)}$$

UNSOLVED PROBLEMS

1. What do you mean by the term error? Discuss all types of error in detail with suitable examples.
2. What are errors? Discuss the various types of errors that occurs, while performing computations. What measures cab be taken to improve accuracy in numerical computation.
3. Discuss various types of errors that occurs while performing numerical computations. Give suitable examples.
4. Round-off the following numbers correct to four significant digits:
   4.26625, 45.46826, 4985561, 0.72135, 0.00042217, 1.6583, 30.6567, 0.858378, 3.14159.
5. The height of an observation tower was estimated to be 47 m. whereas its actual height was 45 m. Calculate the percentage of relative error in the measurement.
6. If the number p is correct to three decimal places, what will be the error?
7. If true value = 10/3 , approximate value = 3.33, find the absolute and relative errors.
8. Round-off the following numbers to two decimal places.
   48.21416, 2.3742, 52.275, 2.375, 2.385, 81.255.
9. Calculate the value of $\sqrt{102} - \sqrt{101}$ correct to four significant digits.
10. If X = 2.536, find the absolute error and relative error when
    (i) X is rounded-off
    (ii) X is truncated to two decimal digits.
11. If $\pi = 22/7$ is approximated as 3.14, find the absolute error, relative error, and percentage of relative error.
12. Given the solution of a problem as X' = 35.25 with the relative error in the solution atmost 2%, find, to four decimal digits, the range of values within which the exact value of the solution must lie.
13. Given that:
    $a = 10.00 \pm 0.05$ $\qquad b = 0.0356 \pm 0.0002$

$c = 15300 \pm 100$ $d = 62000 \pm 500$

Find the maximum value of the absolute error in

(i) $a + b + c + d$ (ii) $a + 5c - d$ (iii) d^3.

14. What do you understand by machine epsilon of a computer? Explain.
15. What do you mean by truncation error? Explain with examples.
16. If $I = 4xy^2z^{-3}$ and errors in x, y, z be 0.001, show that the maximum relative error at $x = y = z = 1$ is 0.006.
17. If $I = 10x^3y^2z^2$ and errors in x, y, z are 0.03, 0.01, 0.02 respectively at $x = 3, y = 1, z = 2$. Calculate the absolute error and percentage relative error in evaluating I.
18. If $I = 4x^2y^3z^{-4}$, find the maximum absolute error and maximum relative error in I when errors in $x = 1, y = 2, z = 3$, respectively, are equal to 0.001, 0.002, 0.003.
19. Write a short note on error in a series approximation.
20. Explain the procedure of adding several numbers of different absolute accuracies.
21. Find the smaller root of the equation $x^2 - 30x + 1 = 0$ correct to three decimal places. State different algorithms. Which algorithm is better and why?
22. Write a short note on Errors in numerical computation.
23. The solution of a problem as $x = 14.65$ and the absolute error of the solution is less than 3%. Find the range of values within which the exact value of the solution lie.
24. The solution of a problem as $x = 15.45$ and the relative error of the solution is at the most than 2.5%. Find to five decimal digits the range of values within which the exact value of the solution lie.
25. The Maclaurin's expansion for e^x is given as

$$e^x = 1 + x + \frac{x^2}{2!} + \frac{x^3}{3!} + \ldots\ldots + \frac{x^{n-1}}{(n-1)!} + R_n(x)$$

Where $R_n(x) = x^n \frac{e^{\xi}}{n!}$, $0 < \xi$

Find the number of terms required for the series such that their sum gives the value of e^x correct to 8 decimal places at $x = 1$.

26. The height of a building was estimated 25 feet. The actual height was 24 feet. Find Absolute error, Relative error and Percentage error.
27. How errors are propagated during computation. Discuss propagation of errors during arithmetic operations.

3.12 MACHINE COMPUTATION

To obtain meaningful results for a given problem using computers, there are five phases:

(i) Select method

(ii) Designing algorithm

(iii) Flow-chart

(iv) Program

(v) Computer execution

A method is defined as a mathematical formula for finding the solution of a given problem. There may be more than one method available to solve the same problem. We have to select the method which suits the given problem. The inherent assumptions and limitations of the method must be studied carefully to solve problem in best way.

When the method has been decided, we must describe a complete and unambiguous set of computational steps to be followed in a particular sequence to obtain the solution. This description is called an **algorithm.** It may be emphasized that the computer is concerned with the algorithm and not with the method. The algorithm tells the computer where from start, what information required, what operations to be performed and in which order, what information to be store and printed, and when to stop.

An algorithm has five important features:

1. **Finiteness:** An algorithm must terminate after a finite number of steps.
2. **Definiteness:** Each step of an algorithm must be clearly defined.
3. **Inputs:** An algorithm must specify the quantities which must be read before the algorithm can begin.
4. **Outputs:** An algorithm must specify the quantities which are to be outputted and their proper place.
5. **Effectiveness:** An algorithm must be effective, which means that all operations are executable.

A **flow-chart** is a graphical representation of a specific sequence of steps (algorithm) to be followed by the computer to produce the solution of a given problem. It makes use of the flow chart symbols to represent the basic operations to be carried out. The various symbols are connected by arrows to indicate the flow of information and processing. While drawing a flow chart, any logical error in the formulation of the problem or application of the algorithm can be easily seen and corrected.

3.13 COMPUTER SOFTWARE

Computer software is to provide a useful computational tool for users. When we write computer software, we requires a good understanding of numerical analysis and art of programming. Good computer software must satisfy certain criteria of **self-starting, accuracy** and **reliability, minimum number of levels, good documentation, ease of use,** and **portability.**

Computer software should be self-starting as far as possible.

A numerical method very often involves parameters whose values are determined by the properties of the problem to be solved. For example, When we have to find the roots of an equation, one or more initial approximations to the root have to be given. The program will be more acceptable if it can be made automatic in the sense that the program will select the initial approximations itself rather than requiring the user to specify them. Accuracy and reliability are measures of the performance of an algorithm on all similar problems. Once an error criterion is fixed, it should produce solutions of all similar problems to that accuracy. The program should be able to prevent and handle most of the exceptional conditions like division by zero, infinite loops, etc.

The structure of the program should avoid many levels. For example, many programs used to find roots of an equation have three levels:

(i) Program calls zero-finder (parameters, function)

(ii) Zero-finder calls function

(iii) Function subprogram

The more number of levels in the program, the more time is wasted in interlinking and transfer of parameters. Documentation that is accurate and easy to use is a very important criteria. The program must have some comment lines or comment paragraphs at various places giving explanation and clarification of the method used and steps involved. Accurate documentation should clarify what kind of problems can be solved using this software, what parameters are to be supplied, what accuracy can be achieved, which method has been used, and other relevant details.

The criterion of portability means that the software should be made independent of the computer being used as far as possible. Since most machines have different hardware configuration, complete independence from the machine may not be possible. However, the aim of writing the computer software should be that the same program should be able to run on any machine with minimum modifications. Machine-dependent constants, for example machine error EPS, must be avoided or automatically generated. A standard dialect of the programming language should be used rather than a local dialect.

Most of the numerical methods are available in the form of **software,** which is a package of thoroughly tested, portable, and self documented subprograms. The general purpose packages contain a number of subroutines for solving a variety of mathematical problems that commonly arise in scientific and engineering computation. The special purpose packages deal with specified problem areas. Many computer installations require one or both types of packages and make it available, on-line, to their users. Most of the software packages are available for PCs also.

General Purpose Packages

IMSL: (International Mathematical and Statistical Library). The IMSL is a general purpose library of over 900 subroutines written in ANSI Fortran for solving a large number of mathematical and statistical problems. NAG: (Numerical Algorithms Group). This package covers the basic areas of mathematical and statistical computation. The package is available in any one of the three languages ANSI Fortran, Algol 60 or Algol 68.

Special Purpose Packages

All the following packages are distributed by IMSL.

BLAS: (Basic Linear Algebra Subroutines). BLAS contains 38 ANSI Fortran subroutines for the methods in numerical linear algebra. The objective is fast computer execution.

B-Splines: A package of subroutines for performing calculations with piecewise polynomials.

DEPACK: (Differential Equations Package). DEPACK contains Fortran subprograms for the integration of initial value problems in ordinary differential equations. This package includes Runge-Kutta methods, variable step, variable order Adams type methods, and backward differentiation methods for stiff problems.

EISPACK: (Matrix Eigensystem Routines). EISPACK contains 51 Fortran subprograms for computing the eigenvalues and/or eigenvectors of a matrix.

ELLPACK: (Elliptic Partial Differential Equations Solver). ELLPACK contains over 30 numerical method modules for solving elliptic partial differential equations in two dimensions with general domains and in three dimensions with rectangular domains. The 5-point discretization is used and the resulting system of equations is solved by Gauss elimination for band matrices and by SOR iterations.

FISHPACK: (Routines for the Helmholtz Problems in Two or Three Dimensions). FISHPACK contains a set of Fortran programs for solving Helmholtz problems in two or three dimensions. There are separate programs for rectangular, polar, spherical and cylindrical coordinates.

FUNPACK: (Special Function Subroutines). The FUNPACK package contains Fortran and assembly language subroutines for evaluating important special functions like exponential integral, elliptic integrals of first and second kind, Bessel functions, Dawson integrals, etc.

ITPACK: (Iterative Methods). ITPACK contains Fortran subprograms for iterative methods for solving linear system of equations. The package is oriented towards the sparse matrices that arise in solving partial differential equations and in other applications.

LINPACK: (Linear Algebra Package). LINPACK contains Fortran subprograms for direct methods for general, symmetric, symmetric positive definite, triangular, and tridiagonal matrices. The package also includes programs for least-squares problems, along with the QR and singular value decompositions of rectangular matrices.

MINPACK: MINPACK is a package of subroutines for solving systems of nonlinear equations and nonlinear least-squares problems. The package also includes programs for minimization and optimization problems.

QUADPACK: QUADPACK contains subroutines for evaluating a definite integral.

Software packages for PCs are also available for most of the areas mentioned above.

3.14 COMMON SOURCES OF ERROR IN NUMERICAL COMPUTATIONS

1. **There is no such thing as "human error"!** This vague phrase does not describe the source of error clearly. Careful description of sources of error allows future experimenters to improve on your techniques. This **long list** of common sources of error is meant to help you identify some of the common sources of error you might encounter while doing experiments. If you find yourself stuck for words when describing sources of error, this list may help. The list goes from the common to the obscure.
2. **Incomplete definition** (may be systematic or random): One reason that it is impossible to make exact measurements is that the measurement is not always clearly defined. For example, if two different people measure the length of the same rope, they would probably get different results because each person may stretch the rope with a different tension. The best way to minimize definition errors is to carefully consider and specify the conditions that could affect the measurement.
3. **Failure to account for a factor** (usually systematic): The most challenging part of designing an experiment is trying to control or account for all possible factors except the one independent variable that is being analyzed. For instance, you may inadvertently ignore air resistance when measuring free-fall acceleration or you may fail to account for the effect of the Earth's magnetic field when measuring the field of a small magnet. The best way to account for these sources of error is to brainstorm with your peers about all the factors that could possibly affect your result. This brainstorm should be done before beginning the experiment so that arrangements can be made to account for the confounding factors before taking data. Sometimes a correction can be applied to a result after taking data, but this is inefficient and not always possible.
4. **Environmental factors** (systematic or random): Be aware of errors introduced by your immediate working environment. You may need to take account for or protect your experiment from vibrations, drafts, changes in temperature, electronic noise or other effects from nearby apparatus.
5. **Instrument resolution** (random): All instruments have finite precision that limits the ability to resolve small measurement differences. For instance, a meter stick cannot distinguish distances to a precision much better than about half of its smallest scale division (0.5 mm in this case). One of the best ways to obtain more precise measurements is to use a null difference method instead of measuring a quantity directly. Null or balance methods involve using instrumentation to measure the difference between two similar quantities, one of which is known very accurately and is adjustable. The adjustable reference quantity is varied until the difference is reduced to zero. The two quantities are then balanced and the magnitude of the unknown quantity can be found by comparison with the reference sample. With this method, problems of source instability are eliminated, and the measuring instrument can be very sensitive and does not even need a scale.
6. **Failure to calibrate or check zero of instrument** (systematic): Whenever possible, the calibration of an instrument should be checked before taking data. If a calibration standard

is not available, the accuracy of the instrument should be checked by comparing with another instrument that is at least as precise, or by consulting the technical data provided by the manufacturer. When making a measurement with a micrometer, electronic balance, or an electrical meter, always check the zero reading first. Re-zero the instrument if possible, or measure the displacement of the zero reading from the true zero and correct any measurements accordingly. It is a good idea to check the zero reading throughout the experiment.

7. **Physical variations** (random): It is always wise to obtain multiple measurements over the entire range being investigated. Doing so often reveals variations that might otherwise go undetected. If desired, these variations may be cause for closer examination, or they may be combined to find an average value.
8. **Parallax** (systematic or random): This error can occur whenever there is some distance between the measuring scale and the indicator used to obtain a measurement. If the observer's eye is not squarely aligned with the pointer and scale, the reading may be too high or low (some analog meters have mirrors to help with this alignment).
9. **Instrument drift** (systematic): Most electronic instruments have readings that drift over time. The amount of drift is generally not a concern, but occasionally this source of error can be significant and should be considered.
10. **Lag time and Hysteresis** (systematic): Some measuring devices require time to reach equilibrium, and taking a measurement before the instrument is stable will result in a measurement that is generally too low. The most common example is taking temperature readings with a thermometer that has not reached thermal equilibrium with its environment.

A similar effect is hysteresis where the instrument readings lag behind and appear to have a "memory" effect as data are taken sequentially moving up or down through a range of values. Hysteresis is most commonly associated with materials that become magnetized when a changing magnetic field is applied.

QUERIES

1. Write a short note on machine computation
2. Discuss various sources of errors.
3. Write down five important features of an algorithm.
4. Why we use computer software in numerical problems?
5. Explain various software's used for numerical problems.

CHAPTER 4

Solution of Transcendental and Polynomial Equations

4.1 INTRODUCTION

Let us consider the equation of the form $f(x) = 0$.

The equation may belong to one of the following types:

(i) Algebraic equations

(ii) Polynomial equations

(iii) Transcendental equations

An expression of the form $f_n(x) = a_0x^n + a_1x^{n-1} + a_2x^{n-2} + ... + a_{n-1}x + a_n$ is known as a **polynomial** in x of degree n, provided $a_0 \neq 0$, where $a_0, a_1, a_2, ..., a_n$ are called constants, independent of x and n is a positive integer. If we put any polynomial equals to zero it becomes an equation e.g. $f_n(x) = 0$ is an equation in x. Any value of x which makes $f_n(x)$ equal to zero is called a zero of the polynomial $f_n(x)$ or root of the equation $f_n(x) = 0$. Every polynomial $f_n(x)$ of degree n has exactly n zeros. Geometrically, the zero of the polynomial $f_n(x)$ is the value of x where the graph of $f_n(x)$ crosses the axis of x. The equation $f_n(x) = 0$ is known as **Algebric equation.** A non-algebraic equation is called **Transcendental equation**. These equations include trigonometric, exponential, or logarithmic functions e.g., $\sin x - 3x = 0$, $ae^{-x} + b\tan x = 0$, $\log_{10}x - x^3$ etc.

In this chapter, we have different methods for the solution of such type of equations. The methods discuss in this chapter are:

The methods for finding roots numerically may broadly be classified into following two types.

1. Direct Methods

2. Iterative Methods (Indirect methods)

Direct methods give the roots of an equation in a finite number of steps whereas iterative methods give the root after certain number of iterations. Now we discuss some major **differences** exist between these two types of methods.

Direct Methods	Iterative Methods
1. These types of methods give roots of an equation in a finite number of steps. e.g. → A quadratic equation $Ax^2 + Bx + C = 0$ where $A \neq 0$ give both the roots x_1, x_2 at same time as $x_{1,2} = (-B \pm \sqrt{B^2-4AC})/2A$.	1. These methods give the root after certain number of iterations.
2. The amount of computation involved can be specified in advance.	2. No such type of computation done in advance.
3. These require no knowledge of an initial approximation.	3. These methods are based on successive approximation they start with a first initial approximation and successively improve it.
4. These methods are independent of the accuracy desired.	4. The accuracy of the solution does not depend on the number of iterations performed.
5. The computation can be completed in a finite number of steps resulting in the exact solution.	5. The computation would require an infinite number of steps to obtain exact solution in the absence of round-off errors.
6. These methods are usually used for dense matrices.	6. These methods are usually used for sparse matrices (that contains very few non-zero elements such matrices are large in size).
7. Round off errors are not negligible.	7. Round off errors are negligible.
8. They can't be expressed as Computational algorithm.	8. They can be expressed as computational algorithm.

4.2 INITIAL APPROXIMATIONS

In iterative methods, we start the process with one or more values and these values generally called as 'initial conditions' or 'initial approximations'. The initial approximation, for any iterative method is to be chosen as close to the required solution as possible. But one question arises as to how we can select the initial approximation when we do not know the solution at all! Answer for this question is that the selection for initial approximation is done by trial and error methods. By trial we start with an initial approximation and for each iteration take the output and after a few iterations we know form the output whether the process is converging or not. If the process is not converging then we change the initial approximation. This process is time consuming but solves the problem.

The initial approximation to a problem can also be calculated based on **graphical** method.

4.2.1 GRAPHICAL METHOD

This method is quite useful in locating approximately the roots of an equation $f(x) = 0$. Frequently, the equation $f(x) = 0$ can be written as $f_1(x) = f_2(x)$ and the roots of $f(x) = 0$ are determined by the intersection of the graphs of the two curves $y = f_1(x)$ and $y = f_2(x)$. We consider the intersection of the graphs $y = f(x)$ and $y = 0$. For example, to solve the equation $f(x) = x^n + ax + b = 0$.

We draw the graphs of $y = x^n$ and $y = -(ax + b)$ and then their point of intersection. In order to find the roots of the equation $e^x \sin x = 1$, we write it as $\sin x = e^{-x}$ and then determine the intersection of $y = \sin x$ with $y = e^{-x}$

We explain the graphical method by considering an example as:

Let $f(x) = x^3 - 2.5x^2 - 2.46x + 3.96 = 0$, it is plotted in the below figure. From the sketch, it is seen that this function has three roots. They are approximately -1.5, 1 and 2.5. Thus it is easy to guess an approximate value for a desired root and refine it using an iterative method.

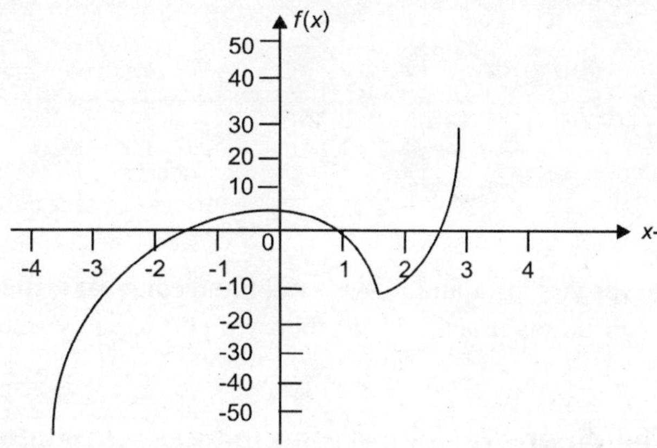

An algorithm to tabulate an arbitrary function is given below. Let x_{min} and x_{max} be the end points of the interval over which $f(x)$ is to be tabulated. Let Δx be the tabulation interval.

Algorithm to tabulate value of *f*(*x*)

1. Read the value of x_{min} , x_{max}, Δx
2. Set $x = x_{min}$
3. while $x \leq x_{max}$ do
   begin
4. Set $y = f(x)$
5. write x and y
6. Set $x = x + \Delta x$
   End
7. Stop

The above example involves a simple function in x if the function is complicated then we use the following approach.

Given $f(x) = 0$, we split $f(x)$, if possible and re-arrange it such that

$$f(x) = g(x) \pm h(x)$$

Where $g(x)$, $h(x)$ are the terms of $f(x) = 0$.

Now, plot $y_1 = g(x)$, $y_2 = h(x)$. The points of intersection of these two curves give the solution (x, y) which is the root of the desired equation. Let us consider an example to illustrate the above analysis.

Consider $f(x) = \sin x + x - \pi/2 = 0$

By splitting the function denote by $y_1 = \sin x$ and $y_2 = x - \pi/2$

The graphs of the two functions are plotted below:

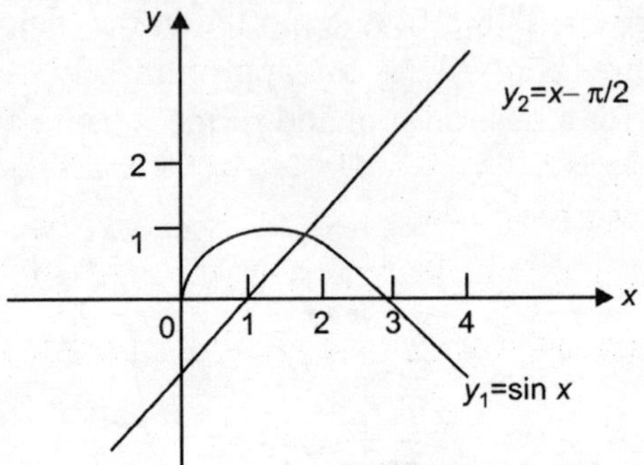

We note that the curves $y_1 = \sin x$ and $y_2 = x - \pi/2$ intersect, near (1.6, 0.8) and hence our initial value in this example can be taken as 2.4 and then we employ any one of the suitable iterative method.

4.2.2 What Happens When We Take Wrong Initial Approximation?

There are number of ways to choose an initial approximation. The choice of an initial approximation itself may lead to convergent or divergent situation. Hence the choice of an initial approximation must be made with care. For each set of iteration values one can observe whether we are converging or not. Situations where an initial approximation x_0 is selected in two different cases leads to converging in the first case and diverging in other case is clear form the following two graphs.

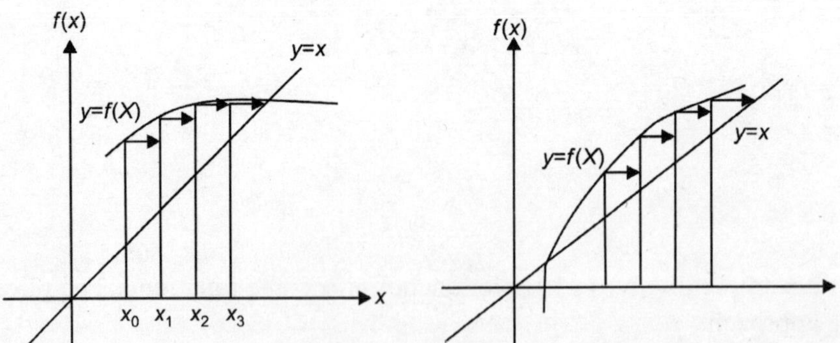

We note that fig (i) represents a convergent process, whereas fig (ii) is divergent. In fig (i) we find the difference between the approximated roots to the exact root is always positive and hence convergence is achieved form one direction. This is known as monotonic convergence. However, there are situations where the value starts oscillating on either side of the true solution. This is generally known as oscillatory convergence.

4.3 WHEN TO TERMINATE AN ITERATIVE PROCESS

We say that the iterative process is continuing till the required accuracy in the solution is achieved. But the new question arises how we can measure the accuracy. Answer is that the prescribed tolerance in the root is epsilon ($\in$).

1. $|x_{i+1} - x_i| \leq \in$ (Absolute error in x)
2. $\left|\dfrac{x_{i+1} - x_i}{x_i}\right| \leq \in$ (Relative error in x), $x \neq 0$
3. $|f(x_{i+1})| \leq \in$ (value of function at root)
4. $|f(x_{i+1}) - f(x_i)| \leq \in$ (Difference in function values)

From the above rules, second rule is more effective than others.

4.4 RATE OF CONVERGENCE OF ITERATIVE METHODS

In numerical analysis, the speed at which a convergent sequence approaches its limit is called the **rate of convergence** i.e. the rate of convergence is measured in terms of the rate at which the relative error decreses between successive approximations or the number of significant digits increases towards the limiting value of the series. On the basis of this we have derived a concept that more the order of rate of convergence more faster will be the method. If we have a method of order ***m*** then it means that in each iteration the number of significant digits incresaes ***m*** times and the error in each step is proportional to the ***m***

the error in each step is proportional to the $\boldsymbol{m}^{\text{th}}$ power of the error in the preecding step. Thus an iterative method is m^{th} order convergent if we have

$$\underset{i \to \infty}{\text{Lt}}\; e_{i+1} \leq P.(e_i)^m$$

Where e_i , e_{i+1} are the errors in successive approximations and P is any finite number.

4.4.1 Types of Convergence

We have mainly two types of convergence:

1. **Linear convergence:** If the error between successive approximations decreases at linear rate then we can say that it is an linear convergence. Linear convergence is also known as convergence of order 1. Linear convergence is known to be slow convergence.
2. **Quadratic Convergence:** In case of quadratic convergence the convergence is of order 2. it is known to be the fast convergence as compared to linear convergence.

In this chapter, we have discussed the methods whose rate of convergence lies between order 1 to 2.

4.4.2 Relationship Between Speed and Rate of Convergence of an Method

There is a simple relationship exists between speed and rate of convergence as if an method has convergence of more order that method will converge fast. From the fast convergence means the method will converge earlier towards the root i.e we have to compute lesser number of iterations.

QUERIES

1. Write the differences between direct and iterative methods.
2. Discuss those parameters, which are helpful for selecting an iterative method.
3. Discuss the rules for stopping an iterative procedure.
4. What do you mean by the term rate of convergence?

4.5 ONE DIMENSIONAL ROOT FINDING ALGORITHM

There are two types of one-dimensional root finding algorithms that we have found in numerical analysis:

1. Root bracketing algorithms
2. Root polishing algorithms

1. Root Bracketing Algorithm

In root bracketing algorithm, a root is bracketed in the interval [a, b] if f(a) and f(b) are of opposite signs. If function is continuous, then at least one root must lie in that interval. Algorithms that proceed by bracketing a root are guaranteed to converge.

Bracketing algorithm begin with a bounded region known to contain a root. The size of this bounded region is reduced until it encloses the root to desired tolerance. Following methods come under the category of root bracketing.

(i) Bisection method

(ii) False position method

2. Root Polishing Algorithm

This technique attempts to improve an initial guess value to the root. These algorithms converge only if started value is "**very close**" to a root and provide rapid convergence. From rapid convergence, we mean that these methods give us the root in quick succession. These methods are normally preferred during solving non- linear equations. Some following methods come under the category of root polishing.

(i) Secant method

(ii) Newton's Raphson method

(iii) Method of successive approximations

4.6 BISECTION METHOD (BOLZANO'S METHOD OR METHOD OF HALF INTERVAL)

This method is one of the simplest iterative methods based on the following theorem.

Theorem: If a function $f(x)$ is continuous in the interval $[a, b]$ and $f(a)$ and $f(b)$ are of opposite signs, then there exists at least one root(say a) between a and b such that $f(a) = 0$. Where a gives the exact root of the function.

The method of bisection is that which is always convergent but the convergence is very slow but sure.

Given $f(x) = 0$ we are interested in finding a root x = a such that $f(a) = 0$, i.e where the curve cuts the x-axis in a plane of x and $y(=f(x))$.

Let us assume that for two values $[x_1, x_2]$ where $f(x_1)$ and $f(x_2)$ give opposite sign values. Bisect the interval $[x_1, x_2]$ and find its mid point i.e. $x_3 = \dfrac{x_1 - x_2}{2}$. Compute $f(x_3)$ and note its sign.

Graphically the analysis is shown in following figure.

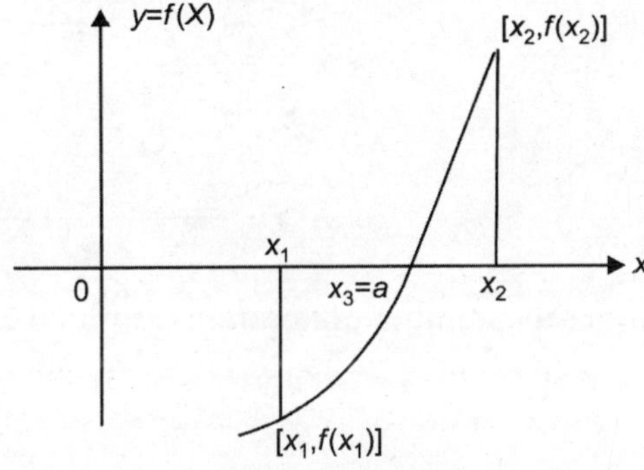

Bisection Method

After notifying the sign following three possibilities can occur:

1. if $f(x_3) = 0$, then the desire root is x_3(i.e. first mid point value).
2. if $f(x_3)$ = -ve, then the root lies between interval $[x_3,x_2]$ and we change the interval $[x_1,x_2]$ to $[x_3,x_2]$ and bisect the new interval $[x_3,x_2]$ which is half of length to the previous interval.
3. if $f(x_3)$ = +ve, then the root lies between interval $[x_1,x_3]$ and we change the interval $[x_1,x_2]$ to $[x_1,x_3]$ and bisect the new interval $[x_1,x_3]$ which is half of length to the previous interval.

This iterative process terminates when the relative error in two successive approximations is less than or nearly equal to the prescribed tolerance. As the interval is bisected by a factor of 2, so after n such bisections we find the root $x = \alpha$ as lying in the interval of length $\dfrac{|x_2 - x_1|}{2^n}$.

Since after n steps, the resultant interval will be of length $\dfrac{|x_2 - x_1|}{2^n}$.

$$\therefore \qquad \frac{|x_2 - x_1|}{2^n} \le \varepsilon$$

or
$$\frac{|x_2 - x_1|}{\in} \leq 2^n$$

$$2^n \geq \frac{|x_2 - x_1|}{\in}$$

By taking logarithms, we have

$$n \log 2 = \log\left(\frac{|x_2 - x_1|}{\in}\right)$$

or
$$n \geq \frac{\left[\log\left(\frac{|x_2 - x_1|}{\in}\right)\right]}{\log 2}$$

or
$$n \geq \frac{\log (x_2 - x_1) - \log \in}{\log 2}$$

The above inequality gives the number of iterations required to achieve accuracy ε. The interval is bisected by a factor of 2^n each time and hence larger number of bisections is not required for convergence.

4.6.1 Criteria For Ending Bisection Process

If ε(epsilon) is the prescribed tolerance (for accuracy) in the root, which may be a very very small value then the iteration cycle terminates when the following condition holds.

$$|x_n - x_{n-1}| \leq \varepsilon$$

i.e. the absolute error becomes less than or equal to ε.

4.6.2 What To Do If Interval Is Not Given

If the starting interval (in which exact root lie) is not given. Give different values to the independent variable (say x) in the function. If there is a change in sign of $f(x)$ either from –ve to +ve or from +ve to –ve for a given set of values, it is implied that a root lies between these two values.

4.6.3 Rate of Convergence of Bisection Method

In the Bisection method, the original interval is divided into half interval in each iteration. If we take mid-points of successive intervals to be the approximations of the root, one half of the current interval is the upper bound to the error.

In Bisection method,

$$e_{i+1} = 0.5\, e_i \quad \text{or} \quad \frac{e_{i+1}}{e_i} = 0.5$$

where e_i and e_{i+1} are the errors in the ith and $(i + 1)^{th}$ iterations, respectively.

Comparing (1) with

$$\lim_{i \to \infty} \left| \frac{e_{i+1}}{e_i^{\,k}} \right| \le A$$

we get $k = 1$ and $A = 0.5$

Thus the Bisection method is I order convergent, or linearly convergent.\

4.6.4 Algorithm For Bisection Method

Suppose we have given a function $f(x) = 0$ whose root lies between the interval $[x_1, x_2]$. Let the root is denoted by α, i denotes the iteration, and epsilon is the prescribed tolerance for root .

Then algorithm for bisection method is follows as:

1. Read x_1, x_2, epsilon
2. Set $i = 1$
3. if$(f(x_1) \times f(x_2) > 0)$
4. write:" The given values are not suitable"
5. do
6. set $x_{mid} = (x_1 + x_2)/2$
7. if$(f(x_{mid}) = = 0)$
8. write: " The root is $= x_{mid}$"
9. if $(f(x_1) \times f(x_{mid}) < 0)$ then
10. set $x_2 = x_{mid}$
11. else
12. set $x_1 = x_{mid}$
13. $i \to i + 1$
14. while $(|x_1 - x_2| \ge \text{epsilon})$ then
15. write : x_{mid}," as the required approximated root"
16. End

4.6.5 Flowchart For Bisection Method

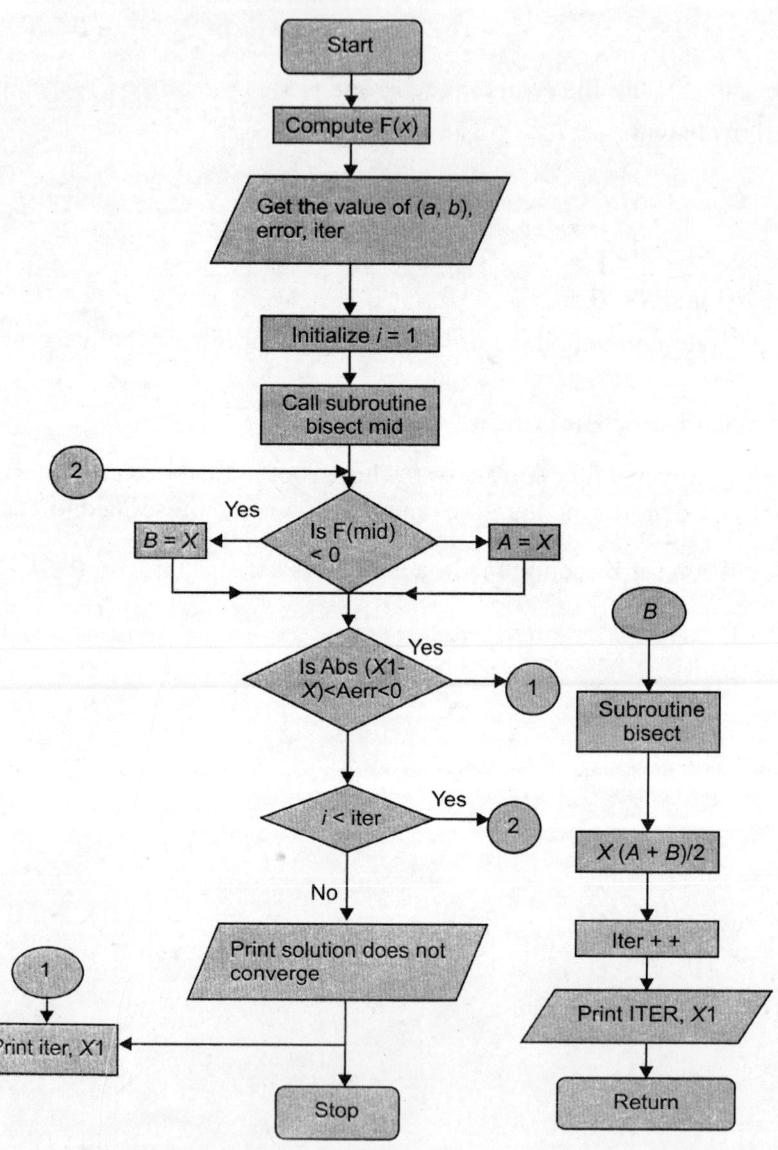

4.6.6 Program For Bisection Method in C

```
#include<stdio.h>
#include<math.h>
#include<conio.h>
#include<process.h>
```

```
#include<string.h>
#define EPS 0.00000005
#define F(x) (x)*log10(x)–1.2
void Bisect();
int count=1,n;
float root=1;
void main()
{
clrscr();
printf("\n BISECTION method \n");
printf("\n Equation to solve is ");
printf("\n\t\t\t x*log(x) – 1.2 = 0\n\n");
printf("Enter the no. of iterations:");
scanf("%d",&n);
Bisect();
getch();
}
void Bisect()
{
float x0,x1,x2;
float f0,f1,f2;
int i=0;
for(x2=1;;x2++)
{
f2=F(x2);
if (f2>0)
{
break;
}
}
for(x1=x2-1;;x2—)
{
f1=F(x1);
if(f1<0)
```

```
{
break;
}
}
//...Printing Result
printf("\n\t\t ITERATIONS\t\t ROOTS\n");
for(;count<=n;count++)
{
x0=(x1+x2)/2.0;
f0=F(x0);
if(f0==0)
{
root=x0;
}
if(f0*f1<0)
{
x2=x0;
}
else
{
x1=x0;
f1=f0;
}
printf("\n\t\t ITERATION %d", count);
printf("\t :\t %f",x0);
if(fabs((x1-x2)/x1) < EPS)
{
printf("\n\t\t Root = %f",x0);
printf("\n\t\t Iterations = %d\n", count);
getch();
exit(0);
}
}
printf("\n\t\t————————————————————————————");
```

```
printf("\n\t\t\t Root = %7.4f",x0);
printf("\n\t\t\t Iterations = %d\n", count-1);
printf("\t\t————————————————————————————————————————");
getch();
}
```

Output:

BISECTION method
Equation to solve is
x* log(x) - 1.2=0
Enter the number of iterations: 30

ITERATIONS	**ROOTS**
ITERATION 1:	2.500000
ITERATION 2:	2.750000
ITERATION 3:	2.625000
ITERATION 4:	2.687500
ITERATION 5:	2.718750
ITERATION 10:	2.741211
ITERATION 11:	2.740723
ITERATION 12:	2.740479
ITERATION 13:	2.740601
ITERATION 14:	2.740662
ITERATION 15:	2.740631
ITERATION 16:	2.740646
ITERATION 17:	2.740639
ITERATION 18:	2.740643
ITERATION 19:	2.740644
ITERATION 20:	2.740645
ITERATION 21:	2.740646
ITERATION 22:	2.740646
ITERATION 23:	2.740646
ITERATION 24:	2.740646
ITERATION 25:	2.740646
ITERATION 26:	2.740646

```
ITERATION 27:          2.740646
ITERATION 28:          2.740646
ITERATION 29:          2.740646
ITERATION 30:          2.740646
--------------------------------------
Iterations = 30        Root = 2.7406
--------------------------------------
```

4.6.7 Program For Bisection Method in C++

```
#include<iostream.h>
#include<conio.h>
#include<math.h>
#include<stdio.h>
#include<iomanip.h>
#define f(x)(x*x*x-4*x-9)
void main()
{
clrscr();
double xmid,x1,x2,epsilon;
int i=1;
cout<<"Enter the given interval values for x1 and x2 =";
cin>>x1>>x2;
cout<<"Enter the prescribed tolerance value 'epsilon'=";
cin>>epsilon;
if(f(x1)*f(x2)>0)
{
cout<<"The given values are not suitable";
getch();
}
do
{
xmid=(x1+x2)/2;
cout<<"After the "<<i<<" iteration"<<" computed xmid= "<<xmid<<endl;
if(f(xmid)==0)
```

```
{
cout<<"The root is = "<<xmid;
getch();
}
if(f(x1)*f(xmid)<0)
x2=xmid;
else
x1=xmid;
i=i+1;
}
while(fabs(x1-x2)>=epsilon);
cout<<"The approximated root = " <<xmid;
getch();
}
```

Output:

```
Enter the given values for x1 and x2 = 2  3
Enter the Prescribed tolerance value 'epsilon' = 0.0001
After the 1 iteration computed xmid = 2.5
After the 2 iteration computed xmid = 2.75
After the 3 iteration computed xmid = 2.625
After the 4 iteration computed xmid = 2.6875
After the 5 iteration computed xmid = 2.71875
After the 6 iteration computed xmid = 2.703125
After the 7 iteration computed xmid = 2.710938
After the 8 iteration computed xmid = 2.707031
After the 9 iteration computed xmid = 2.705078
After the 10 iteration computed xmid = 2.706055
After the 11 iteration computed xmid = 2.706543
After the 12 iteration computed xmid = 2.706299
After the 13 iteration computed xmid = 2.706421
After the 14 iteration computed xmid = 2.706482
The approximated root = 2.706482
```

SOLVED PROBLEMS

Problem 1: Assuming that a root of $x^3 - 9x + 1 = 0$ lies in the interval (2, 4). Find that root by bisection method.

Solution: Let $f(x) = x^3 - 9x + 1$

We have, $f(2) = \text{-ve}$ and $f(4) = \text{+ve}$

∴ root lies between 2 and 4.

Let $x_0 = (2+4)/2 = 3$

Now $f(3) = \text{+ve}$; hence root lies between 2 and 3 as they give opposite sign function values.

$$x_1 = (2+3)/2 = 2.5$$

$f(2.5) = \text{-ve}$

Thus, root lies between 2.5 and 3

$$x_2 = (2.5 + 3)/2 = 2.75$$

$f(2.75) = \text{-ve}$

Thus, root lies between 2.75 and 3

$$x_3 = (2.75+3)/2 = 2.875$$

$f(2.875) = \text{-ve}$

Thus, root lies between 2.875 and 3

$$x_4 = (2.875+3)/2 = 2.9375$$

$f(2.9375) = \text{-ve}$

∴ root lies between 2.9375 and 3

$$x_5 = (2.9375+3)/2 = 2.9688$$

$f(2.9688) = \text{+ve}$

∴ Now root lies between 2.9375 and 2.9688 as they show opposite (sign) function values.

$$x_6 = (2.9375+2.9688)/2 = 2.9532$$

$f(2.9532) = \text{+ve}$

∴ root lies between 2.9375 and 2.9532

$$x_7 = (2.9375+2.9532)/2 = 2.9454$$

$f(2.9454) = \text{+ve}$

∴ root lies between 2.9375 and 2.9454

$$x_8 = (2.9375+2.9454)/2 = 2.9415$$

$f(2.9415) = \text{-ve}$

Thus, root lies between 2.9415 and 2.9454

$$x_9 = (2.9415+2.9454)/2 = 2.9435$$

$f(2.9435) = \text{+ve}$

Thus, root lies between 2.9415 and 2.9435

$$x_{10} = (2.9415+2.9435)/2 = 2.9425$$

$f(2.9425) =$ -ve

Thus, root lies between 2.9425 and 2.9435

$$x_{11} = (2.9425+2.9435)/2 = 2.9430$$

$f(2.9430) =$ +ve

Similarly, $x_{12} = 2.94275$

$$x_{13} = 2.942875$$

At last, we have approximate root is 2.9429 at which given function gives the function value nearly equal to zero.

Problem 2: Find the root of $x - \cos x = 0$ by using bisection method.

Solution: Let $f(x) = x - \cos x$

$f(0) =$ -ve , $f(0.5) =$ -ve and $f(1) =$ +ve

As both $f(0)$ and $f(0.5)$ gives us –ve value now question arises which value we have to taken for starting interval i.e whether we start with (0,1) or (0.5,1), and the answer for this question is that we have to take the interval (0.5,1) as in this interval difference between both values(0.5 and 1) is less due to which number of iterations required also decreases.

Hence, as the root lies between 0.5 and 1

$$x_0 = (0.5+1)/2 = 0.75$$

$$f(0.75) = 0.75 - \cos(0.75) = 0.018311 = +\text{ve}$$

$\therefore$ root lies between 0.5 and 0.75

$$x_1 = (0.5+0.75)/2 = 0.625$$

$$f(0.625) = 0.625 - \cos(0.625) = -0.18596 = -\text{ve}$$

$\therefore$ root lies between 0.625 and 0.75

$$x_2 = (0.625+0.75)/2 = 0.6875$$

$$f(0.6875) = -0.085335 = -\text{ve}$$

$\therefore$ root lies between 0.6875 and 0.75

$$x_3 = (0.6875+0.75)/2 = 0.71875$$

$$f(0.71875) = 0.71875\text{-}\cos(0.71875) = -0.033879 = -\text{ve}$$

$\therefore$ root lies between 0.71875 and 0.75

$$x_4 = (0.71875+0.75)/2 = 0.73438$$

$$f(0.73438) = -0.0078664 = -\text{ve}$$

$\therefore$ root lies between 0.73438 and 0.75

$$x_5 = (0.73438+0.75)/2 = 0.742190$$
$$f(0.742190) = 0.0051999 = +\text{ve}$$

∴ root lies between 0.73438 and 0.742190.

$$x_6 = (0.73438+0.742190)/2 = 0.73829$$
$$f(0.73829) = -0.0013305 = -\text{ve}$$

∴ root lies between 0.73829 and 0.742190

$$x_7 = (0.73829+0.742190)/2 = 0.7402$$
$$f(0.7402) = 0.7402 - \cos(0.7402) = 0.0018663 = +\text{ve}$$

∴ root lies between 0.73829 and 0.7402

$$x_8 = (0.73829+0.7402)/2 = 0.73925$$
$$f(0.73925) = 0.00027593 = +\text{ve}$$

Similarly, $x_9 = 0.7388$

Thus the exact root at which function gives value equal to zero or nearly equal to zero is 0.7388.

Problem 3: Find the negative root of $x^3 - 4x + 9 = 0$ by using bisection method.

Solution: Let
$$f(x) = x^3 - 4x + 9$$
$$f(-x) = -x^3 + 4x + 9$$

The negative root of $f(x) = 0$ is the positive root of $f(-x) = 0$.

∴ we will find the positive root of $f(-x) = 0$, firstly,

i.e.,
$$\phi(x) = x^3 - 4x - 9 = 0$$
$$\phi(2) = -\text{ve and j}(3) = +\text{ve}$$

Hence, root lies between 2 and 3. By following bisection process we conclude the root to be 2.7065 after doing some iterations for $\phi(x) = x^3 - 4x - 9 = 0$.

Hence, the negative root of the given equation $x^3 - 4x + 9 = 0$ is -2.7065.

Problem 4: Compute five iterations of the bisection method to obtain the smallest positive root of equation $f(x) = x^3 - 5x + 1 = 0$.

Solution:
$$f(x) = x^3 - 5x + 1$$

Since
$$f(.2016) = .0001935 \text{ i.e., (+)ve}$$

and
$$f(.2017) = -.0002943 \text{ i.e., (–)ve}$$

Hence, the root lies between .2016 and .2017.

First approximation to the root is
$$x_1 = (.2016 + .2017)/2 = .20165$$

Now
$$f(x_1) = -.00005036 \text{ i.e., (–)ve}$$

Hence, the root lies between .2016 and .20165.

Second approximation to the root is

$$x_2 = (.2016 + .20165)/2 = .201625$$

Now $f(x_2) = .00007159$ i.e., (+)ve

Hence, the root lies between .201625 and .20165.

Third approximation to the root is

$$x_3 = (.201625 + .20165)/2 = .2016375$$

Now $f(x_3) = .00001061$ i.e., (+)ve

Hence, the root lies between .2016375 and .20165.

Fourth approximation to the root is

$$x_4 = (.2016375 + .20165)/2 = .20164375$$

Now $f(x_4) = -.00001987$ i.e., (–)ve

Hence, the root lies between .2016375 and .20164375.

Therefore, Fifth approximation to the root is

$$x_5 = (.2016375 + .20164375)/2 = .201640625$$

Hence, after performing five iterations, the **smallest positive root** of the given equation is **.20164,** correct to **five decimal places.**

Problem 5: Find a real root of $x^3 - x = 1$ between 1 and 2 by bisection method. Compute five iterations.

Solution: Here, $f(x) = x^3 - x - 1$

Since $f(1.324) = -.00306$ i.e., (–)ve

and $f(1.325) = .00120$ i.e., (+)ve

Hence, the root lies between 1.324 and 1.325.

First approximation to the root is

$$x^1 = (1.324 + 1.325)/2 = 1.3245$$

Now $f(x^1) = -.000929$ i.e., (–)ve

Hence, the root lies between 1.3245 and 1.325

Second approximation to the root is

$$x^2 = (1.3245 + 1.325)/2 = 1.32475$$

Now $f(x^2) = .000136$ i.e., (+)ve

Hence, the root lies between 1.3245 and 1.32475.

Third approximation to the root is

$$x^3 = (1.3245 + 1.32475)/2 = 1.324625$$

Now $f(x^3) = -.000396$ i.e., (–)ve

Hence, the root lies between 1.324625 and 1.32475.

Fourth approximation to the root is

$$x^4 = (1.324625 + 1.32475)/2 = 1.3246875$$

Now $f(x^4) = -.0001298$ i.e., (–)ve

Hence, the root lies between 1.3246875 and 1.32475

Therefore, Fifth approximation to the root is

$$x^5 = (1.3246875 + 1.32475)/2 = 1.32471875$$

Hence, the real root of the given equation is **1.324** correct to three decimal places after computing five iterations.

Problem 6: Use bisection method to Compute out the positive square root of 30 correct to 4 decimal places.

Solution: Let $f(x) = x^2 - 30$

Since $f(5.477) = -.00247$ i.e., (–)ve

and $f(5.478) = .00848$ i.e., (+)ve

Hence, the root lies between 5.477 and 5.478

First approximation to the root is

$$x^1 = (5.477 + 5.478)/2 = 5.4775$$

Now $f(x^1) = .003$ i.e., (+)ve

Hence, the root lies between 5.477 and 5.4775

Second approximation to the root is

$$x^2 = (5.477 + 5.4775)/2 = 5.47725$$

Now $f(x^2) = .00026$ i.e., (+)ve

Hence, the root lies between 5.477 and 5.47725

Third approximation to the root is

$$x^3 = (5.477 + 5.47725)/2 = 5.477125$$

Now $f(x^3) = -.0011$ i.e., (–)ve

Hence, the root lies between 5.477125 and 5.47725

Fourth approximation to the root is

$$x^4 = (5.477125 + 5.47725)/2 = 5.4771875$$

Since x^3 and x^4 are the same up to four decimal places, the positive square root of 30, correct to 4 decimal places, is **5.4771.**

UNSOLVED PROBLEMS

1. Find a real root of $e^x = 3x$ by bisection method.
2. Find the smallest positive root of $x^3 - 9x + 1 = 0$, using Bisection method correct to three decimal places.

3. Find the real root lying in interval (1, 2) up to four decimal places for the equation $x^6 - x^4 - x^3 - 1 = 0$ by bisection method.
4. Compute the root of $\tan x + x = 0$ up to two decimal places which lies between 2 and 2.1 using Bisection method.
5. Compute the root of $\log x = \cos x$ correct to two decimal places using Bisection method.
6. Compute the root of $f(x) = \sin 10x + \cos 3x$ using Bisection method. The initial approximations are 4 and 5.
7. Find the real root correct to three decimal places for the following equations:

   (i) $x^3 - x - 2 = 0$ (ii) $x^3 - x^2 - 1 = 0$

   (iii) $x^3 + x^2 - 1 = 0$ (iv) $x^3 - 3x - 5 = 0$.
8. Compute a root of $x^3 - x - 11 = 0$ using Bisection method correct to 3 decimal places which lies between 2 and 3.
9. Find a real root of the equation $x^3 - 2x - 5 = 0$ using Bisection method.
10. Find a positive root of the equation $xe^x = 1$ which lies between 0 and 1.
11. Compute a root of the equation $x^4 + 2x^3 - x - 1 = 0$ in the interval [0, 1].
12. Obtain a root correct to four decimal places for each of these equations using Bisection method.

    (i) $x^3 + x^2 + x + 1 = 0$ (ii) $x^3 - 16 = 0$

    (iii) $x^3 + x - 1 = 0$ (iv) $x^3 - 5x + 3 = 0$.
13. By displaying procedure in tabular form, use Bisection method to compute the root of 36.
14. Find a positive root of the equation $x^3 + 3x - 1 = 0$ by bisection method.
15. Find a real root of $x^3 - 2x - 1 = 0$ which lies between 1 and 2 by using Bisection method correct to 2 decimal places.
16. Compute the approximate value of the root of the equation $3x - \sqrt{1 + \sin x} = 0$ by Bisection method.
17. (i) Explain the Bisection method to calculate the roots of an equation. Write an algorithm and implement it in 'C' and 'C++'.

    (ii) Write a program using C language which implements bisection method to compute the real root of the equation $3x + \sin x - e^x = 0$ in a given interval.
18. Solve $x^3 - 9x + 1 = 0$ for the root between $x = 2$ and $x = 4$ by using Bisection method.
19. The negative root of the smallest magnitude of the equation $f(x) = 3x^3 + 10x^2 + 10x + 7 = 0$ is to be obtained.

    (i) Find an interval of unit length which contains this root.

    (ii) Perform three iterations of the bisection method.
20. The smallest positive root of the equation $f(x) = x^4 - 3x^2 + x - 10 = 0$ is to be obtained.

    (i) Find an interval of unit length which contains this root.

    (ii) Perform three iterations of the bisection method.

4.7 ITERATION METHOD (OR METHOD OF SUCCESSIVE APPROXIMATIONS)

Suppose we want the approximate root of the equation

$$f(x) = 0 \qquad \text{...(i)}$$

Now, write the equation in the form

$$x = g(x) \qquad \text{...(ii)}$$

Assume x_0 to be the starting approximate value to the actual root a of $x = g(x)$.

Setting $x = x_0$ in (ii), we get the first approximation $x_1 = g(x_0)$.

Similarly we obtain a sequence of successive approximations $x_2, x_3, \ldots\ldots$, given by $x_2 = g(x_1)$, $x_3 = g(x_2), \ldots\ldots, x_{i+1} = g(x_i)$ so that each iterate is nearer to the previous iterates move away from the actual root.

Similarly, we will repeat this process until the relative error in the new approximation is within the prescribed tolerance. Method will converge rapidly if we choose initial approximation near to the root otherwise method diverges.

4.7.1 Rate of Convergence of Iteration Method

The formula for iterative method is given by

$$x_{i+1} = g(x_i) \qquad \text{...(i)}$$

Suppose we have a function $f(x) = 0$ whose exact root is r.

Then
$$e_{i+1} = x_{i+1} - r \qquad \text{// error in } (i+1)^{th} \text{ iteration}$$
$$e_i = x_i - r \qquad \text{// error in } I^{th} \text{ iteration}$$

By using these, eq. (i) becomes

$$e_{i+1} + r = g(e_i + r)$$

Expand right side of above equation by using Taylor's series method, we get

$$e_{i+1} + r = g(r) + e_i g'(r) + \frac{e_i^2 g''(r)}{2} + \ldots\ldots$$

Ignore higher powers of e_i's as e_i is very small, so we have left

$$e_{i+1} + r = g(r) + e_i g'(r)$$

As r is the root of $f(x) = 0$. then we have $g(r) = r$

$\therefore$ we have
$$r + e_{i+1} = r + e_i g'(r)$$

$\Rightarrow$
$$e_{i+1} = e_i g'(r)$$

So iteration method has first order convergence, and converges only if $|g'(r)| < 1$.

4.7.2 Algorithm For Iteration Method

Let x_0 be the initial guess value, epsilon be the prescribed tolerance, E_R be the relative error and n be the total number of iterations required for convergence.

1. Read x_0, epsilon, n
2. For $i = 1$ to n in steps of 1 do
3. set x_1=$g(x_0)$
4. Set $E_R = |(x_1-x_0)/x_1|$
5. Set $x_0 = x_1$
6. If $E_R \leq$ epsilon then
7. Write: "Root converges in",i,"iterations"
8. Write: x_2, "is the approximate root"
9. exit
10. endif
11. endfor
12. End

4.7.3 Flowchart for Iteration Method

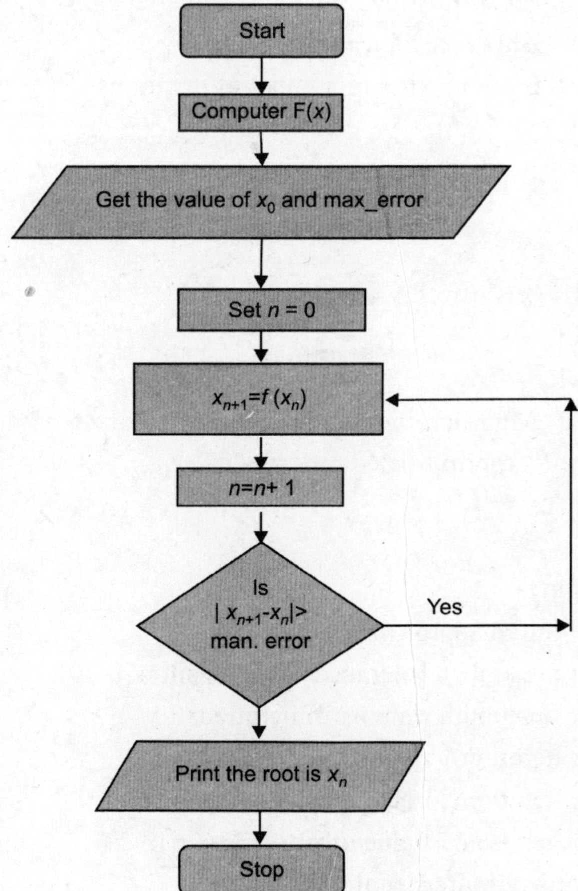

4.7.4 Program For Iteration Method in C

```
#include<conio.h>
#include<stdio.h>
#include<math.h>
#include<iomanip.h>
#define g(x) (pow(2,x)-3)
void main()
{
int i,n;
double x0,epsilon,x1;
clrscr();
printf("Enter initial approximation for x0");
scanf("%ld",&x0);
printf("Enter presribed tolerance value 'epsilon'");
scanf("%ld",&epsilon);
printf("Enter maximum number of iterations:");
scanf("%d",&n);
for(i=1;i<=n;i++)
{
x1=g(x0);
if(fabs((x1-x0)/x1)<=epsilon)
break;
x0=x1;
printf("After iteration number  %d value of x1 = %ld ",i,x1);
}printf("Approximated root = %ld",x1);
getch();
}
```

Output:

```
Enter initial approximation for x0 -3
Enter prescribed tolerance value 'epsilon' 0.001
Enter maximum number of iterations 15
After iteration 1 value of x1 = -2.875
After iteration 2 value of x1 = -2.863687
After iteration 3 value of x1 = -2.862613
```

Approximated root = -2. 862613

4.7.5 Program for Iteration Method in C++

```
#include<iostream.h>
#include<conio.h>
#include<stdio.h>
#include<math.h>
#include<iomanip.h>
#define g(x) (pow(2,x)-3)
void main()
{
int i,n;
double x0,epsilon,x1;
clrscr();
cout<<"Enter initial approximation for x0";
cin>>x0;
cout<<"Enter presribed tolerance value 'epsilon'";
cin>>epsilon;
cout<<"Enter maximum number of iterations:";
cin>>n;
for(i=1;i<=n;i++)
{
x1=g(x0);
if(fabs((x1-x0)/x1)<=epsilon)
break;
x0=x1;
cout<<"After iteration number"<< i <<"value of x1 ="<< x1 <<endl;
}cout<<"Approximated root = "<<x1;
getch();
}
```

Output:

```
Enter initial approximation for x0 -3
Enter prescribed tolerance value 'epsilon' 0.001
Enter maximum number of iterations 15
After iteration 1 value of x1 = -2.875
After iteration 2 value of x1 = -2.863687
```

After iteration 3 value of x1 = -2.862613

Approximated root = -2. 862613

4.7.6 Iteration Method For Non-linear Equations

This method also known as method of successive approximations for solving non-linear equations.

Suppose we have two non-linear equations

$$f_1(x, y) = 0, \quad f_2(x, y) = 0$$

Write these equations into the following form

$$x = g_1(x, y), \quad y = g_2(x, y)$$

Suppose (x_0, y_0) be the initial approximation to the root of these equations, then the next approximation (x_1, y_1) is given by

$$x_1 = g_1(x_0, y_0), \quad y_1 = g_2(x_1, y_0)$$

which gives second approximation to the root.

Similarly, for next approximations we have

$$x_2 = g_2(x_1, y_1), \quad y_2 = g_2(x_2, y_1) \quad \text{Third approximation}$$
$$x_3 = g_3(x_2, y_2), \quad y_3 = g_3(x_3, y_2) \quad \text{Fourth approximation}$$

This process of finding new approximations for x and y will continue until we get actual x and y which are within the prescribed tolerance.

***Note:** - Similar procedure will occur for three or more non-linear equations.*

SOLVED PROBLEMS

Problem 1: Solve the following equations using successive approximation method $x^2 + y^2 = x$; $x^2 - y^2 = y$ correct to two decimal places.

Solution: As we have
$$f_1(x, y) = x^2 + y^2 - x = 0$$
and
$$f_2(x, y) = x^2 - y^2 - y = 0$$

Write these equations in the form as

$$x = x^2 + y^2 = g_1(x, y)$$
and
$$y = x^2 - y^2 = g_2(x, y)$$

Suppose we have the initial approximation $(x_0, y_0) = (0.5, 0.5)$, then the second approximation is given by

$$x_1 = g_1(x_0, y_0) = x_0^2 + y_0^2 = (0.5)^2 + (0.5)^2 = 0.5$$
$$y_1 = g_2(x_1, y_0) = x_1^2 - y_0^2 = (0.5)^2 - (0.5)^2 = 0$$

Thus the second approximation to the root is (0.5, 0)

Now take $(x_1, y_1) = (0.5, 0)$ as the start for next approximation, then the third approximation is given by

$$x_2 = g_1(x_1, y_1) = x_1^2 + y_1^2 = (0.5)^2 + (0)^2 = 0.25$$
$$y_2 = g_2(x_2, y_1) = x_2^2 - y_1^2 = (0.25)^2 - (0)^2 = 0.0625$$

Thus the third approximation to the root is (0.25, 0.0625)

Now take $(x_2, y_2) = (0.25, 0.0625)$ as the start for next approximation, then the fourth approximation is given by

$$x_3 = g_1(x_2, y_2) = x_2^2 + y_2^2$$
$$= (0.25)^2 + (0.0625)^2 = 0.06640$$
$$y_3 = g_2(x_3, y_2) = x_3^2 - y_2^2$$
$$= (0.06640)^2 - (0.0625)^2 = 0.00050271$$

Thus the fourth approximation to the root is (0.06640, 0.00050271)

Now take $(x_3, y_3) = (0.06640, 0.00050271)$ as the start for next approximation, then the fifth approximation is given by

$$x_4 = g_1(x_3, y_3) = x_3^2 + y_3^2$$
$$= (0.06640)^2 + (0.00050271)^2 = 0.00440$$
$$y_4 = g_2(x_4, y_3) = x_4^2 - y_3^2$$
$$= (0.00440)^2 - (0.0.00050271)^2 = 0.0000191$$

As we have seen that the value of x and y comes out to be (0,0) if we take them up to two decimal places (asked in the problem).

So finally the root comes out to be $x = 0$ **and** $y = 0$.

Problem 2: Solve $e^x - 3x = 0$ by the using iteration method.

Solution: Let $f(x) = e^x - 3x = 0$

$$f(0) = 1 = +\text{ve}$$

and $f(1) = e - 3 = -\text{ve}$

$\therefore$ root lies between 0 and 1.

Let $x = \frac{1}{3}e^x = g(x)$

$g'(x) = \frac{1}{3}e^x$ and $g'(0) = 1/3$, $g'(1) < 1$

Hence in the interval (0, 1) we have $|g'(x)| < 1$ which means sequence will converge rapidly if we get $|g'(x)| > 1$ then it shows that the sequence will not converge, we will discuss convergences for various root finding methods in detail later on.

As we have $|g'(x)| < 1$ so **Take $x_0 = 0.6$,**

$$x_1 = \frac{1}{3}e^{0.6} = 0.60737$$

$$x_2 = \frac{1}{3}e^{0.60737} = 0.61187,$$

$$x_3 = \frac{1}{3}e^{0.61187} = 0.61452$$

$$x_4 = \frac{1}{3}e^{0.61452} = 0.61626,$$

$$x_5 = \frac{1}{3}e^{0.61626} = 0.61733$$

$$x_6 = \frac{1}{3}e^{61733} = 0.61799,$$

$$x_7 = \frac{1}{3}e^{0.61799} = 0.61840$$

$$x_8 = \frac{1}{3}e^{0.61840} = 0.61865,$$

$$x_9 = \frac{1}{3}e^{0.61865} = 0.61881$$

$$x_{10} = \frac{1}{3}e^{0.61881} = 0.61891,$$

$$x_{11} = \frac{1}{3}e^{0.61891} = 0.61897$$

$$x_{12} = \frac{1}{3}e^{0.61897} = 0.61900,$$

$$x_{13} = \frac{1}{3}e^{0.61900} = 0.61902$$

As now the value of x_n's get repeated or appear very much near to each other so stop. So correct value of the root of the equation is 0.6190.

Problem 3: Solve $x^3 - 2x - 5 = 0$ for the positive root by using iteration method.

Solution: Let

$$f(x) = x^3 - 2x - 5 = 0$$

$$f(2) = -1 = -\text{ve}$$

and

$$f(3) = 16 = +\text{ve}$$

Hence, root lies between 2 and 3 but closer to 2.

Let

$$x = (2x + 5)^{1/3} = g(x)$$

$$g'(x) = \frac{2}{3}\frac{1}{(2x+5)^{\frac{2}{3}}}$$

As $|g'(x)| < 1$ for all x in (2,3)

Take $x_0 = 2$

$$x_1 = (2'2 + 5)^{1/3} = 9^{1/3} = 2.0801$$
$$x_2 = (9.1602)^{1/3} = 2.0924$$
$$x_3 = (9.1848)^{1/3} = 2.0942$$
$$x_4 = (9.1884)^{1/3} = 2.0945$$
$$x_5 = (9.1890)^{1/3} = 2.0945$$

As the values of respective x_n's gets repeated so stop here.

$\therefore$ Required root is 2.0945

Problem 4: Solve $2^x - x\text{-}3 = 0$ in the interval (-3,-2) by using the method of successive approximation.

Solution: Let
$$x = 2^x\text{-}3 = g(x)$$
$$g'(x) = 2^x.\log 2$$

As $|g'(x)| < 1$ for all x in (-3,-2)

Take $x_0 = \text{-}3$

$$x_1 = 2^{-3} - 3 = -2.875$$
$$x_2 = 2^{-2.875} - 3 = -2.8637$$
$$x_3 = 2^{-2.8637} = -2.8626$$
$$x_4 = 2^{-2.8626} = -2.8625$$
$$x_5 = 2^{-2.8625} = -2.8625$$

$\therefore$ Required root is –2.8625

Problem 5: Use the method of iteration to find a positive root between 0 and 1 of the equation $xe^x = 1$.

Solution: Writing the equation in the form $x = e^{-x}$

we find, $\phi(x) = e^{-x}$ so $\phi'(x) = -e^{-x}$

Hence, $|\phi'(x)| < 1$ for $x < 1$, which assures that iteration is convergent.

Starting with $x_0 = 1$, we find that successive iterates are given by

$$x_1 = 1/e = 0.3678794$$
$$x_2 = e^{-0.3678794} = 0.6922006$$

....

....

....

$$x_{20} = 0.5671477.$$

Problem 6: Find a real root of the equation cos $x = 3x - 1$ correct to 3 decimal places using iteration method.

Solution: We have
$$f(x) = \cos x - 3x + 1 = 0$$

Now, $f(0) = 2$ and $f(\pi/2) = -3\pi/2 + 1 = (-)$ve

A root lies between 0 and $\pi/2$.

Rewriting the given equation as
$$x = 1/3\ (\cos x + 1) = \phi(x)$$

We have
$$\phi'(x) = -\sin x/3$$

and
$$|\phi'(x)| = 1/3\ |\sin x| < 1 \text{ in } (0, \pi/2)$$

Hence the iteration method can be applied and we start with $x_0 = 0$. Then the successive approximations are

$$x_1 = \phi(x_0) = 1/3\ (\cos 0 + 1) = 0.6667$$
$$x_2 = \phi(x_1) = 1/3\ [\cos 0.6667 + 1] = 0.5953$$
$$x_3 = \phi(x_2) = 1/3\ [\cos (0.5953) + 1] = 0.6093$$
$$x_4 = \phi(x_3) = 0.6067$$
$$x_5 = \phi(x_4) = 0.6072$$
$$x_6 = \phi(x_5) = 0.6071.$$

Since x_5 and x_6 are almost the same, the root is 0.607 correct to three decimal places.

Problem 7: Find a real root of $2x - \log_{10} x = 7$ correct to four decimal places using the iteration method.

Solution: We have
$$f(x) = 2x - \log_{10} x - 7$$
$$f(3) = 6 - \log 3 - 7$$
$$= 6 - 0.4771 - 7 = -1.4471$$
$$f(4) = 0.398$$

A root lies between 3 and 4.

Rewriting the given equation as
$$x = 1/2\ (\log_{10} x + 7) = \phi(x),$$

we have
$$\phi'(x) = 1/2\ (1/x \log_{10} e)$$

$|\phi'(x)| < 1$ when $3 < x < 4$ $\qquad (\because \log10\ e = 0.4343)$

Since $|f(4)| < |f(3)|$, the root is near 4.

Hence the iteration method can be applied.

The successive approximations of $x_0 = 3.6$ are

$$x_1 = \phi(x0) = 1/2\ (\log_{10} 3.6 + 7) = 3.77815$$
$$x_2 = \phi(x1) = 1/2\ (\log_{10} 3.77815 + 7) = 3.78863$$

$$x_3 = \phi(x2) = 3.78924$$
$$x_4 = \phi(x3) = 3.78927$$

Since x_3 and x_4 are almost equal, the root is 3.7892, correct to four decimal places.

Problem 8: Find the smallest root of the equation

$$1 - x + \frac{x^2}{(2!)^2} - \frac{x^3}{(3!)^2} + \frac{x^4}{(4!)^2} - \frac{x^5}{(5!)^2} + ... = 0$$

Solution: Writing the given equation as

$$x = 1 + \frac{x^2}{(2!)^2} - \frac{x^3}{(3!)^2} + \frac{x^4}{(4!)^2} - \frac{x^5}{(5!)^2} + ... = \phi(x)$$

and omitting x^2 and higher powers of x, we get $x = 1$ approximately.

Taking $x_0 = 1$, we obtain,

$$x_1 = \phi(x_0) = 1 + \frac{1}{(2!)^2} - \frac{1}{(3!)^2} + \frac{1}{(4!)^2} - \frac{1}{(5!)^2} + ... = 1.2239$$

$$x_2 = \phi(x_1) = 1 + \frac{(1.2239)^2}{(2!)^2} - \frac{(1.2239)^3}{(3!)^2} + \frac{(1.2239)^4}{(4!)^2} - \frac{(1.2239)^5}{(5!)^2} + ... = 1.3263$$

Similarly, $x_3 = \phi(x2) = 1.38$

$x_4 = 1.409$, $x_5 = 1.425$, $x_6 = 1.434$, $x_7 = 1.439$, $x_8 = 1.442$

Values of x_7 and x_8 indicate that the root is 1.44, correct to two decimal places.

Problem 9: Show that the following rearrangement of equation $x^3 + 6x^2 + 10x - 20 = 0$ does not result a convergent sequence of successive approximations by iteration method near $x = 1$, $x = (20 - 6x^2 - x^3)/10$.

Solution: Here, $x = (20 - 6x^2 - x^3)/10 = f(x)$

Hence, $f'(x) = (-12x - 3x^2)/10$

Clearly, $f'(x) < -1$ in neighborhood of $x = 1$. Hence $|f'(x)| > 1$, and neither the method nor the sequence $<x_n>$ converge.

Problem 10: Compute a value of constant k, so that the iteration formula $x = x + k(x^2 - 3)$ may converge at a good rate, given that $x = 3$ is a root.

Solution: Formula $x = f(x)$ where $f(x) = x + k(x^2 - 3)$

will converge if

$$|f'(x)| < 1 \text{ or } -1 < f'(x) < 1$$

i.e., if $-1 < 1 + 2kx < 1$

Moreover, the convergence will be rapid if $f'(a) \approx 0$

i.e., if $1 + 2ka \approx 0$

$$1+2k\sqrt{3}=0 \quad \Rightarrow \quad k=-\frac{1}{2\sqrt{3}}$$

We may take $k = -1/4$ to insure a rapid convergence by this formula.

Problem 11: Starting from $x = 0.12$, solve $x = 0.21 \sin(0.5 + x)$ by using the iteration method.

Solution: Here, $x = 0.21 \sin(0.5 + x)$

Now, First approximation of x is given by

$$x^{(1)} = 0.21 \sin(0.5 + 0.12) = 0.122$$

$$x^{(2)} = 0.21 \sin(0.5 + 0.122) = 0.1224$$

Similarly, $x^{(3)} = 0.12242, \quad x^{(4)} = 0.12242$

Obviously, $x^{(3)} = x^{(4)}$

Hence the required root is 0.12242.

Problem 12: Compute the reciprocal of 41 correct to 4 decimal places by iterative formula $x_{i+1} = x_i(2 - 41x_i)$.

Solution: Iterative formula is

$$x_{i+1} = x_i(2 - 41x_i) \quad \text{...(i)}$$

Putting $i = 0$, $x_1 = x_0(2 - 41x_0)$

Let $x_0 = 0.02$

$$x_1 = (0.02)(2 - 0.82) = 0.024$$

Put $i = 1$ in (i),

$$x_2 = (0.024)\{2 - (41 \times 0.024)\} = 0.0244$$

Put $i = 2$, $x_3 = 0.02439$

$\therefore$ Reciprocal of 41 is 0.0244.

Problem 13: Find the square root of 20 correct to 3 decimal places by using recursion formula $x_{i+1} = 1/2(x_i + 20/x_i)$.

Solution: Put $i = 0$, $x_1 = 1/2\,(x_0 + 20/x_0)$

Let $x_0 = 4.5$

$$x_1 = 1/2\,(4.5 + 20/4.5) = 4.47$$

Put $i = 1$, $x_1 = 4.47,\ x_2 = 1/2\,(4.47 + 20/4.47) = 4.472$

Put $i = 2$, $x_2 = 4.472,\ x_3 = 4.4721$

$\sqrt{20} \approx 4.472$ correct to three decimal places.

Problem 14: Find the cube root of 15 correct to four significant figures by iterative method.

Solution: Let

$$x = (15)^{1/3}$$

$$x^3 - 15 = 0$$

The real root of the above equation lies in (2, 3). The equation may be written as

$$x = (15 + 20x - x^3)/20 = \phi(x)$$

Now, $\phi'(x) = 1 - 3x^2/20 \quad \therefore |\phi'(x)| < 1 \text{ (for } x \approx 2.5)$

Iterative formula is $x_{i+1} = (15 + 20x_i - x_i^3)/20$ ...(i)

Put i = 0, $x_0 = 2.5$, we get $x_1 = 2.47$

Put i = 1 in (i), $x_2 = 2.466$ (where $x_1 = 2.47$)

Similarly, $x_3 = 2.4661$

$\therefore \sqrt[3]{15}$ correct to 3 decimal places is 2.466.

UNSOLVED PROBLEMS

1. Use iteration method to solve $e^{-x} = 10x$.
2. Solve by iterative method, the real root of the equation $3x - \log_{10} x = 6$ correct to four significant figures.
3. Solve by iteration method:
   (i) $1 + \log x = x/2$
   (ii) $\sin x = x + 1/x - 1$
   (iii) $x^3 = x^2 + x + 1$ near 2 (use 5 iterations)
   (iv) $x^3 + x + 1 = 0$
   (v) $x^3 - 2x^2 - 5 = 0$
   (vi) $x^3 - 2x^2 - 4 = 0$.
4. Use the iterative method to find the correct to four significant figures, a real root of each of the following equations:
   (i) $x = 1/(x + 1)^2$ (ii) $x = (5 - x)^{1/3}$
   (iii) $\sin x = 10(x - 1)$ (iv) $x \sin x = 1$
   (v) $e^x = \cot x$ (vi) $1 + x^2 - x^3 = 0$
   (vii) $x^2 - 1 = \sin^2 x$ (viii) $5x^3 - 20x + 3 = 0$.
5. By using iteration method, find $\sqrt{30}$.
6. The equation $f(x) = 0$, where

$$f(x) = 0.1 - x + \frac{x^2}{(2!)^2} - \frac{x^3}{(3!)^2} + \frac{x^4}{(4!)^2} - \ldots$$

   has one root in the interval (0, 1). Calculate this root correct to 5 decimal places.
7. The equation $x^2 + ax + b = 0$ has two real roots, α and β. Show that the iteration method

$$x_{n+1} = -\left(\frac{x_n^2 + b}{a}\right)$$ is convergent near $x = \alpha$ if $2|\alpha| < |\alpha + \beta|$.

8. The equation $x^3-5x^2+4x-3=0$ has one root near $x=4$ which is to be computed by the iteration

$$x_{n+1} = \frac{3+(k-4)x_n+5x_n^2-x_n^3}{k}, \; k \text{ integer}, x_0 = 4$$

(i) Determine which value of k will give the fastest convergence.

(ii) Using this value of k, iterate three times and estimate the error in x^3.

[**Hint:** Put $x_n = \alpha + e_n$, $\alpha = 4 + \delta$, where α is the exact root. Find the error eqn. $ke_{n+1} = (k-12)\, e_n + O(\delta e_n)$]

4.8 REGULA-FALSI (OR METHOD OF FALSE POSITION OR METHOD OF LINEAR INTERPOLATION)

The bisection method guarantees that the iterative process will converge. But it is slow. So, The attempts have been made to speed up the bisection method retains its guaranteed convergence. A new method used to solve this problem is called the **method of false position.**

It is sometimes known as the **method of linear interpolation.**

This is the oldest method for finding the real roots of a equation and closely resembles the bisection method. Regula-Falsi method is faster than the Bisection method. The method is very much similar to the method of bisection.

The method requires an approximate location of the roots such that there is a change in sign of $f(x)$ for say, $f(x_0)$ and $f(x_1)$. This point $[x_0, f(x_0)]$ and $[x_1, f(x_1)]$ can be marked on the curve and a chord is drawn passing through $[x_0, f(x_0)]$ and $[x_1, f(x_1)]$. Let this chord cuts the x-axis at $x = x_2$ (which is the next approximation point).

Graphically the analysis is shown in fig below.

Let α be the desired root that lies between the $[x_0, x_1]$.

Slope of the line joining the points $C(x_2,0)$ and $B(x_1, f(x_1))$ is given by

$$\tan\theta = \frac{f(x_1)-0}{x_1-x_2} = \frac{f(x_1)}{x_1-x_2} \qquad \left[\because \tan\theta = \frac{y_2-y_1}{x_2-x_1}\right]$$

Slope of the line joining the points $A(x_0, f(x_0))$ and $B(x_1, f(x_1)$ is given by

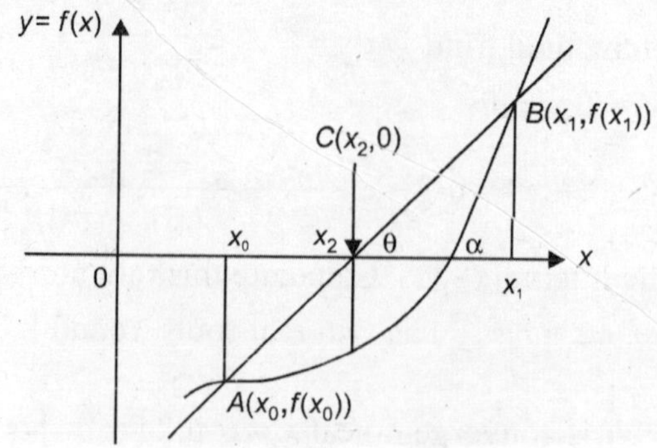

False position method

$$x_{n+1} = \frac{x_{n-1}f(x_n) - x_n f(x_{n-1})}{f(x_n) - f(x_{n-1})}$$

As line *CB* is a part of line *AB*, so we can write

$$\frac{f(x)}{x_1 - x_2} = \frac{f(x_1) - f(x_0)}{x_1 - x_0}$$

After solving for x_2, we have

$$x_2 = \frac{x_0 f(x_1) - x_1 f(x_0)}{f(x_1) - f(x_0)}$$

In general,

$$x_{n+1} = \frac{x_{n-1}f(x_n) - x_n f(x_{n-1})}{f(x_n) - f(x_{n-1})}$$

The difference value $f(x_n) - f(x_{n-1})$ {represents the slope of curve at current interval} plays an important role in proper convergence, i.e. the value should not be small enough that it produces a large error in the root.

Now calculate $f(x_2)$ and note its sign.

After noted the sign following three possibilities can occur while calculating the root.

1. if $f(x_2) = 0$, then the root is x_2.
2. if $f(x_2) = -$ve, then the root lies between interval $[x_2, x_1]$ and we change the interval $[x_0, x_1]$ to $[x_2, x_1]$ and bisect the new interval $[x_2, x_1]$ which is half of length to the previous interval.
3. if $f(x_2) = +$ve, then the root lies between interval $[x_0, x_2]$ and we change the interval $[x_0, x_1]$ to $[x_0, x_2]$ and bisect the new interval $[x_0, x_2]$ which is half of length to the previous interval.

Similarly, we will repeat this process until the relative error between two successive approximations becomes less than or equal to the prescribed tolerance.

4.8.1 Rate of Convergence of Regula-falsi Method

If $< x_n >$ is the sequence of approximations obtained from the formula for False position method is given by

$$x_{n+1} = x_n - \frac{(x_n - x_{n-1})}{f(x_n) - f(x_{n-1})} f(x_n) \quad \text{...(i)}$$

and α is the exact value of the root of the equation $f(x) = 0$, then

Let

$$x_n = \alpha + e_n$$

$$x_n + 1 = \alpha + e_{n+1}$$

where e_n, e_{n+1} are the errors involved in n^{th} and $(n + 1)^{th}$ approximations, respectively.

Clearly, $f(\alpha) = 0$. Hence, (i) gives

$$\alpha + e_{n+1} = \alpha + e_n - \frac{(e_n - e_{n-1})}{f(\alpha + e_n) - f(\alpha + e_{n-1})} \cdot f(\alpha + e_n)$$

or $$e_{n+1} = \frac{e_{n-1} f(\alpha + e_n) - e_n f(\alpha + e_{n-1})}{f(\alpha + e_n) - f(\alpha + e_{n-1})}$$

$$e_{n-1}\left[f(\alpha) + e_n f'(\alpha) + \frac{e_n{}^2}{2!} f''(\alpha) + \ldots\ldots \right]$$

$$= \frac{-e_n\left[f(\alpha) + e_{n-1} f'(\alpha) + \frac{e_{n-1}^2}{2!} f''(\alpha) + \ldots\ldots \right]}{\left[f(\alpha) + e_n f'(\alpha) + \frac{e_n{}^2}{2!} f''(\alpha) + \ldots\ldots \right]} - \left[f(\alpha) + e_{n-1} f'(\alpha) + \frac{e_{n-1}^2}{2!} f''(\alpha) + \ldots\ldots \right]$$

$$= \frac{(e_{n-1} - e_n) f(\alpha) + \frac{e_{n-1} e_n}{2!}(e_n - e_{n-1}) f''(\alpha) + \ldots\ldots}{(e_n - e_{n-1}) f'(\alpha) + \frac{(e_n - e_{n-1})(e_n + e_{n-1})}{2!} f''(\alpha) + \ldots\ldots}$$

$$= \frac{\frac{e_{n-1} e_n}{2} f''(\alpha) + \ldots\ldots}{f'(\alpha) + \left(\frac{e_n + e_{n-1}}{2!} \right) f''(\alpha) + \ldots\ldots} \qquad \mid \therefore f(\alpha) = 0$$

or $$e_{n+1} \approx \frac{e_n e_{n-1}}{2!} \frac{f''(\alpha)}{f'(\alpha)} \qquad \ldots\text{(ii)}$$

(neglecting high powers of e_n, e_{n-1})

Let $e_{n+1} = c\, e_n{}^k$, where c is a constant and $k > 0$.

$\therefore$ $$e_n = c\, e^k_{n-1}$$

or $$e_{n-1} = c^{-1/k} e_n{}^{1/k}$$

From (ii),

$$c\, e_n{}^k = \frac{e_n c^{-1/k}}{2!} e_n{}^{1/k} \cdot \frac{f''(\alpha)}{f'(\alpha)} = \frac{c^{-1/k}}{2!} e_n{}^{1+1/k} \cdot \frac{f''(\alpha)}{f'(\alpha)}$$

Comparing the two sides, we get

$$k = 1 + \frac{1}{k}$$

and $$c = \frac{c^{-1/k}}{2!} \frac{f''(\alpha)}{f'(\alpha)}$$

Now, $k = 1 + 1/k \Rightarrow k^2 - k - 1 = 0 \Rightarrow k = 1.618$

Also,

$$c = c^{-1/k} \cdot \frac{1}{2!} \frac{f''(\alpha)}{f'(\alpha)}$$

$$c^{1+\frac{1}{k}} = c^{1.618} = \frac{1}{2} \frac{f''(\alpha)}{f'(\alpha)}$$

or

$$c = \left[\frac{f''(\alpha)}{2f'(\alpha)} \right]^{0.618}$$

This gives the rate of convergence and $k = 1.618$ gives the order of convergence.

4.8.2 Algorithm For False Position Method

Let $f(x)$ be the given function x_0, x_1 are two initial guess values, epsilon is the prescribed tolerance, and delta is the prescribed lower bound for the slope of $f(x)$.

1. Read x_0, $x1$, epsilon, delta
2. Set $f0 = f(x0)$
3. Set $f1 = f(x1)$
4. if($f0 \times f1 > 0$) then
5. write: "Given initial approximations are not suitable"
6. exit
7. endif
8. do
9. if$|(f1 - f0| <$delta
10. write: "slope too small"
11. exit
12. endif
13. set $x_2 = x_0 - ((x1\text{-}x0) \times f0)/(f1 - f0)$
14. if ($f0 \times f2 < 0$) then
15. set $x_1 = x_2$ and set $f1 = f2$
16. else
17. set $x_0 = x_2$ and set $f0 = f2$
18. endif
19. while(($|x_0 - x_1| >$epsilon) and ($f2 \neq 0$))
20. write: x_2, "as the approximate root"
21. End

4.8.3 Flowchart For False Position Method

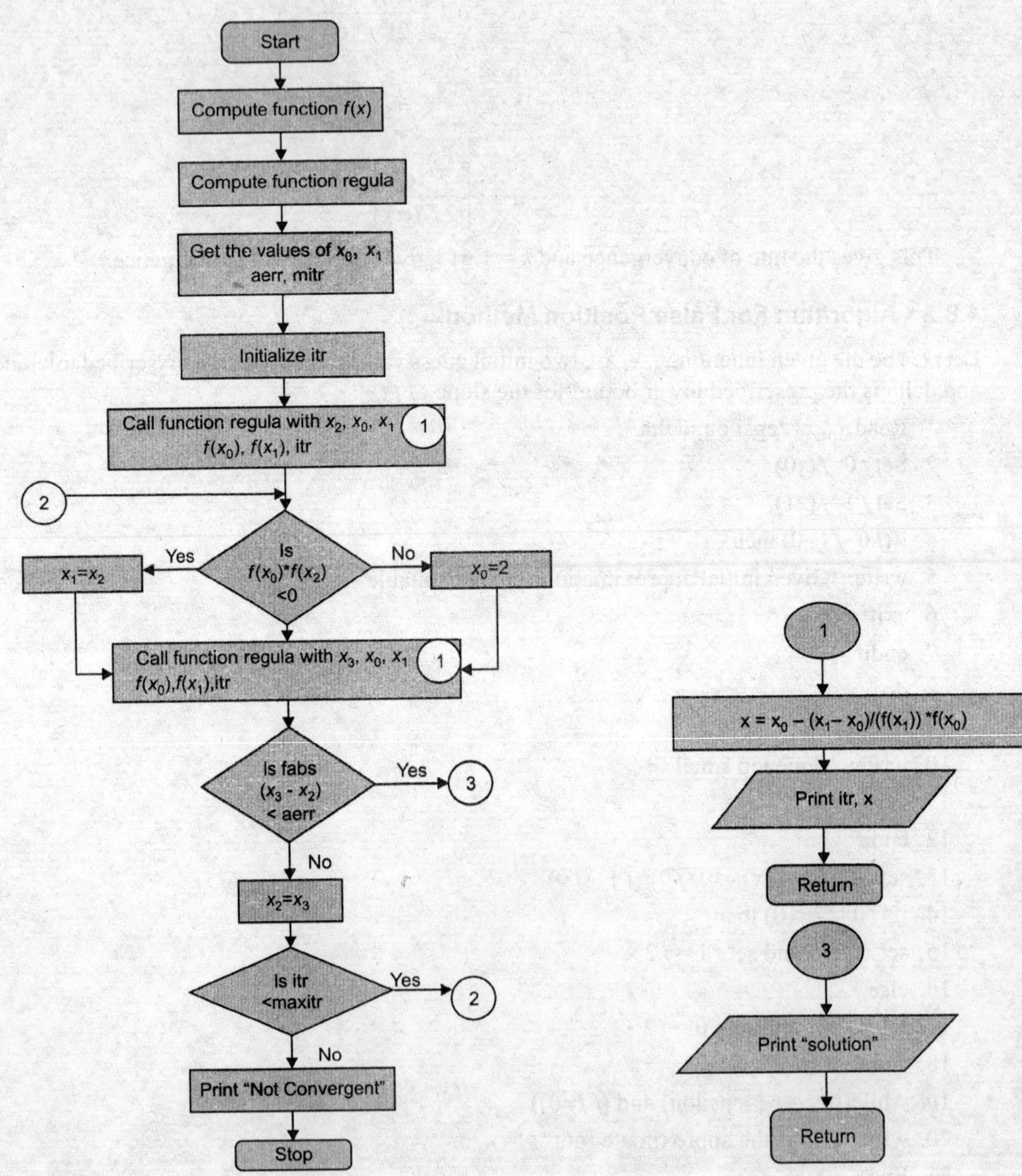

4.8.4 Program For False Position Method in C

```
#include<stdio.h>
#include<conio.h>
#include<stdlib.h>
#include<math.h>
#include<iomanip.h>
#define f(x) (x*x*x+x-1)
void main()
{
int i=1;
float x0,epsilon,x1,x2,delta,temp,f0,f1,f2;
clrscr();
printf("Enter initial approximation for x0");
scanf("%f",&x0);
printf("Enter initial approximation for x1");
scanf("%f",&x1);
printf("Enter prescribed tolerance value 'epsilon'");
scanf("%f",&epsilon);
printf("Enter prescribed lower bound for the slope f(x)");
scanf("%f",&delta);
if(f(x0)*f(x1)>0)
{
printf("\nGiven initial approximations are not suitable\n");
getch();
exit(1);
}
do
{
if(fabs(f(x1)-f(x0))<delta)
{
printf("\nslope too small\n");
exit(1);
}
temp=x2;
```

```
x2=x0-((x1-x0)*f0)/(f1-f0);
printf("\nValue of x2 = %f After iteration number = %d",x2,i);
f2=f(x2);
if(f0*f2<0)
x1=x2;
else
x0=x2;
i++;
}
while((fabs(x0-x1)>epsilon) && (f2!=0));
printf("Approximate root = %f",x2);
printf("\n");
getch();
}
```

Output:

```
Enter initial approximation for x0    0
Enter initial approximation for x1    1
Enter prescribed tolerance value 'epsilon'    0.001
Enter prescribed lower bound for the slope f(x)    0.001
Value of x2 = 0.5 After iteration number = 1
Value of x2 = 0.636364 After iteration number = 2
Value of x2 = 0.671196 After iteration number = 3
Value of x2 = 0.679662 After iteration number = 4
Value of x2 = 0.681690 After iteration number = 5
Value of x2 = 0.68218 After iteration number = 6
Value of x2 = 0.68283 After iteration number = 7
```

4.8.5 Program for False Position Method in C++

```
#include<iostream.h>
#include<conio.h>
#include<stdlib.h>
#include<math.h>
#include<iomanip.h>
#define f(x) (x*x*x+x-1)
```

```
void main()
{
int i=1;
float x0,epsilon,x1,x2,delta,temp,f0,f1,f2;
clrscr();
cout<<"Enter initial approximation for x0";
cin>>x0;
cout<<"Enter initial approximation for x1";
cin>>x1;
cout<<"Enter prescribed tolerance value 'epsilon'";
cin>>epsilon;
cout<<"Enter prescribed lower bound for the slope f(x)";
cin>>delta;
if(f(x0)*f(x1)>0)
{
cout<<"\nGiven initial approximations are not suitable\n";
getch();
exit(1);
}
do
{
if(fabs(f(x1)-f(x0))<delta)
{
cout<<"\nslope too small\n";
exit(1);
}
temp=x2;
x2=x0-((x1-x0)*f0)/(f1-f0);
cout<<"\nValue of x2 = "<<x2<<"After iteration number ="<<i;
f2=f(x2);
if(f0*f2<0)
x1=x2;
else
x0=x2;
```

```
i++;
}
while((fabs(x0-x1)>epsilon) && (f2!=0));
cout<<"Approximate root = "<<x2<<endl;
getch();
}
```

Ouput:

Enter initial approximation for x0 0

Enter initial approximation for x1 1

Enter prescribed tolerance value 'epsilon' 0.001

Enter prescribed lower bound for the slope f(x) 0.001

Value of x2 = 0.5 After iteration number = 1

Value of x2 = 0.636364 After iteration number = 2

Value of x2 = 0.671196 After iteration number = 3

Value of x2 = 0.679662 After iteration number = 4

Value of x2 = 0.681690 After iteration number = 5

Value of x2 = 0.68218 After iteration number = 6

Value of x2 = 0.68283 After iteration number = 7

SOLVED PROBLEMS

Problem 1: Find an approximate root of $x\log_{10}x - 1.2 = 0$ by using False position method.

Solution: Let $f(x) = x\log_{10}x - 1.2$

$f(1) = -1.2 = -\text{ve}$

and $f(2) = 2 \times 0.30103 - 1.2 = -0.597940$

and $f(3) = 3 \times 0.47712 - 1.2 = 0.231364 = +\text{ve}$

Here we take starting interval as $\underset{(2,3)}{(x_0, x_1)}$ as both are nearer to each other due to which we have to calculate less number of iterations.

$$\therefore \quad x_2 = \frac{x_0 f(x_1) - x_1 f(x_0)}{f(x_1) - f(x_0)} = \frac{2f(3) - 3f(2)}{f(3) - f(2)}$$

$$= \frac{2(0.23136) - (-0.59794)}{0.23136 - (-0.59794)} = 2.721014$$

$$f(2.7210) = -0.017104 = -\text{ve}$$

Hence, root lies between 2.721014 and 3.

$$x_3 = \frac{(2.721014 \times 0.231364) - \{3 \times (-0.017104)\}}{0.23136 - (-0.017104)}$$

$$= \frac{0.68084}{0.24846}$$

$$x_3 = 2.740211$$

$$f(2.7402) = -0.00038905 = -\text{ve}$$

Hence, root lies between 2.740211 and 3

$$x_4 = \frac{(2.7402 \times 0.23136) - \{3 \times (-0.00038905)\}}{0.23136 - (-0.00038905)}$$

$$= \frac{0.63514}{0.23175}$$

$$x_4 = 2.740627$$

$$f(2.7406) = 0.00011998 = +\text{ve}$$

Hence, root lies between 2.740211 and 2.740627

$$x_5 = \frac{(2.7402 \times 0.00011998) - \{2.7406 \times (-0.00038905)\}}{0.00011998 - (-0.00038905)}$$

$$= \frac{0.0013950}{0.00050903}$$

$$x_4 = 2.740627$$

Now given equation gives value zero on this value of x.

Hence, the required root is 2.7405

Problem 2: Solve for positive root of $x - \cos x = 0$ by Regula Falsi method.

Solution: Let

$$f(x) = x - \cos x$$

$$f(0) = -1 = -\text{ve}$$

and

$$f(1) = 0.459698 = +\text{ve}$$

$$(x_0, x_1)$$

Hence, required root lies between interval (0, 1)

$$\therefore \quad x_2 = \frac{x_0 f(x_1) - x_1 f(x_0)}{f(x_1) - f(x_0)} = \frac{0 \cdot f(1) - 1 \cdot f(0)}{f(1) - f(0)}$$

$$= \frac{0(0.459698) - 1(-1)}{0.456998 - (-1)} = 0.685073$$

$$f(0.685073) = 0.685073 - \cos(0.685073) = -0.089292$$

Hence, root lies between 0.685078 and 1.

Similarly, for further iterations we have

$$x_3 = 0.73629$$
$$x_4 = 0.738945$$
$$x_5 = 0.73907$$

Hence, the required root is 0.7391

***Note:** Angle should be measured in radians.*

Problem 3: Compute the root of the equation $f(x) = x^3 + x - 1 = 0$, near 1, by using False position method correct to three decimal places.

Solution: Let

$$f(x) = x^3+x-1$$
$$f(0) = -1 = -\text{ve and } f(1) = 1 = +\text{ve}$$
$$(x_0, x_1)$$

Hence, required root lies between interval (0, 1)

$$\therefore \quad x_2 = \frac{x_0 f(x_1) - x_1 f(x_0)}{f(x_1) - f(x_0)} = \frac{0 \cdot f(1) - 1 \cdot f(0)}{f(1) - f(0)}$$

$$= \frac{0(1) - 1(-1)}{1 - (-1)} = 0$$

Now $f(x_2) = f(0.5) = (0.5)^3 + 0.5 - 1 = -0.375 = -\text{ve}$

As $f(x_2) = -\text{ve}$, so root lies between the interval (0.5, 1)

Now $f(x_3) = f(0.636364) = -0.105935 = -\text{ve}$

As $f(x_3) = -\text{ve}$, so root lies between the interval (0.636364, 1)

Similarly by going on this way, the next approximations are

$$x_4 = 0.671196$$
$$x_5 = 0.679662$$
$$x_6 = 0.681690$$
$$x_7 = 0.68218$$
$$x_8 = 0.68245$$

$\therefore$ Root correct to three decimal places is 0.682

Problem 4: By using the method of false position, find the root of the equation $x^6 - x^4 - x^3 - 1 = 0$ up to four decimal places.

Solution: Let

$$f(x) = x^6 - x^4 - x^3 - 1$$
$$f(1.4) = -0.056$$

$$f(1.41) = 0.102$$

Hence, the root lies between 1.4 and 1.41.

Using the method of false position,

$$x_2 = x_0 - \left\{\frac{x_1 - x_0}{f(x_1) - f(x_0)}\right\} f(x_0)$$

$$= 1.4 - \left\{\frac{1.41 - 1.4}{0.102 + 0.056}\right\}(-0.056) \quad \text{Let, } x_0 = 1.4 \text{ and } x_1 = 1.41$$

$$= 1.4 + \left(\frac{0.01}{0.158}\right)(0.056) = 1.4035$$

Now $\qquad f(x_2) = -0.0016$ (–)ve

Hence, the root lies between 1.4035 and 1.41.

Using the method of false position,

$$x_3 = x_2 - \left\{\frac{x_1 - x_2}{f(x_1) - f(x_2)}\right\} f(x_2)$$

Replacing x_0 by x_2

$$= 1.4035 - \left(\frac{1.41 - 1.4035}{0.102 + 0.0016}\right)(-0.0016)$$

$$= 1.4035 + \left(\frac{0.0065}{0.1036}\right)(0.0016) = 1.4036$$

Now $\qquad f(x_3) = -0.00003$ (–)ve

Hence, the root lies between 1.4036 and 1.41.

Using the method of false position,

$$x_4 = x_3 - \left\{\frac{x_1 - x_3}{f(x_1) - f(x_3)}\right\} f(x_3)$$

$$= 1.4036 - \left(\frac{1.41 - 1.4036}{0.102 + 0.00003}\right)(0.00003)$$

$$= 1.4036 + \left(\frac{0.0064}{0.10203}\right)(0.00003) = 1.4036$$

Since $x3$ and $x4$ are approximately the same up to four decimal places, the required root of the given equation is **1.4036.**

Problem 5: Find the root of the equation tan x + tanh x = 0 which lies in the interval (1.6, 3.0) correct to four significant digits using the method of false position.

Solution: Let $f(x) \equiv \tan x + \tanh x = 0$

Since $f(2.35) = -0.03$

and $f(2.37) = 0.009$

Hence, the root lies between 2.35 and 2.37.

Using Regula-Falsi method,

$$x_2 = x_0 - \left\{\frac{x_1 - x_0}{f(x_1) - f(x_0)}\right\} f(x_0)$$

$$= 2.35 - \left\{\frac{2.37 - 2.35}{0.009 + 0.03}\right\}(-0.03) \qquad \text{Let, } x_0 = 2.35 \text{ and } x_1 = 2.37$$

$$= 2.35 + \left\{\frac{0.02}{0.039}\right\}(0.03) = 2.365$$

Now $f(x_2) = -0.00004$ (–)ve

Hence, the root lies between 2.365 and 2.37.

Using Regula-Falsi method,

$$x_3 = x_2 - \left\{\frac{x_1 - x_2}{f(x_1) - f(x_2)}\right\} f(x_2)$$

Replacing x_0 by x_2

$$= 2.365 - \left\{\frac{2.37 - 2.365}{0.009 + 0.00004}\right\}(-0.00004)$$

$$= 2.365 - \left\{\frac{0.005}{0.00904}\right\}(0.00004) = 2.365$$

Hence, the required root is **2.365,** correct to four significant digits.

Problem 6: Find the root of the equation xe^x = cos x in the interval (0, 1) using Regula-Falsi method correct to four decimal places. Write its program in 'C' language.

Solution: Let $f(x) = \cos x - xe^x = 0$ so that

$f(0) = 1,$ $\qquad f(1) = \cos 1 - e = -2.17798$

i.e., the root lies between 0 and 1.

By Regula-Falsi method,

$$x_2 = x_0 - \left\{\frac{x_1 - x_0}{f(x_1) - f(x_0)}\right\} f(x_0)$$

$$= 0 - \frac{1-0}{-3.17798}(1) = 0.31467$$

Now $\qquad f(x_2) = f(0.31467) = 0.51987$

i.e., the root lies between 0.31487 and 1.

Again

$$x_3 = 0.31487 - \frac{(1-0.31487)}{(-2.17798 - 0.51987)}(0.51987)$$

$$= 0.44673$$

Now $\qquad f(x_3) = 0.20356$

The root lies between 0.44673 and 1. Repeating this process, $x_{10} = 0.51775$, corrected as 0.5177 up to 4 decimal places.

\\Method of False Position

```
#include<stdio.h>
#include<conio.h>
#include<math.h>
float f(float x)
{
return cos(x)-x*exp(x);
}
void regula (float *x, float x0,float x1, float fx0, float
fx1,int*itr)
{
*x=x0-((x1-x0)/(fx1-fx0))*fx0;
++(*itr);
printf("Iteration no.%3d x=%7.5f\n",*itr,*x);
}
main()
{
int itr=0,maxitr;
float x0,x1,x2,x3,aerr;
printf("Enter the values for x0,x1, allowed error, max.iteration\n");
scanf("%f%f%f%d",&x0,&x1,&aerr,&maxitr);
regula(&x2,x0,x1,f(x0),f(x1),&itr);
do
{ if(f(x0)*f(x2)<0)
```

```
x1=x2;
else
x0=x2;
regula(&x3,x0,x1,f(x0),f(x1),&itr);
if(fabs(x3-x2)<aerr)
{
printf("After %d iterations,
root=%6.4f\n",itr,x3);
getch();
return (0);
}
x2=x3;
}while (itr<maxitr);
printf("Solution does not converge, iterations not
sufficient\n");
getch();
return(1);
```

Ouput:

```
Enter the values for x0,x1, allowed error, max.iteration
0
1
.00005
20
Iteration number       1 x = 0.31467
Iteration number       2 x = 0.44673
Iteration number       3 x = 0.49402
Iteration number       4 x = 0.50995
Iteration number       5 x = 0.51520
Iteration number       6 x = 0.51692
Iteration number       7 x = 0.51748
Iteration number       8 x = 0.51767
Iteration number       9 x = 0.51773
Iteration number       10 x = 0.51775
After 10 iterations,   root = 0.5177
```

Problem 7: (i) Solve $x^3 - 5x + 3 = 0$ by using Regula-Falsi method.

(ii) Use the method of false position to solve $x^3 - x - 4 = 0$.

Solution: (i) Let $f(x) = x^3 - 5x + 3$

Since $f(.65) = .024625$

and $f(.66) = -.012504$

The root lies between .65 and .66.

Let $x_0 = .65$ and $x_1 = .66$

Using the method of false position,

$$x_2 = x_0 - \left\{\frac{x_1 - x_0}{f(x_1) - f(x_0)}\right\} f(x_0)$$

$$= .65 - \left(\frac{.66 - .65}{-.012504 - .024625}\right)(.024625)$$

$$= .656632282$$

Now $f(x_2) = -.00004392$

Hence, the root lies between .65 and .656632282.

Using the method of false position,

$$x_3 = x_2 - \left\{\frac{x_1 - x_2}{f(x_1) - f(x_2)}\right\} f(x_2)$$

Replacing x_1 by x_2

$$= .65 - \left(\frac{.656632282 - .65}{-.00004392 - .024625}\right)(.024625)$$

$$= .656620474.$$

Since x_2 and x_3 agree up to 4 decimal places, the required root is **.6566,** correct up to four decimal places. Similarly, the other roots of this equation are 1.8342 and – 2.4909.

(ii) Let $f(x) = x^3 - x - 4$

Since $f(1.79) = -.054661$

and $f(1.80) = .032$

The root lies between 1.79 and 1.80

Let $x_0 = 1.79$ and $x_1 = 1.80$

Using the method of false position,

$$x_2 = x_0 - \left\{\frac{x_1 - x_0}{f(x_1) - f(x_0)}\right\} f(x_0)$$

$$= 1.79 - \left\{\frac{(1.80 - 1.79)}{.032 - (-.054661)}\right\}(-.054661)$$

Now, $\quad f(x_2) = -.00012936$

Hence, the root lies between 1.796307 and 1.80.

Using the method of false position,

$$x_3 = x_2 - \left\{\frac{x_1 - x_2}{f(x_1) - f(x_2)}\right\} f(x_2)$$

$$= 1.796307 - \left\{\frac{1.8 - 1.796307}{.032 - (-.00012936)}\right\}(-.00012936)$$

$$= 1.796321.$$

Since $x2$ and $x3$ are the same up to four decimal places, the required root is **1.7963,** correct up to four decimal places.

Problem 8: (i) Apply False-position method to find the smallest positive root of the equation $x - e^{-x} = 0$ correct to three decimal places.

(ii) Find a positive root of $xe^x = 2$ by the method of false position.

Solution: (i) Let $\quad f(x) = x - e^{-x}$

Since $\quad f(.56) = -.01121$

and $\quad f(.58) = .0201$

Hence, the root lies between .56 and .58.

Let $x_0 = .56$ and $x_1 = .58$

Using the method of false position,

$$x_2 = x_0 - \left\{\frac{x_1 - x_0}{f(x_1) - f(x_0)}\right\} f(x_0)$$

$$= .56 - \left(\frac{.58 - .56}{.0201 + .01121}\right)(-.01121)$$

$$= .56716$$

Now $\quad f(x_2) = .00002619$ i.e., (+)ve

Hence, the root lies between .56 and .56716.

Using the method of false position,

$$x_3 = x - \left\{\frac{x_1 - x_2}{f(x_1) - f(x_2)}\right\} f(x_2)$$

Replacing x_1 by x_2

$$= .56 - \left(\frac{.56716 - .56}{.00002619 + .01121}\right)(-.01121)$$

$$= .567143$$

Since x_2 and x_3 agree up to four decimal places, the required root correct to three decimal places is **0.567.**

(ii) Let $f(x) = xe^x - 2$

Since $f(.852) = -.00263$

and $f(.853) = .001715$

The root lies between .852 and .853.

Let $x_0 = .852$ and $x_1 = .853$

Using the method of false position,

$$x_2 = x_0 - \left\{\frac{x_1 - x_0}{f(x_1) - f(x_0)}\right\} f(x_0)$$

$$= .852 - \left(\frac{.852 - .852}{.001715 + (-.00263)}\right)(-.00263)$$

$$= .852605293$$

Now $f(x_2) = -.00000090833$

Hence, the root lies between .852605293 and .853

Using the method of false position,

$$x_3 = x_2 - \left\{\frac{x_1 - x_2}{f(x_1) - f(x_2)}\right\} f(x_2) \qquad \textbf{Replacing } x_0 \textbf{ by } x_2$$

$$= (.852605293) - \left(\frac{.853 - .852605293}{.001715 - (-.00000090833)}\right)(-.00000090833)$$

$$= 0.852605501$$

Since x_2 and x_3 agree up to 6 decimal places, the required root correct to 6 decimal places is **0.852605.**

UNSOLVED PROBLEMS

1. Solve $x^3 - 9x + 1 = 0$ for the root lying between 2 and 4 by the method of false position.
2. Find real cube root of 18 by Regula-Falsi method.
3. Find the smallest positive root correct to three decimal places of the equation

$$\cosh x \cos x = -1.$$

4. Determine the real roots of $f(x) = x^3 - 98$ using False position method within $E_s = 0.1\%$.
5. Write a short note on Regula-Falsi method.
6. Using the False-position method, find x when $x^2 - 9 = 0$. Give computer program using 'C'.
7. Find the real root of the equations

   (i) $x^3 - 4x + 1 = 0$ (ii) $x^3 - x^2 - 2 = 0$

   (iii) $x^3 + x - 3 = 0$ (iv) $x^3 - 5x - 7 = 0$

   by using the method of false-position.

8. Find the real root of the equations

   (i) $x^4 - x^3 - 2x^2 - 6x - 4 = 0$ (ii) $x^6 - x^4 - x^3 - 3 = 0$

   (iii) $xe^x = 3$ (iv) $x^2 - \log_e x - 12 = 0$

   (v) $x = \tan x$ (vi) $3x = \cos x + 1$

   by using the method of false position.

9. (i) Explain Regula-Falsi method by stating at least one advantage over the bisection method.

   (ii) Discuss the method of false position.

10. Solve the following equations by Regula-Falsi method.

    (i) $(5 - x)\, e^x = 5$ near $x = 5$ (ii) $x^3 + x - 1 = 0$ near $x = 1$

    (iii) $2x - \log_{10} x = 7$ lying b/w 3.5 and 4 (iv) $x^3 + x^2 - 3x - 3 = 0$ lying b/w 1 and 2

    (v) $x^3 - 3x + 4 = 0$ b/w -2 and -3

    (vi) $x^4 + x^3 - 7x^2 - x + 5 = 0$ lying b/w 2 and 3.

11. Find the rate of convergence for Regula-Falsi method.

12. Illustrate the false position method by plotting the function on a graph and discuss the speed of convergence to the root. Develop the algorithm for computing the roots using the false-position technique.

13. Find all the roots of $\cos x - x^2 - x = 0$ to 5 decimal places.

14. A root of the equation $f(x) = x - \phi(x) = 0$ can often be determined by combining the iteration method with Regula-Falsi.

    (i) With a given approximate value x_0, we compute $x_1 = \phi(x_0)$, $x_2 = \phi(x_1)$

    (ii) Observing that $f(x_0) = x_0 - x_1$ and $f(x_1) = x_1 - x_2$, we find a better approximation $x2$ using Regula-Falsi on the points $(x_0, x_0 - x_1)$ and $(x_1, x_1 - x_2)$.

    (iii) This last $x2$ is taken as a new x_0 and we start from (i) all over again.

    Compute the smallest root of the equation $x - 5 \log_e x = 0$ with an error less than 0.5×10^{-4} starting with $x_0 = 1.3$.

15. Solve $f(x) = x\tan x + 1 = 0$, whose root lies between the interval [2.5,3] by using Regula-falsi method.

16. Find a real root of the equation $x \log_{10} x = 1.2$ by Regula-falsi method, correct to four decimal places.

4.9 NEWTON- RAPHSON METHOD (METHOD OF TANGENTS)

Most widely used of all methods for finding roots is the Newton-Raphson method. This method can be applied to solve both Algebric and Transcendental equations. We have to choose Proper initial point which is very important in this method. Once we have made proper choice for initial point, it converges faster than all the root finding methods. This method is generally used to improve the result obtained by one of the previous methods.

Formula of the Newton-Raphson may be derived analytically by using the Taylor series approximation of the function $f(x)$.

Suppose we have a function $f(x) = 0$

Let x_0 be the initial point and h (increment or decrement) be the correction value for the root so that

$$f(x_0 + h) = 0$$

Now expand this function by using Taylor series, we have

$$f(x_0 + h) = f(x_0) + hf'(x_0) + \frac{h^2}{2!} f''(x_0) + \frac{h^3}{3!} f'''(x_0) + \ldots\ldots = 0$$

Neglect all the terms of second and higher degree terms, we have

$$f(x_0) + h f'(x_0) = 0$$

$$h = \frac{-f(x_0)}{f'(x_0)}$$

or we can write
$$h = x_1 - x_0 = \frac{-f(x_0)}{f'(x_0)}$$

or
$$x_1 = x_0 - \frac{f(x_0)}{f'(x_0)}$$

In general
$$x_{n+1} = x_n - \frac{f(x_n)}{f'(x_n)}$$

This iterative process terminates when the relative error in two successive approximations is less than or nearly equal to the prescribed tolerance.

4.9.1 Rate of Convergence of Newton Raphson Method

The formula for Newton's method is given by

$$x_{n+1} = x_n - f(x_n)/f'(x_n) \qquad \ldots\text{(i)}$$

suppose we have function $f(x) = 0$, whose exact root is r.

Then
$$e_{n+1} = x_{n+1} - \mathrm{r} \qquad \text{// error in (n+1)}^{\text{th}} \text{ iteration}$$
$$e_n = x_n - \mathrm{r} \qquad \text{// error in n}^{\text{th}} \text{ iteration}$$

By using these, eq. (i) becomes

$$e_{n+1} + \mathrm{r} = e_n + r - \frac{f(e_n + r)}{f'(e_n + r)} \qquad \ldots\text{(ii)}$$

By expanding $f(e_n + r)$ & $f'(e_n + r)$ by Taylor's series & substituting in eq. (ii), we get

$$e_{n+1} = e_n - \frac{[f(r) + e_n f'(r) + (e_n{}^2/2)\cdot f'(r) + \ldots\ldots]}{[f'(r) + e_n f''(r) + (e_n{}^2/2)\cdot f'''(r) + \ldots\ldots]}$$

As y is the root of $f(x) = 0$ so $f(r) = 0$.

$$\therefore \quad e_{n+1} = e_n - \frac{[e_n f'(r) + (e_n^2/2)\cdot f''(r) + ...]}{[f'(r) + e_n f''(r) + (e_n^2/2)\cdot f'''(r) + ...]}$$

Divide num. & den. By $f'(r)$, we have

$$e_{n+1} = e_n - \frac{[e_n + \{e_n^2 \cdot f''(r)\}/\{(2f'(r)\} + ...]}{[1 + \{e_n \cdot f''(r)\}/f'(r) + \{e_n^2 \cdot f'''(r)\}/2f'(r) + ...}$$

Neglect higher powers of e_n's as e_n is very small, we have

$$e_{n+1} = e_n - [e_n + \{e_n^2.f''(r)\}/\{2.f'(r)\}]\,[\,1 + \{e_n f''(r)\}/f'(r)]$$

Expand it by binomial theorem & take first two terms

$$e_{n+1} = e_n - e_n[1 + \{e_n.f''(r)\}/\{2.f'(r)\}]\,[1 - \{e_n f''(r)\}/\{f'(r)\}]$$

After simplification and ignorance of terms involving e_n^2, get

$$e_{n+1} = e_n - e_n[1 - \{e_n.f''(r)\}/\{2.f'(r)\}]$$

$$e_{n+1} = \frac{e_n^2 \cdot f''(r)}{2 \cdot f'(r)}$$

Hence by definition, the order of convergence of Newton-Raphson method is 2, i.e., New Raphson method is **quadratic convergent**. This also shows that subsequent error at each ste proportional to the square of the previous error and as such the convergence is quadratic. Henc at the first iteration we have an answer correct to one decimal place, then it should be correct to places at the second iteration, and to four places at the third iteration. This means that the nun of correct decimal places at each iteration is almost doubled.

$\therefore$ Method converges very rapidly.

4.9.2 Algorithm for Newton-raphson Method

Let $f(x)$ be the given function, $f^1(x)$ be the first derivative of $f(x)$, x_0 be the initial guess va epsilon is the prescribed relative error, E_R be the relative error, delta is the prescribed lower bo for $f^1(x)$ and n be the maximum number of iterations to be allowed.

1. Read x_0, epsilon, delta, n
2. For $i = 1$ to n steps of 1 do
3. Set $y_0 = f(x_0)$
4. Set $dy_0 = f^1(x_0)$
5. if $|dy_0| \le$ delta then
6. write: " Slope too small", x_0
7. exit
8. endif
9. Set $x_1 = x_0 - \{f(x_0)/dy_0)\}$

10. Set $E_R = |(x_1 - x_0)/x_1|$
11. set $x_0 = x_1$
12. if($E_R \leq$ epsilon or $f(x_1)$=0) then
13. write: "Approximate root = ", x_1
14. exit
15. endif
16. End for
17. Write: "Solution doesn't converge in", n, "iterations"
18. End

4.9.3 Flowchart of Newton–Raphson Method

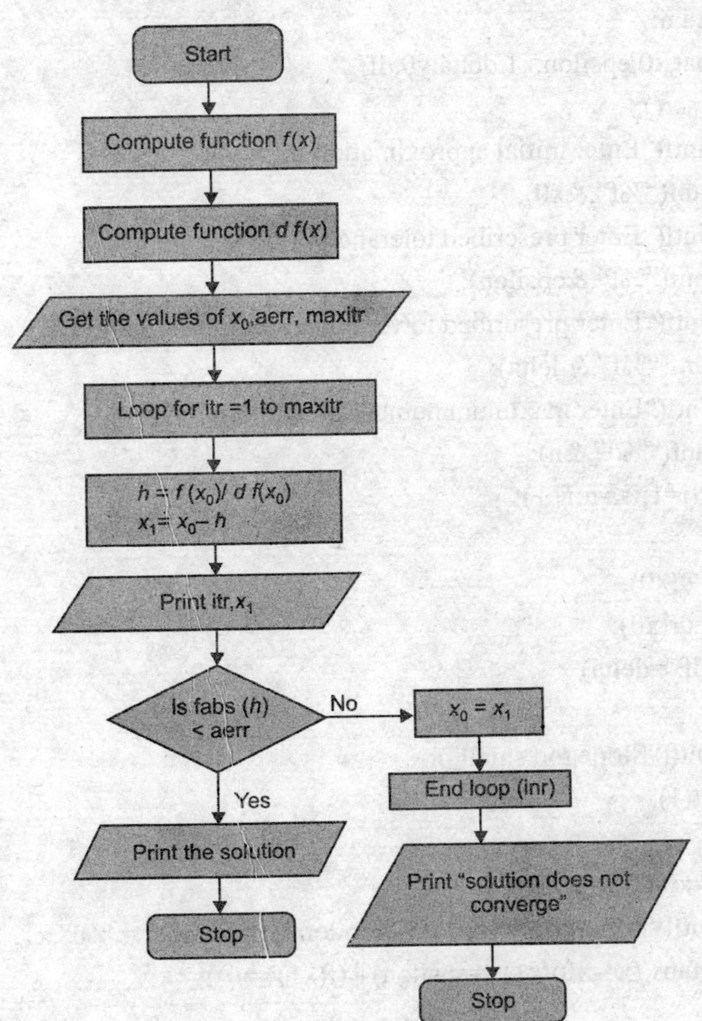

4.9.4 Program for Newton-raphson Method in C

```
#include<conio.h>
#include<stdio.h>
#include<stdlib.h>
#include<math.h>
#include<iomanip.h>
#define f(x) (x*x*x-2x-5)
#define dr(x) (3*x*x-2)
void main()
{
int i,n;
float x0,epsilon,x1,delta,y0,df;
clrscr();
printf("Enter initial approximation for x0");
scanf("%f",&x0);
printf("Enter prescribed tolerance value 'epsilon'");
scanf("%f",&epsilon);
printf("Enter prescribed lower bound for the slope f(x)");
scanf("%f",&delta);
printf("Enter maximum number of iterations");
scanf("%d",&n);
for(i=1;i<=n;i++)
{
y0=f(x0);
df=dr(x0);
if(df<=delta)
{
printf("Slope too small");
exit(1);
}
x1=x0-(y0/df);
printf("\nValue of x1 = %f After iteration number = %d",x1,i);
if((fabs((x1-x0)/x1)<=epsilon) || (f(x1)==0))
{
```

```
printf("\n APPROXIMATE ROOT = %f",x1);
getch();
exit(1);}
x0=x1;
}
printf("Solution doesn't converge in %d iterations",n);
getch();
}
```

Output:

```
Enter initial approximation for x0 2
Enter prescribed tolerance value 'epsilon' 0.001
Enter prescribed lower bound for the slope f(x) 0.001
Enter maximum number of iterations  5
Value of x1 = 2.1 After iteration number = 1
Value of x1 = 2.0.9457 After iteration number = 2
Value of x1 = 2.128934 After iteration number = 3
Value of x1 = 2.09461 After iteration number = 4
Value of x1 = 2.094551 After iteration number = 5
APPROXIMATE ROOT = 2.0946
```

4.9.5 Program for Newton-raphson Method in C++

```
#include<iostream.h>
#include<conio.h>
#include<stdlib.h>
#include<math.h>
#include<iomanip.h>
#define f(x) (x*x*x-2x-5)
#define dr(x) (3*x*x-2)
void main()
{
int i,n;
float x0,epsilon,x1,delta,y0,df;
clrscr();
cout<<"Enter initial approximation for x0";
```

```
cin>>x0;
cout<<"Enter prescribed tolerance value 'epsilon'";
cin>>epsilon;
cout<<"Enter prescribed lower bound for the slope f(x)";
cin>>delta;
cout<<"Enter maximum number of iterations";
cin>>n;
for(i=1;i<=n;i++)
{
y0=f(x0);
df=dr(x0);
if(df<=delta)
{
cout<<"Slope too small";
exit(1);
}
x1=x0-(y0/df);
cout<<"\nValue of x1 = "<<x1<<"After iteration number = "<<i;
if((fabs((x1-x0)/x1)<=epsilon) || (f(x1)==0))
{
cout<<"\n APPROXIMATE ROOT = "<<x1;
getch();
exit(1);}
x0=x1;
}
cout<<"Solution doesn't converge in " << n << "iterations";
getch();
}
```

Output:

Enter initial approximation for x0 2

Enter prescribed tolerance value 'epsilon' 0.001

Enter prescribed lower bound for the slope f(x) 0.001

Enter maximum number of iterations 5

Value of x1 = 2.1 After iteration number = 1

Value of x1 = 2.0.9457 After iteration number = 2
Value of x1 = 2.128934 After iteration number = 3
Value of x1 = 2.09461 After iteration number = 4
Value of x1 = 2.094551 After iteration number = 5
APPROXIMATE ROOT = 2.0946

4.9.6 Limitations of Newton-raphson Method

1. Division by zero may occur if $f'(x_n)$ is zero or very close to zero.
2. if the initial point is too far away from the required root, the process may converge to some other root.
3. A particular value in the iteration sequence may repeat, resulting in an infinite loop. This occurs when the tangent to the curve $f(x)$ at $x = x_{n+1}$ cuts the x-axis again at $x = x_n$.

4.9.7 Newton's Raphson Method for Non Linear Equations

Suppose we have two non-linear equations

$$f_1(x, y) = 0, f_2(x, y) = 0 \qquad \text{...(i)}$$

Suppose (x_0, y_0) be the initial root of these equations, then the next approximation (x_1, y_1) is given by adding increments h and k as

$$x_1 = x_0 + h, y_1 = y_0 + k$$

After substituting these new values in equation (i), we get new pair of equations in the form

$$f_1(x_0 + h, y_0 + k) = 0, f_2(x_0 + h, y_0 + k) = 0 \qquad \text{...(ii)}$$

Now apply Taylor's series on (ii) for expansion of both the equations

$$f_1(x_0 + h, y_0 + k) = f_1(x_0, y_0) + h\frac{\partial f_1(x_0, y_0)}{\partial x_0}\bigg|_{(x_0, y_0)} + k\frac{\partial f_1(x_0, y_0)}{\partial y_0}\bigg|_{(x_0, y_0)} = 0$$

$$f_2(x_0 + h, y_0 + k) = f_2(x_0, y_0) + h\frac{\partial f_2(x_0, y_0)}{\partial x_0}\bigg|_{(x_0, y_0)} + k\frac{\partial f_2(x_0, y_0)}{\partial y_0}\bigg|_{(x_0, y_0)} = 0$$

...(iii)

Solve the equations (iii) for h and k and find the new value for next approximation which is as

$$x_1 = x_0 + h, y_1 = y_0 + k$$

This process of finding new approximations for x and y will continue until we get actual x and y which are within the prescribed tolerance.

***Note:** Similar procedure will occur for three or more non-linear equations.*

SOLVED PROBLEMS

Problem 1: Solve the following system of Non-linear equations by using Newton Raphson method $x^2 + xy + y^2 = 7, x^3 + y^3 = 9$ with (1.5,0.5) as initial approximation.

Solution: As we have, $f_1(x,y) = x^2 + xy + y^2 - 7$ and $f_2(x,y) = x^3 + y^3 - 9$ and $(x_0,y_0) = (1.5,0.5)$

And also we have

$$f_1(x_0+h, y_0+k) = f_1(x_0, y_0) + h\left.\frac{\partial f_1(x_0,y_0)}{\partial x_0}\right|_{(x_0,y_0)} + k\left.\frac{\partial f_1(x_0,y_0)}{\partial y_0}\right|_{(x_0,y_0)} = 0$$

$$f_1(x_0+h, y_0+k) = (x_0^2 + x_0y_0 + y_0^2 - 7) + h(2x_0 + y_0) + k(x_0 + 2y_0) = 0$$

$$f_1(x_0+h, y_0+k) = \{(1.5)^2 + (1.5)(0.5) + (0.5)^2 - 7\} + h\{2(1.5) + (0.5)\} + k\{(1.5) + 2(0.5)\} = 0$$

$$f_1(x_0+h, y_0+k) = 3.5h + 2.5k - 3.75 = 0 \quad \text{...(i)}$$

$$f_2(x_0+h, y_0+k) = f_2(x_0, y_0) + h\left.\frac{\partial f_2(x_0,y_0)}{\partial x_0}\right|_{(x_0,y_0)} + k\left.\frac{\partial f_2(x_0,y_0)}{\partial y_0}\right|_{(x_0,y_0)} = 0$$

$$f_2(x_0+h, y_0+k) = x_0^3 + y_0^3 - 9 + h(3x_0^2) + k(3y_0^2) = 0$$

$$f_2(x_0+h, y_0+k) = \{(1.5)^3 + (0.5)^3 - 9\} + h\{3(1.5)^2\} + k\{3(0.5)^2\} = 0$$

$$f_2(x_0+h, y_0+k) = 6.75h + 0.75k - 5.5 = 0 \quad \text{...(ii)}$$

Solving equations (i) and (ii) for h and k, we have

$$h = 0.7675, k = 0.425$$

Hence, $x_1 = x_0 + h = 1.5 + 0.7675 = 2.2675$

And $y_1 = y_0 + k = 0.5 + 0.425 = 0.925$

∴ the second approximation to the root is **(2.2675,0.925)**

Let $(x_1,y_1) = (2.2675,0.925)$

$$f_1(x_1+h, y_1+k) = (x_1^2 + x_1y_1 + y_1^2 - 7) + h(2x_1 + y_1) + k(x_1 + 2y_1) = 0$$

$$f_1(x_1+h, y_1+k) = 5.46h + 4.1175k + 1.0935 = 0. \quad \text{...(iii)}$$

$$f_2(x_1+h, y_1+k) = x_1^3 + y_1^3 - 9 + h(3x_1^2) + k(3y_1^2) = 0$$

$$f_2(x_1+h, y_1+k) = 15.42h + 2.55k + 3.44 = 0 \quad \text{...(iv)}$$

Solving equations (iii) and (iv) for h and k, we have

$h = -0.228, k = 0.038$

Hence, $x_2 = x_1 + h = 2.2675 + (-0.228) = 2.0395$

And $y_2 = y_1 + k = 0.925 + 0.038 = 0.963$

$\therefore$ the third approximation to the root is **(2.0395,0.963)**

Let $(x_2, y_2) = (2.0395, 0.963)$

$$f_1(x_2 + h, y_2 + k) = (x_2^2 + x_2y_2 + y_2^2 - 7) + h(2x_2 + y_2) + k(x_2 + 2y_2) = 0$$

$$f_1(x_2 + h, y_2 + k) = 5.04h + 3.96k + 0.03 = 0 \quad \text{...(v)}$$

$$f_2(x_2 + h, y_2 + k) = x_2^3 + y_2^3 - 9 + h(3x_2^2) + k(3y_2^2) = 0$$

$$f_2(x_2 + h, y_2 + k) = 12.47h + 2.78 + 0.3 = 0 \quad \text{...(vi)}$$

Solving equations (v) and (vi) for h and k, we have

$h = -0.02$, $k = 0.03$

Hence, $x_3 = x_2 + h = 2.0395 - 0.02 = 2.0195$

$y_3 = y_2 + k = 0.963 + 0.03 = 0.993$

$\therefore$ the fourth approximation to the root is **(2.0195,0.993)**

Similarly, for more refinement we will do one or two further approximations and get the final values as x **= 2 and y = 1.**

Problem 2: Find a real root of the equations $x^2 - y^2 = 4$ and $x^2 + y^2 = 16$ by Newton-Raphson method.

Solution: As we have given the equations

$$x^2 - y^2 = 4 \quad \text{...(i)}$$

$$x^2 + y^2 = 16 \quad \text{...(ii)}$$

To obtain the initial approximations

Add equations (i) & (ii), we have the +ve result

$$x = 2\sqrt{2}$$

Substituting this value of 'x' in any equation gives

$$y = 2\sqrt{2}$$

So the initial approximation point becomes $x_0 = 2\sqrt{2}$, $y_0 = 2\sqrt{2}$

As we have, $f_1(x,y) = x^2 - y^2 - 4$ and $f_2(x,y) = x^2 + y^2 - 16$

and $(x_0, y_0) = (2\sqrt{2}, 2\sqrt{2})$

And also we have

$$f_1(x_0 + h, y_0 + k) = f_1(x_0, y_0) + h\left.\frac{\partial f_1(x_0, y_0)}{\partial x_0}\right|_{(x_0, y_0)} + k\left.\frac{\partial f_1(x_0, y_0)}{\partial y_0}\right|_{(x_0, y_0)} = 0$$

$$f_1(x_0 + h, y_0 + k) = (x_0^2 - y_0^2 - 4) + h(2x_0) + k(-2y_0) = 0$$
$$f_1(x_0 + h, y_0 + k) = \{(2\sqrt{2})^2 - (2\sqrt{2})^2 - 4\} + h\{2(2\sqrt{2})\} - k\{2(2\sqrt{2})\} = 0$$
$$f_1(x_0 + h, y_0 + k) = 4\sqrt{2}h - 4\sqrt{2}k - 4 = 0 \quad ...(iii)$$
$$f_2(x_0 + h, y_0 + k) = f_2(x_0, y_0) + h\frac{\partial f_2(x_0, y_0)}{\partial x_0}\bigg|_{(x_0, y_0)} + k\frac{\partial f_2(x_0, y_0)}{\partial y_0}\bigg|_{(x_0, y_0)} = 0$$
$$f_2(x_0 + h, y_0 + k) = x_0^2 + y_0^2 - 16 + h(2x_0) + k(2y_0) = 0$$
$$f_2(x_0 + h, y_0 + k) = \{(2\sqrt{2})^2 + (2\sqrt{2})^2 - 16\} + h\{2(2\sqrt{2})\} + k\{2(2\sqrt{2})\} = 0$$
$$f_2(x_0 + h, y_0 + k) = 4\sqrt{2}h + 4\sqrt{2}k = 0 \quad ...(iv)$$

Solving equations (iii) and (iv) for h and k, we have

$h = 0.354$, $k = -0.354$

Hence, $x_1 = x_0 + h = 2\sqrt{2} + 0.354$

$x_1 = 2(1.4142) + 0.354 = 3.182$

and $y_1 = y_0 + k = 2\sqrt{2} - 0.354$

$y_1 = 2(1.4142) - 0.354 = 2.474$

∴ the second approximation to the root is **(3.182, 2.474)**

Let $(x_1, y_1) = (3.182, 2.474)$

$$f_1(x_1 + h, y_1 + k) = (x_1^2 - y_1^2 - 4) + h(2x_1) + k(-2y_1) = 0$$
$$= (3.182)^2 - (2.474)^2 - 4 + h\{2(3.182)\} - k\{2(2.474)\} = 0$$
$$f_1(x_1 + h, y_1 + k) = 6.364h - 4.948k + 0.00448 = 0 \quad ...(v)$$
$$f_2(x_1 + h, y_1 + k) = x_1^2 + y_1^2 - 16 + h(2x_1) + k(2y_1) = 0$$
$$= (3.182)^2 + (2.474)^2 - 16 + h\{2(3.182)\} + k\{2(2.474)\} = 0$$
$$= 6.364h + 4.948k - 0.2458 = 0 \quad ...(vi)$$

Solving equations (v) and (vi) for h and k, we have

$h = -0.0204$, $k = -0.0242$

Hence, $x_2 = x_1 + h = 3.182 - 0.0204$

$x_2 = 3.162$

and $y_2 = y_1 + k = 2.474 - 0.0242$

$y_2 = 2.450$

∴ the third approximation to the root is **(3.162, 2.450)**

which gives us the desired root.

Problem 3: Find the real positive root of $3x - \cos x - 1 = 0$ by Newton's method correct decimal places.

Solution: Let $f(x) = 3x - \cos x - 1$

$f(0) = -2 = -ve$ and $f(1) = 3 - \cos(1) - 1 = 2 - 0.540302 = +ve$

Hence, root lies between 0 and 1

As $f'(x) = 3 + \sin x$

Let the initial value be $x_0 = 0.5$

$$f(x_0) = (0.5) = 3 \times (0.5) - \cos(0.5) - 1$$
$$= 1.5 - 0.877582 - 1 = -0.377582 = -ve$$
$$f'(x_0) = f'(0.5) = 3 + \sin(0.5) = 3 + 0.479425 = 3.4794$$

$$\therefore \quad x_1 = x_0 - \frac{f(x_0)}{f'(x_0)}$$
$$= 0.608518$$
$$f(x_1) = f(0.608518) = 0.005058$$
$$f'(x_1) = f(0.608518) = 3.571652$$

$$\therefore \quad x_2 = x_1 - \frac{f(x_1)}{f'(x_1)}$$
$$= 0.608518 - \frac{0.005058}{3.571632}$$
$$= 0.607101$$
$$f(x_2) = f(0.607101) = -0.000002$$
$$f'(x_2) = f'(0.607101) = 3.570488$$

$$\therefore \quad x_3 = x_2 - \frac{f(x_2)}{f'(x_2)}$$
$$= 0.607101 - \frac{(-0.000002)}{3.570488}$$
$$= 0.607101$$

Hence, the required root is 0.607101

Problem 4: Evaluate $\sqrt{5}$ to six decimal places by using Newton's Raphson method.

Solution: Let $x = \sqrt{5}$.

By squaring on both sides, we get

$x^2 = 5 \Rightarrow x^2 - 5 = 0$

So $f(x) = x^2 - 5$

$f(2) = 4 - 5 = -1 = -\text{ve}$ and $f(3) = 9 - 5 = 4 = +\text{ve}$

Hence, the root lies between 2 and 3.

As $$f'(x) = 2x$$

Let the initial value be $x_0 = 2$

$$f(x_0) = (2) = -1$$
$$f'(x_0) = f'(2) = 2 \times 2 = 4$$

$$\therefore \quad x_1 = x_0 - \frac{f(x_0)}{f'(x_0)}$$

$$= 2 - \frac{(-1)}{4} = 2.25$$

$$f(x_1) = f(2.25) = 0.0625$$
$$f'(x_1) = f(2.25) = 4.5$$

$$\therefore \quad x_2 = x_1 - \frac{f(x_1)}{f'(x_1)}$$

$$= 2.25 - \frac{0.0625}{4.5} = 2.236111$$

$$f(x_2) = f(2.236111) = 0.000187$$
$$f'(x_2) = f'(2.236111) = 4.47222$$

$$\therefore \quad x_3 = x_2 - \frac{f(x_2)}{f'(x_2)}$$

$$= 2.236111 - \frac{(0.000187)}{4.47222} = 2.236066$$

$$f(x_3) = f(2.236066) = -0.000035$$
$$f'(x_3) = f'(2.236066) = 4.99996$$

$$\therefore \quad x_4 = x_3 - \frac{f(x_3)}{f'(x_3)}$$

$$= 2.236066 - \frac{(-0.000035)}{4.99996} = 2.236067$$

Hence, the approximate value of $\sqrt{5}$ up to six decimal places is 2.236067.

Problem 5: Solve the equation $x^3 - 2x - 5 = 0$ by using Newton Raphson method, correct to three decimal places.

Solution: Let $$f(x) = x^3 - 2x - 5$$

At $x = 2.25, f(2.25) = -0.359$ and at $x = 2.5, f(2.5) = 5.625$

So root lies between the intervals (2.25, 2.5)

As $$f'(x) = 3x^2 - 2$$

Let the initial approximation of the root $x_0 = 2.25$

$$f'(x_0) = f'(2.25) = 3(2.25)^2 - 2 = 13.1875$$

$$\therefore \quad x_1 = x_0 - \frac{f(x_0)}{f'(x_0)}$$

$$= 2.5 - \frac{(-0.359)}{13.1875} = 2.106635$$

$$f(x_1) = f(2.106635) = (2.106635)^3 - 2(2.106635) - 5$$

$$= 9.3490587 - 4.21327 - 5$$

$$= 0.1357887$$

$$f'(x_1) = f'(2.106635) = 11.313923$$

$$\therefore \quad x_2 = x_1 - \frac{f(x_1)}{f'(x_1)}$$

$$= 2.106635 - \frac{(-0.1357887)}{11.313923} = 2.094598$$

Similarly, by going in the same way the next approximations are

$$x_3 = 2.094552$$

$$x_4 = 2.094552$$

$\therefore$ Correct root up to first three decimal positions is 2.094.

Problem 6: Using Newton's iterative method, find the real root of $x \log_{10} x = 1.2$, correct to five decimal places.

Solution: $$f(x) = x \log_{10} x - 1.2$$

$$\because \quad f(1) = -1.2 = (-)\text{ve}$$

$$f(3) = 3 \log_{10} 3 - 1.2 = (+)\text{ve}$$

So a root of $f(x) = 0$ lies between 1 and 3.

Let us take $x_0 = 2$.

and $$f'(x) = \log_{10} x + \log_{10} e = \log_{10} x + 0.43429$$

Newton's iteration formula gives,

$$x_{n+1} = x_n - \frac{f(x_n)}{f'(x_n)}$$

$$= x_n - \frac{x_n \log_{10} x_n - 12}{\log_{10} x_n + .43429} = \frac{.43429 x_n + 1.2}{\log_{10} x_n + .43429} \quad \text{...(i)}$$

Given $n = 0$, the first approximation is

$$x_1 = \frac{.43429x_0 + 1.2}{\log_{10} 2 + .43429} = 2.81 \qquad (\because x_0 = 2)$$

Similarly, given $n = 1, 2, 3, 4$ in (i), we get

$x_2 = 2.741, x_3 = 2.74064, x_4 = 2.74065, x_5 = 2.74065$

Clearly, $x_4 = x_5$

Hence the required root is **2.74065,** correct to five decimal places.

Problem 7: Evaluate $\sqrt{12}$ to four decimal places by Newton's iterative method.

Solution: Let $x = 12$ so that $x^2 - 12 = 0$ ...(i)

Take $f(x) = x^2 - 12$, Newton's iteration formula gives,

$$x_{n+1} = x_n - \frac{f(x_n)}{f'(x_n)} = x_n - \frac{x_n^{\ 2} - 12}{2x_n} = \frac{1}{2}\left(x_n + \frac{12}{x_n}\right) \quad ...(ii)$$

Now, since
$$f(3) = -3 \text{ (–)ve}$$
$$f(4) = 4 \text{ (+)ve}$$

$\therefore$ The root of (i) lies between 3 and 4.

Given $x_0 = 3.5$, (ii) gives,

$$x_1 = \frac{1}{2}\left(x_0 + \frac{12}{x_0}\right) = \frac{1}{2}\left(3.5 + \frac{12}{3.5}\right) = 3.4643$$

$$x_2 = \frac{1}{2}\left(x_1 + \frac{12}{x_1}\right) = 3.4641$$

$$x_3 = 3.4641$$

Since $x_2 = x_3$ up to 4 decimal places, we have $\sqrt{12}$ **= 3.4641.**

Problem 8: Using Newton's iterative method, find the real root of $x \sin x + \cos x = 0$ which is near $x = \pi$, correct to 3 decimal places.

Solution: We have

$$f(x) = x \sin x + \cos x \text{ and } f'(x) = x \cos x$$

The iteration formula is

$$x_{n+1} = x_n - \frac{x_n \sin x_n + \cos x_n}{x_n \cos x_n}$$

with $x_0 = \pi$

$$x_1 = x_0 - \frac{x_0 \sin x_0 + \cos x_0}{x_0 \cos x_0}$$

$$= \pi - \frac{\pi \sin \pi - \cos \pi}{\pi \cos \pi} = 2.8233$$

Successive iteratives are

$x_2 = 2.7986, x_3 = 2.7984, x_4 = 2.7984$

Since $x_3 = x_4$, the required root is 2.798, correct to three decimal places.

Problem 9: Compute a real root of the equation $x = e^{-x}$ using the Newton-Raphson method.

Solution: We have $\quad f(x) = xe^x - 1$

then, $\quad f'(x) = (1+x)\, e^x$

Let $x_0 = 1$

Then $$x_1 = 1 - \left(\frac{e-1}{2e}\right) = \frac{1}{2}\left(1 + \frac{1}{e}\right) = 0.6839397$$

Now, $f(x_1) = 0.3553424$ and $f'(x_1) = 3.337012$

so that, $$x_2 = 0.6839397 - \frac{0.3553424}{3.337012} = 0.5774545$$

Proceeding in this way, we obtain

$x_3 = 0.5672297, x_4 = 0.5671433$

Hence the required root is **0.5671,** correct to 4 decimal places.

Problem 10: Find to four decimal places, the smallest root of the equation $e^{-x} = \sin x$.

Solution: The given equation is

$$f(x) \equiv e^{-x} - \sin x = 0$$

so that,

$$x_{n+1} = x_n + \frac{e^{-x_n} - \sin x_n}{e^{-x_n} + \cos x_n}$$

Take $x_0 = .6$ then,

$x_1 = .58848, x_2 = .588559$

Hence, the desired value of the root is 0.5885.

Problem 11: (i) Find a positive value of $(17)^{1/3}$, correct to four decimal places, by the Newton-Raphson method.

(ii) Find the cube root of 10.

Solution: (i) The iterative formula is

$$x_{n+1} = \frac{1}{3}\left(2x_n + \frac{a}{x_n^2}\right) \qquad \text{...(i)}$$

Here $a = 17$

Take $x_0 = 2.5$ $\quad\Big|\ \because \sqrt[3]{8} = 2$ and $\sqrt[3]{27} = 3$

Putting $n = 0$ in (i), we get

$$x_1 = \frac{1}{3}\left(2x_0 + \frac{17}{x_0^2}\right) = \frac{1}{3}\left(5 + \frac{17}{6.25}\right) = 2.5733$$

Putting $n = 1$ in (36), we get

$$x_2 = \frac{1}{3}\left(2x_1 + \frac{17}{x_1^2}\right)$$

$$= \frac{1}{3}\left(5.1466 + \frac{17}{6.6220}\right) = 2.5713$$

Again putting $n = 2$ in (36), we get

$$x_3 = \frac{1}{3}\left(2x_2 + \frac{17}{x_2^2}\right)$$

$$= \frac{1}{3}\left(5.1426 + \frac{17}{6.61158}\right) = 2.57128$$

Putting $n = 3$ in (36), we get

$$x_4 = \frac{1}{3}\left(2x_3 + \frac{17}{x_3^2}\right)$$

$$= \frac{1}{3}\left(5.14256 + \frac{17}{6.61148}\right) = 2.57128$$

Since x_3 and x_4 agree to four decimal places, the required root is **2.5713,** correct to four decimal places.

(ii)

$$x_{n+1} = \frac{2x_n^3 + a}{3x_n^2} = \frac{1}{3}\left(2x_n + \frac{a}{x_n^2}\right)$$

Take $x_0 = 2.5$ $\quad\Big|\ \because \sqrt[3]{8} = 2$ and $\sqrt[3]{27} = 3$

$\therefore$

$$x_1 = 2.2 \qquad (n = 0)$$

$$x_2 = 2.155 \qquad (n = 1)$$

$$x_3 = 2.15466 \qquad (n = 2)$$

$$\sqrt[3]{10} \approx 2.15466$$

Problem 12: By applying Newton's formula to find the values of $(30)^{1/5}$.

Solution: To find the pth root of a, we have

$$x_{n+1} = \frac{(p-1)x_n^p + a}{px_n^{p-1}}$$

Here, $a = 30, p = 5$, the first approximation is

$$x_1 = \frac{4x_0^5 + 30}{5x_0^4}$$

Take $x_0 = 1.9$, we get $x_1 = 1.98$

Again, $x_2 = 1.973$

$\therefore$ Value = 1.973 (correct to 3 decimal places).

Problem 13: Using the starting value 2(1 + *i*), solve $x^4 - 5x^3 + 20x^2 - 40x + 60 = 0$ by Newton-Raphson method, given that all the roots of the given equation are complex.

Solution: Let $f(x) = x^4 - 5x^3 + 20x^2 - 40x + 60$

so that, $f'(x) = 4x^3 - 15x^2 + 40x - 40$

Newton-Raphson method gives,

$$x_{n+1} = x_n - \frac{f(x_n)}{f'(x_n)}$$

$$= x_n - \frac{x_n^4 - 5x_n^3 + 20x_n^2 - 40x_n + 60}{4x_n^3 - 15x_n^2 + 40x_n - 40}$$

$$= \frac{3x_n^4 - 10x_n^3 + 20x_n^2 - 60}{4x_n^3 - 15x_n^2 + 40x_n - 40}$$

Put $n = 0$, take $x_0 = 2(1 + i)$ by trial, we get

$$x_1 = 1.92\,(1 + i)$$

Again, $x_2 = 1.915 + 1.908\,i$

Since imaginary roots occur in conjugate pairs roots are $1.915 \pm 1.908\,i$ up to 3 decimal places.

Assuming the other pairs of roots to be $\alpha \pm i\beta$, then

$$\text{Sum} = \begin{pmatrix} \alpha + i\beta + \alpha - i\beta \\ +1.915 + 1.908\,i \\ +1.915 - 1.908\,i \end{pmatrix} = 2\alpha + 3.83 = 5$$

$$\alpha = 0.585$$

Also, the product of the roots = $(\alpha^2 + \beta^2)\,[(1.915)^2 + (1.908)^2] = 60$

$$\beta = 2.805$$

Hence, the other two roots are $0.585 \pm 2.805\,i$.

Problem 14: Using the starting value $x_0 = i$, find a zero of $x^4 + x^3 + 5x^2 + 4x + 4 = 0$.

Solution: By Newton's method

$$x_1 = i - \frac{f(i)}{f'(i)} = i - \frac{3i}{1 + 6i} = .486 + .919\,i$$

Now,

$$x_2 = .486 + .919\,i - \frac{f(.486 + .919\,i)}{f'(.486 + .919\,i)}$$

$$= .486 + .919\,i - \left(\frac{-.292 + .174\,i}{178 + 6.005\,i}\right)$$

$$= -.499 + 0.866\,i$$

The actual root is

$$x = \frac{-1 + i\sqrt{3}}{2}$$

4.10 GENERALIZED NEWTON'S METHOD

Generalized Newton's method is useful when a equation gives multiple roots.

If a is a root of $f(x) = 0$ with multiplicity m, then the iteration formula will be

$$x_{n+1} = x_n - m\frac{f(x_n)}{f'(x_n)}$$

Since a is the root of $f(x) = 0$ with multiplicity m, it implies that α is also a root of $f'(x) = 0$ with multiplicity (m–1) and it is also root of $f''(x) = 0$ with multiplicity (m–2) and so on.

Hence, $x_0 - m\dfrac{f(x_0)}{f'(x_0)},\ x_0 - (m-1)\dfrac{f'(x_0)}{f''(x_0)},\ x_0 - (m-2)\dfrac{f''(x_0)}{f'''(x_0)}, \ldots\ldots$

will all have same value if initial x_0 is chosen very close to the actual root.

Note: *Remember that generalized Newton's method converges more rapidly than the Newton-Raphson method.*

SOLVED PROBLEMS

Problem 1: Find the double root of $x^2 - 2x + 1 = 0$.

Solution: Let $f(x) = x^2 - 2x + 1$

So $f'(x) = 2x - 2$

Let the initial value be $x_0 = 0.9$

Then by Generalised Newton's formula

$$x_0 - 2\frac{f(x_0)}{f'(x_0)} = 0.9 - 2\frac{\{(0.9)^2 - 2(0.9) + 1\}}{2(0.9) - 2}$$

$$= \frac{(-0.18)-2(1.81-1.8)}{(-0.2)} = 1$$

and $$x_0-(2-1)\frac{f'(x_0)}{f''(x_0)} = 0.9-(2-1)\frac{2(0.9)-2}{2}$$

$$= \frac{1.8-(1.8-2)}{2} = 1$$

If α is a double root of $f(x)=0$ then α is a simple root of $f''(x_0)=0$

Since these two values are equal, therefore there is a double root near $x = 1$

If values are very much near about each other, then there also exists double root and we have to refine these values by doing next iteration.

Problem 2: Find the double root of $x^3 - 7x^2 + 16x - 12 = 0$

Solution: Let $$f(x) = x^3 - 7x^2 + 16x - 12$$

So $$f'(x) = 3x^2 - 14x + 16$$

$$f''(x) = 6x - 14$$

Let the initial value be $x_0 = 1.9$

Then by Generalized Newton's formula

Then $$\frac{(-4.94)-0.23}{(-2.6)} = 1.9884615$$

and $$x_0-(2-1)\frac{f'(x_0)}{f''(x_0)} = 1.9-(2-1)\frac{3(1.9)^2-14(1.9)+16}{6(1.9)-14}$$

$$= \frac{(-4.94)-0.23}{(-2.6)} = 1.9884615$$

If α is a double root of $f(x) = 0$ then α is a simple root of $f''(x_0) = 0$

Since these two values are very close, therefore there is a double root near $x = 1.9$

Problem 3: Find the double root of the equation $x^3 - x^2 - x + 1 = 0$.

Solution: Let $$f(x) = x^3 - x^2 - x + 1$$

so that $$f'(x_0) = 3x^2 - 2x - 1$$

$$f''(x_0) = 6x - 2$$

Starting with $x0 = 0.9$, we have

$$x_0 - 2\frac{f(x_0)}{f'(x_0)} = 0.9 - \frac{2\times.019}{(-.37)} = 1.003$$

and $$x_0-(2-1)\frac{f'(x_0)}{f''(x_0)} = 0.9-\frac{(-.37)}{3.4} = 1.009$$

The closeness of these values implies that there is a double root near $x = 1$.

Choosing $x_1 = 1.01$ for the next approximation, we get

$$x_1-2\frac{f(x_0)}{f'(x_0)} = 1.01-2\times\frac{0.0002}{0.0403} = 1.0001$$

and $$x_1-(2-1)\frac{f'(x_0)}{f''(x_0)} = 1.01-\frac{.0403}{4.06} = 1.0001$$

This shows that there is a double root at $x = 1.0001$ which is quite near the actual root $x = 1$.

Problem 4: Show that the equation $f(x) = 1 - xe^{1-x} = 0$ has a double root at $x = 1$. The root is obtained by using the modified Newton- Raphson method with $m = 2$ starting with $x_0 = 0$.

Solution: Since $f(1) = f'(1) = 0$ and $f''(1) \neq 0$, the root $x = 1$ is a double root.

$$x_{n+1} = x_n - 2\left[\frac{x_n^3-7x_n^2+16x_n-12}{3x_n^2-14x_n+16}\right],\quad n = 0, 1, \ldots\ldots$$

Starting with $x_0 = 0$, we get

$$x_1 = .735758882$$
$$x_2 = .978185253$$
$$x_3 = .999842233$$
$$x_4 = 1.000000061$$
$$x_5 = 1.000000061$$

Hence the root correct to six decimal places is 1.000000.

UNSOLVED PROBLEMS

1. By using Newton-Raphson's method, find the root of $x^4 - x - 10 = 0$ which is near to $x = 2$, correct to three decimal places.
2. Compute one positive root of $2x - \log_{10} x = 7$ by the Newton-Raphson method correct to four decimal places.
3. (i) Use the Newton-Raphson method to find a root of the equation $x^3 - 2x - 5 = 0$.

   (ii) Use Newton-Raphson method to find a root of the equation $x^3 - 3x - 5 = 0$.
4. Find the real root of the equations

   (i) $\log x = \cos x$ (ii) $x^2 + 4 \sin x = 0$

   by Newton-Raphson method, correct to three decimal places.

5. Use Newton-Raphson method to obtain a root correct to three decimal places of the following equations:

(i) $\sin x = 1 - x$ (ii) $x^3 - 5x + 3 = 0$

(iii) $x^4 + x^2 - 80 = 0$ (iv) $x^3 + 3x^2 - 3 = 0$

(v) $4(x - \sin x) = 1$ (vi) $x - \cos x = 0$

(vii) $\sin x = x/2$ (viii) $x + \log x = 2$

(ix) $\tan x = x$.

6. Explain the method of Newton-Raphson for computing roots. Apply it for finding x from $x^2 - 25 = 0$. Write a program using 'C'.
7. Write a computer program in 'C' for finding out a real root of eqn. $f(x) = 0$ by the Newton Raphson method.
8. Using the Newton-Raphson method, obtain the formula for N and find 20 correct to 2 decimal places.
9. Obtain the cube root of 120 using the Newton-Raphson method, starting with $x_0 = 4.5$.
10. Develop an algorithm using the Newton-Raphson method to find the fourth root of a positive number N, and find $\sqrt[4]{32}$.
11. Find the cube root of 3 correct to three decimal places by Newton's iterative method.
12. Prove the recurrence formula

$$x_{i+1} = \frac{1}{3}\left(2x_i + \frac{N}{x_i^2}\right)$$

for finding the cube root of N. Find the cube root of 63.

13. Use Newton's formula to prove that the square root of N can be obtained by the recursion formula,

$$x_{i+1} = x_i\left(1 - \frac{x_i^2 - N}{2N}\right)$$

Find the square root of

(a) 26 (b) 29 (c) 35.

14. Show that the iterative formula for finding the reciprocal of n is $x_{i+1} = x_i(2 - nx_i)$, and find the value of 1/31.
15. Determine p, q, and r so that the order of the iterative method

$$x_{n+1} = px_b + \frac{qa}{x_n^2} + \frac{ra^2}{x_n^5}$$

for $a^{1/3}$ becomes as high as possible. **[Hint: $p + q + r = 1, p - 2q - 5r = 0, 3q + 15r = 0$.]**

16. Derive the expression for the Newton-Raphson method to find a root of an equation. Find the order of the convergence of this method.

17. Find all positive roots of the equation

$$\int_0^x e^{-x^2}\,dt - 1 = 0 \text{ with six correct decimals.}$$

18. Given equation

$$2e^{-x} = \frac{1}{x+2} + \frac{1}{x+1}$$

has two roots greater than – 1. Calculate these roots correct to five decimal places.

19. The equation $x = 0.2 + 0.4\sin(x/b)$ where b is a parameter, has one solution near $x = 0.3$. The parameter is known only with some uncertainty: $b = 1.2 \pm 0.05$. Calculate the root with an accuracy reasonable with respect to the uncertainty of b.

20. Find the positive root of the equation

$$e^x = 1 + x + \frac{x^2}{2} + \frac{x^3}{6}e^{0.3x} \text{ correct to 6 decimal places.}$$

4.11 SECANT METHOD (ALSO CALLED CHORD METHOD)

Secant method is quite similar to that of the Regula-Falsi method except for the condition. Secant method is faster as compared to false position method but slower than that of Newton's method. Convergence is not assured in case of secant method. It also overcomes the problem of Newton's method of calculating $f(x)$ in each iteration.

Assume that we have two (x_0, x_1) initial guess values of the root α. Then, according to Newton's Raphson method

$$x_{n+1} = x_n - \frac{f(x_n)}{f'(x_n)}$$

where $f'(x_n)$ represents the slope. So we can write

$$= x_n - \frac{f(x_n)}{[f(x_n) - f(x_{n-1})]/(x_n - x_{n-1})}$$

After simplify, we have result

$$x_{n+1} = \frac{x_{n-1}f(x_n) - x_n f(x_{n-1})}{f(x_n) - f(x_{n-1})}$$

Above formula is used for secant method. If the denominator of the above formula, i.e. $f(x_n) - f(x_{n-1})$ approach towards zero, then the method fails. To avoid this kind of Problem we can write the above formula as

$$x_{n+1} = x_n - \frac{(x_n - x_{n-1})\,f(x_n)}{f(x_n) - f(x_{n-1})}$$

Now, the new derived formula is known as the general formula for secant method. The steps for applying secant method are same as in case of false position method. The Iterative cycle terminates when the relative error in two successive approximations is less than or equal to the prescribed tolerance (epsilon).

4.11.1 Rate of Convergence of Secant Method

The formula for secant method can be written as

$$x_{n+1} = x_n - \frac{(x_n - x_{n-1})\, f(x_n)}{f(x_n) - f(x_{n-1})} \qquad \text{...(i)}$$

Suppose we have a function, $f(x) = 0$ whose exact root is r.

Then $e_{n+1} = x_{n+1} - r$ // error in $(n+1)^{\text{th}}$ iteration

$e_n = x_n - r$ // error in n^{th} iteration

$e_{n-1} = x_{n-1} - r$ // error in $(n-1)^{\text{th}}$ iteration

After substitution & simplification, eq.(i) becomes

$$e_{n+1} = e_n - \frac{(e_n - e_{n-1})\, f(r+e_n)}{f(r+e_n) - f(r+e_{n-1})} \qquad \text{...(ii)}$$

Find $f(r+e_n)$ & $f(r+e_{n-1})$ by expanding them upto third term using Taylor's series, we get

$$f(r+e_n) = f(r) + e_n f'(r) + \frac{e_n^2}{2} f''(r)$$

$$f(r+e_{n-1}) = f(r) + e_{n-1} f'(r) + \frac{e_{n-1}^2}{2} f''(r)$$

After simplification of $f(r+e_n) - f(r+e_{n-1})$, we get

$$f(r+e_n) - f(r+e_{n-1}) = (e_n - e_{n-1}) f'(r) + \frac{e_n^2 - e_{n-1}^2}{2} f''(r)$$

Substitute this in (ii), we get

$$e_{n+1} = e_n - \frac{(e_n - e_{n-1})[f(r) + e_n f'(r) + (e_n^2)/2.f''(r)]}{(e_n - e_{n-1}) f'(r) + (e_n^2 - e_{n-1}^2)/2.f''(r)}$$

After simplification of this, we have

$$e_{n+1} = e_n - \frac{f(r) + e_n f'(r) + (e_n^2)/2.f''(r)}{f'(r) + (e_n + e_{n-1})/2.f''(r)}$$

As r is the root of the $f(x) = 0$ then $f(r) = 0$

So
$$e_{n+1} = e_n - \frac{e_n f'(r) + (e_n^2)/2.f''(r)}{f'(r) + (e_n + e_{n-1})/2.f''(r)}$$

Divide num. & den. By $f'(r)$, we get

$$e_{n+1} = e_n - [e_n + \{(e_n^2)/2\}.\{f''(r)/f'(r)\}]\,[1 - \{(e_n + e_{n-1})/2\}.\{f''(r)/f'(r)\}]$$

$$= e_n - [e_n + \{e_n^2.f''(r)\}/\{2.f'(r)\} - \{e_n(e_n + e_{n-1}).f''(r)\}/\{2.f'(r)\} + ...]$$

$= \{e_n e_{n-1} f''(r)\}/f'(r)$, Neglecting higher e_i's terms

So
$$e_{n+1} = e_n e_{n-1} A \qquad \text{...(iii)}$$

where $A = f''(r)/f'(r)$

To determine order of convergence, the result should be of the type

$$e_{n+1} = k.e_n^{\ m} \qquad ...(iv)$$

From (iv), we have $e_n = k.\ e_{n-1}^{\ m}$, $e_{n-1} = (e_n/k-)^{1/m}$

Substitute these in (iii), we get

$$k.e_n^{\ m} = e_n(e_n/k)^{1/m}.k.A$$

$$k.e_n^{\ m} = e_n^{\ 1+(1/m)}\ .k^{1-(1/m)}.A$$

Equating e_n's coefficients from both sides, we get

$$m = 1 + \frac{1}{m} \quad \Rightarrow \quad m^2 - m - 1 = 0$$

$$\Rightarrow \qquad m = \frac{1 \pm \sqrt{5}}{m} = 1.618 \text{ (by taking +ve value)}$$

Hence, rate of convergence of secent method is **1.618**.

4.11.2 Algorithm for Secant Method

Let $f(x)$ be the given function, x_0, x_1 are two initial guess values, epsilon is the prescribed relative error, delta is the prescribed lower bound for slope of $f(x)$ and n be the maximum number of iterations to be allowed.

1. Read x_0, x_1, epsilon, n, delta
2. for i = 1 to n in steps of 1 do
3. if $|f(x_1) - f(x_0)| \leq$ delta then
4. write: "slope too small"
5. exit
6. end if
7. set $x_2 = \{x_0 f(x_1) - x_1 f(x_0)\}/(f(x_1) - f(x_0))$
8. if $((|x_2 - x_1)/x_2| <$ epsilon) or $(f(x_2)=0)$) then
9. write: "Root is ",x2
10. exit
11. end if
12. set $x_0 = x_1$
13. set $f(x_0) = f(x_1)$
14. set $x_1 = x_2$
15. set $f(x_1) = f(x_2)$
16. end for
17. Write:" Solution does not converge in," n, "iterations"
18. End

4.11.3 Program for Secant Method in C

```
#include<stdio.h>
#include<conio.h>
#include<stdlib.h>
#include<math.h>
#include<iomanip.h>
#define f(x) (x*x*x-9*x+1)
void main()
{
int i,n;
float x0,epsilon,x1,delta,x2;
clrscr();
printf("Enter initial approximation for x0 and x1");
scanf("%f%f ",&x0,&x1);
printf("Enter prescribed tolerance value 'epsilon'");
scanf("%f",&epsilon);
printf("Enter prescribed lower bound for the slope f(x)");
scanf("%f",&delta);
printf("Enter maximum number of iterations");
scanf("%d",&n);
if(f(x0)*f(x1)>0)
{
printf("Initial approximations are not suitable!");
getch();
exit(1);
}
for(i=1;i<=n;i++)
{
if(fabs(f(x0)-f(x1))<delta)
{
printf("Slope too small");
exit(1);
}
x2=(x0*f(x1)-x1*f(x0))/(f(x1)-f(x0));
```

```
printf("\n Value of x2 = %f",x2);
printf("After iteration number = %d",i);
if((fabs((x2-x1)/x2)<epsilon) || (f(x2)==0))
{
printf("\n APPROXIMATE ROOT = %f",x1);
getch();
exit(1);
}
x0=x1;
x1=x2;
}
printf("Solution doesn't converge in %d" ,n);
printf("iterations");
getch();
}
```

Output:

```
Enter initial approximation for x0 and x1   2   4
Enter prescribed tolerance value 'epsilon' 0.001
Enter prescribed lower bound for the slope f(x)  0.001
Enter maximum number of iterations    10
Value of x2 = 2.473684 After iteration number = 1
Value of x2 = 2.739889 After iteration number = 2
Value of x2 = 3.010912 After iteration number = 3
Value of x2 = 2.935228 After iteration number = 4
Value of x2 = 2.942558 After iteration number = 5
Value of x2 = 2.952559 After iteration number = 6
APPROXIMATE ROOT = 2.942558
```

4.11.4 Program for Secant Method in C++

```
#include<iostream.h>
#include<conio.h>
#include<stdlib.h>
#include<math.h>
#include<iomanip.h>
```

```
#define f(x) (x*x*x-9*x+1)
void main()
{
int i,n;
float x0,epsilon,x1,delta,x2;
clrscr();
cout<<"Enter initial approximation for x0 and x1";
cin>>x0>>x1;
cout<<"Enter prescribed tolerance value 'epsilon'";
cin>>epsilon;
cout<<"Enter prescribed lower bound for the slope f(x)";
cin>>delta;
cout<<"Enter maximum number of iterations";
cin>>n;
if(f(x0)*f(x1)>0)
{
cout<<"Initial approximations are not suitable!";
getch();
exit(1);
}
for(i=1;i<=n;i++)
{
if(fabs(f(x0)-f(x1))<delta)
{
cout<<"Slope too small";
exit(1);
}
x2=(x0*f(x1)-x1*f(x0))/(f(x1)-f(x0));
cout<<"\nValue of x2 = "<<x2<<"After iteration number = "<<i;
if((fabs((x2-x1)/x2)<epsilon) || (f(x2)==0))
{
cout<<"\n APPROXIMATE ROOT = "<<x1;
getch();
exit(1);
```

```
}
x0=x1;
x1=x2;
}
cout<<"Solution doesn't converge in " << n << "iterations";
getch();
}
```

Output:

```
Enter initial approximation for x0 and x1  2   4
Enter prescribed tolerance value 'epsilon' 0.001
Enter prescribed lower bound for the slope f(x)  0.001
Enter maximum number of iterations   10
Value of x2 = 2.473684 After iteration number = 1
Value of x2 = 2.739889 After iteration number = 2
Value of x2 = 3.010912 After iteration number = 3
Value of x2 = 2.935228 After iteration number = 4
Value of x2 = 2.942558 After iteration number = 5
Value of x2 = 2.952559 After iteration number = 6
APPROXIMATE ROOT = 2.942558
```

SOLVED PROBLEMS

Problem 1: Compute the Positive root of the equation cos$x$ - xe^x = 0, using Secant Method to three decimal places.

Solution: Let

$$f(x) = \cos x - xe^x$$

$$f(0) = 1 = +\text{ve}$$

and

$$f(1) = -2.17789 = -\text{ve}$$

$$(x_0, x_1)$$

Hence, required root lies between interval (0, 1)

So by using secant method, we have

$$x_2 = \frac{x_0 f(x_1) - x_1 f(x_0)}{f(x_1) - f(x_0)}$$

$$= \frac{0(-2.17789) - 1(1)}{-2.17789 - 1} = 0.31467$$

$$f(x_2) = f(0.31467)$$
$$= \cos(0.31467) - 0.31467(e)^{0.31467}$$
$$= 0.51987 = +\text{ve}$$

Hence, required root now lies between interval (0.31467, 1)

$$\therefore \quad x_3 = \frac{x_1 f(x_2) - x_2 f(x_1)}{f(x_2) - f(x_1)}$$

$$= \frac{1(0.51987) - 0.31467(-2.17789)}{0.51987 - (-2.17789)}$$

$$= \frac{0.51987 + 0.68531}{0.51987 + 2.17789}$$

$$= \frac{1.20518}{2.69776} = 0.44673$$

$$f(x_3) = f(0.44673)$$
$$= \cos(0.44673) - 0.44673(e)^{0.44673}$$
$$= 0.20354 = +\text{ve}$$

Hence, required root lies between interval (0.44673, 1)

Similarly this process will go on until we get the required root.

Here we have, the next approximations

$$x_4 = 0.53171$$
$$x_5 = 0.51690$$
$$x_6 = 0.51775$$
$$x_7 = 0.51776$$

As x_6 and x_7 are very much close to each other so stop here.

Hence required root up to three decimal places is 0.518

Problem 2: Compute a root of the equation $x = e^{-x}$, using Secant Method to three decimal places.

Solution: As $\quad x = e^{-x}$

or $\quad x = 1/e^x$

or $\quad x e^x = 1$ or $x e^x - 1 = 0$

Let $\quad f(x) = x e^x - 1$

since $\quad f(0) = -1 = -\text{ve}$

and $\quad f(1) = e - 1 = +\text{ve}$

$$(\mathbf{x_0, x_1})$$

Hence, required root lies between interval (0, 1)

$$f(x_0) = f(0) = -1$$

and $$f(x_1) = f(1) = e - 1 = 2.718 - 1 = 1.718$$

So by using secant method, we have

$$x_2 = \frac{x_0 f(x_1) - x_1 f(x_0)}{f(x_1) - f(x_0)}$$

$$= \frac{0(1.718) - 1(-1)}{1.718 - (-1)} = 0.3679$$

Now, $$f(x_2) = 0.3679 \times e^{0.3679} - 1$$

$$= 0.5315 - 1 = -0.4685 = -\text{ve}$$

$$(x_1, x_2)$$

Hence, required root now lies between interval (0.3679, 1)

$$\therefore \quad x_3 = \frac{x_1 f(x_2) - x_2 f(x_1)}{f(x_2) - f(x_1)}$$

$$= \frac{(0.3679)(1.718) - 1(-0.4685)}{1.718 - (-0.4685)} = 0.5034$$

Now, $$f(x_3) = 0.5034 \times e^{0.5034} - 1$$

$$= -0.1673 = -\text{ve}$$

$$(x_2, x_3)$$

Hence, required root now lies between interval (0.5034, 1)

$$\therefore \quad x_4 = \frac{x_2 f(x_3) - x_3 f(x_2)}{f(x_3) - f(x_2)}$$

$$= \frac{(0.5034)(1.718) - 1(-0.1673)}{1.718 - (-0.1673)} = 0.5474$$

Similarly, proceeding in this way we will reach at required root which is 0.5671

Problem 3: Find a real root of the equation $f(x) = x^3 - 5x + 1 = 0$ lies in the interval (0, 1). Perform four iterations of the secant method.

Solution: We have, $x_0 = 0, x_1 = 1, f(x_0) = 1, f(x_1) = -3$

By Secant Method,

The first approximation is

$$x_2 = x_1 - \left[\frac{x_1 - x_0}{f(x_1) - f(x_0)}\right] f(x_1) = 0.25$$

$$f(x_2) = -0.234375.$$

The second approximation is

$$x_3 = x_2 - \left[\frac{x_2 - x_1}{f(x_2) - f(x_1)}\right] f(x_2) = 0.186441$$

$$f(x_3) = 0.074276$$

The third approximation is

$$x_4 = x_3 - \left[\frac{x_3 - x_2}{f(x_3) - f(x_2)}\right] f(x_3) = 0.201736$$

$$f(x_4) = -0.000470$$

The fourth approximation is

$$x_5 = x_4 - \left[\frac{x_4 - x_3}{f(x_4) - f(x_3)}\right] f(x_4) = 0.201640$$

$$f(x_5) = 0.201640.$$

Problem 4: Compute the root of the equation $x^2 e^{-x/2} = 1$ in the interval [0, 2] using the secant method. The root should be correct to three decimal places.

Solution: We have,

$x_0 = 1.42, x_1 = 1.43, f(x_0) = -.0086, f(x_1) = .00034.$

By secant method,

The first approximation is

$$x_2 = x_1 - \left[\frac{x_1 - x_0}{f(x_1) - f(x_0)}\right] f(x_1)$$

$$= 1.43 - \left[\frac{1.43 - 1.42}{.00034 + .0086}\right](.00034) = 1.4296$$

$$f(x_2) = -.000011$$

The second approximation is

$$x_3 = x_2 - \left[\frac{x_2 - x_1}{f(x_2) - f(x_1)}\right] f(x_2)$$

$$= 1.4296 - \left[\frac{1.4296 - 1.42}{-.000011 - .0034}\right](-.000011)$$

$$= 1.4292$$

Since x_2 and x_3 agree up to three decimal places, the required root is **1.429.**

UNSOLVED PROBLEMS

1. Write the procedure of the secant method to find a root of a polynomial equation to implement it in 'C' and 'C++'.

2. The equation $x^2 - 2x - 3\cos x = 0$ is given
   (i) Locate the smallest root in magnitude in an interval of length one unit.
   (ii) Hence, find this root correct to 3 decimal points using the secant method.
3. Use the secant method to determine the root of the equation $\cos x - xe^x = 0$.

Now we proceed to discuss some methods useful for obtaining the complex roots of polynomial equations $f(x) = 0$.

4.12 MULLER'S METHOD

Muller's method is an iterative method, which is also used for finding the root but it takes three initial values to start. The method is iterative, converges almost quadratically, and can be used to obtain complex roots.

4.12.1 Analytical Derivation of Muller's Method

Assume that the three initial points are $A[x_0, f(x_0)]$, $B[x_1, f(x_1)]$ and $C[x_2, f(x_2)]$. Consider a parabola

$$y = f(x) = a(x-x_0)^2 + b(x-x_0) + c \quad \text{...(i)}$$

Which passes through these three points and also suppose that this parabola intersect the x-axis at two points, say α and α'. Take the order of x_n's values as $x_2 < x_0 < x_1$.

Assume that $x_1 - x_0 = d_1$, $x_0 - x_2 = d_2$

Substitute the given values of x_n's in equation (i)

At $x = x_0$, $\quad f(x_0) = c \quad$ ...(ii)

At $x = x_1$, $\quad f(x_1) = ad_1^2 + bd_1 + c$

or $\quad ad_1^2 + bd_1 = f(x_1) - f(x_0) \quad$ (as $c = f(x_0)$) $\quad$ ...(iii)

At $x = x_2$, $\quad f(x_2) = ad_2^2 - bd_2 + c$

or $\quad ad_2^2 - bd_2 = f(x_2) - f(x_0) \quad$ ...(iv)

Solving (iii) and (iv) for a and b,

(iii) × d_2 + (iv) × d_1 gives

$$ad_1d_2(d_1+d_2) = d_2 f(x_1) - (d_1 + d_2) f(x_0) + d_1 f(x_2)$$

$$\therefore \quad a = \frac{d_2 f(x_1) - (d_1 + d_2) f(x_0) + d_1 f(x_2)}{d_1 d_2 (d_1 + d_2)} \quad \text{...(v)}$$

From (iii), we have

$$b = \frac{f(x_1) - f(x_0) - ad_1^2}{d_1} \quad \text{...(vi)}$$

Now point of intersection of the parabola $y = a(x-x_0)^2 + b(x-x_0) + c$ with x-axis i.e. $y = 0$ are

$$x - x_0 = \frac{-b \pm \sqrt{b^2 - 4ac}}{2a}$$

After rationalizing and simplification, we have

$$x = x_0 - \frac{2c}{-b \pm \sqrt{b^2 - 4ac}}$$

As from the above result we have seen that x gives two values one for '+' and one for '–' , we have to chose that value of x, which is nearer to x_0.

i.e. if $b > 0$, take the '+' sign in the denominator and if b>0, take '-' sign.

Now for starting the next iteration, choose three values out of these four values as x_2, x_0, x_1, x in such a way that $x > x_0$, neglect x_2, and take x_0, x, x_1, and rename them as x_2, x_0, x_1.

If $x < x_0$, then neglect x_1, and take x_2, x, x_0 and rename them as x_2, x_0, x_1.

And if $x > x_1$, take x_2, x_1, x and rename them as x_2, x_0, x_1. if $x < x_2$, take x, x_2, x_0 and rename them as x_2, x_0, x_1.Proceed in this way till we get the required value.

We will explain these, by using examples.

4.12.2 Algorithm for Muller's Method

1. Start.
2. Input variables xi, xi1, xi2
3. Input absolute error-aerr
4. Repeat Steps 5-12 until |Xn-Xi| < aerr is true
5. Yi = y(Xi)
6. Yil = y(Xi1)
7. Yi2 = y(Xi2)
8. a = A(Xi, Xi1, Xi2, Yi, Yi1, Yi2)
9. b = B(Xi, Xi1, Xi2, Yi, Yi1, Yi2);
10. Xn = approx (Xi, Yi, a, b);
11. Check condition
12. if false
13. exit loop
14. if yes
15. Xi = Xn
16. increment value of i
17. End of loop
18. Print output value
19. End
20. Start of section 1
21. take Xa, Xb, Xc, Ya, Yb, Yc
22. x = ((Yb-Ya)*(Xc-Xa)-(Yc-Ya)*(Xb-Xa))/((Xb-Xa)*(Xc-Xa)*(Xb-Xc))

23. Return value of x
24. End of section 1
25. Start of section 2
26. Take Xa, Xb, Xc, Ya, Yb, Yc
27. c = (((Yc-Ya)*pow((Xb-Xa),2))-((Yb-Ya)*pow((Xc-Xa),2)))/((Xb-Xa)*(Xc-Xa)*(Xb-Xc))
28. Return value of c
29. End of section 2
30. Start of section approx
31. Take x, y, a, b
32. c = sqrt(b*b-4*a*y)
33. If (b + c) > (b-c): t = x-((2*y)/(b + c))
34. Else: t = (x-((2*y)/(b-c)))
35. Return value of t
36. End of section approx

4.12.3 Flow-chart for Muller's Method

See figure on next page

4.12.4 Program for Muller's Method of Finding Roots in C

```
#include <stdio.h>
#include <string.h>
#include <conio.h>
#include <math.h>
#include <process.h>
#include <dos.h>
float y(float);
float A(float,float,float,float,float,float);
float B(float,float,float,float,float,float);
float approx(float,float,float,float);
void main()
{
float a,b,Xi,Xi1,Xi2,Yi,Yi1,Yi2,Xn,aerr;
int i=1;
```

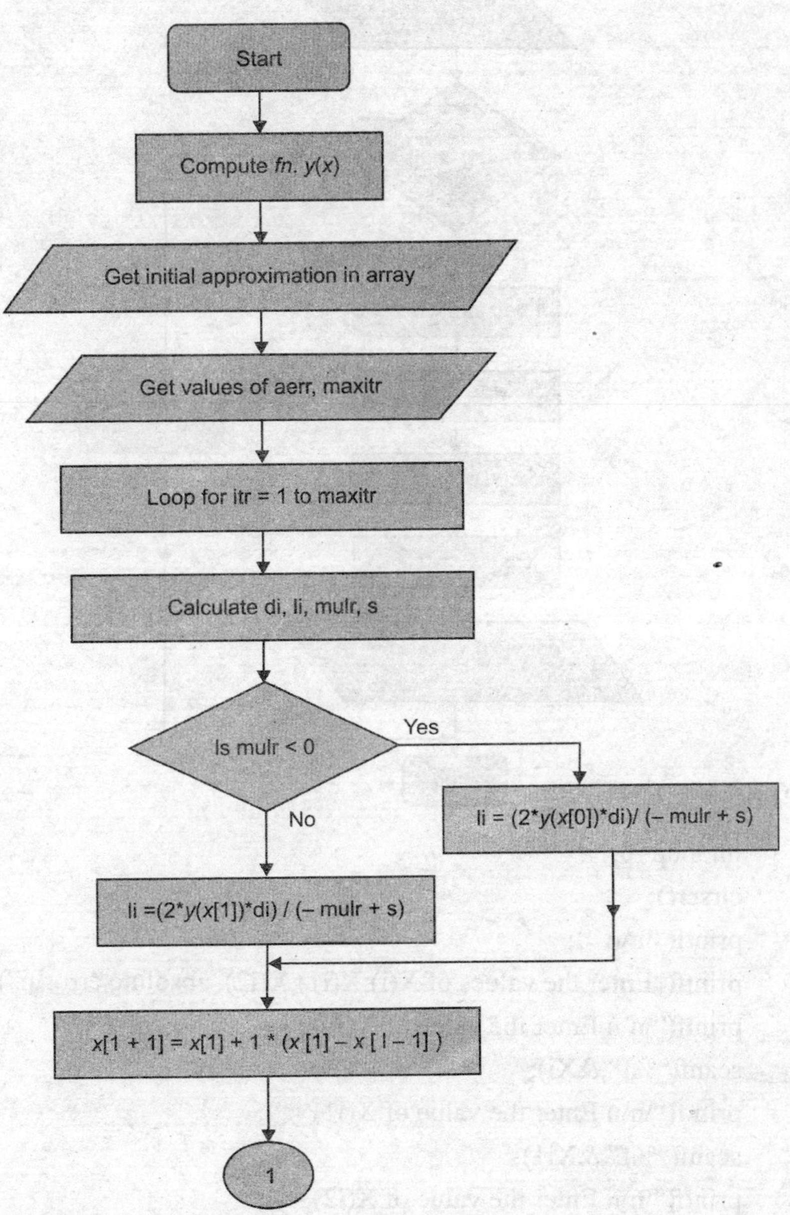
Start
Compute fn. y(x)
Get initial approximation in array
Get values of aerr, maxitr
Loop for itr = 1 to maxitr
Calculate di, li, mulr, s
Is mulr < 0
Yes
No
li = (2*y(x[0])*di)/ (– mulr + s)
li =(2*y(x[1])*di) / (– mulr + s)
x[1 + 1] = x[1] + 1 * (x [1] – x [l – 1])
1

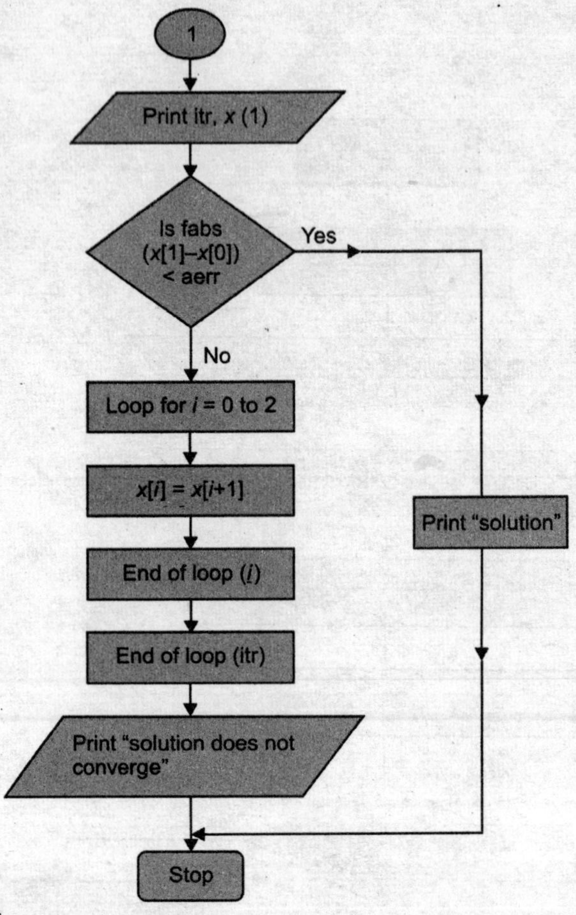


```
int loop=0;
clrscr();
printf("\n\n ");
printf("Enter the values of X(i),X(i1),X(i2), absolute error\n");
printf("\n\n Enter the value of X(i) - ");
scanf("%f",&Xi);
printf("\n\n Enter the value of X(i1) - ");
scanf("%f",&Xi1);
printf("\n\n Enter the value of X(i2) - ");
scanf("%f",&Xi2);
printf("\n\n Enter the value of Absolute Error – ");
scanf("%f",&aerr);
for(loop=0; loop<10;loop++)
```

```
{
delay(200);
printf("...");
}
printf("\n\n\n");
while(1)
{
Yi=y(Xi);
Yi1=y(Xi1);
Yi2=y(Xi2);
a=A(Xi,Xi1,Xi2,Yi,Yi1,Yi2);
b=B(Xi,Xi1,Xi2,Yi,Yi1,Yi2);
Xn=approx(Xi,Yi,a,b);
printf("\n\n After Iteration %d value of x-%f",i,Xn);
if(fabs(Xn-Xi)<aerr)
{
goto jmp;
}
Xi=Xn;
i++;
}
jmp:
printf("\n\n After %d iterations root is-%6.6f\n",i+1,Xn);
printf("\n\n\n Press Enter to Exit");
getch();
}
float y(float x)
{
float t;
t=(x*x*x)-(2*x)-5;
return(t);
}
float A(float Xa;float Xb,float Xc,float Ya,float Yb,
float Yc)
```

```
{
float x;
x=((Yb-Ya)*(Xc-Xa)-(Yc–Ya)*(Xb-Xa))/((Xb-Xa)*(Xc-Xa)
*(Xb-Xc));
return(x);
}
float B(float Xa,float Xb,float Xc,float Ya,float Yb,
float Yc)
{
float c;
c=(((Yc-Ya)*pow((Xb-Xa),2))-((Yb-Ya)*pow((Xc-Xa),2)))
/((Xb-Xa)*(Xc-Xa)*(Xb-Xc));
return(c);
}
float approx(float x,float y,float a,float b)
{
int c;
float t;
c=sqrt(b*b-4*a*y);
if((b+c)>(b-c))
{
t=x-((2*y)/(b+c));
}
else
{
t=(x-((2*y)/(b-c)));
}
return (t);
}
```

Output:

```
Enter the values of X(i),X(i1),X(i2), absolute error
Enter the value of X(i) - 3
Enter the value of X(i1) - 2
Enter the value of X(i2) - 1
```

```
Enter the value of Absolute Error - 0.000001
Processing ................................
After Iteration 1 value of x - 2.085714
After Iteration 2 value of x - 2.094654
After Iteration 3 value of x - 2.094550
After Iteration 4 value of x - 2.094552
After Iteration 5 value of x - 2.094552
After 6 iteration root is - 2.094552
Press Enter to Exit
```

4.12.5 Program for Muller's Method of Finding Roots Using C++

```
#include <iostream.h>
#include <string.h>
#include <conio.h>
#include <math.h>
#include <process.h>
#include <dos.h>
float y(float);
float A(float,float,float,float,float,float);
float B(float,float,float,float,float,float);
float approx(float,float,float,float);
void main()
{
float a,b,Xi,Xi1,Xi2,Yi,Yi1,Yi2,Xn,aerr;
int i=1;
int loop=0;
clrscr();
cout<<"\n\n ";
cout<<"Enter the values of X(i),X(i1),X(i2),absolute error\n";
cout<<"\n\n Enter the value of X(i) - ";
cin>>Xi;
cout<<"\n\n Enter the value of X(i1) - ";
cin>>Xi1;
cout<<"\n\n Enter the value of X(i2) - ";
```

```
cin>>Xi2;
cout<<"\n\n Enter the value of Absolute Error – ";
cin>> aerr;
for(loop=0; loop<10;loop++)
{
delay(200);
cout<<"...";
}
cout<<"\n\n\n";
while(1)
{
Yi=y(Xi);
Yi1=y(Xi1);
Yi2=y(Xi2);
a=A(Xi,Xi1,Xi2,Yi,Yi1,Yi2);
b=B(Xi,Xi1,Xi2,Yi,Yi1,Yi2);
Xn=approx(Xi,Yi,a,b);
cout<"\n\n After Iteration"<<i<<"value of x-"<<Xn;
if(fabs(Xn-Xi)<aerr)
{
goto jmp;
}
Xi=Xn;
i++;
}
jmp:
cout<<"\n\n After"<<i+1<<"iterations root is"<<cout.width(6.6)<<Xn;
cout<<"\n\n\n Press Enter to Exit";
getch();
}
float y(float x)
{
float t;
t=(x*x*x)-(2*x)-5;
```

```
return(t);
}
float A(float Xa;float Xb,float Xc,float Ya,float Yb,
float Yc)
{
float x;
x=((Yb-Ya)*(Xc-Xa)-(Yc–Ya)*(Xb-Xa))/((Xb-Xa)*(Xc-Xa)
*(Xb-Xc));
return(x);
}
float B(float Xa,float Xb,float Xc,float Ya,float Yb,
float Yc)
{
float c;
c=(((Yc-Ya)*pow((Xb-Xa),2))-((Yb-Ya)*pow((Xc-Xa),2)))
/((Xb-Xa)*(Xc-Xa)*(Xb-Xc));
return(c);
}
float approx(float x,float y,float a,float b)
{
int c;
float t;
c=sqrt(b*b-4*a*y);
if((b+c)>(b-c))
{
t=x-((2*y)/(b+c));
}
else
{
t=(x-((2*y)/(b-c)));
}
return (t);
}
```

Output:

Enter the values of X(i),X(i1),X(i2), absolute error

Enter the value of X(i) - 3

Enter the value of X(i1) - 2
Enter the value of X(i2) - 1
Enter the value of Absolute Error - 0.000001
Processing ....
After Iteration 1 value of x - 2.085714
After Iteration 2 value of x - 2.094654
After Iteration 3 value of x - 2.094550
After Iteration 4 value of x - 2.094552
After Iteration 5 value of x - 2.094552
After 6 iteration root is - 2.094552
Press Enter to Exit

SOLVED PROBLEMS

Problem 1: By using Muller's method find root of the equation $f(x) = x^3 - 2x - 5 = 0$. Which lies between 2 and 3?

Solution: As we have given the starting values as 2 and 3.

Let $x_0 = 2$, $x_2 = 1$(that should be less than both the values) and $x_1 = 3$

$\therefore$

$$f(x_0) = f(2) = (2)^3 - 2(2) - 5 = -1$$

$$f(x_1) = f(3) = (3)^3 - 2(3) - 5 = 16$$

$$f(x_2) = f(1) = (1)^3 - 2(1) - 5 = -6$$

$$d_1 = x_1 - x_0 = 3 - 2 = 1,$$

$$d_2 = x_0 - x_2 = 2 - 1 = 1$$

As

$$c = f(x_0) = -1$$

$$a = \frac{d_2 f(x_1) - (d_1 + d_2) f(x_0) + d_1 f(x_2)}{d_1 d_2 (d_1 + d_2)}$$

$$= \frac{16 - (1+1)(-1) + (-6)}{1+1} = 6$$

$$b = \frac{f(x_1) - f(x_0) - a d_1^2}{d_1}$$

$$= \frac{16 + 1 - 6}{1} = 11$$

As $b > 0$, so we will take '+' sign in the denominator of the result as

$$x = x_0 - \frac{2c}{b + \sqrt{b^2 - 4ac}}$$

$$x = 2 - \frac{2(-1)}{11 + \sqrt{121 - 24}}$$

$$= 2 + \frac{2}{11 + 12.0415}$$

$$= 2 + 0.0868$$

$$= 2.0868$$

which is the required root.

Problem 2: Using Muller's method find root of the equation $f(x) = x^3 - x^2 - x - 1 = 0$.

Solution: As starting values are not given, first of all, we have to compute these values as

At $x = 1$, $f(1) = -2$

At $x = 2$, $f(2) = 1$

As at x values 1 and 2, we get opposite sign result, so take the starting values as 1 and 2.

Let $x_0 = 1, x_2 = 0, x_1 = 2$

$\therefore$ $f(x_0) = f(1) = -2$

$f(x_1) = f(2) = 1$

$f(x_2) = f(0) = -1$

$d_1 = x_1 - x_0 = 2 - 1 = 1$

$d_2 = x_0 - x_2 = 1 - 0 = 1$

As $f(x_0) = \text{c}$

$\therefore$ $\text{c} = -2$

$$a = \frac{d_2 f(x_1) - (d_1 + d_2) f(x_0) + d_1 f(x_2)}{d_1 d_2 (d_1 + d_2)}$$

$$= \frac{1 - (1+1)(-2) + (-1)}{(1+1)} = 2$$

$$b = \frac{f(x_1) - f(x_0) - a d_1^2}{d_1}$$

$$= \frac{1 - (-2) - (2)(1)^2}{1} = 1$$

As $b > 0$, so we will take '+' sign in the denominator of the result as

$$x = x_0 - \frac{2c}{b + \sqrt{b^2 - 4ac}}$$

$$x = 1 - \frac{2(-2)}{1 + \sqrt{1+16}}$$

$$= \frac{4.12 + 4}{4.12}$$

$$= 1.97$$

As $x > x_0$ i.e. $1.97 > 1$

Take $x_2 = 0, x_0 = 1, x_1 = 1.97$ for next iteration

$$f(x_0) = -2$$

$$f(x_1) = f(1.97) = (1.97)^3 - (1.97)^2 - (1.97) - 1$$

$$= 7.645 - 3.880 - 1.97 - 1$$

$$= 0.795$$

$$f(x_2) = -1$$

$$d_1 = x_1 - x_0 = 1.97 - 1 = 0.97$$

$$d_2 = x_0 - x_2 = 1 - 0 = 1$$

As $c = f(x_0) = -2$

$$a = \frac{d_2 f(x_1) - (d_1 + d_2) f(x_0) + d_1 f(x_2)}{d_1 d_2 (d_1 + d_2)}$$

$$= \frac{0.795 - (0.97 + 1)(-2) - 0.97}{(0.97 + 1)} = 1.9$$

$$b = \frac{f(x_1) - f(x_0) - a d_1^2}{d_1}$$

$$= \frac{0.795 - (-2) - (1.9)(0.97)^2}{0.97} = 1.05$$

As $b > 0$, so we will take '+' sign in the denominator of the result as

$$x = x_0 - \frac{2c}{b + \sqrt{b^2 - 4ac}}$$

$$x = 1 - \frac{2(-2)}{1.05 + \sqrt{1.1025 + 15.76}}$$

$$= 9.15/5.15 = 1.78$$

Similarly, proceeding in this way we will reach the exact root which is 1.84

UNSOLVED PROBLEMS

1. Use Muller's method to find a root of the equations:
   (i) $x^3 - x - 1 = 0$ (ii) $x^3 - x^2 - x - 1 = 0$
   which lie between 1 and 2.
2. Apply Muller's method to find the root of the equation $\cos x = xe^x$ which lies between 0 and 1.
3. Using Muller's method, find a root of the equations:
   (i) $x^3 - 3x - 5 = 0$ which lie between 2 and 3
   (ii) $\log x = x - 3$ taking $x_0 = 0.25$, $x_1 = 0.5$ and $x_2 = 1$
   (iii) $x^3 - 1/2 = 0$ take $x_0 = 0$, $x_1 = 1$ and $x_2 = 1/2$.
4. Solve by Muller's method: $x^3 + 2x^2 + 10x - 20 = 0$ by taking $x = 0$, $x = 1$, $x = 2$ as initial approximations.

4.13 GRAFFE'S ROOT- SQUARING METHOD

Graffe's root-squaring method is a direct method that gives all the roots simultaneously. This method has a great advantage over the other methods is that it does not require prior information about the approximate values, etc., of the roots. It is applicable to polynomial equations only and is capable of giving all the roots. Here below we discuss the case of the polynomial equation having real and distinct roots.

Suppose we have a polynomial

$$x^n + a_1x^{n-1} + a_2x^{n-2} + ...+a_n = 0$$

where $a_1, a_2, ..., a_n$ are real.

Separating the even terms on one side and the odd terms on the other side and squaring, we get

$$(x^n + a_2x^{n-2} + a_4x^{n-4} + ...)^2 = (a_1x^{n-1} + a_3x^{n-3} + ...)^2$$

After simplification and rearrangement, we have

$$x^{2n} + x^{2n-2}\{2a_2 - a_1^2\} + x^{2n-4}\{a_2^2 + 2a_4 - 2a_1a_3\} +.....+(-1)^n.a_n^2 = 0 \quad ...(i)$$

Replace $x^2 = y$ in eq. (i), we have

$$y^n + c_1.y^{n-1} + c_2.y^{n-2} +.....+ c_n = 0 \quad ...(ii)$$

Where

$$c_1 = -a_1^2 + 2a_2$$
$$c_2 = a_2^2 - 2a_1a_3 + 2a_4$$
$$\vdots$$
$$c_n = (-1)^n.a_n^2$$

The equation (ii), has n roots which a_n squared.

Now again separate the even terms on one side and the odd terms on other side and then square and rearrange and in this fashion the root is squared every time and this procedure is repeated and is finally stopped when the double products can be neglected.

Suppose after k such squaring and rearranging , we have

$$x^n + C_1 . x^{n-1} + C_2 . x^{n-2} + ... + C_n = 0$$

Which has n roots, say $p_1, p_2, ..., p_n$. if the original equation has roots, say $r_1, r_2, ..., r_n$ then we can write

$p_i = r_i^2$, $k = 1, 2, ..., n$

Assume that $\quad |r_1| > |r_2| > ..... > |r_n|$

then $\quad |p_1| > |p_2| ..... > |p_n|$

By using theory of equations, we have

$$C_1 = -\Sigma p_m \approx -p_1$$
$$C_2 = \Sigma p_k \cdot p_j \approx p_1 \cdot p_2$$
$$C_3 = -p_i \cdot p_j p_k \approx -p_1 \cdot p_2 \cdot p_3$$

Hence we have

$$p_1 \approx -C_1$$
$$P_2 \approx -C_2/C_1$$
$$P_3 \approx -C_3/C_2$$

By k such square-root extractions of p_i we obtain all r_i which are the roots of the original equation. The signs of the original roots r_i's are determined by the following analysis:

$$|p_i| = \frac{|C_i|}{|C_{i-1}|} = |r_i^2| = |r_i| 2^k$$

Taking logarithms, we have

$$\log |r_i| = -2^k (\log |C_i| - \log |C_{i-1}|), I = 1, 2, ..., k$$

From this we obtain the absolute value of r_i and by substitution in the original equation we can find the sign. This method is attractive as squaring process increases the separation of the roots which is helpful in finding the closer roots.

SOLVED PROBLEMS

Problem 1: Find all the roots of the polynomial $x^3 - 6x^2 + 11x - 6 = 0$ using Graffe's root squaring method.

Solution: As we have the given equation $x^3 - 6x^2 + 11x - 6 = 0$

After rearranging and squaring the even power and odd power terms, we have

$$(x^3 + 11x)^2 = (6x^2 + 6)^2$$

i.e. $\quad x^6 + 121x^2 + 22x^4 = 36x^2 + 36 + 72x^2$

i.e. $\quad x^6 - 14x^4 + 49x^2 - 36 = 0$

Put $x^2 = y$, we get

$$y^3 - 14\,y^2 + 49\,y - 36 = 0$$

Again rearranging and squaring the even power and odd power terms, we have

$$(y^3 + 49\,y)^2 = (14\,y^2 + 36)^2$$

i.e. $$y^6 + 2401\,y^2 + 98\,y^4 = 196\,y^4 + 1296 + 1008\,y^2$$

i.e. $$y^6 - 98\,y^4 + 1393\,y^2 - 1296 = 0$$

Again put $y^2 = z$, we have

$$z^3 - 98\,z^2 + 1393\,z - 1296 = 0$$

Again rearranging and squaring the even power and odd power terms, we have

$$(z^3 + 1393\,z)^2 = (98\,z^2 + 1296)^2$$

i.e. $$z^6 + 194044\,z^2 + 2786\,z^4 = 9604\,z^4 + 1679616 + 25401\,z^2$$

i.e. $$z^6 - 6818\,z^4 - 1686433\,z^2 - 1679616 = 0$$

put $z^2 = u$, we get

$$u^3 - 6818\,u^2 - 1868433\,u - 1679616 = 0$$

Then

$$|r_1| = (6818)^{1/8} = 3.0144433$$

$$|r_2| = (1868433/6818)^{1/8} = 2.0171006$$

$$|r_3| = (1679616/1868433)^{1/8} = 0.9867714$$

Problem 2: Find all the roots of the equation $x^4 - 3x + 1 = 0$ by Graeffe's method.

Solution: Here $$f(x) = x^4 - 3x + 1 = 0 \qquad ...(i)$$

Now $f(x)$ has two changes in sign i.e., + to – and – to +. Therefore it may have two positive real roots.

Again $f(-x) = x^4 + 3x + 1$. Since no change in sign of $f(-x)$ there is no negative root. But $f(x)$, being of degree four, will have four roots of which two are real positive and the remaining two are complex.

Rewriting (i) as $x^4 + 1 = 3x$.

Squaring and putting $x^2 = y$, we have

$$(y^2 + 1)^2 = 9y$$

Squaring again and putting, $y^2 = z$

$$(z + 1)^4 = 81z$$

i.e., $$z^4 + 4z^3 + 6z^2 - 77z + 1 = 0 \qquad ...(ii)$$

or $$z^4 + 6z^2 + 1 = -z(4z^2 - 77)$$

Squaring once again and putting $z^2 = u$, we get

$$(u^2 + 6u + 1)^2 = u(4u - 77)^2$$

or $$u^4 - 4u^3 + 654u^2 - 5917u + 1 = 0 \qquad ...(iii)$$

If p_1, p_2, p_3, p_4 are the roots of (i) and q_1, q_2, q_3, q_4 are the roots of (iii), then

$$p_1 = (q1)^{1\backslash 8} = (-\lambda)^{1\backslash 8} = (4)^{1\backslash 8} = 1.1892071$$

$$p_2 = (q_2)^{1/8} = \left[-\frac{\lambda_2}{\lambda_1}\right]^{1/8} = \left[\frac{654}{4}\right]^{1/8} = 1.8909921$$

$$p_3 = (q_3)^{1/8} = \left[-\frac{\lambda_3}{\lambda_2}\right]^{1/8} = \left[\frac{5917}{654}\right]^{1/8} = 1.3169384$$

$$p_4 = (q_4)^{1/8} = \left[-\frac{\lambda_4}{\lambda_3}\right]^{1/8} = \left[\frac{1}{5917}\right]^{1/8} = 0.3376659$$

From (ii) and (iii), we observe that the magnitudes of the co-efficients λ_1 and λ_4 have become constant.

p, p_4 are the real roots and p_2, p_3 are complex roots. Let these complex roots be

$$p_2 e^{\pm i\phi_2} = x_2 \pm i\eta_2$$

From (iii), its magnitude is given by

$$p_2^{2(2)3} \approx \frac{\lambda_3}{\lambda_1} = \frac{5917}{4} \qquad \therefore p^2 = 1.5780749$$

also from (i) the sum of the roots = 0, i.e., $p_1 + 2\xi_2 + p_4 = 0$

$$\xi_2 = \frac{1}{2}(p_1 + p_4) = -0.7634365$$

and

$$\eta_2 = \sqrt{p_2^{\ 2} - \xi_2^{\ 2}} = \sqrt{1.9074851} = 1.3811173$$

Hence, the four roots are 1.1892071, 0.3376659, – 0.7634365 ± 1.3811173 *i*.

UNSOLVED PROBLEMS

1. Find all the roots of the following equations by Graeffe's method squaring thrice:
   (i) $x^3 - 4x^2 + 5x - 2 = 0$ (ii) $x^3 - 2x^2 + 5x + 6 = 0$
   (iii) $x^3 - x - 1 = 0$.

4.14 COMPARISON OF NEWTON'S METHOD WITH REGULA-FALSI METHOD

Regula-Falsi is surely convergent while Newton's method is conditionally convergent. But once Newton's method converges, it converges faster. In the Falsi method, we calculate only one more value of the function at each step i.e., $f(x(n))$ while in Newton's method, we require two calculations $f(xn)$ and $f'(xn)$ at each step. Newton's method generally requires fewer iterations but also requires more time for computation at each iteration.

When $f'(x)$ is large near the root the correction to be applied is smaller in the case of Newton's method which is then preferred. If $f'(x)$ is small near the root, the correction to be applied is large and the curve becomes parallel to the x-axis.

In this case the Regula-Falsi method should be applied.

4.15 COMPARISON OF ITERATIVE METHODS

The main basis of comparison of the various iterative methods is done according to following measurements:

1. The rate of convergence
2. The amount of computational effort per iteration.
3. Sensitivity of the method to the starting value and the intermediate values.

We will show the comparison of various iterative methods by using a table as:

Comparison of Iterative Methods

Method	Iterative formulae	Order of Convergence	Functions evaluation per iteration	Reliability of Convergence
Bisection	$x_{n+1} = \frac{(x_n + x_{n-1})}{2}$ x_n and x_{n-1} enclose root	1	1	Guaranteed Convergence
False Position	$x_{n+1} = \frac{x_{n-1}f(x_n) - x_nf(x_{n-1})}{f(x_n) - f(x_{n-1})}$ x_n and x_{n-1} enclose root	1	1	Guaranteed Convergence
Newton-Raphson	$x_{n+1} = x_n - \frac{f(x_n)}{f'(x_n)}$	2	2	Sensitive to start value. Convergence fast if starting point near the root
Secant	$x_{n+1} = x_n - \frac{(x_n - x_{n-1})f(x_n)}{f(x_n) - f(x_{n-1})}$	1.618	1	No guarantee of convergence if not near the root. Economical method
Iteration	$x_i = g(x_{i-1})$	1	1	Easy to program. No guarantee of convergence

From all the methods discussed above we have seen that Newton-Raphson method has more order of convergence as compared to other iterative methods so it will converge fast. We use Newton-Raphson method to improve the result obtained by other methods. Newton-Raphson method is also very sensitive as it may diverge if $f'(x)$ is near to zero anytime during the iterative cycle. The convergence of Bisection method is sure but it is very slow method. Similarly, we can compare other methods by considering the three factors rate of convergence, amount of computational effort required and sensitivity.

QUERIES

1. Write comparison of bisection method with regula-falsi method.
2. Write comparison of Newton-raphson method with regula-falsi method.
3. Write comparison of iteration method with secant method.
4. Write comparison of secant method with regula-falsi method.
5. Give a comparative study of various iterative methods.

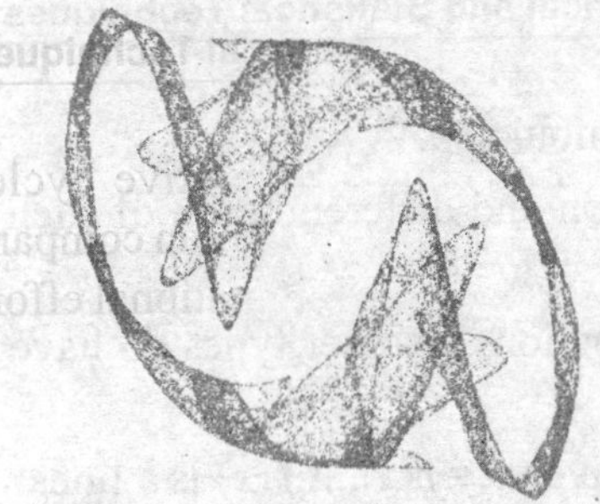

CHAPTER 5

Interpolation

5.1 INTRODUCTION

In the mathematical subfield of numerical analysis, interpolation is a method of constructing new data points within the range of a discrete set of known data points. Interpolation is the technique of filling up intermediate terms of a series. It is an art of reading the values in between the given set of data.

Discrete data sets, or tables of the form

x_1	x_2	x_3	...	x_n
y_1	y_2	y_3	...	y_n

are commonly involved in technical calculations. The source of the data may be experimental observations or numerical computations. In interpolation we construct a curve through the data points. In doing so, we make the implicit assumption that the data points are accurate and distinct.

Suppose $y = f(x)$ be a given function where y is the dependent variable and x is the independent variable. The function gives the values $y_1, y_2, \ldots, y_n$ corresponding to $x_1, x_2, \ldots, x_n$. Then the process of finding the value of y at given values of x is known as ***interpolation***.

Given a set of values in the form of a table, if we are interested to evaluate f(x) for a value of x outside the given range, the problem is that we call as extrapolation. Some of the interpolation formulae can be utilized for ***extrapolation*** also.

It is essential to study the interpolation formulae in view of the important fact that the interpolation formulae are the starting points in the derivations of many methods like numerical differentiation, numerical integration etc. In case we have a large data of information, it is easy to store it in the computer memory and then use any interpolation technique.

In this chapter, we will be mostly concerned with the polynomial interpolation only. Polynomial interpolation is mostly preferred because of the following reasons:

1. They are simple forms of functions, which can be easily manipulated.
2. Computations for definite values of the argument, integration and differentiation of such functions, are easy.
3. Polynomials are free from singularities whereas rational functions or other types do have singularities.

We, discuss in the first place, the techniques of interpolation. Given $y = f(x)$, if $f(x)$ is a linear function (e.g. $y = x + 2$) and it satisfies two points of interpolation, then we have a linear interpolation. The function $f(x)$ may be a second degree equation (i.e. parabola) or any higher order polynomial assumption depending on the type of interpolation. In general, we have a polynomial interpolation.

5.2 FINITE DIFFERENCES

Finite differences deals with the changes that occurs in the value of the function (dependent variable(y)) due to finite changes in the independent variable(x). We are given a set of values (x_a, y_a); $a = 1, 2, 3, \ldots\ldots, n$ of any function like $y = f(x)$. A value of the independent variable x is called argument and the corresponding value of the dependent variable y is called entry. The function $y = f(x)$ is tabulated for the equally spaced values $x = x_0, x_0 + h, x_0 + 2h, \ldots\ldots, x_0 + nh$, giving $y = y_0, y_1, y_2, \ldots\ldots, y_n$. To determine the values of $f(x)$ or $f'(x)$ for some intermediate values of x, the following four types of differences are used:

1. Forward differences
2. Backward differences
3. Divided differences
4. Central differences

We will discuss these finite differences one by one in detail.

5.2.1 Forward Differences

Let $y = f(x)$ be a given function of x and let $y_1, y_2, \ldots y_n$ be the values of y corresponding to $x_1, x_2, \ldots x_n$, the values of x.

In general, the difference between any two consecutive values of x need not be same or equal.

First Order Forward Differences

The differences $y_2 - y_1, y_3 - y_2, y_4 - y_3 \cdots y_n - y_{n-1}$ are called the first order forward Differences of y. These differences are represented as

$$\Delta y_1 = y_2 - y_1$$

$$\Delta y_2 = y_3 - y_2$$

....

....

$$\Delta y_{n-1} = y_n - y_{n-1}$$

Where Δ(delta) is known as forward difference operator, and $\Delta y_1, \Delta y_2, ..., \Delta y_{n-1}$ are known as first order forward differences.

Second Order Forward Differences

The second order forward differences can be calculated as:

$$\Delta^2 y_1 = \Delta y_2 - \Delta y_1$$
$$\Delta^2 y_2 = \Delta y_3 - \Delta y_2$$
$$\cdots\cdots\cdots\cdots$$
$$\cdots\cdots\cdots\cdots$$
$$\Delta^2 y_{n-1} = \Delta y_n - \Delta y_{n-1}$$

where $\Delta^2 y_1, \Delta^2 y_2, \ldots \Delta^2 y_{n-1}$ are known as *second order* forward differences. Similarly, by following this we can calculate higher order forward differences.

Higher Order Forward Differences

The **general formula** for calculating the higher order forward differences is as:

$$\Delta^n y_i = \Delta^{n-1} y_{i+1} - \Delta^{n-1} y_i$$

where superscripted ***n*** gives the order and subscripted ***i*** gives the differences at the point (i^{th} point).

Forward Difference Table

Let $y = f(x)$ be a given function of x and let $y_1, y_2, ..., y_n$ be the values of y corresponding to $x_1, x_2, ..., x_n$, the values of x, which are equally spaced. The forward difference table with the tabulated values is as

Forward Difference Table

x_i	y_i	Δy_i	$\Delta^2 y_i$	$\Delta^3 y_i$
x_1	y_1	$\Delta y_1 = y_2 - y_1$	$\Delta^2 y_1 = \Delta y_2 - \Delta y_1$	$\Delta^3 y_1 = \Delta^2 y_2 - \Delta^2 y_1$
x_2	y_2	$\Delta y_2 = y_3 - y_2$	$\Delta^2 y_2 = \Delta y_3 - \Delta y_2$	
x_3	y_3	$\Delta y_3 = y_4 - y_3$		
x_4	y_4			

5.2.2 Backward Differences

Let $y = f(x)$ be a given function of x and let $y_1, y_2, \ldots y_n$ be the values of y corresponding to $x_1, x_2, \ldots x_n$, the values of x.

First Order Backward Differences

The differences $y_2 - y_1, y_3 - y_2, y_4 - y_3 \cdots y_n - y_{n-1}$ are called the backward differences of y. These differences are represented as

$$\nabla y_2 = y_2 - y_1$$
$$\nabla y_3 = y_3 - y_2$$
$$\dots\dots\dots\dots$$
$$\dots\dots\dots\dots$$
$$\nabla y_n = y_n - y_{n-1}$$

where ∇(delta) is known as backward difference operator, and $\nabla y_2, \nabla y_3, \dots\dots, \nabla y_n$ are known as *first order* backward differences.

Second Order Backward Differences

The second backward differences can be calculated as:

$$\nabla^2 y_2 = \nabla y_2 - \nabla y_1$$
$$\nabla^2 y_3 = \nabla y_3 - \nabla y_2$$
$$\dots\dots\dots\dots$$
$$\dots\dots\dots\dots$$
$$\nabla^2 y_n = \nabla y_n - \nabla y_{n-1}$$

where $\nabla^2 y_2, \nabla^2 y_3, \dots, \nabla^2 y_n$ are known as *second order* backward differences. Similarly, by following this we can calculate higher order backward differences.

Higher Order Backward Differences

The **General formula** for calculating the higher order backward differences is as

$$\nabla^n y_i = \nabla^{n-1} y_i - \nabla^{n-1} y_{i-1}$$

where superscripted ***n*** gives the order and subscripted ***i*** gives the differences at the point (i^{th} point).

Backward Difference Table

Let $y = f(x)$ be a given function of x and let $y_1, y_2, \dots, y_n$ be the values of y corresponding to $x_1, x_2, \dots, x_n$, the values of x, which are equally spaced. The backward difference table with the tabulated values is as

Backward Difference Table

x_i	y_i	Δy_i	$\Delta^2 y_i$	$\Delta^3 y_i$
x_1	y_1			
x_2	y_2	$\nabla y_2 = y_2 - y_1$		
x_3	y_3	$\nabla y_3 = y_3 - y_2$	$\nabla^2 y_3 = \nabla y_3 - \nabla y_2$	
x_4	y_4	$\nabla y_4 = y_4 - y_3$	$\nabla^2 y_4 = \nabla y_4 - \nabla y_3$	$\nabla^3 y_4 = \nabla^2 y_3 - \nabla^2 y_2$

5.2.3 Divided Differences

If $y = f(x)$ be a given function of x and let $y_1, y_2, \ldots y_n$ be the values of y corresponding to $x_1, x_2, \ldots,$ x_n, the values of x.

First Order Divided Differences

The differences $\frac{y_2 - y_1}{x_2 - x_1}, \frac{y_3 - y_2}{x_3 - x_2}, \frac{y_4 - y_3}{x_4 - x_3}, \ldots\ldots, \frac{y_n - y_{n-1}}{x_n - x_{n-1}}$ are called the divided differences of y.

These differences are represented as

$$\Delta_d y_1 = \frac{y_2 - y_1}{x_2 - x_1}$$

$$\Delta_d y_2 = \frac{y_3 - y_2}{x_3 - x_2}$$

$$\Delta_d y_3 = \frac{y_4 - y_3}{x_4 - x_3}$$

....

....

$$\Delta_d y_{n-1} = \frac{y_n - y_{n-1}}{x_n - x_{n-1}}$$

Where Δ_d is called the divided difference operator and $\Delta_d y_1, \Delta_d y_2, \Delta_d y_3, \ldots, \Delta_d y_{n-1}$ are called first order divided differences.

Second Order Divided Differences

The second order divided differences can be calculated as:

$$\Delta^2{}_d y_1 = \frac{\Delta_d y_2 - \Delta_d y_1}{x_3 - x_1}$$

$$\Delta^2{}_d y_2 = \frac{\Delta_d y_3 - \Delta_d y_2}{x_4 - x_2}$$

....

....

$$\Delta^2{}_d y_n = \frac{\Delta_d y_{n+1} - \Delta_d y_n}{x_{n+2} - x_n}$$

where $\Delta^2{}_d y_1, \Delta^2{}_d y_2, \ldots, \Delta^2{}_d y_n$ are known as *second order* divided differences. Similarly, by following this we can calculate higher order divided differences.

Higher Order Divided Differences

The **General formula** for calculating the higher order divided differences is as

$$\Delta^n{}_d y_i = \frac{\Delta^{n-1}{}_d y_{i+1} - \Delta^{n-1}{}_d y_i}{x_{i+n} - x_i}$$

Where superscripted ***n*** gives the order and subscripted ***i*** gives the differences at the point (i^{th} point).

Divided Difference Table

Let $y = f(x)$ be a given function of x and let $y_1, y_2, ..., y_n$ be the values of y corresponding to x_1, x_2, ..., x_n, the values of x, which are unequally spaced. The backward difference table with the tabulated values is as:

Divided Difference Table				
x_i	y_i	Δy_i	$\Delta^2 y_i$	$\Delta^3 y_i$
x_1	y_1	$\Delta_d y_1 = \frac{y_2 - y_1}{x_2 - x_1}$	$\Delta_d{}^2 y_1 = \frac{\Delta_d y_2 - \Delta_d y_1}{x_3 - x_1}$	$\Delta_d{}^3 y_1 = \frac{\Delta_d{}^2 y_2 - \Delta_d{}^2 y_1}{x_4 - x_1}$
x_2	y_2	$\Delta_d y_2 = \frac{y_3 - y_2}{x_3 - x_2}$	$\Delta_d{}^2 y_2 = \frac{\Delta_d y_3 - \Delta_d y_2}{x_4 - x_2}$	
x_3	y_3	$\Delta_d y_3 = \frac{y_4 - y_3}{x_4 - x_3}$		
x_4	y_4			

5.2.4 Central Differences

If we require interpolation near the central part of the given table, then we prefer to use Central difference formulae. Comparatively, central difference formulas do converge faster than other type of formulae. Central differences can be calculated by taking average of the forward and backward differences.

The central difference operator d is defined by the relation

$\delta y_{1/2} = y_1 - y_0, \quad \delta y_{3/2} = y_2 - y_1, ..., \quad \delta y_{n-1/2} = y_n - y_{n-1}$

Similarly, higher order central differences can be calculated.

Central Difference Table

Central differences can be calculated by taking average of the forward and backward differences. Let $y = f(x)$ be a given function. Suppose we take the interval of differencing as h(= constant).

Let $x = a$ be the central point and let us consider the values of x on either side of 'a'.

Let $f(a) = y_0, \quad f(a-h) = y_{-1}, f(a+h) = y_1$

$f(a-2h) = y_{-2}, f(a+2h) = y_2$

$f(a-3h) = y_{-3}, f(a+3h) = y_3$ etc.

Central Difference Table

x	Y	First diff.	Second diff.	Third diff.	Fourth diff.	Fifth diff.
$a-3h$	y_{-3}	$\Delta y_{-3} = \delta_{-5/2}$				
$a-2h$	y_{-2}	$\Delta y_{-2} = \delta_{-3/2}$	$\Delta^2 y_{-3} = \delta^2 y_{-2}$			
$a-h$	y_{-1}	$\Delta y_{-1} = \delta_{-1/2}$	$\Delta^2 y_{-2} = \delta^2 y_{-1}$	$\Delta^3 y_{-3} = \delta^3 y_{-3/2}$	$\Delta^4 y_{-3} = \delta^4 y_{-1}$	$\Delta^5 y_{-3} = \delta^5 y_{-1/2}$
a	y_0	....	$\Delta^2 y_{-1} = \delta^2 y_{-2}$	$\Delta^3 y_{-2} = \delta^3 y_{-1/2}$	$\Delta^4 y_{-2} = \delta^4 y_0$	$\Delta^5 y_{-2} = \delta^5 y_{1/2}$
$a+h$	y_1	$\Delta y_0 = \delta_{1/2}$	$\Delta^2 y_0 = \delta^2 y_1$	$\Delta^3 y_{-1} = \delta^3 y_{1/2}$	$\Delta^4 y_{-1} = \delta^4 y_1$	
$a+2h$	y_2	$\Delta y_1 = \delta_{3/2}$	$\Delta^2 y_1 = \delta^2 y_2$	$\Delta^3 y_0 = \delta^3 y_{3/2}$		
$a+3h$	y_3	$\Delta y_2 = \delta_{5/2}$				

5.3 RELATIONSHIP BETWEEN DIFFERENT OPERATORS

5.3.1 Relation Between $\Delta = E - 1$

Before proving the relation we will discuss about E operator first. Here $Ef(x)$ is known as shifting or displacement or translation operator, and is defined as

$$Ef(x) = f(x+h) \text{ and } E^{-1}f(x) = f(x-h)$$

We know that

$$\begin{aligned} \Delta f(x) &= f(x+h) - f(x) \\ &= Ef(x) - f(x) \\ &= (E-1)f(x) \\ \Delta &= \mathbf{E - 1} \end{aligned}$$

5.3.2 Relation Between $\nabla = 1 - E^{-1}$

We know that

$$\begin{aligned} \nabla f(x) &= f(x) - f(x-h) \\ &= f(x) - E^{-1}f(x) \\ &= (1-E^{-1})f(x) \\ \nabla &= \mathbf{1 - E^{-1}} \end{aligned}$$

Hence

$$\begin{aligned} E^{-1} &= 1 - \nabla \\ E &= (1-\nabla)^{-1} \text{ since } (E^{-1})^{-1} = E \end{aligned}$$

5.3.3 Relation Between $\delta = E^{-1/2}.\Delta = E^{1/2}.\nabla$

$$\begin{aligned} \delta f(x) &= f(x + h/2) - f(x - h/2) \\ &= E^{1/2}f(x) - E^{-1/2}f(x) \\ &= (E^{1/2} - E^{-1/2})f(x) \\ \delta &= E^{1/2} - E^{-1/2} \\ &= E^{-1/2}[E-1] \\ &= E^{-1/2}.\Delta \end{aligned}$$

Also
$$\delta = E^{1/2}[1-E^{-1}]$$
$$= E^{1/2}.\nabla$$
$$\delta = E^{-1/2}.\Delta$$
$$= E^{1/2}.\nabla$$

5.3.4 Relation Between $E = e^{hD}$

$$Df(x) = \frac{df(x)}{dx} = f'(x)$$

By Taylor's theorem

$$f(x+h) = f(x) + \frac{hf'(x)}{1!} + \frac{h^2 f''(x)}{2!} + \ldots\ldots \infty$$

$$Ef(x) = f(x) + \frac{hDf(x)}{1!} + \frac{h^2 D^2 f(x)}{2!} + \ldots\ldots \infty$$

$$= \left[1 + \frac{hD}{1!} + \frac{(hD)^2}{2!} + \frac{(hD)^3}{3!} + \ldots\ldots \infty\right] f(x)$$

$$= e^{hD} f(x)$$

$\therefore$ $$E = e^{hD}$$

$\therefore$ $$E = 1 + \Delta = e^{hD}$$

$$hD = \log E = \log(1+\Delta)$$

$$hD = \Delta - \frac{\Delta^2}{2!} + \frac{\Delta^3}{3!} - \ldots\ldots$$

$$D = \frac{1}{h}\left[\Delta - \frac{\Delta^2}{2!} + \frac{\Delta^3}{3!} - \ldots\ldots\right]$$

SOLVED PROBLEMS

Problem 1: Compute the forward difference table, given values are:

x:	5	10	15	20	25	30
y:	9962	9848	9659	9397	9063	8660

and point out the values of $\Delta^2 y_{10}$, $\Delta^4 y_5$.

Solution: Forward difference table define as:

x	y	Δy	$\Delta^2 y$	$\Delta^3 y$	$\Delta^4 y$
5	9963				
		–114			
10	9848		–75		
		–189		2	
15	9659		–73		–1
		–262		1	
20	9397		–72		2
		–334		3	
25	9063		–69		
		–403			
30	8660				

See the table, $\Delta^2 y_{10} = -73$ and $\Delta^4 y_5 = -1$.

Problem 2: Compute the backward difference table for $y = \log x$ given values are:

x:	10	20	30	40	50
y:	1	1.3010	1.4771	1.6021	1.6990

and find out the values of $\nabla^3 \log 40$ and $\nabla^4 \log 50$.

Solution: Backward difference table define as:

x	y	∇y	$\nabla^2 y$	$\nabla^3 y$	$\nabla^4 y$
10	1				
		0.3010			
20	1.3010		–0.1249		
		0.1761		0.0738	
30	1.4771		–0.0511		–0.0508
		0.1250		0.0230	
40	1.6021		–0.0281		
		0.0969			
50	1.6990				

See the table, $\nabla^3 \log 40 = 0.0738$ and $\nabla^4 \log 50 = -0.0508$.

Problem 3: Calculate:

(i) $\Delta \tan^{-1} x$ **(ii) $\Delta^2 \cos 2x$**

where h(represent as interval of differencing).

Solution:

(i) $\Delta \tan^{-1} x = \tan^{-1}(x+h) - \tan^{-1} x$

$$= \tan^{-1}\left\{\frac{x+h-x}{1+x(x+h)}\right\} = \tan^{-1}\left(\frac{h}{1+hx+x^2}\right)$$

(ii) $\Delta^2 \cos 2x = \Delta[\cos 2(x+h) - \cos 2x]$

$$= [\cos 2(x+2h) - \cos 2(x+h)] - [\cos 2(x+h) - \cos 2x]$$

$$= -2 \sin(2x+3h) \sin h + 2 \sin(2x+h) \sin h$$

$= -2 \sin h\,[2 \cos (2x + 2h) \sin h]$
$= -4 \sin^2 h \cos 2(x + h).$

Problem 4: If $f(x) = \exp(ax)$, calculate $\Delta^n f(x)$.

Solution:

$$\Delta e^{ax} = e^{a(x+h)} - e^{ax} = (e^{ah} - 1)e^{ax}$$
$$\Delta^2 e^{ax} = \Delta(\Delta e^{ax}) = \Delta[(e^{ah} - 1)e^{ax}]$$
$$= (e^{ah} - 1)(e^{ah} - 1)e^{ax} = (e^{ah} - 1)^2 e^{ax}$$

Similarly $\Delta^3 e^{ax} = (e^{ah} - 1)^3 e^{ax}$

....

....

....

$$\Delta^n e^{ax} = (e^{ah} - 1)^n e^{ax}.$$

Problem 5: Prove that

$$\Delta \log f(x) = \log\left[1 + \frac{\Delta f(x)}{f(x)}\right]$$

Solution:

$$\text{L.H.S.} = \log f(x + h) - \log f(x)$$
$$= \log [f(x) + \Delta f(x)] - \log f(x)$$
$$| \because \Delta f(x) = f(x + h) - f(x)$$
$$= \log\left[\frac{f(x) + \Delta f(x)}{f(x)}\right]$$
$$= \log\left[1 + \frac{\Delta f(x)}{f(x)}\right] = \text{R.H.S.}$$

UNSOLVED PROBLEMS

1. Draw a table of differences for the function given below:

   $f(x) = x^3 + 5x - 7$ for $x = -1, 0, 1, 2, 3, 4, 5$

   Compute the table to obtain $f(6)$ and $f(7)$.

2. Draw the difference table

x:	10	15	20	25	30	35
y:	19.97	21.51	22.47	23.52	24.65	25.89.

   Compute the values of $\Delta^2 y_{10}$, Δy_{20}, $\Delta^3 y_{15}$, and $\Delta^5 y_{10}$.

3. Draw the forward difference table for given values

x:	10	20	30	40
y:	1.1	2.0	4.4	7.9.

4. Draw the table of differences for the data given below:

x:	0	1	2	3	4
f(x):	1.0	1.5	2.2	3.1	4.6

Evaluate $\Delta^3 f(2)$.

5.4 NEWTON'S METHOD OF INTERPOLATION

Newton's formula was first discovered by the Scottish mathematician James Gregory (1638–1675) a contemporary of Newton. Gregory and Newton did extensive work on methods of interpolation but now the formula is referred to as Newton's interpolation formula. Newton formula's is used for constructing the interpolation polynomial. It also makes use of divided differences. Newton has derived general forward and backward difference interpolation formulas. Newton's methods of interpolation are divided into three methods. A particular method is use depending on the nature of the problem. There are following three Newton's methods exist for solving suitable problems.

- **Newton's Forward Difference Interpolation Formula**
- **Newton's Backward Difference Interpolation Formula**
- **Newton's Divided Difference Interpolation Formula**

5.4.1 Newton's Forward Difference Interpolation Formula

The **Newton's Forward difference interpolation formula** is used only when the following two possibilities exists together.

(i) The function should be tabulated at equal intervals.

(ii) The point for which the function is to be interpolated should lie in the first or upper half of the table. If both these possibilities exist, then Newton's forward difference formula will give better approximation.

Given function $y(x)$ is tabulated at equally spaced points, in which the difference between the consecutive values is same say h. Let $y(x)$ be a polynomial in x of degree n and is taken in the form:

$$y(x) = a_1 + a_2(x-x_1) + a_3(x-x_1)(x-x_2) + \ldots\ldots\ldots\ldots + a_{n+1}(x-x_1)(x-x_2)\ldots\ldots\ldots(x-x_n) \quad \ldots(1)$$

At $x = x_1$ we have $y_1 = a_1$ (from eq (1))

Similarly at $x = x_2$ we have $y_2 = a_1 + a_2(x_2 - x_1)$

At $x = x_3$ we have $y_1 = a1 + a_2(x_3 - x_1)\ (x_3 - x_2)$

and so on.

As values of x are equally spaced, so we have

$$x_{i+1} - x_i = h$$

and

$$x_{i+p} - x_i = ph$$

So values of y's in terms of h comes

$$y_1 = a_1$$

$$y_2 = a_1 + a_2 h$$

$$y_3 = a_1 + a_2(2h) + a_3(2h)(h)$$

....

....

....

$$y_{n+1} = a_1 + a_2(nh) + a_3(nh)((n-1)h) + \ldots\ldots a_{n+1}(nh)((n-1)h)\ldots\ldots(h)$$

After solving these for a_i's, we get

$$a_1 = y_1$$

$$a_2 = (y_2 - a_1)/h = (y_2 - y_1)/h \qquad \text{(As } a_1 = y_1\text{)}$$

$$a_3 = (y_3 - a_2(2h) - a_1)/2h^2$$

$$= \left[y_3 - \left(\frac{y_2 - y_1}{h} \right) - 2h - y_2 \right] / 2h^2$$

$$= [y_3 - 2y_2 + 2y_1 - y_1] / 2!h^2$$

....

....

....

$$a_{n+1} = [y_{n+1} - ny_n + \ldots\ldots + y_1] / n!h^n$$

By using the forward difference table, we get

$$a_1 = y_1$$

$$a_2 = \frac{\Delta y_1}{h}$$

$$a_3 = \frac{\Delta^2 y_1}{2!h^2}$$

....

....

....

$$a_{n+1} = \frac{\Delta^n y_1}{n!h^n}$$

Substitute these values in eq. (1), we get

$$y(x) = y_1 + \frac{\Delta y_1}{h}(x - x_1) + \frac{\Delta^2 y_1}{2!h^2}(x - x_1)(x - x_2) + \ldots\ldots + \frac{\Delta^n y_1}{n!h^n}(x - x_1)(x - x_2)\ldots\ldots(x - x_n) \quad \ldots(2)$$

Suppose we have the relation $\frac{x-x_1}{h}=u$, then

so $$x-x_1 = hu$$
$$x-x_2 = x-(x_1+h) = hu-h = h(u-1)$$
$$x-x_3 = x-(x_2+h) = h(u-1)-h = h(u-2)$$
....
....
....
$$x-x_n = h(u-(n-1))$$

So after substitution of these above values in eq. (2), we get

$$y(x) = y_1 + \frac{\Delta y_1}{h}hu + \frac{\Delta^2 y_1}{2!h^2}\cdot(hu)\cdot h(u-1) + ... + \frac{\Delta^n y_1}{n!h^n}\cdot(hu)\cdot(uh-h).....h(u-(n-1))$$

$$y(x) = y_1 + \Delta y_1 u + \frac{\Delta^2 y_k}{2!}u(u-1) + ... + \frac{\Delta^{n-k} y_k}{n!}u(u-1).....(u-(n-1))$$

This derivation have assumed that $x_1 < x < x_2$ i.e x lies in the upper half of the table. If however, $x_k < x < x_{k+1}$, then the above formula become

$$y(x) = y_k + \Delta y_k u + \frac{\Delta^2 y_k}{2!}u(u-1) + ... + \frac{\Delta^{n-k} y_k}{(n-k)!}u(u-1).....(u-((n-k)-1))$$

This formula is known as Newton's forward difference interpolation formula.

***Note:** This formula gives better approximation if values are equally spaced and we have to interpolate near the top of the table.*

5.4.1.1 Algorithm for Newton's Forward Difference Interpolation

Algorithm of numerical interpolation using Newton's Forward Difference for the given function. Given two one dimension arrays named x and y each of size 'n' used to store the given values and one two dimensional array d of size (n-1) by (n-1) which is used to store the forward difference table.

1. Begin
2. Read n
3. Do i = 1 to n
4. Read x[i], y[i]

5. End do
6. Read x
7. If ((x < x[1]) or (x > x[n])) then
8. Print "the value lies outside the given range"
9. End if
10. Do j =1 to n-1 by 1 /* Newton's forward difference table
11. Do i = 1 to n –j
12. If (j == 1) then
13. d[i][j] = y[i+1] – y[i]
14. else
15. d[i][j] = d[i+1][j-1] – d[i][j-1]
16. end if
17. end do
18. end do
19. i=2
20. while (x < x[i])
21. i = i + 1
22. end while
23. k= i – 1
24. set u = (x – x[k])/(x[k+1] – x[k])
25. set total = y[k]
26. do I = 1 to (n – k)
27. set prod = 1
28. do j= 0 to (I – 1)
29. prod = prod * (u – j)
30. end do
31. find the factorial of i! and let the value be denoted by 'fact'
32. total = total + (d[k][i] * prod)/fact
33. end do
34. print ' the interpolated value is ", total
35. end

5.4.1.2 Flowchart for Newton's Forward Difference Interpolation

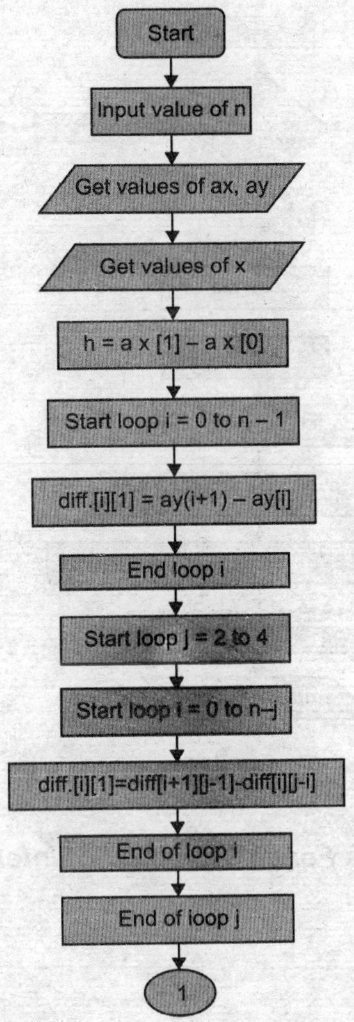

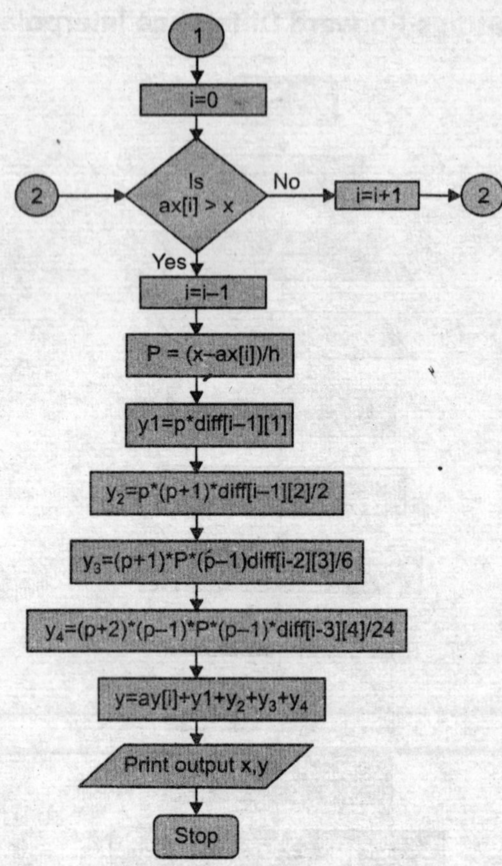

5.4.1.3 *Program for Newton's Forward Method of Interpolation in C*

```
# include <stdio.h>
# include <conio.h>
# include <math.h>
# include <process.h>
# include <string.h>
void main()
{
int n;
int i,j;
float ax[10];
float ay[10];
float x;
```

```
float y = 0;
float h;
float p;
float diff[20][20];
float y1,y2,y3,y4;
clrscr();
printf("\n Enter the number of terms – ");
scanf("%d",&n);
Printf ("\n\n Enter the value in the form of x - ");
for (i=0;i<n;i++)
{
printf("\n\n Enter the value of x%d - ",i+1);
scanf("%f",&ax[i]);
}
printf("\n\n Enter the value in the form of y – ");
for (i=0;i<n;i++)
{
printf ("\n\n Enter the value of y%d – ", i+1);
scanf ("%f",&ay [i]);
}
printf("\nEnter the value of x for");
printf("\nwhich you want the value of y - ");
scanf("%f",&x);
h=ax[1]-ax[0];
for(i=0;i<n-1;i++)
{
diff[i][1]=ay[i+1]-ay[i];
}
for(j=2;j<=4;j++)
{
for(i=0;i<n-j;i++)
{
diff[i][j]=diff[i+1][j-1]-diff[i][j-1];
}
```

```
}
i=0;
do {
i++;
}while(ax[i]<x);
i—;
p=(x–ax[i])/h;
y1=p*diff[i-1][1];
y2=p*(p+1)*diff[i-1][2]/2;
y3=(p+1)*p*(p–1)*diff[i–2][3]/6;
y4=(p+2)*(p+1)*p*(p-1)*diff[i-3][4]/24;
y=ay[i]+y1+y2+y3+y4;
printf("\nwhen x=%6.4f, y=%6.8f ",x,y);
printf("\n\n\n Press Enter to Exit");
getch();
}
```

Output:

Enter the number of terms – 7

Enter the value in the form of x -

Enter the value of x1 - 100

Enter the value of x2 - 150

Enter the value of x3 - 200

Enter the value of x4 - 250

Enter the value of x5 - 300

Enter the value of x6 - 350

Enter the value of x7 - 400

Enter the value in the form of y -

Enter the value of y1 - 10.63

Enter the value of y2 - 13.03

Enter the value of y3 - 15.04

Enter the value of y4 - 16.81

Enter the value of y5 - 18.42

Enter the value of y6 - 19.9

Enter the value of y7 - 21.27

Enter the value of x for which you want the value of y-218

When X=218.0000, Y=15.69701481

Press Enter to Exit

5.4.1.4 *Program for Newton's Forward Method of Interpolation in C++*

```
# include <iostream.h>
# include <conio.h>
# include <math.h>
# include <process.h>
# include <string.h>
void main()
{
int n;
int i,j;
float ax[10];
float ay[10];
float x;
float y = 0;
float h;
float p;
float diff[20][20];
float y1,y2,y3,y4;
clrscr();
cout<<"\n Enter the number of terms – ";
cin>>n;
cout<<"\n\n Enter the value in the form of x - ";
for (i=0;i<n;i++)
{
cout<<"\n\n Enter the value of x - "<<i+1;
cin>>ax[i];
}
cout<<"\n\n Enter the value in the form of y – ");
for (i=0;i<n;i++)
{
cout<<"\n\n Enter the value of y – "<<i+1;
```

```
cin>>ay[i];
}
cout<<"\nEnter the value of x for";
cout<<"\nwhich you want the value of y - ";
cin>>x;
h=ax[1]-ax[0];
for(i=0;i<n-1;i++)
{
diff[i][1]=ay[i+1]-ay[i];
}
for(j=2;j<=4;j++)
{
for(i=0;i<n-j;i++)
{
diff[i][j]=diff[i+1][j-1]-diff[i][j-1];
}
}
i=0;
do {
i++;
}while(ax[i]<x);
i—;
p=(x–ax[i])/h;
y1=p*diff[i-1][1];
y2=p*(p+1)*diff[i-1][2]/2;
y3=(p+1)*p*(p–1)*diff[i–2][3]/6;
y4=(p+2)*(p+1)*p*(p-1)*diff[i-3][4]/24;
y=ay[i]+y1+y2+y3+y4;
cout<<"\nwhen x="<<setprecision(6.4)<<x<<"y=" setprecision (6.8)<<y;
cout<<"\n\n\n Press Enter to Exit";
getch();
}
```

Output:

Enter the number of terms – 7

Enter the value in the form of x -
Enter the value of x1 - 100
Enter the value of x2 - 150
Enter the value of x3 - 200
Enter the value of x4 - 250
Enter the value of x5 - 300
Enter the value of x6 - 350
Enter the value of x7 - 400
Enter the value in the form of y -
Enter the value of y1 - 10.63
Enter the value of y2 - 13.03
Enter the value of y3 - 15.04
Enter the value of y4 - 16.81
Enter the value of y5 - 18.42
Enter the value of y6 - 19.9
Enter the value of y7 - 21.27
Enter the value of x for which you want the value of y-218
When X=218.0000, Y=15.69701481
Press Enter to Exit

SOLVED PROBLEMS

Problem 1: Population of a town was as given below. Compute the population in the year 1955.

Year(x) :	1951	1961	1971	1981	1991
Population(y): (in thousands)	46	66	81	93	101

Solution: Where x = 1955 and interpolation is desired at the beginning of the table so we need Forward Difference table and Newton's forward interpolation formula.

The forward difference table is

x_i	y_i	Δy_i	$\Delta^2 y_i$	$\Delta^3 y_i$	$\Delta^4 y_i$
1951	46	20	–5	2	–3
1961	66	15	–3	–1	
1971	81	12	–4		
1981	93	8			
1991	101				

Newton's forward interpolation formula.

As x =1955 and it lies between $x_1 < x < x_2$ i.e. $1951 < x < 1961$

H is the difference between consecutive values of x which is 10 i.e. $h = 10$

$$y(x) = y_1 + \Delta y_1 u + \frac{\Delta^2 y_1}{2!} u(u-1) + \frac{\Delta^3 y_1}{3!} u(u-1)(u-2) + \frac{\Delta^4 y_1}{4!} u(u-1)(u-2)(u-3)$$

Now $\quad u = \dfrac{x - x_1}{h} = \dfrac{1955 - 1951}{10} = 0.4$

$$y(1955) = 46 + 20(0.4) + \frac{(-5)}{2}(0.4)(0.4-1) + \frac{2}{6}(0.4)(0.4-1)(0.4-2) +$$

$$\frac{(-3)}{24}(0.4)(0.4-1)(0.4-2)(0.4-3)$$

$$= 46 + 8 + 0.6 + 0.128 + 0.1248$$

$$= 54.8528$$

Problem 2: Given values are sin30° = 0.5, sin40° = 0.6428, sin50° = 0.7660, sin60° = 0.8660. Find sin44°, by using suitable Newton's interpolation technique.

Solution: Value of sin44° lies between 2nd and 3rd point, therefore we will start forward difference formula from second value.

The difference table is

x_i	y_i	Δy_i	$\Delta^2 y_i$	$\Delta^3 y_i$
30°	0.50	0.4121	–0.0089	–0.0343
40°	0.6428	0.1332	–0.0432	
50°	0.7760	0.06		
60°	0.8660			

As $x = 44°$, h =10

$$\therefore \qquad u = \frac{x - x_2}{h} = \frac{44° - 40°}{5} = \frac{4}{5} = 0.8$$

The Newton's forward interpolation polynomial is given by

$$y(x) = y_2 + u \cdot \Delta y_2 + u(u-1)\frac{\Delta^2 y_2}{2!}$$

$$= 0.6428 + 0.8(0.1332) + 0.8(0.8\text{-}1)\,\frac{(-0.0432)}{2}$$

$$= 0.6428 + 0.10656 + 0.0034$$

$$= 0.75276$$

Problem 3: Compute the relationship between y and x from the following data.

x :	0	1	2	3	4	5
y:	-5	1	9	25	55	105

Solution: Given values of x are at finite differences, Draw the finite difference table.

The difference table is

x_i	y_i	Δy_i	$\Delta^2 y_i$	$\Delta^3 y_i$	$\Delta^4 y_i$
0	–5	6	2	6	0
1	1	8	8	6	0
2	9	16	14	6	
3	25	30	20		
4	55	50			
5	105				

From the Newton's forward difference formula, we have

$$y(x) = y_1 + \Delta y_1 u + \frac{\Delta^2 y_1}{2!} u(u-1) + \frac{\Delta^3 y_1}{3!} u(u-1)(u-2) + \frac{\Delta^4 y_1}{4!} u(u-1)(u-2)(u-3)$$

Where $u = \dfrac{x-0}{1} = x$

Hence we have

$$f(x) = (-5) + 6x + \frac{2x}{2}(x-1) + \frac{6}{6} x(x-1)(x-2) + 0$$

$$= -5 + 6x + x^2 - x + x^3 - 3x^2 + 2x$$

$$= -5 + 7x - 2x^2 + x^3$$

Problem 4: Given the following values, find log 104.4, log 105 = 2.0211, log 106 = 2.0253, log 107 = 2.0029, log 108 = 2.0334, log 109 = 2.0374.

Solution: It is an **extrapolation problem** and Compute value near the beginning of the table and hence we use Newton's forward interpolation formula.

x_i	$f(x_i) = \log x_i$	Δ	Δ^2	Δ^3	Δ^4
105	2.0211	0.0042	–0.0266	0.0795	–0.1589
106	2.0253	–0.0224	0.0529	–0.0794	
107	2.0029	–0.0305	–0.0265		
108	2.0334	0.0040			
109	2.0374				

Here $u = \dfrac{x - x_1}{h} = \dfrac{104.4 - 105}{1} = -0.6$

From the Newton's forward difference formula, we have

$$y(x) = y_1 + \Delta y_1 u + \frac{\Delta^2 y_1}{2!}u(u-1) + \frac{\Delta^3 y_1}{3!}u(u-1)(u-2) + \frac{\Delta^4 y_1}{4!}u(u-1)(u-2)(u-3)$$

$$= 2.0211 + 0.0042(-0.6) + \frac{(-0.0266)}{2}(-0.6)(-0.6-1) + \frac{(0.0795)}{6}(-0.6)(-0.6-1)(-0.6-2)$$

$$+ \frac{(-0.1589)}{24}(-0.6(-0.6-1)(-0.6-2)(-0.6-3)$$

$$= 1.9132$$

Problem 5: Compute the number of students who obtained marks between 40 and 45 from the data below.

Marks:	30-40	40-50	50-60	60-70	70-80
No. of students:	31	42	51	35	31

Solution: Create the cumulative frequency table, which is as:

Marks less than(x):	40	50	60	70	80
No. of students(y):	31	73	124	159	190

Since the point $y = 40\text{-}45$ lies between first and second point so we use forward difference and it start from first point.

The difference table is

x_i	y_i	Δy_i	$\Delta^2 y_i$	$\Delta^3 y_i$	$\Delta^4 y_i$
40	31	42	9	–25	37
50	73	51	–16	12	
60	124	35	–4		
70	159	31			
80	190				

$$y(<45) = y_1 + \Delta y_1 u + \frac{\Delta^2 y_1}{2!}u(u-1) + \frac{\Delta^3 y_1}{3!}u(u-1)(u-2) + \frac{\Delta^4 y_1}{4!}u(u-1)(u-2)(u-3)$$

Take $x_1 = 40$ and $x = 45$

So $$u = \frac{x - x_1}{h} = \frac{45-40}{10} = 0.5$$

$$y(<45) = 31 + 0.5(42) + (0.5)(0.5-1)\frac{(9)}{2} + (0.5)(0.5-1)(0.5-2)\frac{(-25)}{6} +$$

$$(0.5)(0.5-1)(0.5-2)(0.5-3)\frac{(37)}{24}$$

$y(<45) = 48$(approx.)

Hence the number of students with marks less than 45 is 48 and the number of students having marks less than 40 is 31.

Hence the number of students having marks between 40 and 45 is 48 – 31 = 17.

Problem 6: Given values of $f(x)$ for $x = 0, 1, 2, \ldots\ldots, 6$ are given by

x:	0	1	2	3	4	5	6
f(x):	2	4	10	16	20	24	38

Find the value of $f(3.2)$ using only four of the given values. Choose the four values that you think will give the best approximation.

Solution. Last four values of $f(x)$ for $x = 3, 4, 5, 6$ are taken into consideration so that 3.2 occurs in the beginning of the table.

Here, $a = 3,\quad h = 1,\quad x = 3.2$ $\qquad\therefore a + hu = 3.2$

i.e., $3 + 1 \times u = 3.2$ or $u = 0.2$

The difference table is:

x	f (x)	Δf (x)	Δ²f (x)	Δ³f (x)
3	16			
		4		
4	20		0	
		4		10
5	24		10	
		14		
6	38			

Applying Newton's forward difference formula,

$$f(3.2) = f(3) + u\Delta f(3) + \frac{u(u-1)}{2!}\Delta^2 f(3) + \frac{u(u-1)(u-2)}{3!}\Delta^3 f(3)$$

$$16 + (.2)(4) + \frac{(.2)(.2-1)}{2}(0) + \frac{(.2)(.2-1)(.2-2)}{6}(10)$$

$$= 17.28$$

Problem 7: From the following table, find the value of $e^{0.24}$

x:	0.1	0.2	0.3

Solution: The difference table is:

x	$10^5 y_i$	$10^5 \Delta y$	$10^5 \Delta^2 y$	$10^5 \Delta^3 y$	$10^5 \Delta^4 y_i$
0.1	110517				
		11623			
0.2	122140		1223		
		12846		127	
0.3	134986		1350		17
		14196		144	
0.4	149182		1494		
		15690			
0.5	164872				

Here $h = 0.1$. $\therefore$ $0.24 = 0.1 + 0.1 \times u$ or $u = 1.4$

Newton forward formula is

$$y(.24) = y(.1) + u\Delta y(.1) + \frac{u(u-1)}{2!}\Delta^3 y(.1) + \frac{u(u-1)(u-2)}{3!}\Delta^3 y(.1)$$

$$+\frac{u(u-1)(u-2)(u-3)}{4!}\Delta^4 y(.1)$$

$$\Rightarrow 10^5 y(.24) = 10^5 y(.1) + u10^5 \Delta y(.1) + \frac{u(u-1)}{2!}10^5 \Delta^2 y(.1) + \frac{u(u-1)(u-2)}{3!}\Delta^3 y(.1)$$

$$+\frac{u(u-1)(u-2)}{3!}10^5 \Delta^3 y(.1) + \frac{u(u-1)(u-2)(u-3)}{4!}10^5 \Delta^4 y(.1)$$

$$\Rightarrow 10^5 y(.24) = 10^5 y(.24) = 110517 + (1.4)(11623) + \frac{(1.4)(1.4-1)}{2!}(1223)$$

$$+\frac{(1.4)(1.4-1)(1.4-2)}{3!}(127) + \frac{(1.4)(1.4-1)(1.4-2)(1.4-3)}{4!}(17)$$

$$= 127124.9088$$

$$y(.24) = 1.271249088$$

Hence, $e^{.24} = 1.271249088.$

Problem 8: From the given table of half-yearly premiums for policies maturing at different ages, estimate the premium for policies maturing at age of 46.

Age:	45	50	55	60	65
Premium: (in dollars)	114.81	96.16	83.32	74.48	68.48

Solution: The difference table is:

Age (x)	Premium in dollars (y)	Δy	$\Delta^2 y$	$\Delta^3 y$	$\Delta^4 y_i$
45	114.84				
		−18.68			
50	96.16		5.84		
		−12.84		−1.84	
55	83.32		4		.68
		−8.84		−1.16	
60	74.48		2.84		
		−6			
65	68.48				

Here $h = 5$, $a = 45$, $a + hu = 46$

$$\therefore \quad 45 + 5u = 46 \Rightarrow u = .2$$

By Newton's forward difference formula,

$$y_{46} = y_{45} + u\Delta y_{45} + \frac{u(u-1)}{2!}\Delta^2 y_{45} + \frac{u(u-1)(u-2)}{3!}\Delta^3 y_{45} + \frac{u(u-1)(u-2)(u-3)}{4!}\Delta^4 y_{45}$$

$$= 114.84 + (.2)(-18.86) + \frac{(.2)(.2-1)}{2!}(5.84)$$

$$+ \frac{(.2)(.2-1)(.2-2)}{3!}(-1.84) + \frac{(.2)(.2-1)(.2-2)(.2-3)}{4!}(-.68)$$

$$= 110.525632$$

Hence the premium for policies maturing at the age of 46 is **$ 110.52.**

Problem 9: Compute the cubic polynomial which takes the following values:

x:	0	1	2	3
f(x):	1	2	1	10

Solution: Let us form the difference table:

x	y	Δy	$\Delta^2 y$	$\Delta^3 y$
0	1			
		1		
1	2		−2	
		−1		12
2	1		10	
		9		
3	10			

Here, $h = 1$. Hence, using the formula,

$$x = a + hu$$

and choosing $a = 0$, we get $x = u$

By Newton's forward difference formula,

$$y = y_0 + x\Delta y_0 + \frac{x(x-1)}{2!}\Delta^2 y_0 + \frac{x(x-1)(x-2)}{3!}\Delta^3 y_0$$

$$= 1 + x(1) + \frac{x(x-1)}{2!}(-2) + \frac{x(x-1)(x-2)}{3!}(12)$$

$$= 2x^3 - 7x^2 + 6x + 1$$

Hence, the required cubic polynomial is

$$y = f(x) = 2x^3 - 7x^2 + 6x + 1.$$

Problem 10: The following are the numbers of deaths in four successive ten year age groups. Compute the number of deaths at 45—50 and 50—55.

Age group:	25—35	35—45	45—55	55—65
Deaths:	13229	18139	24225	31496.

Solution: Difference table of cumulative frequencies represented as:

Age upto x	No. of deaths y	$\Delta f(x)$	$\Delta^2 f(x)$	$\Delta^3 f(x)$
35	13229			
		18139		
45	31368		6086	
		24225		1185
55	55593		7271	
		31496		
65	87089			

Here, $h = 10$, $a = 35$, $a + hu = 50$

$$\therefore \quad 35 + 10u = 50 \Rightarrow u = 1.5$$

By Newton's forward difference formula,

$$y_{50} = y_{35} + u\Delta y_{35} + \frac{u(u-1)}{2!}\Delta^2 y_{35} + \frac{u(u-1)(u-2)}{3!}\Delta^3 y_{35}$$

$$= 13229 + (1.5)(18139) + \frac{(1.5)(.5)}{2}(6086) + \frac{(1.5)(.5)(-.5)}{6}(1185)$$

$$= 42645.6875 \approx 42646$$

$\therefore$ Deaths at ages between 45 – 50 are 42646 – 31368 = 11278

and Deaths at ages between 50 – 55 are 55593 – 42646 = 12947.

UNSOLVED PROBLEMS

1. Table gives the distance in nautical miles of the visible horizon for the given heights in feet above the earth's surface.

x:	100	150	200	250	300	350	400
y:	10.63	13.03	15.04	16.81	18.42	19.9	21.27

Use Newton's forward formula to find y when $x = 218$ ft.

2. Given values of $f(x)$ for $x = 0, 1, 2, ..., 6$ are given by

x:	0	1	2	3	4	5	6
f(x):	1	3	11	31	69	131	223

Compute the value of $f(3.4)$, using only four of the given values.

3. Given that:

x:	1	2	3	4	5	6
y(x):	0	1	8	27	64	125

Compute the value of $f(2.5)$.

4. Ordinates $f(x)$ of a normal curve in terms of standard deviation x are given as

x:	1.00	1.02	1.04	1.06	1.08
f(x):	0.2420	0.2371	0.2323	0.2275	0.2227

Find the ordinate for standard deviation $x = 1.025$.

5. Using Newton's formula for interpolation, Compute the population for the year 1905 from the table:

Year:	1891	1901	1911	1921	1931
Population:	98,752	132,285	168,076	195,690	246,050

6. Calculate the number of students from the following data who secured scores not more than 45

Scores range:	30—40	40—50	50—60	60—70	70—80
Number of students:	35	48	70	40	22

7. Find the number of men getting wages between \$ 10 and \$ 15 from the given table:

Wages (in \$):	0—10	10—20	20—30	30—40
Frequency:	9	30	35	42

8. Given below the scores obtained by 492 candidates in a certain examination

Scores	Number of candidates
0—40	210
40—45	43
45—50	54
50—55	74
55—60	32
60—65	79

Find out the number of candidates

(a) who secured scores more than 48 but not more than 50;

(b) who secured scores less than 48 but not less than 45.

9. Use Newton's forward difference formula to obtain the interpolating polynomial $f(x)$, which can satisfying the following data:

x:	1	2	3	4
f(x):	26	18	4	1

If another point $x = 5, f(x) = 26$ is added to the above data, will the interpolating polynomial be the same as before or different. Explain why.

10. The table below gives value of tan x for $.10 \le x \le .30$.

x:	.10	.15	.20	.25	.30
tan x:	.1003	.1511	.2027	.2553	.3093

Calculate tan 0.12 using Newton's forward difference formula

11. (i) Compute the value of $f(22)$ from the following available data:

x:	20	25	30	35	40	45
f(x):	354	332	291	260	231	204

(ii) Find the cubic polynomial which takes the following values:

$y(0) = 1, y(1) = 0, y(2) = 1$ and $y(3) = 10$

Hence or otherwise obtain $y(4)$.

(iii) Use Newton's method to find a polynomial $p(x)$ of lowest possible degree such that $p(n) = 2^n$ for $n = 0, 1, 2, 3, 4$.

5.4.2 Newton's Backward Difference Interpolation Formula

The **Newton's Backward Difference Interpolation formula** is used only when the following two possibilities exists together.

(i) The function should be tabulated at equal intervals.

(ii) The point for which the function is to be interpolated should lie in the second or lower half of the table. If both these possibilities exist, then Newton's Backward Difference formula will give better approximation.

Given function $y(x)$ is tabulated at equally spaced points, in which the difference between the consecutive values is same say h. Let $y(x)$ be a polynomial in x of degree n and is taken in the form:

$$y(x) = a_1 + a_2(x-x_n) + a_3(x-x_n)(x-x_{n-1}) + \ldots\ldots\ldots + a_{n+1}(x-x_n)(x-x_{n-1})\ldots\ldots\ldots\ldots(x-x_1) \quad (1)$$

(from (1))

At $x = x_n$ $\quad y_n = a_1$

Similarly At $x = x_{n-1}$ $\quad y_{n-1} = a_1 + a_2(x_{n-1} - x_n)$

At $x = x_{n-2}$ $\quad y_{n-2} = a_1 + a_2(x_{n-2} - x_n) + a_3(x_{n-2} - x_n)(x_{n-2} - x_{n-1})$

and so on.

As values of x are equally spaced, so we have

$$x_i - x_{i-1} = h \Rightarrow x_{i-1} - x_i = -h, \text{ and } x_{i-m} - x_i = -mh$$

So values of y's in terms of h comes

$$y_n = a_1$$
$$y_{n-1} = a_1 + a_2(-h)$$
$$y_{n-2} = a_1 + a_2(-2h) + a_3(-2h)(-h)$$

....

....

....

$$y_{n+1} = a_1 + a_2(-nh) + a_3(-nh)(-(n-1)h) + \ldots\ldots\ldots + a_{n+1}(-nh)(-(n-1)h)\ldots\ldots(-h)$$

After solving these for a_i's, we get

$$a_1 = y_n$$
$$a_2 = (y_{n-1} - a_1)/h = (y_{n-1} - y_n)/h$$
$$a_3 = (y_{n-2} - a_2(-2h) - a_1)/2h^2$$
$$= \left[y_{n-2} - \frac{(y_{n-1} - y_n)}{h} - (-2h) - u_n \right] / 2h^2$$
$$= [y_{n-2} - 2y_{n-1} + 2y_n - y_n] / 2h^2$$
$$= [y_{n-2} - 2y_{n-1} + y_n] / 2!h^2$$

....

....

$$a_{n+1} = [y_1 - ny_2 + \ldots\ldots + y_n] / n!h^n$$

By using the backward difference table, we get

$$a_1 = y_n$$
$$a_2 = \frac{\nabla y_n}{h}$$
$$a_3 = \frac{\nabla^2 y_n}{2!h^2}$$

....

....

$$a_{n+1} = \frac{\nabla^n y_n}{n!h^n}$$

Substitute these a values in eq. (1), we get

$$y(x) = y_n + \frac{\nabla y_n}{h}(x-x) + \frac{\nabla^2 y_n}{2!h^2}(x-x_n)(x-x_{n-1}) + ... + \frac{\nabla^n y_n}{n!h^n}(x-x_n)(x-x_{n-1})......(x-x_1) \quad ...(2)$$

Suppose, we have the relation $\frac{x-x_n}{h} = u$, then

$x-x_n = hu$

$x-x_{n-1} = x-(x_n-h) = x-x_n+h = hu+h = h(u+1)$

$x-x_{n-2} = x-(x_{n-1}-h) = x-x_{n-1}+h = h(u+1)+h = h(u+2)$

....

....

$x-x_1 = h(u+(n-1))$

So, after substitution of these above values in eq. (2), we get

$$y(x) = y_n + \frac{\nabla y_n}{h}hu + \frac{\nabla^2 y_n}{2!h^2}(hu)\cdot h(u+1) + ... + \frac{\nabla^n y_n}{n!h^n}(hu)\cdot h(u+1)......h(u+(n-1))$$

$$y(x) = y_n + \nabla y_n u + \frac{\nabla^2 y_n}{2!}u(u+1) + ... + \frac{\nabla^n y_n}{n!}u(u+1)......(u+(n-1))$$

we have assumed that $x_{n-1} < x < x_n$ (i.e x lies at the bottom of the table).

If, however, $x_{k-1} < x < x_k$, then the above formula becomes

$$y(x) = y_k + \nabla y_k u + \frac{\nabla^2 y_k}{2!}u(u+1) + ... + \frac{\nabla^{k-1} y_k}{(k-1)!}u(u-1) + ... + (u+((k-1)-1))$$

This formula is known as Newton's Backward difference interpolation formula.

5.4.2.1 *Algorithm for Newton's Backward Difference Interpolation*

Algorithm of Newton's Backward Difference Interpolation for the given function. Given two one dimensional arrays *x* and *y* each of size '*n*' which are used to store the values. One two dimensional array d of size *n*–1 by *n*–1 which is used to store the backward difference table.

1. Begin
2. Read n
3. Do i = 1 to n
4. Read x[i], y[i]
5. End do
6. Read x

7. If ((x < x[1]) or (x > x[n])) then
8. Print "the value lies outside the given range"
9. End if
10. Do j =1 to n-1 by 1 /* newton's backward difference table
11. Do i = j + 1 to n
12. If (j == 1) then
13. d[i][j] = y[i] – y[i - 1]
14. else
15. d[i][j] = d[i][j-1] – d[i - 1][j-1]
16. end if
17. end do
18. end do
19. i=1
20. while (x > x[i])
21. i = i + 1
22. end while
23. k = i
24. h = x[k] – x[k - 1] (where h is the difference between consecutive values of x)
25. set u = (x – x[k])/(x[k] – x[k-1])
26. sum = y[k]
27. do i = 1 to k - 1
28. set prod = 1
29. do j= 0 to (I – 1)
30. prod = prod * (u + j)
31. end do
32. find the factorial of i! and let the value be denoted by 'fact'
33. sum = sum + (d[k][i] * prod)/fact
34. end do
35. print ' the interpolated value is ", sum
36. end

5.4.2.2 Flowchart for Newton's Backward Difference Interpolation

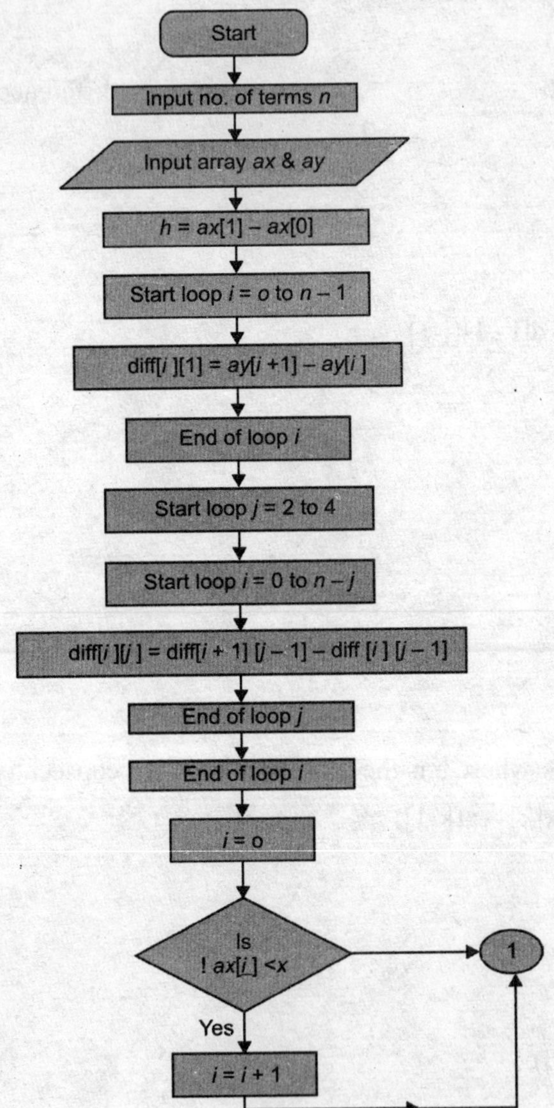

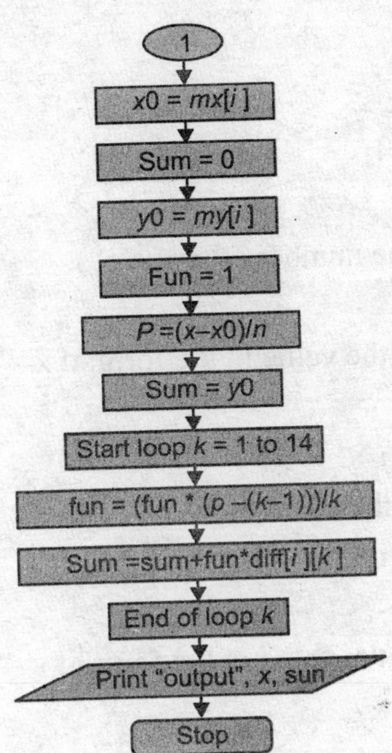

5.4.2.3 Program for Newton's Backward Method of Interpolation in C

```
# include <stdio.h>
# include <conio.h>
# include <math.h>
# include <process.h>
# include <string.h>
void main()
{
int n;
int i,j,k;
float my[10];
float my[10];
float x;
float x0 = 0;
float y0;
float sum;
float h;
float fun;
```

```
float p;
float diff[20][20];
float y1, y2, y3, y4;
clrscr();
printf("\n Enter the number of terms - ");
scanf("%d",&n);
printf("\n\n Enter the value in the form of x - ");
for (i=0;i<n;i++)
{
printf("\n\n Enter the value of x%d - ",i+1);
scanf ("%f",&mx[i]);
}
printf ("\n\n Enter the value in the form of y -");
for (i=0;i<n;i++)
{
printf ("\n\n Enter the value of y%d - ",i+1);
scanf ("%f",&my[i]);
}
printf ("\nEnter the value of x for");
printf("\nwhich you want the value of y - ");
scanf("%f",&x);
h=mx[1]-mx[0];
for(i=0;i<n-1;i++)
{
diff[i][1]=my[i+1]-my[i];
}
for (j=2;j<=4;j++)
{
for (i=0;i<n-j;i++)
{
diff[i][j]=diff[i+1][j-1]-diff[i][j-1];
}
}
i=0;
```

```
while(!mx[i]>x)
{
i++;
}
x0=mx[i];
sum=0;
y0=my[i];
fun=1;
p=(x-x0)/h;
sum=y0;
for (k=1;k<=4;k++)
{
fun=(fun*(p-(k-1)))/k;
sum=sum+fun*diff[i][k];
}
printf ("\nwhen x=%6.4f,y=%6.8f",x,sum);
printf("\n\n\n Press Enter to Exit");
getch( );
}
```

Output:

```
Enter the number of terms-7
Enter the value in the form of x-
Enter the value of x1 - 100
Enter the value of x2 - 150
Enter the value of x3 - 200
Enter the value of x4 - 250
Enter the value of x5 - 300
Enter the value of x6 - 350
Enter the value of x7 - 400
Enter the value in the form of y -
Enter the value of y1 - 10.63
Enter the value of y2 - 13.03
Enter the value of y3 - 15.04
Enter the value of y4 - 16.81
```

```
Enter the value of y5 - 18.42
Enter the value of y6 - 19.90
Enter the value of y7 - 21.27
Enter the value of x for which you want the value of y - 410
When x = 410.0000, y = 21.34462738
Press Enter to Exit
```

5.4.2.4 Program for Newton's Backward Method Of Interpolation in C++

```
# include <iostream.h>
# include <conio.h>
# include <math.h>
# include <process.h>
# include <string.h>
void main()
{
int n;
int i,j,k;
float my[10];
float my[10];
float x;
float x0 = 0;
float y0;
float sum;
float h;
float fun;
float p;
float diff[20][20];
float y1, y2, y3, y4;
clrscr();
cout<<"\n Enter the number of terms - ";
cin>>n;
cout<<"\n\n Enter the value in the form of x - ";
for (i=0;i<n;i++)
{
```

```
cout<<"\n\n Enter the value of x - "<<i+1;
cin>>mx[i];
}
cout<<"\n\n Enter the value in the form of y -";
for (i=0;i<n;i++)
{
cout<<"\n\n Enter the value of y - "<<i+1;
cin>>my[i];
}
cout<<"\nEnter the value of x for";
cout<<"\nwhich you want the value of y - ";
cin>>x;
h=mx[1]-mx[0];
for(i=0;i<n-1;i++)
{
diff[i][1]=my[i+1]-my[i];
}
for (j=2;j<=4;j++)
{
for (i=0;i<n-j;i++)
{
diff[i][j]=diff[i+1][j-1]-diff[i][j-1];
}
}
i=0;
while(!mx[i]>x)
{
i++;
}
x0=mx[i];
sum=0;
y0=my[i];
```

```
fun=1;
p=(x-x0)/h;
sum=y0;
for (k=1;k<=4;k++)
{
fun=(fun*(p-(k-1)))/k;
sum=sum+fun*diff[i][k];
}
cout<<"\n when x="<< setprecision(6.4)<<x<<"y="<< setprecision(6.8)<<sum;
cout<<"\n\n\n Press Enter to Exit";
getch( );
}
```

Output:

```
Enter the number of terms-7
Enter the value in the form of x-
Enter the value of x1 - 100
Enter the value of x2 - 150
Enter the value of x3 - 200
Enter the value of x4 - 250
Enter the value of x5 - 300
Enter the value of x6 - 350
Enter the value of x7 - 400
Enter the value in the form of y -
Enter the value of y1 - 10.63
Enter the value of y2 - 13.03
Enter the value of y3 - 15.04
Enter the value of y4 - 16.81
Enter the value of y5 - 18.42
Enter the value of y6 - 19.90
Enter the value of y7 - 21.27
Enter the value of x for which you want the value of y - 410
When x = 410.0000, y = 21.34462738
Press Enter to Exit
```

SOLVED PROBLEMS

Problem 1: Given the table of value as

x:	2.5	3.0	3.5	4.0	4.5
f(x):	9.75	12.45	15.70	19.52	23.75

Find $y(4.25)$.

Solution: Since the point $x = 4.25$ lies between the fourth and fifth point, therefore we use backward difference at the fifth point.

The difference table is

x_i	y_i	∇y_i	$\nabla^2 y_i$	$\nabla^3 y_i$	$\nabla^4 y_i$
2.5	9.75				
3.0	12.45	2.70			
3.5	15.70	3.25	0.55		
4.0	19.52	3.82	0.57	0.02	
4.5	23.75	4.23	0.41	0.16	0.14

As x lies between $4.0 < x < 4.5$ or $x_4 < x < x_5$, the Newton's backward difference interpolating for the above case becomes

$$y(x) = y_5 + \nabla y_5 + \frac{\nabla^2 y_5}{2!} u(u+1) + \frac{\nabla^3 y_5}{3!} u(u+1)(u+2) + \frac{\nabla^4 y_5}{4!} u(u+1)(u+2)(u+3)$$

$$\text{Now } u = \frac{x - x_5}{h} = \frac{4.25 - 4.5}{0.5} = -0.5$$

$$y(4.25) = 23.75 + 4.23(-0.5) + \frac{0.41}{2}(-0.5)(-0.5+1) + \frac{0.16}{6}(-0.5)(-0.5+1)(-0.5+2) + \frac{0.14}{24}(-0.5)(-0.5+1)(-0.5+2)(-0.5+3)$$

$$= 23.75 - 2.115 - 0.05125 - 0.01 - 0.13125$$

$$= 21.378$$

Problem 2: Population of a town was given below. Compute the population for the year 1925.

Year (x):	1891	1901	1911	1921	1931
Population (y): (in thousands)	46	66	81	93	101

Solution: Here, $a + nh = 1931$, $h = 10$, $a + nh + uh = 1925$

$u = (1925 - 1931)/10 = -0.6$

The difference table is:

x	y	∇y	$\nabla^2 y$	$\nabla^3 y$	$\nabla^4 y$
1891	46				
		20			
1901	66		–5		
		15		2	
1911	81		–3		–3
		12		–1	
1921	93		–4		
		8			
1931	101				

Applying Newton's Backward difference formula, we get

$$y_{1925} = y_{1931} + u\nabla y_{1931} + \frac{u(u+1)}{2!}\nabla^2 y_{1931} + \frac{u(u+1)(u+2)}{3!}\nabla^3 y_{1931} + \frac{u(u+1)(u+2)(u+3)}{4!}\nabla^4 y_{1931}$$

$$= 101 + (-.6)(8) + \frac{(-.6)(.4)}{2!}(-4) + \frac{(-.6)(.4)(1.4)}{3!}(-1) + \frac{(-.6)(.4)(1.4)(2.4)}{4!}(-3)$$

$$= 96.8368 \text{ thousands.}$$

Hence the population for the year 1925 = 96836.8 ≈ 96837.

Problem 3: Population of a town is given below:

Year:	1921	1931	1941	1951	1961	1971
Population: (in Lakhs)	20	24	29	36	46	51

Compute the increase in population during the period 1955 to 1961.

Solution: Here, $a + nh = 1971$, $h = 10$, $a + nh + uh = 1955$

$\therefore$ $1971 + 10u = 1955$ $u = -1.6$

The difference table is:

x	y	∇y	$\nabla^2 y$	$\nabla^3 y$	$\nabla^4 y$	$\nabla^5 y$
1921	20					
		4				
1931	24		1			
		5		1		
1941	29		2		0	
		7		1		–9
1951	36		3		–9	
		10		–8		
1961	46		–5			
		5				
1971	51					

Applying Newton's backward difference formula, we get

$$y_{1955} = y_{1971} + u\nabla y_{1971} + \frac{u(u+1)}{2!}\nabla^2 y_{1971} + \frac{u(u+1)(u+2)}{3!}\nabla^3 y_{1971}$$

$$+\frac{u(u+1)(u+2)(u+3)}{4!}\nabla^4 y_{1971} + \frac{u(u+1)(u+2)(u+3)(u+4)}{5!}\nabla^5 y_{1971}$$

$$= 51+(-1.6)(5)+\frac{(-1.6)(-0.6)}{2!}(-5)+\frac{(-1.6)(-0.6)(0.4)}{6}(-8)$$

$$+\frac{(-1.6)(-0.6)(.4)(1.4)}{24}(-9)+\frac{(-1.6)(-0.6)(0.4)(1.4)(2.4)}{120}(-9)$$

$$= 39.789632$$

Increase in population during period 1955 to 1961 is

$$= 46 - 39.789632$$
$$= 6.210368 \text{ Lakhs}$$
$$= 621036.8 \text{ Lakhs.}$$

Problem 4: From the following table of half-yearly premium for policies maturing at different ages, estimate the premium for a policy maturing at the age of 63:

Age:	45	50	55	60	65
Premium: (in dollars)	114.84	96.16	83.32	74.48	68.48

Solution: The difference table is:

Age x	Premium (in dollars) y	∇y	$\nabla^2 y$	$\nabla^3 y$	$\nabla^4 y$
45	114.84				
		–18.68			
50	96.16		5.84		
		–12.84		–1.84	
55	83.32		4		.68
		–8.84		–1.16	
60	74.48		2.84		
		–6			
65	68.48				

Here $a + nh = 65$, $h = 5$, $\quad a + nh + uh = 63$

$\therefore \quad 65 + 5u = 63 \quad \Rightarrow \quad u = -.4$

By Newton's backward difference formula,

$$y(63) = y(65) + u\nabla y(65) + \frac{u(u+1)}{2!}\nabla^2 y(65) + \frac{u(u+1)(u+2)}{3!}\nabla^3 y(65) + \frac{u(u+1)(u+2)(u+3)}{4!}\nabla^4 y(65)$$

$$= 68.48 + (-.4)(-6) + \frac{(-.4)(-6)}{2}(2.84) + \frac{(-.4)(.6)(1.6)}{6}(-1.16) + \frac{(-.4)(.6)(1.6)(2.6)}{24}(.68)$$

$$= 70.585152$$

Problem 5: Calculate the value of tan 48° 152 from the following table:

$x°$:	45	46	47	48	49	50
tan $x°$:	1.00000	1.03053	1.07237	1.11061	1.15037	1.19175

Solution: Here $a + nh = 50$, $h = 1$, $a + nh + uh = 48° \ 15' = 48.25°$

$$\therefore \quad 50 + u(1) = 48.25 \quad \Rightarrow \quad u = -1.75$$

The difference table is:

$x°$	$10^5 y$	$10^5\nabla y$	$10^5\nabla^2 y$	$10^5\nabla^3 y$	$10^5\nabla^4 y$	$10^5\nabla^5 y$
45	100000					
		3553				
46	103053		131			
		3648		9		
47	107237		140		3	
		3824		12		-5
48	111061		152		-2	
		3976		10		
49	115037		162			
		4138				
50	119175					

$$y_{a+nh+uh} = y_{a+nh} + u\nabla y_{a+nh} + \frac{u(u+1)}{2!}\nabla^2 y_{a+nh} + \frac{u(u+1)(u+2)}{3!}\nabla^3 y_{a+nh} + \frac{u(u+1)(u+2)(u+3)}{4!}\nabla^4 y_{a+nh} + \frac{u(u+1)(u+2)(u+3)(u+4)}{5!}\nabla^5 y_{a+nh}$$

$$10^5 y_{48.25} = 119175 + (-1.75)\times 4138 + \frac{(-1.75)\times(-0.75)}{2}\times 162 + \frac{(-1.75)(-0.75)(0.25)}{3!}\times 10 + \frac{(-1.75)(-.75)(.25)(1.25)}{4!}(-2) + \frac{(-1.75)(-.75)(.25)(1.25)(2.25}{5!}(-5)$$

$$\Rightarrow \qquad 10^5\, y_{48.25} = 112040.2867$$

$$\therefore \qquad y_{48.25} = \tan 48°152 = 1.120402867.$$

UNSOLVED PROBLEMS

1. From the given table find the value of tan 17°

θ°:	0	4	8	12	16	20	24
tan θ°:	0	0.0699	0.1405	0.2126	0.2867	0.3640	0.4402

2. The values of annuities are given for the following ages. Find the value of annuity at the age of $27\frac{1}{2}$.

Age:	25	26	27	28	29
Annuity:	16.195	15.919	15.630	15.326	15.006

3. The table below gives the value of tan x for $0.10 \le x \le 0.30$.

x:	0.10	0.15	0.20	0.25	0.30
y = tan x:	0.1003	0.1511	0.2027	0.2553	0.3093

Find: (*i*) tan 0.50 (*ii*) tan 0.26 (*iii*) tan 0.40.

4. Given:

x:	1	2	3	4	5	6	7	8
f(*x*):	1	8	27	64	125	216	343	512

Find $f(7.5)$ using Newton's Backward difference formula.

5. From the following table of values of x and $f(x)$, determine

(*i*) $f(0.23)$ (*ii*) $f(0.29)$

x:	0.20	0.22	0.24	0.26	0.28	0.30
f (*x*):	1.6596	1.6698	1.6804	1.6912	1.7024	1.7139

6. In an examination, the number of candidates who obtained scores between certain limits are as follows:

Scores	Number of candidates
0—19	41
20—39	62
40—59	65
60—79	50
80—99	17

Estimate the number of candidates who obtained fewer than 70 scores.

7. Compute the value of $f(42)$ from the following available data:

x:	20	25	30	35	40	45
f(*x*):	354	332	291	260	231	204

8. The area A of a circle of diameter d is given for the following values:

d:	80	85	90	95	100
A:	5026	5674	6362	7088	7854

Calculate the area of a circle of diameter 105.

9. Given table, find y, when $x = 1.84$ and 2.4 by Newton's interpolation formula:

x:	1.7	1.8	1.9	2.0	2.1	2.2	2.3
$y \doteq e^x$:	5.474	6.050	6.686	7.389	8.166	9.025	9.974

10. Using Newton's backward difference formula, find the value of $e^{-1.9}$ from the following table of values of e^{-x}:

x:	1	1.25	1.50	1.75	2.00
e^{-x}:	0.3679	0.2865	0.2231	0.1738	0.1353

5.4.3 Newton's Divided Difference Interpolation Formula

The **Newton's Divided Difference Interpolation formula** is used only when the following two possibilities exist together.

(i) The function should be tabulated at unequal intervals.

(ii) The point for which the function is to be interpolated should lie anywhere in the table.

If both these possibilities exist then Newton's divided difference formula will give better approximation. This formula is also suitable when the values are equally spaced.

Given function $y(x)$ is tabulated at unequally spaced points, in which the difference between the consecutive values may or may not be same. Let $y(x)$ be a polynomial in x of degree n and is taken in the form:

$$y(x) = a_1 + a_2(x-x_1) + a_3(x-x_1)(x-x_2) + \dots\dots\dots + a_{n+1}(x-x_1)(x-x_2)\dots\dots(x-x_n) \quad \text{...(1)}$$

At $x = x_1 \quad y_1 = a_1$ (from eq. (1))

Similarly, At $x = x_2 \quad y_2 = a_1 + a_2(x_2 - x_1)$

At $x = x_3 \quad y_3 = a_1 + a_2(x_3 - x_1) + a_3(x_3 - x_1)(x_3 - x_2)$

....

....

....

and so on.

After solving these for a_i's, we get

$$a_1 = y_1$$

$$a_2 = (y_2 - a_1) / (x_2 - x_1) = (y_2 - y_1) / (x_2 - x_1)$$

$$a_3 = (y_3 - a_2(x_3 - x_1) - a_1) / (x_3 - x_1)\,(x_3 - x_2)$$

$$= \left[y3 - \frac{(y_2 - y_1)}{x_2 - x_1}(x_3 - x_1) - y_1 \right] / [(x_3 - x_1)(x_3 - x_2)]$$

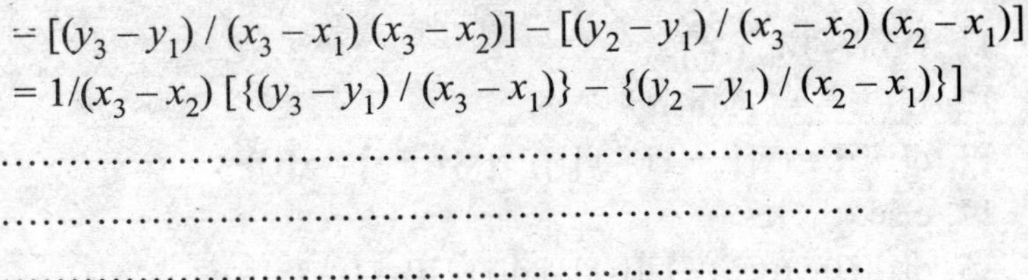

$$= [(y_3 - y_1) / (x_3 - x_1)(x_3 - x_2)] - [(y_2 - y_1) / (x_3 - x_2)(x_2 - x_1)]$$
$$= 1/(x_3 - x_2) [\{(y_3 - y_1) / (x_3 - x_1)\} - \{(y_2 - y_1) / (x_2 - x_1)\}]$$

....

....

....

and so on.

By using the divided difference table, we get

$$a_1 = y_1$$
$$a_2 = \Delta_d y_1$$
$$a_3 = \Delta^2{}_d y_1$$

....

....

....

$$a_{n+1} = \Delta^n{}_d y_1$$

So after substitution of these above values in eq. (1), we get

$$y(x) = y_1 + \Delta_d y_1(x-x_1) + \Delta^2{}_d y_1(x-x_1)(x-x_2) + \ldots\ldots\ldots\ldots + \Delta^n{}_d y_1(x-x_1)(x-x_2)\ldots\ldots\ldots(x-x_n)$$

we have assumed that $x_1 < x < x_2$

If, however, $x_k < x < x_{k+1}$, then the above formula becomes

$$y(x) = y_k + \Delta_d y_k(x-x_k) + \Delta^2{}_d y_k(x-x_k)(x-x_{k+1}) + \ldots\ldots + \Delta^{n-k}{}_d y_k(x-x_k)(x-x_{k+1})\ldots\ldots(x-x_{n-1})$$

This formula is known as Newton's Divided difference interpolation formula.

5.4.3.1 Algorithm for Newton's Divided Difference Interpolation

Algorithm using Newton's Divided Difference Interpolation for the given function. Given two one dimensional arrays *x* and *y* each of size '*n*' which are used to store values. One two dimensional array *d* of size *n*–1 by *n*–1 which is used to store the divided difference table.

1. Begin
2. Read n
3. Do i = 1 to n
4. Read x[i], y[i]
5. End do
6. Read x
7. If ((x < x[1]) or (x > x[n])) then
8. Print "the value lies outside the given range"
9. End if
10. Do j =1 to n-1 by 1   /* divided difference table

```
Do i = 1 to n - j
If ( j  ==  1) then
d[i][j] = (y[i  + 1] – y[i]) / (x[i  + j] – x[i])
else
d[i][j] = (d[i  + 1][j-1] – d[i+1][j-1]) / (x[i  + j] – x[i])
end if
end do
end do
i=1
while (x > x[i])
i = i + 1
end while
k = i  - 1
sum = y[k]
do i = 1 to n
set prod = 1
do j= 0 to (i – 1)
prod = prod * ( x – x[k+j])
end do
sum = sum + (d[k][i] * prod)
end do
print ‘ the interpolated value is “, sum
end
```

SOLVED PROBLEMS

Problem 1: Using Newton's divided difference formula defines $f(x)$ as a polynomial in x from the following data.

x:	–1	0	3	6	7
$f(x)$:	3	–6	39	822	1611

Solution: The divided difference table is

x_i	y_i	$\Delta_d y_i$	$\Delta_d^2 y_i$	$\Delta_d^3 y_i$	$\Delta_d^4 y_i$
–1	3	–9	–6	5	1
0	–6	15	41	13	
3	39	261	132		
6	822	789			
7	1611				

Then the formula leads to

$$f(x) = 3 + (x+1)(-9) + x(x+1)(6) + x(x+1)(x-3)(5) + x(x+1)(x-3)(x-6)$$
$$= x^4 - 3x^3 + 5x^2 - 6$$

UNSOLVED PROBLEMS

1. Given the values:

x:	5	7	11	13	17
f(x):	150	392	1452	2366	5202

Calculate $f(9)$ using Newton's divided difference formula.

2. The observed values of a function are, respectively, 168, 120, 72, and 63 at the four positions 3, 7, 9, and 20 of the independent variable. What is the best estimate you can give for value of the function at the position 6 of the independent variable?

3. Apply Newton's divided difference formula to find the value of $f(8)$ if $f(1) = 3, f(3) = 31, f(6) = 223, f(10) = 1011, f(11) = 1343.$

4. Given that

x:	1	3	4	6	7
y_x:	1	27	81	729	2187

Compute y_5. Why does it differ from 3^5?

5. By applying Newton's divided difference formula to find $f(7)$ if $f(3) = 24$, $f(5) = 120$, $f(8) = 504$, $f(9) = 720$, and $f(12) = 1716$.

6. The following table is given:

x:	0	1	2	5
f(x):	2	3	12	147

What is the form of the function?

5.4.4 Newton's Central Difference Interpolation Formula

In the previous text, we had two Newton's interpolation formula which is used only for interpolation near the beginning and the end of the table. They are not applicable to interpolate near the central value (middle value of the table). To get more accurate results near the middle value of the table, we will obtain a more suitable formula which utilizes differences close to the middle value of the table. Such formulae are named as *central difference interpolation formula.*

5.4.4.1 *Gauss's Forward Interpolation Formula*

By the Newton's Gregory forward interpolation formula, we have

$$y(x) = y_k + \Delta y_k u + \frac{\Delta^2 y_k}{2!}u(u-1) + \frac{\Delta^3 y_k}{3!}u(u-1)(u-2) + \frac{\Delta^4 y_k}{4!}u(u-1)(u-2)(u-3) + \ldots \quad \ldots(1)$$

As we have, $\Delta^2 y_k - \Delta^2 y_{k-1} = \Delta^3 y_{k-1} \Rightarrow \Delta^2 y_k = \Delta^2 y_{k-1} + \Delta^3 y_{k-1}$ ...(2)

$$\Delta^3 y_k = \Delta^3 y_{k-1} + \Delta^4 y_{k-1} \quad ...(3)$$

$$\Delta^4 y_k = \Delta^4 y_{k-1} + \Delta^5 y_{k-1} \quad ...(4)$$

and so on.

Also we have $\Delta^3 y_{k-1} = \Delta^3 y_{k-2} + \Delta^4 y_{k-2}$ ...(5)

$$\Delta^4 y_{k-1} = \Delta^4 y_{k-2} + \Delta^5 y_{k-2} \quad ...(6)$$

and so on.

Substitute the values of (2), (3), (4) in equation (1), we get

$$y(x) = y_k + \Delta y_k u + \frac{\Delta^2 y_{k-1} + \Delta^3 y_{k-1}}{2!} u(u-1) + \frac{\Delta^3 y_{k-1} + \Delta^4 y_{k-1}}{3!} u(u-1)(u-2) + \frac{\Delta^4 y_{k-1} + \Delta^5 y_{k-1}}{4!} u(u-1)(u-2)(u-3) + ...$$

Now by using eq. (5) and (6), we have

$$y(x) = y_k + \Delta y_k u + \frac{\Delta^2 y_{k-1}}{2!} u(u-1) + \frac{\Delta^3 y_{k-1}}{3!} (u+1)u(u-1) + \frac{\Delta^4 y_{k-2}}{4!} (u+1)u(u-1)(u-2) + ...$$

The above formula is known as Gauss's forward interpolation formula.

Algorithm for Gauss's Forward Interpolation Formula

1. Start
2. Input n terms
3. Input the array ax
4. Input the array ay
5. h=ax[1]-ax[0]
6. for i=0;i<n–1;i++
7. diff[i][1]=ay[i+1]-ay[i]
8. End Loop i
9. for j=2;j<=4;j++
10. for i=0;i<n–j;i++
11. diff[i][j]=diff[i+1][j–1]–diff[i][j–1]
12. End Loop i
13. End Loop j
14. i=0
15. Repeat Step 16 until ax[i]<x
16. i=i+1

17. i=i–1;
18. p=(x–ax[i])/h
19. y1=p*diff[i][1]
20. y2=p*(p–1)*diff[i–1][2]/2
21. y3=(p+1)*p*(p-1)*diff[i–2][3]/6
22. y4=(p+1)*p*(p–1)*(p–2)*diff[i–3][4]/24
23. y=ay[i]+y1+y2+y3+y4
24. Print x,y
25. End

Flowchart for Gauss's Forward Interpolation Formula

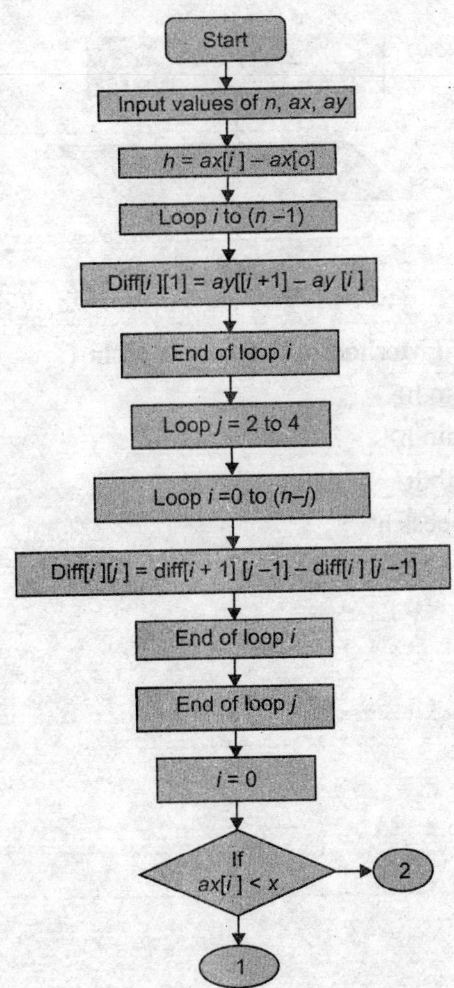

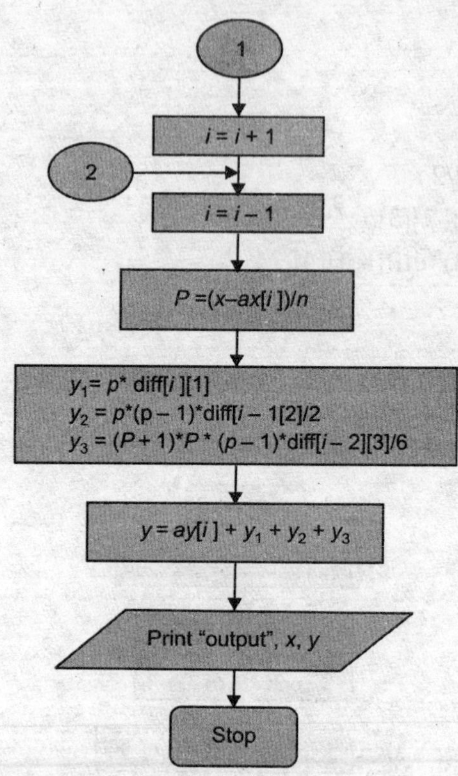

Program for Gauss's Forward Method of Interpolation in C

```
# include <stdio.h>
# include <conio.h>
# include <math.h>
# include <process.h>
# include <string.h>
void main()
{
int n;
int i,j;
float ax[10];
float ax[10];
float x;
float nr,dr;
float y=0;
float h;
```

```
float p;
float diff[20][20];
float y1,y2,y3,y4;
clrscr();
printf("\n Enter the number of terms – ");
scanf("%d",&n);
printf("\n\n Enter the value in the form of x – ");
for (i=0;i<n;i++)
{
printf("\n\n Enter the value of x%d – ",i+i);
scanf("%f",&ax[i]);
}
printf("\n\n Enter the value in the form of y – ");
for(i=0;i<n;i++)
{
printf("\n\n Enter the value of y%d–",i+1);
scanf("%f",&ay[i]);
}
printf("\nEnter the value of x for");
printf("\nwhich you want the value of y–");
scanf ("%f",&x);
h=ax[1]–ax[0];
for(i=0;i<n–1;i++)
{
diff[i][1]=ay[i+1]–ay[i];
}
for(j=2;j<=4;j++)
{
for(i=0;i<n–j;i++)
{
diff[i][j]=diff[i+1][j–1]–diff[i][j–1];
}
}
i=0;
```

```
do {
i++;
}while(ax[i]<x);
i—;
p=(x–ax[i])/h;
y1=p*diff[i][1];
y2=p*(p–1)*diff[i-1][2]/2;
y3=(p+1)*p*(p–1)*diff[i–2][3]/6;
y4=(p+1)*p*(p–1)*(p–2)*diff[i–3][4]/24;
y=ay[i]+y1+y2+y3+y4;
printf("\nwhen x=%6.4f,y=%6.8f ",x,y);
printf("\n\n\n Press Enter to Exit");
getch();
}
```

Output:

Enter the number of terms – 7

Enter the value in the form of x –

Enter the value of x1 – 1.00

Enter the value of x2 – 1.05

Enter the value of x3 – 1.10

Enter the value of x4 – 1.15

Enter the value of x5 – 1.20

Enter the value of x6 – 1.25

Enter the value of x7 – 1.30

Enter the value in the form of y –

Enter the value of y1 – 2.7183

Enter the value of y2 – 2.8577

Enter the value of y3 – 3.0042

Enter the value of y4 – 3.1582

Enter the value of y5 – 3.3201

Enter the value of y6 – 3.4903

Enter the value of y7 – 3.6693

Enter the value of x for
which you want the value of y – 1.17

When $x = 1.17$, y = 3.2221

Press Enter to Exit

Program for Gauss's Forward Method of Interpolation in C++

```
# include <iostream.h>
# include <conio.h>
# include <math.h>
# include <process.h>
# include <string.h>
void main()
{
int n;
int i,j;
float ax[10];
float ax[10];
float x;
float nr,dr;
float y=0;
float h;
float p;
float diff[20][20];
float y1,y2,y3,y4;
clrscr();
cout<<"\n Enter the number of terms – ";
cin>>n;
cout<<"\n\n Enter the value in the form of x – ";
for (i=0;i<n;i++)
{
cout<<"\n\n Enter the value of x – "<<i+i;
cout<<ax[i];
}
cout<<"\n\n Enter the value in the form of y – ";
for(i=0;i<n;i++)
{
cout<<"\n\n Enter the value of y –"<<i+1;
```

```
cin>>ay[i];
}
cout<<"\nEnter the value of x for";
cout<<"\nwhich you want the value of y–";
cin>>x;
h=ax[1]–ax[0];
for(i=0;i<n–1;i++)
{
diff[i][1]=ay[i+1]–ay[i];
}
for(j=2;j<=4;j++)
{
for(i=0;i<n–j;i++)
{
diff[i][j]=diff[i+1][j–1]–diff[i][j–1];
}
}
i=0;
do {
i++;
}while(ax[i]<x);
i—;
p=(x–ax[i])/h;
y1=p*diff[i][1];
y2=p*(p–1)*diff[i-1][2]/2;
y3=(p+1)*p*(p–1)*diff[i–2][3]/6;
y4=(p+1)*p*(p–1)*(p–2)*diff[i–3][4]/24;
y=ay[i]+y1+y2+y3+y4;
cout<<"\nwhen x="<< setprecision(6.4)<<x<<"y=" setprecision(6.8)<<y;
cout<<"\n\n\n Press Enter to Exit";
getch();
}
```

Output:

Enter the number of terms – 7

Enter the value in the form of x –

Enter the value of x1 – 1.00

Enter the value of x2 – 1.05

Enter the value of x3 – 1.10

Enter the value of x4 – 1.15

Enter the value of x5 – 1.20

Enter the value of x6 – 1.25

Enter the value of x7 – 1.30

Enter the value in the form of y –

Enter the value of y1 – 2.7183

Enter the value of y2 – 2.8577

Enter the value of y3 – 3.0042

Enter the value of y4 – 3.1582

Enter the value of y5 – 3.3201

Enter the value of y6 – 3.4903

Enter the value of y7 – 3.6693

Enter the value of x for

which you want the value of y – 1.17

When $x = 1.17$, y = 3.2221

Press Enter to Exit

SOLVED PROBLEMS

Problem 1: By using, Gauss's forward formula to evaluate y_{30}, given that $y_{21} = 18.4708$, $y_{25} = 17.8144$, $y_{29} = 17.1070$, $y_{33} = 16.3432$ and $y_{37} = 15.5154$

Solution: The difference table is

x	u	y_u	Δy_u	$\Delta^2 y_u$	$\Delta^3 y_u$	$\Delta^4 y_u$
21	–2	18.4708	-0.6564	-0.0510	-0.0074	-0.0022
25	–1	17.8144	-0.7074	-0.0564	-0.0076	
29	0	17.1070	-0.7638	-0.0640		
33	1	16.3432	-0.8278			
37	2	15.5154				

By taking $x_0 = 29$, $h = 4$, $x = 30$, we have $u = \dfrac{x - x_0}{h} = \dfrac{30 - 29}{4} = 0.25$

By using Gauss's forward difference formula

$$y(x) = y_0 + \Delta y_0 u + \frac{\Delta^2 y_{-1}}{2!} u(u-1) + \frac{\Delta^3 y_{-1}}{3!} (u+1)u(u-1) + \frac{\Delta^4 y_{-2}}{4!} (u+1)u(u-1)(u-2) + ...$$

$$y(30) = 17.1070 + (0.25)(-0.7638) + \frac{(0.25)(-0.75)}{2} (-0.0564) + \frac{(1.25)(0.25)(-0.75)}{6} (-0.0076)$$

$$+ \frac{(1.25)(0.25)(-0.75)(-1.75}{24}$$

$$= 17.1070 - 0.19095 + 0.00529 + 0.00004$$

$$= 16.9216 \text{(approx.)}$$

Problem 2: Apply here a central difference formula to obtain $f(32)$ given that:

$f(25) = 0.2707 \quad f(35) = 0.3386$

$f(30) = 0.3027 \quad f(40) = 0.3794$.

Solution: Here $a + hu = 32$ and $h = 5$

Take origin at 30 $\therefore a = 30$ then $u = 0.4$

The forward difference table is:

u	x	f(x)	Δf(x)	Δ²f(x)	Δ³f(x)
–1	25	.2707			
			.032		
0	30	.3027		.0039	
			.0359		.0010
1	35	.3386		.0049	
			.0408		
2	40	.3794			

Applying Gauss' forward difference formula, we have

$$f(u) = f(0) + u\Delta f(0) + \frac{u(u-1)}{2!} \Delta^2 f(-1) + \frac{(u+1)u(u-1)}{3!} \Delta^3 f(-1)$$

$$f(.4) = .3027 + (.4)(.0359) + \frac{(.4)(.4-1)}{2!} (.00398) + \frac{(1.4)(.4)(.4-1)}{3!} (.0010)$$

$$= 0.316536.$$

Problem 3: The values of e^{-x} at x = 1.72 to x = 1.76 are given in the following table:

x:	1.72	1.73	1.74	1.75	1.76
e^{-x}:	0.17907	0.17728	0.17552	0.17377	0.17204

Compute the value of $e^{-1.7425}$ using Gauss' forward difference formula.

Solution: Here taking the origin at 1.74 and $h = 0.01$.

$$x = a + uh$$

The difference table is as follows:

u	x	$10^5 f(x)$	$10^5 \Delta f(x)$	$10^5 \Delta^2 f(x)$	$10^5 \Delta^3 f(x)$	$10^5 \Delta^4 f(x)$
–2	1.72	17907				
			–179			
–1	1.73	17728		3		
			–176		–2	
0	1.74	17552		1		3
			–175		1	
1	1.75	17377		2		
			–173			
2	1.76	17204				

Gauss's forward formula is

$$f(u) = f(0) + u\Delta f(0) + \frac{u(u-1)}{2!}\Delta^2 f(-1) + \frac{(u+1)u(u-1)}{3!}\Delta^3 f(-1)$$
$$+ \frac{(u+1)u(u-1)(u-2)}{4!}\Delta^4 f(-1)$$

$$f(.4) = 17552 + (.25)(-175) + \frac{(.25)(-.75)}{2}(1) + \frac{(1.25)(.25)(-.75)}{6}(1)$$
$$+ \frac{(1.25)(.25)(-.75)(-1.75)}{24}(3)$$
$$= 17508.16846$$

$$f(0.25) = e^{-1.7425} = 0.1750816846.$$

UNSOLVED PROBLEMS

1. Use Gauss's forward formula to find the value of $f(x)$ at $x = 3.75$ from the table:

x:	2.5	3.0	3.5	4.0	4.5	5.0
f(x):	24.145	22.043	20.225	18.644	17.262	16.047

2. Given that

x:	25	30	35	40	45
log x:	1.39794	1.47712	1.54407	1.60206	1.65321

Compute the value of log 3.7, using Gauss's forward formula.

3. Calculate the value of $f(41)$ by applying Gauss's forward formula from the following data:

x:	30	35	40	45	50
f(x):	3678.2	2995.1	2400.1	1876.2	1416.3

4. From the following table, calculate the value of $e^{1.17}$ using Gauss forward formula:

x:	1	1.05	1.10	1.15	1.20	1.25	1.30
e^x:	2.7183	2.8577	3.0042	3.1582	3.3201	3.4903	3.6693

5. From the following table Compute y when $x = 1.45$

x:	1.0	1.2	1.4	1.6	1.8	2.0
y:	0.0	– .112	– .016	.336	.992	2.0

5.4.4.2 *Gauss's Backward Interpolation Formula*

By the Newton's Gregory forward interpolation formula, we have

$$y(x) = y_k + \Delta y_k u + \frac{\Delta^2 y_k}{2!}u(u-1) + \frac{\Delta^3 y_k}{3!}u(u-1)(u-2) + \frac{\Delta^4 y_{-2}}{4!}u(u-1)(u-2)(u-3) + \ldots\ldots \quad \ldots(1)$$

As we have,

$$\Delta y_k - \Delta y_{k-1} = \Delta^2 y_{k-1} \Rightarrow \Delta y_k = \Delta y_{k-1} + \Delta^2 y_{k-1} \quad \ldots(2)$$

$$\Delta^2 y_k - \Delta^2 y_{k-1} = \Delta^3 y_{k-1} \Rightarrow \Delta^2 y_k = \Delta^2 y_{k-1} + \Delta^3 y_{k-1} \quad \ldots(3)$$

Similarly,

$$\Delta^3 y_k = \Delta^3 y_{k-1} + \Delta^4 y_{k-1} \quad \ldots(4)$$

anΔ so on.

Also we have

$$\Delta^3 y_{k-1} = \Delta^3 y_{k-2} + \Delta^4 y_{k-2} \quad \ldots(5)$$

$$\Delta^4 y_{k-1} = \Delta^4 y_{k-2} + \Delta^5 y_{k-2} \quad \ldots(6)$$

and so on.

Substitute the values of (2), (3), (4) in equation (1), we get

$$y(x) = y_k + (\Delta y_{k-1}u + \Delta^2 y_{k-1}) + \frac{\Delta^2 y_{k-1} + \Delta^3 y_{k-1}}{2!}u(u-1) + \frac{\Delta^3 y_{k-1} + \Delta^4 y_{k-1}}{3!}u(u-1)(u-2) +$$

$$\frac{\Delta^4 y_{k-1} + \Delta^5 y_{k-1}}{4!}u(u-1)(u-2)(u-3) + \ldots\ldots$$

$$y(x) = y_k + \Delta y_{k-1}u + \frac{\Delta^2 y_{k-1}}{2!}u(u+1) + \frac{\Delta^3 y_{k-1}}{3!}(u+1)(u)(u-1) + \frac{\Delta^4 y_{k-1}}{4!}(u+1)(u)(u-1)(u-2)$$

$$+ \frac{\Delta^5 y_{k-1}}{5!}(u+1)(u)(u-1)(u-2) + \ldots\ldots$$

Now by using eq. (5) and (6), we have

$$y(x) = y_k + \Delta y_{k-1}u + \frac{\Delta^2 y_{k-1}}{2!}u(u+1) + \frac{\Delta^3 y_{k-2} + \Delta^4 y_{k-2}}{3!}(u+1)(u)(u-1) +$$

$$\frac{\Delta^4 y_{k-2} + \Delta^5 y_{k-2}}{4!}(u+1)(u)(u-1)(u-2)$$

Hence,

$$y(x) = y_k + \Delta y_{k-1} u + \frac{\Delta^2 y_{k-1}}{2!} u(u+1) + \frac{\Delta^3 y_{k-2}}{3!} (u+1)(u)(u-1)$$

$$+ \frac{\Delta^4 y_{k-2}}{4!} (u+2)(u+1)(u)(u-1) + \ldots\ldots$$

The above formula is known as Gauss's backward interpolation formula.

Algorithm for Gauss's Backward Formula

1. Start
2. Input n terms
3. Input the array ax
4. Input the array ay
5. h=ax[1]-ax[0]
6. for i=0;i<n-l;i++
7. diff[i][1]=ay[i+1]-ay[i]
8. End Loop i
9. for j=2;j<=4;j++
10. for i=0;i<n–j;i++
11. diff[i][j]=diff[i+1][j–1]–diff[i][j–1]
12. End Loop i
13. End Loop j
14. i=0
15. Repeat Step 16 until ax[i]<x
16. i=i+1
17. i=i–1;
18. p=(x–ax[i])/h
19. y1=p*diff[i-1][1]
20. y2=p*(p+1)*diff[i–1][2]/2
21. y3=(p+1)*p*(p-1)*diff[i–2][3]/6
22. y4=(p+2)*(p+1)*p*(p–1)*diff[i–3][4]/24
23. y=ay[i]+y1+y2+y3+y4
24. Print x,y
25. End

Flowchart for Gauss's Backward Formula

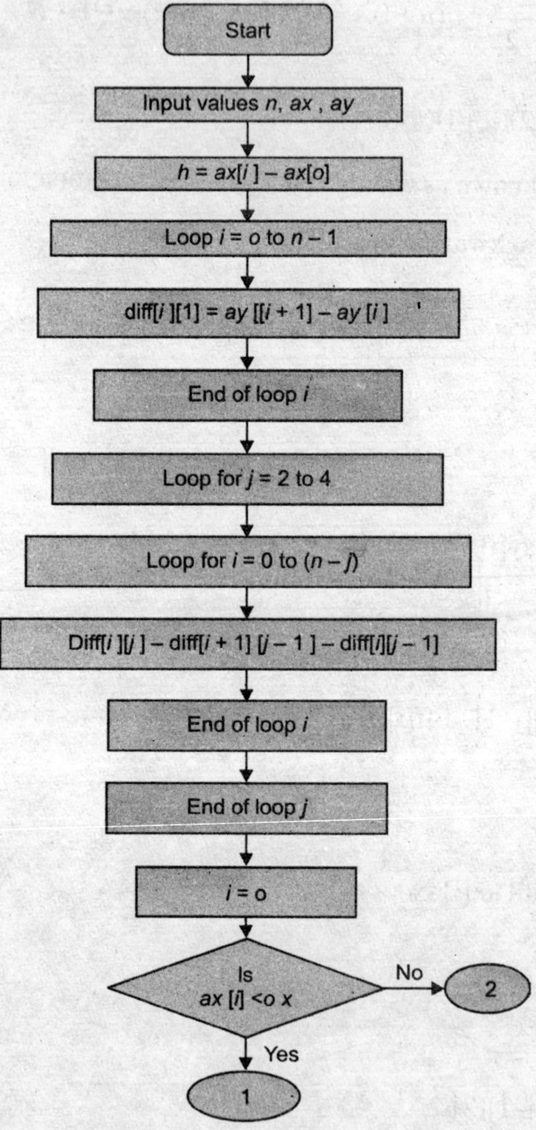

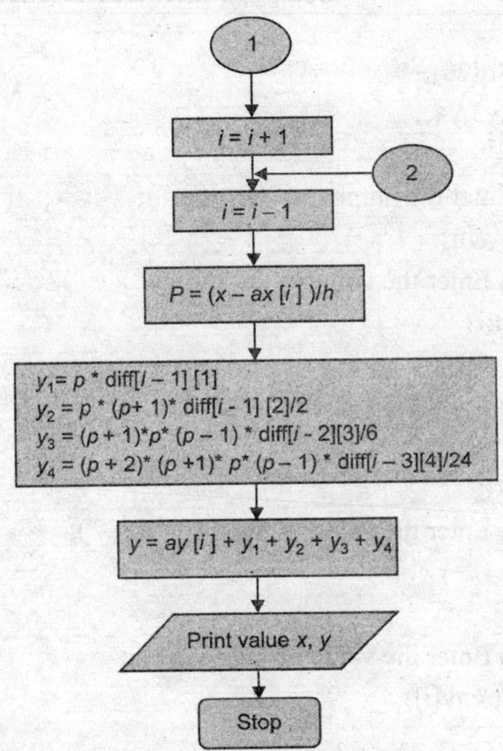

Program for Gauss's Backward Method of Interpolation in C

```
# include <stdio.h>
# include <conio.h>
# include <math.h>
# include <process.h>
# include <string.h>
void main()
{
int n;
int i,j;
float ax[10];
float ay[10];
float x;
float y=0;
float h;
float p;
```

```
float diff[20][20];
float y1,y2,y3,y4;
clrscr();
printf("\n Enter the number of terms – ");
scanf("%d",&n);
printf("\n\n Enter the value in the form of x – ");
for (i=0;i<n;i++)
{
printf("\n\n Enter the value of x%d–",i+1);
scanf("%f",&ax[i]);
}
printf("\n\n Enter the value in the form of y–");
for(i=0;i<n;i++)
{
printf("\n\n Enter the value of y%d–",i+1);
scanf("%f",&ay[i]);
}
printf("\nEnter the value of x for");
printf("\nwhich you want the value of y – ");
scanf("%f",&x);
h=ax[1]–ax[0];
for(i=0;i<n–1;i++)
{
diff[i][1]=ay[i+1]–ay[i];
}
for(j=2;j<=4;j++)
{
for(i=0;i<n–j;i++)
{
diff[i][j]=diff[i+1][j–1]–diff[i][j–1];
}
}
i=0;
do {
i++;
```

```
}while (ax[i]<x);
i––;
p=(x-ax[i])/h;
y1=p*diff[i–1][1];
y2=p*(p+1)*diff[i-1][2]/2;
y3=(p+1)*p*(p–1)*diff[i–2][3]/6;
y4=(p+2)*(p+1)*p*(p–1)*diff[i–3][4]/24;
y=ay[i]+y1+y2+y3+y4;
printf("\nwhen x=%6.1f,y=%6.4f ",x,y);
printf("\n\n\n Press Enter to Exit");
getch();
}
```

Output:

```
Enter the number of terms – 7
Enter the value in the form of x –
Enter the value of x1 – 1.00
Enter the value of x2 – 1.05
Enter the value of x3 – 1.10
Enter the value of x4 – 1.15
Enter the value of x5 – 1.20
Enter the value of x6 – 1.25
Enter the value of x7 – 1.30
Enter the value in the form of y –
Enter the value of y1 – 2.1783
Enter the value of y2 – 2.8577
Enter the value of y3 – 3.0042
Enter the value of y4 – 3.1582
Enter the value of y5 – 3.3201
Enter the value of y6 – 3.4903
Enter the value of y7 – 3.6693
Enter the value of x for
which you want the value of y – 1.35
When x = 1.35, y=3.8483
Press Enter to Exit
```

Program for Gauss's Backward Method of Interpolation in C++

```
# include <iostream.h>
# include <conio.h>
# include <math.h>
# include <process.h>
# include <string.h>
void main()
{
int n;
int i,j;
float ax[10];
float ay[10];
float x;
float y=0;
float h;
float p;
float diff[20][20];
float y1,y2,y3,y4;
clrscr();
cout<<"\n Enter the number of terms – ";
cin>>n;
cout<<"\n\n Enter the value in the form of x – ";
for (i=0;i<n;i++)
{
cout<<"\n\n Enter the value of x–"<<i+1;
cin>>ax[i]);
}
cout<<"\n\n Enter the value in the form of y–";
for(i=0;i<n;i++)
{
cout<<"\n\n Enter the value of y–"<<i+1;
cin>>ay[i];
}
cout<<"\nEnter the value of x for";
```

```
cout<<"\nwhich you want the value of y - ";
cin>>x;
h=ax[1]-ax[0];
for(i=0;i<n-1;i++)
{
diff[i][1]=ay[i+1]-ay[i];
}
for(j=2;j<=4;j++)
{
for(i=0;i<n-j;i++)
{
diff[i][j]=diff[i+1][j-1]-diff[i][j-1];
} }
i=0;
do {
i++;
}while (ax[i]<x);
i--;
p=(x-ax[i])/h;
y1=p*diff[i-1][1];
y2=p*(p+1)*diff[i-1][2]/2;
y3=(p+1)*p*(p-1)*diff[i-2][3]/6;
y4=(p+2)*(p+1)*p*(p-1)*diff[i-3][4]/24;
y=ay[i]+y1+y2+y3+y4;
cout<<"\nwhen x="<< setprecision(6.1)<<x<<"y="<< setprecision(6.4)<<y;
cout<<"\n\n\n Press Enter to Exit";
getch();
}
```

Output:

```
Enter the number of terms - 7
Enter the value in the form of x -
Enter the value of x1 - 1.00
Enter the value of x2 - 1.05
Enter the value of x3 - 1.10
```

Enter the value of x4 – 1.15
Enter the value of x5 – 1.20
Enter the value of x6 – 1.25
Enter the value of x7 – 1.30
Enter the value in the form of y –
Enter the value of y1 – 2.1783
Enter the value of y2 – 2.8577
Enter the value of y3 – 3.0042
Enter the value of y4 – 3.1582
Enter the value of y5 – 3.3201
Enter the value of y6 – 3.4903
Enter the value of y7 – 3.6693
Enter the value of x for
which you want the value of y – 1.35
When x = 1.35, y=3.8483
Press Enter to Exit

SOLVED PROBLEMS

Problem 1: Calculate the value of cos 51° 422 by Gauss's backward formula. Given that

x:	50°	51°	52°	53°	54°
cos x:	0.6428	0.6293	0.6157	0.6018	0.5878.

Solution: Taking the origin at 52° and $h = 1$

$u = (x - a) = 51° \, 422 \; - 52° = - 182 \; = - 0.3°$

Gauss's backward formula is

$$f(u) = f(0) + u\Delta f(-1) + \frac{u(u+1)}{2!}\Delta^2 f(-1) + \frac{(u+1)u(u-1)}{3!}\Delta^3 f(-2)$$

$$+\frac{(u+2)(u+1)u(u-1)}{4!}\Delta^4 f(-2)(39)$$

$$10^4 f(-.3) = 6157 + (-.3)(-136) + \frac{(.7)(-.3)}{2!}(-3) + \frac{(.7)(-.3)(-1.3)}{3!}(-2) + \frac{(1.7)(.7)(-.3)(-1.3)}{4!}(4)$$

$$= 6198.10135$$

The difference table is as below:

u	x	$10^4f(x)$	$10^4\Delta f(x)$	$10^4\Delta^2f(x)$	$10^4\Delta^3f(x)$	$10^4\Delta^4f(x)$
–2	50°	6428				
			–135			
–1	51°	6293		–1		
			[–136]		[–2]	
0	52°	[6157]		[–3]		[4]
			–139		2	
1	53°	6018		–1		
			–140			
2	54°	5878				

$f(-.3) = .619810135$

Hence cos 51°422 $= 0.619810135$.

Problem 2: $f(x)$ is a polynomial of degree four and given that $f(4) = 270, f(5) = 648, \Delta f(5) = 682, \Delta^3 f(4) = 132$.
Calculate the value of $f(5.8)$ using Gauss's backward formula.

Solution:

$$\Delta f(5) = f(6) - f(5)$$

$$f(6) = f(5) + \Delta f(5) = 648 + 682 = 1330$$

$$\Delta^3 f(4) = (E-1)^3 f(4) = f(7) - 3f(6) + 3f(5) - f(4) = 132$$

$$f(7) = 3f(6) - 3f(5) + f(4) + 132$$

$$= 3 \times 1330 - 3 \times 648 + 270 + 132 = 2448.$$

The difference table is (Taking origin at 6):

u	x	$f(x)$	$\Delta f(x)$	$\Delta^2 f(x)$	$\Delta^3 f(x)$
–2	4	270			
			378		
–1	5	648		304	
			[682]		[132]
0	6	[1330]		[436]	
			1118		
1	7	2448			

Here, $a = 6, h = 1,$ $\quad a + hu = 5.8$

$\therefore \quad 6 + u = 5.8 \Rightarrow u = -.2$

Gauss's backward formula is

$$f(-.2) = f(0) + u\Delta f(-1) + \frac{u(u+1)}{2!}\Delta^2 f(-1) + \frac{(u+1)u(u-1)}{3!}\Delta^3 f(-2)$$

$$= 1330 + (-.2)(682) + \frac{(.8)(-.2)}{2}(436) + \frac{(.8)(-.2)(-1.2)}{6}(132) = 1162.944$$

$$\therefore \quad f(5.8) = 1162.944.$$

UNSOLVED PROBLEMS

1. The population of a town in the years 1931, ..., 1971 are as follows:

Year:	1931	1941	1951	1961	1971
Population: (in thousands)	15	20	27	39	52

Calculate the population of the town in 1946 by applying Gauss's backward formula.

2. Apply Gauss's backward formula to find the value of $(1.06)^{19}$ if $(1.06)^{10} = 1.79085$, $(1.06)^{15} = 2.39656$, $(1.06)^{20} = 3.20714$, $(1.06)^{25} = 4.29187$ and $(1.06)^{30} = 5.74349$.

3. Given that

x::	50	51	52	53	54
tan x:	1.1918	1.2349	1.2799	1.3270	1.3764

By using Gauss's backward formula, find the value of tan 51° 422 .

4. Interpolate by means of Gauss's backward formula, the population of a town for the year 1974 given that:

Year:	1939	1949	1959	1969	1979	1989
Population: (in thousands)	12	15	20	27	39	52

5. By applying Gauss's backward formula to compute sin 45° from the following table:

θ°:	20	30	40	50	60	70	80
sin θ:	0.34202	0.502	0.64279	0.76604	0.86603	0.93969	0.98481

5.5 STIRLING'S FORMULA FOR CENTRAL DIFFERENCE

Taking the average of the Gauss forward and backward interpolation formula for central differences, we obtain the Stirling's central difference formula.

The Gauss's forward formula is given by

$$y(x) = y_k + \Delta y_k u + \frac{\Delta^2 y_{k-1}}{2!} u(u-1) + \frac{\Delta^3 y_{k-1}}{3!} (u+1)(u)(u-1) +$$

$$+ \frac{\Delta^4 y_{k-2}}{4!} (u+1)(u)(u-1)(u-2) + \ldots\ldots \quad \ldots(1)$$

and Gauss's backward formula is given by

$$y(x) = y_k + \Delta y_{k-1} u + \frac{\Delta^2 y_{k-1}}{2!} u(u+1) + \frac{\Delta^3 y_{k-2}}{3!} (u+1)(u)(u-1) +$$

$$+ \frac{\Delta^4 y_{k-2}}{4!} (u+2)(u+1)(u)(u-1) + \ldots\ldots$$

Taking the average of the above two formulas:

$$y(x) = y_k + u + \frac{(\Delta y_{k-1} + \Delta y_k)}{2} + \frac{u^2}{2!}\Delta^2 y_{k-1} + \frac{u(u^2-1)}{3!}\frac{(\Delta^3 y_{k-2} + \Delta^3 y_{k-1})}{2}$$

$$+ \frac{u^2(u^2-1)}{4!}\Delta^4 y_{k-2} + \ldots\ldots$$

The above formula is known as Stirling's formula for central formula.

5.5.1 Algorithm for Stirling's Formula

1. Start.
2. Input no. of terms n
3. Input the array ax
4. Input the array ay
5. h = ax[1]-ax[0]
6. for (i = 1;i < n-1; i++)
7. diff [i][1] = ay[i + 1]-ay[i]
8. End of loop i
9. for (j = 2; j < = 4; j++)
10. for (i = 0; i < n-j; i++)
11. diff[i][j] = diff[i + 1][j-1]-diff[i][j-1]
12. End of loop i
13. End of loop j
14. i = 0
15. Repeat step 16 until ax[i] < x
16. i = i + 1
17. i = i-1;
18. p = (x-ax[i])/h
19. y1= p*(diff[i][1] + diff[i-1][1])/2
20. y2 = p*p*diff[i-1][2]/2
21. y3 = p*(p*p-1)*(diff[i-1][3]+diff[i-2][3])/6
22. y4 = p*p*(p*p-1)*diff[i-2][4]/24
23. y = ay[i]+y1 + y2 + y3 + y4
24. Print output value
25. End

5.5.2 Flowchart for Stirling's Formula

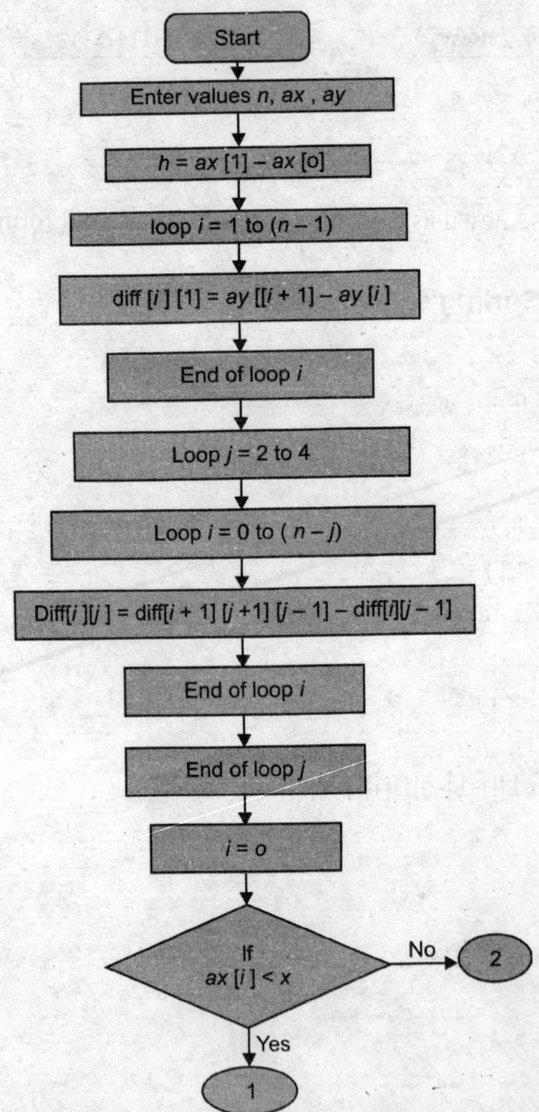

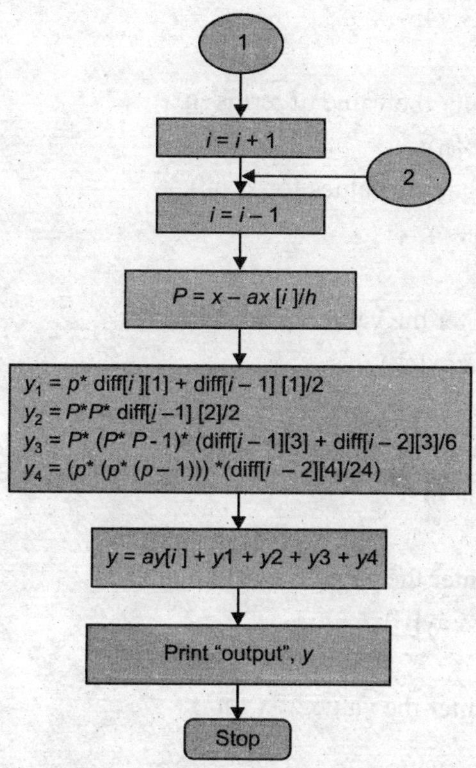

5.5.3 Program for Stirling Method of Interpolation in C

```
#include<stdio.h>
#include<conio.h>
#include<math.h>
#include<process.h>
void main()
{
int n;
int i,j;
float ax[10];
float ax[10];
float h;
float p;
float diff[20][20];
float x,y;
```

```
float y1,y2,y3,y4;
clrscr();
printf("\n Enter the value of terms");
scanf("%d",%n);
printf("\n Enter the values for x \n");
for(i=0;i<n;i++)
{
printf("\n Enter the value for x%d-",i+1);
scanf("%f",&ax[i]);
}
printf("\n Enter the values for y \n");
for(i=0;i<n;i++)
{
printf("\n Enter the value for y%d-",i+1);
scanf("%f",&ay[i]);
}
printf("\n Enter the value of x for");

printf("\n which you want the value of y");
scanf("%f",&x);
h=ax[1]-ax[0];
for(i=0;i<n-1;i++)
{
diff[i][1]=ay[i+1]-ay[i];
}
for(j=2;j<=4;j++)
{
for(i=0;i<n-j;i++)
{
diff[i][j]=diff[i+1][j-1]-diff[i][j-1];
}
}
i=0;
do {
```

```
i++;
}while(ax[i]<x);
i--;
p=(x-ax[i])/h;
y1=p*(diff[i][1]+diff[i-1][1])/2;
y2=p*p*diff[i-1][2]/2;
y3=p*(p*p-1)*(diff[i-1][3]+diff[i-2][3])/6;
y4=p*p*(p*p-1)*diff[i-2][4]/24;
y=ay[i]+y1+y2+y3+y4;
printf("\n\n When x=%6.2f, y=%6.8f",x,y);
getch();
}
```

Output:

```
Enter the value of terms-7
Enter the values for x
Enter the value for x1 - .61
Enter the value for x2 - .62
Enter the value for x3 - .63
Enter the value for x4 - .64
Enter the value for x5 - .65
Enter the value for x6 - .66
Enter the value for x7 - .67
Enter the values for y
Enter the value for y1 - 1.840431
Enter the value for y2 - 1.858928
Enter the value for y3 - 1.877610
Enter the value for y4 - 1.896481
Enter the value for y5 - 1.915541
Enter the value for y6 - 1.934792
Enter the value for y7 - 1.954237
Enter the value of x for
which you want the value of y - 0.6440
When x=0.6440,y=1.90408230
Press Enter to Continue
```

5.5.4 Program for Stirling Method of Interpolation in C++

```
#include<iostream.h>
#include<conio.h>
#include<math.h>
#include<process.h>
void main()
{
int n;
int i,j;
float ax[10];
float ax[10];
float h;
float p;
float diff[20][20];
float x,y;
float y1,y2,y3,y4;
clrscr();
cout<<"\n Enter the value of terms";
cin>>n;
cout<<"\n Enter the values for x \n";
for(i=0;i<n;i++)
{
cout<<"\n Enter the value for x-"<<i+1;
cin>>ax[i];
}
cout<<"\n Enter the values for y \n";
for(i=0;i<n;i++)
{
cout<<"\n Enter the value for y-"<<i+1;
cin>>ay[i];
}
cout<<"\n Enter the value of x for";
cout<<"\n which you want the value of y";
```

```
cin>>x;
h=ax[1]-ax[0];
for(i=0;i<n-1;i++)
{
diff[i][1]=ay[i+1]-ay[i];
}
for(j=2;j<=4;j++)
{
for(i=0;i<n-j;i++)
{
diff[i][j]=diff[i+1][j-1]-diff[i][j-1];
}
}
i=0;
do {
i++;
}while(ax[i]<x);
i—;
p=(x-ax[i])/h;
y1=p*(diff[i][1]+diff[i-1][1])/2;
y2=p*p*diff[i-1][2]/2;
y3=p*(p*p-1)*(diff[i-1][3]+diff[i-2][3])/6;
y4=p*p*(p*p-1)*diff[i-2][4]/24;
y=ay[i]+y1+y2+y3+y4;
cout<<"\n\n When x="<< setprecision(6.2)<<x<<"y="<< setprecision(6.8f)<<y;
getch();
}
```

Output:

```
Enter the value of terms-7
Enter the values for x
Enter the value for x1 - .61
Enter the value for x2 - .62
Enter the value for x3 - .63
Enter the value for x4 - .64
```

Enter the value for x5 - .65

Enter the value for x6 - .66

Enter the value for x7 - .67

Enter the values for y

Enter the value for y1 - 1.840431

Enter the value for y2 - 1.858928

Enter the value for y3 - 1.877610

Enter the value for y4 - 1.896481

Enter the value for y5 - 1.915541

Enter the value for y6 - 1.934792

Enter the value for y7 - 1.954237

Enter the value of x for

which you want the value of y - 0.6440

When x=0.6440,y=1.90408230

Press Enter to Continue

5.5.5 Advantages of Central Difference Interpolation Formula

In general, the coefficients in the central difference interpolation formulae are smaller and rapidly convergent than those in Newton's Gregory formulae. The coefficients in Stirling's formula decrease more rapidly than those in Newton-Gregory formulae. Therefore, it is advisable to use Stirling's central difference formula instead of Newton-Gregory formula.

5.5.6 When to Use Stirling's Formula

If $-\frac{1}{4} \le u \le \frac{1}{4}$, then use Stirling's formula.

SOLVED PROBLEMS

Problem 1: Compute the value of e^x when (*a*). $x = 0.644$ and (*b*). x= 0.638 by using Stirling's formula from the following data:

x:	0.61	0.62	0.63	0.64	0.65	0.66	0.67
$y = e^x$:	1.840341	1.658928	1.877610	1.896481	1.915541	1.934792	1.954237

Solution: The difference table see on next page.

(i) Here $$u = \frac{0.644 - 0.64}{0.001} = 0.4$$

X	y = e^x	Δ	Δ²	Δ³	Δ⁴
0.61	1.840431	0.019497	0.000185	0.000004	–0.00004
0.562	1.658928	0.018682	0.000189	0.00000	0.000002
0.63	1.877610	0.018871	0.000189	0.000002	0.000001
0.64	1.896481	0.019060	0.000191	0.000003	
0.65	1.915541	0.019251	0.000194		
0.66	1.934792	0.019445			
0.67	1.954237				

According to Stirling's formula

$$y(x) = y_4 + u\frac{(\Delta y_3 + \Delta y_4)}{2} + \frac{u^2}{2!}\Delta^2 y_3 + \frac{u(u^2-1)}{3!}\frac{(\Delta^3 y_2 + \Delta^3 y_3)}{2} + \frac{u^2(u^2-1)}{4!}\Delta^4 y_2 + ...$$

$$y(0.644) = 1.896481 + (0.4)\frac{(0.018871 + 0.019060)}{2} + \frac{0.16}{2!}(0.000189) + 0$$

$$= 1.896481 + 0.0075862 + 0.00001512$$

$$= 1.904082$$

(ii) Here $$u = \frac{0.644 - 0.64}{0.001} = 0.4$$

According to Stirling's formula

$$y(x) = y_4 + u\frac{(\Delta y_3 + \Delta y_4)}{2} + \frac{u^2}{2!}\Delta^2 y_3 + \frac{u(u^2-1)}{3!}\frac{(\Delta^3 y_2 + \Delta^3 y_3)}{2} + \frac{u^2(u^2-1)}{4!}\Delta^4 y_2 + ...$$

$$y(0.638) = 1.896481 - (0.2)\frac{(0.018871 + 0.019060)}{2} + \frac{0.04}{2}(0.000189)$$

$$= 1.896481 - 0.0037931 + 0.0000038$$

$$= 1.892692$$

Problem 2: Given that:

θ:	0°	5°	10°	15°	20°	25°	30°
tan θ:	0	0.0875	0.1763	0.2679	0.364	0.4663	0.5774

Calculate the value of tan 16° using Stirling formula.

Solution: Take origin at 15°

$\therefore$ $$a = 15°, h = 5$$

$$a + hu = 16$$

$\Rightarrow$ $$15 + 5u = 16 \quad u = .2$$

The difference table is:

u	θ	$10^4 f(\theta)$	$10^4\Delta f(\theta)$	$10^4\Delta^2 f(\theta)$	$10^4\Delta^3 f(\theta)$	$10^4\Delta^4 f(\theta)$	$10^4\Delta^5 f(\theta)$	$10^4\Delta^6 f(\theta)$
–3	0	0						
			875					
–2	5	875		13				
			888		15			
–1	10	1763		28		2		
			916		17		–2	
0	15	2679		45		0		11
			961		17		9	
1	20	3640		62		9		
			1023		26			
2	25	4663		88				
			1111					
3	30	5774						

Using Stirling's formula,

$$10^4 f(.2) = 2679 + (.2)\left(\frac{961+916}{2}\right) + \frac{(.2)^2}{2!}(45) + \frac{(1.2)(.2)(-.8)}{3!}\left(\frac{17+17}{2}\right) + \frac{(.2)^2\{(.2)^2-1\}}{4!}(0)$$

$$+\frac{(.2)^2\{(.2)^2-1\}}{4!}(0) + \frac{(2.2)(1.2)(.2)(-.8)(-1.8)}{5!}\left\{\frac{9+(-2)}{2}\right\}$$

$$+\frac{(.2)^2\{(.2)^2-1\}\{(.2)^2-4\}}{6!}(11) = 2866.980499$$

$\therefore \quad f(.2) = .2866980499$

Hence tan 16° = 0.2866980499.

Problem 3: By using Stirling's formula to find y_{35}, given $y_{20} = 512, y_{30} = 439, y_{40} = 346$ and y_{50} = 243.

Solution: Let the origin be at 30 and $h = 10$

$$a + hu = 35$$

$$30 + 10u = 35 \quad \Rightarrow \quad u = .5$$

The difference table is as follows:

u	x	y	Δy	$\Delta^2 y$	$\Delta^3 y$
–1	20	512			
			–73		
0	30	439		–20	
			–93		10
1	40	346		–10	
			–103		
2	50	243			

By Stirling's formula,

$$f(.5) = 439 + (.5)\left(\frac{-93-73}{2}\right) + \frac{(.5)^2}{2!}(-20) + \frac{(1.5)(.5)(-.5)}{3!}\left(\frac{10}{2}\right)$$

$$= 394.6875$$

Hence, $y35 = 394.6875$.

UNSOLVED PROBLEMS

1. Use Stirling's central formula to evaluate $y(16)$ given the table:

x:	5	10	15	20	25
y:	0.0875	0.1763	0.2679	0.3640	0.4663

2. Use Stirling's formula to find u_{32} from the data
   $u_{20} = 14.35$, $u_{25} = 13.674$, $u_{30} = 13.257$, $u_{35} = 12.734$, $u_{40} = 12.089$, $u_{45} = 11.309$
3. Compute $f(0.41)$ using Stirling's formula, if
   $f(0.30) = 0.1179, f(0.35) = 0.1368, f(0.40) = 0.1554, f(0.45) = 0.1736, f(0.50) = 0.1915.$
4. Calculate sin (0.197) from the data given below:

x::	0.15	0.17	0.19	0.21	0.23
sin x:	0.14944	0.16918	0.18886	0.20846	0.22798

5. Use Stirling's formula to find u_{32} from the following table:
   $u_{20} = 14.035$ $u_{30} = 13.257$
   $u_{40} = 12.089$ $u_{25} = 13.674$
   $u_{35} = 12.734$ $u_{45} = 11.309.$
6. Apply Stirling's formula to evaluate $y_{12.2}$ from the following table ($y_x = 1 + \log_{10} \sin x$):

x°:	10	11	12	13	14
$10^5\, y_x$:	23967	28060	31788	35209	38368.

7. The following table gives the values of e^x for certain equidistant values of x. Compute the value of e^x when $x = 0.644$ using Stirling's method.

x::	0.61	0.62	0.63	0.64	0.65	0.66	0.67
y = eˣ:	1.840431	1.858928	1.877610	1.896481	1.915541	1.934792	1.954237

5.6 LAGRANGE'S INTERPOLATION

Let $y(x)$ be such that x takes values $x_1, x_2, x_3 \ldots x_n$ which are not necessarily equally spaced. Let us assume a polynomial of the form

$$y(x) = a_1(x-x_2)(x-x_3) + a_2(x-x_1)(x-x_3) + a_3(x-x_1)(x-x_2) \quad \ldots(1)$$

which passes through the points (x_1,y_1), (x_2,y_2) and (x_3,y_3), where all a_i's are constants(unknowns) and are to be determined as:

At $x = x_1$ $\quad y(x_1) = a_1(x_1 - x_2)(x_1 - x_3) \quad \Rightarrow \quad a_1 = y_1 / (x_1 - x_2)(x_1 - x_3)$

At $x = x_2$ $\quad y(x_2) = a_2(x_2 - x_1)(x_2 - x_3) \quad \Rightarrow \quad a_2 = y_2 / (x_2 - x_1)(x_2 - x_3)$

At $x = x_3$ $\quad y(x_3) = a_3(x_3 - x_1)(x_3 - x_2) \quad \Rightarrow \quad a_3 = y_3 / (x_3 - x_1)(x_3 - x_2)$

Substitute these values in equation (1), we get

$$y(x) = y_1 \frac{(x - x_2)(x - x_3)}{(x_1 - x_2)(x_1 - x_3)} + y_2 \frac{(x - x_1)(x - x_3)}{(x_2 - x_1)(x_2 - x_3)} + y_3 \frac{(x - x_1)(x - x_2)}{(x_3 - x_1)(x_3 - x_2)}$$

The above formula in simplified form can be written as

$$y(x) = \sum_{i=1}^{3} y_i \prod_{\substack{j=1 \\ j \neq i}}^{3} \frac{(x - x_i)}{(x - x_j)}$$

This formula is used when polynomial only passing through three points (x_1, y_1), (x_2, y_2) and (x_3, y_3) i.e polynomial is of second order.

If we have a polynomial of degree $(n-1)$ which passes through n points then above formula takes the shape as:

$$y(x) = \sum_{i=1}^{n} y_i \prod_{\substack{j=1 \\ j \neq i}}^{n} \frac{(x - x_i)}{(x - x_j)}$$

The above calculated formula is known as *Lagrange's interpolation formula* and is very simple to operate.

The above process of interpolation needs considerable computation if the number of 'interpolation' required is large in number. It is also seen that introduction of further points, would change all the coefficients. For these reasons, the Lagrange's method is not the most suitable form for interpolation on the computer.

Lagrange's formula is merely a relation between two variables either of which may be taken as the independent variable.

5.6.1 Merits and Demerits of Lagrange's Formula

1. The formula is simple and easy to remember.
2. There is no need to construct the divided difference table and we can directly interpolate the unknown value with the help of given observations.
3. The calculations in the formula are more complicated than in the divided difference formula.
4. The application of the formula is not speedy
5. There is always a chance of commiting some error due to a number of (+)ve and (–)ve sign in the denominator and numerator of each term.
6. The calculations provide no check whether the functional values used are taken correctly or not, whereas the differences used in a difference formula provide a check on the functional values.

5.6.2 Algorithm for Lagrange's Interpolation Formula

1. Start
2. Input no. of terms n
3. Input the array ax
4. Input the array ay
5. for (i=0; i<n; i++)
6. nr=1
7. dr =1
8. for (j=0; j<n; j++)
9. if j !=i
   a. nr=nr*(x-ax[j])
   b.dr*(ax[i]-ax[j])
10. End of Loop j
11. y+=(nr/dr)*ay[i]
12. End of Loop i
13. Print Output values of x, y
14. End

5.6.3 Flowchart for Lagrange's Interpolation Formula

See figure on next page

5.6.4 Program for Lagrange's Method of Interpolation in C

```
# include <stdio.h>
# include <conio.h>
# include <math.h>
# include <process.h>
# include <string.h>
void main()
{
int n;
int i,j;
float ax[100];
float ay[100];
float x=0;
```

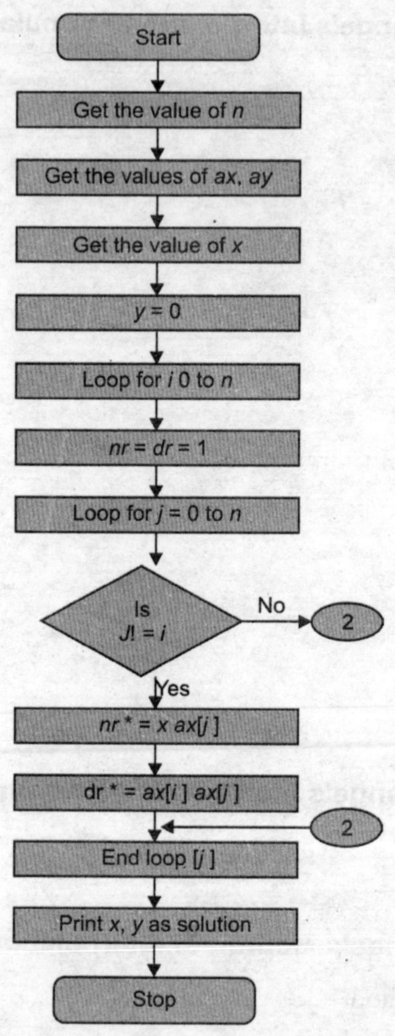


```
float y=0;
float nr;
float dr;
clrscr();
printf("\n Enter the number of terms - ");
scanf("%d",&n);
printf("\n\n Enter the value in the form of x - ");
for (i=0;i<n;i++)
{
printf ("\n\n Enter the value of x%d - ", i+1);
```

```
scanf("%f",&ax[i]);
}
printf("\n\n Enter the value in the form of y - ");
for (i=0;i<n;i++)
{
printf("\n\n Enter the value of y%d - ", i+1);
scanf ("%f",&ay[i]);
}
printf("\n Enter the value of x for ");
printf("\n which you want the value of y - ");
scanf("%f",&x);
for(i=0;i<n;i++)
{
nr=1;
dr=1;
for(j=0;j<n;j++)
{
if(j!=i)
{
nr=nr*(x-ax[j]);
dr=dr*(ax[i]-ax[j]);
}
y+=(nr/dr)*ay[i];
}
}
printf("\n\n When x=%5.2f,y=%5.2f ",x,y);
printf("\n\n\n Press Enter to Exit");
getch();
}
```

Output:

Enter the number of terms - 5

Enter the value in the form of x -

Enter the value of x1- 5

Enter the value of x2 - 7

```
Enter the value of x3 - 11
Enter the value of x4 - 13
Enter the value of x5 - 17
Enter the value in the form of y -
Enter the value of y1 - 150
Enter the value of y2 - 392
Enter the value of y3 - 1452
Enter the value of y4 - 2366
Enter the value of y5 - 5202
Enter the value of x for
Which you want the value of y - 9.0
When x = 9.00, y = 810.00
Press Enter to Exit
```

5.6.5 Program for Lagrange's Method of Interpolation in C++

```
# include <iostream.h>
# include <conio.h>
# include <math.h>
# include <process.h>
# include <string.h>
void main()
{
int n;
int i,j;
float ax[100];
float ay[100];
float x=0;
float y=0;
float nr;
float dr;
clrscr();
cout<<"\n Enter the number of terms - ";
cin>>n;
cout<<"\n\n Enter the value in the form of x - ";
```

```
for (i=0;i<n;i++)
{
cout<<"\n\n Enter the value of x - "<< i+1;
cin>>ax[i];
}
cout<<"\n\n Enter the value in the form of y - ";
for (i=0;i<n;i++)
{
cout<<"\n\n Enter the value of y - "<< i+1;
cin>>ay[i];
}
cout<<"\n Enter the value of x for ";
cout<<"\n which you want the value of y - ";
cin>>x;
for(i=0;i<n;i++)
{
nr=1;
dr=1;
for(j=0;j<n;j++)
{
if(j!=i)
{
nr=nr*(x-ax[j]);
dr=dr*(ax[i]–ax[j]);
}
y+=(nr/dr)*ay[i];
}
}
cout<<"\n\n When x="<< setprecision(5.2)<<x<<"y="<< setprecision(5.2)<<y;
cout<<"\n\n\n Press Enter to Exit";
getch();
}
```

Output:

Enter the number of terms - 5

Enter the value in the form of x -
Enter the value of x1- 5
Enter the value of x2 - 7
Enter the value of x3 - 11
Enter the value of x4 - 13
Enter the value of x5 - 17
Enter the value in the form of y -
Enter the value of y1 - 150
Enter the value of y2 - 392
Enter the value of y3 - 1452
Enter the value of y4 - 2366
Enter the value of y5 - 5202
Enter the value of x for
Which you want the value of y - 9.0
When x = 9.00, y = 810.00
Press Enter to Exit

SOLVED PROBLEMS

Problem 1: By using Lagrange's interpolation formula, find $y(10)$ from the following table:

x:	5	6	9	11
y:	12	13	14	16

Solution: Here $x_0 = 5, x_1 = 6, x_2 = 9, x_3 = 11$

$f(x_0) = 12, f(x_1) = 13, f(x_2) = 14, f(x_3) = 16$

Lagrange's formula is

$$f(x) = \frac{(x-x_1)(x-x_2)(x-x_3)}{(x_0-x_1)(x_0-x_2)(x_0-x_3)} f(x_0)$$

$$+\frac{(x-x_0)(x-x_2)(x-x_3)}{(x_1-x_0)(x_1-x_2)(x_1-x_3)} f(x_1)$$

$$+\frac{(x-x_0)(x-x_1)(x-x_3)}{(x_2-x_0)(x_2-x_1)(x_2-x_3)} f(x_2)$$

$$+\frac{(x-x_0)(x-x_1)(x-x_2)}{(x_3-x_0)(x_3-x_1)(x_3-x_2)} f(x_3)$$

$$f(x) = \frac{(x-6)(x-9)(x-11)}{(5-6)(5-9)(5-11)}(12) + \frac{(x-5)(x-9)(x-11)}{(6-5)(6-9)(6-11)}(13)$$

$$+\frac{(x-5)(x-6)(x-11)}{(9-5)(9-6)(9-11)}(14) + \frac{(x-5)(x-6)(x-9)}{(11-5)(11-6)(11-9)}(16)$$

$$= -\frac{1}{2}(x-6)(x-9)(x-11) + \frac{13}{15}(x-5)(x-9)(x-11)$$

$$-\frac{7}{12}(x-5)(x-6)(x-11) + \frac{4}{15}(x-5)(x-6)(x-9)$$

Putting $x = 10$, we get

$$f(10) = -\frac{1}{2}(10-6)(10-9)(10-11) + \frac{13}{15}(10-5)(10-9)(10-11)$$

$$-\frac{7}{12}(10-5)(10-6)(10-11) + \frac{4}{15}(10-5)(10-6)(10-9)$$

$$= 14.66666667$$

Hence, $y(10) = 14.66666667$

Problem 2: Use Lagrange's formula of interpolation find $y(9.5)$ given

x :	7	8	9	10
y :	3	1	1	9

Solution: Here value of $n = 4$, By expanding Lagrange's interpolation formula we have

$$y(x) = y_1\frac{(x-x_2)(x-x_3)(x-x_4)}{(x_1-x_2)(x_1-x_3)(x_1-x_4)} + y_2\frac{(x-x_1)(x-x_3)(x-x_4)}{(x_2-x_1)(x_2-x_3)(x_2-x_4)}$$

$$+y_3\frac{(x-x_1)(x-x_2)(x-x_4)}{(x_3-x_1)(x_3-x_2)(x_3-x_4)} + y_4\frac{(x-x_1)(x-x_2)(x-x_3)}{(x_4-x_1)(x_4-x_2)(x_4-x_3)}$$

After substituting the values from the given table of data and x = 9.5, we get

$$y(9.5) = 3\frac{(1.5)(0.5)(-0.5)}{(7-8)(7-9)(7-10)} + \frac{(2.5)(0.5)(-0.5)}{(8-7)(8-9)(8-10)}$$

$$+\frac{(2.5)(1.5)(-0.5)}{(9-7)(9-8)(9-10)} + 9\frac{(2.5)(1.5)(0.5)}{(10-7)(10-8)(10-9)}$$

$$= 0.1875 - 0.3125 + 0.9375 + 2.8125$$

$$= 3.625$$

UNSOLVED PROBLEMS

1. If $y(1) = -3, y(3) = 9, y(4) = 30$ and $y(16) = 132$, find the four point Lagrange's interpolation polynomial that takes the same values as y at the given points.
2. Find $\sqrt{155}$ using Lagrange's interpolation formula from the data:

x:	150	152	154	156
$y = \sqrt{x}$:	12.247	12.329	12.410	12.490

3. From the following data, calculate y when $x = 10$ by using Lagrange's interpolation formula.

X :	5	6	9	11
Y :	12	13	14	16

4. Using Lagrange's formula, express y as a function of x.

X :	0	2	3	6
Y :	648	704	729	792

5. Apply Lagrange's formula to find f(5) and f(6) given that
   $f(2) = 4, f(1) = 2, f(3) = 8, f(7) = 128$
   Explain why the result differs from those obtained by completing the series of powers of 2?
6. Values of $f(x)$ for values of x are given as
   $f(1) = 4, f(2) = 5, f(7) = 5, f(8) = 4$
   Compute $f(6)$ and also the value of x for which $f(x)$ is maximum or minimum.
7. Find by Lagrange's formula, the value of
   (i) u_5 if $u_0 = 1, u_3 = 19, u_4 = 49, u_6 = 181$
   (ii) u_4 if $u_3 = 16, u_5 = 36, u_7 = 64, u_8 = 81$ and $u_9 = 100$.
8. Using Lagrange's formula, find the values of
   (i) y_5 if $y_1 = 4, y_3 = 120, y_4 = 340, y_5 = 2544$
   (ii) y_0 if $y_{-30} = 30, y_{-12} = 34, y_3 = 38, y_{18} = 42$.
9. Find the value of tan 33° by Lagrange's formula if
   tan 30° = 0.5774, tan 32° = 0.6249,
   tan 35° = 0.7002, tan 38° = 0.7813.
10. Use Lagrange's formula to find $f(6)$ from the following table:

x :	2	5	7	10	12
f(x):	18	180	448	1210	2028.

11. Apply Lagrange's formula to find $f(15)$, if

x :	10	12	14	16	18	20
f(x):	2420	1942	1497	1109	790	540.

5.7 INVERSE INTERPOLATION

The process of estimating the value of x for the value of y not in the table is called inverse interpolation. Inverse interpolation involves finding the values of x corresponding to a given non-tabulated $f(x)$ value. For inverse interpolation we will consider $x = f^{-1}(y)$.

By interchanging x and y in the above formula we can also write the formula as:

$$X(y) = \sum_{i=1}^{n} x_i \prod_{\substack{j=1 \\ j \neq i}}^{n} \frac{(y - y_i)}{(y_i - y_j)}$$

This formula is known as **Lagrange's Inverse interpolation formula.**

5

SOLVED PROBLEMS

Problem 1: Using inverse interpolation, compute the real root of $x^3 + x - 3 = 0$, close to x 1.2

Solution: The difference table is

X	y $(= x^3 + x - 3)$	Δ	Δ^2	Δ^3	Δ^4
1.0	–1	0.431	0.066	0.006	0.000
1.1	–0.569	0.497	0.072	0.006	
1.2	–0.072	0.569	0.078		
1.3	0.497	0.647			
1.4	1.144				

Clearly the root of the given equation lies between 1.2 and 1.3. Taking x_0 = 1.2 and using Stirling's formula we get

$$0 = -0.072 + x\frac{0.569 + 0.497}{2} + \frac{x^2}{2}(0.072 + \frac{x(x^2 - 1)}{6}\frac{(0.006 + 0.006)}{2} \qquad \text{as } y = 0$$

$$0 = -0.072 + 0.532x + 0.36x^2 + 0.001x^3$$

$$x = \frac{0.072}{0.532} - \frac{0.036}{0.532}x^2 - \frac{0.001}{0.532}x^3$$

$\therefore$ First approximation for $x^{(2)} = \frac{0.072}{0.532} = 0.1353$

Using this in the expression for x above we get

$$x^{(2)} = 0.1353 - 0.067(0.1353)^2 - 0.001879(0.1353)^3$$
$$= 0.1353 - 0.001226 - 0.00000465$$
$$= 0.1353 - 0.00123$$
$$= 0.134$$

Hence the desired root = 1.2 + 0.1(0.134) = 1.2134

UNSOLVED PROBLEMS

1. Using Lagrange's inverse interpolation formula calculate the value of x when y =3 from the following data.

X:	4	7	10	12
Y:	–1	1	2	4

2. Given the following data:

x:	2.2	2.4	2.6	2.8	3.0
y:	4.457	5.466	6.695	8.198	10.018

Find x when $y = 5.0$, using inverse interpolation formula.

5.8 ERROR PROPAGATION IN DIFFERENCE TABLE

Let $y_0, y_1, y_2, \ldots\ldots, y_n$ be the exact values of a function corresponding to $x_0, x_1, x_2, \ldots\ldots, x_n$. Suppose an error e is made in entering the value y_5 in the table. Now, we will see this error e in y_5 affects the entries in the difference columns.

5.8.1 Propagation of Error

x	y	Δ	Δ²	Δ³	Δ⁴	Δ⁵
x_0	y_0					
		Δy_0				
x_1	y_1		$\Delta^2 y_0$			
		Δy_1		$\Delta^3 y_0$		
x_2	y_2		$\Delta^2 y_1$		$\Delta^4 y_0$	
		Δy_2		$\Delta^3 y_1$		$\Delta^5 y+e$
x_3	y_3		$\Delta^2 y_2$		$\Delta^4 y_1+e$	
		Δy_3		$\Delta^3 y_2+e$		$\Delta^5 y_1-5e$
x_4	y_4		$\Delta^2 y_3+e$		$\Delta^4 y_1-e$	
		Δy_4+e		$\Delta^3 y_3-3e$		$\Delta^5 y_2+10e$
x_5	y_5+e		$\Delta^2 y_4-2e$		$\Delta^4 y_2+6e$	
		Δy_5-e		$\Delta^3 y_4+3e$		$\Delta^5 y_3-10e$
x_6	y_6		$\Delta^2 y_5+e$		$\Delta^4 y_4-4e$	
		Δy_6		$\Delta^3 y_5-e$		${}^5 y_4+5e$
x_7	y_7		$\Delta^2 y_6$		$\Delta^4 y_5+e$	
		Δy_7		$\Delta^3 y_6$		$\Delta^5 5_1-e$
x_8	y_8		$\Delta^2 y_7$		$\Delta^4 y_6$	
		Δy_8		$\Delta^3 y_7$		
x_9	y_9		$\Delta^2 y_8$			
		Δy_9				
x_{10}	y_{10}					

From the above table it is clear that:

(i) The error increases with the order of differences.

(ii) The coefficients of the error (e) in any column are the combinational coefficients of $(1+e)^n$

(iii) The algebraic sum of the error in any difference column is zero.

(iv) The maximum error in each column occurs opposite to entry containing the error.

These facts enable us to detect error in a difference table.

SOLVED PROBLEMS

Problem 1: In the given table find and correct the error, by using difference table, in which *y* is a cubic polynomial in *x*.

x :	0	1	2	3	4	5	6	7
y :	25	21	18	18	27	45	76	123

Solution: The difference table is given by

x	y	Δy	$\Delta^2 y$	$\Delta^3 y$
0	25			
		–4		
1	21		1	
		–3		2
2	18		3	
		0		6
3	18		9	
		9		0
4	27		9	
		18		4
5	45		13	
		31		3
6	76		16	
		47		
7	123			

Since y is a polynomial of degree three, so $\Delta^3 y$ must be constant i.e same. The sum of the third order difference being 15, each entry under $\Delta^3 y$ must be $15/5 = 3$. Thus the four entries under $\Delta^3 y$ are in error and can be written as

$3 + (-1), 3 - (-1), 3 + (-1), 3 - (-1)$

We take $e = -1$ and find that the entry corresponding to $x = 3$ is in error. Therefore, $y + e = 18$, this implies that the true values of $y = 18 - e = 18 - (-1) = 19$.

UNSOLVED PROBLEMS

1. If $y = f(x)$ is a polynomial of degree 3 and the following table gives the values of x and y. Find and correct the wrong values of y.

x :	0	1	2	3	4	5	6
y :	4	10	30	75	160	294	490

2. Assuming the following values of y_x belong to a polynomial of degree 5. it is known that $y = f(3)$ is in error. Correct the error.

x :	0	1	2	3	4	5	6
y :	1	2	33	254	1025	3126	7777

3. If following are values of y_x which is polynomial of degree 4.

x :	0	1	2	3	4
y :	1	-1	1	-1	1

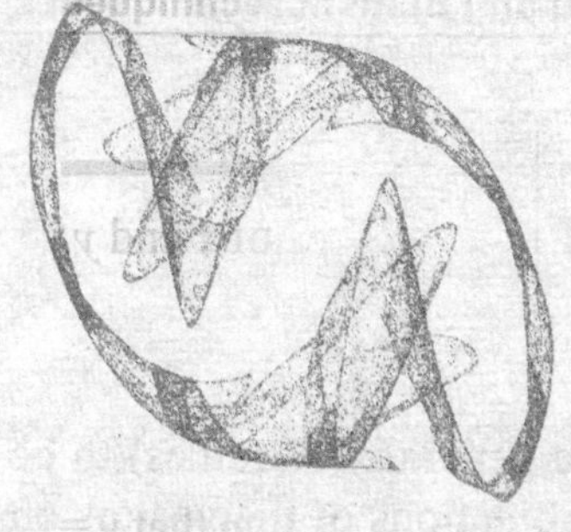

CHAPTER 6

Solution of Linear Equations

6.1 INTRODUCTION

Linear systems of equations are associated with many problems of engineering and science, as well as with applications of social sciences and the quantitative study of business and economic problems.

A general system of *m* linear equations with n unknowns can be written as

$$a_{11}x_1 + a_{12}x_2 + \ldots\ldots + a_{1n}x_n = b_1$$
$$a_{21}x_1 + a_{22}x_2 + \ldots\ldots + a_{2n}x_n = b_2$$
$$a_{31}x_1 + a_{32}x_2 + \ldots\ldots + a_{3n}x_n = b_3$$
$$\ldots\ldots\ldots\ldots\ldots\ldots\ldots\ldots\ldots\ldots\ldots$$
$$\ldots\ldots\ldots\ldots\ldots\ldots\ldots\ldots\ldots\ldots\ldots$$
$$a_{m1}x_1 + a_{m2}x_2 + \ldots\ldots + a_{mn}x_n = b_m$$

Here $x_1, x_2, \ldots, x_n$ are the unknowns, $a_{11}, a_{12}, \ldots, a_{mn}$ are the coefficients of the system, and $b_1, b_2, \ldots, b_m$ are the constant terms.

Often the coefficients and unknowns are real or complex numbers, but integers and rational numbers are also seen, as are polynomials and elements of an abstract algebraic structure.

Where the a_{ij} are the coefficients and b_i are given constant numbers on the right hand side and the quantities x_i are the unknowns to be determined. If all the b's are not equal to zero, the system of linear equation is called **non-homogeneous** and a unique solution of such non-homogeneous equations exist if all the equations are independent. If all the b's are zero, the system of linear equations is known as **homogeneous** and a non-trivial solution of such equations exists only if all such equations are not independent.

6.2 VECTOR EQUATION

One extremely helpful view is that each unknown is a weight for a column vector in a linear combination.

$$x_1\begin{bmatrix} a_{11} \\ a_{21} \\ \cdots \\ a_{m1} \end{bmatrix} + x_2\begin{bmatrix} a_{12} \\ a_{22} \\ \cdots \\ a_{m2} \end{bmatrix} + \ldots + x_n\begin{bmatrix} a_{1n} \\ a_{2n} \\ \cdots \\ a_{mn} \end{bmatrix} = \begin{bmatrix} b_1 \\ b_2 \\ \cdots \\ b_m \end{bmatrix}$$

This allows all the language and theory of vector spaces (or more generally, modules) to be brought to bear. For example, the collection of all possible linear combinations of the vectors on the left-hand side is called their span, and the equations have a solution just when the right-hand vector is within that span. If every vector within that span has exactly one expression as a linear combination of the given left-hand vectors, then any solution is unique. In any event, the span has a basis of linearly independent vectors that do guarantee exactly one expression; and the number of vectors in that basis (its dimension) cannot be larger than m or n, but it can be smaller. This is important because if we have m independent vectors a solution is guaranteed regardless of the right-hand side, and otherwise not guaranteed.

6.3 MATRIX EQUATION

The vector equation is equivalent to a matrix equation of the form

$$Ax = B$$

where A is an $m \times n$ matrix, x is a column vector with n entries, and b is a column vector with m entries.

$$A = \begin{bmatrix} a_{11} & a_{12} & \cdots & a_{1n} \\ a_{12} & a_{22} & \cdots & a_{2n} \\ \cdots & \cdots & \cdots & \cdots \\ a_{m1} & a_{m2} & \cdots & a_{mn} \end{bmatrix}, \quad x = \begin{bmatrix} x_1 \\ x_2 \\ \cdots \\ x_n \end{bmatrix}, \quad b = \begin{bmatrix} b_1 \\ b_2 \\ \cdots \\ b_m \end{bmatrix}$$

The number of vectors in a basis for the span is now expressed as the rank of the matrix.

6.4 SYSTEM OF LINEAR EQUATIONS

In mathematics, a system of linear equations (or linear system) is a collection of linear equations involving the same set of variables. For example,

$$\begin{aligned} 3x + 2y - z &= 1 \\ 2x - 2y + 4z &= -2 \\ -x + \frac{1}{2}y - z &= 0 \end{aligned}$$

is a system of three equations in the three variables x, y, z. A solution to a linear system is an assignment of numbers to the variables such that all the equations are simultaneously satisfied. A solution to the system above is given by

$$x = 1$$
$$y = -2$$
$$z = -2$$

since it makes all three equations valid.

In mathematics, the theory of linear systems is a branch of linear algebra, a subject which is fundamental to modern mathematics. Computational algorithms for finding the solutions are an important part of numerical linear algebra, and such methods play a prominent role in engineering, physics, chemistry, computer science, and economics. A system of non-linear equations can often be approximated by a linear system, a helpful technique when making a mathematical model or computer simulation of a relatively complex system.

6.5 ELEMENTARY PROBLEM

The simplest kind of linear system involves two equations and two variables:

$$2x + 3y = 6$$
$$4x + 9y = 15$$

One method for solving such a system is as follows. First, solve the top equation for x in terms of y:

$$x = 3 - \frac{3}{2}y$$

Now substitute this expression for x into the bottom equation:

$$4\left(3 - \frac{3}{2}y\right) + 9y = 15$$

This results in a single equation involving only the variable y. Solving gives $y = 1$, and substituting this back into the equation for x yields $x = 3/2$. This method generalizes to systems with additional variables (see "elimination of variables" below, or the article on elementary algebra.)

6.6 GENERAL BEHAVIOR

In general, the behavior of a linear system is determined by the relationship between the number of equations and the number of unknowns:

1. Usually, a system with fewer equations than unknowns has infinitely many solutions. Such a system is also known as an underdetermined system.
2. Usually, a system with the same number of equations and unknowns has a single unique solution.
3. Usually, a system with more equations than unknowns has no solution.

In the first case, the dimension of the solution set is usually equal to $n - m$, where n is the number of variables and m is the number of equations.

The following pictures illustrate this trichotomy in the case of two variables:

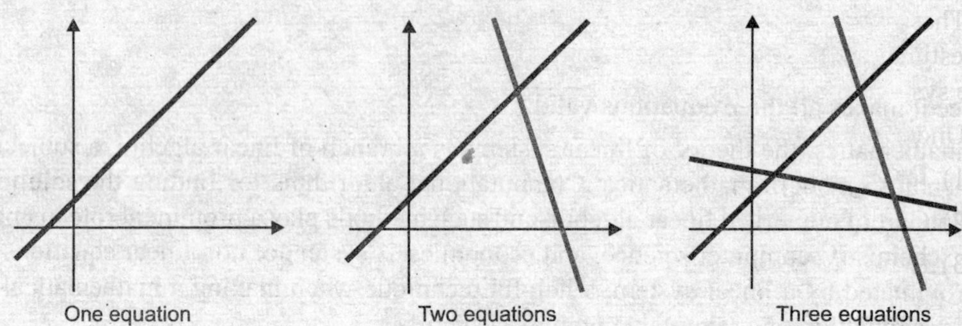

The first system has infinitely many solutions, namely all of the points on the blue line. The second system has a single unique solution, namely the intersection of the two lines. The third system has no solutions, since the three lines share no common point.

Keep in mind that the pictures above show only the most common case. It is possible for a system of two equations and two unknowns to have no solution (if the two lines are parallel), or for a system of three equations and two unknowns to be solvable (if the three lines intersect at a single point). In general, a system of linear equations may behave differently than expected if the equations are linearly dependent, or if two or more of the equations are inconsistent.

6.7 METHODS FOR SOLVING LINEAR EQUATIONS

There are two types of methods involve for solving non-homogeneous system of linear equations.

There are two classes of methods for solving of this type as written below:

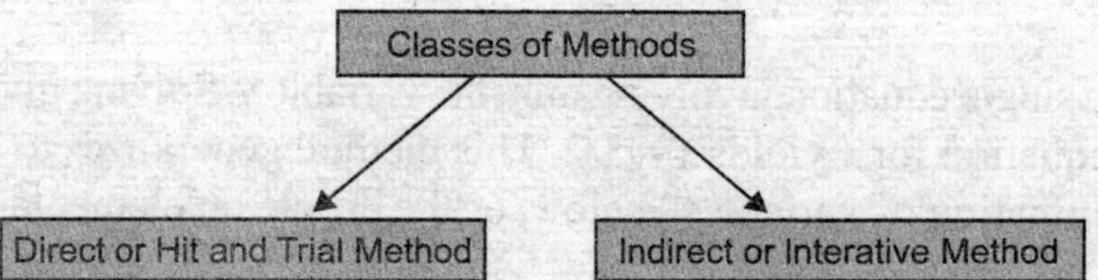

1. Direct Methods

These methods compute the solution in a finite number of steps subject only to round off errors. These methods are guaranteed to succeed and are recommended for general purpose.

Under the category of direct methods we will explain following methods.

(1) Cramer's Rule

(2) Matrix Inversion

(3) Gauss Elimination

(4) Gauss Jordon

(5) Crout's Method

2. Iterative Methods

These methods start with an arbitrary approximation to independent variable x and improve this estimate in an infinite but convergent sequence of steps. These methods are used for solving large system of equations, which have high percentage of zero entries.

Under the category of iterative methods we will explain following methods.

(1) Jacobi's method

(2) Gauss Seidel method

(3) Relaxation method

6.8 DIRECT METHODS

Direct methods give all the roots at the same time. In the next section, we describe five such methods namely, Cramer's rule, Matrix inversion method, Gauss elimination method, Gauss Jordon method, Crout's method.

6.8.1 Cramer's Rule

Cramer's rule is a theorem in linear algebra, which gives the solution of a system of linear equations or corresponding square matrices in terms of determinants. Gabriel Crammer develops this method and hence it is known as Cramer's rule after has name. In order to explain the method we consider a system of three equations as:

$$\begin{aligned} a_1x + b_1y + c_1z &= d_1 \\ a_2x + b_2y + c_2z &= d_2 \\ a_3x + b_3y + c_3z &= d_3 \end{aligned} \qquad \ldots(1)$$

We calculate the determinants Δ, Δ_1, Δ_2 and Δ_3, respectively of coefficient matrix and the coefficient matrices in which first, second and third columns are replaced with the column vector $\begin{bmatrix} d_1 \\ d_2 \\ d_3 \end{bmatrix}$. Thus, we have

$$\Delta = \begin{bmatrix} a_1 & b_1 & c_1 \\ a_2 & b_2 & c_2 \\ a_3 & b_3 & c_3 \end{bmatrix}, \qquad \Delta_1 = \begin{bmatrix} d_1 & b_1 & c_1 \\ d_2 & b_2 & c_2 \\ d_3 & b_3 & c_3 \end{bmatrix}$$

$$\Delta_2 = \begin{bmatrix} a_1 & d_1 & c_1 \\ a_2 & d_2 & c_2 \\ a_3 & d_3 & c_3 \end{bmatrix}, \qquad \Delta_3 = \begin{bmatrix} a_1 & b_1 & d_1 \\ a_2 & b_2 & d_2 \\ a_3 & b_3 & d_3 \end{bmatrix}$$

Then solution of equations (1) is given by Cramer's rule as

$$x = \frac{\Delta_1}{\Delta}, y = \frac{\Delta_2}{\Delta}, z = \frac{\Delta_3}{\Delta}$$

provided $\Delta \neq 0$. If $\Delta = 0$, then Cramer's rule fails. It should be noted that this rule is efficient when we have lesser number of equations.

6.8.1.1 Drawback of The Method

Though the method involves very simple calculation, large amount of computation is required to evaluate large order determinants. For a 10 × 10 system, the Cramer's rule requires:

$\alpha (n + 1)!$ – Number of multiplications $(1 \leq \alpha \leq e{-}1 \cong 2)$

$(2 \times 11)! \cong 70{,}000{,}000$ (approx.) numbers of multiplications

SOLVED PROBLEMS

Problem 1: Use Cramer's rule to solve the equations

$$\begin{aligned} 3x + y + 2z &= 3 \\ 2x - 3y - z &= -3 \\ x - 2y + z &= 4 \end{aligned}$$

Solution:

$$\Delta = \begin{bmatrix} 3 & 1 & 2 \\ 2 & -3 & -1 \\ 1 & -2 & 1 \end{bmatrix} = 8$$

$$\Delta_1 = \begin{bmatrix} 3 & 1 & 2 \\ -3 & -3 & -1 \\ 1 & -2 & 1 \end{bmatrix} = 8$$

$$\Delta_2 = \begin{bmatrix} 3 & 1 & 2 \\ 2 & -3 & -1 \\ 4 & 4 & 1 \end{bmatrix} = 16$$

$$\Delta_3 = \begin{bmatrix} 3 & 1 & 3 \\ 2 & -3 & -3 \\ 1 & -2 & 4 \end{bmatrix} = -8$$

∴ By Cramer's rule, we have

$$x = \frac{\Delta_1}{\Delta} = \frac{8}{8} = 1,$$

$$y = \frac{\Delta_2}{\Delta} = \frac{16}{8} = 2,$$

$$z = \frac{\Delta_3}{\Delta} = \frac{-8}{8} = -1.$$

Hence, the required solution is $x = 1, y = 2, z = -1$.

6.8.2 Matrix Inversion Method

The purpose of this method is to compute the inverse of a matrix. The inverse is the matrix that when multiplied by the original matrix yields the identity matrix (i.e., AA'=I where I has one's on the diagonal and zero's everywhere else).

In practice, the matrix inverse, the determinant, and the solution of a system of a linear equations are all closely related. Each of these can be computed from the LU decomposition of a matrix.

Consider the equations

$$\begin{array}{l} a_{11}x_1 + a_{12}x_2 + \ldots\ldots + a_{1n}x_n = b_1 \\ a_{21}x_1 + a_{22}x_2 + \ldots\ldots + a_{2n}x_n = b_2 \\ a_{31}x_1 + a_{32}x_2 + \ldots\ldots + a_{3n}x_n = b_3 \\ \ldots\ldots\ldots\ldots\ldots\ldots\ldots\ldots\ldots\ldots \\ \ldots\ldots\ldots\ldots\ldots\ldots\ldots\ldots\ldots\ldots \\ a_{n1}x_1 + a_{n2}x_2 + \ldots\ldots + a_{nn}x_n = b_n \end{array} \quad \ldots(1)$$

where a_{ij} and b_{ij} are known constants and x_i's are unknowns.

The system (1) is equivalent to

$$AX = B$$

$$\Rightarrow \quad A^{-1}AX = A^{-1}B$$

$$\Rightarrow \quad 1X = A^{-1}B$$

$$\Rightarrow \quad X = A^{-1}B$$

Now, in matrix form

$$\begin{bmatrix} x_1 \\ x_2 \\ x_3 \\ . \\ . \\ . \\ x_n \end{bmatrix} = \frac{1}{|A|} \begin{bmatrix} c_{11} & c_{21} & \ldots & c_{n1} \\ c_{12} & c_{22} & \ldots & c_{n2} \\ c_{13} & c_{23} & \ldots & c_{n3} \\ \ldots & \ldots & \ldots & \ldots \\ \ldots & \ldots & \ldots & \ldots \\ \ldots & \ldots & \ldots & \ldots \\ c_{1n} & c_{2n} & \ldots & c_{nn} \end{bmatrix} \begin{bmatrix} b_1 \\ b_2 \\ b_3 \\ . \\ . \\ . \\ b_n \end{bmatrix} \quad \ldots(2)$$

where c_{ij} are the cofactors of a_{ij} in A and $|A|$ is the determinant of A. This method fails when A is a singular matrix i.e. $|A| = 0$. This method is also not suitable for larger number of equations because evaluation of A^{-1} by cofactors becomes very cumbersome. We shall now discuss some methods, which can be applied to any number of equations.

SOLVED PROBLEMS

Problem 1: Use Matrix inversion method to solve the equations

$$3x + y + 2z = 3$$
$$2x - 3y - z = -3$$
$$x - 2y + z = 4$$

Solution:

$$A = \begin{vmatrix} \overset{+}{3} & \overset{-}{1} & \overset{+}{2} \\ 2 & -3 & -1 \\ 1 & -2 & 1 \end{vmatrix}$$

$$|A| = 3(-3-2) - 1(2+1) + 2(-4+3)$$
$$= -15 - 3 - 2$$
$$= -20 \neq 0$$

The cofactor of a_{11} in $|A|$ is $c_{11} = (-1)^{1+1} \begin{vmatrix} -3 & -1 \\ -2 & 1 \end{vmatrix} = (-3-2) = -5$

The cofactor of a_{12} in $|A|$ is $c_{12} = (-1)^{1+2} \begin{vmatrix} 2 & -1 \\ 1 & 1 \end{vmatrix} = (-1)(2+1) = -3$

The cofactor of a_{13} in $|A|$ is $c_{13} = (-1)^{1+3} \begin{vmatrix} 2 & -3 \\ 1 & -2 \end{vmatrix} = (-4+3) = -1$

The cofactor of a_{21} in $|A|$ is $c_{21} = (-1)^{2+1} \begin{vmatrix} 1 & 2 \\ -2 & 1 \end{vmatrix} = (-1)(1+4) = -5$

The cofactor of a_{22} in $|A|$ is $c_{22} = (-1)^{2+2} \begin{vmatrix} 3 & 2 \\ 1 & 1 \end{vmatrix} = (3-2) = 1$

The cofactor of a_{23} in $|A|$ is $c_{23} = (-1)^{2+3} \begin{vmatrix} 3 & 1 \\ 1 & -2 \end{vmatrix} = (-1)(-6-1) = 7$

The cofactor of a_{31} in $|A|$ is $c_{31} = (-1)^{3+1} \begin{vmatrix} 1 & 2 \\ -3 & -1 \end{vmatrix} = (-1+6) = 5$

The cofactor of a_{32} in $|A|$ is $c_{32} = (-1)^{3+2} \begin{vmatrix} 3 & 2 \\ 2 & -1 \end{vmatrix} = (-1)(-3-4) = 7$

The cofactor of a_{33} in $|A|$ is $c_{33} = (-1)^{3+3} \begin{vmatrix} 3 & 1 \\ 2 & -3 \end{vmatrix} = (-9-2) = -11$

Now, adjoint of $A = \text{Transpose of} \begin{vmatrix} -5 & -3 & -1 \\ -5 & 1 & 7 \\ 5 & 7 & -11 \end{vmatrix}$

$$\text{Adj. } A = \begin{vmatrix} -5 & -5 & 5 \\ -3 & 1 & 7 \\ -1 & 7 & -11 \end{vmatrix}$$

The inverse of matrix A is given by, $A^{-1} = \dfrac{\text{Adj } A}{|A|}$

$$\therefore \qquad A^{-1} = \frac{-1}{20}\begin{vmatrix} -5 & -5 & 5 \\ -3 & 1 & 7 \\ -1 & 7 & -11 \end{vmatrix}$$

$$= \begin{vmatrix} 1/4 & 1/4 & -1/4 \\ 3/20 & -1/20 & -7/20 \\ 1/20 & -7/20 & 11/20 \end{vmatrix}$$

6.8.3 Gauss Elimination Method

In linear algebra, Gaussian elimination is an efficient algorithm for solving systems of linear equations, finding the rank of a matrix, and calculating the inverse of an invertible square matrix. Gaussian elimination is named after German mathematician and scientist Carl Friedrich Gauss.

This is a direct method that is divided into two parts:

(i) Triangularization (or Elimination)

(ii) Back substitution

In case of triangularization we make an upper triangular system, which then solved through back substitution process.

In order to explain the main concept of the Gauss elimination method, let us consider the following system of three equations in three unknowns:

Consider the equations

$$\begin{aligned} a_{11}x_1 + a_{12}x_2 + \ldots\ldots + a_{1n}x_n &= b_1 \\ a_{21}x_1 + a_{22}x_2 + \ldots\ldots + a_{2n}x_n &= b_2 \\ a_{31}x_1 + a_{32}x_2 + \ldots\ldots + a_{3n}x_n &= b_3 \\ \ldots\ldots\ldots\ldots\ldots\ldots\ldots\ldots\ldots\ldots\ldots\ldots \\ \ldots\ldots\ldots\ldots\ldots\ldots\ldots\ldots\ldots\ldots\ldots\ldots \\ a_{n1}x_1 + a_{n2}x_2 + \ldots\ldots + a_{nn}x_n &= b_n \end{aligned} \qquad \ldots(1)$$

where a_{ij} and b_i are known constants and x_i's are unknowns.

The system (1) is equivalent to

$$AX = B \qquad ...(2)$$

Where A is the matrix formed by coefficients a_{ij}, X is the matrix of unknowns and B is the matrix of b_i elements.

Now we have to reduce the augmented matrix $[A, B]$ to upper triangular matrix.

$$[A, B] = \begin{bmatrix} a_{11} & a_{12} & a_{13} & .... & a_{1n} & b_1 \\ a_{21} & a_{22} & a_{23} & .... & a_{2n} & b_2 \\ .... & .... & .... & .... & .... & .... \\ .... & .... & .... & .... & .... & .... \\ a_{n1} & a_{n2} & a_{n3} & .... & a_{nn} & b_n \end{bmatrix} \qquad ...(3)$$

Now, multiply the first row of (3) (if $a_{11} \neq 0$) by a_{i1}/a_{11} and subtract from the ith row of $[A, B]$, where $\boldsymbol{i = 2,3...n}$. By this, all elements in the first column of $[A, B]$ except a_{11} (pivot element) are made zero. Now (3) becomes

$$\begin{bmatrix} a_{11} & a_{12} & a_{13} & .... & a_{1n} & b_1 \\ 0 & a_{22} & a_{23} & .... & a_{2n} & b_2 \\ .... & .... & .... & .... & .... & .... \\ .... & .... & .... & .... & .... & .... \\ 0 & a_{n2} & a_{n3} & .... & a_{nn} & b_n \end{bmatrix} \qquad ...(4)$$

Now, by considering a_{22} as the pivot, we will make all the elements below a_{22} in the second column of (4) to zeros. That is, multiply second row of (4) by a_{i2}/a_{22} and subtract from the corresponding elements of the ith row **(i = 3,4...n)**. Now all the elements below a_{22} are reduced to zero. Same concept is applied in the next derived matrices. Now (4) reduces to

$$\begin{bmatrix} a_{11} & a_{12} & a_{13} & .... & a_{1n} & b_1 \\ 0 & a_{22} & a_{23} & .... & a_{2n} & b_2 \\ .... & .... & a_{33} & .... & a_{2n} & b_3 \\ .... & .... & .... & .... & .... & .... \\ 0 & a_{n2} & a_{n3} & .... & a_{nn} & b_n \end{bmatrix} \qquad ...(5)$$

Now taking a_{33} as the pivot, using elementary operations, we make all elements below a_{33} as zeros. Continuing the process, till all the elements below the leading diagonal of A are made to zero. This process is known as **triangularization.**

To implement on computer, we take the augmented matrix after elimination as So $[A, B]$ becomes:

$$\begin{bmatrix} a_{11} & a_{12} & a_{13} & .... & a_{1n} & a_1(n+1) \\ 0 & a_{22} & a_{23} & .... & a_{2n} & a_2(n+1) \\ .... & .... & a_{33} & .... & a_{2n} & a_3(n+1) \\ .... & .... & .... & .... & .... & .... \\ 0 & 0 & 0 & .... & a_{nn} & a_n(n+1) \end{bmatrix} \quad ...(6)$$

From (6), the given system of linear equations is equivalent to

$$a_{11}x_1 + a_{12}x_2 + ...+a_{1n}x_n = a_{1(n+1)}$$
$$a_{22}x_2 + ...+a_{2n}x_n = a_{2(n+1)}$$
$$.......................................$$
$$a_{nn}x_n = a_{n(n+1)}$$

Starting from the last equation, we solve for $x_n = a_{n(n+1)} / a_{nn}$. Using this in the previous equation, we get x_{n-1} and so. This is known as back substitution process. Which gives us $x_n, x_{n-2}, x_{n-3}, ...x_1$.

The general formula for back substitution process is derived as

$$x_i = \frac{\left[a_{j(n+1)} - \sum_{j=i+1}^{n} a_{ij}x_j\right]}{a_{ij}} \quad \text{for } i = n-1, n-2, ..., 2, 1.$$

6.8.3.1 Algorithm for Gauss Elimination Method

Suppose A is the augmented matrix of order $n \times (n+1)$ i.e. has n rows and $(n+1)$ columns (including b_i's element column) and matrix A is stored by using a two dimensional array 'a' having size $n \times (n+1)$. Let 'x' be a one dimensional array having size n that stores the solution values of given set of equations.

1. Read n
2. for i =1 to n in steps of 1 do
3. for j = 1 to (n+1) in steps of 1 do
4. read a_{ij}
5. endfor
6. endfor // This part reads the augmented matrix
7. for k = 1 to n-1 in steps of 1 do
8. for i = (k+1) to n in steps of 1 do
9. u ⟵ a_{ik}/a_{kk}
10. for j = k to (n+1) in steps of 1 do
11. a_{ij} ⟵ $a_{ij} - u.a_{kj}$

12. endfor
13. endfor
14. endfor // Triangularization process
15. $x_n = a_{n(n+1)}/a_{nn}$
16. for i = (n-1) to 1 in steps of –1 do
17. sum ⟵ 0
18. for j = (i+1) to n steps of 1 do
19. sum = sum + $a_{ij}x_j$
20. endfor
21. $x_i \longleftarrow (a_{i(n+1)} - \text{sum})/a_{ii}$
22. endfor // Back substitution process
23. for i = 1 to n in steps of 1 do
24. write x_i
25. endfor // Print the results
26. End

6.8.3.2 *Flowchart of Gauss Elimination Method*

See figure on next page

6.8.3.3 *Program for Gauss Elimination Method in C*

```
#include<conio.h>
#include<stdio.h>
#include<math.h>
#include<iomanip.h>
void main()
{
int i,j,k,n;
float a[10][10],x[10],u,sum;
clrscr();
printf("Enter total number of equations");
scanf("%d",&n);
printf("Enter elements of augmented matrix rowwise\n");
for(i=0;i<n;i++)
for(j=0;j<n+1;j++)
```

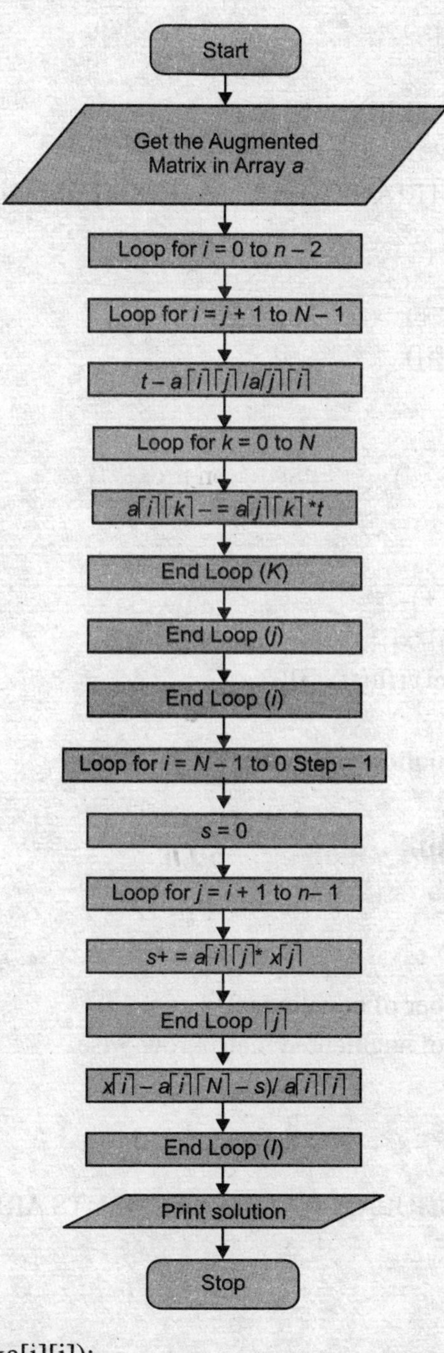


```
scanf("%f",&a[i][j]);
for(k=0;k<n-1;k++)
for(i=k+1;i<n;i++)
{
```

```
u=a[i][k]/a[k][k];
for(j=0;j<n+1;j++)
a[i][j]=a[i][j]-u*a[k][j];
}
printf("UPPER TRIANGULAR MATRIX ELEMENTS ARE AS \n");
for(i=0;i<n;i++)
{
for(j=0;j<n+1;j++)
printf("%f",a[i][j]);
printf("\n");
}
for(i=n-1;i>=0;i--)
{
sum=0;
for(j=i+1;j<n;j++)
sum=sum+a[i][j]*x[j];
x[i]=(a[i][n]-sum)/(float)a[i][i];
}
printf("\nThe solution values are");
for(i=0;i<n;i++)
printf("%f\n",x[i]);
getch();
}
```

Output:

```
Enter total number of equations 3
Enter elements of augmented matrix row-wise
1                    3 -1  5
2                    4 -3  3
2  -3  2  2
UPPER TRIANGULAR MATRIX ELEMENTS ARE AS
2                    3 -1  5
0  -2 -1 -7
0   0  6  18
The solution values are
1
2
3
```

6.8.3.4 *Program for Gauss Elimination Method in C++*

```
#include<iostream.h>
#include<conio.h>
#include<stdio.h>
#include<math.h>
#include<iomanip.h>
void main()
{
int i,j,k,n;
float a[10][10],x[10],u,sum;
clrscr();
cout<<"Enter total number of equations";
cin>>n;
cout<<"Enter elements of augmented matrix rowwise"<<endl;
for(i=0;i<n;i++)
for(j=0;j<n+1;j++)
cin>>a[i][j];
for(k=0;k<n-1;k++)
for(i=k+1;i<n;i++)
{
u=a[i][k]/a[k][k];
for(j=0;j<n+1;j++)
a[i][j]=a[i][j]-u*a[k][j];
}
cout<<"UPPER TRIANGULAR MATRIX ELEMENTS ARE AS"<<endl;
for(i=0;i<n;i++)
{
for(j=0;j<n+1;j++)
cout<<a[i][j]<<endl;
}
for(i=n-1;i>=0;i—)
{
sum=0;
```

```
for(j=i+1;j<n;j++)
sum=sum+a[i][j]*x[j];
x[i]=(a[i][n]-sum)/(float)a[i][i];
}
cout<<"The solution values are"<<endl;
for(i=0;i<n;i++)
cout<<x[i]<<endl;
getch();
}
```

Output:

```
Enter total number of equations 3
Enter elements of augmented matrix row-wise
3           3 -1 5
4           4 -3 3
2 -3 2 2

UPPER TRIANGULAR MATRIX ELEMENTS ARE AS
3           3 -1 5
0 -2 -1 -7
0 0 6 18

The solution values are
1
2
3
```

SOLVED PROBLEMS

Problem 1: Solve the equation by Gauss-Elimination method

$$2x + 3y - z = 5$$
$$4x + 4y - 3z = 3$$
$$2x - 3y + 2z = 2$$

Solution: Write the equations in the form

$$AX = B$$

$$\begin{bmatrix} 2 & 3 & -1 \\ 4 & 4 & -3 \\ 2 & -3 & 2 \end{bmatrix} \begin{bmatrix} x \\ y \\ z \end{bmatrix} = \begin{bmatrix} 5 \\ 3 \\ 2 \end{bmatrix}$$

$\therefore$
$$[A, B] = \begin{bmatrix} 2 & 3 & -1 \\ 4 & 4 & -3 \\ 2 & -3 & 2 \end{bmatrix} \begin{bmatrix} 5 \\ 3 \\ 2 \end{bmatrix}$$

Taking $a_{11} = 2(\neq 0)$ as the pivot, reduce all the elements below that (a_{11} element) to zero. By operating $R_2 \rightarrow R_2 - (a_{21}/a_{11}).R_1$ (for making a_{21} element zero) i.e. $R_2 \rightarrow R_2 - (4/2).R_1$.

$$[A, B] = \begin{bmatrix} 2 & 3 & -1 \\ 0 & -2 & -1 \\ 2 & -3 & 2 \end{bmatrix} \begin{bmatrix} 5 \\ -7 \\ 2 \end{bmatrix}$$

Similarly by operating $R_3 \rightarrow R_3 - (a_{31}/a_{11}).R_1$. (for making a_{31} element zero) i.e. $R_3 \rightarrow R_3 - (2/2).R_1$.

$$[A, B] = \begin{bmatrix} 2 & 3 & -1 \\ 0 & -2 & -1 \\ 0 & -6 & 3 \end{bmatrix} \begin{bmatrix} 5 \\ -7 \\ -3 \end{bmatrix}$$

Taking $a_{22} = -2(\neq 0)$ as the pivot, reduce all the elements below that (a_{22} element) to zero. By operating $R_3 \rightarrow R_3 - (a_{32}/a_{22}).R_2$. (for making a_{32} element zero) i.e. $R_3 \rightarrow R_3 - (6/2).R_2$.

$$[A, B] = \begin{bmatrix} 2 & 3 & -1 \\ 0 & -2 & -1 \\ 0 & 0 & 6 \end{bmatrix} \begin{bmatrix} 5 \\ -7 \\ 18 \end{bmatrix}$$

Hence system of equations becomes

$$2x + 3y - z = 5$$
$$0x - 2y - z = -7$$
$$0x + 0y + 6z = 18$$

By back substitution, we have

$\therefore$
$$x = 1$$
$$y = 2$$
$$z = 3$$

Problem 2: Solve the equation by Gauss-Elimination hod

$$x + y + z + 4w = -6$$
$$x + 7y + z + w = 12$$
$$x + y + 6z + w = -5$$
$$5x + y + z + w = 4$$

Solution: Write the equations in the form

$$AX = B$$

$$\begin{bmatrix} 1 & 1 & 1 & 4 \\ 1 & 7 & 1 & 1 \\ 1 & 1 & 6 & 1 \\ 5 & 1 & 1 & 1 \end{bmatrix} \begin{bmatrix} x \\ y \\ z \\ w \end{bmatrix} = \begin{bmatrix} -6 \\ 12 \\ -5 \\ 4 \end{bmatrix}$$

$$\therefore \qquad [A, B] = \begin{bmatrix} 1 & 1 & & \\ 1 & 7 & & \\ 1 & 1 & & \\ 5 & 1 & & \end{bmatrix} \begin{bmatrix} -6 \\ 12 \\ -5 \\ 4 \end{bmatrix}$$

Taking $a_{11} = 1(\neq 0)$ as the pivot, reduce all the eleme below that (a_{11} element) to zero.

By operating $R_i \to R_i - (a_{i1}/a_{11}).R_1$ where i = , 4

i.e. $R_2 \to R_2 - (1/1).R_1$ (for mak $_{21}$ element zero)

$R_3 \to R_3 - (1/1).R_1$ (for mak $_{31}$ element zero)

$R_4 \to R_4 - (5/1).R_1$ (for mak $_{41}$ element zero)

$$\therefore \qquad [A, B] = \begin{bmatrix} 1 & 1 & & 4 \\ 0 & 6 & & -3 \\ 0 & 0 & & -3 \\ 0 & -4 & & -19 \end{bmatrix} \begin{bmatrix} -6 \\ 18 \\ 1 \\ 34 \end{bmatrix}$$

Taking $a_{22} = 6(\neq 0)$ as the pivot, reduce all the elem below that (a_{22} element) to zero.

By operating $R_i \to R_i - (a_{i2}/a_{22}).R_2$ where i = 4

i.e. $R_3 \to R_3 - (0/6).R_2$ (for mak $_{32}$ element zero)

$R_4 \to R_4 - (-4/6).R_2$ (for mak $_{42}$ element zero)

$$\therefore \qquad [A, B] = \begin{bmatrix} 1 & 1 & & 4 \\ 0 & 6 & & -3 \\ 0 & 0 & & -3 \\ 0 & 0 & & -21 \end{bmatrix} \begin{bmatrix} -6 \\ 18 \\ 1 \\ 46 \end{bmatrix}$$

Taking $a_{33} = 5(\neq 0)$ as the pivot, ... ce all the elements below that (a_{33} element) to zero.

By operating $R_i \to R_i - (a_{i3}/a_{33}$ ... where $i = 4$

i.e. $R_4 \to R_4 - (-4/5)$.. (for making a_{43} element zero)

$$\therefore \quad [A, B] = \begin{bmatrix} 1 & 1 & 1 & 4 \\ 0 & 6 & 0 & -3 \\ 0 & 0 & 5 & -3 \\ 0 & 0 & 0 & -117/5 \end{bmatrix} \begin{bmatrix} -6 \\ 18 \\ 1 \\ 234/5 \end{bmatrix}$$

Hence system of equations beco...

$$x + \ldots + 4w = -6$$

$$0x + 6y \ldots - 3w = 18$$

$$0x + 0y \ldots - 3w = 1$$

$$0x + 0y + 0z - \ldots 7/5)w = 234/5$$

By back substitution, we have

$$w = -2$$

$$z = -1$$

$$y = 2$$

$$x = 1$$

Problem 3: Solve the equation by ... ss-Elimination method

$$3 \ldots y + z = 3$$

$$\ldots y + z = 0$$

$$6\ldots + 4z = 6$$

and also comment on the nature ... lution.

Solution: Write the equations in th... m

$$AX = B$$

$$\begin{bmatrix} 3 & 2 & \ldots \\ 2 & 1 & \ldots \\ 6 & 2 & \ldots \end{bmatrix} \begin{bmatrix} x \\ y \\ z \end{bmatrix} = \begin{bmatrix} 3 \\ 0 \\ 6 \end{bmatrix}$$

$$\therefore \quad [A, B] = \begin{bmatrix} 3 & 2 & 1 \\ 2 & 1 & 1 \\ 6 & 2 & 4 \end{bmatrix} \begin{bmatrix} 3 \\ 0 \\ 6 \end{bmatrix}$$

Taking $a_{11} = 3(\neq 0)$ as the pivot, ... ice all the elements below that (a_{11} element) to zero.

By operating $R_2 \to R_2 - (a_{21}/a$ ... (for making a_{21} element zero)

i.e. $R_2 \to R_2 - (2/3)$..

$$[A, B] = \begin{bmatrix} 3 & 2 & 1 \\ 0 & -1/3 & 1/3 \\ 6 & 2 & 4 \end{bmatrix} \begin{bmatrix} 3 \\ -2 \\ 6 \end{bmatrix}$$

By operating $R_3 \to R_3 - (a_{31}/a_{11}).R_1$ (for making a_{31} element zero)
i.e. $R_3 \to R_3 - (6/3).R_1$

$$[A, B] = \begin{bmatrix} 3 & 2 & 1 \\ 0 & -1/3 & 1/3 \\ 0 & -2 & 2 \end{bmatrix} \begin{bmatrix} 3 \\ -2 \\ 0 \end{bmatrix}$$

Taking $a_{22} = -1/3 (\neq 0)$ as the pivot, reduce all the elements below that (a_{22} element) to zero.
By operating $R_3 \to R_3 - (a_{32}/a_{22}).R_2$ (for making a_{32} element zero)
i.e. $R_3 \to R_3 - (2/(1/3)).R_2$

$$[A, B] = \begin{bmatrix} 3 & 2 & 1 \\ 0 & -1/3 & 1/3 \\ 0 & 0 & 0 \end{bmatrix} \begin{bmatrix} 3 \\ -2 \\ 12 \end{bmatrix}$$

Hence system of equations becomes

$$3x + 2y + z = 3$$
$$0x - (1/3)y + (1/3)z = -2$$
$$0x + 0y + 0z = 12$$

Note: As last equation has no variable that we should use for back substitution, so the given system of equations has no solution, so given system of equations is inconsistent.

UNSOLVED PROBLEMS

1. Apply the Gauss Elimination method and comment on the type of solution.

$$16x + 22y + 4z = -2$$
$$4x - 3y + 2z = 9$$
$$12x + 25y + 2z = -11$$

2. Solve the following system of equations using the Gauss Elimination method:

$$4x_1 + x_2 + 3x_3 = 17$$
$$x_1 + 5x_2 + x_3 = 14$$
$$2x_1 - x_2 + 8x_3 = 12$$

3. Solve the following system of equations using the Gauss Elimination method:

$$20x + y - 2z = 17$$
$$3x + 20y - z = -18$$
$$2x - 3y + 20z = 25$$

4. Solve the following system of equations using the Gauss Elimination method:

$$x + y + z = 6$$
$$3x + 3y + 4z = 20$$
$$2x + y + z = 13$$

5. Solve the following system of equations using the Gauss Elimination method:

$$x + y - z = 2$$
$$2x + 3y + 5z = -3$$
$$3x + 2y - 3z = 6$$

6.8.4 Concept of Pivoting

We have some pivoting techniques that are helpful for us to reduce, as much as possible, the round-off errors, and thus avoid the loss of significant digits due to division by a small number.

For this reason, pivoting techniques play an important role in matrix computations. By using pivoting techniques, we can increase the accuracy in the final value of the determinant. We use pivoting in the cases where the diagonal elements are zero or very small.

PIVOT: A pivot is the largest magnitude element in a column or a row or the principal diagonal of the matrix. The largest element of the whole matrix is known as the pivot of the matrix. For example, consider the matrix

$$A = \begin{bmatrix} 5 & 3 & -8 & 17 \\ -21 & 15 & -4 & 11 \\ 9 & -45 & 13 & 12 \\ 12 & 55 & 22 & 42 \end{bmatrix}$$

Here, -21 is the pivot for the first column; 42 is the pivot for the principal diagonal, 55 is the pivot for matrix *A*.

Now, we will discuss the two most important techniques of pivoting, which are as:

(i) Partial pivoting technique

(ii) Complete pivoting technique

6.8.4.1 *Partial Pivoting Technique*

In partial pivoting, we have to locate the largest magnitude element of first column (formed by coefficients of first variable) and brought as the first pivot by interchanging the first equation with the equation that contains the largest element in magnitude. In second step, we have to locate the largest magnitude element of second column (formed by coefficients of second variable) and brought as the second pivot by interchanging the second equation with the equation that contains the largest element in magnitude.

This process will go on till we have left with a single variable in the equation. This technique is known as partial pivoting technique. Similarly, we can consider a row, and interchange the columns to achieve partial pivoting. So, in simple words we can say that in partial pivoting case we have to choose the column and corresponding to that we have to interchanging the rows to make the largest element on the diagonal position.

6.8.4.2 Complete Pivoting Technique

Complete pivoting technique is different in this we have to choose a leading or trailing principal sub matrix of order i(i varies from 2 to n) then choose from it the pivot, say, a_{pq}, and bring it to the meeting point of the first row and first column of the sub matrix by interchanging both rows and columns.

From both techniques complete pivoting technique is hard to implement and we mostly use partial pivoting technique for solving matrices.

Now, we will discuss both types of pivoting techniques with examples.

6.8.4.3 Advantages and Disadvantages of Partial and Complete Pivoting

First disadvantage of both the technique is that the symmetry or regularity in the matrix is lost. Partial pivoting has an advantage over complete pivoting is that partial pivoting requires less labor in terms of interchanges and search for the pivot than complete pivoting. If the partial pivot turns out to be small in comparison with the elements of matrix, it is better to resort to complete pivoting to avoid rounding errors. Hence, a combination of partial and complete pivoting is expected to be very effective.

SOLVED PROBLEMS

Problem 1: Compute the solution of the following equations by using Gauss elimination method with partial pivoting:

$$\begin{aligned} 5x_1 + x_2 + x_3 + x_4 &= 4 \\ x_1 + 7x_2 + x_3 + x_4 &= 12 \\ x_1 + x_2 + 6x_3 + x_4 &= -5 \\ x_1 + x_2 + x_3 + 4x_4 &= -6 \end{aligned}$$

Solution: The equations in matrix form

$$A = \left[\begin{array}{cccc|c} 5 & 1 & 1 & 1 & 4 \\ 1 & 7 & 1 & 1 & 12 \\ 1 & 1 & 6 & 1 & -5 \\ 1 & 1 & 1 & 4 & -6 \end{array}\right]$$

As the largest magnitude element of first column is 5 (pivot in first column) which is already in its proper place (a_{11}), so no need to interchange the rows here.

Subtracting 1/5 times the first row from the second row, 1/5 times the first row from the third row and 1/5 times the first row from the fourth row, we get

$$A = \left[\begin{array}{cccc|c} 5 & 1 & 1 & 1 & 4 \\ 0 & 34/5 & 4/5 & 4/5 & 56/5 \\ 0 & 4/5 & 29/5 & 4/5 & -29/5 \\ 0 & 4/5 & 19/5 & 19/5 & -34/5 \end{array}\right]$$

As 34/5 is the largest element in magnitude for second column which is also in its proper place (a_{22}), so no need to interchange the rows here.

Subtracting (4/5)/(34/5) times the second row from the third row, (4/5)/(34/5) times the second row from the fourth row, we get

$$A = \left[\begin{array}{cccc|c} 5 & 1 & 1 & 1 & 4 \\ 0 & 34/5 & 4/5 & 4/5 & 56/5 \\ 0 & 0 & 97/17 & 12/17 & -121/17 \\ 0 & 0 & 12/17 & 63/17 & -138/17 \end{array}\right]$$

As 97/17 is the largest element in magnitude for third column which is also in its proper place (a_{33}), so no need to interchange the rows here.

Subtracting (12/17)/(97/17) times the third row from the fourth row, we get

$$A = \left[\begin{array}{cccc|c} 5 & 1 & 1 & 1 & 4 \\ 0 & 34/5 & 4/5 & 4/5 & 56/5 \\ 0 & 0 & 97/17 & 12/17 & -121/17 \\ 0 & 0 & 0 & 351/97 & -702/97 \end{array}\right]$$

So, the above matrix in equation form can be written as

$$5x_1 + x_2 + x_3 + x_4 = 4$$
$$0x_1 + (34/5)x_2 + (4/5)x_3 + (4/5)x_4 = 56/5$$
$$0x_1 + 0x_2 + (97/17)x_3 + (12/17)x_4 = -121/17$$
$$0x_1 + 0x_2 + x_3 + (351/97)x_4 = -702/97$$

By applying back substitution, we have

$$x_1 = 11/17$$
$$x_2 = 30/17$$
$$x_3 = 1$$
$$x_4 = -2$$

Problem 2: Compute the solution of the following equations by using Gauss elimination method with partial pivoting

$$x + y + z = 1$$
$$4x + 3y - z = 6$$
$$3x + 5y + 3z = 4$$

Solution: The equations in matrix form

$$\therefore \quad A = \left[\begin{array}{ccc|c} 1 & 1 & 1 & 1 \\ 4 & 3 & -1 & 6 \\ 3 & 5 & 3 & 4 \end{array}\right]$$

As the largest magnitude element of first column is 4 (pivot in first column) which is not in its proper place (a_{11}), so need to interchange the rows here.

Interchange the first and second row elements, we have

$$\therefore \quad A = \left[\begin{array}{ccc|c} 4 & 3 & -1 & 6 \\ 1 & 1 & 1 & 1 \\ 3 & 5 & 3 & 4 \end{array}\right]$$

Now the pivot element 4(for first column) is in its proper place (a_{11}).

Subtracting 1/4 times the first row from the second row, 3/4 times the first row from the third row, we get

$$\therefore \quad A = \left[\begin{array}{ccc|c} 4 & 3 & -1 & 6 \\ 0 & 1/4 & 5/4 & -1/2 \\ 0 & 11/4 & 15/4 & -1/2 \end{array}\right]$$

As 11/4 is the largest element in magnitude for second column which is not in its proper place (a_{22}), so there is need to interchange the rows here.

Interchanging the second and third row, we have

$$\therefore \quad A = \left[\begin{array}{ccc|c} 4 & 3 & -1 & 6 \\ 0 & 11/4 & 15/4 & -1/2 \\ 0 & 1/4 & 5/4 & -1/2 \end{array}\right]$$

Now 11/4(a_{22} element) is at its proper place.

Subtracting (1/4)/(11/4) times the second row from the third row, we get

$$\therefore \quad A = \left[\begin{array}{ccc|c} 4 & 3 & -1 & 6 \\ 0 & 11/4 & 15/4 & -1/2 \\ 0 & 0 & 10/11 & -5/11 \end{array}\right]$$

So, the above matrix in equation form can be written as

$$4x + 3y - z = 6$$
$$0x + (11/4)y + (15/4)z = -1/2$$
$$0x + 0y + (10/11)z = -5/11$$

By applying back substitution, we have

$$x = 1$$
$$y = 1/2$$
$$z = -1/2$$

Note: The question arises in the second step why we didn't choose 3 as the pivot element for second column the answer is that as we are solving the equations by using Gauss elimination method and in the first phase of the method we have to perform Triangularization if we choose 3 as the pivot we have to interchange the first equation with any other equation which disturbs the Triangularization process.

6

UNSOLVED PROBLEMS

1. What do you mean by pivoting? How it is helpful in triangularization process?
2. Explain both types of pivoting and also its disadvantages and advantages.
3. Compute the determinant of the following matrix by using triangularization algorithm for Complete and partial pivoting.

$$\begin{bmatrix} 1 & -2 & 3 \\ 0 & -1 & 4 \\ -2 & 2 & 0 \end{bmatrix}$$

6.8.5 Gauss Jordan Method

This method is a modification of the above Gauss elimination method. In this, when an unknown is eliminated, it is eliminated form all other equations not only from equations to follow. In this method, each equation is left with only one unknown, which eliminates the use of back substitution.

In this method, the coefficient matrix A of the system $AX = B$ is brought to a diagonal matrix or unit matrix by making the matrix A not only upper triangular but also lower triangular by making all the elements above the leading diagonal of A also as zeros.

The method works as:

In order to explain the essence of the Gauss–Jordan method, let us reconsider the system of linear equations

Consider the equations

$$a_{11}x_1 + a_{12}x_2 + \ldots\ldots + a_{1n}x_n = b_1$$
$$a_{21}x_1 + a_{22}x_2 + \ldots\ldots + a_{2n}x_n = b_2$$

$$a_{31}x_1 + a_{32}x_2 + \ldots\ldots + a_{3n}x_n = b_3$$
$$\ldots\ldots\ldots\ldots\ldots\ldots\ldots\ldots\ldots\ldots \quad \ldots(1)$$
$$\ldots\ldots\ldots\ldots\ldots\ldots\ldots\ldots\ldots\ldots$$
$$a_{n1}x_1 + a_{n2}x_2 + \ldots\ldots + a_{nn}x_n = b_n$$

where a_{ij} and b_i are known constants and x_i's are unknowns.

The system (1) is equivalent to

$$AX = B \quad \ldots(2)$$

Where A is the matrix formed by coefficients a_{ij}, X is the matrix of unknowns and B is the matrix of b_i elements.

Now we have to reduce the augmented matrix $[A, B]$ into upper and lower triangular matrix.

$$[A, B] = \left[\begin{array}{cccccc|c} a_{11} & a_{12} & a_{13} & \ldots & \ldots & a_{1n} & b_1 \\ a_{12} & a_{22} & a_{23} & \ldots & \ldots & a_{2n} & b_2 \\ \ldots & \ldots & \ldots & \ldots & \ldots & \ldots & \ldots \\ \ldots & \ldots & \ldots & \ldots & \ldots & \ldots & \ldots \\ a_{n1} & a_{n2} & a_{n3} & \ldots & \ldots & a_{nn} & b_n \end{array}\right] \quad \ldots(3)$$

Now, multiply the first row of (3) (if $a_{11} \neq 0$) by a_{i1}/a_{11} and subtract from the ith row of $[A, B]$, where $\boldsymbol{i = 2, 3 \ldots n}$. By this, all elements in the first column of $[A, B]$ except a_{11} (pivot element) are made zero. Now (3) becomes

$$\left[\begin{array}{ccccc|c} a_{11} & a_{12} & a_{13} & \ldots & a_{1n} & b_1 \\ 0 & a_{22} & a_{23} & \ldots & a_{2n} & b_2 \\ \ldots & \ldots & \ldots & \ldots & \ldots & \ldots \\ \ldots & \ldots & \ldots & \ldots & \ldots & \ldots \\ 0 & a_{n2} & a_{n3} & \ldots & a_{nn} & b_n \end{array}\right] \quad \ldots(4)$$

Now, by considering a_{22} as the pivot, we will make all the elements below and above a_{22} (element) in the second column of (4) to zeros. That is, multiply second row of (4) by a_{i2}/a_{22} and subtract from the corresponding elements of the ith row $\boldsymbol{(i=1,3,4 \ldots n)}$. Now all the elements below and above a_{22} are reduced to zero. Same concept is applied in the next derived matrices. Now (4) reduces to

$$\left[\begin{array}{ccccc|c} a_{11} & 0 & a_{13} & \ldots & a_{1n} & b_1 \\ 0 & a_{22} & a_{23} & \ldots & a_{2n} & b_2 \\ \ldots & \ldots & a_{33} & \ldots & \ldots & b_3 \\ \ldots & \ldots & \ldots & \ldots & \ldots & \ldots \\ 0 & 0 & a_{n3} & \ldots & a_{nn} & b_n \end{array}\right] \quad \ldots(5)$$

Now taking a_{33} as the pivot, using elementary operations, we make all elements below and above a_{33} to zeros. Continuing the process, till all the elements below and above the leading diagonal of A are made to zero i.e. we have to make the **unit matrix** or **diagonal matrix**. This process is known as **lower and upper triangularization.**

So this process will continue till x_n is eliminated from the first to $(n\text{-}1)^{th}$ equation. The final matrix looks like as:

$$\left[\begin{array}{ccccc|c} a_{11} & 0 & 0 & \dots & 0 & b_1 \\ 0 & a_{22} & 0 & \dots & 0 & b_2 \\ \dots & \dots & a_{33} & \dots & 0 & b_3 \\ \dots & \dots & \dots & \dots & \dots & \dots \\ 0 & 0 & 0 & \dots & a_{nn} & b_n \end{array}\right] \quad \dots(6)$$

Note: All the diagonal elements will come out to be unity.

Now values of the unknowns are given by

$x_1 = b_1,\ x_2 = b_2,\ x_3 = b_3,\ \dots\dots\dots\dots,\ x_n = b_n$

6.8.5.1 Algorithm for Gauss Jordan Method

Suppose A is the augmented matrix of order $n \times (n+1)$ that has n rows and $(n+1)$ columns (including b_i's element column) and matrix A is stored by using a two dimensional array 'a' having size $n \times (n+1)$. Let 'x' be a one dimensional array having size n that stores the solution values of given set of equations. It should me noted that no element in the leading diagonal comes out to be zero.

1. Read n
2. for i =1 to n in steps of 1 do
3. for j = 1 to (n+1) in steps of 1 do
4. Read a_{ij}
5. endfor
6. endfor // This part reads the augmented matrix
7. for k =1 to n in steps of 1 do
8. u ← a_{kk}
9. for j = k to (n+1) in steps of 1 do
10. a_{kj} ← a_{kj}/u
11. endfor
12. for i = 1 to n in steps of 1 do
13. if (i ≠ k) then
14. c ← a_{ik}
15. for j = k to (n+1) in steps of 1 do
16. $a_{ij} = a_{ij} - c.a_{kj}$
17. endfor
18. endif

19. endfor
20. endfor // This part results unit matrix
21. for i = 1 to n in steps of 1 do
22. Write : $a_{i(n+1)}$
23. endfor // Print the results
24. End

6.8.5.2 Flowchart of for Gauss Jordan Method

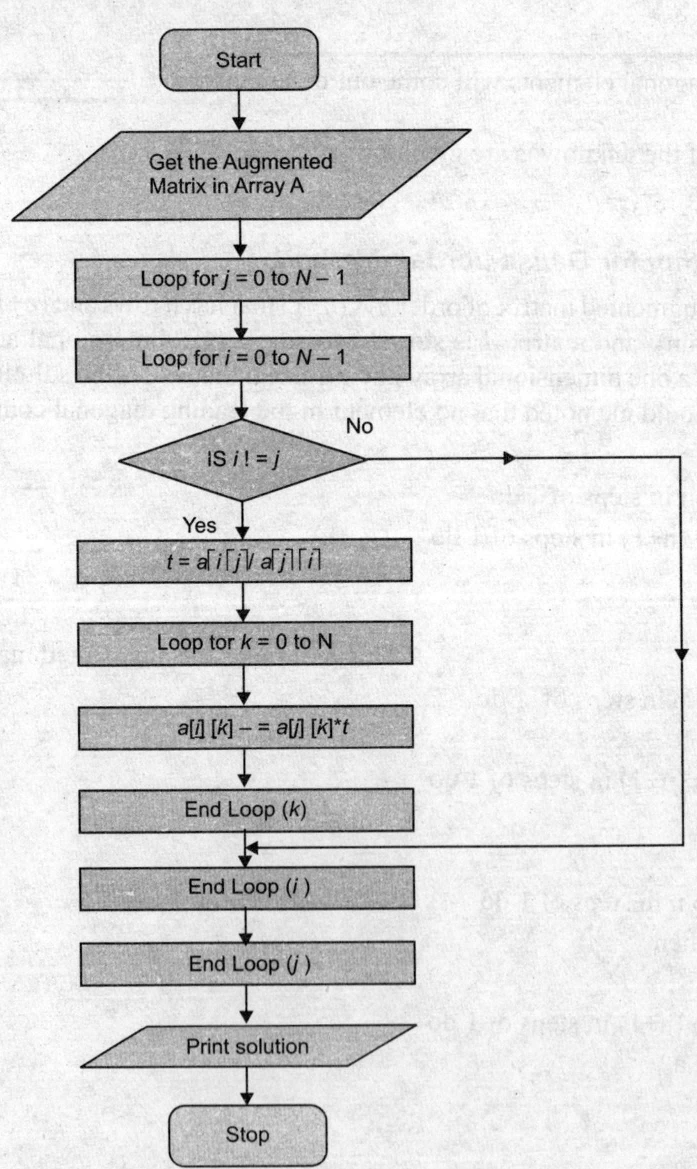

6.8.5.3 Program for Gauss Jordan Method in C

```
#include<conio.h>
#include<stdio.h>
#include<math.h>
#include<iomanip.h>
void main()
{
int i,j,k,n;
float a[10][10],x[10],u,c;
clrscr();
printf("Enter total number of equations");
scanf("%d",&n);
printf("Enter elements of augmented matrix rowwise\n");
for(i=0;i<n;i++)
for(j=0;j<n+1;j++)
scanf("%f",&a[i][j]);
for(k=0;k<n;k++)
{
if(a[k][k]==0)
{
printf("Solution not possible as division by zero");
break;
}
u=a[k][k];
for(j=k;j<n+1;j++)
a[k][j]=a[k][j]/u;
for(i=0;i<n;i++)
{
if(i!=k)
{
c=a[i][k];
for(j=k;j<n+1;j++)
a[i][j]=a[i][j]-c*a[k][j];
}}}
```

```
printf("Resultant values in sequence are\n");
for(i=0;i<n;i++)
{
x[i]=a[i][n];
printf("%f\n",x[i]);
}
getch();
}
```

Output:

```
Enter total number of equations 3
Enter elements of augmented matrix rowwise
1 1 1 1
4 3 -1 6
1 5 3 4
Resultant values in sequence are
0.555556
1.055556
-0.555556
```

6.8.5.4 *Program for Gauss Jordan Method in C++*

```
#include<iostream.h>
#include<conio.h>
#include<stdio.h>
#include<math.h>
#include<iomanip.h>
void main()
{
int i,j,k,n;
float a[10][10],x[10],u,c;
clrscr();
cout<<"Enter total number of equations";
cin>>n;
cout<<"Enter elements of augmented matrix rowwise"<<endl;
for(i=0;i<n;i++)
```

```
for(j=0;j<n+1;j++)
cin>>a[i][j];
for(k=0;k<n;k++)
{
if(a[k][k]==0)
{
cout<<"Solution not possible as division by zero";
break;
}
u=a[k][k];
for(j=k;j<n+1;j++)
a[k][j]=a[k][j]/u;
for(i=0;i<n;i++)
{
if(i!=k)
{
c=a[i][k];
for(j=k;j<n+1;j++)
a[i][j]=a[i][j]-c*a[k][j];
}}}
cout<<"Resultant values in sequence are"<<endl;
for(i=0;i<n;i++)
{
x[i]=a[i][n];
cout<<x[i]<<endl;
}
getch();
}
```

Output:

```
Enter total number of equations 3
Enter elements of augmented matrix rowwise
1 1 1 1
4 3 -1 6
1 5 3 4
```

Resultant values in sequence are

0.555556

1.055556

-0.555556

6.8.5.5 Comparison Between Gauss Elimination and Gauss Jordan Method

For an nth order matrix the total number of multiplications required is given by:

In Gauss elimination method → $n^3/3 + O(n^2)$

In Gauss Jordon method → $n^3/2 + O(n^2)$

For example if n = 10, the number of multiplication required for Gauss-jordon method is approximately 500 whereas for Gauss elimination we need only 333 multiplications

Hence, we note that though Gauss-jordon method seems to be easier but requires about 50 percent more operations that the Gauss elimination method.

Therefore, for large n, the Gauss elimination method is preferred.

SOLVED PROBLEMS

Problem 1: Solve the system of equations by Gauss Jordon method:

$$x + y + z + w = 2$$
$$2x - y + 2z - w = -5$$
$$3x + 2y + 3z + 4w = 7$$
$$x - 2y - 3z + 2w = 5$$

Solution:

$$[A, B] = \left[\begin{array}{cccc|c} 1 & 1 & 1 & 1 & 2 \\ 2 & -1 & 2 & -1 & -5 \\ 3 & 2 & 3 & 4 & 7 \\ 1 & -2 & -3 & 2 & 5 \end{array}\right]$$

Taking $a_{11} = 1(\neq 0)$ as the pivot, reduce all the elements below that (a_{11} element) to zero.

By operating $R_i \to R_i - (a_{i1}/a_{11}).R_1$ where $i = 2, 3, 4$

i.e. $R_2 \to R_2 - (2/1).R_1$ (for making a_{21} element zero)

$R_3 \to R_3 - (3/1).R_1$ (for making a_{31} element zero)

$R_4 \to R_4 - (1/1).R_1$ (for making a_{41} element zero)

$$= \left[\begin{array}{cccc|c} 1 & 1 & 1 & 1 & 2 \\ 0 & -3 & 0 & -3 & -9 \\ 0 & -1 & 0 & 1 & 1 \\ 0 & -3 & -4 & 1 & 3 \end{array}\right]$$

Taking $a_{22} = -3(\neq 0)$ as the pivot, reduce all the elements below and above that (a_{22} element) to zero.

By operating $R_i \rightarrow R_i - (a_{i2}/a_{22}).R_2$ where $i = 1, 3, 4$

i.e. $R_1 \rightarrow R_1 - (-1/3).\ R_2$ (for making a_{12} element zero)

$R_3 \rightarrow R_3 - (1/3).\ R_2$ (for making a_{32} element zero)

$R_4 \rightarrow R_4 - (3/3).\ R_2$ (for making a_{42} element zero)

$$= \left[\begin{array}{cccc|c} 1 & 0 & 1 & 0 & -1 \\ 0 & 1 & 0 & 1 & 3 \\ 0 & 0 & 0 & 2 & 4 \\ 0 & 0 & -4 & 4 & 12 \end{array}\right]$$

As $a_{33} = 0$, so interchange row 3 and 4

$$= \left[\begin{array}{cccc|c} 1 & 0 & 1 & 0 & -1 \\ 0 & 1 & 0 & 1 & 3 \\ 0 & 0 & -4 & 4 & 12 \\ 0 & 0 & 0 & 2 & 4 \end{array}\right]$$

Taking $a_{33} = -4(\neq 0)$ as the pivot, reduce all the elements below and above that (a_{33} element) to zero.

By operating $R_i \rightarrow R_i - (a_{i3}/a_{33}).R_3$ where $i = 1, 2, 4$

i.e. $R_1 \rightarrow R_1 - (-1/4).R_3$ (for making a_{13} element zero)

$R_2 \rightarrow R_2 - (-0/4).R_3$ (for making a_{23} element zero)

$R_4 \rightarrow R_4 - (-0/4).R_3$ (for making a_{43} element zero)

$$= \left[\begin{array}{cccc|c} 1 & 0 & 0 & 1 & 2 \\ 0 & 1 & 0 & 1 & 3 \\ 0 & 0 & -4 & 4 & 12 \\ 0 & 0 & 0 & 2 & 4 \end{array}\right]$$

Taking $a_{44} = 2(\neq 0)$ as the pivot, reduce all the elements above that (a_{44} element) to zero.

By operating $R_i \rightarrow R_i - (a_{i4}/a_{44}).R_4$ where $i = 1, 2, 3$

i.e. $R_1 \rightarrow R_1 - (1/2).R_4$ (for making a_{14} element zero)

$R_2 \rightarrow R_2 - (1/2).R_4$ (for making a_{24} element zero)

$R_3 \rightarrow R_3 - (4/2).R_4$ (for making a_{34} element zero)

$$= \left[\begin{array}{cccc|c} 1 & 0 & 0 & 0 & 0 \\ 0 & 1 & 0 & 0 & 1 \\ 0 & 0 & -4 & 0 & 4 \\ 0 & 0 & 0 & 2 & 4 \end{array}\right]$$

As all the elements above and below the diagonal (1, 1,-4, 2) are comes out to be zero. Hence system of equations becomes

$$\begin{aligned} x + 0y + 0z + 0w &= 0 \\ 0x + y + 0z + 0w &= 1 \\ 0x + 0y - 4z + 0w &= 4 \\ 0x + 0y + 0z + 2w &= 4 \end{aligned}$$

$\therefore \quad x = 0, y = 1, z = -1, w = 2$

Problem 2: Solve the system of equations by Gauss-Jordan method

$$\begin{aligned} x + y + z &= 1 \\ 4x + 3y - z &= 6 \\ x + 5y + 3z &= 4 \end{aligned}$$

Solution: Write the equations in the form

$$AX = B$$

$$\begin{bmatrix} 1 & 1 & 1 \\ 4 & 3 & -1 \\ 1 & 5 & 3 \end{bmatrix} \begin{bmatrix} x \\ y \\ z \end{bmatrix} = \begin{bmatrix} 1 \\ 6 \\ 4 \end{bmatrix}$$

$$\therefore \qquad [A, B] = \left[\begin{array}{ccc|c} 1 & 1 & 1 & 1 \\ 4 & 3 & -1 & 6 \\ 1 & 5 & 3 & 4 \end{array}\right]$$

Taking $a_{11} = 1 (\neq 0)$ as the pivot, reduce all the elements below that (a_{11} element) to zero.

By operating $R_i \to R_i - (a_{i1}/a_{11}).R_1$ where $i = 2, 3$

i.e. $R_2 \to R_2 - (4/1).R_1$ (for making a_{21} element zero)

$R_3 \to R_3 - (1/1).R_1$ (for making a_{31} element zero)

$$\therefore \qquad [A, B] = \left[\begin{array}{ccc|c} 1 & 1 & 1 & 1 \\ 0 & -1 & -5 & 2 \\ 0 & 4 & 2 & 3 \end{array}\right]$$

Taking $a_{22} = -1 (\neq 0)$ as the pivot, reduce all the elements below and above that (a_{22} element) to zero.

By operating $R_i \to R_i - (a_{i2}/a_{22}).R_2$ where $i = 1, 3$

i.e. $R_1 \to R_1 - (-1/1).R_2$ (for making a_{12} element zero)

$R_3 \to R_3 - (-4/1).R_2$ (for making a_{32} element zero)

$$\therefore \quad [A, B] = \left[\begin{array}{ccc|c} 1 & 0 & -4 & 3 \\ 0 & -1 & -5 & 2 \\ 0 & 0 & -18 & 11 \end{array}\right]$$

Taking $a_{33} = -18(\neq 0)$ as the pivot, reduce all the elements above that (a_{33} element) to zero.

By operating $R_i \to R_i - (a_{i3}/a_{33}).R_3$ where $i = 1, 2$

i.e. $R_1 \to R_1 - (2/9).R_3$ (for making a_{13} element zero)

$R_2 \to R_2 - (5/18).R_3$ (for making a_{23} element zero)

$$\therefore \quad [A, B] = \left[\begin{array}{ccc|c} 1 & 0 & 0 & 5/9 \\ 0 & -1 & 0 & -19/18 \\ 0 & 0 & -18 & 11 \end{array}\right]$$

As all the elements above and below the diagonal (1, –1, –18) are comes out to be zero. Hence system of equations becomes

$$x + 0y + 0z = 5/9$$
$$0x - y + 0z = -19/18$$
$$0x + 0y - 18z = 11$$

$\therefore x = 5/9, y = 19/18, z = -11/18$

6.8.6 Crout's Method

Crout's method is also known as LU decomposition method (Choleski method) or compact elimination. Where L represents lower matrix and *U* represents upper matrix. The method is more suitable as compared to Gauss's method. This method reduces the round off errors. This method is also useful for larger number of equations.

Consider the relation

$$AX = B \qquad ...(1)$$

In 3×3 matrix form, relation (1) can be represented as

$$\begin{bmatrix} a_{11} & a_{12} & a_{13} \\ a_{21} & a_{22} & a_{23} \\ a_{31} & a_{32} & a_{33} \end{bmatrix} \begin{bmatrix} x_1 \\ x_2 \\ x_3 \end{bmatrix} = \begin{bmatrix} b_1 \\ b_2 \\ b_3 \end{bmatrix}$$

Now we have to decompose the matrix *A* as the product of two matrices L and U. Lower (*L*) and upper triangular (*U*) matrices are those with zero elements above and below the diagonal.

$$\text{Suppose } L = \begin{bmatrix} l_{11} & 0 & 0 \\ l_{21} & l_{22} & 0 \\ l_{31} & l_{32} & l_{33} \end{bmatrix} \quad \text{and} \quad U = \begin{bmatrix} u_{11} & u_{12} & u_{13} \\ 0 & u_{22} & u_{23} \\ 0 & 0 & u_{33} \end{bmatrix}$$

The main simplification is to write

$$A = L.U$$

Where the diagonal elements of either L or U, but not both, can be chosen unity.

$$\text{Consider} \quad \begin{bmatrix} a_{11} & a_{12} & a_{13} \\ a_{21} & a_{22} & a_{23} \\ a_{31} & a_{32} & a_{33} \end{bmatrix} = \begin{bmatrix} l_{11} & 0 & 0 \\ l_{21} & l_{22} & 0 \\ l_{31} & l_{32} & l_{33} \end{bmatrix} \begin{bmatrix} 1 & u_{12} & u_{13} \\ 0 & 1 & u_{23} \\ 0 & 0 & 1 \end{bmatrix}$$

$$\begin{bmatrix} l_{11} & l_{11}.u_{12} & l_{11}.u_{13} \\ l_{21} & l_{22}\cdot u_{12} + l_{22} & l_{21}\cdot u_{13} + l_{22}\cdot u_{23} \\ l_{31} & l_{31}\cdot u_{12} + l_{32} & l_{31}\cdot u_{13} + l_{32}\cdot u_{23} + l_{33} \end{bmatrix}$$

Now compute the elements of L and U by comparing the above matrix with matrix formed by the elements of A.

(1) The first column elements of L (obtained by equating the first column of A with the first column of LU)

$$l_{11} = a_{11}, l_{21} = a_{21}, l_{31} = a_{31}$$

(2) The first row elements of L (obtained by equating the remaining first row elements of A with the remaining first row elements of LU)

i.e. $$u_{12} = a_{12} / l_{11}, u_{13} = a_{13} / l_{11}$$

(3) The second column elements of U (obtained by equating the remaining first column elements of A and LU)

i.e. $$l_{22} = a_{22} - l_{21}.u_{12}$$

$$l_{32} = a_{32} - l_{31}.u_{12}$$

(4) The second row elements of U (obtained by equating the remaining second row elements of A with the remaining second row elements of LU)

i.e. $$u_{23} = (a_{23} - l_{21}.u_{13})/l_{22}$$

(5) The third column elements of L (obtained by equating the remaining 3rd column of A with that of LU)

i.e. $$l_{33} = a_{33} - l_{31}.u_{13-} - l_{32}.u_{23}$$

Hence all the elements of L and U are determined. Therefore, it is possible to decompose any matrix A into LU by the above process. Now the system $AX = B$ can be written as

$$LUX = B \quad ...(2)$$

Define $$Y = UX = [U]_{3\times3}.[X]_{3\times1}$$

Then (2) is given by, $L.Y = B$

Hence, we have the following two systems:

$$L.Y = B \quad \text{...(3)}$$

$$U.X = Y \quad \text{...(4)}$$

By solving the (3), the values of Y can be obtained and then the values of X can be obtained by solving the system (4).

The number of operations involved in terms of multiplications is of the order of $O(n^2)$ and for $n = 10$ we need about 110 operations as compared to 333 operations of the Gauss's scheme.

6.8.6.1 Flowchart of Crout's Method

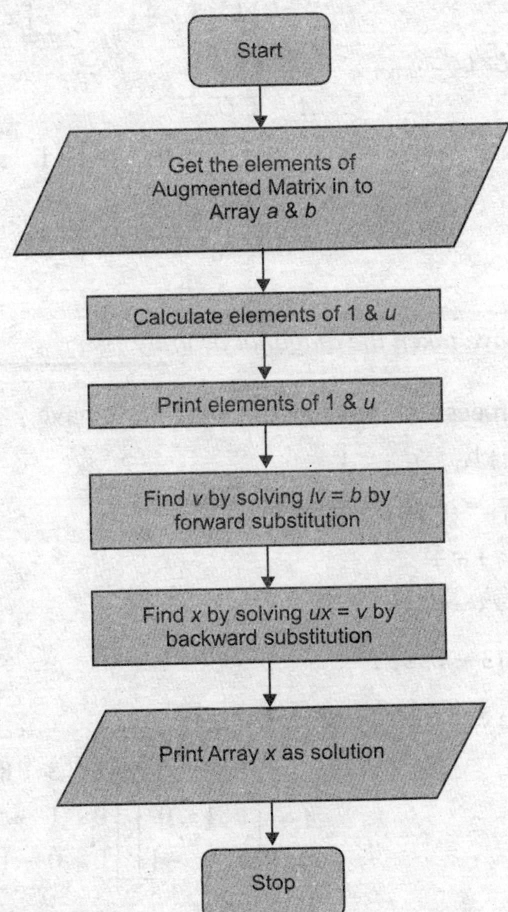

SOLVED PROBLEMS

Problem 1: Solve the system of equations by Crout's method

$$x_1 + 3x_2 + 8x_3 = 4$$
$$x_1 + 4x_2 + 3x_2 = -2$$
$$x_1 + 3x_3 + 4x_3 = 1$$

Solution: The System of equation can be written as

$$AX = B$$

Where $A = \begin{bmatrix} 1 & 3 & 8 \\ 1 & 4 & 3 \\ 1 & 3 & 4 \end{bmatrix}$; $X = \begin{bmatrix} x_1 \\ x_2 \\ x_3 \end{bmatrix}$; $B = \begin{bmatrix} 4 \\ -2 \\ 1 \end{bmatrix}$

Now decompose A into $L \times U$

i.e. $A = L.U$

$$= \begin{bmatrix} l_{11} & 0 & 0 \\ l_{21} & l_{22} & 0 \\ l_{31} & l_{32} & l_{33} \end{bmatrix} \begin{bmatrix} 1 & u_{12} & u_{13} \\ 0 & 1 & u_{23} \\ 0 & 0 & 1 \end{bmatrix}$$

Note that in U matrix we have taken the diagonal as unity.

Now after comparison process of above matrix with A, we have

$l_{11} = a_{11} = 1; l_{21} = a_{21} = 1; l_{31} = a_{31} = 1$

$u_{12} = a_{12}/l_{11} = 3/1 = 3; u_{13} = a_{13}/l_{11} = 8/1 = 8$

$l_{22} = a_{22} - l_{21}.u_{12} = 4 - 1\times3 = 1$

$l_{32} = a_{32} - l_{31}.u_{12} = 3 - 1\times3 = 0$

$u_{23} = (a_{23} - l_{21}.u_{13})/l_{22} = (3 - 1\times8)/1 = -5$

$l_{33} = a_{33} - l_{31}.u_{13} - l_{32}.u_{23} = 4 - 1\times8 - u\times(5/3) = -4$

Hence

$$A = \begin{bmatrix} 1 & 0 & 0 \\ 1 & 1 & 0 \\ 1 & 0 & -4 \end{bmatrix} \cdot \begin{bmatrix} 1 & 3 & 8 \\ 0 & 1 & -5 \\ 0 & 0 & 1 \end{bmatrix} = \text{L.U}$$

Defining $Y = U.X$

We have the two systems

$$LY = B$$

i.e.
$$\begin{bmatrix} 1 & 0 & 0 \\ 1 & 1 & 0 \\ 0 & 0 & -4 \end{bmatrix} \begin{bmatrix} y_1 \\ y_2 \\ y_3 \end{bmatrix} = \begin{bmatrix} 4 \\ -2 \\ 1 \end{bmatrix} \quad ...(1)$$

and
$$UX = Y$$

i.e.
$$\begin{bmatrix} 1 & 3 & 8 \\ 0 & 1 & -5 \\ 0 & 0 & 1 \end{bmatrix} \begin{bmatrix} x_1 \\ x_2 \\ x_3 \end{bmatrix} = \begin{bmatrix} y_1 \\ y_2 \\ y_3 \end{bmatrix} \quad ...(2)$$

From (1), we have

$y_1 = 4, y_2 = -6, y_3 = ¾$

Using these values in (2), and again solve (2) for x values, we have

$x_3 = ¾ ; x_2 = y_2 + 5x_3 = -9/4 ; x_1 = y_1 - 3x_2 - 8x_3 = 19/4$

By physical multiplication of L and U and then comparing the elements of LU with A, the solutions y_1, y_2, y_3 in (1) and hence the solutions x_1, x_2, x_3 in (2) is possible.

UNSOLVED PROBLEMS

1. Solve the following equations by Crout's method.

$$5x_1 + x_2 + x_3 + x_4 = 4$$
$$x_1 + 7x_2 + x_3 + x_4 = 12$$
$$x_1 + x_2 + 6x_3 + x_4 = -5$$
$$x_1 + x_2 + x_3 + 4x_4 = -6$$

6.8.7 *ILL* Condition Systems

ILL conditioning of the system is one of the major causes for inaccurate results of the system of equations. There are some problems which have the property that a very small change in the coefficients or the parameters of the problem produces a very large change in the solution of the problem. Such types of problems are ill-conditioned.

System of equations that have this property is called ill-conditioned system. When we have a large system and if it is ill-conditioned due to the commutative effect of round off we may obtain a totally different solution of the system which is far away from the exact solution. If the system is ill-conditioned it is suggested to follow only the Gauss Elimination method.

Now we will explain this concept with an example.

Let us consider a two 2 × 2 systems of equations:

(a)
$$x + 5y = 17$$
$$1.500x + 7.501y = 25.503$$

(b) $$x + 5y = 17$$
$$1.500x + 7.501 = 25.501$$

Practically the two systems are same except for a very small change in one of coefficients (i.e. from 25.503 to 25.501). The solution for the two systems turns out to:

(a) $x = 2, \quad y = 3$ and

(b) $x = 17, \quad y = 0$

and we note the very large difference in the solutions. *We normally come across such problems when fitting curves by the least squares method.*

How to make ill-conditioned system well conditioned

The system (of linear equations) condition is checked by computing the determinant of the coefficient matrix i.e. if determinant comes out to be small then there is possibility that system of linear equation is ill conditioned.

e.g. Suppose we have the following system of linear equations

$$x + 4y = 13$$
$$1.001x + 4y = 13.001$$

In matrix form. We have

$$\begin{vmatrix} 1 & 4 \\ 1.001 & 4 \end{vmatrix}$$

Determinant $= (1 \times 4 - 4 \times 1.001) = -0.001$(which is a very very small value). That indicates that the given system is ill conditioned.

Now, we will explain an iterative method that will help us to improve the accuracy of the system.

Consider ill conditioned system of simultaneous linear equations of the form

$$a_{11}x_1 + a_{12}x_2 + ...+a_{1n}x_n = b_1$$
$$a_{21}x_1 + a_{22}x_2 + ...+a_{2n}x_n = b_2$$
$$a_{31}x_1 + a_{32}x_2 + ...+a_{3n}x_n = b_3$$
$$..$$
$$..$$
$$a_{n1}x_1 + a_{n2}x_2 + ...+a_{nn}x_n = b_n \quad ...(1)$$

where all a_i's and b_i's are constants.

Let $x_1', x_2', x_3', ..., x_n'$ be the approximate solution values.

Substitute these solution values in system (1), we have

$$a_{11}x_1' + a_{12}x_2' + ...+a_{1n}x_n' = b_1'$$
$$a_{21}x_1' + a_{22}x_2' + ...+a_{2n}x_n' = b_2'$$

$$a_{31}x_1' + a_{32}x_2' + \ldots\ldots + a_{3n}x_n' = b_3'$$
$$\ldots\ldots\ldots\ldots\ldots\ldots\ldots\ldots\ldots\ldots\ldots\ldots$$
$$\ldots\ldots\ldots\ldots\ldots\ldots\ldots\ldots\ldots\ldots\ldots\ldots$$
$$a_{n1}x_1' + a_{n2}x_2' + \ldots\ldots + a_{nn}x_n' = b_n' \qquad \ldots(2)$$

where $b_1', b_2', \ldots\ldots\ldots\ldots, b_n'$ are new values of $b_1, b_2, \ldots\ldots, b_n$

Subtract system (2) from (1), we have

$$a_{11}x_{e1} + a_{12}x_{e2} + \ldots\ldots + a_{1n}x_{en} = c_1$$
$$a_{21}x_{e1} + a_{22}x_{e2} + \ldots\ldots + a_{2n}x_{en} = c_2$$
$$a_{31}x_{e1} + a_{32}x_{e2} + \ldots\ldots + a_{3n}x_{en} = c_3$$
$$\ldots\ldots\ldots\ldots\ldots\ldots\ldots\ldots\ldots\ldots\ldots\ldots$$
$$\ldots\ldots\ldots\ldots\ldots\ldots\ldots\ldots\ldots\ldots\ldots\ldots$$
$$a_{n1}x_{e1} + a_{n2}x_{e2} + \ldots\ldots + a_{nn}x_{en} = c_n \qquad \ldots(3)$$

where $x_{e1} = x_1 - x_1'$, $x_{e2} = x_2 - x_2'$, $\ldots\ldots\ldots\ldots$, $x_{en} = x_n - x_n'$, and $c_i = b_i - b_i'$

Where $i = 1,2,3,\ldots\ldots, n$

Solving system (3) (by using Gauss elimination technique preferably) and compute the values of x_{ei} ; $i = 1,2,3,\ldots\ldots, n$

Hence, we will get a better solution in the form

$x_1 = x_1' + x_{e1}$, $x_2 = x_2' + x_{e2}$, $\ldots\ldots\ldots\ldots$, $x_n = x_n' + x_{en}$

We can repeat this process again and again for improving the accuracy of the system.

6.9 ITERATIVE METHODS

An iterative method is that in which we start from an initial approximation to the actual solution and obtain better and better approximations from a computation cycle repeated time and again for achieving a desired accuracy. Thus in an iterative method, the amount of computation depends on the degree of accuracy required.

For large systems, iterative methods may be faster than the direct methods and even the round off errors in iterative methods is smaller. In fact, iteration is a self-correcting process and any error made at any stage of computation gets automatically corrected in the subsequent steps. Simple iterative method can be devised for systems in which the coefficients of the leading diagonal are large as compared to others. Iterative methods, are particularly preferred over direct methods when coefficient matrix is sparse i.e. matrix has many zeros and to reduce round-off errors. In this section we describe three such methods namely, Jacobi, Gauss-Seidel and Relaxation method.

6.9.1 Jacobi's Method

Jacobi method is an algorithm for determining the solutions of a system of linear equations with largest absolute values in each row and column dominated by the diagonal element. Each diagonal element is solved for, and an approximate value plugged in. The process is then iterated until it converges.

This algorithm is a stripped-down version of the Jacobi transformation method of matrix diagonalization. The method is named after German mathematician Carl Gustav Jakob Jacobi. Let us explain this method in the case of three equations in three unknowns.

Consider the system of equations

$$a_{11}x_1 + a_{12}x_2 + a_{13}x_3 = a_{14}$$
$$a_{21}x_1 + a_{22}x_2 + a_{23}x_3 = a_{24}$$
$$a_{31}x_1 + a_{32}x_2 + a_{33}x_3 = a_{34} \qquad \text{...(1)}$$

Let us assume

$$|a_{11}| \geq |a_{12}| + |a_{13}|$$
$$|a_{22}| \geq |a_{21}| + |a_{23}|$$
$$|a_{33}| \geq |a_{31}| + |a_{32}|$$

If the above condition does not hold then by interchanging the equations we can make the leading diagonal elements dominant.

Initialization step: If you are able to choose the better initial values for the unknowns involved in the system then it is a good practice otherwise assign a zero value to each unknown i.e. take $x_1 = 0, x_2 = 0, x_3 = 0$. The better initial values only affect the number of iterations required for convergence.

Iteration 1: Let us start with the initial values x_1, y_1, z_1 corresponding to x, y, z respectively. Substitute these values(x_1, y_1, z_1) in (1), and compute the first approximations as:

$$x_1 = (a_{14} - a_{12}y_1 - a_{13}z_0)/a_1$$
$$y_1 = (d_2 - a_2x_0 - c_2z_0)/b_2 \qquad \text{...(2)}$$
$$z_1 = (d_3 - a_3x_0 - b_3y_0)/c_3$$

Iteration 2: Substitute these new values x_1, y_1, z_1 for the values x_0, y_0, z_0 respectively in (2), and compute the second approximations as:

$$x_2 = (d_1 - b_1y_1 - c_1z_1)/a_1$$
$$y_2 = (d_2 - a_2x_1 - c_2z_1)/b_2 \qquad \text{...(3)}$$
$$z_2 = (d_3 - a_3x_1 - b_3y_1)/c_3$$

This iterative process will continue until the successive values of each unknown differ only within the prescribed limits.

The method is also applicable for more than three equations in the same way.

SOLVED PROBLEMS

Problem 1: Compute the following system of equations by using Jacobi's method

$$4x + y + 3z = 17$$
$$x + 5y + z = 14$$
$$2x - y + 8z = 12$$

Solution: Here, we have seen that the diagonal elements are dominant. Hence, jacobi's method is very much suitable for the system.

As $|4| \geq |1| + |3|$; $|5| \geq |1| + |1|$; $|8| \geq |2| + |-1|$

Now rewrite the given equations in the form as

$$x = (17 - y - 3z)/4$$
$$y = (14 - x - z)/5$$
$$z = (12 - 2x + y)/8 \quad ...(1)$$

Now start with the initial approximations as, $x_0 = y_0 = z_0 = 0$

Substitute $x_0 = y_0 = z_0 = 0$ on the right sides of (1) in place of x, y, z, we get first approximations as

$$x_1 = 17/4 = 4.25$$
$$y_1 = 14/5 = 2.8$$
$$z_1 = 3/2 = 1.5$$

Substitute x_1, y_1, z_1 on the right sides of (1) in place of x, y, z, we get second approximations as

$$x_2 = (17 - 2.8 - 3\times1.5)/4 = 2.425$$
$$y_2 = (14 - 4.25 - 1.5)/5 = 1.65$$
$$z_2 = (12 - 2\times4.25 + 2.8)/8 = 0.7875$$

Substitute x_2, y_2, z_2 on the right sides of (1) in place of x, y, z, we get third approximations as

$$x_3 = (17 - 1.65 - 3\times0.7875)/4 = 3.25$$
$$y_3 = (14 - 2.425 - 0.7875)/5 = 2.16$$
$$z_3 = (12 - 2\times2.425 + 1.65)/8 = 1.1$$

Substitute x_3, y_3, z_3 on the right sides of (1) in place of x, y, z, we get fourth approximations as

$$x_4 = (17 - 2.16 - 3\times1.1)/4 = 2.885$$
$$y_4 = (14 - 3.25 - 1.1)/5 = 1.93$$
$$z_4 = (12 - 2\times3.25 + 2.16)/8 = 0.96$$

Substitute x_4, y_4, z_4 on the right sides of (1) in place of x, y, z, we get fifth approximations as

$$x_5 = (17 - 1.93 - 3\times0.96)/4 = 3.05$$
$$y_5 = (14 - 2.885 - 0.96)/5 = 2.03$$
$$z_5 = (12 - 2\times2.885 + 1.93)/8 = 1.02$$

Now, we have seen that the successive values of x, y, z, comes out to be very much near to each other so we have to stop here.

Hence, the approximate values are $x = 3, y = 2, z = 1$.

UNSOLVED PROBLEMS

1. Solve the following equations by Gauss Jacobi method.

$$28x + 4y - z = 32$$
$$x + 3y + 10z = 24$$
$$2x + 17y + 4z = 35$$

6.9.2 Gauss Seidel Method

Gauss Seidel method is also an effective iterative method and works very much similar to that of Jacobi's method but the accuracy of Gauss Seidel is much more than that of Jacobi's method. This is the modification to the Jacobi's method. The Gauss–Seidel method, also known as the Liebmann method or the method of successive displacement, is an iterative method used to solve a linear system of equations. It is named after the German mathematicians Carl Friedrich Gauss and Philipp Ludwig von Seidel.

In this method, the previously calculated values (most recent approximations) for some unknowns are used in the computation of the next values (other unknowns) within the same iteration. But this is not in the case of Jacobi's method. This method is effective because in this method we use new values of the unknowns, which are always better than the old values. Though it can be applied to any matrix with non-zero elements on the diagonals, convergence is only guaranteed if the matrix is either diagonally dominant, or symmetric and positive definite.

Let us explain this method in the case of three equations in three unknowns.

Consider the system of equations

$$\begin{aligned} a_1x + b_1y + c_1z &= d_1 \\ a_2x + b_2y + c_2z &= d_2 \\ a_3x + b_3y + c_3z &= d_3 \end{aligned} \quad \text{...(1)}$$

Let us start with the initial values x_0, y_0, z_0 corresponding to x, y, z respectively. Substitute $y = y_0$, $z = z_0$ in **first** equation of (1) and compute first approximation for x as

$$x_1 = (d_1 - b_1y_0 - c_1z_0)/a_1$$

Now substitute $x = x_1$ and $z = z_0$ in the **second** equation of (1) and compute first approximation for y as

$$y_1 = (d_2 - a_2x_1 - c_2z_0)/b_2$$

Now substitute $x = x_1$ and $y = y_1$ in the **third** equation of (1) and compute first approximation for z as

$$z_1 = (d_3 - a_3x_1 - b_3y_1)/c_3$$

Now, we are at the step where we have three new approximations to x, y, z in our hand. Now by using these new approximations (recent) we repeat the whole process again and again until the successive values of each unknown differs only within the prescribed limits.

6.9.2.1 Algorithm of Gauss Seidel Method

Given A an augmented matrix of order $n \times (n+1)$ which is used to store a two dimensional array a of order $n \times (n+1)$. The result is stored in one dimensional array r of order n. The prescribed tolerance is given by **epsilon** and number of permitted iterations is given by max iteration.

1. Begin
2. Read n
3. Do i = 1 to n by 1
4. Do j = 1 to n +1 by 1
5. Read a(i, j)
6. End do
7. End do
8. Do i = 1 to n by 1
9. Read x(i)
10. End do
11. Read ephsilon , max_iteration
12. Do i = 1 to max_iteration
13. max_error = 0
14. Do j = 1 to n by 1
15. Sum =0
16. Do k = 1 to n by 1
17. If (j <> k) then
18. Sum + = a(j , k) * x (k)
19. End if
20. End do
21. Temp = (a(j, n +1) – sum) / a(j, j)
22. Error = abs (temp – x(j)) / temp
23. x(j) = temp
24. if (error > max_error) then
25. max_error = error
26. end if
27. end do
28. if (max_error < = ephislon) then
29. print "the equations converges in", I " iterations"
30. Do i = 1 to n by 1
31. print x(i)
32. End do
33. Exit

34. End if
35. End do
36. Print "the equations does not converges in", max_iterationI "iterations"
37. End

6.9.2.2 Flowchart of Gauss Seidel Method

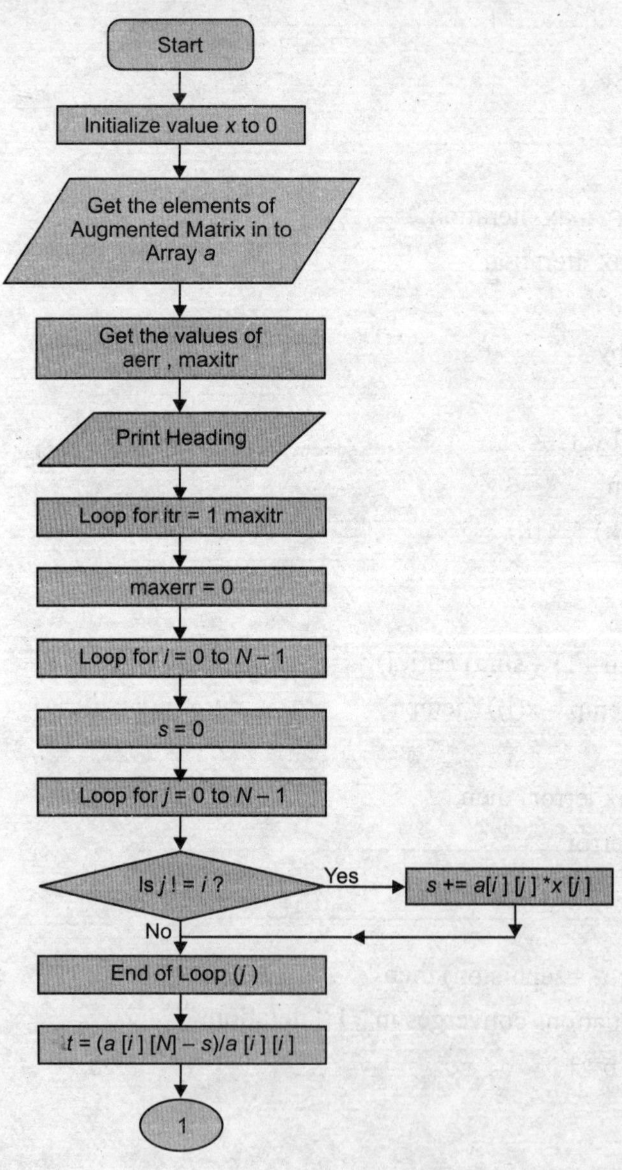

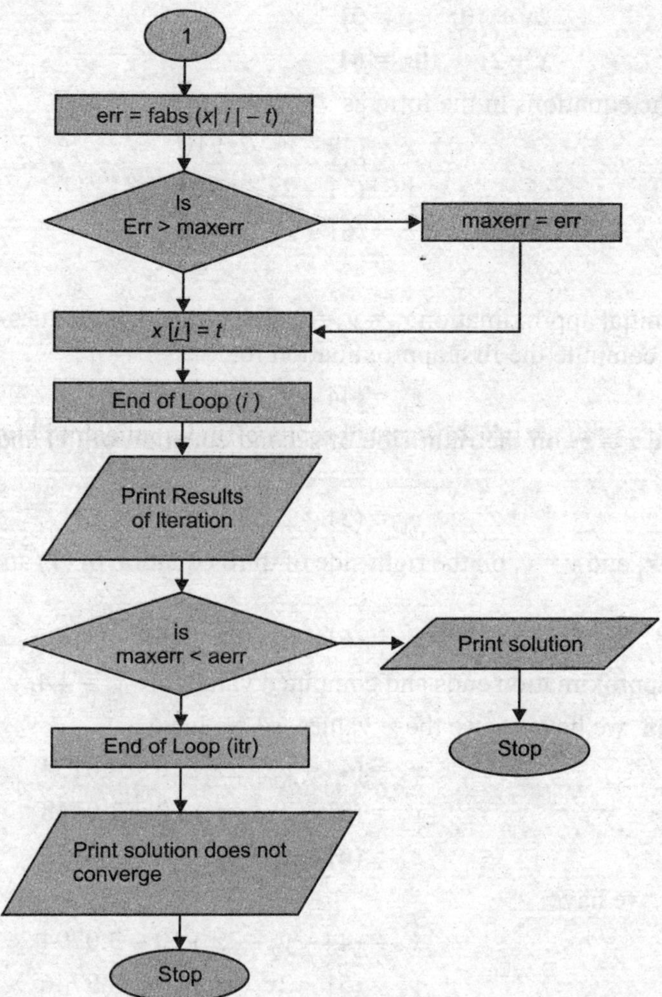

6.9.2.3 Comparison Between Gauss Seidel and Jacobi's Method

The Gauss Seidel method will converge more rapidly as it uses the most recent approximations of the unknowns while proceeding to the next step. The rate of convergence in Gauss Seidel method is approximately two times than that of Gauss-Jacobi method.

The Gauss-Seidel method converges always when the Jacobi's method converges. When the Jacobi's method does not converge, the Gauss Seidel method may converge. Gauss-Seidel method is more efficiently implemented on computer as compared to Jacobi's method.

SOLVED PROBLEMS

Problem 1: Compute the following system of equations by using Gauss-Seidel method

$$10x + y + 2z = 44$$

$$2x + 10y + z = 51$$
$$x + 2y + 10z = 61 \quad \text{...(1)}$$

Solution: Rewrite the equations in the form as

$$x = (44 - y - 2z)/10$$
$$y = (51 - 2x - z)/10$$
$$z = (61 - x - 2y)/10$$

For first iteration

Now, start with the initial approximation $x_0 = y_0 = 0$. Substitute these values on the right side of first equation of (1) and compute the first approximation for x as

$$x_1 = (44 - y_0 - 2z_0)/10 = 4.4$$

Substitute $x = x_1$ and $z = z_0$ on the right side of second equation of (1) and compute the first approximation for y as

$$y_1 = (51 - 2x_1 - z_0)/10 = 4.22$$

Now, Substitute $x = x_1$ and $y = y_1$ on the right side of third equation of (1) and compute the first approximation for z as

$$z_1 = (61 - x_1 - 2y_1)/10 = 4.816$$

Now, here the first approximation ends and computed values are $x_1 = 4.4, y_1 = 4.22, z_1 = 4.816$

For second iteration, we have to use these values, so we have

$$x_2 = (44 - y_1 - 2z_1)/10 = 4.0154$$
$$y_2 = (51 - 2x_2 - z_1)/10 = 3.0148$$
$$z_3 = (61 - x_2 - 2y_2)/10 = 5.0955$$

For third iteration, we have

$$x_3 = (44 - y_2 - 2z_2)/10 = 3.0794$$
$$y_3 = (51 - 2x_3 - z_2)/10 = 3.9746$$
$$z_3 = (61 - x_3 - 2y_3)/10 = 4.9971$$

For fourth iteration, we have

$$x_4 = (44 - y_3 - 2z_3)/10 = 3.0031$$
$$y_4 = (51 - 2x_4 - z_3)/10 = 3.9997$$
$$z_4 = (61 - x_4 - 2y_4)/10 = 4.8001$$

For fifth iteration, we have

$$x_5 = (44 - y_4 - 2z_4)/10 = 3.0400$$
$$y_5 = (51 - 2x_5 - z_4)/10 = 4.0120$$
$$z_5 = (61 - x_5 - 2y_5)/10 = 4.8360$$

For sixth iteration, we have

$$x_6 = (44 - y_5 - 2z_5)/10 = 3.0316$$
$$y_6 = (51 - 2x_6 - z_5)/10 = 4.0101$$
$$z_6 = (61 - x_6 - 2y_6)/10 = 4.9948$$

For seventh iteration, we have

$$x_7 = (44 - y_6 - 2z_6)/10 = 3.0000$$
$$y_7 = (51 - 2x_7 - z_6)/10 = 4.0001$$
$$z_7 = (61 - x_7 - 2y_7)/10 = 5.0000$$

Now, we have seen that the successive values of x, y, z comes out to be very much near to each other so we have to stop here.

Hence, the approximate values are $x = 3$, $y = 4$, $z = 5$.

UNSOLVED PROBLEMS

1. Solve the following equations by Gauss-Seidel method

$$10x - 2y - z - w = 3$$
$$-2x + 10y - z - w = 15$$
$$-x - y + 10z - 2w = 27$$
$$-x - y - 2z + 10w = -9$$

2. Solve the following equations by using Gauss-Seidel method.

$$20x + y - 2z = 17$$
$$3x + 20y - z = -18$$
$$2x - 3y + 20z = 25$$

3. Solve the following equations by using Gauss-Seidel method.

$$2x - y + 2z = 3$$
$$x + 3y + 3z = -1$$
$$x + 2y + 5z = 1$$

   with initial approximation as (0.3, -0.8, 0.3).

4. Solve the following equations by using Gauss-Seidel method.

$$6x + y + z = 105$$
$$4x + 8y + 3z = 155$$
$$5x + 4y - 10z = 65$$

5. Discuss the convergence of the following equations by solving them with Gauss Seidel method.

$$54x + y + z = 110$$
$$2x + 15y + 6z = 72$$
$$-x + 6y + 27z = 85$$

6.9.3 Relaxation Method

Relaxation method also comes under the category of iterative methods which is used to relax the variables by liquidating the residuals. The relaxation method is a method for obtaining numerical approximations to the solutions of systems of equations, including certain types of elliptic partial differential equations, in particular Laplace's equation and its generalization, Poisson's equation.

The function is assumed to be given on the boundary of a shape, and has to be computed on its interior. Here, we start relaxation method with three equations containing three unknowns. The method is also applicable for more than three equations also by following the same procedure.

Consider the system of equations

$$
\begin{aligned}
a_1x + b_1y + c_1z &= d_1 \\
a_2x + b_2y + c_2z &= d_2 \\
a_3x + b_3y + c_3z &= d_3
\end{aligned} \qquad \text{...(1)}
$$

We define the residuals by the following relations.

$$
\begin{aligned}
r_1 &= a_1x + b_1y + c_1z - d_1 \\
r_2 &= a_2x + b_2y + c_2z - d_2 \\
r_3 &= a_3x + b_3y + c_3z - d_3
\end{aligned} \qquad \text{...(2)}
$$

An operation table as below describes the working of the method:

Operation	Changes in					
	x	y	z	r_1	r_2	r_3
R_1	1	0	0	a_1	a_2	a_3
R_2	0	1	0	b_1	b_2	b_3
R_3	0	0	1	c_1	c_2	c_3
		I			A'	

Here operation R_1 increases the value of x by 1, y by 0 and z by 0. Similarly, for R_2 and R_3 respectively. The operation table consists of unit matrix I and transpose of A as A'.

In this method, we have to find such combinational values of x, y, z which makes r_1, r_2, r_3 zero. The method will be understandable more by considering at least one example.

6.9.3.1 Convergence of The Method

The method converges if the diagonal elements are dominant. This method solves the system successfully if at least one inequality holds from the following inequalities.

$$
\begin{aligned}
|a_1| &\geq |b_1| + |c_1| \\
|b_2| &\geq |a_2| + |c_2| \\
|c_3| &\geq |a_3| + |b_3|
\end{aligned}
$$

SOLVED PROBLEMS

Problem 1: Solve the system of equations by using Relaxation method

$$12x + y + z = 31$$
$$2x + 8y - z = 24$$
$$3x + 4y + 10z = 58$$

Solution: As we have seen that the given system is diagonally dominant.

The residuals are given by

$$r_1 = 12x + y + z - 31$$
$$r_2 = 2x + 8y - z - 24$$
$$r_3 = 3x + 4y + 10z = 58 \quad \text{...(1)}$$

Operation Table

	x	y	z	r_1	r_2	r_3
$R_1 \to$	1	0	0	12	2	3
$R_2 \to$	0	1	0	1	8	4
$R_3 \to$	0	0	1	1	–1	10

Take the initial values of x, y, z as 0, 0, 0

After substitution $x = y = z = 0$ in (1), we have

$r_1 = -31, r_2 = -24, r_3 = -58$

Here in the above residuals, the largest residual in magnitude is $r_3 = -58$

In tabular form it can be shown as

x	y	z	r_1	r_2	r_3	
0	0	0	–31	–24	**–58**	...(1)

Now we have to liquidate this residual (-58) by finding a proper multiple of R_3 (as it corresponds to r_3 so that's why we have chosen R_3)

The proper multiple for R_3 is 6 here, which is chosen as

	x	y	z	r_1	r_2	r_3	
6.R_3 gives	0	0	6	6	–6	60	
and	0	0	0	–31	–24	–58	
Add & Check	0	0	6	–25	**–30**	2	...(2)
Residual							

Now we have seen that residual reduces from (–58) to 2. If we choose any other multiple for R_3 then it produces residual more than 2 under r_3.

Similarly, in (2) the largest residual comes out to be (–30). Now we have to liquidate this residual by finding a proper multiple of R_2 (as it corresponds to r_2 so that's why we have chosen R_2)

The proper multiple for R_2 is 4, which is chosen as

	x	y	z	r_1	r_2	r_3	
4.R_2 gives	0	4	6	4	32	16	
and	0	0	0	–25	–30	2	
Add & Check	0	4	0	**–21**	2	18	...(3)
Residual							

Note that here we have not included 6 because operation 4.R_2 only increases y, but not x and z.

Now we have seen that residual reduces from (–30) to 2. if we choose any other multiple for R_2 then it produces residual more than 2 under r_2.

Similarly, in (3) the largest residual comes out to be (–21). Now we have to liquidate this residual by finding a proper multiple of R_1 (as it corresponds to r_1 so that's why we have chosen R_1).

The proper multiple for R_1 is 2 here, which is chosen as

	x	y	z	r_1	r_2	r_3	
2.R_2 gives	2	0	0	24	4	6	
and	0	0	0	–21	2	18	
Add & Check	2	0	0	3	6	**24**	...(4)
Residual							

Note that here we have not included 4 because operation 2.R_1 only increases x, but not y and z.

Similarly, in (4) the largest residual comes out to be (24). Now we have to liquidate this residual by finding a proper multiple of R_3 (as it corresponds to r_3 so that's why we have chosen R_3).

The proper multiple for R_3 is –2 here, which is chosen as

	x	y	z	r_1	r_2	r_3	
–2.R_3 gives	0	0	–2	–2	2	–20	
and	0	0	0	3	6	24	
Add & Check	0	0	–2	1	**8**	4	...(5)
Residual							

Note that here we have not included 2 because operation –2.R_3 only increases z, but not x and y.

Similarly, in (5) the largest residual comes out to be (8). Now we have to liquidate this residual by finding a proper multiple of R_2 (as it corresponds to r_2 so that's why we have chosen R_2).

The proper multiple for R_2 is –1 here, which is chosen as

	x	y	z	r_1	r_2	r_3	
–1.R_2 gives	0	–1	0	–1	–8	–4	
and	0	0	0	1	8	4	
Add & Check	0	–1	0	0	0	**0**	...(6)

Residual

Now, we have seen that all the residuals come out to be zero.

Now add equations from (1) to (6) for finding the relaxed values of x, y, z.

x	y	z	r_1	r_2	r_3	
0	0	0	–31	–24	**–58**	...(1)
0	0	6	–25	**-30**	2	...(2)
0	4	0	**-21**	2	18	...(3)
2	0	0	3	6	**24**	...(4)
0	0	–2	1	**8**	4	...(5)
0	–1	0	0	0	0	...(6)
2	3	4	0	0	0	

Since we have seen that all the residuals comes out to be zero, the solution is $x = 2, y = 3, z = 4$.

Or

We can make direct table for the whole process as:

	x	y	z	r_1	r_2	r_3	
	0	0	0	–31	–24	**–58**	...(1)
$6R_3 \rightarrow$	0	0	6	–25	**-30**	2	...(2)
$4R_2 \rightarrow$	0	4	0	**-21**	2	18	...(3)
$2R_2 \rightarrow$	2	0	0	3	6	**24**	...(4)
$-2R_3 \rightarrow$	0	0	–2	1	**8**	4	...(5)
$-1R_3 \rightarrow$	0	–1	0	0	0	0	...(6)
	2	3	4	0	0	0	

Note: *- This method is also applicable for more than three variables in the same way.*

UNSOLVED PROBLEMS

1. Solve the equations by using Relaxation method.

$$6x + y - z = 14$$
$$x + 5y - z = -18$$
$$2x + y + 9z = 68$$

6.9.4 Matrix Inversion Methods

The inverse of a square matrix is written as A^{-1} so that $AA^{-1} = A^{-1}A = \text{I}$. The inverse of a matrix exists if it is a non-singular matrix. Also, inverse of a matrix, if it exists, is unique. We have various techniques for finding the inverse of a given matrix, of which, the method of obtaining inverse with the help of an adjoint requires a lot of calculations. But here, in this section we shall discuss some other methods which require less of computational labor and can be easily extended to matrices of higher order. The methods are as:

1. Gauss Elimination method
2. Gauss Jordon method
3. Crout's method

Gauss Elimination Method

Consider a non-singular square matrix A of order three whose inverse we have to find. If A^{-1} is the inverse of A, then $AA^{-1} = I$. Where I is the unit matrix of order three. Now, we are interested in finding the elements of A^{-1}.

Let $$A = \begin{bmatrix} a_{11} & a_{12} & a_{13} \\ a_{21} & a_{22} & a_{23} \\ a_{31} & a_{32} & a_{33} \end{bmatrix}, \quad \text{and} \quad A^{-1} = \begin{bmatrix} x_{11} & x_{12} & x_{13} \\ x_{21} & x_{22} & x_{23} \\ x_{31} & x_{32} & x_{33} \end{bmatrix}$$

Then, we have $AA^{-1} = I$

$$\Rightarrow \begin{bmatrix} a_{11} & a_{12} & a_{13} \\ a_{21} & a_{22} & a_{23} \\ a_{31} & a_{32} & a_{33} \end{bmatrix} \begin{bmatrix} x_{11} & x_{12} & x_{13} \\ x_{21} & x_{22} & x_{23} \\ x_{31} & x_{32} & x_{33} \end{bmatrix} = \begin{bmatrix} 1 & 0 & 0 \\ 0 & 1 & 0 \\ 0 & 0 & 1 \end{bmatrix} \quad ...(1)$$

This is equivalent to system of equations

$$AX_1 = I_1, \quad X_1 = \begin{bmatrix} x_{11} \\ x_{21} \\ x_{31} \end{bmatrix}, \quad I_1 = \begin{bmatrix} 1 \\ 0 \\ 0 \end{bmatrix} \quad ...(2)$$

$$AX_2 = I_2, \quad X_2 = \begin{bmatrix} x_{12} \\ x_{22} \\ x_{32} \end{bmatrix}, \quad I_2 = \begin{bmatrix} 0 \\ 1 \\ 0 \end{bmatrix} \quad ...(3)$$

$$AX_3 = I_3, \quad X_3 = \begin{bmatrix} x_{13} \\ x_{23} \\ x_{33} \end{bmatrix}, \quad I_3 = \begin{bmatrix} 0 \\ 0 \\ 1 \end{bmatrix} \quad ...(4)$$

Now, from (2), (3) and (4) we can solve vectors X_1, X_2, X_3 by using Gauss elimination method. The solution set of (2), (3) and (4) gives the corresponding column of the inverse matrix A^{-1}. As we have seen that in (2), (3) and (4) the A matrix (coefficient matrix) are same so we can solve all of them simultaneously. We can also find the inverse of matrices having order more than three with the same procedure.

Now, we will explain this by using an example.

SOLVED PROBLEMS

Problem 1: Find the inverse of the given matrix by using Gauss Elimination method.

$$A = \begin{bmatrix} 0 & 1 & 1 \\ 1 & 2 & 0 \\ 3 & -1 & -4 \end{bmatrix}$$

Solution: The augmented matrix $[A, I]$ can be written as

$$[A, I] = \left[\begin{array}{ccc|ccc} 0 & 1 & 1 & 1 & 0 & 0 \\ 1 & 2 & 0 & 0 & 1 & 0 \\ 3 & -1 & -4 & 0 & 1 & 1 \end{array}\right] \qquad ....(1)$$

Now, we have to reduce the matrix A to an upper triangular matrix by reducing all the elements below the diagonal to zero. First, start with a_{11}, as a_{11} (pivot) to zero. So we will interchange the first and second row. So, reduced system becomes

$$\left[\begin{array}{ccc|ccc} 1 & 2 & 0 & 0 & 1 & 0 \\ 0 & 1 & 1 & 1 & 0 & 0 \\ 3 & -1 & -4 & 0 & 0 & 1 \end{array}\right] \qquad ....(2)$$

Now, for making the matrix (2) an upper triangular we first operate $\boldsymbol{R_3 + (-3).R_1}$ which will make all the elements below a_{11} to be zero. After operation reduced matrix comes out to be

$$\left[\begin{array}{ccc|ccc} 1 & 2 & 0 & 0 & 1 & 0 \\ 0 & 1 & 1 & 1 & 0 & 0 \\ 0 & -7 & -4 & 0 & -3 & 1 \end{array}\right] \qquad ....(3)$$

Now, we will have to make the elements below a_{22} to be zero, for this we operate $\boldsymbol{R_3 + 7.R_2}$ that gives

$$\left[\begin{array}{ccc|ccc} 1 & 2 & 0 & 0 & 1 & 0 \\ 0 & 1 & 1 & 1 & 0 & 0 \\ 0 & 0 & -4 & 0 & -3 & 1 \end{array}\right] \qquad ....(4)$$

As the triangularization process has done, now through back substitution we will find the elements of A^{-1} as

From (4), we have

$$\begin{aligned} x_{11} + 2x_{21} &= 0 \\ x_{21} + x_{31} &= 1 \\ 3x_{31} &= 7 \end{aligned}$$

Which after simplification gives, $\mathbf{x_{11} = 8/3, x_{21} = (-4/3), x_{31} = 7/3.}$

Similarly, we have

$$x_{12} + 2x_{22} = 1$$
$$x_{22} + x_{32} = 0$$
$$3x_{32} = -3$$

Which after simplification gives, $\mathbf{x_{12} = -1, x_{22} = 1, x_{32} = -1.}$

Similarly, we have

$$x_{13} + 2x_{23} = 0$$
$$x_{23} + x_{33} = 0$$
$$3x_{33} = 1$$

Which after simplification gives, $\mathbf{x_{13} = 2/3, x_{23} = (-1/3), x_{33} = 1/3.}$

Hence, A^{-1} comes out to be

$$A^{-1} = \begin{bmatrix} 8/3 & -1 & 2/3 \\ -4/3 & 1 & -1/3 \\ 7/3 & -1 & 1/3 \end{bmatrix}$$

6.9.5 Gauss Jordon Method

Gauss Jordon is also an effective method for finding the inverse of a matrix which is similar to the Gauss elimination method except that the matrix A is directly reduced to unit matrix instead of upper triangular form. In Gauss Jordon method, the two matrices A and I are written side by side and the same transformations are performed on both. As soon as A reduces to I, the unit matrix I reduces to A^{-1}.

Consider a non-singular square matrix A of order three whose inverse we have to find. If A^{-1} is the inverse of A, then $AA^{-1} = I$. Where I is the unit matrix of order three. Now, we are interested in finding the elements of A^{-1}.

Let
$$A = \begin{bmatrix} a_{11} & a_{12} & a_{13} \\ a_{21} & a_{22} & a_{23} \\ a_{31} & a_{32} & a_{33} \end{bmatrix}$$

Now side by side representation of $[A, I]$ for a 3×3 system is as

$$\left[\begin{array}{ccc|ccc} a_{11} & a_{12} & a_{13} & 1 & 0 & 0 \\ a_{21} & a_{22} & a_{23} & 0 & 1 & 0 \\ a_{31} & a_{32} & a_{33} & 0 & 0 & 1 \end{array}\right]$$
$$\qquad\qquad \mathbf{A} \qquad\qquad\qquad \mathbf{I}$$

Now, Gauss Jordon works as

Step 1:

Divide the first row by a_{11} (Here the first row is taken as pivot row whose first element is considered as pivot if it is zero then interchange this row with any other row.) then do the following transformations

$$R_i \rightarrow R_i - a_{i1}R_1 \qquad \text{where } i = 2, 3.$$

By doing this system reduces in the following form

$$\left[\begin{array}{ccc|ccc} 1 & a_{12} & a_{13} & b_{11} & 0 & 0 \\ 0 & a_{22} & a_{23} & b_{21} & 1 & 0 \\ 0 & a_{32} & a_{33} & b_{31} & 0 & 1 \end{array}\right]$$

Step 2:

Divide the second row by a_{22} (Here the second row is taken as pivot row whose second element is considered as pivot if it is zero then interchange this row with any other row.) then do the following transformations

$$R_i \rightarrow R_i - a_{i2}R_2 \qquad \text{where } i = 1, 3.$$

Now we obtain the new matrix as

$$\left[\begin{array}{ccc|ccc} 1 & 0 & a_{13} & b_{11} & b_{12} & 0 \\ 0 & 1 & a_{23} & b_{21} & b_{22} & 0 \\ 0 & 0 & a_{33} & b_{31} & b_{32} & 1 \end{array}\right]$$

Step 3:

Divide the third row by a_{33} (Here the third row is taken as pivot row whose third element is considered as pivot if it is zero then interchange this row with any other row.) then do the following transformations

$$R_i \rightarrow R_i - a_{i3}R_3 \qquad \text{where } i = 1, 1.$$

Now we obtain the new matrix as

$$\left[\begin{array}{ccc|ccc} 1 & 0 & 0 & b_{11} & b_{12} & b_{13} \\ 0 & 1 & 0 & b_{21} & b_{22} & b_{23} \\ 0 & 0 & 1 & b_{31} & b_{32} & b_{33} \end{array}\right]$$

$$I \qquad\qquad A^{-1}$$

The matrix formed by b's elements gives the inverse of A. By using same process we can also compute the inverse of other matrices having order more than three.

Now, we will explain this by using an example.

SOLVED PROBLEMS

Problem 1: Find the inverse of the given matrix by using Gauss Jordon method.

$$A = \begin{bmatrix} 1 & 1 & 3 \\ 1 & 3 & -3 \\ -2 & -4 & -4 \end{bmatrix}$$

Solution: The augmented matrix $[A, I]$ can be written as

$$[A, I] = \left[\begin{array}{ccc|ccc} 1 & 1 & 3 & 1 & 0 & 0 \\ 1 & 3 & -3 & 0 & 1 & 0 \\ -2 & -4 & -4 & 0 & 0 & 1 \end{array}\right] \quad ...(1)$$

Step 1:

Divide the first row by $a_{11} = 1$ (Here the first row is taken as pivot row) and do the transformations

$$R_i \to R_i - a_{i1}R_1 \qquad \text{where } i = 2, 3.$$

i.e. operate $R_2 \to R_2 - R_1$ and $R_3 \to R_3 - (-2).R_1$, so system becomes

$$[A, I] = \left[\begin{array}{ccc|ccc} 1 & 1 & 3 & 1 & 0 & 0 \\ 0 & 2 & -6 & -1 & 1 & 0 \\ 0 & -2 & 2 & 2 & 0 & 1 \end{array}\right]$$

Step 2:

Divide the second row by $a_{22} = 2$ (Here the second row is taken as pivot row)

$$[A, I] = \left[\begin{array}{ccc|ccc} 1 & 1 & 3 & 1 & 0 & 0 \\ 0 & 1 & -3 & -1/2 & 1/2 & 0 \\ 0 & -2 & 2 & 2 & 0 & 1 \end{array}\right]$$

Now perform the transformations

$$R_i \to R_i - a_{i2}R_2 \qquad \text{where } i = 1, 3.$$

i.e. operate $R_1 \to R_1 - R_2$ and $R_3 \to R_3 - (-2).\to_2$, so system becomes

$$[A, I] = \left[\begin{array}{ccc|ccc} 1 & 0 & 6 & 3/2 & -1/2 & 0 \\ 0 & 1 & -3 & -1/2 & 1/2 & 0 \\ 0 & 0 & -4 & 1 & 1 & 1 \end{array}\right]$$

Step 3:

Divide the third row by $a_{33} = -4$ (Here the third row is taken as pivot row)

$$[A, I] = \left[\begin{array}{ccc|ccc} 1 & 0 & 6 & 3/2 & -1/2 & 0 \\ 0 & 1 & -3 & -1/2 & 1/2 & 0 \\ 0 & 0 & 1 & -1/4 & -1/4 & 1/4 \end{array}\right]$$

Now do the transformations

$$R_i \rightarrow R_i - a_{i3}R_3 \qquad \text{where } i = 1, 2.$$

i.e. operate $R_1 \rightarrow R_1 - 6.R_3$ and $R_2 \rightarrow R_2 - (-3).R_3$, so system becomes

$$[A, I] = \left[\begin{array}{ccc|ccc} 1 & 0 & 0 & 3 & 1 & 3/2 \\ 0 & 1 & 0 & -5/4 & -1/4 & -3/4 \\ 0 & 0 & 1 & -1/4 & -1/4 & 1/4 \end{array}\right]$$

Hence, A^{-1} comes out to be

$$A^{-1} = \begin{bmatrix} 3 & 1 & 3/2 \\ -5/4 & -1/4 & -3/4 \\ -1/4 & -1/4 & 1/4 \end{bmatrix}$$

UNSOLVED PROBLEMS

1. Solve the system of equations using the Gauss Jordan method:

$$x + 2y + z = 8$$
$$2x + 3y + 4z = 20$$
$$4x + 3y + 2z = 16$$

2. Solve the system of equations using the Gauss Jordan method:

$$2x + 4y + z = 3$$
$$3x + 2y - 2z = -1$$
$$x - y + z = 6$$

6.9.6 Crout's Method

Crout's method can also be used for finding the inverse of a square matrix A. As we know, in case of Crout's method A is decomposed into $A = LU$ where L and U are lower and upper triangular matrices respectively.

$$A^{-1} = (LU)^{-1} = U^{-1}L^{-1} \qquad \text{...(1)}$$

If L and U are the lower and upper triangulars, then L^{-1} and U^{-1} are also lower and upper triangulars.

As we know that $LL^{-1} = I$ then we can find the lower matrix L^{-1} as L is known. Similarly we can compute U^{-1} from $UU^{-1} = I$ as U is also known.

After knowing both L^{-1} and U^{-1} we can compute A^{-1} as $A^{-1} = U^{-1}L^{-1}$.

Now, we will explain this by using an example.

SOLVED PROBLEMS

Problem 1:Find the inverse of the given matrix by using Crout's method.

$$A = \begin{bmatrix} 1 & -2 & 3 \\ 0 & -1 & 4 \\ -2 & 2 & 0 \end{bmatrix}$$

Solution: As we know $LU=A$ then we have $A^{-1} = U^{-1}L^{-1}$ ...(1)

From $LU = A$, we have

$$\begin{bmatrix} l_{11} & 0 & 0 \\ l_{21} & l_{22} & 0 \\ l_{31} & l_{32} & l_{33} \end{bmatrix} \begin{bmatrix} 1 & u_{12} & u_{13} \\ 0 & 1 & u_{23} \\ 0 & 0 & 1 \end{bmatrix} = \begin{bmatrix} 1 & -2 & 3 \\ 0 & -1 & 4 \\ -2 & 2 & 0 \end{bmatrix}$$

Compute l and u elements by equating as we have done in crout's method.

Now after comparison process of above matrix with A, we have

$l_{11} = a_{11} = 1; l_{21} = a_{21} = 0; l_{31} = a_{31} = -2$

$u_{12} = a_{12}/l_{11} = -2/1 = -2; u_{13} = a_{13}/l_{11} = 3/1 = 3$

$l_{22} = a_{22} - l_{21}.u_{12} = -1 - 0 = -1$

$l_{32} = a_{32} - l_{31}.u_{12} = 2 - 4 = -2$

$u_{23} = (a_{23} - l_{21}.u_{13})/l_{22} = (4 - 0)/(-1) = -4$

$l_{33} = a_{33} - l_{31}.u_{13} - l_{32}.u_{23} = 0 + 6 - 8 = -2$

$$\therefore \quad L = \begin{bmatrix} 1 & 0 & 0 \\ 1 & -1 & 0 \\ -2 & -2 & -2 \end{bmatrix}, \quad \text{and} \quad U = \begin{bmatrix} 1 & -2 & 3 \\ 1 & 1 & -4 \\ 0 & 0 & 1 \end{bmatrix}$$

Now we have to find L^{-1} and U^{-1} by using these L and U matrices.

As $LL^{-1} = I$, so we have

$$\begin{bmatrix} 1 & 0 & 0 \\ 0 & -1 & 0 \\ -2 & -2 & -2 \end{bmatrix} \begin{bmatrix} x_{11} & 0 & 0 \\ x_{21} & x_{22} & 0 \\ x_{31} & x_{32} & x_{33} \end{bmatrix} = \begin{bmatrix} 1 & 0 & 0 \\ 0 & 1 & 0 \\ 0 & 0 & 1 \end{bmatrix}$$

$\therefore \quad x_{11} = 1, x_{22} = -1, x_{33} = -1/2, -x_{21} = 0, x_{11} + x_{21} + x_{31} = 0, x_{22} + x_{32} = 0$

$\therefore \quad x_{31} = -1, x_{32} = 1$

So L^{-1} comes out to be

$$L^{-1} = \begin{bmatrix} 1 & 0 & 0 \\ 0 & -1 & 0 \\ -1 & 1 & -1/2 \end{bmatrix}$$

Similarly as $UU^{-1} = I$, so we have

$$\begin{bmatrix} 1 & -2 & 3 \\ 0 & 1 & -4 \\ 0 & 0 & 1 \end{bmatrix} \begin{bmatrix} 1 & c_{12} & c_{13} \\ 0 & 1 & c_{23} \\ 0 & 0 & 1 \end{bmatrix} = \begin{bmatrix} 1 & 0 & 0 \\ 0 & 1 & 0 \\ 0 & 0 & 1 \end{bmatrix}$$

$\therefore \quad c_{12} - 2 = 0, c_{13} - 2.c_{23} + 3 = 0, c_{23} - 4 = 0,$

$\therefore \quad c_{12} = 2, c_{23} = 4, c_{13} = 5$

So U^{-1} comes out to be

$$U^{-1} = \begin{bmatrix} 1 & 2 & 5 \\ 0 & 1 & 4 \\ 0 & 0 & 1 \end{bmatrix}$$

Now as we know the relation

$$A^{-1} = (LU)^{-1} = U^{-1}L^{-1}$$

$$\text{So } A^{-1} = \begin{bmatrix} 1 & 2 & 5 \\ 0 & 1 & 4 \\ 0 & 1 & 1 \end{bmatrix} \begin{bmatrix} 1 & 0 & 0 \\ 0 & -1 & 0 \\ -1 & 1 & -1/2 \end{bmatrix} = \begin{bmatrix} -4 & 3 & -5/2 \\ -4 & -4 & -2 \\ -1 & 1 & -1/2 \end{bmatrix}$$

UNSOLVED PROBLEMS

1. Find the inverse of given by using Crout's method.

$$\begin{bmatrix} 2 & 1 & -1 \\ 0 & 2 & 1 \\ 5 & 2 & -3 \end{bmatrix}$$

2. Solve using crout's method

(i) $2x + 3y + z = -1$; $5x + y + z = 9$; $3x + 2y + 4z = 11$

(ii) $2x + 3y + z = 9$; $x + 2y + 3z = 6$; $3x + y + 2z = 8$

CHAPTER 7

Numerical Differentiation

7.1 INTRODUCTION

Numerical differentiation is the process of finding or evaluating the derivative or derivatives of a given function at some particular value of the independent variable when the values of the function corresponding to the given value of the independent variable are known. The problem of differentiation is solved by first approximating the function by a polynomial by means of an interpolation formula and then differentiating this formula as many times as desired. If the values of the argument are equally spaced, we represent the function by Newton's formula. Further, if it is required to find the derivative of the function at a point near to the beginning or end of a set of tabular values, we use Newton's forward or backward formula. When it is desired to find derivative at a point near the middle of the table, we should use one of the central difference formulae. If the entries are given for unequal intervals of the argument, Newton's divided difference formula should be used to represent the function. Now in the next context we will discuss these formulae one by one.

7.2 NUMERICAL DIFFERENTIATION

Numerical differentiation is the process of finding the numerical value of a derivative of a given function at a given point. In general, numerical differentiation is more difficult than numerical integration. Numerical differentiation is a technique of numerical analysis to produce an estimate of the derivative of a mathematical function or function subroutine using values from the function and perhaps other knowledge about the function. Numerical differentiation methods are obtained by using one of the following techniques:

1. **Method Based on Interpolation**
2. **Method Based on Finite Difference Operators**

7.2.1 Method Based on Interpolation

Methods based on interpolation uses interpolation formulae for determining the derivatives of the function whereas methods based on finite difference uses finite difference operators for determining the derivatives. Firstly, we will discuss method based on interpolation.

7.2.1.1 *Derivative Using Forward Difference Formula*

This method is applied only when the values of the argument are equally spaced and we have been asked to find the derivative of a function at a point which in the upper half of the table.

As we know that Newton's forward difference interpolating formula is given by

$$y(x) = y_1 + \Delta y_1 + \frac{\Delta^2 y_1}{2!} u(u-1) + \frac{\Delta^3 y1}{3!} u(u-1)(u-2) + \ldots\ldots + \frac{\Delta^n y_1}{n!} u(u-1)(u-2)\ldots(u-(n-1)) \quad \ldots(i)$$

Where Δ(delta) is known as forward difference operator. This is calculated as:

$$\Delta^j y_i = \Delta^{j-1} y_{i+1} - \Delta^{j-1} y_i$$

where i refers to difference at ith point(position) and j refers to order.

i.e. forward differences can be calculated as $\Delta y_1 = y_2 - y_1$, $\Delta y_2 = y_3 - y_2$......are called as first order forward differences.

Similarly, second order forward differences are written as

$$\Delta^2 y_1 = \Delta y_2 - \Delta y_1,$$
$$\Delta^2 y_2 = \Delta y_3 - \Delta y_2 \text{......etc.}$$

Here $x_1 < x < x_2$ (i.e x lies at start or top part of the table)

Where $\frac{x - x_1}{h} = u \Rightarrow x - x_1 = uh$

Differentiate both sides of eq. (i) w.r.t u, we get

$$\frac{dy}{du} = \Delta y_1 + \frac{2u-1}{2!} \cdot \Delta^2 y_1 + \frac{3u^2 - 6u - 2}{3!} \cdot \Delta^3 y_1 + \ldots\ldots$$

As
$$u = \frac{x - x_1}{h}, \quad \frac{du}{dx} = \frac{1}{h}$$

$$\frac{dy}{dx} = \frac{dy}{du} \cdot \frac{du}{dx}$$

$$= \frac{1}{h}\left[\Delta y_1 + \frac{2u-1}{2!}\Delta^2 y_1 + \frac{3u^2 - 6u + 2}{3!}\Delta^3 y_1 + \frac{4u^3 - 18u^2 + 22u - 6}{4!} \cdot \Delta^4 y_1 + \ldots\right] \quad \ldots(ii)$$

At $x = x_1$, $u = 0$,∴ substitute $u = 0$ in (ii), we get

$$(dy/dx)_{x1} = 1/h\left[\Delta y_1 - \frac{1}{2}\Delta^2 y_2 + \frac{1}{3}\Delta^3 y_1 - \Delta^4 y_1 + \ldots\ldots\right] \cdot \frac{1}{h} \quad \ldots(iii)$$

Again diff. (ii) w.r.t x, we get

$$d^2y/dx^2 = d/du(dy/dx) \cdot du/dx$$

$$= \frac{1}{h}\left[\frac{2}{2!}\Delta^2 y_1 + \frac{6u-6}{3!}\Delta^3 y_1 + \frac{12u^2-36u+22}{4!}\Delta^4 y_1 \ldots\ldots\right]\cdot\frac{1}{h}$$

Substituting $u = 0$, as in this case $x = x_1$ so, we get

$$(d^2y/dx^2)_{x1} = \frac{1}{h^2}\left[\Delta^2 y_1 - \Delta^3 y_1 + \frac{11}{12}\Delta^4 y_1 \ldots\ldots\right]$$

Similarly, the higher order derivatives can be calculated.

Algorithm of Newton's Forward Difference Differentiation Method

Algorithm of numerical differentiation using Newton's Forward Difference for the given function. Given two one dimensional arrays x and y each of size 'n' used to store the given values. One two dimensional array dy of size $n - 1$ by $n - 1$.

1. Begin
2. Read n
3. Do i = 1 to n
4. Read x[i], y[i]
5. End do
6. Read x0
7. If ((x0 < x[1]) or (x0 > x[n])) then
8. Print "the value lies outside the given range"
9. End if
10. /* table of newton's forward difference values
11. Do j = 1 to n – 1 by 1
12. Do i = 1 to n – j
13. If (J = = 1) then
14. dy [i][j] = y[i + 1] – y[i]
15. else
16. dy[i][j] = dy[i+1][j – 1] – dy[i][j – 1]
17. end if
18. end do
19. end do
20. i = 1
21. while (x0 > x[i])
22. i = i + 1

23. end while
24. k = i – 1
25. h = x[k + 1] – x[k] (where h is the difference between consecutive values of x)
26. sum = 0
27. do j = 1 to n – k
28. temp = (– 1) $^{j+1}$
29. sum = sum + (temp * dy[k][j]) /j
30. end do
31. derivative = sum/h
32. print "the value of the derivative is" derivative
33. end

7.2.1.2 *Derivative Using Backward Difference Formula*

This method is applied only when the values of the argument are equally spaced and we have been asked to find the derivative of a function at a point which in the lower half of the table.

As we know that Newton's backward difference interpolating formula is given by

$$y(x) = y_n + \nabla y_n u + \frac{\nabla^2 y_n}{2!} u(u+1) + \frac{\nabla^3 y_n}{3!} u(u+1)(u+2) + \ldots\ldots$$

$$+ \frac{\nabla^n y_n}{n!} u(u+1)(u+2)\ldots(u+(n-1)) \qquad \ldots\text{(i)}$$

where ∇(del) is known as backward difference operator. This is calculated as

$$\nabla^j y_i = \nabla^{j-1} y_i - \nabla^{j-1} y_{i-1}$$

where *i* refers to difference at i^{th} point(position) and *j* refers to order. i.e backward differences can be calculated as $\nabla y_2 = y_2 - y_1$, $\nabla y_3 = y_3 - y_2$........

$\nabla y_n = y_n - y_{n-1}$ are called as first order backward differences.

Similarly, second order backward differences are written as

$$\nabla^2 y_3 = \nabla y_3 - \nabla y_2,$$

$$\nabla^2 y_4 = \nabla y_4 - \nabla y_3 \ldots\ldots\ \text{etc.}$$

Here $x_{n-1} < x < x_n$ (i.e *x* lies at end or bottom part of the table)

Where $\frac{x - x_n}{h} = u \Rightarrow x - x_n = uh$

Differentiate (i) w.r.t *u*, we get

$$dy/du = \nabla y_n + \frac{(2u+1)}{2!} \cdot \nabla^2 y_n + \frac{(3u2 + 6u + 2)}{3!} \cdot \nabla^3 y_n + \ldots\ldots$$

As $$u = \frac{x - x_n}{h}, \text{du/dx} = \frac{1}{h}$$

$$\frac{dy}{dx} = \frac{dy}{du} \cdot \frac{du}{dx}$$

$$= \frac{1}{h}\left[\nabla y_n + \frac{2u+1}{2!}\nabla^2 y_n + \frac{(3u^2+6u+2)}{3!}\nabla^3 y_n + \frac{4u^3+18u^2+22u+6}{4!}\nabla^4 y_n + \ldots\right] \quad \text{...(ii)}$$

As at $x = x_n$, $u = 0$, substitute $u = 0$ in (ii), we get

$$(dy/dx)x_n = \frac{1}{h}\left[\nabla y_n + \frac{1}{2}\nabla^2 y_n + \frac{1}{3}\nabla^3 y_n + \frac{1}{4}\cdot\nabla^4 y_n + \ldots\ldots\right]$$

Again differentiate (iii) w.r.t x, we get

$$d^2y/dx^2 = d/du(dy/dx).du/dx$$

$$= \frac{1}{h}\left[\nabla^2 y_n + \frac{6u+6}{3!}\nabla^3 y_1 + \frac{6u^2+18u+11}{4!}\nabla^4 y_n \ldots\ldots\right]\frac{1}{h}$$

Substitute $u = 0$, we get

$$(d^2y/dx^2)x_n = \frac{1}{h^2}\left[\nabla^2 y_n + \nabla^3 y_n + \frac{11}{12}\nabla^4 y_n \ldots\ldots\right]$$

Similarly, we can calculate higher order derivatives.

Algorithm of Newton's Backward Difference Differentiation Method

Algorithm of numerical differentiation using Newton's Backward Difference for the given function. Given two one dimensional arrays x and y each of size 'n' used to store the given values. One two dimensional array dy of size $n - 1$ by $n - 1$.

1. Begin
2. Read n
3. Do i = 1 to n
4. Read x[i], y[i]
5. End do
6. Read x0
7. If ((x0 < x[1]) or (x0 > x[n])) then
8. Print "the value lies outside the given range"
9. End if
10. /* table of newton's backward difference values
11. Do j = 1 to n – 1 by 1
12. Do i = j + 1 to n

```
If (j = = 1) then
dy[i][j] = y[i] – y[i – 1]
i++
else
dy[i][j] = dy[i][j – 1] – dy[i – 1][j – 1]
i++
end if
end do
end do
i = 1
while (x0 > x[i])
i = i + 1
end while
k = i
h = x[k] – x[k – 1]   (where h is the difference between consecutive values of x)
sum = 0
do j = 1 to k – 1
sum = sum + (dy[k][j]) /j
end do
derivative = sum/h
print "the value of the derivative is" derivative
end
```

7.2.1.3 *Derivative Using Divided Difference Formula*

This method is applied only when the values of the argument are unequally spaced. As we know that Newton's divided difference interpolating formula is given by

$$y(x) = y_1 + \Delta_d y_1 (x - x_1) + \Delta_d^2 y_1 (x - x_1)(x - x_2) + \Delta_d^3 y_1 (x - x_1)(x - x_2)(x - x_3) + \ldots\ldots$$

$$\ldots\ldots + \Delta_d^n y_1 (x - x_1)(x - x_2)(x - x_3)\ldots\ldots\ldots (x - x_n) \quad \ldots(1)$$

where Δ_d is known as divided difference operator. This is calculated as

$$\Delta_d y_1 = \frac{y_2 - y_1}{x_2 - x_1},\ \Delta_d y_2 \frac{y_3 = y_2}{x_3 - x_2},\ \Delta_d^2 y_1 = \frac{\Delta_d y_2 - \Delta_d y_1}{x_3 - x_1}$$

$$\Delta_d^2 y_2 = \frac{\Delta_d y_3 - \Delta_d y_2}{x_4 - x_2},\quad \frac{\Delta_d^2 y_2 - \Delta_d^2 y_1}{x_4 - x_1}$$

we have assumed that $x_1 < x < x_2$

if we have $x_k < x < x_{k+1}$, then Equation (1) becomes

$$y(x) = y_k + \Delta_d y_k(x - x_k) + \Delta_d^2 yk(x - x_k)(x - x_{k+1}) + \Delta_d^3 yk(x - x_k)(x - x_{k+1})(x - x_{k+2}) + \ldots\ldots$$

$$\ldots\ldots + \Delta_d^n yk(x - x_k)(x - x_{k+1})(x - x_{k+2})\ldots\ldots\ldots (x - x_n) \quad \ldots(2)$$

Differentiate (2) w.r.t x, we get

$$\frac{dy}{dx} = \Delta_d y_k + \Delta_d^2 y_k\,[(x - x_k)(x - x_{k+1})] + \Delta_d^3 y_k\,[\,(x - x_{k+1})\,(x - x_{k+2}) + (x - x_k)\,(x - x_{k+2})$$

$$+ (x - x_k)(x - x_{k+1})] + \ldots\ldots\ldots$$

Differentiate above equation again w.r.t x

$$\frac{d^2y}{d^2x} = 2\,\Delta_d^2 y_k + \Delta_d^3 y_k\,[(x - x_{k+1})(x - x_{k+2})] + [(x - x_k) + (x - x_{k+2}) + (x - x_k)$$

$$+ (x - x_{k+1}) + \ldots\ldots\ldots$$

$$= 2\,\Delta_d^2 y_k + \Delta_d^3 y_k\,\,[(x - x_k) + (x - x_{k+1}) + (x - x_{k+2})] + \ldots\ldots\ldots$$

Algorithm of Newton's Divided Difference Differentiation Method

Algorithm of numerical differentiation using Newton's Divided Difference for the given function. Given two one dimensional arrays x and y each of size 'n' used to store the given values. One two dimensional array dy of size $n - 1$ by $n - 1$.

1. Begin
2. Read n
3. Do i = 1 to n
4. Read x[i], y[i]
5. End do
6. Read x0
7. If ((x0 < x[1]) or (x0 > x[n])) then
8. Print "the value lies outside the given range"
9. End if
10. /* table of divided difference values
11. Do j = 1 to n – 1 by 1
12. Do i = 1 to n – j
13. If (j = = 1) then
14. dy[i][j] = (y[i + 1] – y[i])/(x[i + 1] – x[i])
15. else

```
dy[i][j] = (dy[i + 1][j – 1] – dy[i+1][j – 1])/(x[i + j] – x[i])
end if
end do
end do
i = 1
while (x0 > x[i])
i = i + 1
end while
k = i – 1
sum = dy[k][1]
do j = 2 to n – k
sum1 = 0
do p = 0 to j – 1
prod = 1
do q = 0 to j – 1
if ( p <> q) then
prod = prod * (x0 – x[k + p])
end if
end do
sum1 = sum1 + prod
end do
sum = sum + dy[k][j] * sum1
end do
print " the derivative is " sum
end
```

7.2.1.4 Derivative Using Stirling Central Difference Formula

This method is applied only when the values of the argument are equally spaced and we are asked to find the derivative of a function at the middle value of the interval.

We know that Stirling formula is

$$y(x) = y(x_0 + h) = y_0 + \frac{(\Delta y_0 + \Delta y_{-1})}{2} + \frac{u^2}{2!}\Delta^2 y_{-1} + \frac{u(u^3 - 1)}{3!}\frac{(\Delta^3 y_{-1} + \Delta^3 y_{-2})}{2}$$

$$\frac{u^2(u^2 - 1)}{4!}\Delta^4 y_{-2} + \ldots\ldots\ldots \qquad \ldots(1)$$

Differentiating (1) w.r.t u, we get

$$\frac{dy}{du} = \frac{(\Delta y_0 + \Delta y_{-1})}{2} + \frac{2u}{2!}\Delta^2 y_{-1} + \frac{3u^2-1}{3!}\frac{(\Delta^3 y_{-1} + \Delta^3 y_{-2})}{2} + \frac{(4u^3-2u)}{4!}\Delta^4 y_{-2} + \ldots\ldots$$

As $$u = \frac{x-x_0}{h}, \frac{du}{dx} = \frac{1}{h}$$

Now $$\frac{dy}{dx} = \frac{dy}{du}\cdot\frac{du}{dx}$$

$$= \frac{1}{h}\left[\frac{1}{2}(\Delta y_0 + \Delta y_{-1}) + u\Delta^2 y_{-1} + \frac{(3u^2-1)}{6}\frac{(\Delta^3 y_{-1} + \Delta^3 y_{-2})}{2} + \frac{(2u^2-u)}{12}\Delta^4 y_{-2} + \ldots\right]$$

As at $x = x_0$, $u = 0$, $\therefore$ after substituting $u = 0$, we get

$$(dy/dx)x_0 = \frac{1}{h}\left[\frac{1}{2}(\Delta y_0 + \Delta y_{-1}) - \frac{1}{6}\frac{(\Delta^3 y_{-1} + \Delta^3 y_{-2})}{2} + \frac{1}{30}\frac{\Delta^5 y_{-2} + \Delta^5 y_{-3}}{2} + \ldots\ldots\right]$$

Similarly $$(d^2y/dx^2)_{x0} = \frac{1}{h^2}\left[\frac{1}{2}(\Delta^2 y_{-1} - \frac{1}{12}\Delta^4 y_{-2} + \frac{1}{90}\Delta^6 y_{-3} + \ldots\ldots\right]$$

Similarly we can find higher order derivatives.

7.2.1.5 *Derivative Using Bessel's Central Difference Formula*

This method is also applied when the values of the argument are equally spaced and we have to ask for finding the derivative of a function at the middle value of the interval. As the values of the arguments are equally spaced, so the Bessel's formula becomes:

$$y(x) = y_0 + u\Delta y_0 + \frac{u(u-1)}{2!}\Delta^2 y_{-1} + \frac{\Delta^2 y_{-1} + \Delta^2 y_0}{2} + \frac{u\left(u-\tfrac{1}{2}\right)u(u-1)}{3!}\Delta^3 y_{-1}$$

$$+ \frac{u(u+1)\,u(u-1)(u-2)}{4!} + \frac{\Delta^4 y_{-2} + \Delta^4 y_{-1}}{2} + \frac{\left(u-\tfrac{1}{2}\right)u(u+1)\,u(u-1)}{5!}\Delta^3 y_{-1} + \ldots\ldots \quad \ldots(1)$$

As $$u = \frac{x-x_0}{h}, \frac{du}{dx} = \frac{1}{h}$$

$$\frac{dy}{dx} = \frac{dy}{du}\cdot\frac{du}{dx} = \frac{1}{h}\frac{dy}{dx} \quad \ldots(2)$$

Differentiate eq. (1) w.r.t u, we get $\frac{dy}{du}$

(Simplification is left as an exercise for the students)

After determine the $\frac{dy}{du}$, substitute the solution into (2).

7.2.2 Method Based on Finite Difference Operators

These methods use finite difference operators for determining the derivatives.

7.2.2.1 *Derivative Using Forward Difference Operator (Δ)*

As we know the relation between *D*(differential operator) and Δ(forward difference operator) is given by

$$1 + \Delta = E = e^{hD}$$

By taking log on both sides, we get

$$hD \log e = \log(1 + \Delta)$$

As $\log e = 1$, so above equation reduces to

$$hD = \log(1 + \Delta)$$

Expand this by using $\log(1+x) = x - \frac{x^2}{2} + \frac{x^3}{3} - \frac{x^4}{4} \ldots\ldots\ldots$

$$\Rightarrow \qquad hD = \Delta - \frac{\Delta^2}{2} + \frac{\Delta^3}{3} - \frac{\Delta^4}{4} \ldots\ldots\ldots$$

$$D = \frac{1}{h}\left[\Delta - \frac{\Delta^2}{2} + \frac{\Delta^3}{3} - \frac{\Delta^4}{4} \ldots\ldots\ldots\right]$$

By squaring, we get

$$D^2 = \frac{1}{h^2}\left[\Delta - \frac{\Delta^2}{2} + \frac{\Delta^3}{3} - \frac{\Delta^4}{4} \ldots\ldots\ldots\right]^2$$

Expand this by using the formula

$$(u + v + w + \ldots\ldots.)^2 = u^2 + v^2 + w^2 + \ldots\ldots. + (2uv + 2uw + 2ux + 2vw + \ldots\ldots\ldots)$$

$$= \frac{1}{h^2}\left[\left(\Delta^2 + \frac{1}{4}\Delta^4 + \frac{1}{9}\Delta^6 + \frac{1}{16}\Delta^8 + \ldots\right) + \left(-\Delta^3 + \frac{2}{3}\Delta^4 - \frac{1}{2}\Delta^5 + \frac{1}{3}\Delta^5 + \frac{2}{5}\Delta^6 + \frac{1}{4}\Delta^6 \ldots\right)\right]$$

After simplification, we get

$$= \frac{1}{h^2}\left[\Delta^2 - \Delta^3 + \frac{11}{12}\Delta^4 - \frac{5}{6}\Delta^5\right]$$

$$D^3 = D^2 \cdot D$$

$$= \frac{1}{h^3}\left[\Delta - \frac{1}{2}\Delta^2 \frac{1}{3}\Delta^3 \ldots\ldots\right]^3$$

$$= \frac{1}{h^3}\left[\Delta^3 - \frac{3}{2}\Delta^4 \ldots\ldots\right]$$

By attaching y_1 to the D, D^2, D^3 equations, we have

$$Dy_1 = (dy/dx)_{x1} = \frac{1}{h}\left[\Delta y_1 - \frac{\Delta^2}{2} y_1 + \frac{\Delta^3}{3} y_1 - \frac{\Delta^4}{4} y_1 \ldots\ldots\right]$$

$$D^2 y_1 = (d^2y/dx^2)_{x1} = \frac{1}{h^2}\left[\Delta^2 y_1 - \Delta^3 y_1 + \frac{11}{12}\Delta^4 y_1 - \frac{5}{6}\Delta^5 y_1 + \ldots\ldots\right]$$

$$D^3 y_1 = (d^3y/dx^3)_{x1} = \frac{1}{h^3}\left[\Delta^3 y_1 - \frac{3}{2}\Delta^4 y_1 + \ldots\ldots\right]$$

Similarly we can find higher order derivatives.

7.2.2.2 *Derivative Using Backward Difference Operator* (∇)

As we know the relation between D (differential operator) and ∇(backward difference operator) is given by

$$1 - \nabla = E^{-1} = e^{-hD}$$

By taking log on both sides, we get

$$-hD \log e = \log(1 - \nabla)$$

As $\log e = 1$, so above equation reduces to

$$-hD = \log(1 - \nabla)$$

Expand this by using $\log(1 - x) = -x - \frac{x^2}{2} + \frac{x^3}{3} - \frac{x^4}{4} \ldots\ldots$

$$-hD = -\left[\nabla + \frac{\nabla^2}{2} + \frac{\nabla^3}{3} + \frac{\nabla^4}{4} + \ldots\ldots\right]$$

$$\therefore \quad D = \frac{1}{h}\left[\nabla - \frac{\nabla^2}{2} + \frac{\nabla^3}{3} - \frac{\nabla^4}{4} + \ldots\ldots\right]$$

After squaring on both sides, we get

$$D^2 = \frac{1}{h^2}\left[\nabla + \frac{\nabla^2}{2} + \frac{\nabla^3}{3} + \frac{\nabla^4}{4} + \ldots\ldots\right]^2$$

$$= \frac{1}{h^2}\left[\left(\nabla^2 + \frac{\nabla^4}{4} + \frac{\nabla^6}{9} + \frac{\nabla^8}{16} + \ldots\ldots\right) + \left(\nabla^3 + \frac{2}{3}\nabla^4 + \frac{1}{2}\nabla^5 + \frac{2}{5}\nabla^6 + \frac{1}{3}\nabla^5 + \ldots\ldots\right)\right]$$

$$= \frac{1}{h^2}\left[\nabla^2 + \nabla^3 + \frac{11}{12}\nabla^4 + \frac{5}{6}\nabla^5 + \ldots\ldots\right]$$

Similarly $$D^3 = \frac{1}{h^3}\left[\nabla + \frac{\nabla^2}{2} + \frac{\nabla^3}{3} + \frac{\nabla^4}{4} + \frac{\nabla^5}{5} + \ldots\ldots\right]$$

$$= \frac{1}{h^3}\left[\nabla^3 + \frac{3}{2}\nabla^4 + \ldots\ldots\right]$$

By attaching y_n to the D, D^2, D^3 equations, we have

$$Dy_n = (dy/dx)_{xn} = \frac{1}{h}\left[\nabla y_n + \frac{\nabla^2}{2}y_n + \frac{\nabla^3}{3}y_n + \frac{\nabla^4}{4}y_n + \ldots\ldots\right]$$

$$D^2y_n = (d^2y/dx^2)_{xn} = \frac{1}{h^2}\left[\nabla^2 + \nabla^3 + \frac{11}{12}\nabla^4 + \frac{5}{6}\nabla^5 + \ldots\ldots\right]$$

$$D^3y_n = (d^3y/dx^3)_{xn} = \frac{1}{h^3}\left[\nabla^3 + \frac{3}{2}\nabla^4 + \ldots\ldots\right]$$

Similarly we can find higher order derivatives.

7.3 ERRORS IN NUMERICAL DIFFERENTIATION

In numerical differentiation, the error in the higher order derivatives occurs due to the fact that, although the tabulated function and its approximating polynomial would agree at the set of data points, their slopes at these points may vary considerably. Numerical differentiation is, therefore, an unsatisfactory process and should be used only in rare cases.

The numerical computation of derivatives involves two types of errors: **truncation errors** and **rounding errors.**

SOLVED PROBLEMS

Problem 1: From the data given below

x:	1	1.2	1.4	1.6	1.8
y:	3.2	4.7	5.6	6.3	7.6

Compute dy/dx and d^2y/dx^2 at $x = 1.1$

Solution: As x values are equally spaced and we have to calculate the first and second order derivatives at a point($x = 1.1$) that lies in the first half (upper half) of the table. So, we will use forward difference formula here.

x	y	Δy_i	$\Delta^2 y_i$	$\Delta^3 y_i$	$\Delta^4 y_i$
1	3.2	1.5	– 0.6	0.4	0.4
1.2	4.7	0.9	– 0.2	0.8	
1.4	5.6	0.7	0.6		
1.6	6.3	1.3			
1.8	7.6				

Here h (interval size) = 0.2

As $x = 1.1$ lies between 1st and 2nd point, i.e between $x = 1.0$ and $x = 1.2$. So we start forward difference formula from **first** point.

By using Newton's forward formula

$$(dy/dx) = \frac{1}{h}\left[\Delta y_1 + \frac{2u-1}{2!}\Delta^2 y_1 + \frac{3u^2-6u+2}{3!}\Delta^3 y_1 + \frac{4u^3-18u+22-6u}{4!}\cdot\Delta^4 y_1\right]$$

where $$u = \frac{x-x_1}{h} = \frac{1.1-1}{0.2} = 0.5$$

As, from the table, values $\Delta y_1 = 1.5, \Delta^2 y_1 = -0.6, \Delta^3 y_1 = 0.4, \Delta^4 y_1 = 0.4$

$$(dy/dx)_{x=1.1} = \frac{1}{0.2}\left[1.5 + \frac{\{2*(0.5)-1\}}{2}\cdot(-0.6) + \frac{\{3(0.5)^2-6*(0.5)+2\}}{6}\cdot(0.4)\right.$$

$$\left. + \frac{\{4+(0.5)^3-18(0.5)^2+22(0.5)-6\}}{24}\cdot(0.4)\right]$$

$$= 5\,[1.5 + 0 - 0.016 + 0.016] = 7.5$$

$$(d^2y/dx^2)_{x=1.1} = \frac{1}{h^2}\left[\frac{2}{2!}\Delta^2 y_1 + \frac{6u-6}{3!}\Delta^3 y_1 + \frac{12u^2-36u+22}{4!}\Delta^4 y_1\right]$$

$$= \frac{1}{0.25}\left[(-0.6) + \frac{\{6*(0.5)-6\}}{6}(0.4) + \frac{\{12(0.5)^2-36(0.5)+22\}}{24}\cdot(0.4)\right]$$

$$= 4\,[-0.6 - 0.2 + 0.116] = -0.684 \text{ (approx.)}$$

Problem 2: Compute dy/dx at $x = 0.1$ from the following table:

x:	0.1	0.2	0.3	0.4
y:	0.9975	0.9900	0.9776	0.9604

Solution: Take a = 0.1. The difference table is:

x	y	Δy	$\Delta^2 y$	$\Delta^3 y$
0.1	0.9975			
		−0.0075		
0.2	0.9900		−0.0049	
		− 0.0124		0.0001
0.3	0.9776		− 0.0048	
		− 0.0172		
0.4	0.9604			

Here $h = 0.1$ and $y_0 = 0.9975$

Here $h = 0.1$ and $y_0 = 0.9975$

$$\left[\frac{dy}{dx}\right]_{x=0.1} = \frac{1}{h}\left[\Delta y_0 - \frac{1}{2}\Delta^2 y_0 + \frac{1}{3}\Delta^3 y_0\right]$$

$$= \frac{1}{0.1}\left[-0.0075 - \frac{1}{2}(-0.0049) + \frac{1}{3}(0.0001)\right]$$

$$= -0.050167.$$

Problem 3: The distance(s) covered as a function of time (t) by a car during 70 km journey is given in the table:

t (minutes):	10	20	30	40	50
s (distance):	13	28	42	52	70

Calculate the speed and acceleration of the car at $t = 25$

Solution: As t values are equally spaced and we have to calculate the speed (ds/dt) and acceleration (d^2s/dt^2) at a point ($t = 25$) that lies in the first half (upper half) of the table. So, we will use forward difference formula here for calculating the speed and acceleration.

T	s	Δs_i	$\Delta^2 s_i$	$\Delta^3 s_i$	$\Delta^4 s_i$
10	13	15	− 1.0	− 3.0	15.0
20	28	14	− 4.0	12.0	
30	42	10	8.0		
40	52	18			
50	70				

Here h(interval size) = 10

As $t = 25$ lies between 2nd and 3rd point, i.e between $t = 20$ and $t = 30$. So we start forward difference formula from **second** point.

∴ By using Newton's forward formula

$$(ds/dt) = \frac{1}{h}\left[\Delta s_2 + \frac{2u-1}{2!}\Delta^2 s_2 + \frac{3u-6u+2}{3!}\Delta^3 s_2\right]$$

where $u = \dfrac{t-t_2}{h} = \dfrac{25-20}{10} = 0.5$

As from the table values $\Delta s_2 = 14$, $\Delta^2 s_2 = -4.0$, $\Delta^3 s_2 = 12.0$

$$(ds/dt)_{t=25} = \frac{1}{10}\left[14 + \frac{\{2*(0.5)-1\}}{2}\cdot(-4.0) + \frac{\{3(0.5)^2 - 6*(0.5)+2\}}{6}\cdot(12.0)\right]$$

$$= 0.1\,[14 + 0 - 0.5]$$

$$= 1.35 \text{ km/m.}$$

$$(d^2s/dt^2)_{t=25} = \frac{1}{h^2}\left[\frac{2}{2!}\Delta^2 y_2 + \frac{6u-6}{3!}\Delta^3 y_2\right]$$

$$= \frac{1}{100}\left[(-4.0) + \frac{\{6*(0.5)^2-6\}}{6}(12.0)\right]$$

$$= 0.01\,[-4-6]$$

$$= -0.1$$

Problem 4: From the data given below

x :	1	2	3	4	5
y :	3.2	4.7	5.6	6.3	7.6

Compute dy/dx and d^2y/dx^2 at $x = 3.5$

Solution: As x values are equally spaced and we have to calculate the first and second order derivatives at a point($x = 3.5$) that lies in the second half (lower half) of the table. So, we will use backward difference formula here.

x	y	∇y_i	$\nabla^2 y_i$	$\nabla^3 y_i$	$\nabla^4 y_i$
1	3.2				
2	4.7	1.5			
3	5.6	0.9	0.6		
4	6.3	0.7	– 0.2	0.4	
5	7.6	1.3	0.6	0.8	0.4

Here h(interval size) = 1.0

As $x = 3.5$ lies between 3rd and 4th point, i.e between $x = 3$ and $x = 4$. So, we start backward difference formula from **fourth** point.

By using Newton's backward formula

$$dy/dx = \frac{1}{h}\left[\nabla y_4 + \frac{(2u-1)}{2!}\cdot\nabla^2 y_4 + \frac{3u^2-6u+2}{3!}\cdot\Delta^3 y_4\right]$$

where $\quad u = \frac{x - x_4}{h} = \frac{3.5-4}{1} = -0.5$

As from the table values $\nabla y_4 = 0.7,\ \nabla^2 y_4 = -0.2,\ \nabla^3 y_4 = 0.4$

$$(dy/dx)_{x=3.5} = \frac{1}{1}\left[0.7 + \frac{\{2*(-0.5)+1\}}{2}\cdot(-0.2) + \frac{\{3(-0.5)^2+6*(-0.5)+2\}}{6}\cdot(0.4)\right]$$

$$= [0.7 + 0 - 0.016] = 0.684$$

$$(d^2y/dx^2)_{x=3.5} = \frac{1}{h^2}\left[\nabla^2 y_4 + \frac{6u-6}{3!}\cdot\nabla^3 y_4\right]$$

$$= 1\left[(-0.2) + \frac{\{6(-0.5)+6\}}{6}\cdot(0.4)\right] = [-0.2+0.2] = 0.$$

Problem 5: The distance covered by an athlete for the 50 meter race is given in the following table:

Time (sec):	0	1	2	3	4	5	6
Distance (meter):	0	2.5	8.5	15.5	24.5	36.5	50

Compute the speed of the athlete at $t = 5$ sec., correct to two decimals.

Solution: Here we are to find derivative at $t = 5$ which is near the end of the table, hence we shall use the formula obtained from Newton's backward difference formula. The backward difference table is as follows:

t	s	∇s	$\nabla^2 s$	$\nabla^3 s$	$\nabla^4 s$	$\nabla^5 s$	$\nabla^6 s$
0	0						
		2.5					
1	2.5		3.5				
		6		–2.5			
2	8.5		1		3.5		
		7		1		–3.5	
3	15.5		2		0		1
		9		1		–2.5	
4	24.5		3		–2.5		
		12		–1.5			
5	36.5		1.5				
		13.5					
6	50						

The speed of the athlete at $t = 5$ sec is given by

$$\left(\frac{ds}{dt}\right)_{t=5} = \frac{1}{h}\left[\Delta s_5 + \frac{1}{2}\Delta^2 s_5 + \frac{1}{3}\Delta^3 s_5 + \frac{1}{4}\Delta^4 s_5 + \frac{1}{5}\Delta^5 s_5\right]$$

$$= \frac{1}{1}\left[12 + \frac{1}{2}(3) + \frac{1}{3}(1) + \frac{1}{4}(0) + \frac{1}{5}(-3.5)\right]$$

$$= 13.1333 \text{ H} \approx 13.13 \text{ metre/sec.}$$

Problem 6: Population of a city (obtained form census) is shown in the following table:

Year(x) :	1950	1960	1970	1980	1990
Population(y) : (in thousands)	17	28	41	55	72

Find the rate of growth of population in 1980 and 1983.

Solution: As x values are equally spaced and we have to calculate the rate of growth (dy/dx) at a point($x = 1980$ and $x = 1983$) that lies in the second half (lower half) of the table. So, we will use backward difference formula here.

x	Y	∇y_i	$\nabla^2 y_i$	$\nabla^3 y_i$	$\nabla^4 y_i$
1950	17				
1960	28	11			
1970	41	13	2		
1980	55	14	1	–1	
1990	72	17	3	2	3

(i) Rate of growth of the population in year 1980

Here h (interval size) = 10

As $x = 1980$ which is the 4th point. So, we start backward difference formula from **fourth** point.

By using Newton's backward formula

$$(dy/dx)_{x4} = \frac{1}{h}\left[\nabla y_4 + \frac{1}{2}\nabla^2 y_4 \frac{1}{3}\nabla^3 y_4\right]$$

As from the table values $\nabla y_4 = 14$, $\nabla^2 y_4 = 1$, $\nabla^3 y_4 = -1$

$$(dy/dx)_{x=1980} = \frac{1}{10}\left[14 + \frac{1}{2}\cdot 1 \frac{1}{10} + \frac{1}{3}\cdot(-1)\right]$$

$$= 0.1[14 + 0.5 - 0.33]$$

$$= 0.1[14.17]$$

$$= 1.417$$

(ii) Rate of growth of the population in year 1983

Here h (interval size) = 10

As $x = 1983$ lies between 4th and 5th point, i.e between $x = 1980$ and $x = 1990$. So we start backward difference formula from **fifth** point.

By using Newton's backward formula

$$dy/dx = \frac{1}{h}\left[\nabla y_5 + \frac{(2u+1)}{2!}\cdot\nabla^2 y_5 + \frac{(3u^2+6u+2)}{3!}\cdot\Delta^3 y_5 + \frac{(4u^3+18u+22u+6)}{4!}\Delta^4 y_5\right]$$

where $$u = \frac{x - x_5}{h} = \frac{1983-1990}{10} = -0.7$$

As from the table values $\nabla y_5 = 17$, $\nabla^2 y_5 = 3$, $\nabla^3 y_5 = 2$, $\nabla^4 y_5 = 3$

$$(dy/dx)_{x=1983} = 0.1\left[17 + \frac{\{2*(-0.7)+1\}}{2}\cdot 3 + \frac{\{3(-0.7)\,2 + 6*(-0.7)+2\}}{6}\cdot 2\right.$$

$$\left. + \frac{\{4(-0.7)3 + 18(-0.7)\,2 + 22(-0.7)+6\}}{24}\cdot 3\right]$$

$$= 0.1[17 - 0.6 - 0.24 - 0.24] = 1.592$$

Problem 7: From the data given below:

x:	3	5	11	19	26
y:	9.7	13.2	22.1	30	42.1

Compute dy/dx and d^2y/dx^2 at $x = 8$.

Solution: As x values are unequally spaced. So, we will use divided difference formula here.

x	y	$\Delta_d y_i$	$\Delta^2_d y_i$	$\Delta^3_d y_i$	$\Delta^4_d y_i$
3	9.7	1.75	−0.03375	−0.00008	0.0001
5	13.2	1.48	−0.0351	0.004	
11	22.1	0.9875	0.0494		
19	30	1.7286			
26	42.1				

As derivative is required at $x = 8$, which lies between 2nd and 3rd point, so we take the difference at second point.

As we know Newton's divided difference formula is

$$dy/dx = \Delta_d y_2 + \Delta^2_d y_2[(x-x_2)+(x-x_3)] + \Delta^3_d y_2[(x-x_2)(x-x_3)+(x-x_2)(x-x_4)+(x-x_3)(x-x_4)]+\ldots$$

After substitution, we get

$$(dy/dx)_{x=8} = 1.48+(-0.0351)[(8-5)+(8-11)] + 0.0040\,[(8-5)(8-11)+(8-5)(8-19) + (8-11)(8-19)]$$

$$= 1.48 - (0.0351)[0] + 0.0040[-9-33+33]$$
$$= 1.48 - 0.036$$
$$= 1.44$$

$(d2y/dx2)\ x = 8$ is calculated same by using double differentiation formula, so it is left as an exercise for the students.

Problem 8: The distance(*s*) covered by a car in a given time (*t*) is given as

Time (in minutes):	10	12	16	17	24
Distance (in km.):	12	15	20	22	32

Compute the speed of the car at $t = 14$ minutes.

Solution: As x values are unequally spaced, so we will use divided difference formula here.

T	s	$\Delta_d s_i$	$\Delta^2_d s_i$	$\Delta^3_d s_i$	$\Delta^4_d s_i$
10	12	1.5	–0.42	0.0274	–0.0035
12	15	1.25	0.15	–0.015	
16	20	2	0		
17	22	2			
22	32				

As derivative is required at t=14, which lies between 2nd and 3rd point, so we take the difference at second point.

As we know Newton's divided difference formula is

$$ds/dt = \Delta_d s_2 + \Delta^2_d s_2[(t-t_2)+(t-t_3)] + \Delta^3_d s_2[(t-t_2)(t-t_3)+(t-t_2)(t-t_4)+(t-t_3)(t-t_4)] + \ldots\ldots.$$

After substitution, we get

$$= 1.25 + 0.15[(14-12) + (14-16)] + (-0.015)[(14-12)(14-16)+(14-12)(14-17)+(14-16)(14-17)]$$
$$= 1.25 + 0.15[0] + (-0.015)[(2).(-2) + (2).(-3) + (-2)(-3)]$$
$$= 1.25 + 0.06$$
$$= 1.31$$

Problem 9: The pressure of a certain gas was recorded at intervals of 0.5 seconds and the results are listed in the following table:

Time(x):	0.35	0.40	0.45	0.50	0.55	0.60	0.65
Pressure(y):	1.521	1.506	1.488	1.467	1.444	1.418	1.389

Estimate the rate of change of Pressure at $x = 0.50$

Solution: As x values are equally spaced and the derivative is required near the middle of the table, so we use Stirling's formula.

t	x	Δx	$\Delta^2 x$	$\Delta^3 x$	$\Delta^4 x$	$\Delta^5 x$	$\Delta^6 x$
0.35	1.521(x_{-3})	−0.015	−0.003	0	0.001	−0.003	0.006
0.4	1.506(x_{-2})	−0.018	−0.003	0.001	−0.002	0.003	
0.45	1.488(x_{-1})	−0.021	−0.002	−0.001	0.001		
0.5	1.467(x_0)	−0.023	−0.003	0			
0.55	1.444(x_1)	−0.026	−0.003				
0.6	1.418(x_2)	−0.029					
0.65	1.389(x_3)						

$$dx/dt = \frac{1}{h}\left[\frac{\Delta x_0 - \Delta x_{-1}}{2} - \frac{1}{6}\frac{\Delta^3 x_{-1} + \Delta^3 x_{-2}}{2} + \frac{1}{30}\frac{\Delta^5 x_{-2} + \Delta^5 x_{-3}}{2} + \dots\dots\right]$$

By taking values from the table, we get

$$(dx/dt)_{t=0.50} = \frac{1}{.05}\left[\frac{(-0.023 - 0.02)}{2} - \frac{1}{6}\frac{(-0.001 + 0.001)}{2} + \frac{1}{30}\frac{(0.003 - 0.003)}{2}\right]$$

$$= \frac{1}{.05}\left[\frac{-0.044}{2} - 0 + 0\right] = -0.44$$

Problem 10: Using Bessel's formula, Compute $f'(7.5)$ from the following table:

x:	7.47	7.48	7.49	7.50	7.51	7.52	7.53
f (x):	0.193	0.195	0.19	0.201	0.203	0.206	0.208

Solution: As *x* values are equally spaced and the derivative is required near the middle of the table, so we use Bessel's formula.

x	U	y_u	Δ	Δ^2	Δ^3	Δ^4	Δ^5	Δ^6
7.47	− 3	0.193	0.002	0.001	− 0.001	0.000	0.003	− 0.010
7.48	− 2	0.195	0.003	0.000	− 0.001	0.003	− 0.007	
7.49	− 1	0.198	0.003	− 0.001	0.002	− 0.004		
7.50	0	0.201	0.002	0.001	− 0.002			
7.51	1	0.203	0.003	− 0.001				
7.52	2	0.206	0.002					
7.53	3	0.208						

We know that Bessel's formula is given by

$$y_u = y_0 + \frac{u(u-1)}{2!}\cdot\frac{\Delta^2 y_{-1} + \Delta^2 y_0}{2} + \frac{u(u-1/2)(u)(u-1)}{3!}\Delta^3 y_{-1}$$

$$+ \frac{(u+1)(u)(u-1)(u-2)}{4!}\cdot\frac{\Delta^4 y_{-2} + \Delta^4 y_{-1}}{2} + \frac{u(u-1/2)(u)(u-1)}{5!}\Delta^5 y_{-2}$$

$$+ \frac{(u+2)(u+1)(u)(u-1)(u-2)(u-3)}{6!}\frac{\Delta^6 y_{-3} + \Delta^6 y_{-2}}{2}$$

As $u = \dfrac{x - x_0}{h}, du/dx = 1/h \quad \therefore dy/dx = \dfrac{dy}{du} \cdot \dfrac{du}{dx} = \dfrac{1}{h} \cdot \dfrac{dy}{du}$

Differentiating (1) w.r.t to u and substituting $u = 0$, we get

$$(dy/dx)_{u=0} = 1/h(dy/du)_{u=0}$$

$$= \frac{1}{h}\left[\Delta y_0 - \frac{1}{4}(\Delta^2 y_{-1} - \Delta^2 y_0)\frac{1}{12}\Delta^3 y_{-1} + \frac{1}{24}(\Delta^4 y_{-2} + \Delta^4 y_{-1}) - \frac{1}{120}\cdot\Delta^5 y_{-2} - \frac{1}{240}\cdot(\Delta^6 y_{-3} + \Delta^6 y_{-2})\right]$$

$$(dy/dx)_{7.5} = \frac{1}{0.01}\left[0.002 - \frac{1}{4}(-0.001 + 0.001) + \frac{1}{12}(0.002) + \frac{1}{24}(-0.004 + 0.003) - \frac{1}{120}(-0.007) - \frac{1}{240}(-0.001 + 0)\right]$$

$$= 0.2 + 0 + 0.01666 - 0.0416 + 0.00583 + 0.00416 = 0.223$$

Problem 11: Compute $f'(4)$ from the following data:

x:	0	2	5	1
f (x):	0	8	125	1

Solution: Though this problem can be solved by Newton's divided difference formula, we are giving here, as an alternative, Lagrange's method. Lagrange's polynomial, in this case, is given by

$$f(x) = \frac{(x-2)(x-5)(x-1)}{(0-2)(0-5)(0-1)}(0) + \frac{(x-0)(x-5)(x-1)}{(2-0)(2-5)(2-1)}(8) + \frac{(x-0)(x-2)(x-1)}{(5-0)(5-2)(5-1)}(125) + \frac{(x-0)(x-2)(x-5)}{(1-0)(1-2)(1-5)}(1)$$

$$= -\frac{4}{3}(x^3 - 6x^2 + 5x) + \frac{25}{12}(x^3 - 3x^2 + 2x) + \frac{1}{4}(x^2 - 7x + 10x) = x^3$$

$\therefore \quad f¢\,(x) = 3x^2$

when $x = 4$, $f'(4)^2 = 3(4)^2 = 48$

UNSOLVED PROBLEMS

1. The distance(s) covered as a function of time (t) by an athlete during his run for the 50 meter race is given in the following table

Time (seconds):	0	1	2	3	4	5	6
Distance (metres):	0	2.5	8.5	15.5	24.5	36.5	50

Compute the speed of the athlete at $t = 5$ seconds.

2. The distance covered by a car in a given time (t) is given in the following table:

Time (minutes):	10	12	14	16	18
Distance (Kms.):	12	15	20	27	37

Compute the acceleration of the car at $t = 13$ minutes.

3. Compute $y'(x_0)$, $y''(x_0)$ at $x_0 = 1.5$ by using the following table:

x:	1.0	1.5	2.0	2.5	3.0
$y = f(x)$:	1.729	1.691	1.505	1.416	1.311

4. Compute first and second derivatives of the function tabulated below at $x = 0.6$

x:	0.4	0.5	0.6	0.7	0.8
y:	1.5836	1.7974	2.0442	2.3275	2.6511

5. Compute $y2'(0)$ and $y'(0)$ from the given table:

x:	0	1	2	3	4	5
y:	4	8	15	7	6	2

6. Compute $y'(1.5)$ and $y'(1.5)$ from the following table:

x:	1.5	2.0	2.5	3	3.5	4
$f(x)$:	3.375	7	13.625	24	38.875	59.

7. Using Stirling's formula for finite differences Compute $y'(1.4)$ and $y''(1.4)$ given the following table:

x :	1.0	1.2	1.4	1.6	1.8
$f(x)$:	2.7183	3.3201	4.0552	4.9530	6.0496

8. Compute $y'(4)$ from the given table:

x:	1	2	4	8	10
y:	0	1	5	21	27.

9. Compute the numerical value of $y'(10°)$ for $y = \sin x$ given that:

$\sin 0° = 0.000$, $\sin 10° = 0.1736$,

$\sin 20° = 0.3420$, $\sin 30° = 0.5000$, $\sin 40° = 0.6428$.

10. Given the following table:

x :	0	1	2	3	4
$f(x)$:	6.9897	7.4036	7.7815	8.1281	8.4510

Compute $y'(2)$ and $y''(2)$ using appropriate central-difference formula.

11. Given the following table of values of x and y:

x :	1.0	1.05	1.10	1.15	1.20	1.25	1.30
y :	1.0	1.0247	1.0488	1.0723	1.0954	1.1180	1.1401

Compute dy/dx and d^2y/dx^2 at $x = 1.25$

12. Compute the first and second derivatives for the function tabulated below at the point $x = 3.0$:

x:	3	3.2	3.4	3.6	3.8	4.0
y:	– 14	– 10.032	– 5.296	0.256	6.672	14.

13. (i) A slider in a machine moves along a fixed straight rod. Its distance x cm along the rod is given below for various values of the time t seconds. Find the velocity of the slider and its acceleration when $t = 0.3$ second.

t:	0	0.1	0.2	0.3	0.4	0.5	0.6
x:	30.13	31.62	32.87	33.64	33.95	33.81	33.24.

(ii) A slider in a machine moves along a fixed straight rod. Its distance x(in cm) along the rod is given at various times t (in secs).

t:	0	0.1	0.2	0.3	0.4	0.5	0.6
x:	30.28	31.43	32.98	33.54	33.97	33.48	32.13

Evaluate dx/dt at $t = .1$ and at $t = .5$.

14. The following data gives the corresponding values of $f(\theta) = \tan\theta$ and θ (in degrees).

θ :	12°	20°	28°	36°	44°
tanθ :	0.213	0.364	0.532	0.727	0.966

Compute the value of $\sec^2\theta$ at 26°.

15. A slider in a machine moves along a fixed straight rod. Its displacement x meter along the rod is given below for various values of time t seconds. Find the velocity of the slider and its acceleration when

(i) $t = 1.1$ sec. (ii) $t = 1.6$ sec.

t:	1.0	1.1	1.2	1.3	1.4	1.5	1.6
x:	7.989	8.403	8.781	9.129	9.451	9.750	10.031

16. Given sin0° = 0.000, sin 10° = 0.1736, sin 20°=0.3420, sin 30° = 0.5000, Sin 40° = 0.6428

(i) Find numerical value of d^2y/dx^2 at $x = 20°$ for $y = \sin x$

(ii) Find numerical value of dy/dx at $x = 10°$ for $y = \sin x$

17. A rod is rotating in a plane about one of its ends. If the following table the angle θ Radians through which the rod has turned for different values of time t seconds. Find its angular velocity and angular acceleration when $t = 0.7$ seconds.

t **seconds :**	0.0	0.2	0.4	0.6	0.8	1.0
θ radians :	0.0	0.12	0.48	1.10	2.0	3.20

18. Compute dy/dx at $x = 1$ form the following table by constructing a central difference table:

x :	0.7	0.8	0.9	1.0	1.1	1.2	1.3
y :	0.644218	0.717356	0.783327	0.841471	0.891207	0.932039	0.963558

19. The following data gives the corresponding values of $f(\theta) = \sec\theta$ and θ(in degrees).

θ:	23°	29°	37°	41°	47°
secθ:	1.09	1.14	1.25	1.33	1.47

Find the value of secθ tanθ at 33°

20. Find the rate of change of pressure with respect to specific volume when v =0.75 based on the following data

Volume(v):	0. 5	0. 75	1. 0	1. 25	1. 5
Pressure(p):	0. 13	0. 42	1. 0	1. 95	3. 38

21. Compute dy/dx at x =0.1 from the following table

X:	0. 1	0. 2	0. 3	0. 4
Y:	0. 9975	0. 9900	0. 9976	0. 9604

22. Compute $f'(1.1)$ and $f''(1.1)$ from the following table:

X:	1.0	1.2	1.4	1.6	1.8	2.0
Y:	0.0	0.1280	0.5540	1.2960	2.4320	4.0

23. From the following table obtain dy/dx and d^2y/dx^2 for $x = 1.2$

X:	1.0	1.2	1.4	1.6	1.8	2	2.2
Y:	2.7183	3.3201	4.0552	4.9530	6.0496	7.3891	9.0250

24. The table below gives the result of an observation. θ is the observed temperature in degrees of a vessel of cooling water, t is the time taken in minutes from the beginning of observations:

T:	1	3	5	7	9
θ:	85.3	74.5	67.0	6.5	54.3

Find the approximate rate of colling at $t = 3$ and t – 3.5

25. Compute $f'(10)$ from the follwoing data by using appropriate Newton's method

X:	3	5	11	27	34
Y:	–13	23	899	17315	35606

CHAPTER 8

Numerical Integration

8.1 INTRODUCTION

In numerical analysis, **numerical integration** constitutes a broad family of algorithms for calculating the numerical value of a definite integral, and by extension, the term is also sometimes used to describe the numerical solution of differential equations. This article focuses on calculation of definite integrals. Numerical integration is the approximate computation of an integral using numerical techniques. The numerical computation of an integral is sometimes called quadrature. Ueberhuber uses the word "**quadrature**" to mean numerical computation of a univariate integral, and "**cubature**" to mean numerical computation of a multiple integral. There are a wide range of methods available for numerical integration.

The Process of determining the value of a definite integral from a set of numerical values of the function is called numerical integration. The process is called numerical quadrature when applied to the integration of a function of a single variable. The process is called numerical cubature when applied to determine double integral of a function of two independent variables.

The basic problem considered by numerical integration is to compute an approximate solution to a definite integral. We study the numerical evaluation of the integrals of the type

$$I = \int_a^b f(x)\, dx \qquad ...(1)$$

Where a and b are finite and $f(x)$ is a continuous function of x for $a \le x \le b$. All other type of integrals like the limits being infinite or integrand having the singularities etc. can be transformed to the above form in most of the cases. If $f(x)$ is a smooth well-behaved function and the limits of integration are fixed, the method of Gaussian quadrature is very effective.

We know that integral (1) represents the area (figure below) bounded by the curve $y = f(x)$, the x-axis and the two ordinates at $x = a$ and $x = b$. By dividing the interval (a, b) into a finite number of equal intervals, finding the areas of these sub-intervals and then summing all such areas and then the area of the required integration is possible.

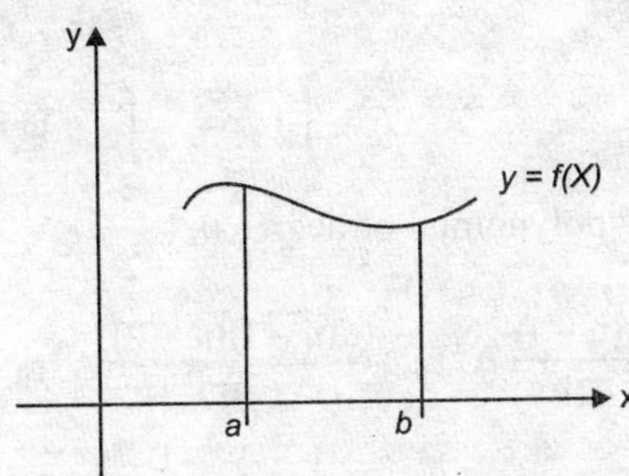

In this approach, we can distinguish two types:

(a) Selecting the equidistant intervals, we can derive some formulae. The simplest two formulae of this class are the trapezoidal and Simpson's rule. Because of the simplicity, these formulae are quite often used.

(b) Select the maximum possible accuracy to be achieved and then the intervals are selected to yield this accuracy. Generally, formulae derived with this approach are twice as accurate as the formulae of (a). These formulae are known as Gaussian quadrature ('numerical integration' is often called numerical Quadrature).

The trapezoidal and Simpson's formulae are special cases of general Newton-Cotes quadrature formulae. The use of Newton-Cotes and other formulae of higher order than Simpson's rule is not necessary in most of the engineering applications.

8.2 NEWTON -COTES INTEGRATION FORMULA

8.2.1 A General Quadrature Formula (Newton-cotes Formula)

The Newton-Cotes formulas are an extremely useful and straightforward family of numerical integration techniques. The Newton-Cotes formulas, the most commonly used numerical integration methods, approximate the integration of a complicated function by replacing the function with many polynomials across the integration interval. The integration of the original function can then be obtained by summing up all polynomials whose "areas" are calculated by the weighting coefficients and the values of the function at the nodal points.

Let $$I = \int_a^b y\,dx$$

where $y = f(x)$ is given for certain equidistant values of x say $x_0, x_0 + h, x_0 + 2h, \ldots\ldots\ldots$

Let the range (a, b) be divided into n equal parts, each of width h so that $b - a = nh$.

Let $x_0 = a, x_1 = x_0 + h = a + h, x_2 = a + 2h, \ldots\ldots, x_n = a + nh = b$, we have assumed that the there are corresponding $n + 1$ ordinates $y_0, y_1, \ldots\ldots, y_n$.

$$\therefore \qquad I = \int_a^b y dx = \int_{x_0}^{x_0+nh} f(x)dx = \int_{x_0}^{x_0+nh} p_n(x)dx$$

Where $P_n(x)$ is interpolating polynomial of degree n.

$$= \int_0^n \left(y_0 + u\Delta y_0 + \frac{u(u-1)}{2!}\Delta^2 y_0 + \frac{u(u-1)(u-2)}{3!}\Delta^3 y_0 + \ldots\ldots \right) \text{hdu} \qquad \ldots(1)$$

Since dx = hdu, and when $x = x_0$, $u = 0$ and when $x = x_0 + nh$, $u = n$

$$= h\int_0^n \left(y_0 + u\Delta y_0 + \frac{(u^2-u)}{2!}\Delta^2 y_0 + \frac{(u^3-3u+2u)}{3!}\Delta^3 y_0 + \ldots\ldots \right) du$$

$$= h\left[y_0(u) + \frac{u^2}{2}\Delta y_0 + \frac{(u^3/3 - u^2/2)}{2}\Delta^2 y_0 + \frac{1}{6}(u^3/4 - u^3 + u^2)\Delta^3 y_0 + \ldots\ldots \right]_0^n$$

$$\int_{x_0}^{x_n} y dx = h\left[ny_0 + \frac{n^2}{2}\Delta y_0 + \frac{1}{2}(n^3/3 - n^2/2)\Delta^2 y_0 + \frac{1}{6}(n^4/4 - n^3 + n^2)\Delta^3 y_0 + \ldots\ldots \right] \qquad \ldots(2)$$

The equation (2), called **Newton-Cotes quadrature formula**. A number of formulae can be deduced from this by substituting n = 1,2,3,...... and 6, respectively, in formula (2).

8.2.2 TRAPEZOIDAL RULE (n = 1)

Trapezoidal rule is one of the simplest method of finding the integral of a function. Trapezoidal rule suits best if the integrand is a linear function.

Consider the function $f(x)$, whose graph between $x = a$ and $x = b$ is shown in figure. An approximation to the area under the curve is obtained by dividing the interval $[a, b]$ into n strips of width h each, and approximating the area of each strip by that of a trapezoid as shown by shaded area.

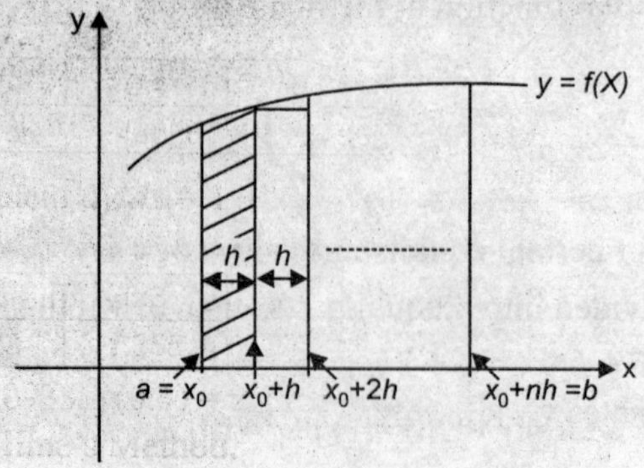

The general quadrature formula is given by

$$\int_{x_0}^{x_0+nh} f(x)dx$$

$$= h\left[ny_0 + \frac{n^2}{2}\Delta y_0 + \frac{1}{2}(n^3/3 - n^2/2)\,\Delta^2 y_0 + \frac{1}{6}(n^4/4 - n^3 + n^2)\,\Delta^3 y_0 + \ldots\ldots\right] \quad \ldots(1)$$

By substituting $n = 1$, in equation (1), we have

$$\int_{x_0}^{x_0+nh} f(x)dx = h\left[1.y_0 + \frac{1}{2}\Delta y_0\right]$$

Since other differences do not exist so if $n = 1$.

$$= h\left[y_0 + \frac{1}{2}(y_1 + y_0)\right]$$

$$= \frac{h}{2}(y_0 + y_1) \quad \ldots(2)$$

This is the area of one strip. In this way we calculate the area of all the strips and sum it.

$$\int_{x_0}^{x_n} f(x)dx = \int_{x_0}^{x_0+nh} f(x)dx$$

$$= \int_{x_0}^{x_0+h} f(x)dx + \int_{x_0+h}^{x_0+2h} f(x)dx + \ldots\ldots \int_{x_0+(n-1)h}^{x_n+nh} f(x)dx$$

$$= \frac{h}{2}(y_0 + y_1)\frac{h}{2}(y_1 + y_2) + \ldots\ldots\ldots\frac{h}{2}(y_{n-1} + y_n)$$

$$= \frac{h}{2}[y_0 + y_n) + 2(y_1 + y_2 + y_3 + \ldots\ldots\ldots\ldots + y_{n-1})$$

$$= \frac{h}{2}[(\text{sum of first \& last ordinate}) + 2(\text{sum of the remaining ordinates})]$$

This is known as **Trapezoidal rule.** And this method gives the accurate results when the given curve is a Linear curve.

The truncation error for this method is greater than any other method, because of its simplicity it is worth discussing the order of truncation error involved in the trapezoidal rule.

8.2.2.1 Truncation Error in Trapezoidal Rule

Mainly the area between the curve and trapezoid produces the error. In the neighborhood of $x = x_0$, we can expand $y = f(x)$ by Taylor's series in powers of $x - x_0$. That is

$$y(x) = y_0 + \frac{(x-x_0)}{1!} y_0' + \frac{(x-x_0)^2}{2!} y_0'' + \ldots\ldots \qquad \ldots(1)$$

where $y'_0 = [y'(x)]\ x = x_0$

$$\int_{x_0}^{x_1} y dx = \int_{x_0}^{x_1} \left[y_0 + \frac{(x-x_0)}{1!} y_0' + \frac{(x-x_0)^2}{2!} y_0'' + \ldots\ldots \right] dx$$

$$= \left[y_0 x + \frac{(x-x_0)^2}{2!} y_0' + \frac{(x-x_0)^3}{3!} y_0'' + \ldots\ldots \right]_{x_0}^{x_1}$$

$$= y_0(x_1 - x_0) + \frac{(x_1-x_0)^2}{2!} y_0' + \frac{(x_1-x_0)^3}{3!} y_0'' + \ldots\ldots$$

$$= hy_0 + \frac{h^2}{2!} y_0' + \frac{h^3}{3!} y_0'' + \ldots\ldots\ldots \qquad \ldots(2)$$

If h is the equal interval width

Also $\int_{x0}^{x_1} y dx \approx \frac{h}{2}(y_0 + y_1) =$ area of the first trapezium $= A_0$ ...(3)

Putting $x = x_1$ in (1), we get

$$y(x_1) = y_1 = y_0 + \frac{(x_1-x_0)}{1!} y_0' + \frac{(x_1-x_0)^2}{2!} y_0'' + \ldots\ldots$$

i.e.,

$$y_1 = y_0 + \frac{h}{1!} y_0' + \frac{h}{2!} y_0'' + \ldots\ldots\ldots \qquad \ldots(4)$$

$$A_0 \approx \frac{h}{2}\left[y_0 + y_0 + \frac{h}{1!} y_0' + \frac{h^2}{2!} y_0'' + \ldots\ldots\ldots \right]$$

$$\approx hy_0 + \frac{h^2}{2} y_0' + \frac{h^3}{4} y_0'' + \ldots\ldots$$

Subtracting A_0 value from (2), we get

$$\int_{x0}^{x_1} y dx - A_0 = h^3 y_0'' (1/3! - 1/4) + \ldots\ldots$$

$$= -\frac{1}{12} h^3 y_0'' + \ldots\ldots$$

$\therefore$ The error in the first interval (x_0, x_1) is $-\frac{1}{12}h^3 y_0''$ (neglecting other terms).

Similarly error in the ith interval= $-\frac{1}{12}h^3 y_{i-1}'' + ...$

$\therefore$ Total cumulative error (approx.),

$$E = -\frac{1}{12}h^3(y_0'' + y_1'' + y_2'' + ...y_{n-1}'')$$

$|E| < \frac{nh^3}{12} \cdot M$ where M is the maximum value of $|y_0''|, |y_1''|, |y_2''|, ...$

$< \frac{(b-a)h^2}{12} \cdot M$ if the interval is (a, b) and $h = \frac{(b-a)}{n}$

Hence, the error in the trapezoidal rule is of the order of $\boldsymbol{h^2}$.

8.2.2.2 Algorithm of Trapezoidal Rule

Algorithm of numerical integration using Trapezoidal method when the points are given for the function. Given one dimensional arrays x and y each of size '$n + 1$' used to store the given values.

1. Begin
2. Read n /* n is the number of intervals
3. Do I = 1 to n + 1
4. Read x[i], y[i]
5. End do
6. h = x[2] – x[1] (h is the interval size between two consecutive values of x)
7. sum = y[1] + y[n + 1]
8. do I = 2 to n by 1
9. sum + = 2 * y[i]
10. end do
11. integrated_value = (h/2) * sum
12. print "the area under the curve is" integrated_value
13. End

Algorithm of numerical integration using Trapezoidal method when the function '*f*' is given. The function to be integrated is divided into '*n*' equal size intervals between the limits *a* (lower limit) and *b* (upper limit).

1. Begin
2. Read n /* n is the number of intervals */

3. Read a, b
4. h = (b – a) / n
5. Do I = 1 to n – 2 by 1
6. x2 = (a + I * h)
7. sum + = 2 * f(x2)
8. End do
9. integrated_value = (h/2) * sum
10. print "the area under the curve is" integrated_value
11. End

8.2.2.3 Flow-chart for Trapezoidal Rule

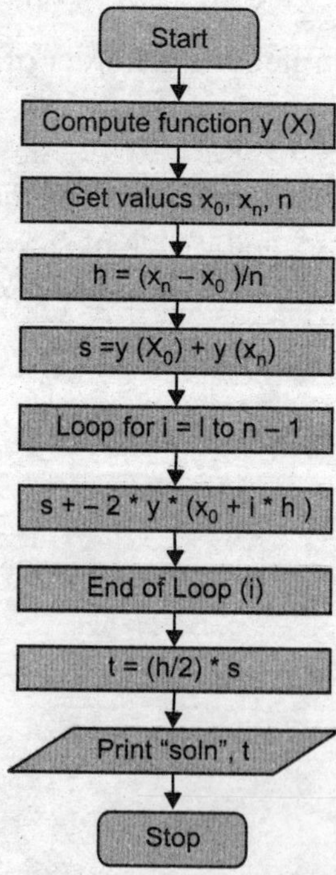

8.2.2.4 Program for Trapezoidal Method of Numerical Integration in C

```
# include <stdio.h>
# include <conio.h>
# include <math.h>
# include <process.h>
# include <string.h>
```

```
float fun(float);
void main()
{
float result=1;
float a,b;
float h,sum;
int i,j;
int n;
clrscr();
printf("\n\n Enter the range - ");
printf("\n\n Lower Limit a - ");
scanf("%f" ,&a);
printf("\n\n Upper Limit b - ");
scanf("%f" ,&b);
printf("\n\n Enter number of subintervals - ");
scanf("%d" ,&n);
h=(b-a)/n;
sum=0;
sum=fun(a)+fun(b);
for(i=1;i<n;i++)
{
sum+=2*fun(a+i);
}
result=sum*h/2;
printf("n\n\n\n Value of the integral is %6.4f\t",result);
printf("\n\n\n Press Enter to Exit");
getch();
}
float fun(float x)
{
float temp;
temp = 1/(1+(x*x));
return temp;
}
```

Output:

```
Enter the range -
Lower Limit a - 0
Upper Limit b - 6
Enter number of subintervals - 6
Value of the integral is 1.4108
Press Enter to Exit
```

8.2.2.5 Program for Trapezoidal Method of Numerical Integration in C++

```
# include <iostream.h>
# include <conio.h>
# include <math.h>
# include <process.h>
# include <string.h>
float fun(float);
void main()
{
float result=1;
float a,b;
float h,sum;
int i,j;
int n;
clrscr();
cout<<"\n\n Enter the range - ";
cout<<"\n\n Lower Limit a - ";
cin>>a;
cout<<"\n\n Upper Limit b - ";
cin>>b;
cout<<"\n\n Enter number of subintervals - ";
cin>>n;
h=(b-a)/n;
sum=0;
sum=fun(a)+fun(b);
for(i=1;i<n;i++)
{
sum+=2*fun(a+i);
}
```

```
result=sum*h/2;
cout<<"\n\n\n\n Value of the integral is"<<cout.width(6.4)<<result<<"f\t";
cout<<"\n\n\n Press Enter to Exit";
getch();
}
float fun(float x)
{
float temp;
temp = 1/(1+(x*x));
return temp;
}
```

Output:

```
Enter the range -
Lower Limit a - 0
Upper Limit b - 6
Enter number of subintervals - 6
Value of the integral is 1.4108
Press Enter to Exit
```

8.2.3 Simpson's One-third Rule ($N = 2$)

By connecting successive points on a curve by straight lines, the area under a curve is obtained, using trapezoidal rule. In the case of Simpson's rule (1/3rd rule), we connect the successive three sets of points on a curve by second-degree parabola and then sum the areas under the parabolas to obtain the approximate area under the curve.

Consider the function $f(x)$, whose graph between $x = a$ and $x = b$ is shown in fig. An approximation to the area under the curve is obtained by dividing the interval $[a, b]$ into n strips of width h each, and approximating the area of two strips by that of the area under the parabolas as shown by shaded area.

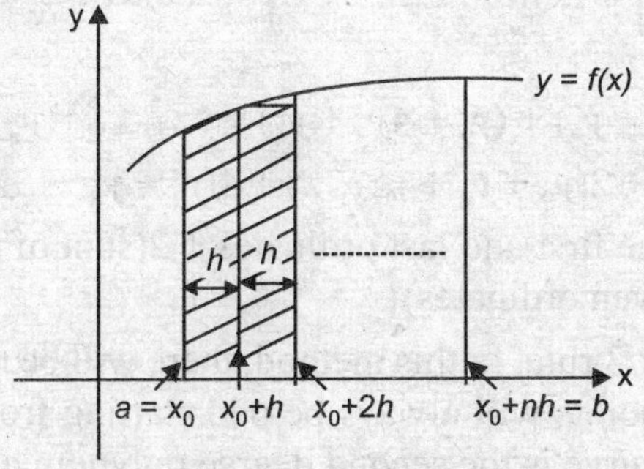

The general quadrature formula is given by

$$\int_{x_0}^{x_0+nh} f(x)dx \approx \left[ny_0 + \frac{n^2}{2}\Delta y_0 + \frac{1}{2}(n^3/3 - n^2/2)\Delta^2 y_0 + \frac{1}{6}(n^4/4 - n^3 + n^2)\Delta^3 y_0 + \ldots\ldots\right] \quad \ldots(1)$$

By substituting $n = 2$, in equation (1), we have

$$\int_{x_0}^{x_2} f(x)dx = h\,[2y_0 + 2\Delta y_0 + 1/2\,(8/3 - 2)\Delta^2 y_0\,]$$

(since other terms vanish)

$$= h\,[2y_0 + 2(y_1 - y_0) + 1/3(E-1)^2 y_0\,]$$
$$= h\,[2y_0 + 2y_1 - 2y_0 + 1/3(y_2 - 2y_1 + y_0)]$$
$$= h\,[1/3y_0 + 4/3y_1 + 1/3y_2]$$
$$= h/3(y_0 + 4y_1 + y_2)$$

Similarly calculating area under different strips as:

$$\int_{x_2}^{x_4} f(x)dx = h/3(y_2 + 4y_3 + y_4)$$

$$\int_{x_i}^{x_{i+2}} f(x)dx = h/3(y_i + 4y_{i+1} + y_{i+2})$$

if n is an even integer, last integral will be

$$\int_{x_{n-2}}^{x_n} f(x)dx = h/3(y_{n-2} + 4y_{n-1} + y_n)$$

By adding all these integrals, we have

$$\int_{x_0}^{x_n} f(x)dx = \int_{x_0}^{x_2} f(x)dx + \int_{x_2}^{x_4} f(x)dx + \ldots\ldots\ldots + \int_{x_{n-2}}^{x_n} f(x)dx$$

$$= h/3[(y_0 + 4y_1 + y_2) + (y_2 + 4y_3 + y_4) + \ldots\ldots\ldots + (y_{n-2} + 4y_{n-1} + y_n)]$$
$$= h/3[(y_0 + y_n) + 2(y_2 + y_4 + \ldots\ldots\ldots) + 4(y_1 + y_3 + \ldots\ldots\ldots)]$$

= h/3[sum of the first and last ordinates + 2(sum of remaining odd ordinates) + 4(sum of even ordinates)]

This is known as Simpson's 1/3 rule. In this method, there will be minimum of two intervals or three points and the number of points will always be odd starting from 3. This method gives the accurate results when the given curve is the second degree polynomial.

8.2.3.1 Truncation Error in Simpson's One-third Rule

By expanding $y = f(x)$ by using Taylor's series in the neighborhood of $x = x_0$, we have

$$y = y_0 + \frac{(x-x_0)}{1!} y_0' + \frac{(x-x_0)^2}{2!} y_0'' + \ldots\ldots \quad \ldots(1)$$

$$\int_{x0}^{x_2} y\,dx = \int_{x0}^{x_2} \left[y_0 + \frac{(x-x_0)}{1!} y_0' + \frac{(x-x_0)^2}{2!} y_0'' + \ldots\ldots \right] dx$$

$$= \left[y_0 x + \frac{(x-x_0)^2}{2!} y_0' + \frac{(x-x_0)^3}{3!} y_0'' + \ldots\ldots \right]_{x0}^{x2}$$

$$= y_0(x_2 - x_0) + \frac{(x_2-x_0)^2}{2!} y_0' + \frac{(x_2-x_0)^3}{3!} y_0''$$

$$= 2hy_0 + \frac{4h^2}{2!} y_0' + \frac{8h^3}{3!} y_0'' + \frac{16h^4}{4!} y_0''' + \ldots\ldots\ldots\ldots$$

$$= 2hy_0 + 2h^2 y_0' + \frac{4h^3}{3} y_0'' \frac{2h^4}{3} y_0''' + \frac{4h^5}{15} y'''' + \ldots\ldots\ldots\ldots \quad \ldots(2)$$

$$A_1 = \text{area} = \int_{x0}^{x_2} y\,dx = h/3(y_0 + 4y_1 + y_2), \text{ by Simpson's rule} \quad \ldots(3)$$

By substituting $x = x_1$ in (1), we get

$$y_1 = y_0 + \frac{(x_1-x_0)}{1!} y_0' + \frac{(x_1-x_0)^2}{2!} y_0'' + \ldots\ldots$$

$$= y_0 + \frac{2h}{1!} y_0' + \frac{4h^2}{2!} y_0'' + \ldots\ldots\ldots\ldots \quad \ldots(4)$$

By substituting $x = x_2$ in (1), we get

$$y_2 = y_0 + \frac{2h}{1!} y_0' + \frac{4h^2}{2!} y_0'' + \ldots\ldots\ldots\ldots \quad \ldots(5)$$

Substituting (4) and (5) in (3), we get

$$A_1 = 2hy_0 + 2h^2 y_0' + \frac{4h^3}{3} y_0'' + \frac{2h^4}{3} y_0''' + \frac{5h^5}{18} y'''' + \ldots\ldots\ldots\ldots \quad \ldots(6)$$

By operating (2) – (6), we get

$$\int_{x0}^{x_2} ydx - A_1 = (4/15 - 5/18)h^5\ y_0'''' + ...$$

$$= \frac{-h^5}{90}\ y_0'''' + ...$$

Leaving the higher order terms involving h^6 and higher powers of h.

As we see that the main part of error in (x_0, x_2) is

$$= \frac{-h^5}{90}\ y_0''''$$

Similarly, error in the interval (x_2, x_4) is

$$= \frac{-h^5}{90}\ y_2'''' \quad \text{and so on.}$$

Hence, total error in all the intervals is given by

$$E = \frac{-h^5}{90}\ y_0'''' + y_2'''' + ...$$

$\therefore\ |E| < \dfrac{nh^5}{90}M$ is the maximum value of y_0'''', y_2'''',......, y_{2n-2}'''' since (x_{2n}, y_{2n})

is the last paired value because we require odd number of ordinates to apply Simpson's 1/3 rule (i.e. $2n$ intervals).

If the interval is (a, b) then $b - a = h(2n)$, by using this,

$$|E| < \frac{(b-a)h^4}{180} \cdot M$$

Hence the error in Simpson's 1/3 rule is of the order h^4.

8.2.3.2 Algorithm of Simpson's One-third Rule

Algorithm of numerical integration using Simpson's 1/3 method when the points are given for the function. Given one dimensional arrays x and y each of size '$n + 1$' used to store the given values.

1. Begin
2. Read n /* n is the number of intervals */
3. Do i = 1 to n + 1
4. Read x[i], y[i]
5. End do

6. h = x[2] – x[1] (h is the interval size between two consecutive values of x)
7. sum = y[1] + y[n + 1]
8. sum1 = 0
9. sum2 = 0
10. Do j = 2 to n - 1 by 2
11. sum1 + = y[j]
12. End do
13. Do k = 3 to n by 2
14. sum2 + = y[k]
15. End do
16. integrated_value = (h/2) * (sum + 2 * sum2 + 4 * sum1)
17. print "the area under the curve is" integrated_value
18. End

Algorithm of numerical integration using Simpson's 1/3 method when the function 'f' is given. The function to be integrated is divided into 'n' equal size intervals between the limits a(lower limit) and b(upper limit).

1. Begin
2. Read n
3. Read a, b
4. h = (b – a) / n
5. sum = f(a) + f(b)
6. sum1 = 0
7. sum2 = 0
8. Do I = 1 to n– 2 by 1
9. x1 = f(a + I * h)
10. Sum1 + = 4 * f(x1)
11. End do
12. Do I = 2 to n– 1 by 1
13. x1 = f(a + I * h)
14. Sum2 + = 2 * f(x1)
15. End do
16. integrated_value = (h/2) * (sum + sum2 + sum1)
17. print "the area under the curve is" integrated_value
18. End

8.2.3.3 Flow-chart of Simpson's One-third Rule

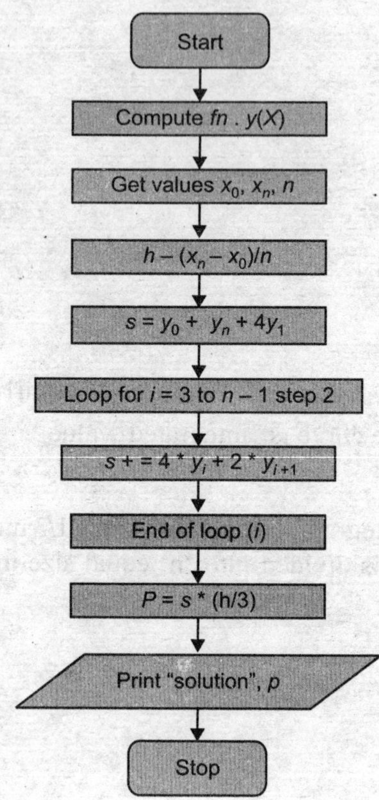

8.2.3.4 Program for Simpson's One-third Rule of Numerical Integration In C

```
# include <stdio.h>
# include <conio.h>
# include <math.h>
# include <process.h>
# include <string.h>
float fun(float);
void main()
{
float result=1;
float a,b;
float h,sum;
int i,j;
```

```
int n;
clrscr();
printf("\n\n Enter the range - ");
printf("\n\n Lower Limit a - ");
scanf("%f" ,&a);
printf("\n\n Upper Limit b - ");
scanf("%f" ,&b);
printf("\n\n Enter number of subintervals - ");
scanf("%d",&n);
h=(b-a)/n;
sum=0;
sum=fun(a)+4*fun(a+h)fun(b);
for(i=3;i<n;i+=2)
{
sum+=2*fun(a+(i-1)*h)+4*fun(a+i*h);
}
result=sum*h/3;
printf("\n\n\n\n Value of the integral is %6.4f\t",result);
printf("\n\n\n Press Enter to Exit");
getch();
}
float fun(float x)
{
float temp;
temp=1/(1+(x*x));
return temp;
}
```

Output:

```
Enter the range -
Lower Limit a - 0
Upper Limit b - 6
Enter number of subintervals - 6
Value of the integral is 1.3662
Press Enter to Exit
```

8

8.2.3.5 Program for Simpson's One-third Rule of Numerical Integration in C++

```
# include <iostream.h>
# include <conio.h>
# include <math.h>
# include <process.h>
# include <string.h>
float fun(float);
void main()
{
float result=1;
float a,b;
float h,sum;
int i,j;
int n;
clrscr();
cout<<"\n\n Enter the range - ";
cout<<"\n\n Lower Limit a - ";
cin>>a;
cout<<"\n\n Upper Limit b - ";
cin>>b;
cout<<"\n\n Enter number of subintervals - ";
cin>>n;
h=(b-a)/n;
sum=0;
sum=fun(a)+4*fun(a+h)fun(b);
for(i=3;i<n;i+=2)
{
sum+=2*fun(a+(i-1)*h)+4*fun(a+i*h);
}
result=sum*h/3;
cout<<"\n\n\n\n Value of the integral is "<<cout.width(6.4)<<result<<"f\t";
cout<<"\n\n\n Press Enter to Exit";
getch();
}
```

```
float fun(float x)
{
float temp;
temp=1/(1+(x*x));
return temp;
}
```

Output:

Enter the range -

Lower Limit a - 0

Upper Limit b - 6

Enter number of subintervals - 6

Value of the integral is 1.3662

Press Enter to Exit

8.2.4 Simpson's Three-eighth Rule (n = 3)

Simpson's 3/8th formula gives correct value of the integral if the function is cubic. Consider the function $f(x)$, whose graph between $x = a$ and $x = b$ is shown in figure. An approximation to the area under the curve is obtained by dividing the interval $[a, b]$ into n strips of width h each, and approximating the area of three strips by that of the area under the parabolas as shown by shaded area.

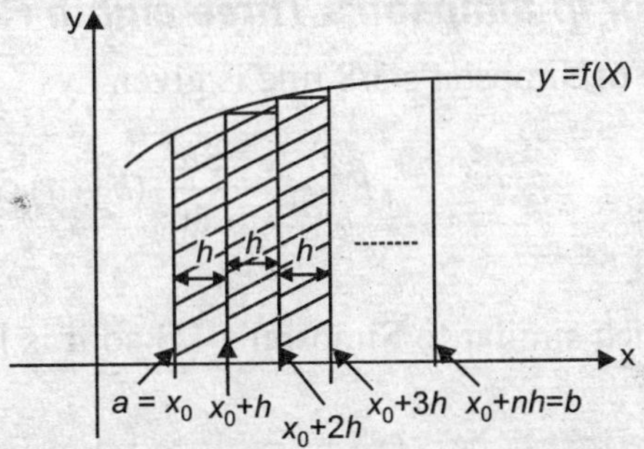

The general quadrature formula is given by

$$\int_{x_0}^{x_0+nh} f(x)dx \approx h\left[ny_0 + \frac{n^2}{2}\Delta y_0 + \frac{1}{2}(n^3/3 - n^2/2)\,\Delta^2 y_0 + \frac{1}{6}(n^4/4 - n^3 + n^2)\,\Delta^3 y_0 + \ldots\ldots\right] \quad ...(1)$$

By substituting $n = 3$, in equation (1), we have

$$\int_{x_0}^{x_3} f(x)dx = h\left[3y_0 + \frac{9}{2}\Delta y_0 + \frac{1}{2}(9/2)\,\Delta^2 y_0 + \frac{1}{6}(81/4 - 27 + 9)\,\Delta^3 y_0\right]$$

$$= h[3y_0 + 9/2(y_1 - y_0) + 9/4(E-1)^2 y_0 + 3/8(E-1)^3 y_0]$$

$$= h[3y_0 + \frac{9}{2}y_1 - \frac{9}{2}y_0 + 9/4(y_2 - 2y_1 + y_0) + 3/8(y_3 - 3y_2 + 3y_1 - y_0)]$$

$$= 3h/8(y_0 + 3y_1 + 3y_2 + y_3)$$

This is the area under the curve between the points $x = x_0$ to $x = x_3$. This covers three intervals of width h each as shown through shaded area. If we use this process repetitively $n/3$ times, we can get the area under the curve for n such intervals.

As we have seen that n is a multiple of 3, so we can write:

$$\int_{x_0}^{x_0+nh} f(x)dx = \int_{x_0}^{x_0+3h} f(x)dx + \int_{x_0+3h}^{x_0+6h} f(x)dx + \ldots\ldots + \int_{x_0+(n-3)h}^{x_0+nh} f(x)dx$$

$$= 3h/8\,[(y_0 + 3y_1 + 3y_2 + y_3) + (y_3 + 3y_4 + 3y_5 + y_6) + \ldots\ldots + (y_{n-3} + 3y_{n-2} + 3y_{n-1} + y_n)]$$

$$= 3h/8\,[(y_0 + y_n) + 3(y_1 + y_2 + y_4 + y_5 + \ldots\ldots + y_{n-1}) + 2(y_3 + y_6 + y_9 + \ldots\ldots + y_n)]$$

This is known as Simpson's 3/8 rule. This method requires minimum of four points or three intervals and the number of intervals is always a multiple of 3.

8.2.4.1 *Truncation Error in Simpson's Three-eighth Rule*

The truncation error in case of Simpson's 3/8 rule is given by

$$|E| < -\frac{3h^5}{80}(b-a)\cdot M$$

Which is of the order h^5.

As the proof is very much similar to Simpson's 1/3 so it is left as an exercise for the readers to prove.

8.2.4.2 Algorithm Of Simpson's Three-eighth Rule

Algorithm of numerical integration using Simpson's 3/8 method when the points are given for the function. Given one dimensional arrays x and y each of size '$n + 1$' used to store the given values.

1. Begin
2. Read n /* n is the number of intervals */
3. Do i = 1 to n + 1
4. Read x[i], y[i]
5. End do

6. h = x[2] – x[1] (h is the interval size between two consecutive values of x)
7. sum = y[1] + y[n + 1]
8. sum1 = 0
9. sum2 = 0
10. Do j = 2 to n - 1 by 3
11. sum1 + = y[j] + y[j + 1]
12. End do
13. Do k = 4 to n - 2 by 3
14. sum2 + = y[k]
15. End do
16. integrated_value = (h/2) * (sum + 2 * sum2 + 3 * sum1)
17. print "the area under the curve is" integrated_value
18. End

Algorithm of numerical integration using Simpson's 3/8 method when the function '*f*' is given. The function to be integrated is divided into '*n*' equal size intervals between the limits *a* (lower limit) and *b* (upper limit).

1. Begin
2. Read n
3. Read a, b
4. h = (b – a) / n
5. sum1 = f(a) + f(b)
6. sum2 = 0
7. sum3 = 0
8. Do i = 1 to n– 3 by 1
9. x1 = f(a + i * h)
10. x2 = f(a + (i + 1) *h)
11. x3 = f(a + (i + 2) * h)
12. Sum2 + = f(x1) + f(x2)
13. Sum3 + = f(x3)
14. End do
15. integrated_value = (h/2) * (sum + 3 *sum2 + 2 * sum3)
16. print "the area under the curve is" integrated_value
17. End

8.2.4.3 Flow-chart of Simpson's Three-eighth Rule

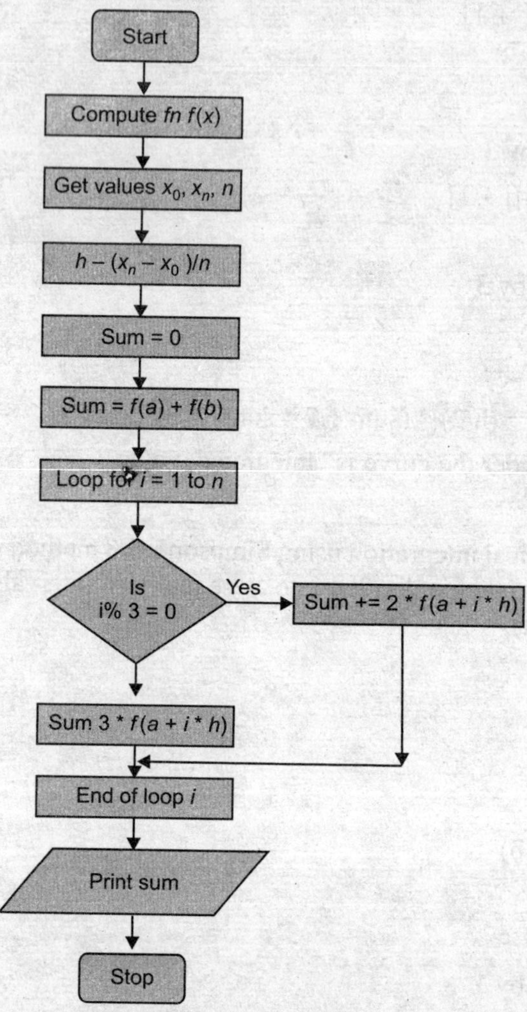

8.2.4.4 Program For Simpson's Three-eighth Rule of Numerical Integration In C

```
# include <stdio.h>
# include <conio.h>
# include <math.h>
# include <process.h>
# include <string.h>
float fun(float);
void main()
```

```
{
float result=1;
float a,b;
float h,sum;
int i,j;
int n;
clrscr();
printf("\n\n Enter the range - ");
printf("\n\n Lower Limit a - ");
scanf("%f" ,&a);
printf("\n\n Upper Limit b - ");
scanf("%f" ,&b);
printf("\n\n Enter number of subintervals - ");
scanf("%d" ,&n);
h=(b-a)/n;
sum=0;
sum=fun(a)+fun(b);
for(i=1;i<n;i++)
{
if(i%3==0)
{
sum+=2*fun(a+i*h)
}
else
{
sum+=3*fun(a+(i)*h);
}
}
result=sum*3*h/8;
printf("\n\n\n\n Value of the integral is %6.4f\t",result);
printf("\n\n\n Press Enter to Exit");
getch();
}
float fun(float x)
```

```
{
float temp;
temp=1/(1+(x*x));
return temp;
}
OUTPUT:
Enter the range -
Lower Limit a - 0
Upper Limit b - 6
Enter number of subintervals - 6
Value of the integral is 1.3571
Press Enter to Exit
```

8.2.4.5 Program For Simpson's Three-eighth Rule of Numerical Integration In C++

```
# include <iostream.h>
# include <conio.h>
# include <math.h>
# include <process.h>
# include <string.h>
float fun(float);
void main()
{
float result=1;
float a,b;
float h,sum;
int i,j;
int n;
clrscr();
cout<<"\n\n Enter the range - ";
cout<<"\n\n Lower Limit a - ";
cin>>a;
cout<<"\n\n Upper Limit b - ";
cin>>b;
cout<<"\n\n Enter number of subintervals - ";
```

```
cin>>n;
h=(b-a)/n;
sum=0;
sum=fun(a)+fun(b);
for(i=1;i<n;i++)
{
if(i%3==0)
{
sum+=2*fun(a+i*h)
}
else
{
sum+=3*fun(a+(i)*h);
}
}
result=sum*3*h/8;
cout<<"\n\n\n\n Value of the integral is "<<cout.width(6.4)<<result<<"f\t";
cout<<"\n\n\n Press Enter to Exit";
getch();
}
float fun(float x)
{
float temp;
temp=1/(1+(x*x));
return temp;
}
OUTPUT:
Enter the range -
Lower Limit a - 0
Upper Limit b - 6
Enter number of subintervals - 6
Value of the integral is 1.3571
Press Enter to Exit
```

8.2.5 Boole's Rule (n =4)

Boole's rule is suitable for finding the integral value if the polynomial is of degree 4.

The general quadrature formula is given by

$$\int_{x_0}^{x_0+nh} f(x)dx \approx \left[ny_0 + \frac{n^2}{2}\Delta y_0 + \frac{1}{2}(n^3/3 - n^2/2)\Delta^2 y_0 + \frac{1}{6}(n^4/4 - n^3 + n^2)\Delta^3 y_0 + \ldots\ldots\right] \quad (1)$$

By substituting $n = 4$ in equation (1), we have

$$\int_{x_0}^{x_4} f(x)dx = 4h\left(y_0 + 2\Delta y_0 + \frac{5}{3}\Delta^2 y_0 + \frac{2}{3}\Delta^3 y_0 + \frac{7}{90}\Delta^4 y_0\right)$$

$$= \frac{2h}{45}(7y_0 + 32y_1 + 12y_2 + 32y_3 + 7y_4$$

As n is a multiple of 4, so we can write

$$\int_{x_0}^{x_0+nh} f(x)dx = \int_{x_0}^{x_0+4h} f(x)dx + \int_{x_0+4h}^{x_0+8h} f(x)dx + \ldots\ldots + \int_{x_0+(n-4)h}^{x_0+nh} f(x)dx$$

$$= 2h/45[7y_0 + 32(y_1 + y_3 + y_5 + \ldots.) + 12(y_2 + y_6 + y_{10} + \ldots) + 14(y_4 + y_8 + y_{12} + \ldots\ldots.) + 7y_n]$$

This is known as Boole's rule.

8.2.5.1 Truncation Error in Boole's Rule

The truncation error in case of Boole's rule is given by

$$|E| < -\frac{8h^7}{945}(b-a)\cdot M$$

This is of the order $h6$.

8.2.6 Weddle's Rule (n = 6)

This rule is suitable for calculating integral value if the polynomial is of order 6.

The general quadrature formula is given by

$$\int_{x_0}^{x_0+nh} f(x)dx \approx h\left[ny_0 + \frac{n^2}{2}\Delta y_0 + \frac{1}{2}(n^3/3 - n^2/2)\Delta^2 y_0 + \frac{1}{6}(n^4/4 - n^3 + n^2)\Delta^3 y_0 + \ldots\ldots\right]$$

...(1)

By substituting $n = 6$, in equation (1), we have

$$\int_{x_0}^{x_0+6h} f(x)dx = h\,[6y_0 + 18\Delta y_0 + \frac{1}{2}(72-18)\Delta^2 y_0 + \frac{1}{6}(324-216+36)\Delta^3 y_0 + ...]$$

$$= h\,[6y_0 + 18\Delta y_0 + 27\Delta^2 y_0 + 24\Delta^3 y_0 + \frac{123}{110}\Delta^4 y_0 + \frac{33}{10}\Delta^5 y_0 + \frac{41}{140}\Delta^6 y_0]$$

Replace the term $\frac{41}{140}\Delta^6 y_0$ by $\frac{42}{140}\Delta^6 y_0$

By changing this, the error introduced is $\frac{h}{140}\Delta^6 y_0$ which is negligible when h and $\Delta^6 y_0$ are small.

Using $\Delta = E - 1$ and replacing all differences in terms of y's, we get

$$\int_{x_0}^{x_0+6h} f(x)dx = 3h/10\,[y_0 + 5y_1 + y_2 + 6y_3 + y_4 + 5y_5 + y_6]$$

So, $$\int_{x_0}^{x_0+nh} f(x)dx = \int_{x_0}^{x_0+6h} f(x)dx + \int_{x_0+6h}^{x_0+12h} f(x)dx + ... + \int_{x_0+(n-6)h}^{x_0+nh} f(x)dx$$

$$\int_{x_0}^{x_0+nh} f(x)dx = 3h/10[(y_0+5y_1+y_2+6y_3+y_4+5y_5) + (2y_6+5y_7+y_8+6y_9+y_{10}+5y_{11}) + ...$$

$$......... + (2y_{n-6} + 5y_{n-5} + y_{n-4} + 6y_{n-3} + y_{n-2} + 5y_{n-1} + y_n)]$$

$$= 3h/10\,[y_0 + 5(y_1 + y_5 + y_7 + y_{11} + ...) + (y_2 + y_4 + y_8 + y_{10} + ...) + 6(y_3 + y_9 + y_{15} + ...) + 2(y_6 + y_{12} + y_{18} + ...) + y_n]$$

This is called Weddle's Rule.

8.2.6.1 Truncation Error in Weddle's Rule

The truncation error in case of Weddle's rule is given by

$$|E| < -\frac{h^7}{140}(b-a)\cdot M$$

This is of the order $h7$.

SOLVED PROBLEMS

Problem 1: Evaluate $\int_0^6 dx/(1+x^2)$. Using:

(i) Trapezoidal rule

(ii) Simpson's 1/3rd rule

(iii) Simpson's 3/8 rule

(iv) Weddle's rule and comment the results.

Solution: We divide the interval (0, 6) into six parts with $h = 1$.

The values of x and $y = 1/1+x^2$

x :	0	1	2	3	4	5	6
y :	1	0.5	0.2	0.1	0.0588	0.0385	0.027

(i) The formula for Trapezoidal rule with $n = 6$ becomes

$$I = h/2\,[(y_0 + y_6) + 2(y_1 + y_2 + y_3 + y_4 + y_5)]$$

$$= \frac{1}{2}\,[(1 + 0.027) + 2(0.5 + 0.2 + 0.1 + 0.0588 + 0.0385)]$$

$$= 1.4108$$

(ii) The formula for Simpson's 1/3rd rule with $n = 6$ becomes

$$I = h/3\,[(y_0 + y_6) + 2(y_2 + y_4) + 4(y_1 + y_3 + y_5)]$$

$$= \frac{1}{3}\,[(1 + 0.027) + 2(0.2 + 0.0588) + 4(0.5 + 0.1 + 0.0385)]$$

$$= 1.3662$$

(iii) The formula for Simpson's 3/8 rule with $n = 6$ becomes

$$I = 3h/8\,[(y_0 + y_6) + 3(y_1 + y_2 + y_4 + y_5) + 2(y_3)]$$

$$= 3/8\,[(1 + 0.027) + 3(0.5 + 0.2 + 0.0588) + 0.0385) + 2(0.1)]$$

$$= 1.3571$$

(iv) The formula for Weddle's rule with $n = 6$ becomes

$$I = 3h/10\,[y_0 + 5(y_1 + y_5) + (y_2 + y_4) + 6(y_3) + y_6]$$

$$= 3/10\,[1 + 5(0.5 + 0.0385) + (0.2 + 0.0588) + 6(0.1) + 2(0.027)]$$

$$= 1.3735$$

Also by direct method

$$\int_0^6 dx/1+x^2 = |\tan^{-1}|_0^6 = \tan^{-1}6 = 1.4056$$

This shows that the values of the integral found by Trapezoidal rule is the near to the actual value.

Problem 2: Evaluate $\int_0^1 dx/(1+x^2)$, using Trapezoidal rule with $h = 0.2$. Hence, obtain an approximate value of p. Can you use other formulae in this case?

Solution: Let $y(x) = 1/(1+x^2)$

Interval is (1–0) = 1 because the value of y are calculated as points taking $h = 0.2$

x :	0	0.2	0.4	0.6	0.8	1.0
y :	1	0.96154	0.86207	0.73529	060976	0.50000

(i) By Trapezoidal rule, we have

$$I = h/2\,[(y_0 + y_5) + 2(y_1 + y_2 + y_3 + y_4)]$$
$$= 0.2/2\,[(1 + 0.5) + 2(0.96154 + 0.86207 + 0.73529 + 0.60976)]$$
$$= 0.783732$$

By actual integration, we have

$$\int_0^1 dx/1 + x^2 = \Big|_0^1 \tan^{-1} \Big| = \pi/4$$

$\therefore$ $$\pi/4 = 0.783732$$

Hence $\pi = 3.13493$(approx.)

Problem 3: Use Trapezoidal rule to evaluate $\int_0^1 x^3 dx$ considering five subintervals.

Solution: Dividing the interval (0, 1) into 5 equal parts, each of width $h = (1-0)/5$

x:	0	0.2	0.4	0.6	0.8	1.0
f(x):	0	0.008	0.064	0.216	0.512	1.000
	y_0	y_1	y_2	y_3	y_4	y_5

By Trapezoidal rule, we have = 0.2, the values of $f(x) = x^3$ are given below:

$$\int_0^1 x^3 dx = \frac{h}{2}[(y_0 + y_5) + 2(y_1 + y_2 + y_3 + y_4)]$$

$$= \frac{0.2}{2}[(0 + 1) + 2(0.008 + 0.064 + 0.216 + 0.512)]$$

$$= 0.1 \times 2.6 = 0.26.$$

Problem 4: Evaluate $\int_0^6 \frac{dx}{1+x^2}$. By using:

(i) Simpson's one-third rule

(ii) Simpson's three-eighth rule

(iii) Trapezoidal rule

(iv) Weddle's rule.

Solution: Divide the interval (0, 6) into six parts each of width $h = 1$.

The values of $f(x) = 1/(1 + x^2)$ are given below:

x:	0	1	2	3	4	5	6
$f(x)$:	1	0.5	0.2	0.1	$\frac{1}{17}$	$\frac{1}{26}$	$\frac{1}{37}$
	y_0	y_1	y_2	y_3	y_4	y_5	y_6

(i) By Simpson's one-third rule,

$$\int_0^6 \frac{dx}{1+x^2} = \frac{h}{3}[(y_0 + y_6) + 4(y_1 + y_3 + y_5) + 2(y_2 + y_4)]$$

$$= \frac{1}{3}\left[\left(1+\frac{1}{37}\right) + 4\left(0.5 + 0.1 + \frac{1}{26}\right) + 2\left(0.2 + \frac{1}{17}\right)\right]$$

$$= 1.366173413.$$

(ii) By Simpson's three-eighth rule,

$$\int_0^6 \frac{dx}{1+x^2} = \frac{3h}{8}[(y_0 + y_6) + 3(y_1 + y_2 + y_4 + y_5) + 2y_3]$$

$$= \frac{3}{8}\left[\left(1+\frac{1}{37}\right) + 3\left(.5 + .2 + \frac{1}{17} + \frac{1}{26}\right) + 2\,(.1)\right]$$

$$= 1.357080836.$$

(iii) By Trapezoidal rule,

$$\int_0^6 \frac{dx}{1+x^2} = \frac{h}{2}[(y_0 + y_6) + 3(y_1 + y_2 + y_3 + y_4 + y_5)]$$

$$= \frac{1}{2}\left[\left(1+\frac{1}{37}\right) + 2\left(.5 + .2 + \frac{1}{17} + \frac{1}{26}\right)\right]$$

$$= 1.410798581.$$

(iv) By Weddle's rule,

$$\int_0^6 \frac{dx}{1+x^2} = \frac{3h}{10}[(y_0 + 5y_1 + y_2 + 6y_3 + y_4 + 5y_5 + y_6)$$

$$= \frac{3}{10}\left[1 + 5(.5) + .2 + .6(.1)\frac{1}{17} + 5\left(\frac{1}{26}\right)\frac{1}{37}\right]$$

$$= 1.373447475..$$

Problem 5: By dividing the range into equal parts, evaluate $\int_0^{\pi} \sin x\, dx$ by Trapezoidal rule and Simpson's rule. Verify your answer with integration.

Solution: Range = π – 0 = π

Hence $h = \pi/10$

We tabulate below the values of y at different x values by dividing the given interval into 10 equal intervals.

x :	0	π/10	2π/10	3π/10	4π/10	5π/10	6π/10	7π/10	8π/10	9π/10	π
y :	0.0	0.3090	0.5878	0.8090	0.9511	1.0	0.9511	0.8090	0.5878	0.3090	0

(i) By Trapezoidal rule, we have

$$I = \pi/20\ [(0+0)+2(0.3090 + 0.5878 + 0.8090 + 0.9511 + 1.0 + 0.9511 + 0.8090 + 0.5878 + 0.3090)]$$
$$= 1.9843(\text{approx.})$$

(ii) By Simpson's 1/3rd rule (as there are 11 ordinates), we have

$$I = \pi/30[(0 + 0) + 2(0.5878 + 0.9511 + 0.9511 + 0.5878) + 4(0.3090 + 0.8090 + 1 + 0.3090)]$$
$$= 2.00091$$

We can't use Simpson's 3/8th rule or weddle's rule here.

(iii) By direct method, we have

$$I = |-\cos x|_0^{\pi} = 2$$

Hence, we have seen that the value given by direct method is very much close to Simpson's rule (1/3) so we can say that Simpson's rule is more accurate than the Trapezoidal rule.

Problem 6: A solid of revolution is formed by rotating about the x-axis, the area between the lines $x = 0$ and $x = 1$ and a curve through the points with following coordinates.

x :	0.0	0.25	0.50	0.75	1.0
y :	1.0	0.9896	0.9589	0.9089	0.8415

Estimate the volume of the solid formed by using Simpson's 1/3rd rule.

Solution: Here $h = 0.25$

Therefore, the required volume of the solid generated, by Simpson's rule is given by

$$V = \int_0^1 \pi y^2\, dx = \pi h/3\left[\left(y_0^2 + y_4^2\right)\ 4\left(y_1^2 + y_3^2\right) + 2y_2^2\right]$$

$$= \frac{0.25(3.1416)}{3}[\{1 + (0.8415)2\} + 4\{(0.9896)2 + (0.9089)2\} + 2(0.9589)2] = 2.8192.$$

This is the required volume.

Problem 7: A river is 80 metres wide. The depth 'd' in meters at a distance x meters from one bank is given by the following table. Calculate the area of cross-section of the river using Simpson's rule.

x :	0	1	20	30	40	50	60	70	80
y:	0	4	7	9	12	15	14	8	3

Solution: Here $h = 10$, Area of cross section is given by $\int_0^{80} ydx$

$$A = 10/3\ [(0+3)+2(7+12+14)+4(4+9+15+8)]$$
$$= 710 \text{ sq. meters.}$$

Problem 8: A reservoir discharging water through sluices at a depth H below the water surface area A for various values of H as given below:

H(ft.):	10	11	12	13	14	15	16
A(sq.ft):	950	1070	1200	1350	1530	1890	2010

If t denotes time in minutes, the rate of fall of the surface is given by $dH/dt = -48$ H/A. Estimate the time taken for the water level to fall from 16 to 10 ft. above the sluices by Weddle's rule.

Solution: Here $h = 1$

A(sq.ft) :	950	1070	1200	1350	1530	1890	2010
A/48H:	1.979	2.027	2.083	2.163	2.277	2.625	2.617

Given $\quad dH/dt = -48$ H/A i.e $t = \int_{16}^{10} \text{A dH/48H}$

According to Weddle's rule, we have

$$t = 3h/10\ [y_0 + 5(y_1 + y_5) + (y_2 + y_4) + 6y_3 + 2y_6]$$
$$= 0.3\ [1.979 + 5(2.027 + 2.625) + (2.083 + 2.277) + 6(2.163) + 2(2.617)]$$
$$= 13.56 \text{ min.}$$

Problem 9: Evaluate $\int_{0.6}^{2} ydx$, where y is given by the following table:

x :	0.6	0.8	1.0	1.2	1.4	1.6	1.8	2.0
y :	1.23	1.58	2.03	4.32	6.25	8.36	10.23	12.45.

Solution: Here the number of subintervals is 7, which is neither even nor a multiple of 3. Also, this number is neither a multiple of 4 nor a multiple of 6, hence using Trapezoidal rule, we get

$$\int_{0.6}^{2} y dx = \frac{h}{2}[(y_0 + y_7) + 2(y_1 + y_2 + y_3 + y_4 + y_5 + y_6)]$$

$$= \frac{0.2}{2}[(1.23 + 12.45) + 2(1.58 + 2.03 + 4.32 + 6.25 + 8.36 + 10.23)] \quad | \text{Here } h = 0.2$$

$$= 7.922.$$

Problem 10: Find $\int_{1}^{11} f(x)dx$ where $f(x)$ is given by the following table, using a suitable integration formula.

x :	1	2	3	4	5	6	7	8	9	10	11
f(x) :	543	512	501	489	453	400	352	310	250	172	95

Solution: Since the number of subintervals is 10 (even) hence we shall use Simpson's 1/3rd rule.

$$\int_{1}^{11} f(x)dx = \frac{h}{3}[y_0 + y_{10}) + 4(y_1 + y_3 + y_5 + y_7 + y_9) + 2(y_2 + y_4 + y_6 + y_8)]$$

$$= 1/3\ [(543 + 95) + 4(512 + 489 + 400 + 310 + 172) + 2(501 + 453 + 352 + 250)]$$

$$= 1/3\ [638 + 7532 + 3112] = 3760.67.$$

Problem 11: Compute from the following table, the area bounded by the curve and the x-axis from $x = 7.47$ to $x = 7.52$.

x:	7.47	7.48	7.49	7.50	7.51	7.52
f (x):	1.93	1.95	1.98	2.01	2.03	2.06

Solution: We know that

$$\text{Area} = \int_{7.47}^{7.52} f(x)dx$$

with $h = 0.01$, the trapezoidal rule gives,

$$\text{Area} = .01/2\ [(1.93 + 2.06) + 2(1.95 + 1.98 + 2.01 + 2.03)]$$

$$= 0.09965.$$

Problem 12: Use Simpson's rule for evaluating $\int_{-0.6}^{0.3} f(x)dx$ from the table given below:

x:	– 0.6	– 0.5	– 0.4	– 0.3	– 0.2	– 0.1	0	.1	.2	.3
f (x) :	4	2	5	3	– 2	1	6	4	2	8

Solution: Since the number of subintervals is 9(a multiple of 3), we will use Simpson's 3/8th rule here.

$$\int_{-0.6}^{0.3} f(x)dx = 3(.1)/8\ [(4 + 8) + 3\{2 + 5 + (-2) + 1 + 4 + 2\} + 2(3 + 6)] = 2.475.$$

Problem 13: Evaluate $\int_1^2 e^{-1/2x} dx$ using four intervals.

Solution: The table of values is:

x :	1	1.25	1.5	1.75	2
$y = e^{-x/2}$:	.60653	.53526	.47237	.41686	.36788
	y_0	y_1	y_2	y_3	y_4

Since we have four (even) subintervals here, we will use Simpson's 1/3rd rule.

$$\int_1^2 e^{-\frac{1}{2}x}\, dx = \frac{h}{3}[(y_0 + y_4) + 4(y_1 + y_3) + 2y_2]$$

$$= .25/3[(.60653 + .36788) + 4(.53526) + .41686) + 2(.47237)]$$

$$= 0.4773025.$$

Problem 14: A river is 80 m wide. The depth 'y' of the river at a distance 'x' from one bank is given by the following table:

x :	0	10	20	30	40	50	60	70	80
y :	0	4	7	9	12	15	14	8	3

Compute the approximate area of cross-section of the river using

(i) Boole's rule. **(ii) Simpson's 1/3rd rule.**

Solution: The required area of the cross-section of the river = $\int_0^{80} y\, dx$

Here the number of sub intervals is 8.

(i) By Boole's rule,

$$\int_0^{80} y\, dx = \frac{2h}{45} + [7y_0 + 32y_1 + 12y_2 + 32y_3 + 7y_4 + 7y_4 + 32y_5 + 12y_6 + 32y_7 + 7y_8]$$

$$= \frac{2h}{45} + [(7(0) + 32(4) + 12(7) + 32(9) + 7(12) + 7(12)$$

$$+ 32(15) + 12(14) + 32(8) + 7(3)]$$

$$= 708$$

Hence the required area of the cross-section of the river = 708 sq. m.

(ii) By Simpson's 1/3rd rule3

$$\int_0^{80} y\,dx = \frac{h}{3}[(y_0 + y_8) + 4(y_1 + y_3 + y_5 + y_7) + 2(y_2 + y_4 + y_6)]$$

$$= \frac{10}{3}[(0+3) + 4(4+9+15+8) + 2(7+12+14)]$$

$$= 710$$

Hence the required area of the cross-section of the river = 710 sq. m.

UNSOLVED PROBLEMS

1. Evaluate $\int_{-3}^{3} x^4 dx$ by using (i). Trapezoidal rule (ii). Simpson's rule. Verify your results by actual integration.
2. The velocity v(km/min) of a train with initial velocity 1km/min, is given at fixed intervals of time t in minutes as follows:

t:	0	0.1	0.2	0.3	0.4	0.5	0.6	0.7	0.8	0.9	1.0
v:	1	1.2	1.4	1.6	1.8	2.0	2.2	2.4	2.6	2.8	3.0

Estimate approximately the distance covered by a train in 1.0 minute by the Trapezoidal rule.

3. Evaluate the integral $I = \int_4^{5.2} \log_e x\,dx$ by using Trapezoidal, Simpson's and Weddle's rule.
4. Evaluate $\int_0^1 \log e^x\,dx$ by Simpson's one-third rule correct to five decimal places, by choice of h.
5. A curve passes through the points (1, 2), (1.5, 2.4), (2.0, 2.7), (2.5, 2.8),(3, 3),(3.5, 2.6) and (4.0, 2.1). Obtain the area bounded by the curve, the x axis and $x = 1$

[**Hint:** Area = $\int_1^4 y\,dx$ and Volume = $\pi\int_1^4 y^2\,dx$]

6. Evaluate $\int_0^{\pi/2} e^{\sin x}\,dx$ taking $h = \pi/6$
7. Evaluate $\int_0^{1/2} dx/\sqrt{1-x^2}$ by Weddle's rule talkig $n = 6$

8. Evaluate $\int_0^4 e^x\, dx$ by Simpson's rule, given that $e = 2.72$, $e^2 = 7.39$, $e^3 = 20.09$, $e^4 = 54.6$ and compare it with the actual value.

9. Evaluate $\int_1^2 1/x\, dx$ by Simpson's 1/3 rd rule with four strips and determine the error by direct integration.

10. Evaluate the values of dy/dx and d^2y/dx^2 at $x = 2.01$ from the table below:

X :	1.96	1.98	2.00	2.02	2.04
Y :	0.7825	0.7739	0.7652	0.7563	0.7473

11. Find the value cos1.74 from the data:

X :	1.70	1.74	1.78	1.82	1.86
Sinx :	0.9916	0.9857	0.9781	0.9691	0.9584

12. Calculate the angular velocity and angular acceleration at the instant $t = 0.06$ from the table below:

θ :	0.052	0.105	0.168	0.242	0.327	0.408	0.489
T:	0	0.02	0.04	0.06	0.08	0.010	0.12

13. Evaluate $\int_1^2 e^{-x/2}\, dx$ using four intervals. Compare the result with the actual integration.

14. Using trapezoidal method, find the area between the x-axis, the lines $x = 1$ and $x = 4$ and the curve through the following points:

X:	1	1. 5	2	2 . 5	3	3. 5	4
Y:	2	2. 3	2. 9	3. 6	4. 5	5. 2	5. 8

15. Compute $\int_1^2 dx/x$ using simpson's 1/3 method by dividing [1,2] into 6 equal parts. Also obtain approximately the value of $\log_e 2$.

16. By using simpson's 1/3 method calculate the value of $\int_0^{\pi/2} (\sin x)^{1/2}\, dx$

17. Evaluate $\int_0^1 dx/\,(1+x)$ by taking 6 intervals using simpson's 1/3.

18. Evaluate $\int_{0.1}^1 (1+x^3)dx$ using simpson's 3/8 method with a step size of 0.1. Compare the result with the exact value.

19. Integrate $f(x) = 0.3 + 1.6x + 0.027x^2$ between the limits $x = 0$ an $dx = 0.9$.

20. Find $\int_0^6 e^x dx/(1+x)$ approximately using simpson's 3/8 rule on integration.

21. Find $\int_1^{11} f(x)dx$, where $f(x)$ is given by the following table, use the suitable integration formula.

X :	1	2	3	4	5	6	7	8	9	10	11
F(x) :	543	512	501	489	453	400	352	310	250	172	95

8.3 ROMBERG'S INTEGRATION

Romberg's method of integration is basically Richardson's extrapolation procedure or sometimes referred as "the deferred approach to the limit". This method is a relatively simple modification of the trapezoidal rule for a better approximation to the value of an integral.

For an interval of size h, let the error in the trapezoidal rule be kh^2 where k is a constant.

Suppose we evaluate $I = \int_{x_0}^{x_n} y\,dx$, taking two different values of h, say h_1 and h_2, then

$$I = I_1 + E_1 = I_1 + kh_1^2 \quad ...(1)$$

$$I = I_2 + E_2 = I_2 + kh_2^2 \quad ...(2)$$

Where I_1 and I_2 are the values of I got by two different values of h, by trapezoidal rule and E_1 E_2 are the corresponding errors.

$$\therefore \quad I_1 + kh_1{}^2 = I_2 + kh_2{}^2$$

$$\therefore \quad I = \frac{I_1 - I_2}{h_2^2 - h_1^2}$$

Substitute k in (1), we get

$$I = I_1 + [(I_1 - I_2)/h_2{}^2 - h_1{}^2].h_1{}^2$$

$$I = \frac{I_1 h_2^2 - I_2 h_1^2}{h_2^2 - h_1^2}$$

This I is a better result than either I_1 or I_2.

If $h_1 = h$ and $h_2 = \frac{1}{2}h$, then we get

$$I = \frac{I_1((1/4)\cdot h^2) - I_2 h^2}{((1/4)h^2 - h^2} = \frac{4I_2 - I_1}{3}$$

i.e
$$I(h.h/2) = \frac{4I(h/2) - I(h)}{3} \quad ...(3)$$

We can also write
$$I = I_2 + \frac{1}{3}(I_2 - I_1) \quad ...(4)$$

We got this result by applying Trapezoidal rule twice. By applying the trapezoidal rule many times, every time halving h, we get a sequence of results A_1, A_2, A_3, we apply the formula given by (3), to each of adjacent pairs and get the resultants $B_1, B_2, B_3, \ldots\ldots\ldots$(which are improved values). Again applying (3), to each pairs $B_1, B_2, B_3, \ldots\ldots\ldots$ we get another sequence of better results C_1, $C_2, C_3, \ldots\ldots\ldots$ continuing in this way, we proceed until we get two successive values which are very much closer to each other. It is also called Richardson's method.

We use the trapezoidal rule several times successively halving h and apply step (3) to each pair of values as per the following scheme.

$I(h)$			
	$I(h, h/2)$		
$I(h/2)$		$I(h, h/2, h/4)$	
	$I(h/2, h/4)$		$I(h, h/2, h/4, h/8)$
$I(h/4)$		$I(h/2, h/4, h/8)$	
	$I(h/4, h/8)$		
$I(h/8)$			

SOLVED PROBLEMS

Problem 1: Evaluate $\int_0^1 dx/1 + x^2$ using Romberg's method. Hence, obtain an approximate value for p.

Solution: By taking $h = 0.5$, tabulate the values of $y = 1/1+x^2$

x:	0	0.5	1
y:	1.0	0.80	0.50

$\therefore$
$$I = 0.5/2\ [1.5 + 2(0.8)] = 0.775$$

By taking $h = 0.25$, we have the table

x:	0	0.25	0.5	0.75	1
y:	1	0.9412	0.8	0.64	0.5

$\therefore$
$$I = 0.25/2\ [1.5 + 2(0.9412 + 0.8 + 0.64)]$$
$$= 0.78280$$

By taking $h = 0.125$, the tabular values are

x :	0	0.125	0.25	0.375	0.5	0.625	0.75	0.875	1
y :	1	0.9846	0.9412	0.8767	0.8	0.7191	0.64	0.5664	0.5

$$I = 0.125/2\ [(1 + 0.5) + 2(0.9846 + 0.9412 + 0.8767 + 0.8 + 0.7191 + 0.64 + 0.5664)]$$
$$= 0.784750$$

The different values got by Trapezoidal rule for various h's are

0.77500 0.78280 0.78475

Applying the formula $I = I_2 + 1/3(I_2 - I_1)$ we will get two improved values, namely 0.7828 + 1/ 3(0.7828 – 0.7750)

And 0.78475 + 1/3(0.78475 – 0.78280) i.e 0.7854 and 0.7854

As both values are equal so finally $I = 0.7854$.

By actual integration,

$$\int dx/(1+x^2) = \tan^{-1}1 = 0.7844875$$

The difference comes only in the third digit.

$$I = \tan^{-1}1 = p/4$$

$$\therefore \qquad \pi/4 = 0.7854$$

$$\text{and} \qquad \pi = 3.1416$$

UNSOLVED PROBLEMS

1. Using Romberg's method to compute $\int_0^1 dx/(1+x^2)$ correct to 4 decimal places.

2. Compute $\int_0^{\pi/2} \sin x\, dx$ significant to five digits by Romberg's Integration method.

8.4 GAUSS'S QUADRATURE FORMULA

\In the previous section, we have discussed Trapezoidal, Simpson's rule for calculating the integral value of the function; these methods are suitable when the range of integration is divided into intervals of equal size. But what to do if the range of integration is not divided into intervals of equal size, to solve this problem Gauss introduces a new formula known as Gauss's Quadrature formula that is suitable for unequal spacing between the interval values. This formula also produces less amount of error and produces better result.

Suppose we have an integral of the form

$$I = \int_a^b f(x)\,dx \qquad \text{...(1)}$$

Gauss formula requires that the integral should be in the form

$$\int_{-1}^{1} f(x)\,dx \qquad \text{...(2)}$$

Suppose we have a new variable y which is related to x linearly as

$$x = p + qy \qquad \text{...(3)}$$

Suppose $x = a$(lower limit) corresponds to $y = -1$ and

$x = b$(upper limit) corresponds to $y = 1$, thus

$$I = \int_a^b f(x)\,dx$$

As $\quad x = p + qy$ then $dx = qdy$

So

$$I = \int_{-1}^{1} f(p+qy)\cdot qdy$$

$$= q\int_{-1}^{1} f(p+qy)dy \qquad \text{...(4)}$$

From (3), we have

At $x = a$, $y = -1$ so $\quad a = p + q(-1)$ or $a = p - q$

At $x = b$, $y = 1$ so $\quad b = p + q$

After solving above equations for p & q, we get

$$p = \frac{a+b}{2}, q\frac{b-a}{2}$$

$$\therefore \quad x = [a + b + (b-a)y]$$

As from (4),

$$I = \int_a^b f(x)dx$$

$$= \frac{b-a}{2}\int_{-1}^{1} f\left[\frac{a+b+(b-a)y}{2}\right]dy$$

$$= \frac{b-a}{2}\int_{-1}^{1} g(y)dy,$$

where $g(y) = f\left[\frac{a+b+(b-a)y}{2}\right]$

or

$$I = \frac{b-a}{2}\sum_{i=1}^{n} R_i g(y_i)$$

where $R_1, R_2, \ldots\ldots, R_n$ are constants known as weights and $y_1, y_2, \ldots\ldots, y_n$ are values of y where we determined $g(y)$ and n denotes number of points for determining the functions.

The values of R_i's and y_i's are determined in table below for upto $n = 10$ points. All these parameters are standard.

Number of	Weights(R_i) Points (n)	y_i(co-ordintes)	Order of truncation error
2	$R_1 = 1.0$ $R_2 = 1.0$	$y_1 = -0.5773503$ $y_2 = -y_1$	$O(f^4(y_0))$
3	$R_1 = 0.5555556$ $R_2 = 0.8888889$ $R_3 = R_1$	$y_1 = -0.7745967$ $y_2 = 0$ $y_3 = -y_1$	$O(f^6(y_0))$
4	$R_1 = 0.3478549$ $R_2 = 0.6521452$ $R_3 = R_2$ $R_4 = R_1$	$y_1 = -0.8611363$ $y_2 = -0.3399810$ $y_3 = -y_2$ $y_4 = -y_1$	$O(f^8(y_0))$
5	$R_1 = 0.2369269$ $R_2 = 0.4786287$ $R_3 = 0.5688889$ $R_4 = R_2$ $R_5 = R_1$	$y_1 = -0.9061799$ $y_2 = -0.5384693$ $y_3 = 0$ $y_4 = -y_2$ $y_5 = -y_1$	$O(f^{10}(y_0))$
6	$R_1 = 0.1713245$ $R_2 = 0.3607616$ $R_3 = 0.4679139$ $R_4 = R_3$ $R_5 = R_2$ $R_6 = R_1$	$y_1 = -0.9324695$ $y_2 = -0.6612094$ $y_3 = -0.2386192$ $y_4 = -y_3$ $y_5 = -y_2$ $y_6 = -y_1$	$O(f^{12}(y_0))$
7	$R_1 = 0.1294850$ $R_2 = 0.2797054$ $R_3 = 0.3818301$ $R_4 = 0.4179591$	$y_1 = -0.9491079$ $y_2 = -0.7415312$ $y_3 = -0.4058452$ $y_4 = 0$	$O(f^{14}(y_0))$

	$R_5 = R_3$ $R_6 = R_2$ $R_7 = R_1$	$y_5 = -y_3$ $y6 = -y_2$ $y_7 = -y_1$	
8	$R_1 = 0.1012285$ $R_2 = 0.2223810$ $R_3 = 0.3137066$ $R_4 = 0.3626838$ $R_5 = R_4$ $R_6 = R_3$ $R_7 = R_2$ $R_8 = R_1$	$y_1 = -0.9602899$ $y_2 = -0.7966665$ $y_3 = -0.5255324$ $y_4 = -0.1834346$ $y_5 = -y_4$ $y_6 = -y_3$ $y_7 = -y_2$ $y_8 = -y_1$	$O(f^{16}(y_0))$
9	$R_1 = 0.0812744$ $R_2 = 0.1806482$ $R_3 = 0.2606107$ $R_4 = 0.3123471$ $R_5 = 0.3302394$ $R_6 = R_4$ $R_7 = R_3$ $R_8 = R_2$ $R_9 = R_1$	$y_1 = -0.9681602$ $y_2 = -0.8360311$ $y_3 = -0.6133714$ $y_4 = -0.3242534$ $y_5 = 0$ $y_6 = -y_4$ $y_7 = -y_3$ $y_8 = -y_2$ $y_9 = -y_1$	$O(f^{18}(y_0))$
10	$R_1 = 0.0666713$ $R_2 = 0.1494513$ $R_3 = 0.2190864$ $R_4 = 0.2692667$ $R_5 = 0.2955242$ $R_6 = R_5$ $R_7 = R_4$ $R_8 = R_3$ $R_9 = R_2$ $R_{10} = R_1$	$y_1 = -0.9379065$ $y_2 = -0.8650633$ $y_3 = -0.6794096$ $y_4 = -0.4333954$ $y_5 = -0.1488743$ $y_6 = -y_5$ $y_7 = -y_4$ $y_8 = -y_3$ $y_9 = -y_2$ $y_{10} = -y_1$	$O(f^{20}(y_0))$

Where $-1 \le y_0 \le 1$.

8.4.1 How to Use Gauss's Quadrature Formula

To use the Gauss's quadrature formula we have to follows three steps as

Step1: Transform the independent variable

Step2: Choose the appropriate value of *n*.

Step3: Substitute R_i's and y_i's and solve the formula.

8.4.2 Discussion on Different Methods

(i) For a given accuracy, Simpson's rule(1/3) requires about half as much effort as the trapezoidal rule and the Gauss's quadrature requires about half as much effort as that of Simpson's rule.

(ii) For experimental data, selecting the abscissas properly, the Gauss's quadrature formula can be used, whereas, the Simpson's rule can be applied if the data are at equal intervals.

(iii) In view of simplicity of the Simpson's method, in implementation, this rule is most commonly use for practical computation.

(iv) A combination of the trapezoidal rule with Gauss's quadrature gives a more accurate result.

SOLVED PROBLEMS

Problem 1: Evaluate $\int_2^4 (x^2 + 2x)dy$ buy using Gauss's Quadrature formula.

Solution: Here $f(x) = x^2 + 4x$ and $a = 2$(lower limit) and $b = 4$(upper limit)

Substitute $x = p + qy$, $dx = qdy$

Transform the interval [2,4] to interval [–1, 1]

Substitute $x = p + qy$

When $x = 2, y = -1$, we have $p - q = 2$

When $x = 4, y = 1$, $p + q = 4$

for p & q, we have

$\therefore$ $x = p + qy = 3 + y$

and $dx = dy$

$$I = \int_2^4 f(x)dx = \int_{-1}^{1} (3 + y)dy + \int_{-1}^{1} g(y)dy$$

By taking $n = 2$, Gauss's formula takes the form

$$I = [R_1 g(y_1) + R_2 g(y_2)] \quad ...(1)$$

$$g(y) = f(3 + y) = (3 + y)^2 + 2(3 + y) = y^2 + 8y + 15$$

$$R_1 = 1.0, R_2 = 1.0$$

$$y_1 = -0.5773503$$

$$g(y_1) = y_1^2 + 8y_1 + 15$$

$$= (-0.5773503)^2 + 8(-0.5773503) + 15$$

$$= 10.714531$$

$$y_2 = -y_1, g(y_2) = (-y_1)^2 + 8(-y_1) + 15 = y_1^2 - 8y_1 + 15$$
$$= 19.952135$$

∴ From (1), we have

$$I = [1.0(10.714531) + 1.0(19.952135)]$$
$$= 30.666667$$

UNSOLVED PROBLEMS

1. Evaluate $\int_{0.2}^{1.5} e^{-x} dx$ using Gauss's Quadrature formula.

8.5 DOUBLE INTEGRATION

We can solve double integral of the form $\int_c^d \int_a^b f(x, y) dx\, dy$ by performing two

Successive integration in x and y directions by taking one variable at a time. In the next section, we will discuss how to solve double integral by using Trapezoidal and Simpson's 1/3rd rule.

8.5.1 Trapezoidal Rule for Double Integration

Suppose we have an integral of the form

$$I = \int_c^d \int_a^b f(x, y) dx\, dy \quad ...(1)$$

Assume that we have divided the interval $[a, b]$ into n sub-intervals each of width h, and the interval $[c, d]$ into m sub-intervals each of width k, so we have

$$x_i = x_0 + ih$$

Where $x_0 = a$ (lower limit for first integral)

and $x_n = b$(upper limit for first integral)

$i = 1,2,\ldots\ldots\ldots\ldots n.$

and $y_j = y_0 + jk$

Where $y_0 = a$(lower limit for second integral) &

$y_n =$ b(upper limit for second integral)

$j = 1,2,\ldots\ldots\ldots\ldots m.$

Equation (1) can also be written as

$$I = \int_c^d \left[\int_a^b f(x, y) dx \right] dy \quad ...(2)$$

After applying Trapezoidal rule in x direction, we get resulting integral in the form

$$I = \int_c^d h/2\{f(x_0, y) + 2f(x_1, y) + 2f(x_2, y) + \ldots\ldots 2f(x_{n-1}, y) + f(x_n, y)\}]dy$$

Now applying Trapezoidal rule in y direction, we have

$$I = h/2[k/2\ \{f(x_0, y_0) + 2f(x_0, y_1) + 2f(x_0, y_2) + \ldots + 2f(x_0, y_{m-1}) + f(x_0, y_m)\} +$$
$$k/2\ \{f(x_1, y_0) + 2f(x_1, y_1) + 2f(x_1, y_2) + \ldots + 2f(x_1, y_{m-1}) + f(x_1, y_m)\} +$$
$$k/2\ \{f(x_2, y_0) + 2f(x_2, y_1) + 2f(x_2, y_2) + \ldots + 2f(x_2, y_{m-1}) + f(x_2, y_m)\} +$$
$$\ldots\ldots\ldots\ldots\ldots\ldots\ldots\ldots\ldots\ldots$$
$$\ldots\ldots\ldots\ldots\ldots\ldots\ldots\ldots\ldots\ldots$$
$$k/2\ \{f(x_n, y_0) + 2f(x_n, y_1) + 2f(x_n, y_2) + \ldots + 2f(x_n, y_{m-1}) + f(x_n, y_m)\}]$$

$$I = hk/4\ [\ \{f(x_0, y_0) + 2f(x_0, y_1) + 2f(x_0, y_2) + \ldots + 2f(x_0, y_{m-1}) + f(x_0, y_m)\} +$$
$$\{f(x_1, y_0) + 2f(x_1, y_1) + 2f(x_1, y_2) + \ldots + 2f(x_1, y_{m-1}) + f(x_1, y_m)\} +$$
$$\{f(x_2, y_0) + 2f(x_2, y_1) + 2f(x_2, y_2) + \ldots + 2f(x_2, y_{m-1}) + f(x_2, y_m)\} +$$
$$\ldots\ldots\ldots\ldots\ldots\ldots\ldots\ldots\ldots\ldots$$
$$\ldots\ldots\ldots\ldots\ldots\ldots\ldots\ldots\ldots\ldots$$
$$\{f(x_n, y_0) + 2f(x_n, y_1) + 2f(x_n, y_2) + \ldots + 2f(x_n y_{m-1}) + f(x_n, y_m)\}]$$

This is the formula for Trapezoidal rule for double integration.

8.5.2 Simpson's 1/3rd Rule for Double Integration

Suppose we have an integral of the form

$$I = \int_c^d \int_a^b f(x, y)dx\,dy \qquad \ldots(1)$$

Assume that we have divided the interval $[a, b]$ into n sub-intervals each of width, and the interval $[c, d]$ into m sub-intervals each of width k, here n and m are even. So we have

$$x_i = x_0 + ih$$

Where $x_0 = a$(lower limit for first integral) and

$x_n = b$(upper limit for first integral)

$$i = 1, 2, \ldots\ldots n.$$

and

$$y_j = y_0 + jk$$

Where $y_0 = a$(lower limit for second integral) and

$y_n = b$(upper limit for second integral)

$$j = 1, 2, \ldots\ldots m.$$

Equation (1) can also be written as

$$I = \int_c^d \left[\int_a^b f(x, y)dx \right] dy \qquad \text{... (2)}$$

After applying Simpson's 1/3rd rule in x direction, we get resulting integral in the form

$$I = \int_c^d h/3\{f(x_0, y) + 4f(x_1, y) + 2f(x_2, y) + ...4f(x_{n-1}, y) + f(x_n, y)\}]dy$$

Now applying Simpson's 1/3rd rule in y direction, we have

$$\begin{aligned} I = \; & h/3[k/3 \; \{f(x_0, y_0) + 4f(x_0, y_1) + 2f(x_0, y_2) + ...+4f(x_0, y_{m-1}) + f(x_0, y_m)\} + \\ & 4k/3 \; \{f(x_1, y_0) + 4f(x_1, y_1) + 2f(x_1, y_2) + ...+4f(x_1, y_{m-1}) + f(x_1, y_m)\} + \\ & 2k/3 \; \{f(x_2, y_0) + 4f(x_2, y_1) + 2f(x_2, y_2) + ...+4f(x_2, y_{m-1}) + f(x_2, y_m)\} + \\ & \ldots\ldots\ldots\ldots\ldots\ldots\ldots\ldots\ldots\ldots \\ & \ldots\ldots\ldots\ldots\ldots\ldots\ldots\ldots\ldots\ldots \\ & 4k/3 \; \{f(x_{n-1}, y_0) + 4f(x_{n-1}, y_1) + 2f(x_{n-1}, y_2) + ...+4f(x_{n-1}, y_{m-1}) + f(x_{n-1}, y_m)\} + \\ & k/3 \; \{f(x_n, y_0) + 4f(x_n, y_1) + 2f(x_n, y_2) + ...+4f(x_n, y_{m-1}) + f(x_n, y_m)\}] \end{aligned}$$

$$\begin{aligned} I = \; & hk/9 \; [\; \{f(x_0, y_0) + 2f(x_0, y_1) + 2f(x_0, y_2) + ...+2f(x_0, y_{m-1}) + f(x_0, y_m)\} + \\ & 4\{f(x_1, y_0) + 2f(x_1, y_1) + 2f(x_1, y_2) + ...+2f(x_1, y_{m-1}) + f(x_1, y_m)\} + \\ & 2\{f(x_2, y_0) + 2f(x_2, y_1) + 2f(x_2, y_2) + ...+2f(x_2, y_{m-1}) + f(x_2, y_m)\} + \\ & \ldots\ldots\ldots\ldots\ldots\ldots\ldots\ldots\ldots\ldots \\ & \ldots\ldots\ldots\ldots\ldots\ldots\ldots\ldots\ldots\ldots \\ & 4\{f(x_{n-1}, y_0) + 2f(x_{n-1}, y_1) + 2f(x_{n-1}, y_2) + ...+2f(x_{n-1}, y_{m-1}) + f(x_{n-1}, y_m)\} \\ & \{f(x_n, y_0) + 2f(x_n, y_1) + 2f(x_n, y_2) + ...+2f(x_n y_{m-1}) + f(x_n, y_m)\}] \end{aligned}$$

This is the formula for Simpson's 1/3rd rule for double integration.

SOLVED PROBLEMS

Problem 1: Evaluate $I = \int_0^{2.6} \int_4^{4.4} f(x, y)dxdy$ **where** $f(x, y) = 1/xy$ **by taking** $h = 0.2$ **and** $k = 0.3$

Solution: we take $h = 0.2$, $k = 0.3$ so that $m = n = 2$. Then the values of the integrand $f(x, y) = 1/xy$ are given as:

y \ X	4	4.2	4.4
2	0.125	0.119	0.1136
2.3	0.1087	0.1035	0.0988
2.6	0.0961	0.0915	0.0874

(i) Trapezoidal Rule

$I = hk/4\,[\{f(4, 2) + 2f(4, 2.3) + f(4, 2.6)\} + 2\{f(4.2, 2) + 2f(4.2, 2.3) + f(4.2, 2.6)\}$
$+ \{f(4.4, 2) + 2f(4.4, 2.3) + f(4.4, 2.6)\}]$

$= \dfrac{0.06}{4}[\{(0.125)+2(0.1087) + 0.0961\} + 2\{(0.1190) + 2(0.1035) + (0.0915)\}$
$+ \{(0.1136) + 2(0.0988) + (0.0874)\}]$

$= 0.015[0.4385 + 0.835 + 0.3986]$

$= 0.015[1.6721]$

$= 0.0250815$

(ii) Simpson's rule

$I = hk/9\,[\{f(4, 2)+4(4, 2.3)+(4, 2.6)\} + 4\{f(4.2, 2) + 4f(4.2, 2.3) + f(4.2, 2.6)\}$
$+ \{f(4.4, 2) + 4f(4.4, 2.3) + f(4.4, 2.6)\}]$

$= \dfrac{0.06}{9}[\{(0.125)+4(0.1087)+(0.0961)\}+4\{(0.1190)+4(0.1035) + (0.0915)\}$
$+ \{(0.1136) + 4(0.0988) + (0.0874)\}]$

$= 0.006[0.6559 + 2.498 + 0.5962]$

$= 0.006[3.7501]$

$= 0.0225006$

Problem 2: Evaluate the Integral $I = \int_0^1 \int_0^1 e^{x+y}\,dydx$ by using (i) Trapezoidal rule (ii). Simpson's rule

Solution: we take $h = k = 0.5$ and $f(x, y) = e^{x+y}$. Then the values of $f(x, y)$ are given as

y \ X	0	0.5	1.0
0	1	1.6487	2.7183
0.5	1.6487	2.7183	4.4817
1	2.7183	4.4817	7.3891

(i) Trapezoidal Rule

$$I = \frac{0.25}{4}[1.0 + 4(1.6487) + 6(2.7183) + 4(4.4817) + 7.3891)]$$

$= 12.3050 / 4$

$= 3.762$

(ii) Simpson's 1/3 rule

$$I = \frac{0.25}{9}[1.0+2.7183+7.3891+2.7183+4(1.6487+4.4817+4.4817+ 1.6487) +$$

16(2.7183)]

$= 26.59042 / 9$

$= 2.9545$

The exact value of the double integral is 2.9525 and thus the result obtained by Simpson's rule is sixty times more accurate than given by Trapezoidal rule.

UNSOLVED PROBLEMS

1. Evaluate $\int_0^1 \int_0^1 xe^y \, dydx$ using Trapezoidal rule ($h = k = 0.5$).

2. Evaluate $\int_0^{2.6} \int_0^{3.2} dxdy/(x + y)$, using Simpson's rule.

3. Using Trapezoidal and Simpson's rule, to evaluate $\int_0^1 \int_0^1 e^{x+y}$, taking two sub-intervals.

CHAPTER 9

Solution of Ordinary Differential Equations

9.1 INTRODUCTION

An ordinary differential equation (or ODE) is a relation that contains functions of only one independent variable, and one or more of its derivatives with respect to that variable. A differential equation is a mathematical equation for an unknown function of one or several variables that relates the values of the function itself and its derivatives of various orders. Differential equations play a prominent role in engineering, physics, economics and other disciplines. Many analytical techniques exist for solving such equations, but these methods can be applied to solve only a selected class of differential equations. However, a majority of differential equations appearing in physical problems cannot be solved analytically. Thus it becomes imperative to discuss their solution by numerical methods. In this chapter, we will discuss various methods that deal with different types of differential equations.

It must be noted that even the differential equations which are solvable by analytical methods can be solved numerically as well.

9.1.1 Differential Equations

The equations that involve variables and their derivatives are called differential equations. A simple differential equation is that which have two variables x and y, and one derivative. dy/dx is as $dy/dx = x + y$

9.1.2 Ordinary and Partial Derivatives

Suppose $y = f(x)$ be a given function where x is an independent variable, then dy/dx is called ordinary derivative.

Similarly, suppose u is a function of x and y i.e $u = f(x,y)$ then $\partial u/\partial x$ is called partial derivative of u w.r.t x while keeping y constant and $\partial u/\partial y$ is called the partial derivative of u w.r.t y while keeping x constant.

9.1.3 Ordinary and Partial Differential Equations

The differential equations that involve ordinary derivatives are called ordinary differential equations. Some examples of O.D.E (ordinary differential equation) are $dy/dx = x + y$, $dy/dx = x^2 + y^2$, $d^2y/d^2x + y = 0$.

Similarly the differential equations that involve partial derivatives are called partial differential equations. Some examples of partial differential equations are

$$\partial u/\partial x = x.\,\partial u/\partial y,\ x.\,\partial z/\partial x + y.\,\partial z/\partial y = nz$$

9.1.4 Order and Degree of a Differential Equation

The order of a differential equation is the order of highest derivative in the equation. e.g $dy/dx = 5x$, $d^2y/dx^2 = -y$ are of order 1 and 2 respectively.

Degree of a differential equation is the degree of highest order derivative.

$$\text{e.g } (d^2y/dx^2)^2 + 2x.dy/dx + 3y = 0.$$

Here the degree of differential is 2 as higher order derivative d^2y/dx^2 have degree 2.

9.1.5 Initial and Boundary Value Problems

The general solution of the nth order ordinary differential equations contains n independent arbitrary constants. In order to determine these, n conditions are prescribed at one point, these are called initial conditions. The differential equation with the initial condition at one point is called initial value problem.

If n conditions are prescribed at more than one point, then these are called boundary value problems. The differential equation with condition at more than one point, is called the boundary value problem.

Boundary value problems arise in several branches of physics. Problems involving the wave equation, such as the determination of normal modes, are often stated as boundary value problems.

To be useful in applications, a boundary value problem should be well posed.

9.1.6 Numerical Solution of Differential Equation

Consider the differential equation

$$dy/dx = f(x, y),\ y(x_0) = y_0$$

The solution means the computation of values of y for various values of x, usually at an equal intervals e.g for $x = 0.1$ to 1.0. A mathematical solution means finding an explicit formula for y in terms of finite number of elementary functions of x e.g Polynomial , Trignometric or exponential functions. A numerical solution is frequently possible when explicit mathematical solution is not.

The numerical methods for computing the solutions of initial value problem may be classified into the following categories:

(i) Single step methods

(ii) Multistep methods

Now first we will discuss single step methods one by one and after that we will discuss multistep methods.

9.2 SINGLE STEP METHOD

These methods use information from only one preceding point i.e to estimate the value of y_n, we need the conditions at the previous point y_{n-1} only. These are direct methods. In this case solution is not iterated. The main drawback of single step methods is that it is difficult to estimate the error. Under single step methods, we will study Picard's method, Taylor's series method, Euler & Runge-kutte methods.

9.2.1 Taylor's Series Method

The Taylor series is a representation of a function as an infinite sum of terms calculated from the values of its derivatives at a single point. It may be regarded as the limit of the Taylor polynomials. Taylor series are named after the English mathematician Brook Taylor. Taylor-Series Method is a straightforward adaptation of classic calculus to develop the solution as an infinite series. The catch is that a computer usually cannot be programmed to construct the terms and one doesn't know how many terms should be used.

As this method comes under the category of single step methods, so we have required one initial condition.

Consider the differential equation

$$dy/dx = f(x, y) \qquad \ldots(1)$$

with initial condition $y(x_1) = y_1$

Let $y = f(x)$ be a continuously differentiable function. Which satisfies equation (1).

By expanding y in terms of Taylor's series around the point $x = x_1$, we have

$$y = y(x) = y(x_1) + (x - x_1)\, y'(x_1) + \frac{(x - x_1)^2}{2!}\, y''(x_1) + \ldots\ldots\ldots \quad (2).$$

Where $y'(x_1), y''(x_1), \ldots\ldots\ldots$, are the differential coefficients at $x = x_1$ and can be computed from equation (2) as:

Given that $\qquad y' = dy/dx = f(x, y)$

Similarly $\qquad y'' = \frac{d}{dx}(y')$ [as y is a function of x]

$$= \frac{d}{dx}[y'(x)]$$

$$= \frac{d}{dx}[f(x,y)]$$

$$= \frac{\partial}{\partial x} f(x, y) + \frac{\partial}{\partial y} f(x, y) \cdot \frac{dy}{dx}$$

$$= \frac{\partial}{\partial x} f(x, y) + \frac{\partial}{\partial y} f(x, y). f(x, y)$$

[as $dy/dx = f(x, y)$]

If we denote $f_x(x, y) = \frac{\partial}{\partial x} f(x, y)$ and $f_y(x, y) = \frac{\partial}{\partial y} f(x, y)$

then $y''(x) = f_x(x, y) + f_y(x, y) \cdot f(x, y)$

At $x = x_1 + h = x_2$ equation (2) becomes

$$y_2 = y(x_1 + h) = y(x_1) + hy'(x_1) + \frac{h^2}{2!} y''(x_1) + \ldots\ldots\ldots\ldots\ldots$$

So $$y_2 = y(x_1 + h) = y(x_1) + hf(x_1, y_1) + \frac{h^2}{2!} [f_x(x_1, y_1) + f_y(x_1, y_1) \cdot f(x_1, y_1)] + \ldots\ldots$$

Where $y(x_1) = y_1$.

Similarly $$y_3 = y_2(x_2 + h) = y_2 + h f(x_2, y_2) + \frac{h^2}{2!} [f_x(x_2, y_2) + f_y(x_2, y_2) \cdot f(x_2, y_2)] + \ldots\ldots\ldots$$

In general,

$$y_{n+1} = y_n + h f(x_n, y_n) + \frac{h^2}{2!} [f_x(x_n, y_n) + f_y(x_n, y_n) \cdot f(x_n, y_n)] + \ldots\ldots\ldots$$

This formula is known as Taylor's series formula.

The error produced in case of Taylor's series is of the order $O(h^3)$.

This method is generally impractical for computational work.

SOLVED PROBLEMS

Problem 1: Compute the solution $y(0.1)$ of $dy/dx = -2xy^2$, given $y(0) = 1$ with $h = 0.1$, using Taylor's series method of order four.

Solution: The given differential equation is

$$dy/dx = -2xy^2$$

The Taylor's series of order four is given as

$$y(x) = y(x_1) + (x - x_1)y'(x_1) + \frac{(x - x_1)^2}{2!} y''(x_1) + \frac{(x - x_1)^3}{3!} y'''(x_1) + \frac{(x - x_1)^4}{4!} y''''(x_1) + \ldots. \quad ...(1)$$

Now, $y(x_1) = 1$ when $x_1 = 1$.

$$y'(x) = dy/dx = -2xy^2$$
$$y'(x_1) = 2\cdot(0)\cdot(1)^2 = 0$$
$$y''(x) = -2y^2 - 4xyy'$$
$$y''(x_1) = -2(1)^2 - 4(0) \times 1 \times 0 = -2$$
$$y'''(x) = -8yy' - 4x(y')^2 - 4xy(y'')$$
$$y'''(x_1) = -8(1)(0) - 4(0)(0)^2 - 4(0)(1)(-2) = 0$$
$$y''''(x) = -8(y')^2 - 8y(y'') - 4(y')^2 - 4x(y')^3 - 4yy'' - 44xy'y'' - 4xyy'''$$
$$y''''(x_1) = -8(0)^2 - 8(1)(-2) - 4(0)^2 - 4(0)(0)^3 - 4(1)(-2) - 4(0)(0)(-2) - 4(0)(1)(0) = 24$$

Substitute these values in eq.(1), we get

$$y(x) = 1 + x \times 0 + (x^2/2) \times (-2) + (x^3/6) \times (0.x^4/24) \times 24 = 1 - x^2 + x^4$$

Now, $y(0.1) = 1 - (0.1)^2 + (0.1)^4 = 0.9901$

Problem 2: Use Taylor's series method to solve $dy/dx = x + y; y(1) = 0$ numerically up to $x = 1.2$ with $h = 0.1$. Compare the final result with the value of explicit solution.

Solution: Here, $x_0 = 1, y_0 = 0$

$$y' = x + y \text{ i.e., } y_0' = x_0 + y_0 = 1$$
$$y'' = 1 + y' \text{ i.e., } y_0'' = 1 + y_0' = 2$$
$$y''' = y'' \text{ i.e., } y_0''' = y_0'' = 2$$
$$y^{(iv)} = y''' \text{ i.e., } y_0^{(iv)} = 2$$
$$y^{(v)} = y^{(iv)} \text{ i.e., } y_0^{(v)} = 2$$

By Taylor's series, we have

$$y_1 = y_0 + hy_0' + \frac{h^2}{2!} + y_0'' + \frac{h^3}{3!} y_0''' + \frac{h^4}{4!} y_0^{(iv)} + \ldots\ldots\ldots$$

$$y(1 + h) = 0 + (0.1)1 + \frac{(0.1)^2}{2!} 2 + \frac{(0.1)^3}{3!} + 2 + \frac{(0.1)^4}{4!} + 2 \ldots\ldots\ldots$$

$$y(1.1) = 0.1103081 = 0.110 \text{ (app.)}$$

Also, $x_1 = x_0 + h = 1.1$

Again, $y' = x_1 + y_1 = 1.1 + 0.11 = 1.21$

$$y_1'' = 1 + y_1' = 1 + 1.21 = 2.21$$
$$y_1''' = y_1'' = 2.21$$
$$y_1^{(iv)} = 2.21$$

$$y_1^{(v)} = 2.21$$

Now,

$$(y + 1.1h) = y_1 + hy_1' + \frac{h^2}{2!} + y_1'' + \frac{h^3}{3!} y_1''' + ...$$

$$0.11 + (0.1)(1.21) + \frac{(0.1)^2}{2}(2.21) + ...$$

$$y(1.2) = 0.232 \text{ (app.)}$$

The analytical solution of the given differential equation is

$$y = -x - 1 + 2e^{x-1}$$

when $x = 1.2$, we get

$$y = -1.2 - 1 + 2e^{0.2} = 0.242.$$

Problem 3: By using Taylor's series, find the solution of the differential equation $xy' = x - y$, $y(2) = 2$ at $x = 2.1$ correct to five decimal places.

Solution: Here $x_0 = 2, y_0 = 2$

Also,

$$y' = 1 - \frac{y}{x} \qquad y_0' = 0$$

$$y'' = -\frac{y'}{x} + \frac{y}{x^2} \qquad y_0'' = -0 + \frac{2}{4} = \frac{1}{2}$$

$$y''' = -\frac{y''}{x} + \frac{2y'}{x^2} - \frac{2y}{x^3} \qquad y_0'' = \frac{-3}{4}$$

$$y^{(iv)} = -\frac{y'''}{x} + \frac{3y''}{x^2} - \frac{6y'}{x^3} + \frac{6y}{x^4} \qquad y_0^{(iv)} = \frac{3}{1} \text{ and so on}$$

Putting these values in Taylor's series, we get

$$y(2 + h) = 2 + \frac{h^2}{4} - \frac{h^3}{8} + \frac{h^4}{16} + ...$$

Put $h = 0.1$, we get

$$y(2.1) = 2.00238 \text{ (correct to 5 decimal places).}$$

Problem 4: Solve the simultaneous equations

$$\mathbf{y' = 1 + xyz, y(0) = 0}$$

$$\mathbf{z' = x + y + z, z(0) = 1.}$$

Solution: Differentiating the given equations

$$y'' = yz + xy'z + xyz',$$

$$y''' = 2y'z + 2yz' + 2xy'z' + xy''z + xyz''$$

$$z'' = 1 + y' + z', \; z''' = y' + z'$$

with $x = 0, y = 0, z = 1$; we get $y' = 1, y'' = 0, y''' = 2$

Also $z' = 1, z'' = 3, z''' = 3$

Hence, $$y(x) = x + x^3/3$$

and $$z(x) = 1 + x + 3/2\, x^2 + 1/2\, x^3.$$

UNSOLVED PROBLEMS

1. From the Taylor's series for $y(x)$, find $y(0.1)$ correct to four decimal places if $y(x)$ satisfies
$$y' = x - y^2 \text{ and } y(0) = 1$$
2. Use Taylor's series to solve the differential equation
$$dy/dx = 1 + xy, \; y(0) = 1 \text{ at } x = 0.1$$
correct to three decimal places
3. Given $dy/dx - 1 = xy$ and $y(0) = 1$, obtain the Taylor's series for $y(x)$ and compute $y(0.1)$ correct to four decimal places.
4. Compute y for $x = 0.1$ and 0.2 correct to four decimal places given: $y' = y - x, y(0) = 2$.
5. Compute the solution by using Taylor's method, $y' = x^2 + y^2, y(0) = 1$ compute $y(0.1)$.
6. Compute the solution by Taylor's method: $y' = y - 2x/y; y(0) = 1$. Also compute $y(0.1)$.
7. Using Taylor series method, solve $dy/dx = x^2 - y, y(0) = 1$ at $x = 0.1, 0.2, 0.3$ and 0.4. Compare the values with exact solution.
8. Compute $dy/dx = x + z, dz/dx = x - y^2$ with $y(0) = 2, z(0) = 1$ to get $y(0.1), y(0.2), z(0.1)$ and $z(0.2)$ approximately by Taylor's algorithm.

9.2.2 Picard's Method

Picard's method, sometimes called the method of successive approximations, gives a means of proving the existence of solutions to differential equations. Emile Picard, a French mathematician, developed the method in the early 20th century. It has proven to be so powerful that it has replaced the Cauchy-Lipschitz method that was previously employed for such endeavors. Picard's method is utilized in similar situations as those that employ the Taylor series method. It is a method that converts the differential equation into an equation involving integrals. Some differential equations are difficult to solve, but Picard's method provides a numerical process by which a solution can be approximated. The method consists of constructing a sequence of functions that will approach the desired solution upon successive iteration. It is similar to the Taylor series method in that successive iterations also approach the desired solution to a differential equation.

Picard's method allows us to find a series solution about some fixed point. The number of terms or iterations that is required to reach the desired solution depends on how far from the chosen point the solution must apply. The closer the chosen point to the unknown point, the fewer terms that are needed. It can be shown that the series is convergent and provides a solution to the differential equation of interest although the number of terms will depend upon how rapidly the series converges as well.

This is also single step method. Consider the differential equation

$$dy/dx = f(x, y) \qquad \ldots(1)$$

or

$$dy = f(x, y)dx.$$

with initial condition $y(x_0) = y_0$.

Integrating equation (1) between corresponding limits for x and y, we have

$$\int_{y_0}^{y} dy = \int_{x_0}^{x} f(x, y)dx$$

$$y - y_0 = \int_{x_0}^{x} f(x, y)dx$$

$$y = y_0 + \int_{x_0}^{x} f(x, y)dx \qquad \ldots(2)$$

The term $\int_{x_0}^{x} f(x, y)dx$ involved in eq. (2) can be solved by method of successive approximation.

For the first approximation y_1 to the solution, we put $y = y_0$ in $f(x,y)$ and obtain

$$y_1 = y_0 + \int_{x_0}^{x} f(x_0, y_0)dx$$

Now substitute this y_1 in place of y in eq. (2) for getting next approximation y_2 as

$$y_2 = y_0 + \int_{x_0}^{x} f(x_1, y_1)dx$$

....

....

Similarly, in general

$$y_{n+1} = y_0 + \int_{x_0}^{x} f(x, y_n)dx$$

This process is stopped when the two corresponding values of y say y_n and y_{n-1} are same to the desired degree of accuracy.

The above formula is known as Picard's iteration formula and is applied to equations where successive integration can be obtained easily.

SOLVED PROBLEMS

Problem 1: Compute a solution of $dy/dx = 1 + xy$ by using Picard's method upto third order approximation, when $y = 0$ at $x = 0$.

Solution: As $f(x, y) = 1 + xy$ and $y_0 = 0$ when $x_0 = 0$, then the integral equation

$$y = y_0 + \int_{x_0}^{x} f(x,y)dx$$

becomes

$$y = 0 + \int_{x_0}^{x} (1+xy)dx \Rightarrow y = \int_{x_0}^{x} (1+xy)dx \qquad \text{... (1)}$$

Now, substitute $y = 0$ in place of y in (1), we get

$$y_1 = 0 + \int_{0}^{x} (1+0)dx = x$$

Now, substitute new value of $y = x$ in $(1+xy)$ of (1), we get

$$y_2 = 0 + \int_{0}^{x} (1+x^2)dx = x + x^3/3$$

Now, substitute new value of $y = x + x^3/3$ in $(1+ xy)$ of (1), we get

$$y_3 = 0 + \int_{0}^{x} [1 + x(x + x^3/3)]dx = \int_{0}^{x} (1 + x^2 + x^4/3)dx$$

Hence,

$$y_3 = x + x^3/3 + x^5/15$$

Problem 2: Use Picard's method to obtain y for $x = 0.1$. Given that: $dy/dx = 3x + y^2$; $y = 1$ at $x = 0$.

Solution: Here $f(x, y) = 3x + y2$, $x_0 = 0$, $y_0 = 1$

First approximation,

$$y^{(1)} = y_0 + \int_{x_0}^{x} f(x, y_0)dx$$

$$= 1 + \int_{0}^{x} (3x+1)dx$$

$$= 1 + x + \frac{3}{2}x^2$$

Second approximation,

$$y^{(2)} = 1 + x + \frac{5}{2}x^2 + \frac{4}{3}x^3 + \frac{3}{4}x^4 + \frac{9}{20}x^5$$

Third approximation,

$$y^{(3)} = 1 + x + \frac{5}{2}x^2 + 2x^3 + \frac{23}{12}x^4 + \frac{25}{12}x^5$$

when $x = 0.1$, we have $y^{(1)} = 1.115$, $y^{(2)} = 1.1264$, $y^{(3)} = 1.12721$

Thus, $y = 1.127$ when $x = 0.1$.

Problem 3: If $\frac{dy}{dx} = \frac{y - x}{y + x}$, Compute the value of y at $x = 0.1$ using Picard's method. Given that $y(0) = 1$.

Solution: First approximation,

$$y^{(1)} = y_0 + \int_0^x \frac{y_0 - x}{y_0 + x}\,dx = 1 + \int_0^x \left(\frac{1 - x}{1 + x}\right) dx$$

$$= 1 + \int_0^x \left(\frac{2}{1 + x} - 1\right) dx = 1 - x + 2\log(1 + x)$$

Second approximation,

$$y^{(2)} = 1 + x - 2\int_0^x \frac{x\,dx}{1 + 2\log(1 + x)}$$

which is difficult to integrate.

Thus, when, $x = 0.1$, $y^{(1)} = 1 - 0.1 + 2\log(1.1) = 0.9828$

Here in this example, only I approximation can be obtained and so it gives the approximate value of y for $x = 0.1$.

UNSOLVED PROBLEMS

1. Compute $dy/dx = x^2 + y^2$, $y(0) = 1$ by Picard's method
2. Given $dy/dx = xy$ with $y(1) = 5$. Find the solution correct to three decimal position in the interval [1.1.5] using step size $h = 0.1$
3. For the differential equation $dy/dx = x - y^2$, $y(0) = 0$ Calculate $y(0.2)$ by Picard's method to third approximations and round-off the value at the 4th place of decimals.
4. Compute $y(0.2)$ if $dy/dx = \log(x + y)$; $y(0) = 1$. Use Picard's method.
5. Use Picard's method to obtain the solution of $dy/dx = x^2 + y^2$ for $x = 0.1$ correct to four decimal places, given that $y = 0$ when $x = 0$.
6. Compute an approximate value of y when $x = 0.1$ if $dy/dx = x - y^2$ and $y = 1$ at $x = 0$ using Picard's method.

9.2.3 Euler's Method

The Euler method, named after Leonhard Euler, is a first-order numerical procedure for solving ordinary differential equations (ODEs) with a given initial value. It is the most basic kind of explicit method for numerical integration for ordinary differential equations The method is a single step method and does not require the evaluation of any derivatives.

Consider the differential equation

$$dy/dx = f(x, y) \qquad \ldots(1)$$

or

$$dy = f(x, y)dx.$$

with initial condition $y(x_0) = y_0$.

Integrating (1), we get

$$y - y_0 = y_0 + \int_{x_0}^{x} f(x,y)dx$$

$$y = y_0 + \int_{x_0}^{x} f(x,y)dx \qquad \ldots(2)$$

Suppose we want to find the solution of equation (1) for values of y at $x = x_r = x_0 + rh$, Where $r = 1,2,3$..................

Then from equation (2), we have

$$y_1 = y_0 + \int_{x_0}^{x_1} f(x,y)dx \qquad \ldots(3)$$

Assuming that $f(x, y) = f(x_0, y_0)$ in $x_0 \le x \le x_1$, the equation (3) yields to

$$y_1 = y_0 + h\,f(x_0, y_0)$$

Similarly for range $x_1 \le x \le x_2$, we have

$$y_2 = y_1 + \int_{x_1}^{x_2} f(x,y)dx = y_1 + hf(x_1 + y_1)$$

In general, we have

$$y_{n+1} = y_n + hf(x_n, y_n),\ n = 0,1,2,3,\ldots\ldots\ldots\ldots$$

This process is very slow and to obtain reasonable accuracy with Euler's method, we need to take a smaller value for h because of this restriction on h, the method is unsuitable for practical use and a modification of it, known as modified Euler's method, which gives more accurate results.

The order of Euler's method is $O(h^2)$ because it produces a truncation error as $\frac{h^2}{2}y''(n)$.

9.2.3.1 *Algorithm of Euler's Method*

1. Function of euler's $F(x, y) = (x-y)/(x+y)$
2. Input values $x0$, y0, h, xn

3. n =(($xn - x0$)/h) +1
4. For i = 1 to n
5. $y = y0 + h*F(x0, y0)$
6. $x = x + h$
7. Print $x0$, $y0$
8. If $x < xn$ then

   $x0 = x$

   $y0 = y$

   ELSE
9. Next i
10. Stop

9.2.3.2 Flow-chart of Euler's Method

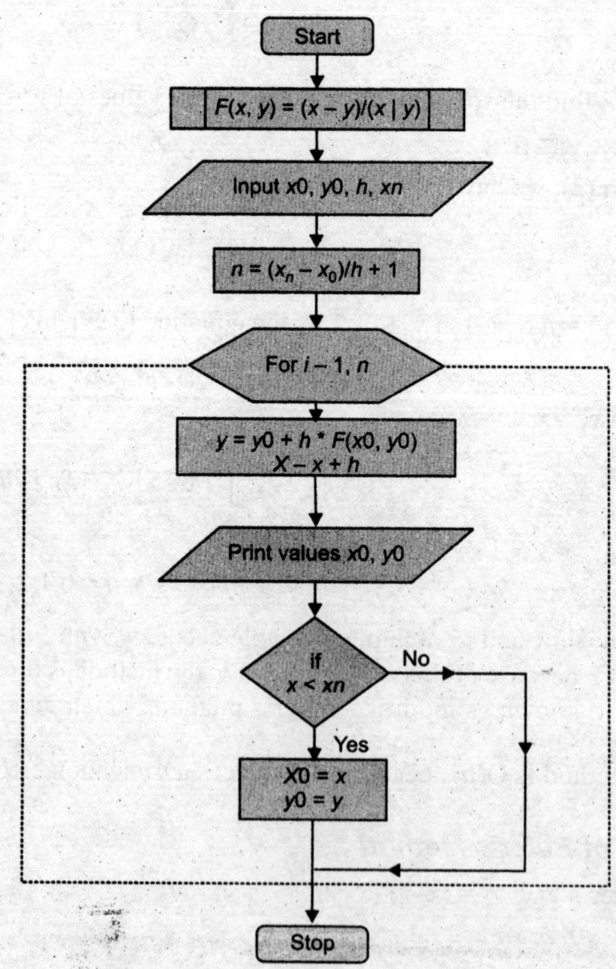

9.2.3.3 *Program for Euler's Method in C*

```
#include<stdio.h>
#include<conio.h>
#define F(x,y) (x–y)/(x+y)
main ( )
{
int i,n;
float x0,y0,h,xn,x,y;
printf("\n Enter the values: x0,y0,h,xn: \n");
scanf ("%f%f%f%f",&x0,&y0,&h,&xn);
n=(xn–x0)/h+1;
for (i=1;i<=n;i++)
{
y=y0+h*F(x0,y0);
x= x0+h;
printf("\n X=%f Y=%f",x0,y0);
if(x<xn)
{
x0= x;
y0=y;
}
}
getch();
}
```

Output:

```
Enter the values: x0,y0,h,xn:
0 1  0.02  0.1
X=0.000000 Y=1.000000
X=0.020000 Y=0.980000
X=0.040000 Y=0.960800
X=0.060000 Y=0.942399
X=0.080000 Y=0.924793
X=0.100000 Y=0.907978
```

9.2.3.4 Program for Euler's Method in C++

```
#include<iostream.h>
#include<conio.h>
#define F(x,y) (x–y)/(x+y)
main ( )
{
int i,n;
float x0,y0,h,xn,x,y;
cout<<"\n Enter the values: x0,y0,h,xn: \n";
cin>>x0>>y0>>h>>xn;
n=(xn–x0)/h+1;
for (i=1;i<=n;i++)
{
y=y0+h*F(x0,y0);
x= x0+h;
cout<<"\n X="<<x0<<"Y="<<y0;
if(x<xn)
{
x0= x;
y0=y;
}
}
getch();
}
```

Output:

```
Enter the values: x0,y0,h,xn:
0 1  0.02  0.1
X=0.000000 Y=1.000000
X=0.020000 Y=0.980000
X=0.040000 Y=0.960800
X=0.060000 Y=0.942399
X=0.080000 Y=0.924793
X=0.100000 Y=0.907978
```

SOLVED PROBLEMS

Problem 1: Given that $dy/dx = xy$ and $y = 5$ when $x = 1$, find an approximate value of $y(1.5)$ with step size 0.1, using Euler's method.

Solution: The Euler's formula is given by the relation

$$y_{n+1} = y_n + hf(x_n, y_n)$$

The first approximate solution for $f(x, y) = xy$ is given as $x_0 = 1$, $y_0 = 5$ and step size is given as $h = 0.1$

For second approximate solution, $y_1 = y_0 + hf(x_0, y_0)$

$$y_1 = y_0 + (0.1)x_0y_0$$

$$y_1 = 5 + (0.1)(1)(5) = 5.5$$

and $x_1 = x_0 + h = 1 + 0.1 = 1.1$

So the point of second solution is $(x_1, y_1) = (1.1, 5.5)$

For third approximate solution, $y_2 = y_1 + hf(x_1, y_1)$

$$y_2 = y_1 + (0.1)x_1y_1$$

$$y_2 = 5.5 + (0.1)(1.1)(5.5) = 6.105$$

and $x_2 = x_1 + h = 1.1 + 0.1 = 1.2$

So the point of third solution is $(x_2, y_2) = (1.2, 6.105)$

For fourth approximate solution, $y_3 = y_2 + hf(x_2, y_2)$

$$y_3 = y_2 + (0.1)x_2y_2$$

$$y_3 = 6.105 + (0.1)(1.2)(6.105) = 6.838$$

and $x_3 = x_2 + h = 1.2 + 0.1 = 1.3$

So the point of fourth solution is $(x_3, y_3) = (1.3, 6.838)$

For fifth approximate solution, $y_4 = y_3 + hf(x_3, y_3)$

$$y_4 = y_3 + (0.1)x_3y_3$$

$$y_4 = 6.838 + (0.1)(1.3)(6.838) = 7.727$$

and $x_4 = x_3 + h = 1.3 + 0.1 = 1.4$

So the point of fifth solution is $(x_4, y_4) = (1.4, 7.727)$

For sixth approximate solution, $y_5 = y_4 + hf(x_4, y_4)$

$$y_5 = y_4 + (0.1)x_4y_4$$

$$y_5 = 7.727 + (0.1)(1.4)(7.727) = 8.809$$

and $x_5 = x_4 + h$

$$= 1.4 + 0.1 = 1.5$$

So the point of sixth solution is $(x_5, y_5) = (1.5, 8.809)$

Thus, the required value of y at $x = 1.5$ is 8.809.

Problem 2: Solve the equation $dy/dx = 1 - y$ with the initial condition $x = 0, y = 0$ using Euler's algorithm and tabulate the solutions at $x = 0.1, 0.2, 0.3$.

Solution. Here, $f(x, y) = 1 - y$

Taking $h = 0.1, x_0 = 0, y_0 = 0$, we obtain

$$y_1 = y_0 + hf(x_0, y_0)$$
$$= 0 + (.1)(1 - 0) = .1$$
$$y(0.1) = 0.1$$

Again,
$$y_2 = y_1 + hf(x_1, y_1)$$
$$= 0.1 + (0.1)(1 - .1)$$
$$= 0.1 + .09 = .19$$
$$y(0.2) = 0.19$$

Again,
$$y_3 = y_2 + hf(x_2, y_2)$$
$$= .19 + (.1)(1 - .19)$$
$$= .19 + (.1)(.81) = .271$$
$$y(0.3) = .271$$

Tabulated values are

x	y(x)
0	0
0.1	0.1
0.2	0.19
0.3	0.271

UNSOLVED PROBLEMS

1. Using Euler's method, find an approximate value of y corresponding to $x = 2$,given that $dy/dx = x + 2y$ and $y = 1$ when $x = 1$.
2. Use Euler's method $dy/dx = x + y, y(0) = 0$, take $h = 0.2$ and compute $y(0.4)$ and $y(0.6)$
3. Apply Euler's method to the initial value problem $dy/dx = x + y, y(0) = 0$ at $x = 0$ to $x = 1.0$ taking $h = 0.2$.
4. Use Euler's method with $h = 0.1$ to solve the differential equation $dy/dx = x^2 + y^2, y(0) = 1$ in the range $x = 0$ to $x = 0.3$.

9.2.4 Modified Euler's Method

As we have seen that the simplest Euler's method uses the slope at starting point of the solution curve to obtain the next point of solution curve but in case of Modified Euler's method we find slopes at both end points of the interval and then find its average. This increases speed as well as accuracy.

Suppose we have the interval which is bounded by the point $x = x_n$ and $x = x_{n+1}$ then its average is given by

$$\frac{1}{2}[y'(x_n) + y'(x_{n+1})] \qquad \text{[As the slope is given by } dy/dx = y'\text{]}$$

According to Euler's method, we have

$$y_{n+1} = y_n + \frac{h}{2}[y'(x_n) + y(x_{n+1})]$$

As $\quad y'(x_n) = f(x_n, y_n)$

$$\therefore \quad y_{n+1} = y_n + \frac{h}{2}\,[f(x_n, y_n) + f(x_{n+1}, y_{n+1})]$$

Where $x_{n+1} = x_n + h$ and $y_{n+1} = y_n + hf(x_n, y_n)$

This is improved value of y_{n+1} at x_{n+1}. The order of modified Euler's method is h^3 because it produces a truncation error as $\frac{h^3}{2}\,y'''(n)$.

9.2.4.1 *Algorithm of Modified Euler's Method*

1. Function of modified euler's F(x)=(x–y)/(x+y)
2. Input values of x(1),y(1),h,xn
3. yp=y(1)+h*F(x(1),y(1))
4. itr=(xn–x(1))/h
5. Print the value x(1),y(1)
6. For i=1,itr
7. x(i+1)= x(i)+h
8. For n=1 to 50
9. yc(n+1)=y(i)+(h/2*(F(x(i),y(i))+F(x(i+1),yp))
10. Print value of n,yc(n+1)
11. p=yc (n+1)-yp
12. If abs(p)<.0001 then
    goto Step 14
    ELSE
    yp=yc(n+1)
13. Next n
14. y(i+1)=yc(n+1)
15. print x(i+1),yp
16. Next i
17. Stop

9.2.4.2 Flow-chart of Modified Euler's Method

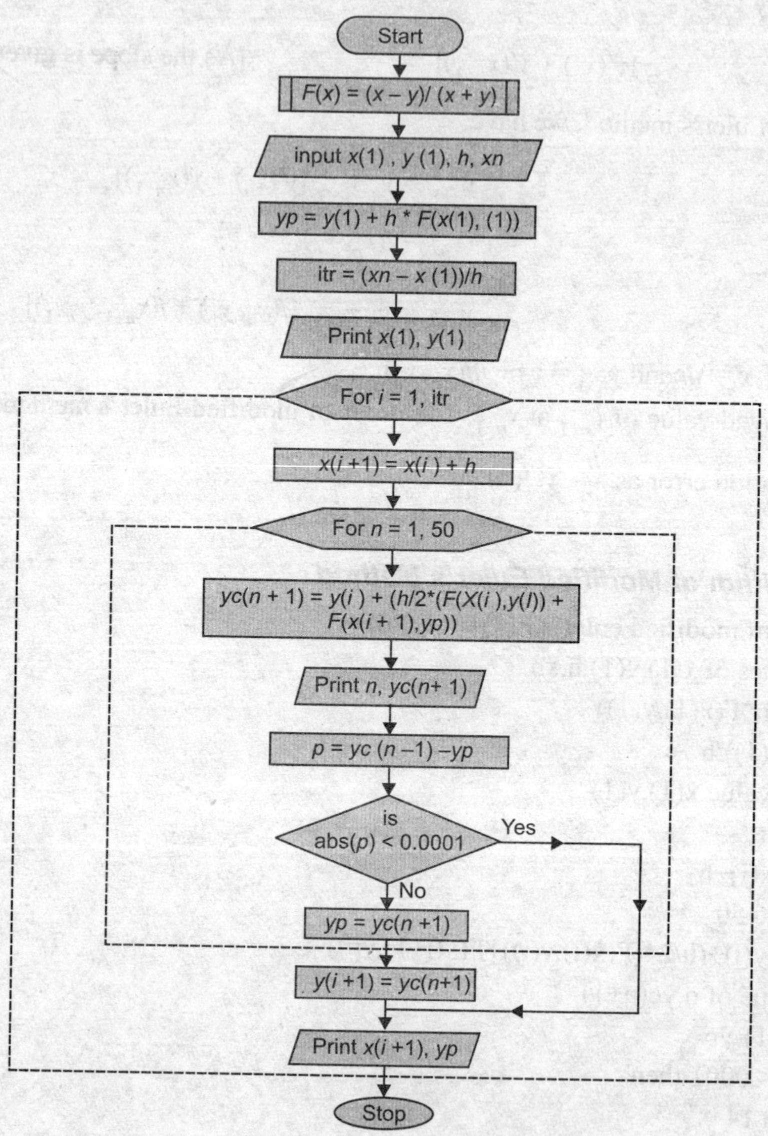

9.2.4.3 PROGRAM OF MODIFIED EULER'S METHOD IN C

```
#include<stdio.h>
#include<conio.h>
#include<math.h>
#define F(x,y) (x–y)/(x+y)
```

```
main ()
{
int i,n,itr ;
float x[5],y[50],yc[50],h,yp,p,xn;
printf("\n Enter the values: x[1],y[1],h,xn:\n");
scanf("%f%f%f%f",&x[1],&y[1],&h,&xn);
yp=y[1]+h*F(x[1],y[1]);
itr=(xn–x[1])/h;
printf("\n\n X=%f Y=%f\n",x[1],y[1];
for (i=1;i<=itr;i++)
{
x[i+1]=  x[i]+h;
for (n=1;n<=50;n++)
{
yc[n+1]=y[i]+(h/2.0)*(F(x[i],y[i])+F(x[i+1],yp));
printf("\nN=%d Y=%f",n,yc[n+1]);
p=yc[n+1]–yp;
if(fabs (p)<0.0001)
goto next;
else
yp=yc[n+1];
}
next:
y[i+1]=yc[n+1];
printf("\n\n X=%f Y=%f\n",x[i+1], yp);
}
getch();
}
```

Output:

```
Enter the values: x[1],y[1],h,xn:
0 1 0.02 0.06
X=0.000000 Y=1.000000
N=1 Y=0.980400
N=2 Y=0.980400
```

```
X=0.020000 Y=0.980400
N=1 Y=0.961584
N=2 Y=0.961598
X=0.040000 Y=0.961584
N=1 Y=0.943572
N=2 Y=0.943593
X=0.060000 Y=0.943572
```

9.2.4.4 Program of Modified Euler's Method in C++

```
#include<iostream.h>
#include<conio.h>
#include<math.h>
#define F(x,y) (x–y)/(x+y)
main ()
{
int i,n,itr ;
float x[5],y[50],yc[50],h,yp,p,xn;
cout<<"\n Enter the values: x[1],y[1],h,xn:\n";
cin>>x[1]>>y[1]>>h>>xn;
yp=y[1]+h*F(x[1],y[1]);
itr=(xn–x[1])/h;
cout<<"\n\n X="<<x[1]<<"Y="<<y[1];
for (i=1;i<=itr;i++)
{
x[i+1]=  x[i]+h;
for (n=1;n<=50;n++)
{
yc[n+1]=y[i]+(h/2.0)*(F(x[i],y[i])+F(x[i+1],yp));
cout<<"\nN="<<n<<"Y="<<yc[n+1];
p=yc[n+1]-yp;
if(fabs (p)<0.0001)
goto next;
else
yp=yc[n+1];
```

```
}
next:
y[i+1]=yc[n+1];
cout<<"\n\n X="<< x[i+1]<<"Y="<< yp;
}
getch();
}
```

Output:

```
Enter the values: x[1],y[1],h,xn:
0 1 0.02 0.06
X=0.000000 Y=1.000000
N=1 Y=0.980400
N=2 Y=0.980400
X=0.020000 Y=0.980400
N=1 Y=0.961584
N=2 Y=0.961598
X=0.040000 Y=0.961584
N=1 Y=0.943572
N=2 Y=0.943593
X=0.060000 Y=0.943572
```

SOLVED PROBLEMS

Problem 1: By using modified Euler's method find y at $x = 0.1$ and y at $x = 0.2$ given that $dy/dx = (y - 2x/y)$, $y(0) = 1$.

Solution: As we have given $x_0 = 0$ and $y_0 = 1$ and let $h = 0.1$

By Modified Euler's method

$$y_{n+1} = y_n + \frac{h}{2}[f(x_n, y_n) + f(x_{n+1}, y_{n+1})]$$

or we can write

$$y_{n+1} = y_n + \frac{h}{2}[f(x_n, y_n) + f(x_n + h, y_n + h\, f(x_n, y_n))] \qquad \ldots(1)$$

$$\therefore \qquad y_1 = y_0 + \frac{h}{2}[f(x_0, y_0) + f(x_0 + h, y_0 + h\, f(x_0, y_0))] \qquad \ldots(2)$$

$$f(x_0, y_0) = y_0 - \frac{2x_0}{y_0} = 1 - 0 = 1$$

$$f(x_0 + h, y_0 + hf(x_0, y_0)) = f\{(0+0.1), 1+0.1(1)\} = f(0.1, 1.1)$$

$$= 1.1 - \frac{2 \times (0.1)}{1.1} = 0.91818$$

Now $$y(0.1) = y_1 = 1 + \frac{0.1}{2}[1+0.91818] = 1.095909$$

Similarly, $$y_2 = y(0.2) = y_1 + \frac{h}{2}[f(x_1, y_1) + f(x_1+h, y_1 + hf(x_1,y_1))] \quad ...(3)$$

$$f(x_1, y_1) = y_1 - \frac{2x_1}{y_1} = 1.095909 - \frac{2 \times 0.1}{1.095909} = 0.913412$$

$$f(x_1+h, y_1 + hf(x_1, y_1)) = f(0.2, 1.095909 + (0.1)(0.913412))$$

$$= f(0.2, 1.18732) = 1.8732 - \frac{2 \times 0.2}{1.18732} = 0.8504268$$

So $$y_2 = 1.095909 + \frac{0.1}{2}[0.913412 + 0.850427]$$

$$= 1.1841009$$

Problem 2: Given $dy/dx = x - y^2$; $y(.2) = 0.2$, find $y(.4)$ by modified Euler's method correct to 3 decimal places, taking $h = 0.2$.

Solution: Here, $f(x, y) = x - y^2$; $x_0 = 0.2$, $y_0 = .02$ and $h = 0.2$

Let $x_1 = 0.4$ then we are to find $y_1 = y(0.4)$

We have $$f(x_0, y_0) = x_0 - y_0$$

$$2 = 0.2 - (.02)2 = 0.2 - .0004 = 0.1996$$

$$y_1^{(1)} = y_0 + hf(x_0, y_0) = .02 + (.2)(.1996) = .060$$

$$f\{x_1, y_1^{(1)}\} = x_1 - \{y_1^{(1)}\}2 = .4 - (.06)2 = .3964$$

$$y_1^{(2)} = y_0 + h/2[f(x_0, y_0) + f\{x_1, y_1^{(1)}\}]$$

$$= .02 + .2/2[.1996 + .3964] = .0796 = .080$$

Now, $$f\{x_1, y_1^{(2)}\} = x_1 - [y_1^{(2)}]2 = .4 - (.08)2 = .3936$$

$$y_1^{(3)} = y_0 + h/2[f(x_0, y_0) + f\{x_1, y_1^{(2)}\}]$$

$$= .02 + .2/2[.1996 + .3936] = .07932 = .079$$

$$f\{x_1, y_1^{(3)}\} = x_1 - [y_1^{(3)}]2 = .4 - (.079)2 = .3938$$

$$y_1^{(4)} = y_0 + h/2[f(x_0, y_0) + f\{x_1, y\, y_1^{(3)}\}]$$

$$= .02 + .2/2[.1996 + .3938] = .0793 = .079$$

Since $y_1^{(3)} = y_1^{(4)}$

hence $y_1 = .079.$

UNSOLVED PROBLEMS

1. Find numerically $y' = y + e^x$, $y(0) = 0$ for $x = 0.2, 0.4$ by Modified Euler's method.
2. Using Modified Euler's method find y at $x = 0.1$ and y at $x = 0.2$ given $dy/dx = y - (2x/y)$, $y(0) = 1$
3. Find y for $x = 0.2$ and $x = 0.5$ using modified Euler's method, given that $dy/dx = \log_e(x+y)$; $y(0) = 1$.
4. Taking $h = .05$, determine the value of y at $x = 0.1$ by Euler's modified method, given that, $dy/dx = x^2 + y$; $y(0) = 1$
5. Given $dy/dx = x^2 + y$, $y(0) = 1$, find $y(.02)$, $y(.04)$ and $y(.06)$ using Euler's modified method.

9.3 RUNGE-KUTTA'S METHODS

The Runge – Kutta methods are an important family of implicit and explicit iterative methods for the approximation of solutions of ordinary differential equations. These techniques were developed around 1900 by the German mathematicians C. Runge and M.W. Kutta. Runge Kutta methods only uses function values at some specific points. These methods have the following properties:

(i) They are single step methods.

(ii) They agree with the Taylor's series through terms in h^p, where p is different for different methods.

(iii) Derivatives are not needed.

In this family of methods we will discuss Runge-kutta second order and Runge-kutta fourth order method. In second and fourth order methods the interval $[x_1, x_n]$ is divided into subintervals and a weighted average of derivatives (slopes) at these intervals is used to determine the value of the dependent variable.

9.3.1 Runge- Kutta Second Order Method

Runge-kutta second order method is also known as Heun's method. Consider the differential equation $dy/dx = f(x, y)$ with initial condition $y(x_1) = y_1$ ...(1)

Draw a straight line from (x_1, y_1) with a slope $s_1 = f(x_1, y_1)$. Suppose this line cut the vertical line through $x_1 + h$ at (x_2, y_2') where $x_2 = x_1 + h$ and $y_2' = y_1 + s_1 h$.

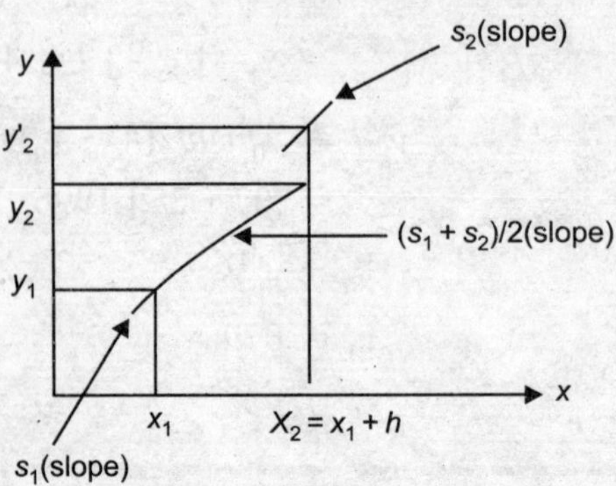

Determine the slope (s_2) dy/dx of the solution curve at the point (x_2, y_2') as $f(x_2, y_2')$. Here $y_2' = y_1 + s_1 h$. Now draw a straight line from (x_1, y_1) with a slope $(s_1 + s_2)/2$. The point where this straight line cuts the vertical line at x_1+h is the approximate solution of the differential equation at x_1+h. Thus we have

$$y_2 = y_1 + \frac{h}{2}(s_1 + s_2) \qquad \text{...(2)}$$

In general the $(n + 1)^{\text{th}}$ point is obtained from n^{th} point using the formula

$$y_{n+1} = y_n + \frac{h}{2}(s_n + s_{n+1}) \qquad \text{...(3)}$$

Where $s_n = f(x_n, y_n)$ and $s_{n+1} = f(x_{n+1}, y_n + s_n h)$

Show that Runge-Kutta second order methods are identical to Taylor's series method

As from equation (3), we have

$$\begin{aligned} y_{n+1} &= y_n + \frac{h}{2}(s_n + s_{n+1}) \\ &= y_n + \frac{h}{2}[f(x_n, y_n) + f(x_n + h, y_n + s_n h)] \end{aligned}$$

Expand this by using Taylor's series, we have

$$y_{n+1} = y_n + \frac{h}{2}[f(x_n, y_n) + f(x_n, y_n) + hf_x(x_n, y_n) + s_n hf_y(x_n, y_n) + \ldots\ldots]$$

$$y_{n+1} = y_n + \frac{h}{2}[f(x_n, y_n) + f(x_n, y_n) + hf_x(x_n, y_n) + f(x_n, y_n) \cdot hf_y(x_n, y_n) + \ldots\ldots]$$

$$y_{n+1} = y_n + h f(x_n, y_n) + \frac{h^2}{2!}[f_x(x_n, y_n) + f_y(x_n, y_n) \cdot f(x_n, y_n)] + \ldots\ldots\ldots$$

Which is identical to Taylor's series method.

9.3.2 Runge-kutta Fourth Order Method

Runge-Kutta 4th order method is a numerical technique used to solve ordinary differential equation of the form

$$\frac{dy}{dx} = f(x, y), y(0) = y_0$$

So only first order ordinary differential equations can be solved by using the Runge-Kutta 4th order method.

Runge-kutta's fourth order methods are better than Runge-kutta's second order methods as error in this case is of order $O(h^5)$ per step whereas in case of second order methods it is of order $O(h^3)$. So, if more precision is required then fourth order methods are useful.

In Runge-kutta's fourth order methods, the slope (i.e. derivatives) at four points including the starting point is computed, and then the weighted average of these slopes is computed for finding the value of the dependent variable (y).

Then by using this point we will compute the next points. This process is repeated till we find the solution in the desired interval.

Geometrically, suppose we have given the initial point $A = (x_i, y_i)$ on the solution curve. Draw a straight line through it with slope $s_1 = f(x_i, y_i)$. Let this straight line cuts the vertical line erected at $x_i + h/2$ at point B.

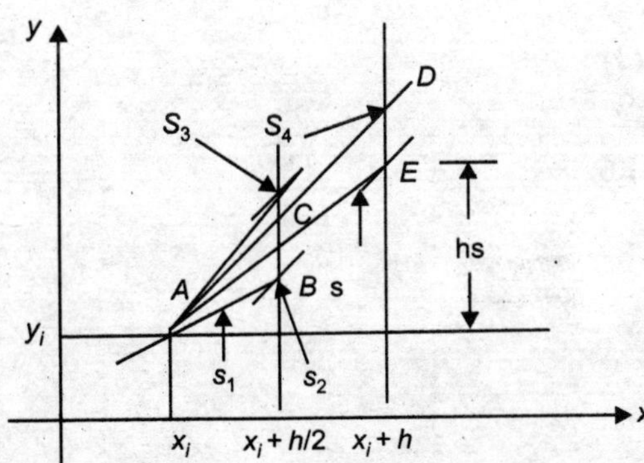

Compute the slope s_2 of the solution curve at B. This slope s_2 equals $f(x_i + h/2, y_i + s_1h/2)$. From the point A draw a line with slope s_2. Let this cut the vertical line at $x_i + h/2$ at point C. Find the slope s_3 at C. This slope s_3 equals $f(x_i + h/2, y_i + s_2h/2)$. From point A draw a straight line again with slope s_3. Let this straight line cut the vertical line erected at $x_i + h$ at D. compute the slope s_4 of the solution curve at D. this slope s_4 equals $f(x_i + h, y_i + s_3h)$. Now find the weighted average(s) of these four slopes which is as:

$$s = \frac{(s_1 + 2s_2 + 2s_3 + s_4)}{6}$$

Now again draw a line from A with slope s. Let it cut the vertical line erected at $x_i + h$ at E. the point E is taken as the solution of the differential equation at $x_i + h$.

The formula may be summarized as:

$$y_{i+1} = \frac{h}{6}(s_1 + 2s_2 + 2s_3 + s_4)$$

Where

$$s_1 = f(x_i, y_i)$$
$$s_2 = f(x_i + h/2, y_i + s_1 h/2)$$
$$s_3 = f(x_i + h/2, y_i + s_2 h/2)$$
$$s_4 = f(x_i + h, y_i + s_3 h)$$

9.3.3 Algorithm of Runge-kutta Method

1. Function of runge-kutta F(x)=(x-y)/(x+y)
2. Input value of x0,y0,h,xn
3. n=(xn-x0)/h
4. x= x0
5. y=y0
6. For i=0 to n
7. k1=h*F(x,y)
8. k2=h*F(x+h/2,y+k1/2)
9. k3=h*F(x+h/2,y+k2/2)
10. k4=h*F(x+h,y+k3)
11. k=(k1+(k2+k3)2+k4)/6
12. Print value of x,y
13. x= x+h
14. y=y+k
15. Next i
16. Stop

9.3.4 Flow-chart of Runge-kutta Method

See figure on next page

9.3.5 Program of Runge-kutta Method in C

```
#include<stdio.h>
#include<conio.h>
#define F(x,y) (x-y)/(x+y)
main()
```

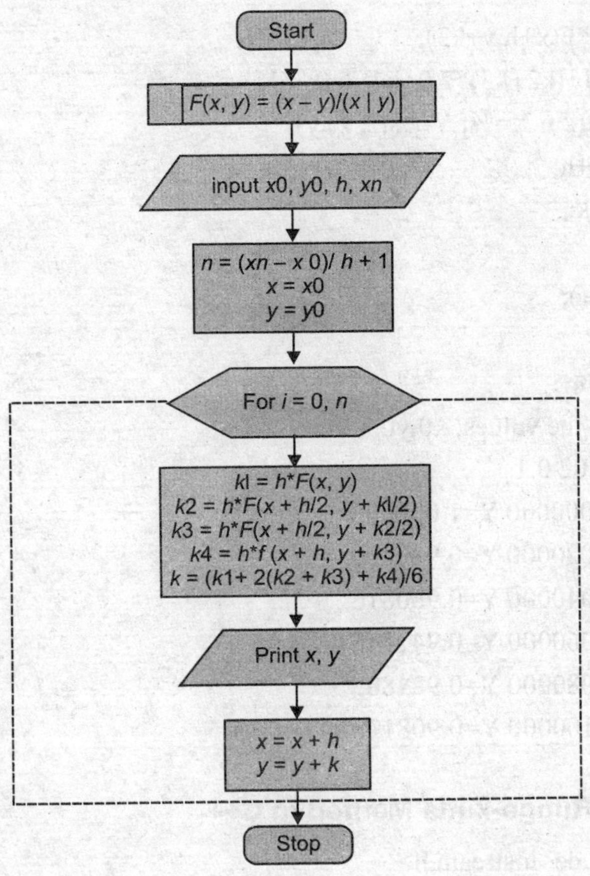


```
{
int i,n;
float x0,y0,h,xn,k1,k2,k3,k4,x,y,k;
printf("\n Enter the values: x0,y0,h,xn:\n");
scanf("%f%f%f%f", &x0,&y0,&h,&xn);
n=(xn-x0)/h;
x= x0;
y=y0;
for(i=0;i<=n;i++)
{
k1=h*F(x,y);
k2=h*F(x+h/2.0,y+k1/2.0);
k3=h*F(x+h/2.0,y+k2/2.0);
```

9

```
k4=h*F(x+h,y+k3);
k=(k1+(k2+k3)*2.0+k4)/6.0;
printf("\n X=%f Y=%f", x, y);
x= x+h;
y=y+k;
}
getch();
}
```

Output:

```
Enter the values: x0,y0,h,xn:
0 1 0.02 0.1
X=0.000000 Y=1.000000
X=0.020000 Y=0.980000
X=0.040000 Y=0.960816
X=0.060000 Y=0.942446
X=0.080000 Y=0.924885
X=0.100000 Y=0.908128
```

9.3.6 Program of Runge-kutta Method in C++

```
#include<iostream.h>
#include<conio.h>
#define F(x,y) (x-y)/(x+y)
main()
{
int i,n;
float x0,y0,h,xn,k1,k2,k3,k4,x,y,k;
cout<<"\n Enter the values: x0,y0,h,xn:\n";
cin>>x0>>y0>>h>>xn;
n=(xn-x0)/h;
x= x0;
y=y0;
for(i=0;i<=n;i++)
{
```

```
k1=h*F(x,y);
k2=h*F(x+h/2.0,y+k1/2.0);
k3=h*F(x+h/2.0,y+k2/2.0);
k4=h*F(x+h,y+k3);
k=(k1+(k2+k3)*2.0+k4)/6.0;
cout<<"\n X="<<x<<"Y="<<y;
x= x+h;
y=y+k;
}
getch();
}
```

Output:

```
Enter the values: x0,y0,h,xn:
0 1 0.02 0.1
X=0.000000 Y=1.000000
X=0.020000 Y=0.980000
X=0.040000 Y=0.960816
X=0.060000 Y=0.942446
X=0.080000 Y=0.924885
X=0.100000 Y=0.908128
```

SOLVED PROBLEMS

Problem 1: Find $y(0.1)$ and $y(0.2)$ in steps of 0.1 by applying Runge-Kutta second order method, given that $dy/dx = y - x$ with $y(0) = 2$.

Solution: Here $x_1 = 0, y_1 = 2$ and $h = 0.1$

Given, $f(x, y) = y - x$

$$s_1 = f(x_1, y_1) = y_1 - x_1 = 2 - 0 = 2$$

$$s_2 = f(x_1 + h, y_1 + hf(x_1, y_1))$$

$$= f(0 + 0.1, 2 + 0.1(2)) = f(0.1, 2.2) = 2.2 - 0.1$$

$$= 2.1$$

The second order Runge-Kutta formula is given by

$$y_2 = y_1 + \frac{h}{2}[s_1 + s_2]$$

$$y_2 = 5 + \frac{0.1}{2}[2 + 2.1]$$
$$= 5 + 0.205$$
$$= 5.205$$

and $\quad x_2 = x_1 + h = 0 + 0.1 = 0.1$

Now,
$$s_1 = f(x_2, y_2) = y_2 - x_2$$
$$= 5.205 - 0.1 = 5.105$$
$$s_2 = f(x_2 + h, y_2 + hf(x_2, y_2))$$
$$= f(0.1+0.1,\ 5.205 + 0.1(5.105)) = f(0.2,\ 5.7155)$$
$$= 5.7155 - 0.2$$
$$= 5.5155$$

$$y_3 = y_2 + \frac{h}{2}[s_1 + s_2]$$
$$= 5.205 + \frac{0.1}{2}[5.105 + 5.5155]$$
$$= 5.205 + 0.531025$$
$$= 5.736025$$

Hence we have $y(0.1) = 5.205$ and $y(0.2) = 5.736025$

Problem 2: Using Runge-kutta's fourth order method, to find an approximate value of y for $x = 0.2$ and 0.4, if $dy/dx = (y^2 - x^2)/(y^2 + x^2)$, given that $y = 1$ when $x = 0$.

Solution: Here $x_1 = 0$, $y_1 = 1$ and h = 0.2

Given, $\quad f(x, y) = (y^2 - x^2)/\ (y^2 + x^2)$

The fourth order Runge – Kutta formula is given by

$$y_2 = y_1 + \frac{h}{6}[s_1 + 2s_2 + 2s_3 + s_4]$$
$$s_1 = f(x_1, y_1) = (y_1^2 - x_1^2)/(y_1^2 + x_1^2) = 1$$
$$s_2 = f(x_1 + h/2, y_1 + s_1 h/2) = f(0.1,\ 1.1) = 0.9836$$
$$s_3 = f(x_1 + h/2,\ y_1 + s_2 h/2) = f(0.1,\ 1.09836) = 0.9835$$
$$s_4 = f(x_1 + h,\ y_1 + s_3 h) = f(0.2,\ 1.1967) = 0.9955$$

Substitute all slope values in above formula for y_2, we have

$$y(0.2) = 1 + 0.2\ [1 + 2 \times 0.9836 + 2 \times 0.9835 + 0.9455]$$
$$= 1.196$$

Now we have to compute $y(0.4)$ by using $x_2 = 0.2, y_2 = 1.196$ and h = 0.2

So,
$$s_1 = f(x_2, y_2) = (y_2^2 - x_2^2)/(y_2^2 + x_2^2) = 0.9455$$
$$s_2 = f(x_2 + h/2, y_2 + s_1 h/2) = f(0.3, 1.2906) = 0.8975$$
$$s_3 = f(x_2 + h/2, y_2 + s_2 h/2) = f(0.3, 1.2858) = 0.8965$$
$$s_4 = f(x_2 + h, y_2 + s_3 h) = f(0.4, 1.3753) = 0.844$$

Now
$$y_3 = y_2 + \frac{h}{6}[s_1 + 2s_2 + 2s_3 + s_4]$$

Substitute all slope values in above formula for y_3, we have

$$y(0.4) = 1.196 + \frac{0.2}{6}[0.9455 + 2 \times 0.8975 + 2 \times 0.8965 + 0.844] = 1.3752$$

Problem 3: A ball at 1200 K is allowed to cool down in air at an ambient temperature of 300 K. Assuming heat is lost only due to radiation, the differential equation for the temperature of the ball is given by

$$\frac{d\theta}{dt} = -2.2067 \times 10^{-12}\left(\theta^4 - 81 \times 10^8\right), \theta(0) = 1200\text{ K}$$

where q is in *K* and *t* in seconds. Find the temperature at $t = 480$ seconds using Runge-Kutta 4th order method. Assume a step size of $h = 240$ seconds.

Solution:
$$\frac{d\theta}{dt} = -2.2067 \times 10^{-12}\left(\theta^4 - 81 \times 10^8\right)$$
$$f(t, \theta) = -2.2067 \times 10^{-12}(\theta^4 - 81 \times 10^8)$$
$$\theta^{i+1} = \theta_i + \frac{1}{6}(k_1 + 2k_2 + 2k_3 + k_4)h$$

For $i = 0$, $t_0 = 0$, $\theta_0 = 1200$ K
$$k1 = f(t_0, \theta_0)$$
$$= f(0,1200)$$
$$= -2.2067 \times 10^{-12}(1200^4 - 81 \times 10^8)$$
$$= -4.5579$$
$$k_2 = f\left(t_0 + \frac{1}{2}h, \theta_0 + \frac{1}{2}k_1 h\right)$$
$$= f\left(0 + \frac{1}{2}(240), 1200 + \frac{1}{2}(4.5579) \times 240\right)$$

$$= f(120{,}653.05)$$
$$= -2.2067 \times 10^{-12}\,(653.05^4 - 81 \times 10^8)$$
$$= -\,0.38347$$

$$k_3 = f\left(t_0 + \frac{1}{2}h, \theta_0 + \frac{1}{2}k_2 h\right)$$
$$= f\left(0 + \frac{1}{2}(240), 1200 + \frac{1}{2}(-0.38347) \times 240\right)$$
$$= f(120{,}1154.0)$$
$$= -2.2067 \times 10^{-12}\,(1154.0^4 - 81 \times 10^8)$$
$$= -\,3.8954$$

$$k_4 = f(t_0 + h, \theta_0 + k_3 h)$$
$$= f(0 + 240{,}1200 + (-3.894) \times 240)$$
$$= f(240{,}265.10)$$
$$= -2.2067 \times 10^{-12} * (265.10^4 - 81 \times 10^8)$$
$$= 0.0069750$$

$$\theta_1 = \theta_0 + \frac{1}{6}(k_1 + 2k_2 + 2k_3 + k_4)h$$
$$= 1200 + \frac{1}{6}(-4.5579 + 2\,(-0.38347) + 2(-3.8954) + (0.069750)\;240$$
$$= 1200 + (-2.1848) \times 240$$
$$= 675.65 \text{ K}$$

θ_1 is the approximate temperature at

$$t = t_1$$
$$= t_0 + h$$
$$= 0 + 240 = 240$$

$$\theta_1 = \theta(240)$$
$$\approx 675.65 \text{ K}$$

For $i = 1,\ t_1 = 240, \theta_1 = 675.65$ K

$$k_1 = f(t_1, \theta_1)$$
$$= f(240{,}675.65)$$

$$= -2.2067 \times 10^{-12}(675.65^4 - 81 \times 10^8)$$

$$= -0.44199$$

$$k_2 = f\left(t_1 + \frac{1}{2}h, \theta_1 + \frac{1}{2}k_1 h\right)$$

$$= f\left(240 + \frac{1}{2}(240), 675.65 + \frac{1}{2}(-0.44199)\,240\right)$$

$$= f(360, 622.61)$$

$$= -2.2067 \times 10^{-12}(622.61^4 - 81 \times 10^8)$$

$$= -0.31372$$

$$k_3 = f\left(t_1 + \frac{1}{2}h, \theta_1 + \frac{1}{2}k_2 h\right)$$

$$= f\left(240 + \frac{1}{2}(240), 675.65 + \frac{1}{2}(-0.31372) \times 240\right)$$

$$= f(360, 638.00)$$

$$= -2.2067 \times 10^{-12}(638.00^4 - 81 \times 10^8)$$

$$= -0.34775$$

$$k_4 = f(t_1 + h, \theta_1 + k_3 h)$$

$$= f(240 + 240, 675.65 + (-0.34775) \times 240)$$

$$= f(480, 592.19)$$

$$= 2.2067 \times 10^{-12}(592.19^4 - 81 \times 10^8)$$

$$= -0.25351$$

$$\theta_2 = \theta_1 + \frac{1}{6}(k_1 + 2k_2 + 2k_3 + k_4)h$$

$$= 675.65 + \frac{1}{6}(-0.44199 + 2(-0.31372) + 2(-0.34775) + (-0.25351)) \times 240$$

$$= 675.65 + \frac{1}{6}(-2.0184) \times 240$$

$$= 594.91 \text{ K}$$

θ_2 is the approximate temperature at

$$t = t_1$$

$$= t_0 + h$$
$$= 240 + 240$$
$$= 480$$
$$\theta_2 = \theta(480)$$
$$\approx 594.91 \text{ K}$$

Figure 1 compares the exact solution with the numerical solution using the Runge-Kutta 4th order method with different step sizes.

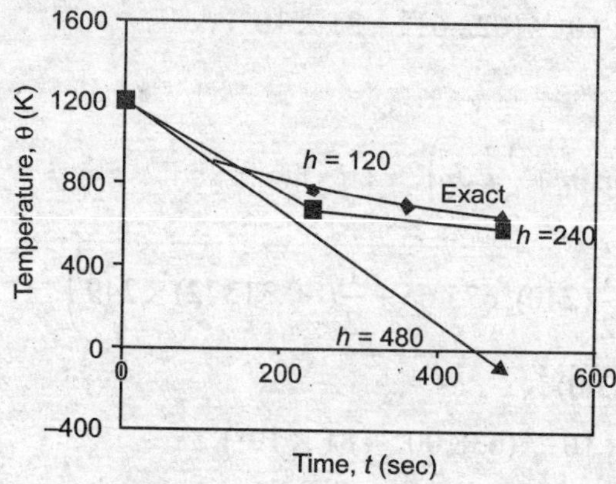

Figure 1: Comparison of Runge-Kutta 4th order method with exact solution for different step sizes.

Table 1 and Figure 2 show the effect of step size on the value of the calculated temperature at $t = 480$ seconds.

Value of temperature at time, t = 480 s for different step sizes			
step size, h	**$\theta(480)$**	**E_t**	**$\lvert E_t \rvert$ %**
480	–90.278	737.85	113.94
240	594.91	52.660	8.1319
120	646.16	1.4122	0.21807
60	647.54	0.033626	0.0051926
30	647.57	0.00086900	0.00013419

In Figure 3, we are comparing the exact results with Euler's method (Runge-Kutta 1st order method), Heun's method (Runge-Kutta 2nd order method), and Runge-Kutta 4th order method.

The formula described in this chapter was developed by Runge. This formula is same as Simpson's 1/3 rule, if $f(x, y)$ were only a function of x. There are other versions of the 4th order method just like there are several versions of the second order methods. The formula developed by Kutta is

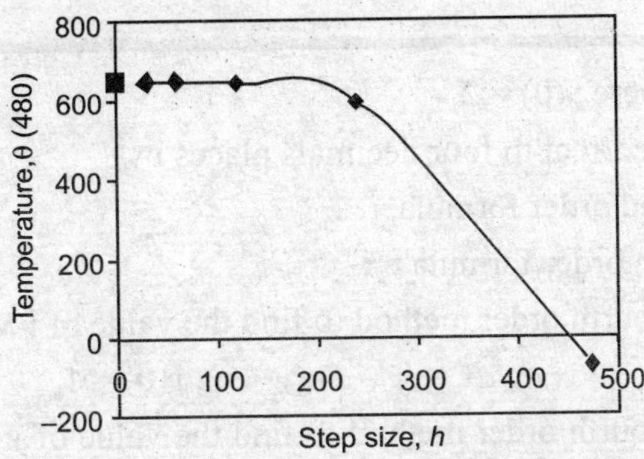

Figure 2: Effect of step size in Runge-Kutta 4th order method

$$y_{i+1} = y_i + \frac{1}{8} k_1 + 3k_2 + 3k_3 + k_4)\, h \qquad ...(6)$$

where $$k_1 = f(x_i, y_i) \qquad ...(7a)$$

$$k_2 = f\left(x_i + \frac{1}{3}h,\, y_i + \frac{1}{3}hk_1\right) \qquad ...(7b)$$

$$k_3 = f\left(x_i + \frac{2}{3}h,\, y_i - \frac{1}{3}hk_1 + hk_2\right) \qquad ...(7c)$$

$$k_4 = f\left(x_i + h,\, y_i + hk_1 - hk_2 + hk_3\right) \qquad ...(7d)$$

This formula is the same as the Simpson's 3/8 rule, if $f(x, y)$ is only a function of x.

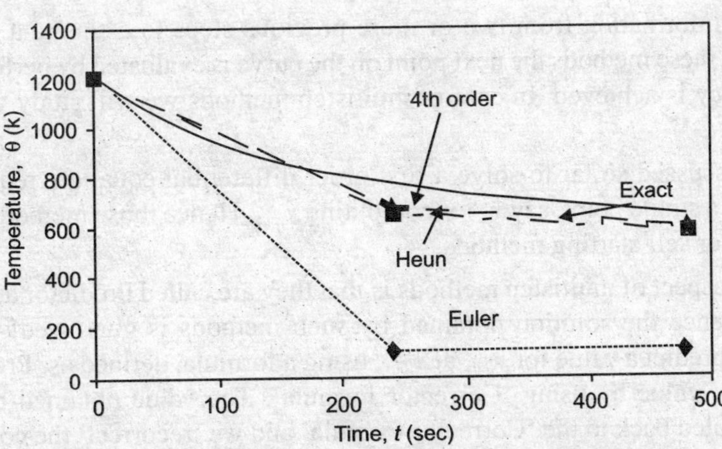

Figure 3: Comparison of Runge-Kutta methods of 1st (Euler), 2nd, and 4th order.

UNSOLVED PROBLEMS

1. Given $dy/dx = y - x$ where $y(0) = 2$
   Find $y(0.1)$ and $y(0.2)$ correct to four decimals places by
   (i) Runge-Kutta second order formula
   (ii) Runge-Kutta fourth order formula
2. Use the Runge-Kutta fourth order method to find the value of y when $x = 1$ given that
   $$dy/dx = (y - x)/(y + x),\ y(0) = 1$$
3. Use the Runge-Kutta fourth order method to find the value of y when $x = 0.2$ and $x = 0.4$ given that $dy/dx = (y^2 - x^2)/(y^2 + x^2)$, $y(0) = 1$
4. Solve the initial value problem
   $$du/dt = -2tu^2,\ u(0) = 1$$
   with $h = 0.2$ on the interval [0,0.4] using the fourth order Runge-Kutta method.
5. Apply the fourth order Runge-Kutta method to find $y(0.2)$ given that
   $$y' = x + y,\ y(0) = 1$$
6. Use the Runge-Kutta Method to approximate y when $x = 0.1$ given that $x = 0$ when $y = 1$ and $dy/dx = x + y$.
7. Apply the Runge-Kutta Fourth Order Method to solve $10\ dy/dx = x^2 + y^2$; $y\ (0) = 1$ for $0 < x \le 0.4$ and $h = 0.1$.
8. Use Runge-Kutta Fourth Order Formula to find $y(1.4)$ if $y(1) = 2$ and $dy/dx = xy$. Take $h = 0.2$.

9.4 MULTISTEP METHOD

These methods use information from two or more previous steps to estimate a value. These are iterative methods. In these methods the next point on the curve is evaluated by performing iterations till sufficient accuracy is achieved. In case of multistep methods we will study various Predictor corrector's methods.

The methods discussed so far to solve a first order differential equation, require information about the solution at a single point, say x_n for computing x_{n+1}. Hence those methods were termed as single step methods or self starting methods.

Most important aspect of multistep methods is that they are called Predictor and Corrector and iterative type and hence the solution obtained by such methods is guarnteed! As per the title, 'Predictor', we first predict a value for y_{n+1} at x_{n+1}, using a formula, defined as 'Predictor formula'. Then we correct this value by using 'Corrector formula'. The value obtained by this corrector formula is re-substituted back to the 'Corrector formula' and we 'recorrect' the corrected solution. This recorrection process is continued till the difference between two succeeding stages of usage of 'corrector formula' is less than a small quantity.

Now the question arises why we call predictor corrector methods as multistep methods the answer to this question is that in multistep method, to compute y_{n+1} we need the functional values at $y_n, y_{n-1}, \ldots\ldots\ldots$ i.e. evaluation of y_{n+1} depends on the multistep points at $y_n, y_{n-1}, \ldots\ldots\ldots$

Now we will discuss three such Predictor Corrector methods as:

(1) Euler's Predictor Corrector method

(2) Adams Bashforth method

(3) Milne's method

9.4.1 Euler's Predictor Corrector Method

The simplex Euler's method & Modified Euler's method *y*ields Euler's Predictor Corrector method, which consists of two steps:

Step 1:

Predict y_{n+1} by using the formula

$$= y_n + hf(x_n, y_n)$$

[Same as obtained in simple Euler's method]

The above formula is also known as Euler's Predictor formula that is useful for predicting the next value of *y*.

y^p_{n+1}

Step 2:

Correct y_{n+1} by using formula

$$y^c_{n+1} = y_n + h/2[f(x_n, y_n) + f(x_{n+1}, y^p_{n+1})$$

[Same as obtained in Modified Euler's method]

The above formula is known as Euler's Corrector formula that is useful for correcting the value of *y* produced by Predictor's formula. The values are corrected until we get two values, which are very-very closer to each other.

SOLVED PROBLEMS

Problem 1: By using Euler's Predictor Corrector method, to find a solution of the differential equation *dy*/*dx* = *xy* in the range [1,1.6] using step size 0.2 for the boundary of the boundary conditions *y* = 5 at *x* = 1.

Solution: Here $x_1 = 1, y_1 = 5, h = 0.2$

Given, $$f(x, y) = xy$$

Now first predict the value for y_2 through Euler's Predictor formula as

$$y^p_2 = y_1 + hf(x_1, y_1) = 5 + 0.2(1 \times 5) = 6$$

and $$x_2 = x_1 + h = 1+0.2 = 1.2$$

Now correct this predicted value of $y_2 = 6$ by using Euler's corrector formula as

$$y_2^c = y_1 + \frac{h}{2}[f(x_1, y_1) + f(x_1, y_1^p)]$$

$$= 5 + \frac{0.2}{2}[1 \times 5 + 1.2 \times 6] = 6.22$$

Now start from (1.2, 6.22), we first predict the value of y_3 using Euler's Predictor's formuls as

$$y_3^p = y_2 + h f(x_2, y_2)$$

$$= 5 + 0.2(1.2 \times 6.22) = 7.7128$$

and

$$x_3 = x_2 + h$$

$$= 1.2 + 0.2 = 1.4$$

Now correct this predicted value of $y_3 = 7.7128$ by using Euler's corrector formula as

$$y_3^c = y_2 + \frac{h}{2}[f(x_2, y_2) + f(x_2, y_2^p)]$$

$$= 6.22 + \frac{0.2}{2}[1.2 \times 6.22 + 1.4 \times 7.7128]$$

$$= 8.0462$$

Now start from (1.4,8.0462), we first predict the value of y_4 using Euler's Predictor's formuls as

$$y_4^p = y_3 + h f(x_3, y_3)$$

$$= 8.0462 + 0.2(1.4 \times 8.0462) = 10.2991$$

and

$$x_4 = x_3 + h = 1.4 + 0.2 = 1.6$$

Now correct this predicted value of $y_4 = 10.2991$ by using Euler's corrector formula as

$$y_4^c = y_3 + \frac{h}{2}[f(x_3, y_3) + f(x_3, y_3^p)]$$

$$= 10.2991 + \frac{0.2}{2}[1.4 \times 8.0462 + 1.6 \times 10.2991]$$

$$= 10.8205$$

UNSOLVED PROBLEMS

1. Find $y(0.8)$ given that $y' = (y - x)/(y + x)$, $y(0) = 1$ taking $h = 0.2$ Using Predictor Corrector methods.
2. Use Predictor-Corrector method for tabulating a solution of $10\,y' = x^2 + y^2$, $y(0) = 1$ for the range $0.5 \le x \le 1.0$

9.4.2 Adams Bashforth Method

Adams Bashforth is one of the most important method to solve the differential equation. To start with, this method requires information at past four solution points to extrapolate the solution at next point, three points out of four can be computed or obtained from the methods used for in the chapter.

For the derivation of the mathematical expression for Adams Bashforth method, we take the first order differential equation as

$$dy/dx = f(x, y)$$

or $$dy = f(x, y)dx \quad \ldots(1)$$

Integrate equation (1) within the limits $x = x_k$ and $x = x_{k+1}$, we have

$$\int_{x_k}^{x_{k+1}} dy = \int_{x_k}^{x_{k+1}} f(x, y)\, dx$$

$$\Rightarrow \qquad y_{k+1} = y_k + \int_{x_k}^{x_{k+1}} f(x, y)\, dx$$

Now to compute integral on the right side of (2), we approximate the function $f(x, y)$ as a polynomial in x by making it a fit through the past four points (x_{k-3}, y_{k-3}), (x_{k-2}, y_{k-2}), (x_{k-1}, y_{k-1}), (x_k, y_k).

By using Newton's backward difference interpolating polynomial, we have

$$f(x, y) = f_k + u \cdot \nabla f_k \frac{u(u+1)}{2} \nabla^2 f_k + \frac{u(u+1)(u+2)}{6} \nabla^3 f_k \quad \ldots(3)$$

where $u = \dfrac{x - x_k}{h}$ and ∇f_k, $\nabla^2 f_k$, $\nabla^3 f_k$ are backward differences at point $x = x_k$

Substitute (3) in (2), we have

$$y_{k+1} = y_k + \int_{x_k}^{x_{k+1}} \left[f_k + u \cdot \nabla f_k \frac{u(u+1)}{2} \nabla^2 f_k + \frac{u(u+1)(u+2)}{6} \nabla^3 f_k \right]$$

Since $u = \dfrac{x - x_k}{h}$, therefore $dx = hdu$

When $x = x_k$, $u = 0$

$x = x_{k+1}$, $u = 1$ (as $x_{k-1} - x_k = h$)

By substituting these values, we get

$$y_{k+1} = y_k + h \int_0^1 \left[f_k + u \cdot \nabla f_k \frac{u(u+1)}{2} \nabla^2 f_k + \frac{u(u+1)(u+2)}{6} \nabla^3 f_k \right] du$$

$$= y_k + h[f_k + (1/2)\nabla f_k + (5/12)\nabla^2 f_k + (3/8)\nabla^3 f_k]$$

Now $$\nabla f_k = f_k - f_{k-1}$$

$$\nabla^2 f_k = \nabla f_k - \nabla f_{k-1} = f_k - f_{k-1} - (f_{k-1} - f_{k-2}) = f_k - 2f_{k-1} + f_{k-2}$$
$$\nabla^3 f_k = \nabla^2 f_k - \nabla^2 f_{k-1} = f_k - 3f_{k-1} + 3f_{k-2} - f_{k-3}$$

Substituting these values, we get

$$y_{k+1} = y_k + h[f_k + (1/2)\{f_k - f_{k-1}\} + (5/12)\{f_k - 2f_{k-1} + f_{k-2}\} + (3/8)\{f_k - 3f_{k-1} + 3f_{k-2} - f_{k-3}]$$

After simplification, we have

$$y_{k+1}^p = y_k + \frac{h}{24}[55f_k - 59f_{k-1} + 37f_{k-2} - 9f_{k-3}]$$

This is known as Adams-Bashforth's Predictor formula. The subscript p here indicates the predicted value of y_{k+1}

In a similar way, we can develop Adams-Bashforth's Corrector formula. Now consider Newton's backward difference interpolating polynomial that passes through the points (x_{k-2}, y_{k-2}), (x_{k-1}, y_{k-1}), (x_k, y_k), (x_{k+1}, y_{k+1}^p) as

$$f(x, y) = f_{k+1} + u \cdot \nabla f_{k+1} \frac{u(u+1)}{2} \nabla^2 f_{k+1} + \frac{u(u+1)(u+2)}{6} \nabla^3 f_{k+1} \qquad (4)$$

where $u = \dfrac{x - x_{k+1}}{h}$ and ∇f_{k+1}, $\nabla^2 f_{k+1}$, $\nabla^3 f_{k+1}$ are backward differences at $x = x_{k+1}$

Substituting (4) in (2), we get

$$y_{k+1} = y_k + \int_{x_k}^{x_{k+1}} \left[f_{k+1} + u \cdot \nabla f_{k+1} \frac{u(u+1)}{2} \nabla^2 f_{k+1} + \frac{u(u+1)(u+2)}{6} \nabla^3 f_{k+1} \right] dx$$

Since $u = \dfrac{x - x_{k+1}}{h}$, therefore $dx = hdu$

When $x = x_k$, $u = -1$

$x = x_{k+1}$, $u = 0$

By substituting these values, we get

$$y_{k+1} = y_{k+1} + h \int_{-1}^{0} \left[f_{k+1} + u \cdot \nabla f_{k+1} \frac{u(u+1)}{2} \nabla^2 f_{k+1} + \frac{u(u+1)(u+2)}{6} \nabla^3 f_{k+1} \right] du$$

$$= y_k + h[f_{k+1} - (1/2)\nabla f_{k+1} - (1/12)\nabla^2 f_{k+1} - (1/24)\nabla^3 f_{k+1}]$$

Now

$$\nabla f_{k+1} = f_{k+1} - f_k$$
$$\nabla^2 f_{k+1} = f_{k+1} - 2f_k + f_{k-1}$$
$$\nabla^3 f_{k+1} = f_{k+1} - 3fk + 3f_{k-1} - f_{k-2}$$

Substituting these values, we get

$$y_{k+1} = y_k + h[f_{k+1} - (1/2)\{f_{k+1} - f_k\} - (1/12)\{f_{k+1} - 2f_k + f_{k-1}\}$$
$$- (1/24)\{f_{k+1} - 3f_k + 3f_{k-1} - f_{k-2}]$$

After simplification, we have

$$y^c_{k+1} = y_k + \frac{h}{24}[9f_{k+1}{}^{P} + 19f_k - 5f_{k-1} + f_{k-2}]$$

This is known as Adams-Moultan's Corrector formula. The subscript c here indicates the corrected value of y_{k+1}. By using this formula repeatedly we can improve the accuracy of computed y_{k+1}.

SOLVED PROBLEMS

Problem 1: By using Adams Bashforth method, find $y(0.4)$ of the differential equation $dy/dx = (½)xy$ and $y(0) = 1, y(0.1) = 1.01, y(0.2) = 1.022, y(0.3) = 1.023$

Solution: Given $h = 0.1, f(x, y) = (½)xy$

$x_0 = 0, x_1 = 0.1, x_2 = 0.2, x_3 = 0.3, x_4 = 0.4$

$y_0 = 1, y_1 = 1.01, y_2 = 1.022, y_3 = 1.023, y_4 = ?$

Starting values of Adams Bashforth method are

k	x_k	y_k	$f(x_k, y_k) = (½)x_k\, y_k$
0	0.0	1	0
1	0.1	1.01	0.0505
2	0.2	1.022	0.1022
3	0.3	1.023	0.1535

To obtain $y(0.4)$, we use the Adams-Bashforth's Predictor formula as

$$y^p_{0.4} = y_3 + \frac{h}{24}[55f_3 - 59f_2 + 37f_1 - 9f_0]$$

$$= 1.023 + \frac{0.1}{24}[55(1.1535) - 59(0.1022) + 37(0.0505) - 9(0)]$$

$$= 1.0408$$

$$\therefore \quad f^p_{0.4} = f(0.4, 1.0408)$$

$$= (½)(0.4)(1.0408)$$

$$= 0.2082$$

To correct the predicted value $y(0.4)$, we use the Adams-Moultan's Corrector formula as

$$y^c_{0.4} = y_3 + \frac{h}{24}[9\, f^p_{0.4} + 19f_3 - 5f_2 + f_1]$$

$$= 1.023 + \frac{0.1}{24}[9(0.2082) + 19(0.1535) - 5(0.1022) + 0.0505] = 1.0410$$

Hence $y(0.4) = 1.0410$

This value can more be improved by repeatedly using Corrector formula.

Problem 2: Using Runge-kutta fourth order method to find starting solutions to extrapolate the solution and apply Adam's method to find a solution of the differential equation $dy/dx = x - y^2$ at $x = 0.4$ in the range [0,1] using step size 0.1 for the boundary conditions $y = 1$ at $x = 0$.

Solution: Here $x_1 = 0, y_1 = 1$ and $h = 0.1$

Given, $$f(x, y) = x - y^2$$

To determine $y(0.1)$

The fourth order Runge-Kutta formula is given by

$$y_1 = y_0 + \frac{h}{6}[s_1 + 2s_2 + 2s_3 + s_4]$$

$$s_1 = f(x_1, y_1) = f(0,1) = -1$$

$$s_2 = f(x_1 + h/2, y_1 + s_1 h/2)$$

$$= f(0.05, 0.95) = -0.8525$$

$$s_3 = f(x_1 + h/2, y_1 + s_2 h/2)$$

$$= f(0.05, 0.9574) = -0.867$$

$$s_4 = f(x_1 + h, y_1 + s_3 h)$$

$$= f(0.1, 0.9137) = -0.7341$$

Substitute all slope values in above formula for y_1, we have

$$y(0.1) = 1 + \frac{0.1}{6}[-1 + 2(-0.8525) + 2(-0.867) + (-0.7341)] = 0.9117$$

To determine $y(0.2)$

Here, $x_2 = 0.1, y_2 = 0.9117$ and $h = 0.1$

The fourth order Runge-Kutta formula is given by

$$y_2 = y_1 + \frac{h}{6}[s_1 + 2s_2 + 2s_3 + s_4]$$

$$s_1 = f(x_2, y_2) = f(0.1, 0.9117) = -0.731$$

$$s_2 = f(x_2 + h/2, y_2 + s_1 h/2) = f(0.15, 0.8751) = -0.616$$

$$s_3 = f(x_2 + h/2, y_2 + s_2 h/2) = f(0.15, 0.8809) = -0.626$$

$$s_4 = f(x_2 + h, y_2 + s_3 h) = f(0.2, 0.8491) = -0.521$$

Substitute all slope values in above formula for y_2, we have

$$y(0.2) = 0.9117 + \frac{0.1}{6}[-0.731 + 2(-0.616) + 2(-0.626) + (-0.521)] = 0.8494$$

To determine $y(0.3)$

Here $x_3 = 0.2$, $y_3 = 0.8494$ and $h = 0.1$

The fourth order Runge-Kutta formula is given by

$$y_3 = y_2 + \frac{h}{6}[s_1 + 2s_2 + 2s_3 + s_4]$$

$$s_1 = f(x_3, y_3)$$

$$= f(0.2, 0.8494) = -0.521$$

$$s_2 = f(x_3 + h/2, y_3 + s_1 h/2)$$

$$= f(0.25, 0.8233) = -0.428$$

$$s_3 = f(x_3 + h/2, y_3 + s_2 h/2)$$

$$= f(0.25, 0.828) = -0.436$$

$$s_4 = f(x_3 + h, y_3 + s_3 h)$$

$$= f(0.3, 0.8058) = -0.349$$

Substitute all slope values in above formula for y_3, we have

$$y(0.3) = 0.8489 + \frac{0.1}{6}[-0.521 + 2(-0.428) + 2(-0.436) + (-0.349)] = 0.8061$$

k	x_k	y_k	$f(x_k, y_k) = (½)x_k\ y_k$
0	0.0	1	–1
1	0.1	0.9117	–0.7312
2	0.2	0.8494	–0.5215
3	0.3	0.8061	–0.3498

To obtain $y(0.4)$, we use the Adams-Bashforth's Predictor formula as

$$y_{0.4}^{p} = y_3 + \frac{h}{24}[55f_3 - 59f_2 + 37f_1 - 9f_0]$$

$$= 0.8061 + \frac{0.1}{24}[55(-0.3498) - 59(-0.5215) + 37(-0.7312) - 9(-1)]$$

$$= 0.7789$$

$$\therefore \quad f_{0.4}^{p} = f(0.4, 0.7789)$$

$$= (0.4) - (0.7789)^2$$

$$= -0.2067$$

To correct the predicted value $y(0.4)$, we use the Adams–Moultan's Corrector formula as

$$y_{0.4}^{c} = y_3 + \frac{h}{24}\,[9\,f_{0.4}^{p} + 19f_3 - 5f_2 + f_1]$$

$$= 0.8061 + \frac{0.1}{24}\,[9(-0.2067) + 19(-0.3498) - 5(-0.5215) + (-0.7312)]$$

$$= 0.7785$$

Hence $y(0.4) = 0.7785$

This value can more be improved by repeatedly using Corrector formula.

UNSOLVED PROBLEMS

1. Obtain $y(0.6)$ given $dy/dx = x + y$, $y(0) = 1$ with $h = 0.2$ by Adam's Bashforth method.
2. Using Adams–Bashforth Formula, find $y(0.4)$ and $y(0.5)$ if y satisfies the differential equation $dy/dx = 3e^x + 2y$ with $y(0) = 0$. Compute y at $x = 0.1, 0.2, 0.3$ by means of Runge-Kutta Method.
3. Determine $y(0.4)$ given the equation $dy/dx = 1/2xy$ using Adams–Moulton Method, given that $y(0) = 1$, $y(0.1) = 1.0025$, $y(0.2) = 1.0101$, $y(0.3) = 1.0228$.
4. Using Adams–Bashforth Predictor–Corrector Method, find $y(1.4)$ given that $x^2y^2 + xy = 1$; $y(1) = 1$, $y(1.1) = 0.996$, $y(1.2) = 0.986$, $y(1.3) = 0.972$

9.4.3 Milne-simpson Method

The Milne's method is very much similar to Adams-Bashforth's method that uses the information at past four points to extrapolate the solution at the next point. So in Milne's method, in addition to starting solution point, three more solution points are computed by using any method discusses so far.

For the derivation of the mathematical expression for Milne's method, we take the first order differential equation as

$$dy/dx = f(x, y)$$

or

$$dy = f(x, y)dx \qquad \ldots(1)$$

Integrate equation (1) within the limits $x = x_{k-3}$ and $x = x_{k+1}$, we have

$$\int_{x_{k-3}}^{x_{k+1}} dy = \int_{x_{k-3}}^{x_{k+1}} f(x, y)\,dx$$

$$\Rightarrow \qquad y_{k+1} = y_{k-3} + \int_{x_{k-3}}^{x_{k+1}} f(x, y)dx \qquad \ldots(2)$$

Now to compute integral on the right side of (2), we approximate the function $f(x, y)$ as a polynomial in x by making it a fit through the past four points (x_{k-3}, y_{k-3}), (x_{k-2}, y_{k-2}), (x_{k-1}, y_{k-1}), (x_k, y_k).

By using Newton's forward difference interpolating polynomial, we have

$$f(x, y) = f_{k-3} + u \cdot \Delta f_{k-3} \frac{u(u-1)}{3} \Delta^2 f_{k-3} + \frac{u(u-1)(u-2)}{6} \Delta^3 f_{k-3} \quad \text{... (3)}$$

where $u = \dfrac{x - x_{k-3}}{h}$ and Δfk_{-3}, $\Delta^2 f_{k-3}$, $\Delta^3 f_{k-3}$ are forward differences at point $x = x_{k-3}$

Substitute (3) in (2), we have

$$y_{k+1} = y_{k-3} + \int_{x_{k-3}}^{x_{k+1}} \left[f_{k-3} + u \cdot \Delta f_{k-3} \frac{u(u-1)}{2} \Delta^2 f_{k-3} + \frac{u(u-1)(u-2)}{6} \Delta^3 f_{k-3} \right] dx$$

Since $u = \dfrac{x - x_{k-3}}{h}$, therefore $dx = hdu$

When $\quad x = x_k, \qquad u = 0$

$\qquad x = x_{k+1}, \qquad u = 1 \qquad$ (as $x_{k-1} - x_k = h$)

Since $u = \dfrac{x - x_k}{h}$, therefore $dx = hdu$

When $\quad x = x_{k-3}, \qquad u = 0$

$\qquad x = x_{k+1}, \qquad u = 4$

(as $x_{k+1} - x_{k-3} = x_{k+1} - x_k + (x_k - x_{k-1}) + (x_{k-1} - x_{k-2}) + (x_{k-2} - x_{k-3}) = h + h + h + h = 4h$)

By substituting these values, we get

$$y_{k+1} = y_k + h \int_0^4 \left[f_{k-3} + u \cdot \Delta f_{k-3} \frac{u(u-1)}{2} \Delta^2 f_{k-3} + \frac{u(u-1)(u-2)}{6} \Delta^3 f_{k-3} \right] du$$

$$= y_{k-3} + h[f_{k-3} + 8\Delta f_{k-3} + (20/3)^2 f_{k-3} + (8/3)\Delta^3 f_{k-3}]$$

Now $\Delta f_k = f_{k+1} - f_k$ and $\Delta f_{k-3} = f_{k-2} - f_{k-3}$

$$\Delta^2 f_{k-3} = \Delta f_{k-2} - \Delta f_{k-3} = (f_{k-1} - f_{k-2}) - (f_{k-2} - f_{k-3}) = f_{k-1} - 2f_{k-2} + f_{k-3}$$

$$\Delta^3 f_{k-3} = \Delta^2 f_{k-2} - \Delta^2 f_{k-3} = (f_k - 2f_{k-1} + f_{k-2}) - (f_{k-1} - 2f_{k-2} + f_{k-3})$$

$$= fk - 3f_{k-1} + 3f_{k-2} - fk_{-3}$$

Substituting these values, we get

$$y_{k+1} = y_{k-3} + h[4f_{k-3} + 8\{f_{k-2} - f_{k-3}\} + (20/3)\{f_{k-1} - 2f_{k-2} + f_{k-3}\}$$
$$+ (8/3)\{fk - 3f_{k-1} + 3f_{k-2} - f_{k-3}]$$

After simplification, we have

$$y^p_{k+1} = y_{k-3} + \frac{4h}{3}[2f_{k-2} - f_{k-1} + 2fk]$$

This is known as Milne's Predictor formula. The subscript p here indicates the predicted value of y_{k+1}

In a similar way, we can develop Milne's Corrector formula. Now consider Newton's forward difference interpolating polynomial that passes through the points (x_{k-1}, y_{k-1}), (x_k, y_k), (x_{k+1}, y^p_{k+1}) as

$$f(x,y) = f_{k-1} + u.\Delta f_{k-1} + \frac{u(u-1)}{2}\Delta^2 f_{k-1} \quad ...(4)$$

to determine the integral $\int_{x_{k-1}}^{x_{k+1}} f(x,y)dx$

for obtaining the value of y_{k+1} as

$$y_{k+1} = y_{k-1} + \int_{x_{k-1}}^{x_{k+1}} f(x,y)dx$$

According to Simpson's 1/3rd rule, we have

$$\int_{x_{k-1}}^{x_{k+1}} f(x,y)dx = h/3[f_{k-1} + 4fk + f^p_{k+1}]$$

$$f^c_{k+1} = y_{k-1} + \frac{h}{3}[f_{k-1} + 4fk + f^p_{k+1}]$$

This is known as Adams-Moultan's Corrector formula. The subscript c here indicates the corrected value of y_{k+1}. By using this formula repeatedly we can improve the accuracy of computed y_{k+1}.

9.4.3.1 Algorithm of Milne's Predictor-corrector Method

1. Function f (x,y)= x + y
2. Input value of xn
3. For i = 0 to 3
4. Input x(i), y(i)
5. Next i
6. h = x(1)–x(0)
7. n =(xn – x(0))/h
8. For i = 3 to n
9. x(i+1)= x(i)+h
10. f = F(x(i),y(i))
11. f1 = F(x(i–1), y(i–1))

12. f2 = F(x(i–2), y(i–2))
13. yp = y(i–3) + 4h/3(2f2 – f1 + 2f)
14. yc = y(i–1)+ h/3(f1 + 4f + F(x(i+1),yp))
15. If abs (yp–yc) < 0.0005 then
    y(i + 1)= yc
    print value of x(i + 1), y(i + 1)
    ELSE
    yp = yc
16. Next i
17. Stop

9.4.3.2 *Flow-chart of Milne's Method*

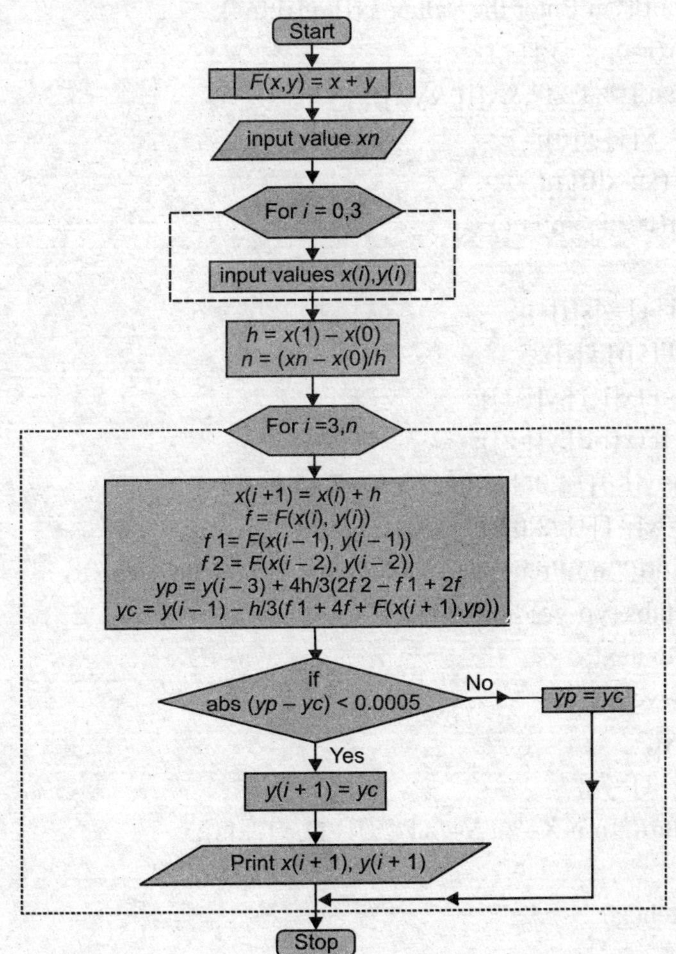

9.4.3.3 Program of Milne's Method in C

```
#include<stdio.h>
#include<conio.h>
#include<math.h>
#define F(x,y) x+y
main()
{
int i,n;
float x[20],y[20],h,f,f1,f2,yp,yc,xn;
printf("\n Enter the value: xn: "};
scanf{"%f",&xn);
printf("\n Enter the value: x{i], y[i]:\n"};
for(i=0;i<=3;i++)
scanf("%f%f",&x[i],&y[i]);
h= x[1]-x[0];
n=(xn-x[0]/h;
for(i=3;i<=n;i++)
{
x[i+1]= x[i]+h;
f=F[x[i],y[i]);
f1=F(x[i-1],y[i-1]);
f2=F(x[i-2],y[i-2]);
yp=y[i-3]+4.0*h/3.0*(2.0*f2-f1+2.0*f);
yc=y[i-1]+h/3.0*(f1+4.0*f+F(x[i+1],yp));
printf("\n\nPredicated Y=%f Correctd Y=%f", yp,yc);
If(fabs (yp-yc)<0.00005)
goto next;
yp=yc;
next;
y[i+1]=yc;
printf("\n\n X=%f Y=%f", x[i+1], y[i+1]);
}
getch();
}
```

Output:

```
Enter the value: xn: 1
Enter the value: x[i], y[i]:
0.0 0.0
0.2 0.02
0.4 0.0906
0.6 0.2214
Predicted Y=0.423147 Corrected Y=0.429650
X=0.800000 Y=0.429650
Predicted Y=0.721307 Corrected Y=0.718820
X=1.000000 Y=0.718820
```

9.4.3.4 *Program of Milne's Method in C++*

```
#include<iostream.h>
#include<conio.h>
#include<math.h>
#define F(x,y) x+y
main()
{
int i,n;
float x[20],y[20],h,f,f1,f2,yp,yc,xn;
cout<<"\n Enter the value: xn: ";
cin>>xn;
cout<<"\n Enter the value: x[i], y[i]:\n";
for(i=0;i<=3;i++)
cin>>x[i]>>y[i]);
h= x[1]-x[0];
n=(xn-x[0]/h;
for(i=3;i<=n;i++)
{
x[i+1]= x[i]+h;
f=F[x[i],y[i]);
```

```
f1=F(x[i-1],y[i-1]);
f2=F(x[i-2],y[i-2]);
yp=y[i-3]+4.0*h/3.0*(2.0*f2-f1+2.0*f);
yc=y[i-1]+h/3.0*(f1+4.0*f+F(x[i+1],yp));
cout<<"\n\nPredicated Y="<<yp<<"Correctd Y="<<yc;
If(fabs (yp-yc)<0.00005)
goto next;
yp=yc;
next;
y[i+1]=yc;
cout<<"\n\n X="<<x[i+1]<<"Y="<< y[i+1];
}
getch();
}
```

Output:

Enter the value: *xn*: 1

Enter the value: *x*[i], *y*[i]:

0.0 0.0

0.2 0.02

0.4 0.0906

0.6 0.2214

Predicted Y=0.423147 Corrected Y=0.429650

X=0.800000 Y=0.429650

Predicted Y=0.721307 Corrected Y=0.718820

X=1.000000 Y=0.718820

SOLVED PROBLEMS

Problem 1: Given $dy/dx = (½)(1+x^2)y^2$ and $y(0) = 1$, $y(0.1) = 1.06$, $y(0.2) = 1.12$, $y(0.3) = 1.21$, evaluate $y(0.4)$ by Milne-Simpson's method.

Solution: Given $h = 0.1$ and $f(x, y) = (½)(1+x^2)y^2$

$x_0 = 0, x_1 = 0.1, x_2 = 0.2, x_3 = 0.3, x_4 = 0.4$

$y_0 = 1, y_1 = 1.06, y_2 = 1.12, y_3 = 1.21, y_4 = ?$

Starting values of Milne's method are

k	x_k	y_k	$f(x_k, y_k) = (½)(1+x_k^2)y_k^2$
0	0.0	1	0.5
1	0.1	1.06	0.5674
2	0.2	1.12	0.6522
3	0.3	1.21	0.7979

To obtain $y(0.4)$, we use the Milne's Predictor formula as

$$y_{0.4}^p = y_0 + \frac{4h}{3}[2f_1 - f_2 + 2f_3]$$

$$= 1 + \frac{4(0.1)}{3}[2(0.5674) - 0.6522 + 2(0.7979)]$$

$$= 1.2771$$

$$\therefore \quad f_{0.4}^p = f(0.4, 1.2771)$$

$$= (½)(1+x_4^2)\,y_4^2$$

$$= (½)(1+0.16)(1.2771)^2$$

$$= 0.9460$$

To correct the predicted value $y(0.4)$, we use the Simpson's Corrector formula as

$$y_{0.4}^c = y_2 + \frac{h}{3}[f_2 + 4f_3 + f_{0.4}^p]$$

$$= 1.12 + \frac{0.1}{3}[0.6522 + 4(0.7979) + 0.9460]$$

$$= 1.2797$$

Hence $y(0.4) = 1.2797$

This value can more be improved by repeatedly using Corrector formula.

Problem 2: Find $y(2)$ if $y(x)$ is the solution of $dy/dx = 1/2\,(x + y)$ where $y(0) = 2$, $y(0.5) = 2.636$, $y(1) = 3.595$, $y(1.5) = 4.968$

Solution: Let $x_0 = 0$, $x_1 = 0.5$, $x_2 = 1$, $x_3 = 1.5$ then we are given y_0, y_1, y_2, y_3 and we require y_4 corresponding to $x_4 = 2$.

By Predictor Formula, we get

$$y_4 = y_0 + 4h/3\,(2y_1' - y_2' + 2y_3') \quad ...(1)$$

we have, $y' = 1/2(x + y)$

$$y_1' = 1/2\,(x_1 + y_1) = 1.568$$

Similarly
$$y_2' = 2.2975,\ y_3' = 3.234$$

from (1),
$$y_4 = 2 + 4\ (0.5)/3\ [3.136 - 2.2975 + 6.468] = 6.871$$
$$y_4' = 1/2\ (x_4 + y_4) = 4.4355$$

This is corrected by

$$y_4^{(1)} = y_2 + h/3\ (\ y_2' + 4\ y_3' \ +\ y_4'\ )$$
$$= 3.595 + 0.5/3\ [2.2975 + 12.936 + 4.4355]$$
$$= 6.87317$$

Now,
$$(y_4^{(1)})' = 1/2[x_4 + y_4^{(1)}] = 1/2\ (2 + 6.87317) = 4.43659$$

Again by the Corrector Formula, we get the second corrected value i.e., $y_{2.00}$.

$$y_4^{(2)} = y_2 + h/3[\ y_2' \ + 4\ y_3' \ +\ (y_4^{(1)})'\ ]$$
$$= 3.595 + 0.5/3\ [2.2975 + 12.936 + 4.43659]$$
$$= 6.87335$$

UNSOLVED PROBLEMS

1. Compute $y(0.6)$ by Milne's method given that $y¢ = x + y, y(0) = 1$ with $h = 0.2$. Obtain the required data by Taylor's series method.
2. Using Milne's method find $y(4.4)$ given $5xy¢ + y^2 - 2 = 0$ given $y(4) = 1, y(4.1) = 1.0049$, $y(4.2) = 1.0097$ and $y(4.3) = 1.0143$
3. Apply Milne's Method to solve the differential equation $dy/dx = -xy^2$ at $x = 0.8$, given that $y(0) = 2$, $y(0.2) = 1.923$, $y(0.4) = 1.724$, $y(0.6) = 1.471$
4. Solve $10\ dy/dx = x^2 + y^2, y(0) = 1$ and compute $y(0.4)$ and $y(0.5)$ by Milne's Method given

x:	0.1	0.2	0.3
y:	1.0101	1.0206	1.0317

5. Solve the differential equation $dy/dx = y - x^2$ by Milne's Method and compute y at $x = 0.80$ when:

x:	0	0.2	0.4	0.6
y:	1	1.12186	1.46820	1.73790

6. Solve the initial value problem $dy/dx = 1 + xy^2, y(0) = 1$ for $x = 0.4, 0.5$ by using Milne's Method. It is given that,

x:	0.1	0.2	0.3
y:	1.105	1.223	1.355

7. Given: $y(0) = 2, y(0.2) = 2.0933, y(0.4) = 2.1755, y(0.6) = 2.2493$, find $y(0.8)$ and $y(1.0)$ by solving $dy/dx = 1/x + y$ by Milne's Method.

9.5 BOUNDARY VALUE PROBLEMS

For solving an ordinary differential equation of nth order, n conditions must be given. If those conditions are prescribed at one point only, then differential together with the conditions constitute an initial value Problems of nth order. However if the conditions are prescribed at two or more points the differential equation together with the conditions constitute a boundary value problem of the nth order.

To be useful in applications, a boundary value problem should be well posed. In this section we will discuss two methods that handles boundary value problems.

9.5.1 Finite Difference Method

Finite difference method for solving two-boundary value problem is an effective method of solving the boundary value problems in a simple manner. In this method, we have to replace the derivatives appearing in the differential equation and boundary conditions by their finite difference approximations. After the replacement process, we get resulting linear system of equations, which are solved by a standard procedure. Now, to obtain the appropriate finite difference approximations to the derivatives, we proceed as follows.

If $y(x)$ and its derivatives are single valued continuous functions of x than by using Taylor's series expansion, we have

$$y(x+h) = y(x) + hy'(x) + \frac{h^2}{2!}y''(x) + \frac{h^3}{3!}y'''(x) + \ldots\ldots\ldots \quad \ldots(1)$$

Solving this for $y'(x)$, we have

$$y'(x) = \frac{[y(x+h) - y(x)]}{h} - \frac{h}{2}y''(x) - \ldots\ldots\ldots\ldots$$

i.e. $$y'(x) = \frac{1}{h}[y(x+h) - y(x)] + O(h) \quad \ldots(2)$$

Which is the forward difference approximation of $y'(x)$ with an error of the order h.

$O(h)$ stands for terms containing powers of h.

Similarly expansion of $y(x-h)$ by using Taylor's series, we have

$$y(x-h) = y(x) - hy'(x) + \frac{h^2}{2!}y''(x) - \frac{h^3}{3!}y'''(x) + \ldots\ldots\ldots \quad \ldots(3)$$

Solving (3) for $y'(x)$, we have

i.e. $$y'(x) = \frac{1}{h}[y(x) - y(x-h)] + O(h) \quad (4)$$

Which is the backward difference approximation of $y'(x)$ with an error h.

Now subtract (3) from (1), we get

$$y'(x) = \frac{[y(x+h) - y(x-h)]}{2h} \, O(h^2) \qquad ...(5)$$

$O(h^2)$ stands for terms containing powers of h^2, h^3........etc.

Which is the central difference approximation of $y'(x)$ with an error of order h^2. This central difference approximation gives us better than the forward and backward difference approximations as it generates an error of order h^2, which is better than h hence it should be, preferred.

Now by adding (1) and (3), we have

$$y''(x) = \frac{[y(x+h) - 2y(x) + (x-h)]}{h^2} \, O(h^2) \qquad ...(6)$$

which is the central difference approximation of $y''(x)$. Similarly, by following the same procedure we can derive central difference approximations for higher order also.

To solve boundary value problems defined by a differential equation and corresponding boundary conditions, we divide the range $[x_0, x_n]$ into n equal subintervals of width h so that

$$x_i = x_0 + ih, \qquad i = 1,2,3,........,n$$

and the corresponding values of y at these points are denoted by

$$y(x_i) = y_i = y(x_0 + ih), \quad i = 1,2,3,..........,n$$

Hence the working expressions for the central difference approximations to the first four derivatives of y_i are as:

$$y_i' = \frac{[y_{i+1} - y_{i-1}]}{2h}$$

$$y_i'' = \frac{[y_{i+1} - 2y_i + y_{i-1}]}{h^2}$$

$$y_i''' = \frac{[y_{i+2} - 2y_{i+1} + y_{i-2}]}{2h^3}$$

$$y_i^{(iv)} = \frac{[y_{i+2} - 4y_{i+1} + 6y_i - 4y_{i-1} + y_{i-2}]}{h^4}$$

The accuracy of this method depends on h, if we take h as a small value it gives us better results but the problem arises is that number of equations evolved also increases.

SOLVED PROBLEMS

Problem 1: Compute values of y at the pivotal points of the interval (0,1) if y satisfies the boundary value problem

$$y^{iv} = 81x^2 - 81y, \; y(0) = y(1) = y''(0) = y''(1) = 0 \qquad \text{(Take } n = 3\text{)}$$

Solution: The given differential equation is

$$y^{iv} = 81x^2 - 81y \quad \text{...(1)}$$

Divide the interval (0, 1) into three equal subintervals so that $h = 1/3$, and the pivotal points are

$$x_0 = 0, x_1 = 1/3, x_2 = 2/3, x_3 = 1$$

Given that, $y(0) = y(1) = y''(0) = y''(1) = 0$

$\Rightarrow$ $y_0 = 0, y_3 = 0,\ y_0'' = 0,\ y_3'' = 0$

By using finite difference method, we have

$$y_i^{(iv)} = \frac{[y_{i+2} - 4y_{i+1} + 6y_i - 4y_{i-1} + y_{i-2}]}{h^4}$$

Substitute this $y_i^{(iv)}$ in equation (1) ,we have

$$\frac{1}{h^4}(y_{i+2} - 4y_{i+1} + 6y_i - 4y_{i-1} + y_{i-2}) + 81y_i = 81\,x_i^2$$

Substituting $h = 1/3$ here, we have

$$81(y_{i+2} - 4y_{i+1} + 6y_i - 4y_{i-1} + y_{i-2}) = 81\,x_i^2 - 81y_i$$

$\Rightarrow$ $y_{i+2} - 4y_{i+1} + 7y_i - 4y_{i-1} + y_{i-2} = x_i^2$, where $i = 1,2$

Substituting $i = 1,2$ and $y_0 = 0$, $y_3 = 0$, we get

$$y_3 - 7y_1 - 4y_0 + y_{-1} = (x_1^2)$$

$\Rightarrow$ $-4y_2 + 7y_1 + y_{-1} = 1/9$ ...(1)

and $y_4 - 4y_3 + 7y_2 - 4y_1 + y_0 = x_2^2$

As $x_1 = 1/3$ and $x_2 = 2/3$

$\therefore$ $y_4 + 7y_2 - 4y_1 = 4/9$ ...(2)

As we have,

$$y_i'' = \frac{1}{h^2}(y_{i+1} - 2y_i + y_{i-1})$$

$i = 0,$ $y''_0 = 9(y_1 - 2y_0 + y_{-1})$

As $y_0 = y''_0 = 0$

$\therefore$ $y_{-1} = -y$ ...(3)

At $i = 3,$ $y''_3 = 9(y_4 - 2y_3 + y_2)$

As $y_3 = y''_3 = 0$

$\therefore$ $y_4 = -y_2$ ...(4)

From (1) and (3), we get

$$6y_1 - 4y_2 = 1/9 \quad \text{...(5)}$$

From (2) and (4), we get

$$-4y_1 + 6y_2 = 4/9 \quad \text{...(6)}$$

After solving (5) and (6), we get

$$y_1 = 0.1222,\ y_2 = 0.1556$$

Hence $y(1/3) = 0.1222$, $y(2/3) = 0.1556$

9.5.2 Shooting Method

Shooting method is the simplest method of solving second order boundary value problem. This method can be applied to both linear and non-linear problems and required good initial guesses for slope.

The main steps involved in this method are:

(1) Transformation of the boundary value problem into an initial value problem

(2) Solution of the initial value problem by Taylor's series method, or Runge-Kutta method etc. and

(3) Solution of the given boundary value problem

Consider the second order boundary value problem defined by

$$y''(x) = f(x) =, y(0) = 0, y(1) = 1 \quad \text{...(1)}$$

Suppose the true value of $y'(0)$ be m. Take m_0, m_1 be two initial guesses of m. Let $y(m_0,1)$ and $y(m_1,1)$ be the corresponding values of $y(1)$ obtained by initial values method.

By using linear interpolation, we obtain a better approximation m_2 for m, given by

$$\frac{m_2 - m_0}{y(1) - y(m_0,1)} = \frac{m_2 - m_0}{y(m_1,1) - y(m_0,1)} \quad \text{...(2)}$$

$$\Rightarrow \quad m_2 = m_0 + (m_1 - m_0)\cdot\frac{y(1) - y(m_0,1)}{y(m_1,1) - y(m_0,1)} \quad \text{.... (3)}$$

We now solve that initial value problem

$$y''(x) = f(x), y(0) = 0 \text{ and } y'(0) = m_2 \quad \text{...(4)}$$

Further find the value of y at $x = 1$ and denote it by $y(m_2,1)$

We again use linear interpolation with $(m_1, y(m_1,1))$ and $(m_2, y(m_2,1))$ to obtain a better approximation m_3 form m and so on. This process will repeated until convergence is obtained, i.e. until the value of $y(m_i, 1)$ agrees with $y(1)$ to the desired accuracy.

The method will be difficult to apply to higher order boundary values problems.

SOLVED PROBLEMS

Problem 1: Solve the boundary value problem $y''(x) = y(x)$, $y(0) = 0$ and $y(1) = 1.1752$ by shooting method, taking $m_0 = 0$ and $m_1 = 0.9$

Solution: The given equation is $y''(x) = y(x)$ ...(1)

with the boundary conditions $y(0) = 0$ and $y(1) = 1.1752$ ...(2)

By using Taylor's Series, we have

$$y(x) = y(0) + xy'(0) + \frac{x^2}{2!}\, y''(0) + \frac{x^3}{3!} y'''(0) + \frac{x^4}{4!} y^{iv}(0) + \ldots\ldots\ldots \quad \ldots(3)$$

From (1) and (2), we get

$$y''(00) = y(0) = 0,\ y'''(0) = y'(0)$$

$$y^{iv}(0) = y''(0) = 0,\ y^{v}(0) = y'''(0)$$

$$y^{vi}(0) = y^{iv}(0) = y''(0) = 0,\ y^{vii}(0) = y^{v}(0)$$

$$y^{viii}(0) = y^{vi}(0) = 0$$

$\therefore$ $$y^{viii}(0) = y^{vi}(0) = y^{iv}(0) = y''(0) = 0$$

and $$y'(0) = y'''(0) = y^{v}(0) = y^{vii}(0)$$

Using these values in (3), we get

$$y(x) = xy'(0) + \frac{x^3}{3!} y'(0) + \frac{x^5}{5!} y'(0) + \frac{x^7}{7!} y'(0) + \ldots\ldots\ldots$$

$$= y'(0)[x + (x^3/6) + (x^5/120) + (x^7/5040) + (x^9/362800) + \ldots\ldots\ldots]$$

Substituting $x = 1$, we get

$$y(1) = y'(0)[1 + (1/6) + (1/120) + (1/5040) + (1/362800) + \ldots\ldots]$$

$$= 1.1752y'(0)$$

Taking $m_0 = 0.8$ and $m_1 = 0.9$, we have

$$y(m_0,1) = (0.8)(1.1752) = 0.9402$$

$$y(m_1,1) = (0.9)(1.1752) = 1.0578$$

if m_2 is better approximation of m_0 and m_1 then

$$m_2 = \frac{y(1) - y(m_{0,1})}{y(m_1,1) - y(m_{0,1})}$$

$$= 0.8 + (0.9 - 0.8)0 \cdot \frac{(1.752 - 0.9402)}{1.0578 - 0.9402}$$

$$= 0.8 + 0.1998$$

$$= 0.9998$$

which is closer to the exact value of $y'(0) = 1$. We now solve the initial value problem $y''(x) = y(x)$, $y(0) = 0$, m_2. Again using Taylor's series solution, we get $y(m_2, 1) = 1.174$ which is closer to the exact value $y(1) = 1.175$

The problem cal also be solve by using Runge-Kutta fourth order method.

UNSOLVED PROBLEMS

1. To solve the equation $y'' + y + 1 = 0$ with the boundary conditions $y(0) = 0$ and $y(1) = 0$
2. Solve the boundary value problem $y''(x) = y(x), y(0) = 0, y(1) = 1$ by using Shooting method.
3. Given the boundary value problem

$$y'' = 3x + 4y, y(0) = 1 \text{ and } y(1) = 1$$

Obtain its solution in the range 0 £ x £ 1 with $h = 0.25$ using finite difference method.

4. Solve the boundary value problem $y'' = y(x); y(0) = 0 , y(1) = 1$ by the Shooting method.

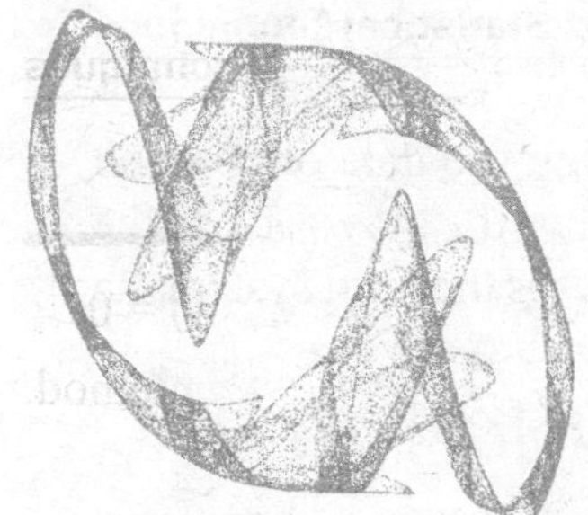

CHAPTER 10

Curve Fitting

10.1 INTRODUCTION

Curve fitting is finding a line or curve that passes most closely through a given set of points, minimizing the distances between the line and the actual points and providing the function whose curve is the line. Curve fitting is the process of constructing a curve, or mathematical function, that has the best fit to a series of data points, possibly subject to constraints. Curve fitting can involve either interpolation, where an exact fit to the data is required, or smoothing, in which a "smooth" function is constructed that approximately fits the data.

The idea of curve fitting is to find a mathematical model that fits your data. We assume that you have theoretical reasons for picking a function of a certain form. The curve fit finds the specific coefficients (parameters) which make that function match your data as closely as possible. Some people try to use curve fitting to find which of thousands of functions fit their data. Field data is often accompanied by noise. Even though all control parameters (independent variables) remain constant, the resultant outcomes (dependent variables) vary. A process of quantitatively estimating the trend of the outcomes, also known as regression or curve fitting, therefore becomes necessary.

The curve fitting process fits equations of approximating curves to the raw field data. Nevertheless, for a given set of data, the fitting curves of a given type are generally NOT unique. Thus, a curve with a minimal deviation from all data points is desired. This best-fitting curve can be obtained by the method of least squares.

The fitting of a curve to a given data is very important both from the point of view theoretical and practical Statistics. In theoretical Statistics, we shall find that the study of regression and correlation (whether total, partial or multiple) can be regarded as fitting of linear curves (straight lines, planes, hyper-planes) to the given bivariate or multivariate frequency or probability distributions. In practical statistics, it enables us to get a close functional relation between x and y. The relation is, in general, expressed by a polynomial, but other types of relationship, algebraic, exponential or logarithmic can also be fitted by using the **principle of least squares**.

As a mathematical equation is fitted to be experimental data by plotting the data on a graph paper and then passing a straight line through the data points. This method has the drawback as the straight line is not unique. So a new method known as **Method of least squares** is the most systematic procedure to fit a unique curve through given points.

10.2 LEAST SQUATRE METHOD

Least squares can be interpreted as a method of fitting data. The best fit, between modeled and observed data, in the least-squares sense is that instance of the model for which the sum of squared residuals has its least value, a residual being the difference between an observed value and the value given by the model. The method was first described by Carl Friedrich Gauss around 1794. Least squares corresponds to the maximum likelihood criterion if the experimental errors have a normal distribution and can also be derived as a method of moments estimator. Regression analysis is available in most statistical software packages.

Least square provides a unique set of values to the constants and hence suggests a curve of best fit to the given data.

Suppose we have (x_i, y_i) be a point where $i = 1,2,3........,n$ be the n sets of observations & let $y = f(x)$ be a relation between $x$and y. Let (x_i, y_i) be represented by the point P_i.

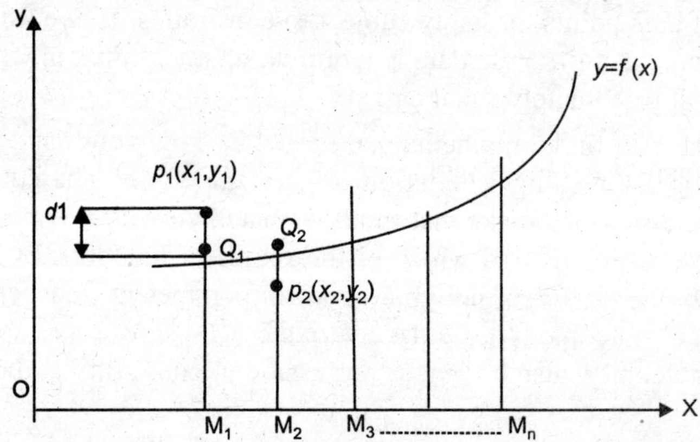

Let the ordinate at P_i meet $y = f(x)$ at Q_i and the x-axis at M_i. Therefore

$$M_iQ_i = f(x_i) \text{ and } M_iP_i = y_i$$

and

$$Q_iP_i = M_iP_i - M_iQ_i = y_i - f(x_i)$$

or $d_i = y_i - f(x_i)$ is called the residual(error) at $x = x_i$. As some of the d_i's may be positive and some may be negative.

$$S = \sum_{i=1}^{n} d_i^2 = \sum_{i=1}^{n} [y_i - f(x_i)]^2$$

is the sum of the squares of the residuals(errors).

S will be either minimum or maximum if the partial derivative if the partial derivative $\partial S/\partial a =$ 0, where a is one of the unknowns assumed in the approximation $y = f(x)$.

If $S = 0$ i.e each $d_i = 0$ then all the n points P_i's will lie on $y = f(x)$

if not we will choose $f(x)$ such that S is minimum i.e the best fitting curve to the set of points is that for which E(sum of residuals) is minimum. This principle is known as principle of least squares.

We will consider some of the best fitting curves of the type in next texts.

10.3 FITTING A STRAIGHT LINE (USING LEAST SQUARES)

Let (x_i, y_i) be a point where $i = 1,2,3\ldots,n$ be the n sets of observations & let the related relation be

$$y = f(x) = ax + b \qquad \text{...(1)}$$

Where a is known as the slope and b is the y intercept of the line (i) also known as regression line of y on $x$in which a, b represents the regression coefficients. We have to select a and b such that the straight line is the best fit to the data.

As by method of least squares sum of squares of the residuals is given by

$$S = \sum_{i=1}^{n} d_i^2$$

$$= \sum_{i=1}^{n} [y_i - f(x_i)]^2 = \sum_{i=1}^{n} [y_i - ax_i - b]^2$$

By principles of least squares, E is minimum,

$$\therefore \qquad \partial S/\partial a = 0 \text{ and } \partial S/\partial b = 0$$

i.e $$\partial S/\partial a = 2\sum_{i=1}^{n} [y_i - ax_i - b]\,(-1)$$

and $$\partial S/\partial b = 2\sum_{i=1}^{n} [y_i - ax_i - b]\,(-x_i)$$

Acc. to the principle of least squares, substituting $\partial S/\partial a = 0$ and $\partial S/\partial b = 0$,we get

$$\sum_{i=1}^{n} [y_i - ax_i - b] = 0 \qquad \text{...(2)}$$

$$\sum_{i=1}^{n} [y_i x_i - ax_i^2 - bx_i] = 0 \qquad \text{...(3)}$$

After eliminating suffices and simplification of (ii) and (iii) we get the normal equations as

$$nb + a\,\Sigma x = \Sigma y$$

$$b\,\Sigma x + a\,\Sigma x^2 = \Sigma xy$$

After solving these equations by using Cramer's rule, we get

$$a = \frac{\begin{vmatrix} n & \Sigma y \\ \Sigma x & \Sigma xy \end{vmatrix}}{\begin{vmatrix} n & \Sigma x \\ \Sigma x & \Sigma x^2 \end{vmatrix}} \text{ and } b = \frac{\begin{vmatrix} \Sigma n & \Sigma x \\ \Sigma xy & \Sigma x^2 \end{vmatrix}}{\begin{vmatrix} n & \Sigma x \\ \Sigma x & \Sigma x^2 \end{vmatrix}}$$

$$a = \frac{n\Sigma xy - \Sigma x\Sigma y}{n\Sigma y^2 - (\Sigma y)^2},$$

$$b = \frac{\Sigma y\Sigma x^2 - \Sigma x\Sigma xy}{n\Sigma x^2 - (\Sigma x)^2} \text{ or } \Sigma y/n - a\Sigma x/n$$

Similarly, we can obtain regression line of x on y

$$x = f(y) = ay + b$$

For this regression line, we take

$$a = \frac{n\Sigma xy - \Sigma y\Sigma x}{n\Sigma y^2 - (\Sigma y)^2},$$

$$b = \frac{\Sigma x\Sigma y^2 - \Sigma y\Sigma xy}{n\Sigma y^2 - (\Sigma y)^2} \text{ or } \Sigma x/n - a\Sigma y/n$$

i.e simply replace x with y and y with x.

Both these regression lines will intersect to each other at the point ($\Sigma x/n$, $\Sigma y/n$).

10.3.1 Algorithm for Fitting a Straight Line of The Form *Y* = *A* + *Bx* for a Given Set of Data Points

1. Start
2. Input no. of value observations
3. Input array ax
4. Input array ay
5. for i=0 to no. of observations
6. sum1+=x[i]
7. sum2+=y[i]
8. xy[i]=x[i]*y[i];
9. sum3= sum3+xy[i]
10. End of Loop i
11. for i = 0 to no. of observations
12. x2[i]=x[i]*x[i]
13. sum4= sum4+x2[i]

14. End of Loop i
15. temp1=(sum2*sum4)-(sum3*sum1)
16. a=temp1/((observations *sum4)-(sum1*sum1))
17. b=(sum2- observations *a)/sum1
18. Print output value of a,b
19. Print "line is: y = a+bx"
20. End

10.3.2 Flow-chart for Fitting a Straight Line $Y = A + Bx$ for a Given Set of Data Points

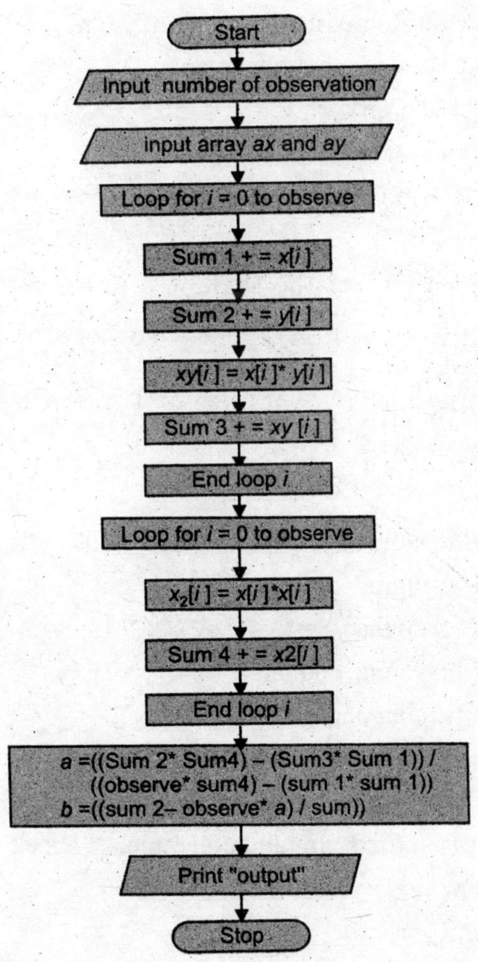

10.3.3 Program: Program to Fit a Linear Regression Line Y on *X* Using Least Square Method

```
#include<stdio.h>
#include<conio.h>
#include<process.h>
void main()
{
int i,n;
float x[100],y[100],sx=0,sy=0,sx2=0,sxy=0,a0,a1,denominator;
clrscr();
printf("\n\t Regression line  Y on X\n\n");
printf(" How many items in data?");
scanf("%d",&n);
printf(" Enter the data as Xi Yi\n");
for(i=1;i<=n;i++)
{
scanf("%f %f",&x[i],&y[i]);
sx+=x[i];
sy+=y[i];
sx2+=x[i]*x[i];
sxy+=x[i]*y[i];
}
printf("\n Number of observation =%d",n);
printf("\n sigma x  =%.4f",sx);
printf("\n sigma y   =%.4f",sy);
printf("\n sigma x square =%.4f",sx2);
printf("\n sigma xy        =%.4f",sxy);
denom=(n*sx2-sx*sx);
if (denom==0)
{ printf("\n Error: Invalid code, please check");
getch();
exit(1);
}
a0=(sy*sx2-sx*sxy)/ denominator;
```

```
a1=(n*sxy-sx*sy)/ denominator;
printf("\n Value of a0 =%.4f",a0);
printf("\n value of a1 =%.4f",a1);
getch();
}
```

10.3.4 Program: Program to Fit a Linear Regression Line X on Y Using Least Square Method

```
#include<stdio.h>
#include<conio.h>
#include<process.h>
void main()
{
int i,n;
float x[100],y[100],sx=0,sy=0,sy2=0,sxy=0,b0,b1, denominator;
clrscr();
printf("\n\t Regression line  X on Y\n\n");
printf(" How many items in data?");
scanf("%d"2,&n);
printf(" Enter the data as Xi Yi\n");
for(i=1;i<=n;i++)
{
scanf("%f %f",&x[i],&y[i]);
sx+=x[i];
sy+=y[i];
sy2+=x[i]*x[i];
sxy+=x[i]*y[i];
}
printf("\n Number of observation =%d",n);
printf("\n sigma x   =%.4f",sx);
printf("\n sigma y   =%.4f",sy);
printf("\n sigma x square =%.4f",sy2);
printf("\n sigma xy       =%.4f",sxy);
denom=(n*sy2-sy*sy);
if (denom==0)
```

```
{
printf("\n Error: Invalid code, please check");
getch();
exit(1);
}
b0=(sx*sy2-sx*sxy)/ denominator;
b1=(n*sxy-sx*sy)/ denominator;
printf("\n Value of b0 =%.4f",b0);
printf("\n value of b1 =%.4f",b1);
getch();
}
```

10.4 FITTING A POLYNOMIAL

We use polynomial models to estimate and predict the shape of response values over a range of input parameter values. Polynomial models are a great tool for determining which input factors drive responses and in what direction. These are also the most common models used for analysis of designed experiments.

Let (x_i, y_i) be a point where $i = 1, 2, 3........,n$ be the n sets of observations

Suppose we want to fit a polynomial of degree m as

$$y = f(x) = a_0 + a_1x + a_2x^2 + ...+ a_mx^m \quad ...(1)$$

Where $a_0,\ a_1,\ a_2,..........,\ a_m$ are the constants that are to be computed for giving the best polynomial fit.

The sum of squares of the residuals is given by

$$E = \sum_{i=1}^{n} d_i^2$$

$$= \sum_{i=1}^{n} [y_i - f(x_i)]^2$$

$$= \sum_{i=1}^{n} [y_i - (a_0 + a_1x_i + a_2x_i^2 + ...+ a_mx_i^m)]^2$$

By principles of least squares, E is minimum.

$$\therefore \quad \partial E/\partial a_0 = 0, \partial E/\partial a_1 = 0, ..., \partial E/\partial a_m = 0$$

$$\partial E/\partial a_0 = \sum_{i=1}^{n} [y_i - (a_0 + a_1x_i + a_2x_i^2 + ...+ a_mx_i^m)](-1) = 0$$

$$\partial E/\partial a_1 = \sum_{i=1}^{n}[y_i - (a_0 + a_1x_i + a_2x_i^2 + ... + a_mx_i^m)](-x_i) = 0$$

$$\partial E/\partial a_m = \sum_{i=1}^{n}[y_i - (a_0 + a_1x_i + a_2x_i^2 + ... + a_mx_i^m)](-x_i^m) = 0$$

After eliminating suffices and simplification of the above equations we get the normal equations as:

$$na_0 + a_1\Sigma x + a_2\Sigma x^2 + ... + a_m x^m = \Sigma y$$

$$a_0\Sigma x + a_1\Sigma x^2 + a_2\Sigma x^3 .... + a_m\Sigma x^{m+1} = \Sigma xy$$

....

....

$$a_0\Sigma x^m + a_1\Sigma x^{m+1} + a_2\Sigma x^{m+3} .... + a_m\Sigma x^{m+m} = \Sigma x^m y$$

These equations can be written in matrix form as

$$\begin{bmatrix} n & \Sigma x & \Sigma x^2 & .... & \Sigma x^m \\ \Sigma x & \Sigma x^2 & \Sigma x^3 & .... & \Sigma x^{m+1} \\ ...... & ..... & .... & ...... & .... \\ ...... & ..... & ...... & .... & ...... \\ \Sigma x^m & \Sigma x^{m+1} & \Sigma x^{m+3} & .... & \Sigma x^{2m} \end{bmatrix} \begin{bmatrix} a_0 \\ a_1 \\ ..... \\ ..... \\ a_m \end{bmatrix} = \begin{bmatrix} \Sigma y \\ \Sigma xy \\ ..... \\ ..... \\ \Sigma x^m y \end{bmatrix}$$

$$AX = B$$

The elements of X and B matrices can be calculated by the following formula

$$X_{ij} = \sum_{k=1}^{n} X_k^{i+j-2}$$

Where $i = 1, 2, 3,........., m$ and $j = 1,2,3,..............., m$

$$B_i = \sum_{k=1}^{n} X_k^{i-1} y_k$$

Where $i = 1,2,3, ..., m$.

SOLVED PROBLEMS

Problem 1: By the method of least squares, find the straight line that best fits the following data:

x :	1	2	3	4	5
y :	14	27	40	55	68

Solution: Let the straight line of best fit be

$$y = a + bx$$

Normal equations are $\Sigma y = ma + b\,\Sigma x$ ...(1)

and $\Sigma xy = a\,\Sigma x + b\,\Sigma x^2$ ...(2)

Here $m = 5$

The table is as below:

x	y	xy	x^2
1	14	14	1
2	27	54	4
3	40	120	9
4	55	220	16
5	68	340	25
$\Sigma x = 15$	$\Sigma y = 240$	$\Sigma xy = 748$	$\Sigma x^2 = 55$

Substituting in (1) and (2), we get

$$204 = 5a + 15b$$

$$748 = 15a + 55b$$

Solving, we get $a = 0$, $b = 13.6$

Hence required straight line is $y = 13.6\,x$

Problem 2: Fit a straight line to the following data:

x:	0	1	2	3	4
y:	1	1.8	3.3	4.5	6.3

Solution: Let the straight line obtained from the given data be $y = a + bx$ then the normal equations are

$$\Sigma y = ma + b\,\Sigma x \quad ...(3)$$

$$\Sigma xy = a\,\Sigma x + b\,\Sigma x^2 \quad ...(4)$$

Here $m = 5$

x	y	xy	x^2
0	1	0	0
1	1.8	1.8	1
2	3.3	6.6	4
3	4.5	13.5	9
4	6.3	25.2	16
$\Sigma x = 10$	$\Sigma y = 16.9$	$\Sigma xy = 47.1$	$\Sigma x^2 = 30$

From (3) and (4), $16.9 = 5a + 10b$

and $47.1 = 10a + 30b$

Solving, we get $a = 0.72$, $b = 1.33$

Required line is $y = 0.72 + 1.33\,x.$

Problem 3: Fit a straight line to the following data:

x:	71	68	73	69	67	65	66	67
y:	69	72	70	70	68	67	68	64.

Solution: Let the equation of the straight line to be fitted be

$$y = a + bx$$

Normal equations are

$$\Sigma y = ma + b\,\Sigma x \quad ...(5)$$

and

$$\Sigma xy = a\,\Sigma x + b\,\Sigma x^2 \quad ...(6)$$

Here $m = 8$. Table is as below:

x	y	xy	x^2
71	69	4899	5041
68	72	4896	4624
73	70	5110	5329
69	70	4830	4761
67	68	4556	4489
65	67	4355	4225
66	68	4488	4356
67	64	4288	4489
$\Sigma x = 546$	$\Sigma y = 548$	$\Sigma xy = 37422$	$\Sigma x^2 = 37314$

Substituting these values in equations (5) and (6), we get

$$548 = 8a + 546b$$

$$37422 = 546a + 37314b$$

Solving, we get

$$a = 39.5454, \quad b = 0.4242$$

Hence the required line of best fit is

$$y = 39.5454 + 0.4242\,x.$$

Problem 4: Given the data below:

x:	0.2	0.4	0.6	0.8	1.0
y:	7.2	7.6	8.0	8.4	8.8

Prove that regression lines (y on x and x on y) intersect at the point ($\Sigma x/n$, $\Sigma y/n$).

Solution:

i	x_i	y_i	x_i^2	y_i^2	$x_i y_i$
1	0.2	7.2	0.04	51.84	1.44
2	0.4	7.6	0.16	57.76	3.04
3	0.6	8.0	0.36	64.00	4.8
4	0.8	8.4	0.64	70.56	6.72
5	1.0	8.8	1.0	77.44	8.8
$n = 5$	$x = 3$	$\Sigma y = 40$	$\Sigma x^2 = 2.2$	$\Sigma y^2 = 321.6$	$\Sigma xy = 24.8$

As we know regression line of y on x is

$$y = ax + b \qquad ...(1)$$

Where

$$a = \frac{n\Sigma xy - \Sigma y \Sigma x}{n\Sigma y^2 - (\Sigma y)^2},\ b = \frac{\Sigma x \Sigma y^2 - \Sigma y \Sigma xy}{n\Sigma y^2 - (\Sigma y)^2}$$

After substitution a and b becomes

$$a = 2,\ b = 6.8$$

$\therefore$ regression line of y on x becomes

$$y = 2x + 6.8 \qquad ...(2)$$

Similarly take regression line x on y be

$$x = ay + b \qquad ...(3)$$

$$a = \frac{n\Sigma xy - \Sigma y \Sigma x}{n\Sigma y^2 - (\Sigma y)^2},\ b = \frac{\Sigma x \Sigma y^2 - \Sigma y \Sigma xy}{n\Sigma y^2 - (\Sigma y)^2}$$

After substitution a and b are

$$a = 0.5,\ b = -3.4$$

$\therefore$ Regression line x on y becomes

$$x = 0.5y - 3.4 \qquad ...(4)$$

This can be checked by substituting ($\Sigma x/n = 3/5 = 0.6$, $\Sigma y/n = 40/5 = 8$) in calculated regression lines. If both regression lines satisfied on these points then we can say that both intersect each other at point ($\Sigma x/n$, $\Sigma y/n$).

By substituting $x = 0.6$ in equation (2), we get

$$y = 2 \times 0.6 + 6.8 = 8$$

Which matches $\Sigma y/n = 8$

Similarly for equation (iv) substitute $y = 8$, we get

$$x = 0.5 \times 8 - 3.4 = 0.6$$

Which matches $\Sigma x/n = 0.6$

Hence we conclude that both regression lines pass through common point ($\Sigma x/n$, $\Sigma y/n$)

Problem 5: For the given data determine the regression line of y on x and x on y. Show that both regression lines pass through ($\bar{x}$, y) where $\bar{x}$ and y are arithmetic means of x and y respectively.

X:	1.2	1.4	1.6	1.8	2.0	2.4
Y:	4.2	6.3	8.4	11.6	15.0	24.2

Solution: Calculate the values of Σx, Σy, Σxy, Σx^2, Σy^2 as shown below

X	Y	XY	X^2	Y^2
1.2	4.2	5.04	1.44	17.64
1.4	6.3	8.82	1.96	39.69
1.6	8.4	13.44	2.56	70.56
1.8	11.6	20.88	3.24	134.56
2.0	15.00	30.00	4.00	225.00
2.4	24.2	58.08	5.76	585.64
$\Sigma x = 10.4$	$\Sigma y = 69.7$	$\Sigma xy = 136.26$	$\Sigma X^2 = 18.96$	$\Sigma Y^2 = 1073.09$

Regression line of y on x:

Let it be $Y = a + bx$

Substituting the values calculated in the above table in the following formula, we get

$$b = \frac{n\Sigma xy - \Sigma x \Sigma y}{n\Sigma X^2 - (\Sigma x)^2}$$

$$b = \frac{6(136.26) - (10.4)(69.7)}{6(18.96) - (10.4)2}$$

$$b = \frac{817.56 - 724.88}{113.76 - 108.16} = \frac{92.68}{5.6}$$

$$b = 16.55$$

$$a = \frac{\Sigma y}{n} - \frac{b\Sigma x}{n}$$

$$= \frac{69.7}{6} - \frac{(16.55)(10.4)}{6}$$

$$= 11.616 - 28.686$$

$$a = -17.07$$

So regression line y on x is given by

$$Y = 16.55\,x - 17.07 \quad (1)$$

Regression line of x on y:

Let it be $X = a + bY$

Substituting the values calculated in the above table in the following formula, we get

$$b = \frac{n\sum xy - \sum y \sum x}{n\sum Y^2 - (\sum y)^2}$$

$$b = \frac{6(136.26) - (69.7)(10.4)}{6(1073.09) - (69.7)^2}$$

$$b = \frac{817.56 - 724.88}{6438.54 - 4858.09} = \frac{92.68}{1580.45}$$

$$b = 0.0586$$

$$a = \frac{\sum x}{n} - \frac{b\sum y}{n}$$

$$= \frac{10.4}{6} - \frac{(0.0586)(69.7)}{6}$$

$$= 1.7333 - 0.6807$$

$$a = 1.0526$$

So regression line x on y is given by

$$X = 1.0526 + 0.0586y \qquad ...(2)$$

Now arithmetic means are calculated as

$$\overline{x} = \frac{\sum x}{n} = \frac{10.4}{6} = 1.7333 \text{ and}$$

$$\overline{y} = \frac{\sum y}{n} = \frac{69.7}{6} = 11.616$$

Now we would check that these two lines intersect at (1.7333,11.616)

We first check for regression line y on x i.e.

$$Y = 16.55x - 17.07$$

Substitute x =1.7333 on R.H.S of equation (1)

We get $$y = 16.55(1.7333) - 17.07$$

$$= 11.616$$

Which means that regression line of y on x passes through (1.7333,11.616).

Now we check for regression line x on y i.e.

$$X = 1.0526 + 0.0586y$$

Substitute y =11.616 on R.H.S of equation (2)

We get $$X = 1.0526 + 0.0586(11.616) = 1.7333$$

Which means that regression line of x on y passes through (1.7333, 11.616).

So both the regression lines pass through the common point (1.7333, 11.616) i.e ($\overline{x}$, y)

Problem 6: Given the data

X:	0.1	0.2	0.3	0.4	0.5	0.6
Y:	5.1	5.3	5.6	5.7	5.9	6.1

Determine both regression lines and also prove that they pass through ($\Sigma x/n$, $\Sigma y/n$).

Solution: Calculate the values of Σx, Σy, Σxy, Σx^2, Σy^2 as shown below

X	Y	XY	X^2	Y^2
0.1	5.1	0.51	0.01	26.01
0.2	5.3	1.06	0.04	28.09
0.3	5.6	1.68	0.09	31.36
0.4	5.7	2.28	0.16	32.49
0.5	5.9	2.95	0.25	34.81
0.6	6.3	3.78	0.36	39.69
$\Sigma x = 2.1$	$\Sigma y = 33.9$	$\Sigma xy = 12.26$	$\Sigma X^2 = 0.91$	$\Sigma Y^2 = 192.45$

Regression line of *y* on *x*:

Let it be

$$Y = a + bx$$

Substituting the values calculated in the above table in the following formula, we get

$$b = \frac{n\,\Sigma xy - \Sigma x \Sigma y}{n \Sigma X^2 - (\Sigma x)^2}$$

$$b = \frac{6(12.26) - (2.1)(33.9)}{6(0.91) - (2.1)2}$$

$$b = 2.26$$

$$a = \frac{\Sigma y}{n} - \frac{b\,\Sigma x}{n}$$

$$= \frac{33.9}{6} - \frac{(2.26)(2.1)}{6}$$

$$= 5.65 - 0.791$$

$$a = 4.859$$

So regression line *y* on *x* is given by

$$Y = 2.26\,x + 4.86 \qquad ...(1)$$

Regression line of *x* on *y* :

Let it be

$$X = a + bY$$

Substituting the values calculated in the above table in the following formula, we get

$$b = \frac{n\sum xy - \sum y \sum x}{n\sum y^2 - (\sum y)^2}$$

$$b = \frac{6(12.26) - (33.9)(2.1)}{6(192.45) - (33.9)^2}$$

$$b = 0.43$$

$$a = \frac{\sum x}{n} - \frac{b\sum y}{n}$$

$$= \frac{2.1}{6} - \frac{(0.43)(33.9)}{6}$$

$$= 0.35 - 2.429$$

$$a = -2.09$$

So regression line x on y is given by

$$X = 0.43y - 2.09 \qquad ...(2)$$

Now we would see that they intersect at ($\Sigma x/n$, $\Sigma y/n$) i.e (0.35, 5.65)

We first check for regression line y on x i.e.

$$Y = 2.26\,x + 4.86$$

Substitute $x = 0.35$ on R.H.S of equation (1)

We get $Y = 2.26\,(0.35) + 4.86$

$$= 5.65$$

Which means that regression line of y on x passes through (0.35, 5.65)

Now we check for regression line x on y i.e.

$$X = 0.43y - 2.09$$

Substitute y =5.65 on R.H.S of equation (2)

We get $X = 0.43(5.65) - 2.09$

$$= 0.35$$

Which means that regression line of x on y passes through (0.35, 5.65)

So both the regression lines pass through the common point (0.35, 5.65).

UNSOLVED PROBLEMS

1. Fit a straight line to the given data regarding x as the independent variable:

x :	1	2	3	4	6	8
y :	2.4	3.1	3.5	4.2	5.0	6.0

2. Find the best values of a and b so that $y = a + bx$ fits the given data:

x :	0	1	2	3	4
y :	1.0	2.9	4.8	6.7	8.6

3. Fit a straight line approximate to the data:

x :	1	2	3	4
y :	3	7	13	21

4. Discuss the method of least squares. How it is helpful while fitting any curve or explain the role of method of least squares in curve fitting.

5. From the data given below, find out

(a) Regression equation of y on x and find the value of y when $x = 10$

(b) Regression equation of x on y and find the value of x when $y = 10$

x :	1	2	3	4	5
y :	2	4	5	3	6

Also prove that both these regression lines intersect at $(\Sigma x/n, \Sigma y/n)$

6. The weight of a calf taken at weekly intervals is given below. Fit a straight line using the method of least squares and calculate the average rate of growth per week.

Age :	1	2	3	4	5	6	7	8	9	10
Weight :	52.5	58.7	65	70.2	75.4	81.1	87.2	95.5	102.2	108.4

7. Find the least square line for the data points

$(-1, 10), (0, 9), (1, 7), (2, 5), (3, 4), (4, 3), (5, 0)$ and $(6, -1)$.

8. Find the least square line $y = a + bx$ for the data:

x_i :	−2	−1	0	1	2
y_i :	1	2	3	3	4

10.5 FITTING OF NON-LINEAR CURVES

In this text we will discuss four types of non-linear curves

1. Geometric Curve
2. Exponential Curve
3. Hyperbola
4. Trigonometric Curve

10.5.1 Fitting of a Geometric Curve

The simple geometric curve is given by the equation

$$y = f(x) = ax^b \quad ...(1)$$

Where a, b are real numbers.

By taking log on both sides of equation (1), we get

$$\log y = \log(ax^b)$$

$$\log y = \log a + \log x^b$$

[by using $\log(m.n) = \log m + \log n$]

$$= \log a + b\log x \qquad \text{[by using } \log(m^n) = n\log m\text{]} \qquad ...(2)$$

Now substitute $\log y = Y$, $\log a = A$, $\log x = X$ in (2), we get

$$Y = A + bX \qquad ...(3)$$

As equation (3) is in linear form so the normal equations obtained are

$$nA + b\Sigma X = \Sigma Y$$

$$A\Sigma Y + b\Sigma X^2 = \Sigma XY$$

or

$$nA + b(\Sigma\log x) = \Sigma\log y$$

$$(\Sigma\log x)A + (\Sigma\log x)^2 \cdot b = \Sigma\log x \log y$$

After simplification of these equations, we get

$$b = \frac{n\Sigma\log x \log y - \Sigma\log x\Sigma\log y}{n\Sigma(\log x^2) - (\Sigma\log x)^2} \text{ and}$$

$$A = \frac{\Sigma\log y\,(\log x)^2 - \Sigma\log x\Sigma\log x \log y}{n\Sigma(\log x^2) - (\Sigma\log x)^2}$$

a in the curve can be computed by $\log a = c$ or $e^A = a$

10.5.2 Fitting an Exponential Curve

The exponential curve is given by the equation

$$y = f(x) = ae^{bx} \qquad ..(1)$$

Where a, b are real numbers.

By taking log on both sides of (1), we get

$$\log y = \log(ae^{bx})$$

$$= \log a + \log e^{bx}$$

$$= \log a + bx \log e$$

$$= \log a + bx \qquad (\text{as } \log e = 1) \qquad ...(2)$$

Now substitute $\log y = Y$, $\log a = A$ in (2)

i.e

$$Y = A + bx \qquad ...(3)$$

as equation (iii) is in linear form, so the normal equations obtained from (3) becomes

$$nA + b\Sigma x = EY$$

$$A\Sigma x + b\Sigma x^2 = \Sigma xY$$

or $$nA + b\Sigma x = \Sigma \log y$$
$$A\Sigma x + b\Sigma x^2 = \Sigma x \log y$$

After simplification of these equations, we get

$$b = \frac{n\Sigma xk\log y - \Sigma x\Sigma \log y}{n\Sigma x^2 - (\Sigma x)^2}$$

$$A = \frac{\Sigma \log y \Sigma x^2 - \Sigma x\Sigma x\log y}{n\Sigma x^2 - (\Sigma x)^2}$$

a in the curve can be computed by $\log a = c$ or $e^A = a$.

10.5.3 Fitting a Hyperbola

A hyperbola is given by the equation

$$y = f(x) = 1/(a + bx) \quad ...(1)$$

or $$1/y = a + bx$$

By substituting $1/y = Y$,we get

$$Y = a + bx \quad ...(2)$$

As equation (ii) is in linear form so the normal equations obtain form (ii) becomes

$$na + b\Sigma x = \Sigma Y$$
$$a\Sigma x + b\Sigma x^2 = \Sigma xY$$

or $$na + b\Sigma x = \Sigma(1/y)$$
$$a\Sigma x + b\Sigma x^2 = \Sigma x(1/y)$$

After simplification of these equations, we get

$$b = \frac{n\Sigma(1/y) - \Sigma x\Sigma(1/y)}{n\Sigma x^2 - (\Sigma x)^2}$$

$$a = \frac{\Sigma(1/y)\Sigma x^2 - \Sigma x\Sigma x(1/y)}{n\Sigma x^2 - (\Sigma x)^2}$$

10.5.4 Fitting a Trigonometric Curve

The equation of a trigonometric curve is given as

$$y(t) = a \sin(rt + \theta) \quad ...(1)$$

where a, θ are unknown quantities and r is known (assumed) quantity.

Or by using $\sin(A + B) = \sin A\cos B + \cos A\sin B$ on (i), we get

$$y(t) = a \sin(rt) \cos\theta + a \cos(rt) \sin\theta$$

By substituting $A = a \cos\theta$ and $B = a \sin\theta$, we get

$$y(t) = A \sin(rt) + B \cos(rt)$$

By method of least squares the sum of the residuals (errors) is given by

$$E = \sum_{i=1}^{n} d_i^2$$

$$= \sum_{i=1}^{n} (y_i - y(t_i))^2$$

$$= \sum_{i=1}^{n} (y_i - (A\sin(rt_i) + B\cos(rt_i)))^2$$

$$= \sum_{i=1}^{n} (y_i - A\sin(rt_i) - B\cos(rt_i)))^2$$

By principle of least squares, E is minimum

$\therefore$ $\partial E/\partial A = 0$ and $\partial E/\partial B = 0$

i.e $$\partial E/\partial A = 2\sum_{i=1}^{n} (y_i - A\sin(rt_i) - B\cos(rt_i))\,(-\sin(rt_i))$$

$$= \sqrt{A^2 + B^2} \quad \text{...(2)}$$

$$\partial E/\partial B = 2\sum_{i=1}^{n} (y_i\ A\sin(rt_i) - B\cos(rt_i)\,(-\cos(rt_i)) = 0$$

After eliminating suffices and simplification of (2) and (3), we get the normal equations as

$$A\Sigma\sin^2(rt) + B\Sigma\sin(rt)\cos(rt) = \Sigma y\sin(rt)$$

$$A\Sigma\sin(rt)\cos(rt) + B\Sigma\cos^2(rt) = \Sigma y\cos(rt)$$

After solving these equations

$$B = \frac{\Sigma\sin^2(rt)\Sigma y\cos(rt) - \Sigma\sin(rt)\cos(rt)\Sigma y\sin(rt)}{\Sigma\sin^2(rt)\Sigma\cos^2(rt) - (\Sigma\sin(rt)\cos(rt))^2}$$

$$A = \frac{\Sigma\cos^2(rt)\Sigma y\sin(rt) - \Sigma y\cos(rt)\Sigma\sin(rt)\cos(rt)}{\Sigma\sin^2(rt)\Sigma\cos^2(rt) - (\Sigma\sin(rt)\cos(rt))^2}$$

The magnitude of a and θ can be computed as

$$a = \sqrt{A^2 + B^2},\ \theta = \tan^{-1}(B/A)$$

Program: Program to Fit a Non Linear Trend $Y = AX^B$.

```
#include<conio.h>
#include<stdio.h>
```

```
#include<math.h>
#include<process.h>
void main()
{
float d, b, a,a0,n,i,x[30],y[30],xly[30],xx[30],sum1=0,sum2=0,sum3=0,sum4=0;
clrscr();
printf(" How the value of n ");
scanf("%f",&n);
printf("\n Ente the x series \n");
for(i=1;i<=n;i++)
{
scanf("%f",&x[i]);
}
printf("\n Ente the Y series \n");
for(i=1;i<=n;i++)
{
scanf("%f",&y[i]);
}

for(i=1;i<=n;i++)
{
xly[i]=log(x[i]) * log(y[i]);
xx[i]=log(x[i]) * log(x[i]);
sum1=sum1+log(x[i]);
sum2=sum2+log(y[i]);
sum3=sum3+xly[i];
sum4=sum4+xx[i];
}
d = (n * sum4) - ( sum1 * sum1);
if (d ==0)
{
printf("\n Curve cannot be fitted");
exit(1);
}
```

```
b = ((n * sum3) - (sum2 * sum1))/ d;
a0= (sum2/n) - (b* sum1)/n;
a=exp(a0);
printf("Value of a =%f",a);
printf("Value of b =%f",b);
getch();
}
```

Program: Program to Fit a Non Linear Trend Y = AE^B.

```
#include<conio.h>
#include<stdio.h>
#include<math.h>
#include<process.h>
void main()
{
float d, b, a,a0,n,i,x[30],y[30],xly[30],xx[30],sum1=0,sum2=0,sum3=0,sum4=0;
clrscr();
printf(" How the value of n ");
scanf("%f",&n);
printf("\n Ente the x series  \n");
for(i=1;i<=n;i++)
{
scanf("%f",&x[i]);
}
printf("\n Ente the Y series  \n");
for(i=1;i<=n;i++)
{
scanf("%f",&y[i]);
}

for(i=1;i<=n;i++)
{
xly[i]=x[i] * log(y[i]);
xx[i]=x[i] * x[i];
```

```
sum1=sum1+x[i];
sum2=sum2+log(y[i]);
sum3=sum3+xly[i];
sum4=sum4+xx[i];
}
d = (n * sum4) - ( sum1 * sum1);
if (d ==0)
{
printf("\n Curve cannot be fitted");
exit(1);
}
b = ((n * sum3) - (sum2 * sum1))/ d;
a0= (sum2/n) - (b* sum1)/n;
a=exp(a0);
printf("Value of a =%f",a);
printf("Value of b =%f",b);
getch();
}
```

Program: Program to fit a non linear trend $Y=AB^X$.

```
#include<stdio.h>
#include<conio.h>
#include<math.h>
main()
int i,n;
float x[100],y[100],sx=0,sx2=0,sly=0,sxly=0,a0,a1,a,b,denom;
clrscr();
printf("\n How many itemsin data ?");
scanf("%d",&n);
printf(" Enter the data as Xi Yi\n");
for(i=1;i<=n;i++)
{
scanf("%f %f",&x[i],&y[i]);
sx+=x[i];
```

```
sx2+=x[i]*x[i];
sly+= log(y[i]);
sxly+=x[i]*log(y[i]);
}
denom=(n*sx2-sx*sx);
a0=(sly*sx2-sx*sxly)/denom;
a=exp(a0);
a1=(n*sxly-sx*sly)/denom;
b=exp(a1);
printf("\n Value of a=%9.4f",a);
printf("\n Value of b=%9.4f",b);
getche();
}
```

10.6 HOW TO SOLVE EQUATION THAT CONTAINS THREE UNKNOWN VARIABLES SAY A,B,C IN EQUATIONS FORM

$$y = a + bx + cx^2 \text{ (parabola)}$$
$$y = ax^b + c$$
$$y = ab^x + c$$
$$y = ae^{bx} + c.$$

These equations are solved by firstly converted into equations that contained two unknowns and then solved by previously discussed methods.

10.6.1 For Equation $Y = A + Bx + Cx^2$ (Parabola)

We shall reduce this equation to a linear graph from which two unknowns can be found.

Assume that a particular point (x_1, y_1) satisfies the given equation. That is

$$y_1 = a + bx_1 + x_1^2$$

Hence,
$$y - y_1 = b(x - x_1) + c(x^2 - x_1^2)$$

i.e.
$$\frac{y - y_1}{x - x_1} = b + c(x + x_1)$$

$$\frac{y - y_1}{x - x_1} = Y \text{ and } x + x_1 = X$$

Hence the equation becomes,

$$Y = b + cX \text{ which is linear equation in } X \text{ and } Y.$$

Now solve this equation by using previous method that is used for solving equation of parabola that contains two variables.

10.6.2 For Equation $Y = AX^B + C$

First, we will get the value of c so that the points $(x_1, y_1), (x_2, y_2), (x_3, y_3)$ be three points on the curve where x_1, x_2, x_3 are in geometric progression (i.e., $x_2^2 = x_1x_3$).

Substituting these points on the curve,

$$y_1 = ax_1^b + c$$

$$\therefore\ y_1 - c = ax_1^b;\quad y_2 - c = ax_2^b;\quad y_3 - c = ax_3^b$$

Hence $$(y_1 - c)(y_3 - c) = a^2(x_1x_3)^b = a^2(x_2^2)^b = (ax_2^b)^2 = (y_2 - c)^2$$

i.e. $$y_1y_3 - c(y_1 + y_3) + c^2 = y_2^2 + c^2 - 2cy_2$$

$$c = \frac{y_1y_3 - y_2^2}{y_1 + y_3 - 2y_2}$$

Using this value of c in $y - c = ax^b$ we have only two unknowns left out.

$$\log_{10}(y - c) = \log_{10} a + b\log_{10} x$$

$$Y = A + bX,$$

where $Y = \log_{10}(y - c)$, $A = \log_{10} a$ and $X = \log_{10} x$

This being linear is X and Y, we get the equation and also A and b. Knowing A, we get a.

Hence the equation $y = ax^b + c$.

10.6.3 For Equation $Y = AB^X + C$

If the curve is of the form $y = ab^x + c$, we select three points $(x_1, y_1), (x_2, y_2), (x_3, y_3)$ on this curve so that x_1, x_2, x_3 are in arithmetic progression i.e., $x_1 + x_3 = 2x_2$

Substituting these points on the curve,

$$y_1 = ab_1^x;\quad y_2 - c = ab_2^x;\quad y_3 - c = ab_3^x$$

$$(y_1 - c)(y_3 - c) = a^2 b_1^x + x_3 = a^2\ b_2^{2x} = (ab_2^x)^2 = (y_2 - c)^2$$

$$\therefore\quad c = \frac{y_1y_3 - y_2^2}{y_1 + y_3 - 2y_2}$$

if this c is found out, $y - c$ is known.

Now $$y - c = ab^x$$

$$\therefore\quad \log_{10}(y - c) = \log a + x \text{ log10 b}$$

i.e $$Y = A + Bx$$

This is linear in x and Y. Hence A, B can be found out. Therefore a, b can be got using A, B. Thus the curve is got.

10.6.4 For Equation of the form $Y = AE^{BX} + C$

This can be solved by splitting the curve into following form

$$y = a(e^b)^x + c = a \cdot k^x + c \text{ where } k = e^b.$$

Now solve this by using procedure of the curve

$$y = ab^x + c$$

10.7 CORRELATION ANALYSIS

Correlation analysis is the study of knowing the degree of association between different variables belonging to the same variable.

The formula for computing Correlation Coefficient (r) is given as

$$r = \frac{n\Sigma xy - \Sigma x \Sigma y}{[n\Sigma x^2 - (\Sigma x)^2][n\Sigma y^2 - (\Sigma y)^2}$$

This formula computes the correlation between two variables only.

SOLVED PROBLEMS

Problem 1: Fit a parabola of the pattern $y = a + bx + cx^2$ to the data given below

X:	2	4	6	8	10	12
Y:	1.5	3.6	4.9	6.5	7.2	9.2

Solution: The normal equations for the given parabola are

$$an + b\Sigma x + c\Sigma x^2 = \Sigma y \quad \ldots(1)$$

$$a\Sigma x + b\Sigma x^2 + c\Sigma x^3 = \Sigma xy \quad \ldots(2)$$

$$a\Sigma x^2 + b\Sigma x^3 + c\Sigma x^4 = \Sigma x^2 y \quad \ldots(3)$$

I	x_i	y_i	x_i^2	x_i^3	x_i^4	$x_i y_i$	$x_i^2 y_i$
1	2	1.5	4	8	16	3.0	6.0
2	4	3.6	14	64	256	14.4	57.6
3	6	4.9	36	216	1296	29.4	176.4
4	8	6.5	64	512	4096	52.0	416.0
5	10	7.2	100	1000	10000	72.0	720.0
6	12	9.2	144	1728	20736	110.4	1324.8
$n = 6$	$\Sigma x = 42$	$\Sigma y = 32.9$	$\Sigma x^2 = 364$	$\Sigma x^3 = 3528$	$\Sigma x^4 = 36400$	$\Sigma xy = 281.2$	$\Sigma x^2 y = 2700.8$

After substitution in (1), (2) and (3), we get

$$6a + 42bx + 364c = 32.9$$
$$42a + 364b + 3528c = 281.2$$
$$364a + 3528b + 36400c = 2700.8$$

After solving, we get

$$a = -0.028$$
$$b = 0.90$$
$$c = -0.013$$

Hence parabola becomes

$$y = -0.028 + 0.90x - 0.013x^2$$

Problem 2: Fit a second degree parabola to the following data, taking y as dependent variable.

x :	1	2	3	4	5	6	7	8	9
y :	2	6	7	8	10	11	11	10	9

Solution: Let $X = x - x_m = x - 5$ and $Y = y - y_m = y - 7$

Where x_m, y_m are assumed means of x and y respectively.

Let $Y = aX^2 + bX + c$ be the best fit.

x	y	X	Y	X^2	X^3	X^4	XY	X^2Y
1	2	–4	–5	16	–64	256	20	–80
2	6	–3	–1	9	–27	81	3	–9
3	7	–2	0	4	–8	16	0	0
4	8	–1	1	1	–1	1	–1	1
5	10	0	3	0	0	0	0	0
6	11	1	4	1	1	1	4	4
7	11	2	4	4	8	16	8	16
8	10	3	3	9	27	81	9	27
9	9	4	2	16	64	256	8	32
Total		0	11	60	0	708	–51	–9

The normal equations are

$$708a + 60c = -9$$
$$60b = 51$$
$$60a + 9c = 11$$

After solving these, we get $a = -0.2673$, $b = 0.85$, $c = 3.0042$

Hence required parabola becomes

$$Y = -0.2673X^2 + 0.85X + 3.0042$$

$\therefore$ $$y-7 = -0.2673(x-5)^2 + 0.85(x-5) + 3.0042$$

i.e. $$y = -0.2673x^2 + 3.523x - 0.9283$$

Problem 3: Given a table for the function $y = ae^{bx}$ as

x:	20	40	60	90
y:	2	6	10	15

Obtain the equation of best fit.

Solution: The normal equations for the curve $y = ae^{bx}$ are

$$nA + b\Sigma x = \Sigma \log y \quad ...(1)$$

$$A\Sigma x + b\Sigma x^2 = \Sigma x \log y \quad ...(2)$$

Where $$A = \log a$$

i	x_i	y_i	$\log y_i$	x_i^2	$x_i \log y_i$
1	20	2	0.30103	400	0.0206
2	40	6	0.77815	1600	31.126
3	60	10	1.00	3600	60.00
4	90	15	1.17609	8100	105.8481
$n = 4$	$\Sigma x = 210$	$\Sigma y = 33$	$\Sigma \log y = 3.25527$	$\Sigma x^2 = 13700$	$\Sigma x \log y = 202.9947$

After substitution these value in normal equations, we get

$$4A + 210b = 3.25527$$

$$210A + 13700 = 202.9947$$

After solving these equations for A, b, we get

$$b = 0.012, \quad A = 0.183$$

As $\log a = A$ $\therefore$ $a = \text{antilog } A \Rightarrow \text{antilog } (0.183) = 1.20$

Hence exponential curve becomes $y = 1.20e^{0.012x}$

Problem 4: Given the data below

Y:	2	10	26	61
X:	600	500	400	350

Fit a curve of type $y = ae^{bx}$. Find the best possible values of a and b.

Solution: The normal equations for the curve $y = ae^{bx}$ are

$$nc + b\,\Sigma x = \Sigma \log y$$

$$c\,\Sigma x + b\,\Sigma x^2 = \Sigma x \log y$$

So on solving we get

$$b = \frac{n\Sigma x \log y - \Sigma \log y \Sigma x}{n\Sigma x^2 - (\Sigma x)^2} \quad .... (1)$$

$$c = \frac{\Sigma \log y}{n} - \frac{b\Sigma x}{n} \quad ...(2)$$

where $$c = \log a$$

Calculate the values of Σx, $\Sigma \log y$, $\Sigma x \log y$, Σx^2 as shown below

X	Y	Log Y	X^2	X log Y
600	2	0.30103	360000	180.618
500	10	1	250000	500
400	26	1.414973	160000	565.9892
350	61	1.78533	122500	624.8655
$\Sigma x = 1850$	$\Sigma y = 99$	$\Sigma \log Y = 4.501333$	$\Sigma X^2 = 892500$	$\Sigma X \log Y = 1871.473$

Substitute these values in (1) and (2)

$$b = \frac{4(1871.473) - 1850(4.501333)}{4(892500) - (1850)^2}$$

$$= \frac{7485.892 - 8327.46605}{3570000 - 3422500}$$

$$= \frac{-841.5740}{147500} = -0.0057056$$

$$c = \frac{4.501333}{4} + \frac{(0.0057056)(1850)}{4}$$

$$= \frac{4.501333}{4} + \frac{10.55536}{4} = \frac{15.0566}{4} = 3.76417$$

$$c = \log a$$

a = antilog (c) = antilog(3.76417)= 43.12777

The required values of a and b are 43.12777, – 0.0057056 and relation between x and y is

$$Y = 43.12777\, e^{-0.0057056\,x}$$

Problem 5: Given the data below

x:	10	20	30	40	50	60	70	80
y:	1.06	1.33	1.52	1.68	1.81	1.91	2.01	2.11

Fit a curve of type $y = a\,x^b$. Find the best possible values of a and b.

Solution: The normal equations for the curve $y = ax^b$ are

$$nc + b\,\Sigma X = \Sigma Y$$

$$c\,\Sigma X + b\,\Sigma X^2 = \Sigma XY$$

So on solving we get

$$b = \frac{n\,\Sigma XY - \Sigma X \Sigma Y}{n\,\Sigma X^2 - (\Sigma X)^2} \quad ...(1)$$

$$c = \frac{\Sigma Y}{n} - \frac{b\Sigma X}{n} \quad ...(2)$$

where $c = \log a$, $Y = \log y$ and $X = \log x$

Calculate the values of ΣX, ΣY, ΣXY, ΣX^2 as shown below

x	**y**	**X = Log x**	**Y = log y**	**XY = Log x log y**	**X² = log² x**
10	1.06	1	0.0253	0.0253	1
20	1.33	1.301	0.1238	0.16106	1.6926
30	1.52	1.4771	0.1818	0.2685	2.1818
40	1.68	1.602	0.2253	0.3609	2.5664
50	1.81	1.6989	0.2576	0.4376	2.88623
60	1.91	1.7781	0.281	0.4996	3.1616
70	2.01	1.845	0.3031	0.5592	3.404
80	2.11	1.903	0.3242	0.6169	3.6214
		Σlog x = 12.6051	**Σlog y = 1.7221**	**Σlog x log y = 2.92906**	**Σlog² x = 20.514**

Substitute these values in (1) and (2)

$$b = \frac{8(2.92906) - 12.6051\,(1.7221)}{8(20.514) - (12.6051)^2}$$

$$= \frac{23.4324 - 21.7072}{164.112 - 158.8885} = \frac{1.7252}{5.2235} = 0.3302$$

$$c = \frac{1.7221}{8} - \frac{(0.3302)(\,12.6051)}{8}$$

$$c = \frac{1.7221}{8} - \frac{4.1622}{8} = -2.4401 = -0.30501$$

$$c = \log a$$

$$a = \text{antilog}\ (c) = \text{antilog}(-\,0.30501) = 0.4954$$

The required values of a and b are 0.4954, 0.3302 and relation between x and y is

$$Y = 0.4954\, x^{0.3302}$$

Problem 6: Given the data below

x:	1.2	1.4	1.6	1.8	2.0	2.4
y:	4.2	6.1	8.5	11.5	14.9	23.5

Fit a curve of type $y = ax^b$. Find the best possible values of a and b.

Solution: The normal equations for the curve $y = ax^b$ are

$$nc + b\,\Sigma X = \Sigma Y$$

$$c\,\Sigma X + b\,\Sigma X^2 = \Sigma XY$$

So on solving we get:

$$b = \frac{n\Sigma XY - \Sigma X \Sigma Y}{n\Sigma X^2 - (\Sigma X)^2} \quad ...(1)$$

$$c = \frac{\Sigma Y}{n} - \frac{b\Sigma X}{n} \quad ...(2)$$

where $c = \log a$, $Y = \log y$ and $X = \log x$

Calculate the values of $\Sigma X, \Sigma Y, \Sigma XY, \Sigma X^2$ as shown below:

x	y	X = Log x	Y = log y	XY = Log x log y	$X^2 = \log^2 x$
1.2	4.2	0.0791	0.6232	0.0492	0.0062
1.4	6.1	0.1461	0.7853	0.1147	0.0213
1.6	8.5	0.2041	0.9294	0.1896	0.0416
1.8	11.5	0.2552	1.0606	0.2706	0.0651
2	14.9	0.301	1.1731	0.3531	0.0906
2.4	23.5	0.3802	1.371	0.5212	0.1445
		Σlog x = 1.3657	Σlog y = 5.9427	Σlog x log y = 1.4985	$\Sigma\log^2 x$ = 0.3694

Substitute these values in (1) and (2)

$$b = \frac{6(1.4985) - 1.3657\,(5.9427)}{6(0.3694) - (1.3657)^2}$$

$$= \frac{8.991 - 8.1159}{2.2164 - 1.8651} = \frac{0.8751}{0.3513} = 2.49103$$

$$c = \frac{5.9427}{6} - \frac{(2.49103)(1.3657)}{6}$$

$$c = \frac{5.9427}{6} - \frac{3.40199}{6}$$

$$= 0.99045 - 0.5669 = 0.42355$$

$$c = \log a$$

$$a = \text{antilog}\,(c) = \text{antilog}(0.42355) = 2.6518$$

The required values of a and b are 2.6518, 2.49103 and relation between x and y is

$$Y = 2.6518x^{2.49103}$$

Problem 7: Given the data below

x :	1	2	3	4	5	6
y :	0.5	2.1	4.5	7.5	12.6	19.8

Fit a geometric curve of the form y = ax^b

Solution: The normal equations for the geometric curve $y = ax^b$ are

$$nA + b\Sigma X = \Sigma Y \quad ...(1)$$

$$A\Sigma X + b\Sigma X^2 = \Sigma XY \quad ...(2)$$

Where $Y = \log y$, $A = \log a$, $X = \log x$

i	x_i	y_i	$X_i = \log x_i$	$Y_i = \log y_i$	x_i^2	X_iY_i
1	1	0.5	0	–0.301	0	0
2	2	2.1	0.301	0.3222	0.0906	0.09698
3	3	4.5	0.4771	0.6532	0.2276	0.31164
4	4	7.5	0.602	0.875	0.3624	0.52675
5	5	12.6	0.699	1.1003	0.4886	0.7691
6	6	19.8	0.7781	1.2967	0.6054	1.00896
$n = 6$	$\Sigma x = 21$	$\Sigma y = 47$	$\Sigma X = 2.8572$	$\Sigma Y = 3.9464$	$\Sigma X^2 = 1.7746$	$\Sigma XY = 2.71343$

After substitution these values in normal equations, we get

$$6A + 2.8572b = 3.9464$$

$$2.8572A + 1.7746b = 2.71343$$

After solving for A and b, we get

$$A = 0.2947, b = 2$$

As A = log a ∴ a = antilog(A) ⇒ a = antilog(0.2947) ⇒ a = 1.3427

Hence geometric curve becomes $y = 1.3427x^2$

Problem 8: The observations in the following table fit a law of the form $y = 1/(a + b \log x)$. Fit the given equation by method of least squares.

x :	2	4	5	7	9	12
y :	0.5	1	2	4	6	9

Solution:

As

$$y = 1/(a + b \log x) \quad ...(1)$$

Substitute $Y = 1/y$ in (1)

$$Y = a + b \log x$$

Substitute $\log x = X$

∴

$$Y = a + bx$$

As this in linear form the normal equations for this are

$$na + b\Sigma X = \Sigma Y \quad ...(2)$$

$$a\Sigma X + b\Sigma X^2 = \Sigma XY \quad ...(3)$$

i	x_i	y_i	X_i=log x	Y = 1/y	x_i^2	y_i^2
1	2	0.5	0.301	2	0.0906	0.602
2	4	1	0.602	1	0.3624	0.602
3	5	2	0.699	0.5	0.4886	0.3495
4	7	4	0.845	0.25	0.714	0.2112
5	9	6	0.9542	0.16	0.9104	0.1526
6	12	9	1.0791	0.11	1.1644	0.1187
n = 6	Σx = 39	Σy = 22.5	ΣX = 4.4803	ΣY = 4.02	ΣX^2 = 1.1644	ΣXY = 2.036

Substitute these values in (2) and (3), we get

$$6a + 4.4803b = 4.02$$

$$4.4803a + 1.1644b = 2.036$$

After solving these for a and b, we get

$$a = 0.3439, \quad b = 0.4367$$

∴ Required hyperbola becomes

$$y = 1 / (0.3439 + 0.4367x)$$

Problem 9: The data given below will fit a formula of the type $y = a + bx + cx^2$. Find the required equation or formula.

x :	1.5	5	8.8	13.7	16.7
y :	7	23	41	54	69

Solution: Take the point (1.5, 7) as the particular point on the curve

$$y = a + bx + cx^2 \quad ...(1)$$

∴

$$7 = a + (1.5)b + c(1.5)^2 \quad ...(2)$$

Subtract (2) from (1), we get

$$y - 7 = b(x - 1.5) + c[x^2 - (1.5)^2]$$

$$\frac{y-7}{x-1.5} = b + c(x + 1.5)$$

i.e

$$Y = b + cX$$

where $Y = \dfrac{y-7}{x-1.5}$ and $X = x + 1.5$

Normal equations for $Y = b + cX$ becomes

$$nb + c\Sigma X = \Sigma Y \quad ...(3)$$

$$b\Sigma X + c\Sigma X^2 = \Sigma XY \quad ...(4)$$

As we have taken the starting point as (1.5,7) so there is no need to involve this point in the table.

i	X = x + 1.5	Y = (y–7)/(x–1.5)	X^2	XY
1	7.5	4.57	56.25	34.275
2	10.3	4.6575	106.09	47.972
3	15.2	3.85	231.04	58.52
4	18.2	4.08	331.24	74.256
n = 4	ΣX = 54.2	ΣY = 17.1575	ΣX^2 = 724.58	ΣXY = 215.023

Substitute the corresponding values in (3) and (4), we get

$$4b + 54.2c = 17.1575 \quad \text{...(5)}$$

$$54.2b + 724.58c = 215.023 \quad \text{...(6)}$$

After simplification, we get

$$c = 1.776, \quad b = -19.775$$

From (2),

$$7 = a - 19.775*1.5 + 1.776*2.25$$

$$a = 32.6665$$

Substitute these values of a, b and c in (1), we get

$$y = 32.6665 - 19.775x + 1.776x^2$$

Problem 10: A curve given by the equation of the form $y = ax^b + c$ is to be fitted to the following data. Determine the values of a, b and c.

x :	0.2	0.5	1.0	2.0	3.0	4.0
y :	3.2	3.2	4.1	8.1	13.7	22.6

Solution: Let us search for three points whose abscissas are in G.P.

$\therefore x_1 = 0.5, x_2 = 1.0, x_3 = 2.0$ since $x_1x_3 = x_2^2$

$$\therefore \quad c = \frac{y_1y_3 - y_2^2}{y_1 + y_3 - 2y_2} = \frac{3.2 * 8.1 - (4.1)^2}{3.2 + 8.1 - 2(4.1)} \quad \text{...(1)}$$

$$\log(y - 2.9387) = \log a + b \log x$$

$$Y = A + bX \quad \text{...(2)}$$

$X = \log x$, $Y = \log(y - 2.9387)$, $A = \log a$

Normal equations for $Y = A + bX$ becomes

$$nA + b\Sigma X = \Sigma Y \quad \text{...(3)}$$

$$A\Sigma X + b\Sigma X^2 = \Sigma XY \quad \text{...(4)}$$

i	$y-2.9387$	$Y = \log(y-2.9387)$	$X = \log x$	X^2	XY
1	0.2613	–0.5829	0.6989	0.4884	0.4073
2	0.2613	–0.5829	–0.3010	0.0906	0.1754
3	1.1613	0.0649	0	0	0
4	5.1613	0.7128	0.3010	0.0906	0.2145
5	10.7613	1.0319	0.4771	0.2276	0.4923
6	19.6613	1.2936	0.6021	0.3625	0.7789
$n = 6$		$\Sigma Y = 1.9374$	$\Sigma X = 0.3803$	$\Sigma X^2 = 1.2597$	$\Sigma XY = 2.0684$

After substituting the values in (3) and (4), we get

$$6A + 0.3803b = 1.9374$$

$$0.3803A + 1.2597b = 2.0684$$

After simplification, we get

$$A = 0.2231 \text{ and } b = 1.5746$$

As $A = \log a$

$\Rightarrow$ $a = \text{antilog } (A) = 1.6714$

So final equation becomes

$$y = 1.6714x^{1.5746} + 2.9387$$

UNSOLVED PROBLEMS

1. By using the following table values

x:	0.5	0.7	0.9	1.2
y:	0.4	0.3	0.2	0.1

Fit a best curve of the form, $y = a + b/x + c/x^2$

2. By using the following table values

x:	500	1000	2000	4000	6000
y:	0.20	0.33	0.38	0.45	0.51

Fit a best curve of the form, $y = a + b\sqrt{x}$

3. Fit an equation of the form $y = ae^{bx}$ to the following data by the method of least squares:

x:	1	2	3	4
y:	1.65	2.7	4.5	7.35

4. Fit the exponential curve $y = ae^{bx}$ to the following data, e being Napierian base, 2.71828

x:	0	2	4
y:	5.012	10	31.62

5. Using the method of least squares, fit the non-linear curve of the form $y = ae^{bx}$ to the following data:

x:	0	2	4
y:	5.012	10	31.62

6. Fit a curve of the form $y = ax^b$ to the data given below:

x:	1	2	3	4	5
y:	7.1	27.8	62.1	110	161

7. Given the data below:

x:	1	2	3	4	5	6
y:	0.5	2.1	4.5	7.5	12.6	19.8

Fit a geometric curve.

8. Fit an exponential curve of the form $y = ab^x$ to the following data:

x:	1	2	3	4	5	6	7	8
y:	1	1.2	1.8	2.5	3.6	4.7	6.6	9.1

9. Fit a curve $y = ax^b$ to the following data:

x:	1	2	3	4	5	6
y:	2.98	4.26	5.21	6.1	6.8	7.5

10. The following number relate to the flow of water over a triangular notch.

H:	1.2	1.4	1.6	1.8	2.0	2.4
Q:	4.2	6.1	8.5	11.5	14.9	23.5

If the law is $Q = CH^n$, find C and H.

11. Using the following table, fit a curve of the form $y = ax^b$. Find a and b.

x:	10	20	30	40	50	60	70	80
y:	1.06	1.33	1.52	1.68	1.81	1.91	2.01	2.11

12. A curve given by the equation of the form $y = ax^b + c$ is to be fitted to the following data. Determine the values of a, b and c.

x:	2	4	6	8	10
y:	5	8.6	13.9	25.3	35.6

13. Fit a parabola of the pattern $y = a + bx + cx^2$ to the data given below:

x:	5	10	15	20	25	30
y:	1.2	3.7	5.1	6.4	8.3	10.2

14. The values of x and y are related by the law $y = ab^x + c$. Corresponding values of x and y are given below. Find the best values of a, b and c.

x:	2	4	6	8	10	12
y:	12	20	35	63	110	202

15. Fit a curve of the form $y = a + bx^n$ given the data:

x :	20	40	60	80	100
y :	5	8.6	13.9	25.3	35.6

16. Fit a curve of the form $y = ab^{-x}$ form the data given below:

x :	0.1	0.2	0.4	0.6	1.0	1.5	2.0
y :	50	36	28	19	16	11	7

17. Obtain a curve of the form $y = ae^{bt} + c$ to the data given below:

t:	11	8	6	2	1
v:	10.99	11.34	11.65	12.46	12.71

18. Fit a second degree parabola to the following data:

x:	1.0	1.5	2.0	2.5	3.0	3.5	4.0
y:	1.1	1.3	1.6	2.0	2.7	3.4	4.1

19. The observations in the following table fit a law of the form $y = a \sin(4t + \theta)$. Fit the given by method of least squares.

t :	0.2	0.4	0.5	0.6	0.8
y :	0.3	0.4	0.6	0.7	0.9

20. Fit a curve of type $y = A \cos(wx + \theta)$ for the data given below. Assume $w = 1.0$

x:	0	0.2	0.4	0.6	0.8	1.0	1.2
y:	0.05	0.57	1.17	1.68	2.13	2.52	2.79

21. The pressure and volume of a gas are related by the equation $pv^r = k$, r and k being constants. Fit this equation to the following set of observations.

P(kg/cm²) :	0.5	1.0	1.5	2.0	2.5	3.0
V(litres) :	1.62	1.0	0.75	0.62	0.52	0.46

22. Find the curve $y = ax + bx^2$ that best fits the following data:

X :	1	2	3	4	5
Y :	1.8	5.1	8.9	14.1	19.8

CHAPTER 11

Introduction to Statistics

11.1 INTRODUCTION

Statistics is a mathematical science pertaining to the collection, analysis, interpretation or explanation, and presentation of data. Statisticians improve the quality of data with the design of experiments and survey sampling. Statistics also provides tools for prediction and forecasting using data and statistical models. Statistics is applicable to a wide variety of academic disciplines, including natural and social sciences, government, and business.Statistics is the scientific application of mathematical principles to the collection, analysis, and presentation of numerical data. Statisticians contribute to scientific inquiry by applying their mathematical and statistical knowledge to the design of surveys and experiments; the collection, processing, and analysis of data; and the interpretation of the results. Statisticians may apply their knowledge of statistical methods to a variety of subject areas, such as biology, economics, engineering, medicine, public health, psychology, marketing, education, and sports. Many economic, social, political, and military decisions cannot be made without statistical techniques, such as the design of experiments to gain federal approval of a newly manufactured drug. Statistics is a set of tools used to organize and analyze data. Data must either be numeric in origin or transformed by researchers into numbers.

Statistical methods can be used to summarize or describe a collection of data; this is called **descriptive statistics**. This is useful in research, when communicating the results of experiments. In addition, patterns in the data may be modeled in a way that accounts for randomness and uncertainty in the observations, and are then used to draw inferences about the process or population being studied; this is called **inferential statistics**. Inference is a vital element of scientific advance, since it provides a prediction (based in data) for where a theory logically leads. To further prove the guiding theory, these predictions are tested as well, as part of the scientific method. If the inference holds true, then the descriptive statistics of the new data increase the soundness of that hypothesis. Descriptive statistics and inferential statistics (**predictive statistics**) together comprise applied statistics. There is also a discipline called **mathematical statistics**, which is concerned with the theoretical basis of the subject.

The word statistics can either be singular or plural. In its singular form, statistics refers to the mathematical science discussed in this article. In its plural form, statistics is the plural of the word statistic, which refers to a quantity calculated from a set of data.

In applying statistics to a scientific, industrial, or societal problem, one begins with a process or population to be studied. This might be a population of people in a country, of crystal grains in a rock, or of goods manufactured by a particular factory during a given period. It may instead be a process observed at various times; data collected about this kind of "population" constitute what is called a time series.

For practical reasons, rather than compiling data about an entire population, one usually instead studies a chosen subset of the population, called a sample. Data are collected about the sample in an observational or experimental setting. The data are then subjected to statistical analysis, which serves two related purposes: description and inference.

1. **Descriptive Statistics**
2. **Predictive Statistics (or Inferential Statistics)**

1. **Descriptive Statistics:** Descriptive statistics can be used to summarize the data, either numerically or graphically, to describe the sample. Basic examples of numerical descriptors include the mean and standard deviation. Graphical summarizations include various kinds of charts and graphs.
2. **Inferential Statistics:** Inferential statistics is used to model patterns in the data, accounting for randomness and drawing inferences about the larger population. These inferences may take the form of answers to yes/no questions (hypothesis testing), estimates of numerical characteristics (estimation), forecasting of future observations, descriptions of association (correlation), or modeling of relationships (regression). Other modeling techniques include ANOVA, time series, and data mining.

The concept of correlation is particularly noteworthy. Statistical analysis of a data set may reveal that two variables (that is, two properties of the population under consideration) tend to vary together, as if they are connected. For example, a study of annual income and age of death among people might find that poor people tend to have shorter lives than affluent people. The two variables are said to be correlated. However, one cannot immediately infer the existence of a causal relationship between the two variables; see correlation does not imply causation. The correlated phenomena could be caused by a third, previously unconsidered phenomenon, called a lurking variable.

If the sample is representative of the population, then inferences and conclusions made from the sample can be extended to the population as a whole. A major problem lies in determining the extent to which the chosen sample is representative. Statistics offers methods to estimate and correct for randomness in the sample and in the data collection procedure, as well as methods for designing robust experiments in the first place; see experimental design.

The fundamental mathematical concept employed in understanding such randomness is probability. Mathematical statistics (also called statistical theory) is the branch of applied mathematics that uses probability theory and analysis to examine the theoretical basis of statistics.

The use of any statistical method is valid only when the system or population under consideration satisfies the basic mathematical assumptions of the method. Misuse of statistics can produce subtle but serious errors in description and interpretation — subtle in that even experienced professionals sometimes make such errors, and serious in that they may affect social policy, medical practice and the reliability of structures such as bridges and nuclear power plants.

Even when statistics is correctly applied, the results can be difficult to interpret for a non-expert. For example, the statistical significance of a trend in the data — which measures the extent to which the trend could be caused by random variation in the sample — may not agree with one's intuitive sense of its significance. The set of basic statistical skills (and skepticism) needed by people to deal with information in their everyday lives is referred to as statistical literacy. Statistics are used to describe the characteristics of groups. These characteristics are referred to as variables. Data is gathered and recorded for each variable. Descriptive statistics can then be used to reveal the distribution of the data in each variable. Statistics is a discipline, which is concerned with:

- designing experiments and other data collection,
- summarizing information to aid understanding,
- drawing conclusions from data, and
- estimating the present or predicting the future.

The two statements at the beginning illustrate some of these points. In making predictions, Statistics uses the companion subject of Probability, which models chance mathematically and enables calculations of chance in complicated cases.

Today, statistics has become an important tool in the work of many academic disciplines such as medicine, psychology, education, sociology, engineering and physics, just to name a few. Statistics is also important in many aspects of society such as business, industry and government. Because of the increasing use of statistics in so many areas of our lives, it has become very desirable to understand and practise statistical thinking. This is important even if you do not use statistical methods directly.

Statistics is also frequently used for purposes of prediction. Prediction is based on the concept of generalizability: if enough data is compiled about a particular context (e.g., students studying writing in a specific set of classrooms), the patterns revealed through analysis of the data collected about that context can be generalized (or predicted to occur in) similar contexts. The prediction of what will happen in a similar context is probabilistic. That is, the researcher is not certain that the same things will happen in other contexts; instead, the researcher can only reasonably expect that the same things will happen.

Prediction is a method employed by individuals throughout daily life. For instance, if writing students begin class every day for the first half of the semester with a five-minute free writing exercise, then they will likely come to class the first day of the second half of the semester prepared to again free write for the first five minutes of class. The students will have made a prediction about the class content based on their previous experiences in the class: Because they began all previous class sessions with free writing, it would be probable that their next class session will begin the same way. Statistics is used to perform the same function; the difference is that precise probabilities are determined in terms of the percentage chance that an outcome will occur, complete with a range of error. Prediction is a primary goal of inferential statistics.

11.2 SOME BASIC DEFINITIONS USED IN STATISTICS

11.2.1 Constant

A quantity which can be assuming only one value is called a constant. It is usually denoted by the first letters of alphabets a, b, c.

For Example: Value of pi =3.14159 and value of e = 2.71828….

11.2.2 Variable

A quantity which can vary from one individual or object to and other is called a variable. It is usually denoted by the last letters of alphabets x, y, z.

For Example: Heights and Weights of students, Income, Temperature, No. of Children in a family etc…

11.2.3 Continuous Variable

A variable which can assume each and every value within a given range is called a continuous variable. It can occur in decimals.

For Example: Heights and Weights of students, Speed of a bus, the age of a Shopkeeper, the life time of a T.V etc…

11.2.4 Continuous Data

Data which can be described by a continuous variable is called continuous data.

For Example: Weights of 50 students in a class.

11.2.5 Discrete Variable

A variable which can assume only some specific values within a given range is called discrete variable. It cannot occur in decimals. It can occur in whole numbers.

For Example: Number of students in a class, number of flowers on the tree, number of houses in a street, number of chairs in a room etc…

11.2.6 Discrete Data

Data which can be described by a discrete variable is called discrete data.

For Example: Number of students in a college.

11.2.7 Quantitative Variable

A characteristic which varies only in magnitude from on individual to another is called quantitative variable. It can be measurable.

For Example: Wages, Prices, Heights, Weights etc…

11.2.8 QUALITATIVE VARIABLE

A characteristic which varies only in quality from one individual to another is called qualitative variable. It cannot be measured.

For Example: Beauty, Marital Status, Rich, Poor, Smell etc…

11.3 STATISTICS

Bowley defines Statistics as ***"Numerical statement of facts in any department of enquiry placed in relation to each other."***

Also Bowely himself defines Statistics in the following three different ways:

- Statistics may be called the science of counting.
- Statistics may rightly be called the science of averages.
- Statistics is the science of the measurements of Social organism, regarded as a whole in all its manifestations.

Webster defines Statistics as *"classified facts representing the conditions of the people in a state .... especially those facts which can be stated in number or in any other tabular or classified arrangement."*

Professor Horace Secrist define as follows: *"By Statistics we mean aggregates of Facts affected to a marked extent by multiply of cause numerically expressed enumerated or estimated according to reasonable standards of accuracy, collected in a systematic manner for a pre-determined purpose and placed in relation to each other."*

Boddingtons defines: *"Statistics is the science of estimates and probabilities."*

Another mathematician King defines: *"The science of statistics is the methods of judging collective, natural or social phenomenon from the results obtained from the analysis or enumeration or collection of estimates."*

Lovet also said about the Statistics as: *"Statistics is the science which deals with collection, classification and tabulation of numerical facts as the basis for explanation description and comparison of Phenomenon."*

At end we can say the best definition is coated by the ***Croxton*** *and* ***Cowden****. They said: "It is the science which deals with the collection, analysis and interpretation of numerical data."*

11.4 HISTORY OF STATISTICS

The word *statistics* ultimately derives from the New Latin term *statisticum collegium* ("council of state") and the Italian word *statista* ("statesman" or "politician"). The German *Statistik*, first introduced by Gottfried Achenwall (1749), originally designated the analysis of data about the state, signifying the "science of state" (then called *political arithmetic* in English). It acquired the meaning of the collection and classification of data generally in the early 19th century. It was introduced into English by Sir John Sinclair.

Thus, the original principal purpose of *Statistik* was data to be used by governmental and (often centralized) administrative bodies. The collection of data about states and localities continues, largely through national and international statistical services. In particular, censuses provide regular information about the population.

The mathematical methods of statistics emerged from probability theory, which can be dated to the correspondence of Pierre de Fermat and Blaise Pascal (1654). Christiaan Huygens (1657) gave the earliest known scientific treatment of the subject. Jakob Bernoulli's *Ars Conjectandi* (posthumous, 1713) and Abraham de Moivre's *Doctrine of Chances* (1718) treated the subject as a branch of mathematics.[1] In the modern era, the work of Kolmogorov has been instrumental in formulating the fundamental model of Probability Theory, which is used throughout statistics.

The theory of errors may be traced back to Roger Cotes's *Opera Miscellanea* (posthumous, 1722), but a memoir prepared by Thomas Simpson in 1755 (printed 1756) first applied the theory to the discussion of errors of observation. The reprint (1757) of this memoir lays down the axioms that positive and negative errors are equally probable, and that there are certain assignable limits within which all errors may be supposed to fall; continuous errors are discussed and a probability curve is given.

Pierre-Simon Laplace (1774) made the first attempt to deduce a rule for the combination of observations from the principles of the theory of probabilities. He represented the law of probability of errors by a curve. He deduced a formula for the mean of three observations. He also gave (1781) a formula for the law of facility of error (a term due to Lagrange, 1774), but one which led to unmanageable equations. Daniel Bernoulli (1778) introduced the principle of the maximum product of the probabilities of a system of concurrent errors.

The method of least squares, which was used to minimize errors in data measurement, was published independently by Adrien-Marie Legendre (1805), Robert Adrain (1808), and Carl Friedrich Gauss (1809). Gauss had used the method in his famous 1801 prediction of the location of the dwarf planet Ceres. Further proofs were given by Laplace (1810, 1812), Gauss (1823), James Ivory (1825, 1826), Hagen (1837), Friedrich Bessel (1838), W. F. Donkin (1844, 1856), John Herschel (1850), and Morgan Crofton (1870).

Other contributors were Ellis (1844), De Morgan (1864), Glaisher (1872), and Giovanni Schiaparelli (1875). Peters's (1856) formula for r, the probable error of a single observation, is well known.

In the nineteenth century authors on the general theory included Laplace, Sylvestre Lacroix (1816), Littrow (1833), Richard Dedekind (1860), Helmert (1872), Hermann Laurent (1873), Liagre, Didion, and Karl Pearson. Augustus De Morgan and George Boole improved the exposition of the theory.

Adolphe Quetelet (1796-1874), another important founder of statistics, introduced the notion of the "average man" (*l'homme moyen*) as a means of understanding complex social phenomena such as crime rates, marriage rates, or suicide rates.

11.4.1 Where Statistics Stands Today

During the 20th century, the creation of precise instruments for public health concerns (epidemiology, biostatistics, etc.) and economic and social purposes (unemployment rate, econometry, etc.) necessitated substantial advances in statistical practices: the Western welfare states developed after World War I had to possess specific knowledge of the "population".

Today the use of statistics has broadened far beyond its origins as a service to a state or government. Individuals and organizations use statistics to understand data and make informed decisions throughout the natural and social sciences, medicine, business, and other areas. Statistics is generally regarded not as a subfield of mathematics but as a distinct, albeit allied, field. Many universities maintain separate mathematics and statistics departments. Statistics is also taught in departments as diverse as psychology, education, and public health.

11.5 IMPORTANCE OF STATISTICS IN DIFFERENT FIELDS

Statistics plays a vital role in every fields of human activity. Statistics has important role in determining the existing position of per capita income, unemployment, population growth rate, housing, schooling medical facilities etc…in a country. Now statistics holds a central position in almost every field like Industry, Commerce, Trade, Physics, Chemistry, Economics, Mathematics, Biology, Botany, Psychology, Astronomy etc…, so application of statistics is very wide. Now we discuss some important fields in which statistics is commonly applied.

11.5.1 Business

Statistics play an important role in business. A successful businessman must be very quick and accurate in decision making. He knows that what his customers wants, he should therefore, know what to produce and sell and in what quantities. Statistics helps businessman to plan production according to the taste of the costumers, the quality of the products can also be checked more efficiently by using statistical methods. So all the activities of the businessman based on statistical information. He can make correct decision about the location of business, marketing of the products, financial resources etc…

11.5.2 Economics

Statistics play an important role in economics. Economics largely depends upon statistics. National income accounts are multipurpose indicators for the economists and administrators. Statistical methods are used for preparation of these accounts. In economics research statistical methods are used for collecting and analysis the data and testing hypothesis. The relationship between supply and demands is studies by statistical methods, the imports and exports, the inflation rate, the per capita income are the problems which require good knowledge of statistics.

11.5.3 Mathematics

Statistical plays a central role in almost all natural and social sciences. The methods of natural sciences are most reliable but conclusions draw from them are only probable, because they are based on incomplete evidence. Statistical helps in describing these measurements more precisely. Statistics is branch of applied mathematics. The large number of statistical methods like probability averages, dispersions, estimation etc... is used in mathematics and different techniques of pure mathematics like integration, differentiation and algebra are used in statistics.

11.5.4 Banking

Statistics play an important role in banking. The banks make use of statistics for a number of purposes. The banks work on the principle that all the people who deposit their money with the banks do not withdraw it at the same time. The bank earns profits out of these deposits by lending to others on interest. The bankers use statistical approaches based on probability to estimate the numbers of depositors and their claims for a certain day.

11.5.5 State Management

Statistics is essential for a country. Different policies of the government are based on statistics. Statistical data are now widely used in taking all administrative decisions. Suppose if the government wants to revise the pay scales of employees in view of an increase in the living cost, statistical methods will be used to determine the rise in the cost of living. Preparation of federal and provincial government budgets mainly depends upon statistics because it helps in estimating the expected expenditures and revenue from different sources. So statistics are the eyes of administration of the state.

11.5.6 Accounting and Auditing

Accounting is impossible without exactness. But for decision making purpose, so much precision is not essential the decision may be taken on the basis of approximation, know as statistics. The correction of the values of current asserts is made on the basis of the purchasing power of money or the current value of it.

In auditing sampling techniques are commonly used. An auditor determines the sample size of the book to be audited on the basis of error.

11.5.7 Natural and Social Sciences

Statistics plays a vital role in almost all the natural and social sciences. Statistical methods are commonly used for analyzing the experiments results, testing their significance in Biology, Physics, Chemistry, Mathematics, Meteorology, Research chambers of commerce, Sociology, Business, Public Administration, Communication and Information Technology etc...

11.5.8 Astronomy

Astronomy is one of the oldest branch of statistical study, it deals with the measurement of distance, sizes, masses and densities of heavenly bodies by means of observations. During these measurements errors are unavoidable so most probable measurements are founded by using statistical methods.

Example: This distance of moon from the earth is measured. Since old days the astronomers have been statistical methods like method of least squares for finding the movements of stars.

11.6 VARIOUS CHARACTERISTICS OF STATISTICS

1. Statistics are aggregates of facts.
2. Statistics are numerically expressed.
3. Statistics are affected to a marked extent by multiplicity of causes.
4. Statistics are enumerated or estimated according to a reasonable standard of accuracy.
5. Statistics are collected for a predetermine purpose.
6. Statistics are collected in a systemic manner.
7. Statistics must be comparable to each other.

11.7 USES OF STATISTICS

1. Statistics helps in providing a better understanding and exact description of a phenomenon of nature.
2. Statistical helps in proper and efficient planning of a statistical inquiry in any field of study.
3. St atistical helps in collecting an appropriate quantitative data.
4. Statistics helps in presenting complex data in a suitable tabular, diagrammatic and graphic form for an easy and clear comprehension of the data.
5. Statistics helps in understanding the nature and pattern of variability of a phenomenon through quantitative obersevations.
6. Statistics helps in drawing valid inference, along with a measure of their reliability about the population parameters from the sample data.

11.8 VARIOUS LIMITATIONS OF STATISTICS

1. Statistics laws are true on average. Statistics are aggregates of facts. So single observation is not a statistics, it deals with groups and aggregates only.
2. Statistical methods are best applicable on quantitative data.
3. Statistical cannot be applied to heterogeneous data.
4. It sufficient care is not exercised in collecting, analyzing and interpretation the data, statistical results might be misleading.
5. Only a person who has an expert knowledge of statistics can handle statistical data efficiently.
6. Some errors are possible in statistical decisions. Particularly the inferential statistics involves certain errors. We do not know whether an error has been committed or not.

11.9 STATISTICS — AN EXPERIMENT AND OBSERVATION

A common goal for a statistical research project is to investigate causality, and in particular to draw a conclusion on the effect of changes in the values of predictors or independent variables on response or dependent variables. There are two major types of causal statistical studies, experimental studies and observational studies. In both types of studies, the effect of differences of an independent variable (or variables) on the behavior of the dependent variable are observed. The difference between the two types is in how the study is actually conducted. Each can be very effective.

An experimental study involves taking measurements of the system under study, manipulating the system, and then taking additional measurements using the same procedure to determine if the manipulation may have modified the values of the measurements. In contrast, an observational study does not involve experimental manipulation. Instead data are gathered and correlations between predictors and the response are investigated.

An example of an experimental study is the famous Hawthorne studies which attempted to test changes to the working environment at the Hawthorne plant of the Western Electric Company. The researchers were interested in whether increased illumination would increase the productivity of the assembly line workers. The researchers first measured productivity in the plant then modified the illumination in an area of the plant to see if changes in illumination would affect productivity. As it turns out, productivity improved under all the experimental conditions (see Hawthorne effect). However, the study is today heavily criticized for errors in experimental procedures, specifically the lack of a control group and blindedness.

An example of an observational study is a study which explores the correlation between smoking and lung cancer. This type of study typically uses a survey to collect observations about the area of interest and then perform statistical analysis. In this case, the researchers would collect observations of both smokers and non-smokers and then look at the number of cases of lung cancer in each group.

The basic steps for an experiment are to:

1. plan the research including determining information sources, research subject selection, and ethical considerations for the proposed research and method,
2. design the experiment concentrating on the system model and the interaction of independent and dependent variables,
3. summarize a collection of observations to feature their commonality by suppressing details (descriptive statistics),
4. reach consensus about what the observations tell us about the world we observe (statistical inference),
5. document and present the results of the study.

11.10 VARIOUS LEVELS OF MEASUREMENT IN STATISTICS

There are four types of measurements or levels of measurement or measurement scales used in statistics:

1. Nominal
2. Ordinal
3. Interval
4. Ratio

They have different degrees of usefulness in statistical research. Ratio measurements have both a zero value defined and the distances between different measurements defined; they provide the greatest flexibility in statistical methods that can be used for analyzing the data. Interval measurements have meaningful distances between measurements defined, but have no meaningful zero value defined (as in the case with IQ measurements or with temperature measurements in Fahrenheit). Ordinal measurements have imprecise differences between consecutive values, but have a meaningful order to those values. Nominal measurements have no meaningful rank order among values.

Since variables conforming only to nominal or ordinal measurements cannot be reasonably measured numerically, sometimes they are called together as categorical variables, whereas ratio and interval measurements are grouped together as quantitative or continuous variables due to their numerical nature.

11.11 SOME STATISTICAL TESTS AND PROCEDURES

1. **Analysis of Variance (Anova):** Analysis of variance (ANOVA) is a collection of statistical models, and their associated procedures, in which the observed variance is partitioned into components due to different explanatory variables. In its simplest form ANOVA gives a statistical test of whether the means of several groups are all equal, and therefore generalizes Student's two-sample t-test to more than two groups.
2. **Chi-square Test:** A chi-square test (also chi-squared or ÷2 test) is any statistical hypothesis test in which the sampling distribution of the test statistic is a chi-square distribution when the null hypothesis is true, or any in which this is asymptotically true, meaning that the sampling distribution (if the null hypothesis is true) can be made to approximate a chi-square distribution as closely as desired by making the sample size large enough.
3. **Correlation:** Correlation (often measured as a correlation coefficient, ñ) indicates the strength and direction of a linear relationship between two random variables. In general statistical usage, correlation or co-relation refers to the departure of two random variables from independence. In this broad sense there are several coefficients, measuring the degree of correlation, adapted to the nature of the data.
4. **Factor Analysis:** Factor analysis is a statistical method used to describe variability among observed variables in terms of fewer unobserved variables called factors. The observed variables are modeled as linear combinations of the factors, plus "error" terms. The information gained about the interdependencies can be used later to reduce the set of variables in a dataset. Factor analysis originated in psychometrics, and is used in behavioral sciences, social sciences, marketing, product management, operations research, and other applied sciences that deal with large quantities of data.

5. **Mann-whitney U:** the Mann–Whitney U test (also called the Mann–Whitney–Wilcoxon (MWW), Wilcoxon rank-sum test, or Wilcoxon–Mann–Whitney test) is a non-parametric test for assessing whether two independent samples of observations come from the same distribution. It is one of the best-known non-parametric significance tests. It was proposed initially by Frank Wilcoxon in 1945, for equal sample sizes, and extended to arbitrary sample sizes and in other ways by H. B. Mann and Whitney (1947).

6. **Pearson Product-moment Correlation Coefficient:** the Pearson product-moment correlation coefficient (sometimes referred to as the PMCC, and typically denoted by r) is a measure of the correlation (linear dependence) between two variables X and Y, giving a value between +1 and "1 inclusive. It is widely used in the sciences as a measure of the strength of linear dependence between two variables. It was first introduced by Francis Galton in the 1880s, and is named after Karl Pearson. The correlation coefficient is sometimes called "Pearson's r."

7. **Regression Analysis:** Regression analysis refers to techniques for modeling and analyzing several variables, when the focus is on the relationship between a dependent variable and one or more independent variables. Regression analysis helps us understand how the typical value of the dependent variable changes when any one of the independent variables is varied, while the other independent variables are held fixed. Regression analysis estimates the conditional expectation of the dependent variable given the independent variables — that is, the average value of the dependent variable when the independent variables are held fixed. Less commonly, the focus is on a quantile, or other location parameter of the conditional distribution of the dependent variable given the independent variables. In all cases, the estimation target is a function of the independent variables called the regression function. In regression analysis, it is also of interest to characterize the variation of the dependent variable around the regression function, which can be described by a probability distribution.Regression analysis is widely used for prediction (including forecasting of time-series data). Regression models for prediction are often useful even when the assumptions are moderately violated, although they may not perform optimally.

8. **Spearman's Rank Correlation Coefficient:** Spearman's rank correlation coefficient or Spearman's rho, named after Charles Spearman and often denoted by the Greek letter ρ (rho) or as r_s, is a non-parametric measure of correlation – that is, it assesses how well an arbitrary monotonic function could describe the relationship between two variables, without making any other assumptions about the particular nature of the relationship between the variables.

9. **Student's T-test:** A t-test is any statistical hypothesis test in which the test statistic follows a Student's t distribution if the null hypothesis is true. It is most commonly applied when the test statistic would follow a normal distribution if the value of a scaling term in the test statistic were known. When the scaling term is unknown and is replaced by an estimate based on the data, the test statistic (under certain conditions) follows a Student's t-distribution.

10. **Time Series Analysis:** Signal processing, and many other fields, a time series is a sequence of data points, measured typically at successive times, spaced at (often uniform) time intervals. Time series analysis comprises methods that attempt to understand such time series, often either to understand the underlying context of the data points (Where did they come from? What generated them?), or to make forecasts (predictions). Time series forecasting is the use of a model to forecast future events based on known past events: to forecast future data points before they are measured. A standard example in econometrics is the opening price of a share of stock based on its past performance.

11.12 VARIOUS SPECIALIZED DISCIPLINE

Some fields of inquiry use applied statistics so extensively that they have specialized terminology.

These disciplines include:

1. Actuarial science
2. Biostatistics
3. Business statistics
4. Data mining (applying statistics and pattern recognition to discover knowledge from data)
5. Economic statistics (Econometrics)
6. Engineering statistics
7. Geography and Geographic Information Systems, more specifically in Spatial analysis
8. Demography
9. Psychological statistics
10. Quality
11. Social statistics (for all the *social* sciences)
12. Statistical literacy
13. Statistical surveys
14. Process analysis and chemometrics (for analysis of data from analytical chemistry and chemical engineering)
15. Reliability engineering
16. Image processing
17. Statistics in various sports, particularly baseball and cricket
18. In Nano technology

Statistics form a key basis tool in business and manufacturing as well. It is used to understand measurement systems variability, control processes (as in statistical process control or SPC), for summarizing data, and to make data-driven decisions. In these roles it is a key tool, and perhaps the only reliable tool.

11.13 STATISTICAL PACKAGES (SOFTWARES)

The rapid and sustained increases in computing power starting from the second half of the 20th century have had a substantial impact on the practice of statistical science. Early statistical models were almost always from the class of linear models, but powerful computers, coupled with suitable numerical algorithms, caused a resurgence of interest in nonlinear models (especially neural networks and decision trees) and the creation of new types, such as generalised linear models and multilevel models. The popular statistical softwares are:

1. SPSS
2. SAS
3. WINKS SDA
4. EXCEL etc.

The computer revolution has implications for the future of statistics, with a new emphasis on "experimental" and "empirical" statistics and a large number of statistical packages.

11.14 STATISTICAL COMPUTING

The rapid and sustained increases in computing power starting from the second half of the 20th century have had a substantial impact on the practice of statistical science. Early statistical models were almost always from the class of linear models, but powerful computers, coupled with suitable numerical algorithms, caused a resurgence of interest in nonlinear models (especially neural networks and decision trees) and the creation of new types, such as generalised linear models and multilevel models.

Increased computing power has also led to the growing popularity of computationally-intensive methods based on resampling, such as permutation tests and the bootstrap, while techniques such as Gibbs sampling have made Bayesian methods more feasible. The computer revolution has implications for the future of statistics, with a new emphasis on "experimental" and "empirical" statistics. A large number of both general and special purpose statistical packages are now available to practitioners.

QUERIES

1. What do you mean by Statistics?
2. Define Statistics?
3. Explain historical development of statistics.
4. What is the role of Statistics today?
5. What are the various applications of Statistics?
6. What are the various characteristics of statistics?
7. What are the various uses of statistics?

8. What are the various limitations of Statistics?
9. How can you apply an experiment on Statistics?
10. What is the level of measurement?
11. What are the various Statistical tests?
12. What is the role of Statistics in different discipline?
13. Name the various software used for Statistics?
14. Why statistics is used?

CHAPTER 12

Measures of Central Tendency

12.1 INTRODUCTION

The term central tendency refers to the "middle" value or perhaps a typical vale of the data and is measured using mean, median and mode. Each of these measures calculated differently, and one that is best to use depends on the situation. The objective of statistical analysis is to find various numerical measures, which describe the characteristics of a frequency distribution. One of such measure is measure of central tendency. The measures of central tendency are the measures, which represent a huge amount of numerical data by a single numerical value, which is capable of representing the characteristics of the whole distribution.

According to Prof. R.A. Fisher, *"The inherent inability of the human mind to grasp in its entirety a large body of numerical data compels us to seek relatively few constants that will adequately describe the data."*

12.2 MEASURES OF CENTRAL TENDENCY

Measure of central tendency is defined as "***method of finding the average value of any statistical series***".

Here "***measure***" means methods and the "***central tendency***" means the average of a series. It is based on quantitative information. It is also called as ***statistical averages***.

In mathematics, an **average, central tendency** of a data set is a measure of the "middle" or "expected" value of the data set.

There are many different descriptive statistics that can be chosen as a measurement of the central tendency of the data items. These include arithmetic mean, the median and the mode. Other statistical measures such as the standard deviation and the range are called measures of spread and describe how spread out the data is.

An average is a single value that is meant to typify a list of values. If all the numbers in the list are the same, then this number should be used. If the numbers are not the same, an easy way to get a representative value from a list is to randomly pick any number from the list. However, the word 'average' is usually reserved for more sophisticated methods that are generally found to be more useful. In the latter case, the average is calculated by combining the values from the set in a specific way and computing a single number as being the average of the set.

The most common method is the arithmetic mean but there are many other types of central tendency, such as median, mode etc.

There are many different descriptive statistics that can be chosen as a measurement of the central tendency. The most common method, and the one generally referred as *the average*, is the arithmetic mean. According to A.L. Bowley, ***"Statistics may rightly be called as the science of averages."*** This average is also known as central value. Through statistical analysis we condense the whole data to a single value, which represents the data in true sense.

Thus averages are the values which lie between two extreme values; and it is the reason that these values are known as "Measures of Central Tendency" or "Measures of Location". These measures are also known as Averages.

Definitions :

Measures of Central Tendency or Averages have been defined differently by different statisticians.

According to the Professor Bowley, *" Statistical constants which enables us to comprehend in a single effort the significance of the whole."*

In the words of Croxton and Cowden, *"An average value is a single value within the range of the data that is used to represent all the values in the series. Since an average is somewhere within the range of the data it is sometimes called a measure of central value."*

According to the Murry R. Speigal, *"Average is a value which is typical or representative of a set of data."*

According to the Simpson and Kafka, *"A measure of central tendency is a typical value around which other figures congregate."*

According to the Clark, *"Average is an attempt to find one single figure to describe whole of figures."*

Ya-Luo-Chou, a Chinese statistician says about the statistics, *" An average is a typical value in the sense that it is sometimes employed to represent all the individual values in a series of a variable."*

According to A.E. Waugh, *"An average stands for the whole group of which it forms a part, yet represents the whole."*

Thus the basic purpose of the statistical analysis is to determine the value which represents the whole series. This value is termed as central value or an average. In other words, average is a statistical measure representing a group of individual values in simple and comprehensive manner.

12.2.1 Why We Need Measure of Central Tendency

Central Tendency is used for the computations due to the following reasons:

1. It is very simple to compute.
2. It is very easy for a person to learn the central tendency formulas.
3. It is very difficult to enquire the facts collected by statisticians and to satisfy the query done by any person or layman without using these measures.
4. It is very difficult to remove confusion about the particular topic until the collected facts are computed through the central tendency.
5. Collected facts can be computed through central tendency and presented for the further processing.
6. It is very easy to use and apply for the research work.
7. Only the values computed through the central tendency give a clear picture for the conclusion.
8. Comparisons can be done very easily and a decision can be taken.

12.2.2 Main Objectives of Measure of Central Tendency

Following are the major objectives to enhance the performance of central tendency:

1. It removes the complexity from the data.
2. Comparitive study becomes easier.
3. To present a huge data in the simple form.
4. Final results can be used for further purposes.
5. More statistical methods and analysis can be applied on the computed data.

12.2.3 Features or Characteristics of a Measure of Central Tendency

There are different types of characteristics associated with different types of central tendencies. Some of the major characteristics are discussed as follows:

1. It should be rigidly defined i.e. its definition should be clear and unambiguous.
2. It should be simple to understand and calculate.
3. It should be based on all the experimental observations.
4. It should be capable of further algebraic treatment.
5. It should be least affected by small fluctuations of sampling.
6. It should not be affected by the extreme observations, i.e. very small or very large observations.

12.2.4 Advantages of Measure of Central Tendency

Following are the advantages of measure of central tendency:

12

1. Complexity of the data removed by the central tendency. In other words, it simplifies the complexity of data. As data collected from various resources and sources is complex, the job of central tendency is to compute and present in the simple form.
2. It increases the understandability about the data.
3. It makes the comparative study in simpler form.
4. It enhances the efficiency for the processing of data.
5. The presentation of data in a single form is easier to understand and use for further processing and analyzing some more statistical methods.
6. By using the central tendency, it is very easy to remember the single figures.
7. Processing of computation becomes fast for further operations.

12.2.5 Disadvantages of Measure of Central Tendency

Following are the disadvantages of measure of central tendency:

1. Results fluctuate if value is not within the range.
2. If a value does not belong to the series, then it will raise the given figure and results will change.
4. Central tendency gives dangerous and wrong results, if major changes occur in the data.
5. In some cases, results are different and useless.
6. Some times, it does not give central idea.

12.2.6 Types of Central Tendency or Averages

Average most often refers to the arithmetic mean, but is actually ambiguous and may be used to also refer to the mode, median, or midrange. You should always clarify which average is being used, preferably by using a more specific term. Averages give us information about a typical element of a data set. They are **measures of central tendency**. There are different types of central tendencies or averages used for same purpose in different ways.

Definition: *"A single value which can represent the whole set of data is called an average". If the average tends to lie or indicating the center of the distribution is called measure of central tendency or sometimes they locate the general position of the data, so they are also called measure of location.*

Mainly two types of averages are used in the Statistics.

These are discussed in detail as follows:

1. Mathematical Averages
2. Positional Averages

1. **Mathematical Averages:** Such types of averages are Mathematical Averages because their values are computed by using some mathematical equations or formulas. Further Mathematical Averages are subdivided into three categories as given below:
   (i) Arithmetic Mean
   (ii) Geometric Mean
   (iii) Harmonic Mean
2. **Positional Averages:** Such type of averages are also called as Positional Averages or Positional Point because their values are either computed from their central point from its position in series or we can say determined by locating their positions in the series of the variables. Further positional averages are sub divided into three categories written as below:
   (i) Median
   (ii) Mode
   (iii) Quantile

12.2.7 Objectives and Functions of Averages or Central Value

1. The human mind cannot retain all details of large number of activities and their inter-relations, so averages are a must. An average represents all the features of a group; hence the results about the whole group can be deduced from it.
2. An average gives us simple and brief description of the main features of the whole data.
3. The measures of central tendency or averages reduce the data to a single value which is highly useful for making comparative studies.
4. Averages help to develop a business in case of a firm or help the economy of a country to develop. For example, in case of an aviation company, the management will be interested to know about the average number of persons boarding plane on the desired certain route. In such a case, a finance minister or finance secretary would apply some economic measures to increase per capital income if he feels that it is lower as compared to other developed countries per capital income.
5. Other statistical devices such as mean deviation co-efficient of variation, co-relation, analysis of time series and index numbers are also based on the averages; and hence the use of averages becomes compulsory.

12.2.8 Essentials of a Good Average

1. Only those averages, where all the data are used give best result, whereas the averages which use less data are not representative of the whole group. Averages are not satisfactory where whole of the data are not used.
2. No term should affect the average too much. If one or two very small or very large items unduly affect the average then the average cannot be really typical of the entire group. Thus extreme terms may distort the average and reduce its usefulness.

3. There should be no confusion about the meaning or description of an average. It must have a rigid or to the point definition. The best thing is that the average should be defined by algebraic formula and the person prejudiced or biased of the investigator should not lead to ill-defining of the average.
4. Measures of central tendency are used in many other techniques of statistical analysis like measures of Dispersion, Correlation etc. This purpose can be solved only if measures of central tendency or average are capable of further algebraic treatment.
5. That average is considered a good average which can be found by arithmetic as well as by graphic method.
6. A good average will be least affected by sampling fluctuations. If a few samples are taken from the same universe, the average should be such as has the least variation in values derived in the individual samples. The results obtained will be considered to be the true representative of the universe in this case.
7. We will not call an average good one if it is affected by skewness present in the distribution.
8. A popular average which is known to common people will be more useful as a simple person will be able to understand it. Otherwise its use will be limited to higher section of people only.
9. A good average should have a clear and stable definition.

   There are different kinds of calculations for central tendency, the kind of calculation that should be used depends on the type of data (or level of measurement) and purpose for which the central tendency is being calculated.

QUERIES

1. Define measure of central tendency. Discuss the desirable characteristics of central tendency.
2. Discuss the measures of central tendency? Write its merits and demerits?
3. What are the various goals of central tendency?
4. What are the various features or characteristics of an ideal Average?
5. What are the various pros and cons of Central Tendency?
6. Write a critical note on the measure of central tendency.
7. Define and explain relative uses of the various measures of central tendency?
8. What are the various mathematical averages? Explain all?
9. What are the objectives and functions of averages or central value?
10. What are the essentials of a good average?

12.3 ARITHMETIC MEAN

In statistics, the **arithmetic mean** (or simply the **mean**) of a list of numbers is the sum of the entire list divided by the number of items in the list. If the list is a statistical population, then the mean of that population is called a **population mean.** If the list is a statistical sample, we call the resulting statistic a **sample mean**.

"Arithmetic mean is quotient of sum of the given values and number of the given values".

The mean is the most commonly-used type of average and is often referred to simply as the *average*. The term "mean" or "arithmetic mean" is preferred in mathematics and statistics to distinguish it from other averages such as the median and the mode.

It is denoted by $\bar{X}$ (pronounced as "x-bar"). **Mean** most often refers to the **arithmetic mean**, but is also ambiguous. Unless specified otherwise, we will assume arithmetic mean whenever the term mean is used.

Definition: *"The Arithmetic Mean is obtained by summing all the elements of the data set and dividing by the number of elements."*

Let us suppose, $x_1, x_2, x_3, x_4, \ldots, x_n$ are the *n* elements, then the arithmetic mean is computed as:

$$\bar{X} = \frac{\Sigma X}{n}$$

Application of this formula to the following data:

42 35 36 45 38 38 39 39 39 40 32 40 37

yields the following results:

$$\bar{X} = \frac{\Sigma X}{N} = \frac{500}{13} = 38.29$$

Use of means as a way of describing a set of scores is fairly common; for example, batting average, bowling average, grade point average, and average points scored per game are all means. Note the use of the word "average" in all of the above terms. In most cases when the term "average" is used, it refers to the mean, although not necessarily. When a politician uses the term "average income", for example, he or she may be referring to the mean, median, or mode.

The mean is sensitive to extreme scores. For example, the mean of the following data is 39.23 somewhat larger than the preceding example by just changing the last value.

42 35 36 45 38 38 39 39 39 40 32 40 47

12.3.1 Methods of Computation

There are three methods to find mean for a frequency distribution. These methods are as follows:

1. Direct method
2. Short-cut method
3. Step deviation method

1. Direct Method

Some times it is also called hit method. It is computed directly. The computation of this method changes depending upon the type of the series, i.e. whether the series is Individual, discrete or continuous.

- *In Case of Individual Series:*

$$\bar{X} = \frac{\Sigma X}{n}$$

where n is the total number of observations.

X is the given set of observations.

- *In Case of Discrete Series:*

$$\bar{X} = \frac{\Sigma fX}{N}$$

where $N = \Sigma f$ Sum of all the frequencies.

X is the given set of observations.

- *In Case of Continuous Series:*

$$\bar{X} = \frac{\Sigma fm}{N}$$

where $N = \Sigma f$ Sum of all the frequencies.

m are the mid points of the classes.

2. Shortcut Method

This method is also called deviation method. Another name of this method is assumptive method. Here we assume a mean from the middle position. Further from the assumed mean, a deviation is computed on the whole series and then all the related deviations are added to the assumed mean. That is the final result. The computation of this method changes according to the given series.

- *In Case of Individual Series:*

$$\bar{X} = A + \frac{\Sigma d}{n}$$

where n is the total number of observations.

d is the deviation of X from the assumed mean, say A.

- *In Case of Discrete Series:*

$$\bar{X} = A + \frac{\Sigma fd}{N}$$

where $N = \Sigma f$ is the sum of all the frequencies.

d is the deviation of X from the assumed mean, say A.

- *In Case of Continuous Series:*

$$\bar{X} = A + \frac{\Sigma fd}{N}$$

where $N = \sum f$ is the sum of all the frequencies.

m are the mid points of the classes.

d is the deviation of m from the assumed mean, say A.

3. Step Deviation Method

This method is also called incremented method. It is the incremental form of short-cut method, i.e. one step advance of the short-cut method. This method is best if the deviation has big value. In this method all the figures of the deviations are reduced by dividing them all by a common factor. The computation of this method changes according to the given series.

- *In Case of Individual Series:*

$$\overline{X} = A + \frac{\sum d'}{n} \times c$$

where n is the total number of observations

d' is the deviation of X from assumed mean A

c is the common factor

- *In Case of Discrete Series:*

$$\overline{X} = A + \frac{\sum fd'}{N} \times c$$

where N is the total of the frequencies

d' is the deviation of X from the assumed mean A

c is the common factor

- *In Case of Continuous Series:*

$$\overline{X} = A + \frac{\sum fd'}{N} \times c$$

where N is the total of the frequencies

d' is the deviation of midpoints m from the assumed mean A

c is the common factor

SOLVED PROBLEMS

Problem 1: Find the arithmetic mean (average) of the following numbers: 9, 3, 7, 3, 8, 10, and 2.

Solution: Add up all the numbers. Then divide by 7 because there are 7 different numbers. 0

$$\frac{9+3+7+3+8+10+2}{7} = \frac{42}{7} = 6$$

Problem 2: Find the arithmetic mean of –4, 3, 18, 0, 0, and – 10.

Solution: Sum the numbers. Divide by 6 because there are 6 numbers.

The answer is 7/6, or 1.167

Problem 3: The one-sided train fare of five selected BS students is recorded as follows($) : 10, 5, 15, 8 and 12. Calculate arithmetic mean of the following data.

Solution: Let train fare is indicated by x, then

x($)
10
5
15
8
12
$\Sigma x = 50$

Arithmetic mean of $X = \bar{X} = \frac{\Sigma x}{n}$, we decide to use above-mentioned formula. Form the given data, we have $\Sigma x = 50$ and $n = 5$. Placing these two quantities in above formula, we get the arithmetic mean for given data.

$$\bar{X} = \frac{50}{5} = \$10$$

Problem 4: Given the following frequency distribution of first year students of a particular college.

Age (Years):	13	14	15	16	17
Number of Students:	2	5	13	7	3

Solution: The given distribution belongs to a grouped data and the variable involved is ages of first year students. While the number of students Represent frequencies.

Ages (Years)	Number of Students	
x	f	fx
13	2	26
14	5	70
15	13	195
16	7	112
17	3	51
Total	$\Sigma f = 30$	$\Sigma fx = 545$

Now we will find the Arithmetic Mean as $\bar{X} = \frac{\Sigma fx}{\Sigma f} = \frac{454}{30} = 15.13$ years.

Problem 5: You are given the following data:

Year:	2001	2002	2003	2004	2005
Profit:	50	30	40	70	60

Compute the arithmetic mean of profits per year with different ways.

Solution:

1. Using direct method:

Year	Profit 'X'
2001	50
2002	30
2003	40
2004	70
2005	60
$n = 5$	$\Sigma X = 250$

We get, $$\bar{X} = \frac{\Sigma X}{n} = \frac{250}{5} = 50$$

So the arithmetic mean or average profit is 50.

2. Using Short-cut method:

Year	Profit 'X'	$d = (X–A)$
2001	50	10
2002	30	–10
2003	40	0
2004	70	20
2005	60	10
		$\Sigma d = 50$

Here Assumed mean $A = 40$ So we get,

$$\bar{X} = A + \frac{\Sigma d}{n} = 40 + \frac{50}{5} = 50$$

So the arithmetic mean or average profit is 50.

3. Using Step Deviation method:

Year	Profit 'X'	$d = (X–A)$	d' ($c = 10$)
2001	50	10	1
2002	30	–10	–1
2003	40	0	0
2004	70	30	3
2005	60	20	2
			$\Sigma d' = 5$

Let us take the assumed mean $A = 40$

Here $N = 5$

So we get,

$$\overline{X} = A + \frac{\Sigma d'}{n} \times c = 40 + \frac{5}{5} \times 10 = 50$$

So the arithmetic mean or average profit is 50.

III. Having Continuous Series:

Problem 6: The following data shows distance covered by 100 persons to perform their routine jobs.

Distance (Km):	0-10	10-20	20-30	30-40
Number of Persons:	10	20	40	30

Calculate Arithmetic Mean by Step-Deviation Method; also explain why it is better than direct method in this particular case.

Solution: The given distribution belongs to a grouped data and the variable involved is ages of "distance covered". While the "number of persons" Represent frequencies.

Distance Covered in (Km)	Number of Persons f	Mid Points x	$u = \left(\frac{x-5}{10}\right)$	fu
0-10	10	5	-1	-10
10-20	20	15	0	0
20-30	40	25	1	40
30-40	30	35	2	60
Total	$\Sigma f = 100$			$\Sigma fu = 90$

Now we will find the Arithmetic Mean as $\overline{X} = A + \frac{\Sigma fu}{\Sigma f} \times h$

Where

$A = 15, \Sigma fu = 90, \Sigma f = 100$ and $h = 10$

$$\overline{X} = 15 + \frac{90}{100} \times 10 = 24 \text{ Km}$$

Explanation:

Here from the mid points (x) it is very much clear that each mid point is multiple of 5 and there is also a gap of 10 from mid point to mid point, i.e. class size or interval (h). Keeping in view this, we should prefer to take method of Step-Deviation instead of Direct Method.

Problem 7: The following frequency distribution showing the marks obtained by 50 students in statistics at a certain college. Find the arithmetic mean using (1) Direct Method (2) Short-Cut Method (3) Step-Deviation.

Marks	20-29	30-39	40-49	50-59	60-69	70-79	80-89
Frequency	1	5	12	15	9	6	2

Solution:

			Direct Method	Short-Cut Method		Step-Deviation Method	
Marks	f	x	Fx	$D = x - A$	fD	$u = \frac{x-A}{h}$	fu
20-29	1	24.5	24.5	–30	–30	–3	–3
30-39	5	34.5	172.5	–20	–100	–2	–10
40-49	12	44.5	534.5	–10	–120	–1	–12
50-59	15	54.5	817.5	0	0	0	0
60-69	9	64.5	580.5	10	90	1	9
70-79	6	74.5	447.5	20	120	2	12
80-89	2	84.5	169.5	30	60	3	6
Total	50		274.5		20		2

(1) Direct Method:

$$\bar{X} = \frac{\Sigma fx}{\Sigma f} = \frac{2745}{50} 54.9 \text{ or } 55 \text{ marks}$$

(2) Short-Cut Method:

$$\bar{X} = A + \frac{\Sigma fdD}{\Sigma f} \quad \text{Where } A = 54.5$$

$$= 54.5 + \frac{20}{50} = 54.5 + 0.4 = 54.9$$

(3) Step-Deviation Method:

$$\bar{X} = A + \frac{\Sigma fu}{\Sigma f} \times h$$

Where $A = 54.5 \quad h = 10$

$$= 54.5 + \frac{2}{50} \times 10$$

$$= 54.5 + 0.4$$

$$= 54.9 \text{ Marks}$$

Problem 8: You are given the following data of students in a college:

Attendance:	0 – 10	10 – 20	20 – 30	30 – 40	40 – 50
No. of Students :	3	5	10	15	20

Compute the arithmetic mean of attendance per student with different ways.

12

Solution:

1. Using direct method:

Attendance	No. of Students 'f'	Mid Value — m	fm
0-10	2	5	10
10-20	8	15	120
20-30	10	25	250
30-40	15	35	525
40-50	15	45	675
	Σf or $N = 50$		$\Sigma fm = 1580$

We get,

$$\bar{X} = \frac{\Sigma fm}{N} = \frac{1580}{50} = 31.60 \cong 32$$

Hence the average attendance of student is 32 approx.

2. Using Short-cut method:

Attendance	No. of Students 'f'	Mid Values — m	$d = m - A$	fd
0-10	2	5	–20	–40
10-20	8	15	–10	–80
20-30	10	25	0	0
30-40	15	35``	10	150
40-50	15	45	20	300
	Σf or $N = 50$			$\Sigma fd = 330$

Here Assumed mean $A = 25$

So we get,

$$\bar{X} = A + \frac{\Sigma fd}{N} = 25 + \frac{330}{50} = 31.60 \cong 32$$

Hence the average attendance of students is 32 approx.

3. Using Step deviation Method:

Attendance	No. of Students 'f'	Mid Values — m	$d' = \frac{m-25}{10}$	fd'
0-10	2	5	–2	–4
10-20	8	15	–1	–8
20-30	10	25	0	0
30-40	15	35	1	15
40-50	15	45	2	30
	Σf or $N = 50$			$\Sigma fd' = 33$

So we get,

$$\bar{X} = A + \frac{\sum fd'}{N} \times c = 25 + \frac{33}{50} \times 10 = 31.60 \cong 32$$

PROGRAMS

1. Program to Find the Mean of Individual Series Using Direct Method.

```
#include<stdio.h>
#include<conio.h>
void main( )
{
float x[100],s=0.0,am;
int i,n;
clrscr( );
printf(" How many items in the individual distribution ?");
scanf("%d", &n);
printf("\n Enter the data \n");
for(i=1;i<=n;i++)
{scanf("%f",&x[i]);
s+=x[i];
}
am=s/n;
printf("\n The arithmetic mean is %7.4f",am);
getch();
}
```

Output:

```
How many items in the individual distribution? 5
Enter the data
5
20
10
30
45
The arithmetic mean is 22.000
```

2. Program to Find the Mean of Discrete Series Using Direct Method.

```
#include<stdio.h>
#include<conio.h>
void main( )
{
float x[100], sfx=0.0,am;
 int i,n,f[100],sf = 0;
clrscr( );
printf(" How many items in the discrete series?");
scanf("%d", &n);
printf("\n Enter the data: \n");
for(i=1;i<=n;i++)
scanf("%f",&x[i]);
printf("\n Enter the frequencies: \n");
for(i=1;i<=n;i++)
scanf("%f",&f[i]);
for(i=1;i<=n;i++)
{
sfx=sfx+x[i]*f[i];sf=sf+f[i];
}
am=sfx/sf;
printf("\n The arithmetic mean is %8.3f",am);
getche( );
}
```

Output:

```
How many items in the discrete series? 5
Enter the data
2000
2500
3000
3500
4000
Enter the frequencies:
5
```

```
10
15
8
2
The arithmetic mean is 2900.000
```

3. Program to Find the Mean of Continuous Series Using Direct Method.

```
#include<stdio.h>
#include<conio.h>
void main( )
{
float x[100],l[100],u[100],sfm=0.0,am;
int i,n,f[100],sf=0;
clrscr( );
printf(" How many classes in the continuous series?");
scanf("%d", &n);
printf("\n Enter the lower and upper data: \n");
for(i=1;i<=n;i++)
scanf("%f %f",&l[i],&u[i]);
printf("\n Enter the frequencies: \n");
for(i=1;i<=n;i++)
scanf("%d",&f[i]);
for(i=1;i<=n;i++)
{
m[i]=(l[i]+u[i])/2;
sfm=sfm+m[i]*f[i];sf=sf+f[i];
}
am=sfm/sf;
printf("\n The arithmetic mean is %8.2f",am);
getch( );
}
```

Output:

```
How many classes in the continuous series? 5
Enter the lower and upper data:
0 10
```

```
10 20
20 30
30 40
40 50
Enter the frequencies:
2
8
10
15
15
The arithmetic mean is 31.90
```

4. Program to Find the Mean of Individual Series Using Shortcut Method (or Indirect Method).

```
#include<stdio.h>
#include<conio.h>
void main( )
{
float x[100],sd=0,am,a;
int i,n;
clrscr( );
printf(" How many items in the individual distribution ?");
scanf("%d", &n);
printf("\n Enter the value of assumed mean");
scanf("%f",&a);
printf(" Enter the data \n");
for(i=1;i<=n;i++)
{
scanf("%f",&x[i]);
sd+=x[i]-a;
}
am=a+sd/n;
printf("\n The arithmetic mean is %7.3f",am);
getch();
}
```

```
OUTPUT:
How many items in the individual distribution? 5
Enter the value of assumed mean 50
Enter the data
50
30
40
70
60
The arithmetic mean is 50.000
```

5. Program to Find the Mean of Discrete Series Using Indirect Method or Shortcut Method.

```
#include<stdio.h>
#include<conio.h>
void main( )
{
float a,x[100],sfd=0.0,am;
int i,n,f[100],sf=0;
clrscr( );
printf("\n How many items in the discrete series?");
scanf("%d", &n);
printf("\n Enter the value of assumed mean");
scanf("%f",&a);
printf("\n Enter the values of X \n");
for(i=1;i<=n;i++)
scanf("%f",&x[i]);
printf("\n Enter the frequencies: \n");
for(i=1;i<=n;i++)
scanf("%d",&f[i]);
for(i=1;i<=n;i++)
{
sfd=sfd+(x[i]-a)*f[i];
sf=sf+f[i];
}
am=a+sfd/sf;
```

12

```
printf("\n The arithmetic mean is :  %8.3f",am);
getch( );
}
```

Output:

How many items in the discrete series? 5
Enter the value of assumed mean 2500
Enter values of X:
2000
2500
3000
3500
4000
Enter the frequencies:
5
10
15
8
2
The arithmetic mean is 2900.000

6. Program to Find the Mean of Continuous Series Using Indirect Method (or Short-cut Method).

```
#include<stdio.h>
#include<conio.h>
void main( )
{
float a,m[100],l[100],u[100], sfd=0.0,am;
int i,n,f[100],sf=0;
clrscr( );
printf(" How many classes in the continuous series?");
scanf("%d", &n);
printf("\n Enter the value of assumed mean");
scanf("%f",&a);
printf("\n Enter the lower and upper data: \n");
for(i=1;i<=n;i++)
```

```
scanf("%f %f",&l[i],&u[i]);
printf("\n Enter the frequencies: \n");
for(i=1;i<=n;i++)
scanf("%d",&f[i]);
for(i=1;i<=n;i++)
{
m[i]=(l[i]+u[i])/2;
sfd=sfd+(m[i]-a)*f[i];sf=sf+f[i];
}
am=a + sfd/sf;
printf("\n The arithmetic mean is %8.2f",am);
getch( );
}
```

Output:

```
How many classes in the continuous series? 5
Enter the value of assumed mean 25
Enter the lower and upper data:
0 10
10 20
20 30
30 40
40 50
Enter the frequencies:
2
8
10
15
15
The arithmetic mean is 31.60
```

7. Program to Find the AM of Individual Series Using Step Deviation Method.

```
#include<stdio.h>
#include<conio.h>
void main( )
{
```

```
float x[100],sd=0,am,a,d,h;
int i,n;
clrscr( );
printf(" How many items in the individual distribution ?");
scanf("%d", &n);
printf("\n Enter the value of assumed mean");
scanf("%f",&a);
printf(" Enter the value of step deviation h ");
scanf("%f",&h);
printf(" Enter the data \n")
for(i=1;i<=n;i++)
{
scanf("%f",&x[i]);
sd+=(x[i]-a)/h;
}
am=a+h*sd/n;
printf("\n The arithmetic mean of individual series is %7.2f",am);
getch( );
}
```

Output:

```
How many items in the individual distribution? 5
Enter the value of assumed mean 40
Enter the value of step deviation h 10
Enter the data
50
30
40
70
60
The arithmetic mean of individual series is 50.00
```

8. Program to Find the AM of Discrete Series Using Step Deviation Method.

```
#include<stdio.h>
#include<conio.h>
void main( )
```

```
{
float x[100],sfd=0,am,a,c;
int i,n,f[100],sf=0;
clrscr( );
printf(" How many items in the distribution ?");
scanf("%d",&n);
printf("\n Enter the value of assumed mean");
scanf("%f",&a);
printf(" Enter the value of step deviation c ");
scanf("%f",&c);
printf(" Enter the data \n")
for(i=1;i<=n;i++)
{
scanf("%f",&x[i]);
sfd+=((x[i]-a)/c)*f[i];
}
printf("\n Enter the frequencies: \n");
for(i=1;i<=n;i++)
{scanf("%d",&f[i]);
sf=sf+ f[i];
}
am=a+c*sfd/sf;
printf("\n The arithmetic mean of discrete series is %7.2f",am);
getche( );
}
```

Output:

```
How many items in the individual distribution? 5
Enter the value of assumed mean 2500
Enter the value of step deviation c 100
Enter the data
2000
2500
3000
3500
```

```
4000
Enter the frequencies:
5
10
15
8
2
The arithmetic mean of discrete series is 2900.00
```

9. Program to Find the Mean of Continuous Series Using Step Deviation Method.

```
#include<stdio.h>
#include<conio.h>
void main( )
{
float a,m[100],l[100],u[100], sfd=0.0,am,c;
int i,n,f[100],sf=0;
clrscr( );
printf(" How many classes in the continuous series?");
scanf("%d", &n);
printf("\n Enter the value of assumed mean");
scanf("%f",&a);
printf("\n Enter the step deviation c");
scanf("%f", &c);
printf("\n Enter the lower and upper data: \n");
for(i=1;i<=n;i++)
scanf("%f %f",&l[i],&u[i]);
printf("\n Enter the frequencies: \n");
for(i=1;i<=n;i++)
scanf("%d",&f[i]);
for(i=1;i<=n;i++)
{
m[i]=(l[i]+u[i])/2;
sfd=sfd+((m[i]-a)/c)*f[i];sf=sf+f[i];
}
am=a + (sfd/sf)*c;
```

```
printf("\n The arithmetic mean is %8.2f",am);
getch( );
}
```

Output:

```
How many classes in the continuous series? 5
Enter the value of assumed mean 25
Enter the lower and upper data:
0 10
10 20
20 30
30 40
40 50
Enter the frequencies:
2
8
10
15
15
The arithmetic mean is 31.60
```

12.3.2 Properties of Arithmetic Mean

Some properties of arithmetic mean

1. If x_1 and x_2 are the means of the two groups computed from the values n_1 and n_2 then the mean x is given by the formula

$$x = n_1x_1 + n_2x_2/n_1 + n_2$$

2. If each observation in the data is replaced by x, the sum total of all the observations remains unchanged.

$$x = x_1, x_2, x_3, x_4........, x_n/n$$

So $$x_1, x_2, x_3, x_4........, x_n = nx$$

Replacing each observation by x, we get

$$x + x + x........+ x = nx$$

3. If every value of the variable x is either increased, decreased, divided or multiplied by a constant, the observations so obtained also increases, decreases, gets multiplied or gets divided respectively by the same constant.
4. Algebraic sum of the deviation of a set of values from their arithmetic mean is 0.

SOLVED PROBLEMS

Problem 1: Find the sum of the deviations of the variate values 3, 4, 6, 8, 14 from their mean.

Solution: Mean of 3, 4, 6, 8, 14 is

$$\bar{x} = \frac{3+4+6+8+14}{5} = \frac{35}{5} = 7$$

x_i	$x_i - \bar{x} = x_i - 7$
3	3–7 = –4
4	4–7 = –3
6	6–7 = –1
8	8–7 = 1
14	14–7 = 7
	$\Sigma(x_i - \bar{x} = 0)$

Hence the sum of the deviations about mean = 0

Problem 2: The average marks of three batches of students having 70, 50 and 30 students respectively are 50, 55 and 45. Find the average marks of all the 150 students, taken together.

Solution:

Let x be the average marks of all 150 students taken together.

Batch - I	**Batch - II**	**Batch - III**
$\bar{X}_1 = 50$	$\bar{X}_2 = 55$	$\bar{X}_3 = 45$
$n_1 = 70$	$n_2 = 50$	$n_3 = 30$

$$\bar{X} = \frac{n_1\bar{X}_1 + n_2\bar{X}_2 + n_3\bar{X}_3}{n_1 + n_2 + n_3}$$

$$= \frac{70 \times 50 + 50 \times 55 + 30 \times 45}{70 + 50 + 30} = \frac{7600}{150}$$

$$\bar{X} = 50.67 \text{ Marks}$$

Problem 3: Find the missing frequencies in the following data, where mean salary of the employees is 38 and total number of employees is 70:

Salary :	0-10	10-20	20-30	30-40	40-50	50-60
Frequency employees :	4	10	12	?	20	?

Solution: Let the missing frequencies be a and b.

Class Interval	Mid points m	Frequency f	fm
0-10	5	4	20
10-20	15	10	150
20-30	25	12	300
30-40	35	a	$35a$
40-50	45	20	900
50-60	55	b	$55b$
		$\Sigma f = 70$	$\Sigma fm = 1370 + 35a + 55b$

Putting the values in

$$\bar{X} = \frac{\Sigma fm}{N}$$

$$\Rightarrow \quad 38 = \frac{1370 + 35a + 55b}{70}$$

$$\Rightarrow \quad 2660 = 1370 + 35a + 55b \qquad \text{... (1)}$$

Also $\Sigma f = 46 + a + b$

$$70 + a + b \Rightarrow a + b = 24 \qquad \text{... (2)}$$

Solving (1) and (2) simultaneously, we get $a = 1$, $b = 23$

Problem 4: Find the class intervals for the following distribution, if the arithmetic mean of the original distribution is 35.84 and d are the deviations of the mid points from the assumed mean.

d:	–21	–14	–7	0	7	14	21
f:	2	12	19	29	20	13	5

Solution:

Let A is the assumed mean from which the deviations of mid points are calculated.

f	$d = m - A$	fd	Mid points m	Class Intervals
2	–21	–42	14	10.5 – 17.5
12	–14	–168	21	17.5 – 24.5
19	–7	–133	28	24.5 – 31.5
29	0	0	35	31.5 – 38.5
20	7	140	42	38.5 – 45.5
13	14	182	49	45.5 – 52.5
5	21	105	56	52.5 – 59.5
	$N = 100$	$\Sigma fd = 84$		

By substituting the values in

$$\bar{X} = A + \frac{\Sigma fd}{N}$$

$$\Rightarrow \quad 35.84 = A + \frac{84}{100}$$

$$\Rightarrow \quad A = 35$$

Thus the mid value of the class corresponding to the value $d = 0$ is $m = 35$. Since the magnitude of the class interval is 7, the corresponding class interval is obtained on adding and subtracting 7/2 = 3.5 from 35 and is given by (35 – 3.5, 35 + 3.5), i.e. 31.5 – 38.5. The other class intervals are shown in the last column of the above table.

Problem 5: The mean of 1000 observations was found to be 180.40. Later on, it was discovered that two observations were wrongly entered as 297 and 165. Their correct values were 197 and 185. Find the correct arithmetic mean.

Solution:

We are given that $\bar{X} = 180.40$, $n = 1000$

$$\bar{X} \text{ (Incorrect)} = \frac{\Sigma X}{n} = 180.40$$

$$\Rightarrow \quad \Sigma x \text{ (Incorrect)} = 1{,}80{,}400$$

$$\frac{180320}{1000} = 180.32$$

$$\Sigma x \text{ (Correct)} = 1{,}80{,}400 - \text{(Sum of wrong observations)}$$
$$+ \text{(Sum of correct observations)}$$
$$= 180400 - (297 + 165) + (197 + 185)$$
$$= 180320$$

$\therefore$ Corrected mean =

Problem 6: From the following distribution, calculate the arithmetic mean:

Marks	No of students
Below 30	16
Below 40	36
Below 50	61
Below 60	76
Below 70	87
Below 80	95
Below 90	100

Solution:

It is a case of open-end classes. Here we will assume that lower limit of first class is 20. Then we calculate the arithmetic mean as:

Marks	Mid value m	Less than c.f.	f	$d' = \frac{m-45}{10}$	fd'
20 - 30	25	16	16	−2	−32
30 - 40	35	36	20	−1	−20
40 - 50	45	61	25	0	0
50 - 60	55	76	15	1	15
60 - 70	65	87	11	2	22
70 - 80	75	95	8	3	24
80-90	85	100	5	4	20
			$\Sigma f = 100$		$\Sigma fd' = 29$

$$\bar{X} = A + \frac{\Sigma fd'}{N} \times c$$

$$= 45 + \frac{29}{100} \times 10 = 47.9$$

12.3.3 Weighted Arithmetic Mean

Computation of arithmetic mean is based on this assumption that all the observations in the given series are of equal importance. But there are many real life situations where some observations are given more importance as compared to others. In such cases arithmetic mean is not a true representative of the given series. Here we apply weighted arithmetic mean. In weighted arithmetic mean, weights are assigned to different observations based on their relative importance. Suppose that $W_1, W_2, W_3, \ldots\ldots W_n$ are the weights associated with the observations $X_1, X_2, X_3, \ldots\ldots X_n$ respectively, then weighted arithmetic mean represented by $\bar{X}_w$ and is given by:

$$\bar{X}_w = \frac{\Sigma WX}{\Sigma W}$$

There are three different ways to compute the weighted arithmetic mean:

1. Direct method
2. Indirect or shortcut method
3. Step deviation method

Here we will discuss only direct method and the other two will be used similarly.

SOLVED PROBLEMS

Problem 1: A student obtained 40, 50, 60, 80, and 45 marks in the subjects of Math, Statistics, Physics, Chemistry and Biology respectively. Assuming weights 5, 2, 4, 3, and 1 respectively for the above mentioned subjects. Find Weighted Arithmetic Mean per subject.

Solution:

Subjects	Marks Obtained x	Weight w	wx
Math	40	5	200
Statistics	50	2	100
Physics	60	4	240
Chemistry	80	3	240
Biology	45	1	45
Total		$\Sigma w = 15$	$\Sigma wx = 825$

Now we will find weighted arithmetic mean as:

$$\bar{X}_w = \frac{\sum wx}{\sum w} = \frac{825}{15} = 55 \text{ marks/subject.}$$

Problem 2: You are given the following data of students in a college:

%age of Attendance :	40	60	70	50	30
No. of students :	3	5	2	4	2

Compute the weighted arithmetic mean of attendance with direct method.

Solution:

1. Using direct method:

Here number of students will act as the weights associated with the percentage attendance.

%age of Attendance X	No. of students W	WX
40	3	120
60	5	300
70	2	140
50	4	200
30	2	60
	$\Sigma W = 16$	$\Sigma WX = 820$

$$\bar{X}_w = \frac{\sum WX}{\sum W}$$

$$= \frac{880}{16} = 51.25$$

It means the weighted arithmetic mean of the percentage of attendance is approx 51.

Problem 3: Compare the performance of two universities on the basis of the following data. Find which of them is better and why?

Name of Exams	University A (Number of students)		University B (Number of students)	
	Appeared	Passed	Appeared	Passed
M.C.A.	300	250	1000	800
M.B.A.	500	450	1200	950
M. Sc.	2000	1560	1000	700
M. Tech.	1200	750	800	500
Total	4000	2950	4000	2950

Solution:

Name of Exams	University A (Number of students)			University B (Number of students)		
	Appeared W_A	Passed	Pass %age X_A	Appeared W_B	Passed	Pass %age X_B
M.C.A.	300	250	83.33	1000	800	80
M.B.A.	500	450	90	1200	950	79.17
M. Sc.	2000	1560	75	1000	700	70
M. Tech.	1200	750	62.5	800	500	62.5
Total	4000	2950	73.75	4000	2950	73.75

To compare their performance, we should use weighted mean in the two cases.

$$\bar{X}_w(A) = \frac{300(83.33) + 500(90) + 2000(75) + 1200(62.5)}{4000} = 73.75$$

$$\bar{X}_w(B) = \frac{1000(80) + 1200(79.17) + 1000(70) + 800(62.5)}{4000} = 73.75$$

So, on the basis of two weighted means, we conlude that both the universities are equally good by considering their pass percentage.

Problem 4: From the following data of income distribution calculate the arithmetic mean. It is given that (i) the total income of persons in the highest group is Rs. 435, and (ii) none is earning less than Rs. 20.

Income (Rs.)	No. of persons	Income (Rs.)	No. of persons
Below 30	16	Below 70	87
Below 40	36	Below 80	95
Below 50	61	80 and over	5
Below 60	76		

Solution: The open class "Income below 30" includes the persons with income less than Rs. 30. But since we are given that none is earning less than Rs. 20, this class will be 20-30. Moreover, we are given the cumulative frequency distribution which has to be converted into the ordinary frequency distribution as given in the following table:

Income	Mid-value (X)	No. of persons (f)	fX
20-30	25	16	400
30-40	35	36-16 = 20	700
40-50	45	61-36 = 20	1100
50-60	55	76-61 = 15	825
60-70	65	87-76 = 11	715
70-80	75	95-87 = 8	600
80 and over	—	5	435*
		$\Sigma f = 100$	$\Sigma fx = 4,800$

$$\text{Arithmetic Mean } = \frac{\Sigma fx}{\Sigma f} = \frac{4,800}{100} = \text{Rs.}48.$$

Problem 5: The average daily wage of all workers in a factory is Rs. 444. If the average daily wages paid to male and female workers are Rs. 480 and Rs. 360 respectively. find the percentage of male and female workers employed by the factory.

Solution: Let n_1 and n_2 denote respectively the number of male and female workers in the factory and $\bar{X}_1$ and $\bar{X}_2$ denote respectively their average daily salary (in Rupees). Let $\bar{X}$ denote the average salary of all the workers in the factory. Then we are given that:

$$\bar{X}_1 = 480, \quad \bar{X}_2 = 360 \quad \text{and} \quad \bar{X} = 444$$

We have $\bar{X} = \dfrac{n_1\bar{X}_1 + n_2\bar{X}_2}{n_1 + n_2}$ $\Rightarrow$ $(n_1 + n_2)\,\bar{X} = n_1\bar{X}_1 + n_2\bar{X}_2$

$\Rightarrow$ $444\,(n_1 + n_2) = 480n_1 + 360n_2$ $\Rightarrow$ $(480 - 444)n_1 = (444 - 360)n_2$

$\therefore$ $36n_1 = 84n_2$ $\Rightarrow$ $\dfrac{n_1}{n_2} = \dfrac{84}{36} = \dfrac{7}{3}$

Hence, the male workers in the factory are : $\dfrac{7}{7+3} \times 100 = \dfrac{7}{10} \times 100 = 70\%$

and the female workers in the factory are : $\dfrac{3}{7+3} \times 100 = \dfrac{3}{10} \times 100 = 30\%$

Problem 6: The arithmetic mean height of 50 students of a college is 5' – 8". The height of 30 of there is given in the frequency distribution below. Find the arithmetic mean height of the remaining 20 students.

Height in inches :	5' – 4"	5' – 6"	5' – 8"	5' – 10"	6' – 10"
Frequency :	4	12	4	8	2

* *It given that total income in the highest group is Rs. 435.*

Solution: Let the variable X denote the height of the students in inches.

Computation of Mean Height of 30 Students

Height in inches (X)	Frequency (f)	$d = \frac{X-68}{2}$	fd
64	4	–2	–8
66	12	–1	–12
68	4	0	0
70	8	1	8
72	2	2	4

$$\bar{X}_1 = \text{Mean height (in inches) of } n_1 = 30 \text{ students}$$

$$= A + \frac{h\Sigma fd}{\Sigma f} = 68 + \frac{2\times(-8)}{30}$$

$$= \frac{2040-16}{30} = \frac{2024}{30} \text{ inches} \qquad \ldots(*)$$

Let $\bar{X}_2$ denote the mean height of the remaining $n_2 = 50 - 30 = 20$ students. If $\bar{X}$ is the mean height of the 50 students, then we are given that:

$$\bar{X} - 5'\,8{:} = 68''$$

$$\bar{X} = \frac{n_1\bar{X}_1 + n_2\bar{X}_2}{n_1+n_2} \quad \Rightarrow \quad 68 = \frac{30\times\left(\frac{2024}{30}\right)+20\bar{X}_2}{50} \qquad \text{[Using (*)]}$$

$$\therefore \quad 2024 + 20\bar{X}_2 = 68\times 50 = 3400 \quad \Rightarrow \quad \bar{X}_2 = \frac{3400-2024}{20} = \frac{1376}{20} = 68.8'' = 5' - 8.8''.$$

Problem 7: A candidate obtained the following percentages of marks in an examination: English 60; Hindi 75; Mathematics 63; Physics 59; Chemistry 55. Find the candidate's weighted arithmetic mean if weights 1, 2, 1, 3, 3 respectively are allotted to the subjects.

Solution: Let the variable X denote the percentage of marks in the examination.

Computation of Weighted Mean

Subject	Marks (%) (X)	Weight (W)	WX
English	60	1	60
Hindi	75	2	150
Mathematics	63	1	63
Physics	59	3	177
Chemistry	55	3	165

$$\therefore \quad \text{Weighted Arithmetic Mean (in \%)} = \frac{\Sigma WX}{\Sigma W} = \frac{615}{10} = 61.5.$$

Program to Find the Weighted Arithmetic Mean of Individual Series Using Direct Method.

```
#include<stdio.h>
#include<conio.h>
void main( )
{
float x[100],w[100],sw=0.0,swx=0.0,wam;
int i,n;
clrscr( );
printf(" How many items in the individual distribution ?");
scanf("%d", &n);
printf("\n Enter the data \n");
for(i=1;i<=n;i++)
{scanf("%f",&x[i]);
scanf("%f",&w[i]);
sw+=w[i];
swx=swx+w[i]*x[i];
}
wam=swx/sw;
printf("\n The weighted arithmetic mean is %7.2f",wam);
getch( );
}
```

Output:

```
How many items in the individual distribution? 5
Enter the data
40    3
60    5
70    2
50    4
30  2
The weighted arithmetic mean is 51.25
```

12.3.4 Merits of Arithmetic Mean

Following are the merits of the Arithmetic mean:

1. It is rigidly defined.

2. It is easy to calculate and understand.
3. It is based upon all the given observations.
4. It does not require any type of rearrangement of the series either in ascending or in descending or in grouping order.
5. It is least affected by the fluctuations in sampling.

12.3.5 Demerits of Arithmetic Mean

Following are the major demerits of the Arithmetic mean:

1. It is very much affected by the values of extreme items of the series.
2. It fails to give a representative value of the series in some cases.
3. It cannot be computed in case of any missing observation.
4. Some times its results are abrupt, i.e. result is not present in the series.
5. So me times it gives absurd and impracticable values.
6. It cannot be computed in case of open-end classes.
7. It cannot be computed for a qualitative data.
8. It fails to provide a representative item where the distribution of the series is not normal, i.e. highly asymmetrical or skewed distribution.
9. Some times it leads to wrong conclusions on the basis of computed results.
10. It cannot be computed either by graphically or by inspection method.
11. It is highly affected by abnormal values.
12. The loss of even a single observation makes it impossible to compute the aritmetic mean correctly.

UNSOLVED PROBLEMS

1. Define arithmetic mean.
2. What are the different ways to compute the A.M.?
3. Differentiate between direct and indirect method to compute A.M..
4. How can you find mean of an individual series?
5. How can you find mean of a discrete series?
6. How can you find mean of a continuous series?
7. Discuss the main properties, merits and demerits of arithmetic mean.
8. Explain the concept of weighted arithmetic mean.
9. Find arithmetic mean of the following data:

Mid values :	2.5	5.5	7.5	12.5	27.5	32.5
No. of Students :	7	10	11	4	3	2

10. If the mean of the following data is 20.6, find the missing frequency.

X:	10	15	20	25	30
F:	3	10	?	7	5

11. Calculate weighted mean of the following data by using the direct and shortcut method:

Items:	76	74	72	60	53
Weights:	2	3	4	6	7

12. Calculate arithmetic mean for the following data.

Marks:	0-5	5-10	10-15	15-20	20-25
No. of Students:	7	3	8	2	5

13. Calculate arithmetic mean for the following data.

X:	10-20	20-30	30-40	40-50	50-60	60-70
f :	10	28	30	42	65	120

14. (a) A certain number of salesmen was appointed in different territories and the following data were compiled from their reports:

Sales ('000 Rs.):	4-8	8-12	12-16	16-20	20-24	24-48	28-32	32-36	36-40
No. of salesmen:	11	13	16	14	—	9	17	6	4

If the average sales is believed to be Rs. 19,920, find the missing information.

(b) From the following data, find the missing frequency when mean is 15.38.

Size:	10	12	14	16	18	20
Frequency:	3	7	?	20	8	5

15. (a) The average salary of 49 out of 50 employees in a firm is Rs. 100. The salary of the 50th employee is Rs. 97.50 more than the average salary of all the 50 workers. Find the mean salary of all the employees in the firm.

(b) The mean of 99 items is 55. The value of 100th item is 99 more than the mean of 100 items. What is the value of 100th item.

12.4 GEOMETRIC MEAN

Geometric mean is a kind of average of a set of numbers that is different from the arithmetic average. The geometric mean is well defined only for sets of positive real numbers. This is calculated by multiplying all the numbers (call the number of numbers "n"), and taking the nth root of the total.

"Geometric mean is the nth positive root of the product of "n" positive given values"

A common example of where the geometric mean is the correct choice is when averaging growth rates.

FORMULA

Geometric Mean

$$\text{Geometric Mean} = ((X_1)(X_2)(X_3)........(X_N))^{1/N}$$

or
$$\text{G.M.} = \sqrt[n]{X_1 \times X_2 \times X_3 \times .... \times X_n} \quad ...(1)$$

where

X = Individual score

N = Sample size (Number of scores)

If we are dealing with two numbers say 4 and 16, then their G.M. is $\sqrt{4 \times 16} = \sqrt{64} = 8$

But if the number of observations is large, then the calculation of *nth* root is very difficult. In such cases the calculations can be done with the help of logarithms. Taking logs on both sides in (1):

$$\log(\text{G.M.}) = \frac{1}{n}\log(X_1 X_2 .... X_n)$$

$$= \frac{1}{n}(\log X_1 + \log X_2 + ... + \log X_n)$$

$$= \frac{1}{n}\Sigma \log X$$

$$\therefore \quad \text{G.M} = \text{Antilog}\left[\frac{1}{n}\Sigma \log X\right]$$

Geometric Mean Example: To find the Geometric Mean of 1, 2, 3, 4 ,5.

Step 1: $N = 5$, the total number of values. Find $1/N$.

$$1/N = 0.2$$

Step 2: Now find Geometric Mean using the formula.

$$((1)(2)(3)(4)(5))^{0.2} = (120)^{0.2}$$

So, Geometric Mean = 2.60517

This example will guide you to calculate the geometric mean manually.

12.4.1 Methods of Computation

In Case of Individual Series:

$$\text{G.M.=Antilog}\left[\frac{1}{n}\Sigma \log X\right]$$

where n is total number of observations

In Case of Discrete Series:

$$\text{G.M.} = \text{Antilog}\left[\frac{1}{N}\Sigma f \log X\right]$$

where N is the total of frequencies

In Case of Continuous Series:

$$\text{G.M.}=\text{Antilog}\left[\frac{1}{N}\Sigma f \log m\right]$$

where N is the total of frequencies and m are the mid points of classes.

12.4.2 Properties of Geometric Mean

The main properties of geometric mean are:

1. The geometric mean is less than arithmetic mean, G.M < A.M
2. The product of the items remains unchanged if each item is replaced by the geometric mean.
3. The geometric mean of the ratio of corresponding observations in two series is equal to the ratios their geometric means.
4. The geometric mean of the products of corresponding items in two series is equal to the product of their geometric mean.
5. The sum of the deviations of the logarithm of the original values above and below the logarithm of the G.M. is equal.
6. In case of any negative value in the series, GM can not be computed.

   The GM raised to the nth power equals to product of the values, i.e.

$$(\text{G.M.})^n = X_1 \times X_2 \times X_3 \times \text{.........} \times X_n$$

7. If there are two G.M.s, i.e. G_1 and G_2 having n_1 and n_2 observations belonging to two different groups A and B respectively, then by combining the two groups, a new GM for $n_1 + n_2$ observations is :

$$\text{G.M.} = \text{Antilog}\left[\frac{n_1 \log G_1 + n_2 \log G_2}{n_1 + n_2}\right]$$

12.4.3 Uses of Geometric Mean

Geometric mean is very useful in finding the average ratios, percentages and rates of increase between two periods. G.M. is the appropriate average to be used for computing the average rate of growth of population, average increase in the rate of profits, sales, production etc. It is very useful to apply in those situations in which it is more desirable to give more weightage to smaller items and small weightage to larger items.

SOLVED PROBLEMS

Problem 1: Find the Geometric Mean of the values 10, 5, 15, 8, 12.

Solution: Here $x_1 = 10, x_2 = 5 x_3 = 15 x_4 = 8 x_5 = 12$ and $n = 5$

$$\text{G.M. of } X = \bar{X} = \sqrt[5]{10 \times 5 \times 15 \times 8 \times 12}$$

$$\bar{X} = \sqrt[5]{72000} = (72000)^{\frac{1}{5}} = 9.36$$

Problem 2: Find the Geometric Mean of the following Data

X:	13	14	15	16	17
f:	2	5	13	7	3

Solution: We may write it as given below:

Here, $x_1 = 13, x_2 = 14, x_3 = 15, x_4 = 16, x_5 = 17$

$f_1 = 2, f_2 = 5, f_3 = 13, f_4 = 7, f_5 = 3$

$n = \Sigma f = f_1 + f_2 + f_3 + f_4 + f_5 = 2 + 5 + 13 + 7 + 3 = 30$

Using the formula of geometric mean for grouped data, geometric mean in this case will become:

$$\text{G.M of } X = \bar{X} = \sqrt[n]{x_1^{f_1} \cdot x_2^{f_2} \cdot x_3^{f_3} \cdot x_4^{f_4} \cdot x_5^{f_5}}$$

$$\bar{X} = \sqrt[30]{(13)^2 \cdot (14)^5 \cdot (15)^{13} \cdot (16)^7 \cdot (17)^3}$$

$$\bar{X} = \sqrt[30]{2.33292 \times 10^{35}} = (2.33292 \times 10^{35})^{\frac{1}{30}}$$

$$\bar{X} = 15.0984 \approx 15.10$$

The method explained above for the calculation of geometric mean is useful when the numbers of values in given data are small in number and the facility of electronic calculator is available. When a set of data contains large number of values then we need an alternative way for computing geometric mean. The modified or alternative way of computing geometric mean is given as under:

For Ungrouped Data

$$\text{G.M of } X = \bar{X} = \text{Antilog}\left(\frac{\Sigma \log x}{n}\right)$$

For Grouped Data

$$\text{G.M of } X = \bar{X} = \text{Antilog}\left(\frac{\Sigma f \log x}{\Sigma f}\right)$$

Problem 3: Find the Geometric Mean of the values 10, 5, 15, 8, 12

x	log x
10	1.0000
5	0.6990
15	1.1761
8	0.9031
12	1.0792
Total	Σlog x = 4.8573

$$\text{G.M of } X = \bar{X} = \text{Antilog}\left(\frac{\Sigma \log x}{n}\right)$$

$$\bar{X} = \text{Antilog}\left(\frac{4.8573}{5}\right)$$

$$\bar{X} = \text{Antilog } (0.9715)$$

$$\bar{X} = 9.36$$

Problem 4: Find the Geometric Mean for the following distribution of students' marks:

Marks :	0-30	30-50	50-80	80-100
No. of Students :	20	30	40	10

Solution:

Marks	No. of Students *f*	Mid Points *x*	*f* log *x*
0-30	25	15	20 log 15 = 23.5218
30-50	30	40	30 log 40 = 48.0168
50-80	40	65	40 log 65 = 72.5165
80-100	10	90	10 log 90 = 19.5424
Total	Σ*f* = 100		Σ*f* log *x* = 163.6425

$$\text{G.M of } X = \bar{X} = \text{Antilog}\left(\frac{\Sigma f \log x}{\Sigma f}\right)$$

$$\bar{X} = \text{Antilog}\left(\frac{163.6425}{100}\right)$$

$$\bar{X} = \text{Antilog } (1.6364)$$

$$\bar{X} = 43.29$$

Problem 5: Find the geometric mean of the following data: 1296, 16, 81, 2401.

Solution: As the GM formula is:

$$\text{G.M.} = \sqrt[n]{X_1 \times X_2 \times X_3 \times \ldots\ldots \times X_n}$$

Here $n = 4$, $x_1 = 1296$, $x_2 = 16$, $x_3 = 81$ and $x_4 = 2401$

By applying all the above values in the given geometric formula,

We get, $$G = \sqrt[4]{1296 \times 16 \times 81 \times 2401} = 6 \times 2 \times 3 \times 7 = 252.$$

Problem 6: Find the geometric mean of the following data: 5, 10, 20, 25.

Solution:

x	log *x*
5	0.699
10	1.0000
20	1.301
25	1.3979
	Σlog *X* = 4.3979

By using the geometric mean formula, we get (Here $n = 4$)

$$\text{G..M} = \text{Antilog}\left[\frac{1}{n}\Sigma \log X\right]$$

$$\text{G..M} = \text{Antilog}\frac{4.3979}{4} = \text{Antilog}1.0994 = 12.5718$$

Problem 7: Find the geometric mean of the grace marks for the following data:

Grace Marks :	5	10	15	20	25
No. of Students :	10	5	20	5	15

Solution:

Grace Marks 'X'	No. of Student 'f'	log *X*	*f* log *X*
5	10	0.699	6.99
10	5	1	5
15	20	1.1761	23.522
20	5	1.301	6.505
25	15	1.3979	20.9685
	***N* = 55**		**Σ*f* log *X* = 62.9855**

12

By using the geometric mean formula, we get

$$\text{G.M} = \text{Antilog}\left[\frac{1}{N}\Sigma f \log X\right]$$

$$= \text{Antilog}\,\frac{62.9855}{55} = 13.9668$$

Problem 8: Find the geometric mean of the prices for the following data:

Qty:	5	10	15	20	5
Price:	10-20	20-30	30-40	40-50	50-60

Solution:

Price 'X'	Qty 'f'	Mid Values 'm'	log *m*	*f* log *m*
10-20	5	15	1.1761	5.8805
20-30	10	25	1.3979	13.979
30-40	15	35	1.5441	23.161
40-50	20	45	1.6532	33.064
50-60	5	55	1.7404	8.702
	N = 55			Σ*f* log *m* = 84.7865

By using the geometric mean formula, we get

$$\text{GM} = \text{Antilog}\left[\frac{1}{N}\Sigma f \log m\right]$$

$$= \text{Antilog}\,\frac{84.7865}{55} = 34.7936$$

12.4.4 Weighted Geometric Mean

Weighted Geometric mean is applied in those cases where all observations in the given series are not given equal importance. Depending upon their relative importance, different weights are assigned to different observations and accordingly weighted geometric mean is calculated.

For a given set of n observations say $X = (X_1, X_2, \ldots, Xn)$ if the corresponding weights are given by $W = (W_1, W_2, \ldots, Wn)$ then the weighted geometric mean is calculated as

$$\text{G.M.}\,(W) = \text{Antilog}\left[\frac{\Sigma W \log X}{\Sigma W}\right]$$

Note that if all the weights are equal, the weighted geometric mean is the same as the unweighted geometric mean.

Problem 9: Find the Weighted Geometric Mean of the following data:

W:	5	2	10	2	6
X:	10	5	20	5	15

Solution:

X	*W*	log *X*	*W* log *X*
10	5	1	5
5	2	0.699	1.398
20	10	1.301	13.01
5	2	0.699	1.398
15	6	1.1761	7.0566
	$\Sigma W = 25$		$\Sigma W \log X = 27.86$

By using the Weighted Geometric Mean formula, we get

$$\text{G.M. }(W) = \text{Antilog}\left[\frac{\Sigma W \log X}{\Sigma W}\right]$$

$$= \text{Antilog}\left[\frac{27.86}{25}\right] = \text{Antilog} 1.114 = 13.001$$

PROGRAMS

1. Program to Find the GM of Individual Series.

```
#include<stdio.h>
#include<conio.h>
#include<math.h>
void main( )
{
float x[100],slx=0,gm;
int i,n;
clrscr( );
printf(" How many data items ?");
scanf("%d",&n);
printf(" Enter the data items \n");
for(i=1;i<=n;i++)
{
```

12

```
scanf("%f",&x[i])
slx+=log(x[i]);
}
gm=exp(slx/n);
printf("\n Geometric mean =%7.4f",gm);
getch( );
}
```

Output:

```
How many data items? 4
Enter the data items
5
10
20
25
Geometric mean = 12.5718
```

2. Program to Find the GM of Discrete Series.

```
#include<stdio.h>
#include<conio.h>
#include<math.h>
void main( )
{
float x[100],sflx=0,gm;
int i,n,sf=0,N,f[100];
clrscr( );
printf(" How many data items?");
scanf("%d",&n);
printf(" Enter the data items as Xi  and Fi \n");
for(i=1;i<=n;i++)
{
scanf("%f %d",&x[i],&f[i]);
sf+=f[i];
sflx=sflx+f[i]*log(x[i]);
}
```

```
N=sf;
gm=exp(sflx/N);
printf("\n Value of N=%d",sf);
printf(" Geometric mean =%7.4f",gm);
getch( );
}
```

Output:

```
How many data items? 5
Enter the data items as Xi and Fi
5     10
10    5
15    20
20    5
25  15
Value of N=55
Geometric mean = 13.9668
```

3. Program to Find the GM of Continuous Series.

```
#include<stdio.h>
#include<conio.h>
#include<math.h>
void main( )
{
float a[100],b[100],x[100],sflx=0,gm;
int i,n,sf=0,N,f[100];
clrscr( );
printf(" How many data items?");
scanf("%d",&n);
printf(" Enter the data items as lower value Ai, upper value Bi, frequency Fi \n");
for(i=1;i<=n;i++)
{
scanf("%f %f %d", &a[i],&b[i],&f[i]);
x[i]=(a[i]+b[i])/2;
sflx+=f[i]*log(x[i]);
```

```
sf+=f[i];
}
N=sf;
gm=exp(sflx/N);
printf("\n Value of N =%d",sf);
printf("\n Geometric mean =%7.4f",gm);
getch( );
}
```

Output:

```
How many data items? 5
Enter the data items as lower value Ai, upper value Bi, frequency Fi
10      20   5
20      30  10
30      40  15
40  50   20
50  60    5
Value of N = 55
Geometric mean = 34.7936
```

12.4.5 Merits of Geometric Mean

Following are the some merits of the Geometric mean:

1. It is rigidly defined and its value is a precise figure.
2. It is based on all observations.
3. It is capable of further algebraic treatment.
4. It is not much affected by fluctuation of sampling.
5. It is not affected by extreme values.

12.4.6 Demerits of Geometric Mean

Following are the some demerits of the Geometric mean:

1. It cannot be calculated if any of the observation is zero or negative.
2. Its calculation is rather difficult.
3. It is not easy to understand.
4. It may not coincide with any of the observations.

UNSOLVED PROBLEMS

1. When we use geometric mean?
2. What are the various properties of geometric mean?
3. What do you know about weighted geometric mean?
4. What are the merits and demerits of G.M?
5. Give the relationship of arithmetic mean with the geometric mean?
6. Compute Geometric mean of the following series

   100 150 200 250 300 350 400

7. Compute Geometric mean of the following data:

X:	10	15	20	25	30
f:	2	3	5	4	6

8. Calculate Geometric mean for following series.

Marks:	0-10	10-20	20-30	30-40	40-50
No of Students :	5	8	15	20	25

9. Find the missing frequency of the following data if mean is 28.

Wage rate:	0-10	10-20	20-30	30-40	40-50	50-60
No of Workers :	12	8	25	17	?	6

10. Find combined Geometric mean of the following three:

	A	B	C
No of items :	5	7	8
GM :	5.52	8.52	6.75

12.5 HARMONIC MEAN

Harmonic mean is another measure of central tendency and also based on mathematic footing like arithmetic mean and geometric mean. Like arithmetic mean and geometric mean, harmonic mean is also useful for quantitative data. Harmonic mean is defined in following terms:

Definition: *Harmonic mean is quotient of "number of the given values" and "sum of the reciprocals of the given values".*

For a given set of *n* observations, say $X = (X_1, X_2, \ldots, Xn)$, the harmonic mean is given by

$$\text{H.M.} = \frac{n}{\Sigma \frac{1}{X}}$$

i.e.the harmonic mean is the number of variables, divided by the sum of the reciprocals of the variables. It is an average which is useful for sets of numbers which are defined in relation to some unit, for example speed (distance per unit of time).

12.5.1 Methods of Computation

In Case of Individual Series:

$$\text{H.M.} = \frac{n}{\Sigma \frac{1}{X}}$$

where n is the total number of observations

In Case of Discrete Series:

$$\text{H.M.} = \frac{N}{\Sigma \frac{f}{X}}$$

where N is the total frequency

In Case of Continuous Series:

$$\text{H.M.} = \frac{N}{\Sigma \frac{f}{m}}$$

where N is the total frequency and m are the mid points.

12.5.2 Uses of Harmonic Mean

Harmonic mean is very useful in finding the average rates like speed (distance per unit time), items produced per day, etc.

SOLVED PROBLEMS

Problem 1: Consider the following data and find Harmonic Mean *X*: 1, 2, 3, 4, 5

Solution:

As we know that H.M is given by

$$\text{H.M.} = \frac{n}{\Sigma \frac{1}{X}}$$

$$= \frac{5}{2.2833} = 2.189$$

Problem 2: Calculate the harmonic mean of the numbers: 13.5, 14.5, 14.8, 15.2 and 16.1

Solution: The harmonic mean is calculated as below:

X	$\frac{1}{x}$
13.2	0.0758
14.2	0.0724
14.8	0.0676
15.2	0.0658
16.1	0.0621
Total	$\Sigma\left(\frac{1}{x}\right) = 0.3417$

$$\text{H.M of } X = \bar{X} = \frac{n}{\Sigma\left(\frac{1}{x}\right)}$$

$$\text{H.M of } X = \bar{X} = \frac{5}{0.3417} = 14.63$$

Problem 3: Given the following frequency distribution of first year students of a particular college. Calculate the Harmonic Mean.

Age (years) :	13	14	15	16	17
Number of Students :	2	5	13	7	3

Solution: The given distribution belongs to a grouped data and the variable involved is ages of first year students. While the number of students Represent frequencies.

Ages (Years)	Number of Students f	$\frac{1}{x}$
13	2	0.1538
14	5	0.3571
15	13	0.8667
16	7	0.4375
17	3	0.1765
Total	$\Sigma f = 30$	$\Sigma\left(\frac{1}{x}\right) = 1.9916$

Now we will find the Harmonic Mean as

$$\bar{X} = \frac{\Sigma f}{\Sigma\left(\frac{f}{x}\right)}$$

$$= \frac{30}{1.9916} = 15.0631 \approx 15 \text{ years.}$$

Problem 4: Consider the following data and find HM of wages of workers

Wages (in Rs. 000) X	Number of workers
20	8
21	10
22	11
23	1
24	20
25	25

Solution:

Wages X	Number of workers f	$\frac{1}{X}$	$\frac{f}{X}$
20	8	0.05	0.4
21	10	0.047619	0.47619
22	11	0.045455	0.5
23	1	0.043478	0.043478
24	20	0.041667	0.83334
25	25	0.04	1
	$N = 75$		$\Sigma\frac{f}{X} = 3.253008$

$$\text{H.M} = \frac{N}{\Sigma \frac{f}{X}} = \frac{75}{3.253008} = 23.055$$

Problem 5: Consider the following data and find the harmonic mean:

Classes	Frequency
2-4	10
4-6	20
6-8	30
8-10	40
Total	100

Solution:

Class	f	m	$1/m$	f/m
2-4	10	3	0.333333	3.333333
4-6	20	5	0.2	4
6-8	30	7	0.142857	4.285714
8-10	40	9	0.111111	4.444444
	$N = 100$			16.0635

$$H.M = \frac{N}{\Sigma \frac{f}{m}}$$

$$= \frac{100}{16.06349}$$

$$= 6.22529$$

Problem 6: An airplane flies around a square. The sides of which measure 100 Km. each. The aero plane covers at a speed of 100 Km. Per hour first side, at 200 km. Per hour the second side, at 300 km. Per hour the third side sand at 400 km. Per hour the fourth side. Use the correct mean to find the average speed around the square.

Solution: To find the average speed, harmonic mean of 4 speeds should be calculated which is given by

$$H.M = \frac{4}{\frac{1}{100} + \frac{1}{200} + \frac{1}{300} + \frac{1}{400}}$$

$$= 192 \text{ Km/hour}$$

Problem 7: Calculate the harmonic mean for the given below:

Marks :	30-39	40-49	50-59	60-69	70-79	80-89	90-99
f :	2	3	11	20	32	25	7

Solution: The necessary calculations are given below

Marks	x	f	$\frac{f}{X}$
30-39	34.5	2	0.0580
40-49	44.5	3	0.0674
50-59	54.5	11	0.2018
60-69	64.5	20	0.3101
70-79	74.5	32	0.4295
80-89	84.5	25	0.2959
90-99	94.5	7	0.0741
Total		$\Sigma f = 100$	$\Sigma\left(\frac{1}{x}\right) = 1.4368$

Now we will find the Harmonic Mean as

$$\bar{X} = \frac{\Sigma f}{\Sigma\left(\frac{f}{x}\right)} = \frac{100}{1.4368} = 69.60$$

12

PROGRAMS

1. Program to Find the HM for Individual Data.

```
#include<stdio.h>
#include<conio.h>
#include<math.h>
main( )
{
float x[100],sr=0,hm;
int i,n;
clrscr( );
printf(" How many data items you have?");
scanf("%d",&n);
printf(" Enter the data as xi \n");
for(i=1;i<=n;i++)
{
scanf("%f",&x[i]);
sr+=1/x[i];
}
hm=n/sr;
printf("\n Harmonic mean is =%f",hm);
getch( );
}
```

Output:

```
How many data items you have?6
Enter the data as xi
34
27
45
55
22
34
Harmonic mean is =33.017983
```

2. Program to Find the HM for Discrete Series.

```
#include<stdio.h>
#include<conio.h>
#include<math.h>
main( )
{
float x[100],s=0,hm;
int i,n,f[100],sf=0;
clrscr( );
printf(" How many data items you have? ");
scanf("%d", &n);
printf(" Enter data as Xi Fi\n");
for(i=1;i<=n;i++)
{
scanf(" %f %d",&x[i],&f[i]);
sf+=f[i];
s+=f[i]/x[i];
}
hm=sf/s;
printf("\n Harmonic Mean is =%f",hm);
getche( );
}
```

Output:

```
How many data items you have?6
Enter data as Xi Fi
20      8
21      10
22      11
23      1
24      20
25  25
Harmonic Mean is = 23.055000
```

3. Program to Find the HM for Grouped Data.

```
#include<stdio.h>
#include<conio.h>
#include<math.h>
main( )
{
float a[100],b[100],x[100], ,s=0,hm;
int i,n, f[100],sf=0;
clrscr( );
printf(" How many data items you have? ");
scanf("%d", &n);
printf(" Enter data as lower value Ai, upper value Bi and fi : \n");
for(i=1;i<=n;i++)
{
scanf(" %f %f %d",&a[i],&b[i],&f[i]);
x[i]=(a[i]+b[i])/2;
sf+=f[i];
s+=f[i]/x[i];
}
hm=sf/s;
printf("\n Harmonic Mean is =%f",hm);
getch( );
}
```

Output:

```
How many data items you have? 4
Enter data as lower value Ai, upper value Bi and fi:

2  4  10
4  6  20
6  8  30
8  10 40
Harmonic Mean is =6.225290
```

12.5.3 Weighted Harmonic Mean

For a given set of n observations say $X = (X_1, X_2, \ldots, Xn)$ if the corresponding weights are given by $W = (W_1, W_2, \ldots, Wn)$then the weighted harmonic mean is calculated as

$$H.M_{w}. = \frac{\Sigma W}{\Sigma \frac{W}{X}}$$

Note that if all the weights are equal, the weighted harmonic mean is the same as the unweighted harmonic mean.

Problem 7: Consider the following data and find the weighted harmonic mean:

Items	Weights
2	5
3	10
5	15
7	20
8	25
13	30

Solution:

Item *X*	Weight *W*	*W/X*
2	5	2.5
3	10	3.333333
5	15	3
7	20	2.857143
8	25	3.125
13	30	2.307692
Total	105	17.12317

As we know that weighted harmonic mean is given as:

$$H.M_{w}. = \frac{\Sigma W}{\Sigma \frac{W}{X}} = \frac{105}{17.12317} = 6.132043$$

Problem 8: A cyclist covers his first 3 kms. at an average speed of 8 kms. Per hour, another 2 kms. at 9 kms. Per hour and the last 2 kms. at 4 kms. Per hour. Find the average speed for the entire journey.

Solution: Since the cyclist is traveling different distances with different speeds, we will apply weighted harmonic mean to find the average speed by taking the distances traveled as the weights. If the cyclist traveled equal distances, then we would apply simple harmonic mean to find the average speed.

Distances W	Speed X	$\frac{1}{X}$	$\frac{W}{X}$
3	8	0.125	0.375
2	9	0.111	0.222
2	4	0.25	0.50
$\Sigma W = 7$			$\Sigma \frac{W}{X} = 1.097$

$$\text{H.M}_{w}. = \frac{\Sigma W}{\Sigma \frac{W}{X}}$$

$$= \frac{7}{1.097} = 6.38 \text{ Kms. Per hour}$$

12.5.4 Merits of Harmonic Mean

The harmonic mean has the following advantages:

1. It is rigidly defined.
2. It is based on all the observations of a series.
3. It is not affected very much by sampling fluctuations.
4. It cannot be computed, if we ignore any observation from the series.
5. It is capable of further algebraic treatment.
6. It gives the greater weightage to the smaller observations of a series.
7. It is very useful in finding the average of rates and ratios where time factor is variable and the act being performed remains constant.

12.5.5 Demerits of Harmonic Mean

The harmonic mean has the following disadvantages:

1. It is not easy to understand by a layman or non-mathematical person.
2. Its computation is very complex and difficult as it involves finding of the reciprocals of the numbers.
3. Its results are not accurate.
4. Its efficiency decreases in the case of huge amount of data.
5. Its algebraic treatment is very much limited and not far and wide as that of the arithmetic mean and geometric mean.
6. It has great impact on the results if the extreme observations are omitted.
7. It cannot be calculated, if any of the observations is zero.
8. Its reliability is lesser as compared to AM and GM.

12.5.6 Relation Between Arithmetic Mean, Geometric Mean and Harmonic Mean

The arithmetic mean (A.M.), the geometric Mean (G.M.) and the harmonic mean (H.M.) of a series of n observations are connected by the relation:

$$\textbf{A.M.} \geq \textbf{G.M.} \geq \textbf{H.M.}$$

Equality holds if and only if all the *n* observations are equal.

UNSOLVED PROBLEMS

1. Define harmonic mean and discuss its merits and demerits.
2 Explain the main applications of arithmetic mean, Geometric mean and harmonic mean. Explain with examples.
3. Prove that Geometric mean is greater than harmonic mean.
4. What do you know about weighted harmonic mean?
5. What are the various properties of harmonic mean?
6. Explain Relation between Arithmetic Mean, Geometric Mean and Harmonic Mean
7. Find AM, GM, and HM of the income:

Family:	*A*	*B*	*C*	*D*	*E*	*F*	*G*	*H*
Income:	70	30	20	10	8	200	42	150

8. From the following series, calculate harmonic mean:

   10 20 30 40 50 100

9. Find the Harmonic mean for the following data:

Marks:	0-10	10-20	20-30	30-40	40-50	50-60
No of Students:	6	10	15	16	24	30

10. Find out the missing frequency of the following series if its Harmonic mean is 20.08.

Marks:	10	20	30	40	50
Frequency:	10	30	15	?	20

12.6 MEDIAN

In statistics, a **median** is a value of the varaible that divides the given set of data into two equal parts. The *median* of a finite list of numbers can be found by arranging all the observations from lowest value to highest value and picking the middle one. If there are an even number of observations, the median is not unique, so one often takes the mean of the two middle values.

At most half of the given data have values less than the *median* and at most half have values greater than the median. If observations of a variable are arrenged in ordered series, the median value corresponds to the middle observation in that ordered series or list. The median value corresponds to a cumulative percentage of 50% (i.e., 50% of the values are below the median and 50% of the values are above the median).

Median is another measure of central tendency quite commonly used in statistics and probability. In a generic idea, a **median** is described as the number separating the higher half of a sample, a population, or a probability distribution, from the lower half. The **median** of a finite list of numbers can be found by arranging all the observations from lowest value to highest value and picking the middle one.

Definitions:

Median is defined as "It is the value of the middle term or item of an arranged series". Arrangement may be in ascending or descending order.

***Croxton and Cowdon** quoated the median as " It is the value of a variable which divides a series in two halves. One half of the items are equal to the second half of items."*

*According to the **Secrist** " It is the value of the item actual or estimated, when series is arranged in order of magnitude, which divides the distribution in two equal parts".*

METHODS OF COMPUTATION OF MEDIAN

Median can be computed in two ways. One is tabular method and second method is Graphical method.

12.6.1 Tabular Method

In Case of Individual Series:

Following steps are being followed to calculate median in individual series:

1. Arrange the data in ascending or descending order.
2. If N (Number of observations) is odd then

$$\text{Median} = \text{Value of } ((N+1)/2)^{\text{th}} \text{ item}$$

3. If N is even then

$$\text{Median} = \text{Value of } \left(\frac{\left(\frac{N}{2}\right)^{\text{th}} \text{item} + \left(\frac{N}{2}+1\right)^{\text{th}} \text{item}}{2} \right)$$

In Case of Discrete Series:

Following steps are being followed to calculate median in individual series:

1. Arrange the values in increasing order.
2. Prepare the 'less than' cumulative frequency (c.f.) distribution.

3. Find $N/2$. Here N is the total frequency.
4. Mark the cumulative frequency just greater than $N/2$.
5. The value of the variable corresponding to the marked c.f. gives median.

In Case of Continuous Series:

Following steps are being followed to calculate median in continuous series:

1. Arrange the classes in increasing order.
2. Prepare the less than cumulative frequency (c.f.) distribution.
3. Find $N/2$. Here N is the total frequency.
4. Mark the cumulative frequency just greater than $N/2$.
5. The class corresponding to the marked c.f. gives the median class and the value of median lies in this class.

The value of median is now calculated by using the following formula:

$$\text{Median} = l + \frac{\left(\frac{N}{2} - c\right)}{f} \times h$$

where

l is the lower limit of median class

f is the frequency of the median class

c is the cumulative frequency of the class just preceding the median class.

N is the total frequency

h is the class width or size of the median class

SOLVED PROBLEMS

Median in Individual Series

Problem 1: Find Median for the following data:

***X*: 1 2 3 4 5 6 7**

Solution:

Here $N = 7$

$\text{Median} = ((N+1)/2)^{\text{th}}$ item

$\text{Median} = ((7+1)/2) = 8/2 = 4$

Problem 2: Calculate the median for the following data:

26.1, 25.6, 25.7, 25.2, 25.0 and 24.7

Solution:

Here $N = 6$

Arrange the data in ascending order.

S.No.	X
1	24.7
2	25.0
3	25.2
4	25.6
5	25.7
6	26.1

$$\text{Median} = \text{Value of}\left(\frac{\left(\frac{6}{2}\right)^{\text{th}}\text{ item} + \left(\frac{6}{2}+1\right)^{\text{th}}\text{ item}}{2}\right)$$

$$= \text{Value of}\left(\frac{3^{\text{rd}}\text{ item} + 4^{\text{th}}\text{ item}}{2}\right)$$

$$= \frac{25.2 + 25.6}{2} = 25.4$$

Median in Discrete Series

Problem 3: You are given the following data. Find the median.

X:	1	2	3	4	5
f:	3	5	9	6	2

Solution:

X	f	c.f.
1	3	3
2	5	8
3	9	17
4	6	23
5	2	25

Here $\frac{N}{2} = \frac{25}{2} = 12.5$

The c.f. just greater than 12.5 is 17 and value of X corresponding to this c.f. is 3.

Hence median = 3

Median in Continuous Series

Problem 4: You are given the following data. Find the median.

X:	0-10	10-20	20-30	30-40	40-50
f:	3	8	17	12	5

Solution:

X	f	c.f.
0-10	3	3
10-20	8	11
20-30	17	28
30-40	12	40
40-50	5	45

Here $$\frac{N}{2} = \frac{45}{2} = 22.5$$

The c.f. just greater than 22.5 is 28 that lies in the class 20-30. Hence the median class is 20-30.

$l = 20,\ h = 10,\ f = 17,\ N = 45,\ c = 11$

Putting the values in

$$\text{Median} = l + \frac{\left(\frac{N}{2} - c\right)}{f} \times h$$

$$= 20 + \frac{22.5 - 11}{17} \times 10 = 26.76$$

Problem 5: Calculate the median for the following distribution:

Weight (Kg)	Number of bundlles
410-419	14
420-429	20
430-439	42
440-449	54
450-459	45
460-469	18
470-479	7

Solution: First of all convert the inclusive class intervals into exclusive class interval by making an adjustment. Add 0.5 to the upper limit and subtract 0.5 from the lower limit of each class.

Weight(Kg.)	No. of bundles f	c.f.
409.5-419.5	14	14
419.5-429.5	20	34
429.5-439.5	42	76
439.5-449.5	54	130
449.5-459.5	45	175
459.5-469.5	18	193
469.5-479.5	7	200

Here $\dfrac{N}{2} = 100$

So c.f. just greater than 100 is 130. Hence median class is 439.5–449.5

$l = 439.5, c = 76, f = 54, h = 10$

By substituting the values in

$$\text{Median} = l + \frac{\left(\frac{N}{2} - c\right)}{f} \times h$$

$$= 439.5 + \frac{100 - 76}{54} \times 10 = 443.94$$

Problem 6: Find the missing frequency from the following distribution, It is given that median is 24:

Marks :	0-10	10-20	20-30	30-40	40-50
No of students :	5	25	?	18	7

Solution: Let the missing frequency is X..

Marks	No. of students f	c.f.
0-10	5	5
10-20	25	30
20-30	X	30+X
30-40	18	48+X
40-50	7	55+X

It is given that median = 24. So median class is 20-30.

$L = 20, \dfrac{N}{2} = \dfrac{55 + X}{2}, c = 30, f = X, h = 10$

Putting the values in

$$\text{Median} = l + \frac{\left(\frac{N}{2} - c\right)}{f} \times h$$

$$24 = 20 + \frac{\frac{55 + X}{2} - 30}{X} \times 10$$

$\Rightarrow$ $$4X = 5X - 25$$

$\Rightarrow$ $$X = 25$$

Hence the missing frequency is 25.

Problem 7: The following table shows the age distribution of persons in a particular region.

Age (years)	No. of persons (in thousands)	Age (years)	No. of persons (in thousands)
Below 10	2	Below 50	14
Below 20	5	Below 60	15
Below 30	9	Below 70	15.5
Below 40	12	70 and over	15.6

(i) Find the median age.

(ii) Why is the median a more suitable measure of central tendency than the mean in this case?

Solution:

Age (years)	No. of persons in '000 (f)	c.f. (less than)
0-10		
10-20	25–2 = 3	2
20-30	9–5 = 4	5
30-40	12–9 = 3	9
40-50	14–12 = 2	12
50-60	15 – 41 = 1	14
60-70	15.5 – 15 = 0.5	15
70 and over	15.6 – 15.5 = 0.1	15.5
	$N = \Sigma f = 15.6$	15.6

(i) First of all we shall convert the given distribution into the continuous frequency distribution as given in the adjoining table and then compute the median.

Here $\frac{N}{2} = \frac{15.6}{2} = 7.8$. Cumulative frequency (c.f.) greater than 7.8 is 9. Thus the corresponding class 20-30 is the median class. Hence, using the median formula (5.13), we get

$$\text{Median} = 20 + \frac{10}{4}(7.8 - 5)20 + \frac{5}{2} \times 2.8$$

Hence, median age is 27 years.

(ii) In this case median is a more measure of central tendency than mean because the last class viz, 70 and over is open end class and as such we cannot obtain the class mark for this class and hence arithmetic mean cannot be computed.

Problem 8: Find the missing frequency from the following of daily sales of shops, given that\the median sale of shops is Rs. 2,400.

Sales in hundred Rs.:	0-10	10-20	20-30	30-40	40-50
No. of shops :	5	25	—	18	7

Solution: Let the missing frequency be '*a*'.

Calculations For Median		
Sales in hundred Rs.	No. of shops (f)	Cumulative frequency (c.f.)
0-10	5	5
10-20	25	30
20-30	a	$30 + a$
30-40	18	$48 + a$
40-50	7	$N = 55 + a$

Since median sales is Rs. 2,400 (24 hundred), 20-30 is the median class. Using median formula, we get

$$24 = 20 + \frac{10}{a}\left(\frac{55+a}{2} - 30\right) \Rightarrow \left(\frac{55+a-60}{2}\right) = \frac{5(a-5)}{a}$$

$$\therefore \quad 4a = 5a - 25 \quad \Rightarrow \quad a = 25.$$

PROGRAMS

1. Program to Find the Median of Individual Series.

```
#include<stdio.h>
#include<conio.h>
void main( )
{
float x[100],s=0.0,am,median;
int i,n;
clrscr( );
printf(" How many items in the individual distribution ?");
scanf("%d", &n);
printf("\n Enter the data \n");
for(i=1;i<=n;i++)
```

```
scanf("%f",&x[i]);
if(n%2==0)
median=(x[n/2]+x[n/2+1])/2.0;
else
median=x[(n+1)/2];
printf("\n The MEDIAN is %7.4f",median);
getch( );
}
```

Output:

```
How many items in the individual distribution? 5
Enter the data
5
20
10
30
45
The MEDIAN is 20.0000
```

2. Program to Find the Median of Discrete Series.

```
#include<stdio.h>
#include<conio.h>
void main( )
{
float x[100],tempx,tempf,median,p;
int i,n,j,k,sf=0,f[100],cf[100];
clrscr( );
printf(" How many items in the discrete series?");
scanf("%d", &n);
printf("\n Enter the data: \n");
for(i=1;i<=n;i++)
scanf("%f",&x[i]);
printf("\n Enter the frequencies: \n");
for(i=1;i<=n;i++)
{
```

12

```
scanf("%f",&f[i]);
sf+=f[i];
}
for(k=1;k<=n-1;k++)
{
for(j=k;j<n;j++)
{
if(x[j]>x[j+1])
{
tempx=x[j];
tempf=f[j];
x[j]=x[j+1];
f[j]=f[j+1];
x[j+1]=tempx;
f[j+1]=tempf;
}
}
}cf[1]=f[1];
for(i=2;i<=n;i++)
cf[i]=cf[i-1]+f[i];p=(float)(sf)/2;
i=0;
while(cf[i]<p)
i++;
median=x[i];
printf("\n The MEDIAN is %8.3f",median);
getche( );
}
```

Output:

How many items in the discrete series? 5

Enter the data

50

10

20

40

30

Enter the frequencies:

3

2

3

2

3

The MEDIAN is 30.000

3. Program to Find the Median of Continuous Series.

```
#include<stdio.h>
#include<conio.h>
void main( )
{
float x[100],f[100],l[100],u[100],median,p;
int i,n,cf[100],f[100],sf=0;
clrscr( );
printf(" How many classes in the continuous series?");
scanf("%d", &n);
printf("\n Enter the lower and upper data: \n");
for(i=1;i<=n;i++)
scanf("%f %f",&l[i],&u[i]);
printf("\n Enter the frequencies: \n");
for(i=1;i<=n;i++)
{
scanf("%f",&f[i]);
sf=sf+f[i];
}
cf[1]=f[1];
for(i=2;i<=n;i++)
cf[i]=cf[i-1]+f[i];p=(float)sf/2;
i=0;
while(cf[i]<p)
i++;
```

```
median=l[i]+((p-cf[i-1])/ f[i])* (u[i]- l[i]);
printf("\n The MEDIAN is %8.2f",median);
getch( );
}
```

Output:

```
How many classes in the continuous series? 5
Enter the lower and upper data:
10 20
20 30
30 40
40 50
50 60
Enter the frequencies:
2
1
4
2
1
The MEDIAN is 35.00
```

12.6.2 Graphical Method to Find the Median

Median can be easily calculated graphically with the help of a curve called the cumulative frequency curve or Ogive We can use 'less than ogive' or 'more than ogive' or both to calculate median. The procedure involves the following steps:

Using 'Less than ogive':

1. Prepare less than cumulative frequency of the given distribution.
2. Take the values of the variables (in discrete series) or class intervals (in continuous series) along the X-axis and the cumulative frequency along Y-axis.
3. Plot cumulative frequency against the corresponding value of the variable or against the upper limit of the corresponding class on Y-axis.
4. The smooth curve obtained by joining the points so obtained with the help of a free hand drawing is called less than cumulative frequency curve or 'less than ogive'.
5. Find a point corresponding to the $N/2$ on the Y-axis and locate the corresponding value on X-axis. That point represents median.

Using 'More than ogive':

1. Prepare more than cumulative frequency of the given distribution.
2. Take the values of the variables (in discrete series) or class intervals (in continuous series) along the *X*-axis and the cumulative frequency along *Y*-axis.
3. Plot cumulative frequency against the corresponding value of the variable (or against the lower limit of the corresponding class in continuous series).
4. The smooth curve obtained by joining the points so obtained with the help of a free hand drawing is called 'more than cumulative frequency curve' or 'more than ogive'.
5. Find a point corresponding to the *N*/2 on the *Y*-axis and locate the corresponding value on *X*-axis. That point represents the value of median.

Note: *We can also use both 'less than ogive' and 'more than ogive' to find median. For that purpose we draw both less than and more than ogive on the same graph paper. One curve will be rising and the other will be falling in nature. Find the point of intersection of the two curves. Value on* X-*axis corresponding to this point of intersection gives the value of median.*

12.6.3 Merits of Median

The median has the following advantages:

1. It is rigidly defined and easy to compute.
2. It can be computed or located by inspection.
3. It does not require all the observations of the data for its determination.
4. It is not affected by the extreme values of a series.
5. It can be determined graphically.
6. It can be determined easily in open-end series without any class limit.
7. It can be computed for irregular data.
8. It can be computed for qualitative data.
9. It can be determined straightway in case of a series of unequal class intervals without converting the series into equal class intervals as it is done in case of mode.
10. It is affected by large number of observations. It is best for smaller data.
11. It is considered suitable for computation of the mean deviation, as the sum of the deviations taken from it is the minimum.
12. It gives a value, which very much exists in the series, and is a round figure in most of the cases.

12.6.4 Demerits of Median

The median has the following disadvantages:

1. It is not unique always.
2. Median is not based on all the values or items or observations of the series.
3. It is affected by fluctuation in sampling.
4. It is not capable of further mathematical and algebraic treatment like mean, geometric mean, harmonic mean and quadratic mean.
5. In case of very small number of observations, it will give erroneous result.
6. It has the requirement to arrange the series either in ascending or descending order.
7. In case of frequency distribution it needs the arrangementof the series either in ascending order or in descending order, which will waste the time in computation.
8. Some times it gives results, which never exists in the series.
9. In some cases it gives fractional and impracticable results.
10. Its processing becomes slow in case of huge amount of data.

UNSOLVED PROBLEMS

1. Define median. Write its merits and demerits.
2. How can you compute the median?
3. What are the various properties of median?
4. Compare the mean and median with an example.
5. Find out the median for the following series:

   10 20 30 40 50 60 70 80 90 100

6. From the given data, find value of median of wages:

Wages in (Rs) :	60-70	50-60	40-50	30-40	20-30
No of Workers :	2	9	11	13	15

7. Find mean and median from the data given below:

Marks obtained:	0-10	10-20	20-30	30-40	40-50
No. of students:	12	18	27	17	6

8. Calculate the median for the following statistical distribution:

Value:	Less than 100	100-200	200-300	300-400	400 and above	Total
Frequency:	50	90	158	68	134	500

9. Calculate arithmetic mean and median from the following series:

Income (Rs.):	0-5	5-10	10-15	15-20	20-25	25-30
Frequency:	5	7	10	8	6	4

10. Calculate the mean and median from the following distribution:

Class.:	10-20	20-30	30-40	40-50	50-60	60-70	70-80	80-90
Frequency:	4	12	40	41	27	13	9	4

12.7 MODE

In statistics, the **mode** is the value that occurs the most frequently in a data set or a probability distribution. In some fields, notably education, sample data are often called **scores**, and the sample mode is known as the **modal score.**

Like the statistical mean and the median, the mode is a way of capturing important information about a random variable or a population in a single quantity. The mode is in general different from the mean and median, and may be very different for strongly skewed distributions.

The mode is not necessarily unique, since the same maximum frequency may be attained at different values. The most ambiguous case occurs in uniform distributions, wherein all values are equally likely.

It is defined as the value around which the items are most heavily concentrated. In other words, it is the value of the variable, which occurs maximum number of times.

Mode has the following major features:

1. It represents the most typical value of a series, which almost coincides with the existing items.
2. It is not affected by the extreme values of the items but by the extreme frequencies of the values.

12.7.1 Methods of Computation

The Mode is determined by means of location by using any of the following four methods:

1. Inspection Method
2. Grouping Method
3. Graphical Method
4. Empirical relation Method

1. Inspection Method

In this method the value of modal class (or mode) is determined by an inspection of the distribution. Here the observed value or class occurring maximum number of times is the mode or modal class. In other words most frequently occurred value is mode. This is called inspection of values or observations. This method fails in the following situations:

1. If the maximum frequency corresponds to two or more adjacent values, i.e. it fails in bimodal series (a series having two values of mode).
2. If the maximum frequency is present either in the very beginning or at the end of the distribution.
3. If there are irregularities in the distribution i. e. the frequencies of the variable increase or decrease in an irregular way.

In such situations, Grouping method or empirical formula method is applied.

2. Method of Grouping

This method involves the following steps:

1. Put the X values or the classes of values in the first column.
2. Put the corresponding frequencies of X values in the second column.
3. Frequency column is the columns number 1 and so on continues the numbering to the next columns. Circle the largest frequency in the column
4. In the column number 2 add the two's frequencies in groups and circle the largest frequency.
5. Again in the third column, group the frequencies in two's beginning with the second frequency and again circle the largest.
6. Further in the fourth column, group the frequencies in three's beginning with the first frequency.
7. Similarly in the fifth column group the frequencies in three's beginning with the second frequency.
8. Similarly in the sixth column group the frequencies in three's beginning with the third frequency.
9. After completing the grouping table, prepare an Analysis table. Analysis table is prepared to locate the mode or modal class against which the maximum frequency occurs.

Mode in the Case of Continuous Series

In this case, modal class is determined either by inspection method or by grouping method as discussed above. Exact value of the mode can be computed by using the following formula:

$$M_0 = l + \frac{h|f_1 - f_0|}{|f_1 - f_0| + |f_1 - f_2|}$$

Where M_0 is the mode or modal value.

l is the lower limit of the modal class.

h is the interval or width of the modal class

f_1 is the frequency of the modal class

f_0 is the frequency of the class preceding to the modal class

f_2 is the frequency of the class succeeding to the modal class

Before applying the above formula series must be arranged in the ascending order. The series should have equal class intervals.

Note: *This method fails if the series is bimodal or multi-modal in nature. Then we have to apply empirical formula to find the unique value of mode.*

3. Graphical Method

This method is best in the case of continuous frequency distribution. Here first of all draw a histogram with the given class interval and their frequencies.

The rectangle shown in the histogram having the highest height is the modal class (when the maximum frequency does not correspond to more than one class interval). To find the exact value of mode in the modal class, draw two lines diagonally starting from each of its upper corner and ending with the nearest corner of their respective adjacent bar.

Further draw a perpendicular from the intersecting point to the base axis, i.e. X-axis. The point where the perpendicular touches the X-axis gives the value of mode.

4. Empirical Relation Method

This method was designed by the Prof. Karl Pearson. This method is used when all the above discussed methods fail to compute the mode or modal class. In this method he found a relationship in Mean, Median and Mode.

From his experiment he derived a formula, which is also called Empirical formula or Empirical Relation to find the mode.

$$\text{Mode} = 3\ \text{Median} - 2\ \text{Mean}$$

This formula is especially useful to determine the value of mode in case it is ill defined, e.g. in case of bimodal or multimodal distributions.

12.7.2 Merits of Mode

Following are the advantages of the Mode:

1. It is very easy to compute.
2. Mode gives most representative value of a series.
3. The extreme items or observations do not affect it.
4. It can be computed directly from an open-end series without estimating the two extreme class limits.
5. It is best for data having ordering or ranking or classification or categorization.
6. It has no impact of values, i.e. it needs only frequencies.
7. It can be computed or analyzed graphically.
8. To determine its value, Histogram or Frequency Polygon can be used.
9. It is reliable for analyzing skewness of a distribution.

12.7.3 Demerits of Mode

Following are the disadvantages of the Mode:

1. It gives different results in some cases so it is not unique always.
2. It is not based on all the observations.

3. It is not capable of mathematical treatment like mean and median.
4. In case of a continuous and bimodal series its computation becomes difficult.
5. Some times it needs number of trial in continuous series.
6. It is strongly affected by sampling fluctuations.
7. It cannot be determined from a series with unequal class intervals.
8. Some times it creates contradiction.

SOLVED PROBLEMS

Problem 1: Find the mode for the following:

1, 2, 3, 5, 1, 2, 4, 5, 4, 5, 2, 5, 6, 5, 7, 2, 4, 5, 1, 4, 5

Solution: Make the frequency distribution of the given data as:

X	f
1	3
2	4
3	1
4	4
5	7
6	1
7	1

By Inspection method, we know that on counting the instances of the observations, the maximum instances are that of 7 which corresponds to $X = 5$. Hence the Mode of this series is 5.

Problem 2: Find the value of mode for the following distribution:

X:	0-10	10-20	20-30	30-40	40-50	50-60
f:	3	8	17	12	6	4

Solution:

X	f
0-10	3
10-20	8
20-30	17
30-40	12
40-50	6
50-60	4

By inspection method, we find that the maximum number of observations concentrates on the 20-30 class, hence it is the modal class.

Here, $l = 20$, $f_1 = 17$, $f_0 = 8$, $f_2 = 12$, $h = 10$

Putting these values in

$$M_0 = l + \frac{h|f_1 - f_0|}{|f_1 - f_0| + |f_1 - f_2|}$$

$$= 20 + \frac{|17-8|}{|17-8|+|17-12|} \times 10 = 20.64$$

Problem 3: Find the value of mode for the following distribution by using grouping method:

X:	0-10	10-20	20-30	30-40	40-50	50-60
f:	3	8	17	12	6	4

Solution:

Grouping Table

X	f	C1 = 1	C2 = 1+2	C3 = 2+3	C4 = 1+2+3	C5 = 2+3+4	C6 = 3+4+5
0-10	3	3					
			11				
10-20	8	8			28		
20-30	17	17	29	25		37	
30-40	12	12					35
				18			
40-50	6	6			22		
			10				
50-60	4	4					

Analysis Table

X	C1	C2	C3	C4	C5	C6	Total
0-10				Y			1
10-20			Y	Y	Y		3
20-30	Y	Y	Y	Y	Y	Y	6
30-40		Y			Y	Y	2
40-50						Y	1
50-60							0

Since total frequency for the class 20-30 is maximum, hence the mode lies in this class.

Accordingly $l = 20$, $f_1 = 17$, $f_0 = 8$, $f_2 = 12$, $h = 10$

Putting these values in

$$M_0 = l + \frac{h|f_1 - f_0|}{|f_1 - f_0| + |f_1 - f_2|} = 20 + \frac{|17-8|}{|17-8|+|17-12|} \times 10 = 20.64$$

Problem 4: You are given the following data:

Score:	10-20	20-30	30-40	40-50	50-60
No. of Players:	4	8	20	16	2

Find the value of mode graphically.

Solution: First of all arrange the data in continuous series as in the below table:

Score	No. of Players f
10-20	4
20-30	8
30-40	20
40-50	16
50-60	2
	$N = 50$

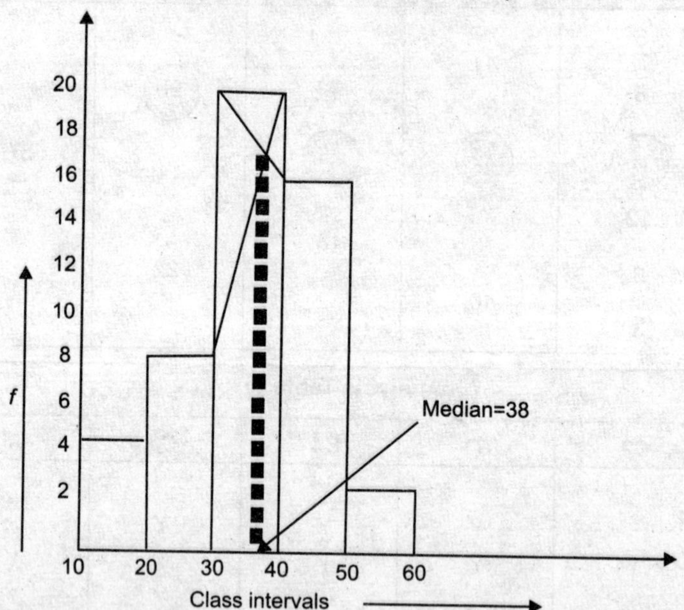

From the above figure, the mode is 33 approx.

Problem 5: Following data is given to you:

Score:	10	20	30	40	50	60	70	80	90
Frequency:	22	11	23	21	21	16	19	5	20

Calculate mode for the above distribution.

Solution:

Grouping Table

Score X	Frequency f	Grouping in Two's		Grouping in Three's		
	Column 1	Column 2	Column 3	Column 4	Column 5	Column 6
10	22	32		56		
20	11		34		55	
30	**23**	**44**				**65**
40	21		**42**	**58**		
50	21	37			**56**	
60	16		35			40
70	19	24		44		
80	5		25			
90	20					

Analysis Table

Column Number	Maximum Frequency in a column	30	40	50	60	70	80	90
1	23	1						
2	44	1	1					
3	42		1	1				
4	58		1	1	1			
5	56			1	1	1		
6	65	1	1	1				
Total Frequencies		**3**	**4**	**4**	**2**	**1**		

As from the above analysis, the highest value of frequency corresponds to two values of X, i.e. 40 and 50. So the given series is a bimodal series. Grouping table method will not give us the unique value of mode. So we have to apply empirical formula method to find the unique value of mode. For that purpose we require the values of mean and median, which are calculated below:

X	f	c.f.	$d' = (X-50)/10$	fd'
10	22	22	–4	–88
20	11	33	–3	–33
30	23	56	–2	–46
40	21	77	–1	–21
50	21	98	0	0
60	16	114	1	16
70	19	133	2	38
80	5	138	3	15
90	20	158	4	80
				$\Sigma fd' = -39$

Median:

$$N/2 = 79$$

So c.f. just greater than 79 is 98 and the value of X corresponding to this c.f. is 50.

Hence median $= 50$

Mean:

$$\bar{X} = A + \frac{\Sigma fd'}{N} \times c$$

$$= 50 + \frac{-39}{158} \times 10 = 47.531$$

Now Mode $= 3$ median $- 2$ mean

$$= 3(50) - 2(47.531) = 54.938$$

PROGRAMS

1. Program to Find the Mode of Individual Series, Using Empirical Formula.

```
#include<stdio.h>
#include<conio.h>
void main( )
{
float x[100],s=0.0,am,median,temp,mode;
int i,n;
clrscr( );
printf(" How many items in the individual distribution ?");
scanf("%d", &n);
printf("\n Enter the data \n");
```

```
for(i=1;i<=n;i++)
{scanf("%f",&x[i]);
s+=x[i];
}
am=s/n;
for(k=1;k<=n-1;k++)
{
for(j=k;j<n;j++)
{
if(x[j]>x[j+1])
{
temp=x[j];
x[j]=x[j+1];
x[j+1]=temp;
}
}
}
if(n%2==0)
median=(x[n/2]+x[n/2+1])/2.0;
else
median=x[(n+1)/2];
mode=3*median-2*am;
printf("\n The MODE is %7.4f",mode);
getch( );
}
```

Output:

```
How many items in the individual distribution? 5
Enter the data
5
20
10
30
45
The MODE is 16.0000
```

2. Program to Find the Mode of Discrete Series, Using Empirical Formula.

```
#include<stdio.h>
#include<conio.h>
void main( )
{
float x[100],sfx=0.0,am,tempx,tempf,median,p,mode;
int i,n,j,k,f[100],sf=0,cf[100];
clrscr( );
printf(" How many items in the discrete series?");
scanf("%d", &n);
printf("\n Enter the data: \n");
for(i=1;i<=n;i++)
scanf("%f",&x[i]);
printf("\n Enter the frequencies: \n");
for(i=1;i<=n;i++)
scanf("%d",&f[i]);
for(k=1;k<=n-1;k++)
{
for(j=k;j<n;j++)
{
if(x[j]>x[j+1])
{
tempx=x[j];
tempf=f[j];
x[j]=x[j+1];
f[j]=f[j+1];
x[j+1]=tempx;
f[j+1]=tempf;
}
}
}
for(i=1;i<=n;i++)
{
sfx=sfx+x[i]*f[i];sf=sf+f[i];
```

```
}
am=sfx/sf;
cf[1]=f[1];
for(i=2;i<n;i++)
cf[i]=cf[i-1]+f[i];p=(float)(sf)/2;
i=0;
while(cf[i]<p)
i++;
median=x[i];
mode=3*median-2*am;
printf("\n The MODE is %7.4f",mode);
getche( );
}
```

Output:

```
How many items in the discrete series? 5
Enter the data
50
10
20
40
30
Enter the frequencies:
3
2
3
2
3
The MODE is 28.462
```

3. Program to Find the Mode of Continuous Series, Using Empirical Formula.

```
#include<stdio.h>
#include<conio.h>
void main( )
{
float x[100],l[100],u[100],sfx=0.0,am,median,mode;
```

```
int i,n,f[100],cf[100],sf=0;
clrscr( );
printf(" How many classes in the continuous series?");
scanf("%d", &n);
printf("\n Enter the lower and upper data: \n");
for(i=1;i<=n;i++)
scanf("%f%f",&l[i],&u[i]);
printf("\n Enter the frequencies: \n");
for(i=1;i<=n;i++)
scanf("%f",&f[i]);
for(i=1;i<=n;i++)
{
x[i]=(l[i]+u[i])/2;
sfx=sfx+x[i]*f[i];sf=sf+f[i];
}
am=sfx/sf;
cf[1]=f[1];
for(i=2;i<n;i++)
cf[i]=cf[i-1]+f[i];}
sf=cf[n];
p=(float)sf/2;
i=0;
while(cf[i]<p)
i++;
median=l[i]+((p-cf[i-1])/ f[i])* (u[i]- l[i]);
mode=3*median-2*am;
printf("\n The MODE is %7.4f",mode);
getch( );
}
```

Output:

How many classes in the continuous series? 5

Enter the lower and upper data:

10 20

20 30

30 40

40 50

50 60

Enter the frequencies:

2

1

4

2

1

The MODE is **35.00**

UNSOVED PROBLEMS

1. Define mode. Explain its merits and demerits.
2. How can you compute a mode?
3. What are the various merits and demerits of mode?
4. What is the relationship between mean, median and mode?
5. Find the mean and mode from the following data:

% marks:	10-19	20-29	30-39	40-49	50-59	60-69	70-79
Students:	8	19	29	36	25	13	4

6. Calculate mean, median, mode for the following data:

Earnings (Rs.):	66-67	67-68	68-69	69-70	70-71	71-22
No. of Persons:	15	24	40	20	14	11

7. An employer decides to offer a cash gift of 5% of the average weekly wage in his factory to every employee. Calculate it taking average to be (i) mode (ii) median.

Monthly wage ('00 Rs.):	20-30	30-40	40-50	50-60	60-70	70-80	80-90
No. of Employees:	28	32	45	60	56	40	20

8. Calculate mode for the following series:

5 10 11 12 13 15 17 19 20 30

9. Calculate mode for the following data:

Weight (kg):	10-20	20-30	30-40	40-50	50-60	60-70	70-80
No of : Persons	2	4	6	8	20	30	40

10. From the following data, obtain the value of mode.

Marks :	0-10	10-20	20-30	30-40	40-50
No of students :	8	10	10	5	7

11. Calculate mode from the following data:

X	f
0-10	20
10-20	25
20-30	30
30-40	35
40-50	40
50-60	45

12. Calculate Mean, Median, and Mode for the following data:

Weight :	10-20	20-30	30-40	40-50	50-60	60-70	70-80	80-90
c.f. :	2	4	10	60	80	90	100	120

13. Calculate Mean, Median, and Mode for the following data:

Central value:	5	10	15	20	30	40	50	60
Frequency:	5	9	13	15	20	15	8	3

12.8 MORE ABOUT POSITIONAL AVERAGES

Quantiles are the points taken at regular intervals from the cumulative distribution in order to divide the whole in a number of equal parts. Dividing ordered data into q essentially equal-sized data subsets is the motivation for q-quantiles; the quantiles are the data values marking the boundaries between consecutive subsets.

Some quantiles have special names:

- The 100-quantiles are called percentiles.
- The 20-quantiles are called duo-deciles.
- The 10-quantiles are called deciles.
- The 9-quantiles are called noniles.
- The 5-quantiles are called quintiles.
- The 4-quantiles are called quartiles.

Out of all the quantiles, quartiles, deciles and percentiles are widely used. Here we will discuss only these quantiles.

12.9 QUARTILE

In descriptive statistics, a **quartile** is any of the three values which divide the sorted data set into four equal parts, so that each part represents 1/4th of the given data values. The total observations are broken into four equal parts by the points: 25%, 50%, and 75%. In other words, these three points divides the series or data into four equal parts. It is denoted as Q_1, Q_2, and $Q3$ respectively.

Thus

- **first quartile** (designated as Q_1) = **lower quartile** = cuts off lowest 25% of data = *25th percentile*
- **second quartile** (designated as Q_2) = *median* = cuts data set in half = *50th percentile*
- **third quartile** (designated as Q_3) = **upper quartile** = cuts off highest 25% of data, or lowest 75% = *75th percentile*

12.9.1 Methods of Computing Quartiles

Calculation of quartiles is done on the same way like the calculation of median. The following steps are involved in the calculation of quartiles:

In Case of Individual Series:

Following steps are being followed to calculate quartiles in individual series:

1. Arrange the data in ascending or descending order.
2. Find $N/4^{th}$ item, and that gives the value of Q_1, where N (it should be the multiple of 4) is the total number of observations.
3. Q_3 is determined by finding $3N/4^{th}$ item, where N is the total number of observations.

In Case of Discrete Series:

Following steps are being followed to calculate quartiles in individual series:

1. Arrange the data in ascending order
2. Prepare the 'less than' cumulative frequency (c.f.) distribution.
3. Find $N/4$ (for Q_1) or $3N/4$ (for Q_3)
4. Mark the cumulative frequencies just greater than $N/4$ (for Q_1) or $3N/4$ (for Q_3)
5. The corresponding values of the variable give the respective quartiles.

In Case of Continuous Series:

Following steps are being followed to calculate quartiles in continuous series:

1. Arrange the data in ascending order
2. Prepare the less than cumulative frequency (c.f.) distribution.
3. Find $N/4$ (for Q_1) or $3N/4$(for Q_3).
4. Mark the cumulative frequency just greater than $N/4$(for Q_1)or $3N/4$(for Q_3).
5. The corresponding classes are known as quartile classes and the value of quartile lies in these respective classes.

The values of quartile are now calculated by using the following formulae:

$$Q_1 = l + \frac{\left(\frac{N}{4} - c\right)}{f} \times h \quad \text{and} \quad Q_3 = l + \frac{\left(\frac{3N}{4} - c\right)}{f} \times h$$

l is the lower limit of the respective quartile class

f is the frequency of the respective quartile class

c is the cumulative frequency of the class preceding the respective quartile class.

N is the total frequency

h is the class size of the respective quartile class

Note: *In addition to the above methods, quartiles can be calculated graphically just like median.*

12.10 DECILES

Deciles are the nine points in a set of data which are 1/10, 2/10, …, 9/10 as in the same way for quartiles. It divides the series in 10 equal parts. These are denoted as $D_1, D_2, D_3, D_4, D_5, D_6, D_7, D_8, D_9$. So a decile is any of the 9 values that divide the sorted data into 10 equal parts, such that each part represents 1/10th of the given distribution. Deciles split ordered data into tenths. Thus

- The 1st decile cuts off lowest 10% of data, i. e., the 10th percentile.
- The 5th decile cuts off lowest 50% of data, i. e., the 50th percentile, 2nd quartile, or median.
- The 9th decile cuts off lowest 90% of data, i. e., the 90th percentile.

12.10.1 Methods of Computing Deciles

In Case of Individual Series:

Following steps are being followed to calculate k-th decile say D_k in individual series:

1. Arrange the data in ascending or descending order.
2. Find $kN/10^{th}$ item, that gives the value of D_k, where *N* is the total number of observations.

In Case of Discrete Series:

Following steps are being followed to calculate kth decile say D_k in discrete series:

1. Arrange the data values in increasing order.
2. Prepare the 'less than' cumulative frequency (c.f.) distribution.
3. Find (*kN*/10).
4. Mark the cumulative frequency just greater than kN/10.
5. The corresponding value of the variable gives D_k, i.e. the *k-th* decile.

In Case of Continuous Series:

Following steps are being followed to calculate k-th decile say D_k in continuous series:

1. Arrange the classes in increasing order.
2. Prepare the less than cumulative frequency (c.f.) distribution.
3. Find $(kN/10)$.
4. Mark the cumulative frequency just greater than $(kN/10)$.
5. The corresponding class is known as k-th decile class and the value of k-th decile lies in this class.

The value of k-th decile is now calculated by using the following formula:

$$D_k = l + \frac{\left(\frac{kN}{10} - c\right)}{f} \times h$$

where

l is the lower limit of marked decile class

f is the frequency of the marked decile class

c is the cumulative frequency of the class preceding the marked decile class.

N is the total frequency

h is the class size of the marked decile class

In addition to the above methods, deciles can be calculated graphically just like median.

12.11 PERCENTILES

A percentile is a measure that tells us what percent of the total frequency scored at or below that measure. So the percentiles are the set of 99 points that divides the series into 100 equal parts. These are denoted by $P_1, P_2, \ldots, P_{99}$. Percentiles split a set of ordered data into hundredths. For example, 70% of the data should fall below the 70th percentile. A percentile is the value of a variable below which a certain percent of observations fall. So the 20th percentile is the value (or score) below which 20 percent of the observations may be fou

nd. Note that the 25th percentile is the first quartile and the 50th percentile is the median.

Relation between percentile, decile and quartile:

- $P_{25} = Q_1$
- $P_{50} = D_5 = Q_2 =$ value of median
- $P_{75} = Q_3$
- $P_{10} = D_1$
- $P_{20} = D_2$
- $P_{30} = D_3$
- $P_{40} = D_4$

- $P_{60} = D_6$
- $P_{70} = D_7$
- $P_{80} = D_8$
- $P_{90} = D_9$

12.11.1 METHODS OF COMPUTING PERCENTILES

In Case of Individual Series:

Following steps are being followed to calculate k-th percentile, say P_k, in individual series:

1. Arrange the data in ascending or descending order.
2. Find $kN/100^{th}$ item, which is equal to P_k, where N(it should be a multiple of 100) is the total number of observations.

In Case of Discrete Series:

Following steps are being followed to calculate k-th percentile, say P_k, in discrete series:

1. Arrange the data in increasing order
2. Prepare the 'less than' cumulative frequency (c.f.) distribution.
3. Find ($kN/100$).
4. Mark the cumulative frequency just greater than $kN/100$.
5. The corresponding value of the variable gives P_k.

In Case of Continuous Series:

Following steps are being followed to calculate kth percentile say P_k in continuous series:

1. Arrange the data in increasing order
2. Prepare the less than cumulative frequency (c.f.) distribution.
3. Find ($kN/100$).
4. Mark the cumulative frequency just greater than ($kN/100$).
5. The corresponding class is known as k-th percentile class and the value of k-th percentile lies in this class.

The value of k-th percentile is now calculated by using the following formula:

$$P_k = l + \frac{\left(\frac{kN}{100} - c\right)}{f} \times h$$

where

l is the lower limit of respective percentile class

f is the frequency of the respective percentile class

c is the cumulative frequency of the class preceding the respective percentile class.

N is the total frequency

h is the class size of the respective percentile class

In addition to the above methods, percentiles can be calculated graphically just like median.

SOLVED PROBLEMS

Problem 1: Calculate quartiles, 6th decile and 70th percentile from the following data:

Wages in Rs. 000	No. of workers
0-10	5
10-20	8
20-30	7
30-40	12
40-50	28
50-60	20
60-70	10
70-80	10

Solution:

Wages	No. of workers *f*	Less than c.f.
0-10	5	5
10-20	8	13
20-30	7	20
30-40	12	32
40-50	28	60
50-60	20	80
60-70	10	90
70-80	10	100

For Quartiles:

$$\frac{N}{4} = \frac{100}{4} = 25 \quad \text{and} \quad \frac{3N}{4} = \frac{300}{4} = 75$$

Mark the c.f. just greater than 2, i.e 32. Hence the corresponding class is 30-40.

$$Q_1 = l + \frac{\left(\frac{N}{4} - c\right)}{f} \times h$$

$$= 30 + \frac{(25 - 20)}{12} \times 10 = 34.16$$

Mark the c.f. just greater than 75, i.e. 80. Hence the corresponding class is 50-60.

$$Q_3 = l + \frac{\left(\frac{3N}{4} - c\right)}{f} \times h$$

$$= 50 + \frac{(75-60)}{20} \times 10 = 57.5$$

For 6th decile:

$$\frac{6N}{10} = \frac{600}{10} = 60$$

Mark the c.f. just greater than 60, i.e. 80. Hence the corresponding class is 50-60.

$$D_6 = l + \frac{\left(\frac{6N}{10} - c\right)}{f} \times h$$

$$= 50 + \frac{(60-60)}{20} \times 10 = 50$$

For 7th percentile:

$$\frac{70N}{100} = \frac{7000}{100} = 70$$

Mark the c.f. just greater than 70, i.e. 80. Hence the corresponding class is 50-60.

$$P_{70} = l + \frac{\left(\frac{70N}{100} - c\right)}{f} \times h$$

$$= 50 + \frac{(70-60)}{20} \times 10 = 55$$

UNSOLVED PROBLEMS

1. What do you know about Quartile and Inter Quartile?
2. Write a short note on following:

   (i) Decile (ii) Percentile (iii) Septile

   (iv) Octile (v) Hexile (vi) Quantile
3. Determine the median and the two extreme quartiles from the following data:

   X 19 12 11 15 17 16 13 16
4. You are provided the data given below, find out the quartiles.

   4 9 2 10 18 7 6 3

5. From the given data relating to the size of shoes determine the following for the sizes:

(i) Median (ii) Upper Quartile (iii) 4th deciles

Size of the shoe:	3	4	5	6	7	8	9	10
Frequency:	7	6	1	6	5	8	14	15

6. From the given data, find the median and two extreme quartiles of the output.

Output :	10-14	15-19	20-24	25-29	30-34	35-39
No. of Workers:	2	5	6	15	12	3

7. Determine the value of median, 9th deciles, and 80 percentile for the marks from the following data:

Marks:	10-14	15-19	20-24	25-29	30-34	35-39
No. of Boys:	12	15	16	14	32	30

8. From the following data, locate the median and quartiles of the income.

Income	No. of persons
Below Rs. 30	50
Rs. 30 and below Rs. 40	100
Rs. 40 and below 50	65
Rs. 50 and below Rs. 60	200
Rs. 60 and below 70	50
Rs. 70 and below Rs 80	25

9. You are provided the data given below; ascertain the value of median and upper quartile.

Marks:	70%	60%	50%	40%	30%	20%
No. of students:	17	16	11	15	50	40

10. Find the median and the first and the third quartiles of the daily wages.

Daily wages:	30-32	32-34	34-36	36-38	38-40	40-42
No. of Workers:	30	15	16	14	32	30

11. You are provided the following data. Draw a cumulative frequency graph from the following distribution and locate the two extreme quartiles of the marks.

Marks:	10	11	12	13	14	15	16	17	18	19	20	30
No. of Students:	38	35	82	70	15	16	20	60	30	40	75	26

12.12 FREQUENCY CHARTS

12.12.1 Variable

A quantity which can vary from one individual to another is called a **variable**. It is also called a **variate.** Wages, barometer readings, rainfall records, heights, and weights are the common examples of variables. Quantities which can take any numerical value within a certain range are called

continuous variables. For example, the height of a child at various ages is a continuous variable since, as the child grows from 120 cm to 150 cm, his height assumes all possible values within the limit. Quantities which are incapable of taking all possible values are called **discontinuous** or **discrete variables.** For example, the number of rooms in a house can take only the integral values such as 2, 3, 4, etc.

12.12.2 Frequency Distributions

The scores of 50 students in mathematics are arranged below according to their roll numbers, the maximum scores being 100. 19, 70, 75, 15, 0, 23, 59, 56, 27, 89, 91, 22, 21, 22, 50, 89, 56, 73, 56, 89, 75, 65, 85, 22, 3, 12, 41, 87, 82, 72, 50, 22, 87, 50, 89, 28, 89, 50, 40, 36, 40, 30,28, 87, 81, 90, 22, 15, 30, 35. The data given in the crude form (or raw form) is called **ungrouped data.**

If the data is arranged in ascending or descending order of magnitude, it is said to be arranged in an array. Let us now arrange it in the intervals 0–10, 10–20, 20–30, 30–40, 40–50, 50–60, 60–70, 70–80, 80–90, 90–100. This is arranged by a method called the **tally method.** In this we consider every observation and put it in the suitable class by drawing a vertical line. After every 4 vertical lines, we cross it for the 5th entry and then a little space is left and the next vertical line is drawn.

Scores (Class-interval)	Number of Students	Frequecy (*f*)	Cumulative frequencies
0-10	||	2	2
10-20	|| ||	4	6
20-30	~~||||~~ ~~||||~~	10	16
30-40	|| ||	4	20
40-50	|||	3	23
50-60	~~||||~~ |||	8	31
60-70	|	1	32
70-80	~~||||~~	5	37
80-90	~~||||~~ ~~||||~~ |	11	48
90-100	||	2	50
Total		$\Sigma f = 50$	

Simply a **frequency distribution.** The groups are called the **classes** and the boundary ends 0, 10, 20, ... etc. are called **class limits.** In the class limits 10-20, 10 is the **lower limit** and 20 is the **upper limit.** The difference between the upper and lower limits of a class is called its magnitude or **class-interval.** The number of observations falling within a particular class is called its **frequency** or **class frequency.**

The frequency of the class 80-90 is 11. The variate value which lies mid-way between the upper and lower limits is called mid-value or mid-point of that class. The mid-points of these are respectively 5, 15, 25, 35, ... The **cumulative frequency** corresponding to a class is the total of all the frequencies up to and including that class. Thus the cumulative frequency of the class 10-20 is 2 + 4, *i.e.,* 6 the cumulative frequency of the class 20-50 is 6 + 10, *i.e.,* 16, and so on.

While preparing the frequency distribution the following points must be remembered:

1. The class-intervals should be of equal width as far as possible A comparison of different distributions is facilitated if the class interval is used for all. The class-interval should be an integer as far as possible.
2. The number of classes should never be fewer than 6 and not more than 30. With a smaller number of classes, the accuracy may be lost, and with a larger number of classes, the computations become tedious.
3. The observation corresponding to the common point of two classes should always be put in the higher class. For example, a number corresponding to the value 30 is to be put up in the class 30-40 and not in 20-30.

The following forms of the above table may also be used:

Cumulative Frequency			
Scores	**Number of Students**	**Scores**	**Number of Students**
Under 10	2	above 90	2
Under 20	6	above 80	13
Under 30	16	above 70	18
Under 40	20	above 60	19
Under 50	23	above 50	27
Under 60	31	above 40	30
Under 70	32	above 30	34
Under 80	37	above 20	44
Under 90	48	above 10	48
Under 100	50	above 0	50

12.13 GRAPHICAL REPRESENTATION OF A FREQUENCY DISTRIBUTION

Representation of frequency distribution by means of a diagram makes the unwieldy data intelligible and conveys to the eye the general run of the observations. The graphs and diagrams have a more lasting effect on the brain. It is always easier to compare data through graphs and diagrams. Forecasting also becomes easier with the help of graphs. Graphs help us in interpolation of values of the variables.

However there are certain disadvantages as well. Graphs do not give measurements of the variables as accurate as those given by tables. The numerical value can be obtained to any number of decimal places in a table, but from graphs it can not be found to 2nd or 3rd places of decimals. Another disadvantage is that it is very difficult to have a proper selection of scale. The facts may be misrepresented by differences in scale.

12.13.1 Types of Graphs and Diagrams

Generally the following types of graphs are used in representing frequency distributions:

(1) Histograms, (2) Frequency Polygon, (3) Frequency Curve, (4) Cumulative Frequency Curve or the Ogive, (5) Historigrams, (6) Bar Diagrams, (7) Area Diagrams, (8) Circles or Pie Diagrams, (9) Prisms, (10) Cartograms and Map Diagrams, (11) Pictograms.

1. Histograms

To draw the histograms of a given grouped frequency distribution, mark off along a horizontal base line all the class-intervals on a suitable scale. With the class-intervals as bases, draw rectangles with the areas proportional to the frequencies of the respective class-intervals. For equal class-intervals, the heights of the rectangles will be proportional to the frequencies. If the classintervals are not equal, the heights of the rectangles will be proportional to the ratios of the frequencies to the width of the corresponding classes. A diagram with all these rectangles is a **Histogram.**

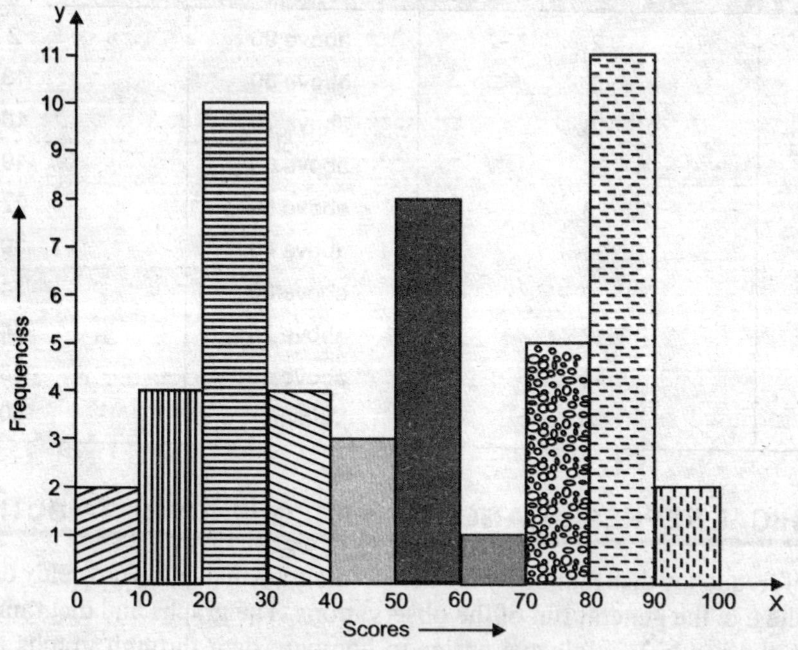

(Histogram for the previous table)

Histograms are also useful when the class-intervals are not of the same width. They are appropriate to cases in which the frequency changes rapidly.

2. Frequency Polygon

If the various points are obtained by plotting the central values of the class intervals as *x* co-ordinates and the respective frequencies as the *y* co-ordinates, and these points are joined by straight lines taken in order, they form a polygon called **Frequency Polygon.**

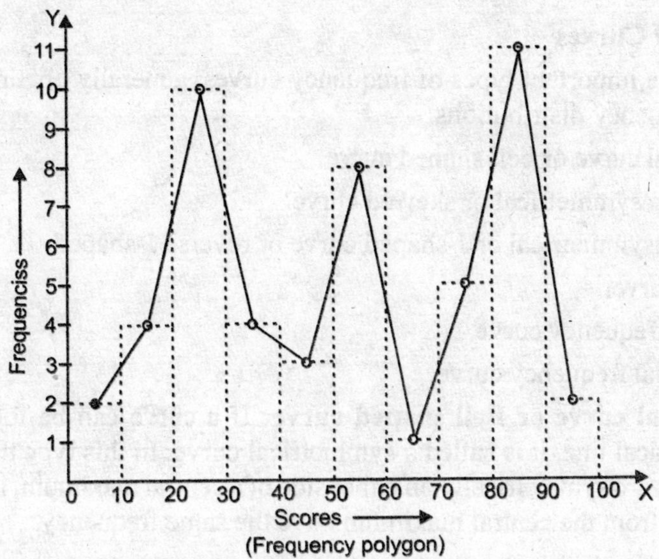

(Frequency polygon)

In a frequency polygon the variables or individuals of each class are assumed to be concentrated at the mid-point of the class-interval. Here in this diagram dotted is the **Histogram** and a polygon with lines as sides is the **Frequency Polygon.**

3. Frequency Curve

If through the vertices of a frequency polygon a smooth freehand curve is drawn, we get the **Frequency Curve.** This is done usually when the class-intervals are of small widths.

4. Cumulative Frequency Curve or the Ogive

If from a cumulative frequency table, the upper limits of the class taken as x co-ordinates and the cumulative frequencies as the y co-ordinates and the points are plotted, then these points when joined by a freehand smooth curve give the **Cumulative Frequency Curve or the Ogive.**

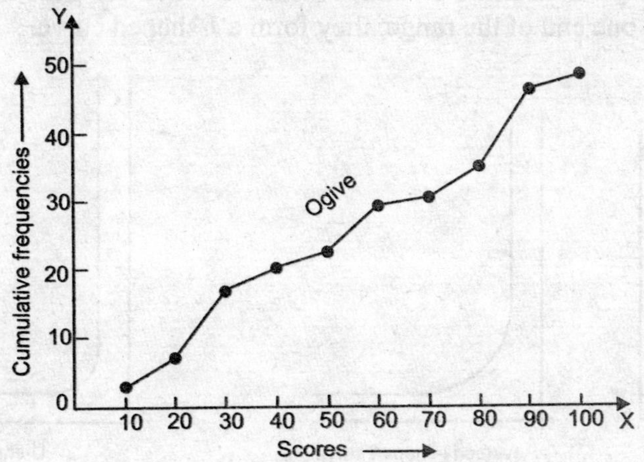

5. Types of Frequency Curves

Following are some important types of frequency curves, generally obtained in the graphical representations of frequency distributions:

(a) Symmetrical curve or bell shaped curve.

(b) Moderately asymmetrical or skewed curve.

(c) Extremely asymmetrical or J-shaped curve or reverse J-shaped.

(d) U-shaped curve.

(e) A bimodal frequency curve.

(f) A multimodal frequency curve.

(a) Symmetrical curve or Bell shaped curve: If a curve can be folded symmetrically along a vertical line, it is called a symmetrical curve. In this type the class frequencies decrease to zero symmetrically on either side of a central maximum, i.e., the observations equidistant from the central maximum have the same frequency.

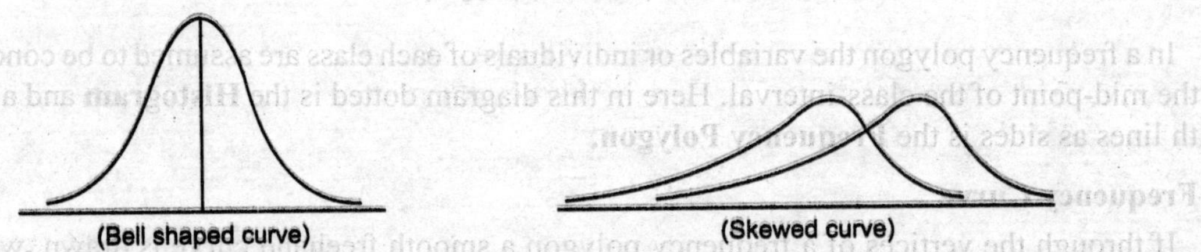

(Bell shaped curve) (Skewed curve)

(b) Moderately asymmetrical or skewed curve: If there is no symmetry in the curve, it is called a **Skew Curve.** In this case the class frequencies decrease with greater rapidity on one side of the maximum than on the other. In this curve one tail is always longer than the other. If the long tail is to the to be a positive side, it is said to be a positive skew curve, if long tail is to the negative side, it is said to be a negative skew curve.

(c) Extremely asymmetrical or J-shaped curve: When the class frequencies run up to a maximum at one end of the range, they form a J-shaped curve.

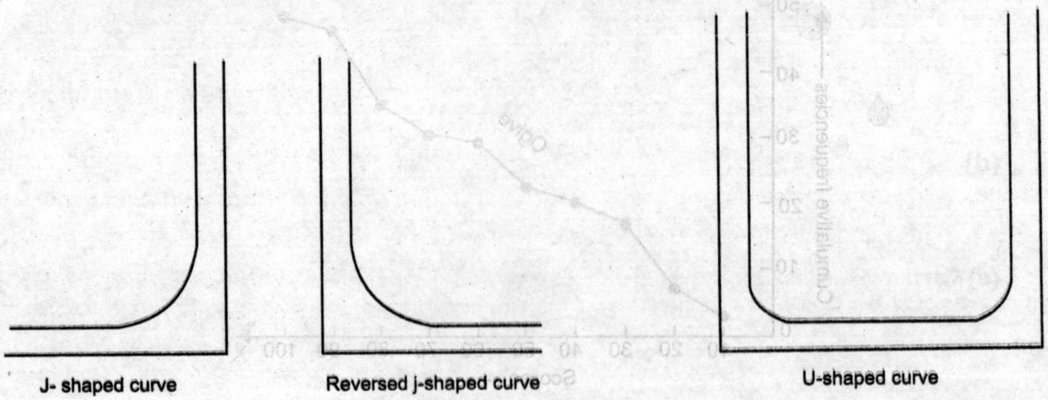

J- shaped curve Reversed j-shaped curve U-shaped curve

(d) **U-shaped curve:** In this curve, the maximum frequency is at the ends of the range and a maximum towards the center.

(e) A Bimodal curve has two maxima.

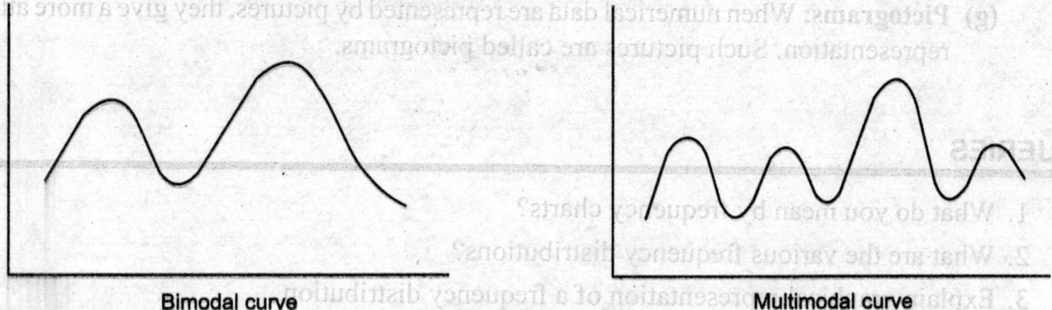

(f) A multimodal curve has more than two maxima.

6. Diagrams

(a) **Bar diagrams:** Bar diagrams are used to compare the simple magnitude of different items. In bar diagrams, equal bases on a horizontal or vertical line are selected and rectangles are constructed with the length proportional to the given data. The width of bars is an arbitrary factor. The distance between two bars should be taken at about one-half of the width of a bar.

(b) **Area diagrams:** When the difference between two quantities to be compared is large, bars do not show the comparison so clearly. In such cases, squares or circle are used.

(c) **Circle or Pie-diagrams:** When circles are drawn to represent an area equivalent to the figures, they are said to form pie-diagrams or circlesdiagrams. In case of circles, the square roots of magnitudes are proportional to the radii.

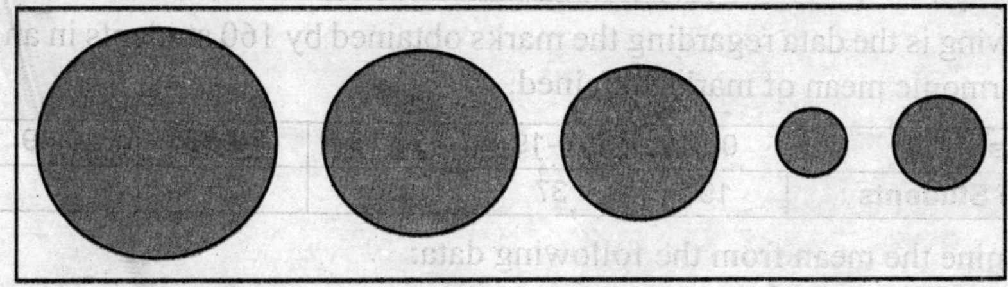

(d) **Subdivided Pie-diagram:** Subdivided Pie-diagrams are used when comparison of the component parts is done with another and the total. The total value is equated to 360° and then the angles corresponding to the component parts are calculated.

(e) **Prisms and Cubes:** When the ratio between the two quantities to be compared is very great so that even area diagrams are not suitable, the data can be represented by spheres, prisms, or cubes. Cubes are in common use. Cubes are constructed on sides which are taken in the ratio of cube roots of the given quantities.

(f) **Cartograms or map diagrams:** Cartograms or map diagrams are most suitable for geographical data. Rainfalls and temperature in different parts of the country are shown with dots or shades in a particular map.

(g) **Pictograms:** When numerical data are represented by pictures, they give a more attractive representation. Such pictures are called pictograms.

QUERIES

1. What do you mean by frequency charts?
2. What are the various frequency distributions?
3. Explain graphical representation of a frequency distribution.
4. Explain various types of graphs and diagrams
5. What are the various types of frequency curves?
6. Explain various types of diagrams.

UNSOLVED PROBLEMS

1. Calculate weighted mean for the following data:

Items :	40	50	55	60	70	80
Weight :	2	3	6	7	2	7

2. Calculate weighted Geometric mean of the following:

Items :	120	125	130	135	140	145	150	155	160
Weights:	3	4	5	5	3	5	2	1	1

3. Following is the data regarding the marks obtained by 160 students in an examination. Find the Harmonic mean of marks obtained.

Marks :	0-9	10-19	20-29	30-39	40-49
No of Students :	19	37	60	25	20

4. Determine the mean from the following data:

X:	100	150	160	168	167	166	144	165
f:	120	130	176	167	155	168	140	159

5. You are provided the following data. Find out the mean.

X:	17	70	60	68	60	66	44	65
f:	71	60	76	67	55	68	40	59

6. From the given data, find out the missing frequencies when median is 30 and the total no. of girls is 100.

Marks :	0-10	10-20	20-30	30-40	40-50	50-60
No. of girls :	7	?	11	?	17	20

7. You are provided the following data, determine the Median and lower quartile and upper quartile by graphical method.

Weights :	10-14	15-19	20-24	25-29	30-34	35-39	40-49
No. of Boys :	12	15	16	14	32	30	34

8. Locate the median by graphic method from the following distribution:

Marks :	0-10	10-20	20-30	30-40	40-50	50-60
No. of students :	1	5	6	4	3	0

9. Locate the mode from the following series using a suitable method.

Marks :	10	20	30	40	50	60	70
No. of students :	38	14	7	30	18	6	20

10. Find out the mean weight of the persons of the following:

Weight in kg :	179	170	160	168	167	166	144	165
No. of persons :	17	16	76	67	15	16	14	15

11. You are given the following data. Find the A.M.

Marks :	0-5	50-15	15-45	45-55	55-95
No. of students :	7	3	8	7	6

12. From the following data, determine the arithmetic mean. of income:

Income :	279	270	260	268	367	266	244	365
No. of students :	71	16	17	16	15	168	140	159

13. From the following data, determine the arithmetic mean:

X :	179	170	160	168	167	166	144	165
f :	171	160	176	167	155	168	140	159

14. Find out the missing frequency from the following distribution for which the arithmetic mean is 16.

Wages :	41.5	44.5	47.5	50.5	53.5
No. of workers :	31	58	60	—	27

15. Find out the missing frequency from the following distribution for which mean is 46.2.

Students:	0-5	5-10	10-15	15-20	20-25
No. of workers :	31	58	60	—	27

16. From the following data, find out the geometric mean (G.M).

X:	179	170	160	168	167	166	144	165
f:	171	160	176	167	155	168	140	159

17. From the following data of income of 8 persons, find the arithmetic mean by using
(i) Direct method (ii) Shortcut method

Income in 000 Rs.	300	100	200	350	400	280	380	450

18. From the following data, calculate the average marks secured by students, using appropriate method.

Marks :	65	80	100	109	102	112	114
No. of Students :	110	120	128	140	180	190	123

19. From the following data, determine the average income by the step deviation method.

Income :	700	600	400	300	900	1100	1200
Person :	2	8	18	16	11	6	2

20. Determine the Arithmetic Mean from the given observations:

X:	10	11	12	13	14	15	16	17	18	19	20	21
f:	9	35	70	105	116	107	88	46	40	30	38	5

21. Determine the arithmetic mean of the wages from the series as given below:

Wages :	11-13	13-15	15-17	16-17	17-18	18-19	19-20
No. of Workers :	3	7	2	6	8	3	5

22. From the following series, find out arithmetic mean.

X:	179	170	160	168	167	166	144	165
f:	171	160	176	167	155	168	140	159

23. Find the arithmetic mean from the given data:

X:	100	102	105	106	101	110	225	112	115
f:	90	92	94	95	92	100	101	110	112

24. Find the Arithmetic mean of marks obtained from the following series:

Marks :	0	10	20	30	40	50	60	70
No. of students:	12	60	76	67	55	68	25	1

25. Compute the geometric mean for the following data:
.7894 .03456 .0040 .6465 .0007 .00975 .8754 .854

26. You are given the following data. Find out the Geometric mean.
179 170 160 168 167 166 144 165

27. Compute the Geometric Mean of the following stock prices:

Stock prices(Rs) :	60-79.9	80-99.9	100-119.9	120-139.9	140-159.9
Frequency :	7	6	8	16	5

28. From the following data relating to the marks, locate the value of mode by appropriate method:

Marks :	30-32	32-34	34-36	36-38	38-40	40-42
No. of Students :	4	10	16	14	32	30

29. You are given the following data. Find out the value of Mode by graphic method.

Marks :	0-10	10-20	20-30	30-40	40-50
No. of Boys :	6	12	7	11	21

30. From the following frequency distribution, determine the value of Mode by a suitable method.

Output :	10-14	15-19	20-24	25-29	30-34	35-39
No. of Workers :	2	5	6	15	12	3

31. You are given the following data, from this find the value of mode by appropriate method.

Size of items :	0-10	0-20	0-30	0-40	0-50	0-60
Frequency :	10	14	15	17	19	21

32. From the following data, determine the value of mode using appropriate method:

Central value :	10	20	30	40	50	60	70
Frequency :	12	30	80	100	111	135	200

33. You are provided the data given below, from this find the value of mode.

Marks :	10-14	15-19	20-24	25-29	30-34	35-39
Frequency :	38	15	16	17	18	33

34. You are provided the data given below. Find out the mode.

Marks less than :	10	20	30	40	50	60
Students :	32	5	6	15	20	23

35. Determine the mean, mode, and median from the following data:

Central value :	5	15	25	35	45	55
Frequency :	6	9	13	20	15	30

36. Calculate the harmonic mean for the following series:

X :	19	12	10	18	17	16	13	16

37. You are provided the data as given below. For this find the Harmonic mean.

Values :	1	0.17	0.6	0.8	0.7	0.66	0.44	0.65
Weights :	7	6	1	6	5	8	14	15

38. Compute the weighted Harmonic mean of the income for the following series:

Income in Rs Lac :	1	0.3	1.5	4.5	170	0.1	0.44	0.16
Persons :	8	15	2	10	10	16	14	19

39. You are given the following data. Find out the harmonic mean of marks.

Marks :	0-10	10-20	20-30	30-40	40-50	50-60
No. of girls :	7	10	11	15	17	20

40. You are provided the data as given below, calculate the harmonic mean of the output per worker.

Output :	10-14	15-19	20-24	25-29	30-34	35-39
No. of Workers :	2	5	6	15	12	3

CHAPTER 13

Measures of Dispersion

13.1 INTRODUCTION

A modern student of statistics is mainly interested in the study of variability and uncertainty. In this section we shall discuss variability and its measures and uncertainty will be discussed in probability. We live in a changing world. Changes are taking place in every sphere of life. A man of statistics does not show much interest in those things which are constant. The total area of the earth may not be very important to a research minded person but the area under different crops, area covered by forests, area covered by residential and commercial buildings are figures of great importance because these figures keep on changing form time to time and from place to place. Very large number of experts is engaged in the study of changing phenomenon. Experts working in different countries of the world keep a watch on forces which are responsible for bringing changes in the fields of human interest. The agricultural, industrial and mineral production and their transportation from one part to the other parts of the world are the matters of great interest to the economists, statisticians, and other experts. The changes in human population, the changes in standard living, and changes in literacy rate and the changes in price attract the experts to make detailed studies about them and then correlate these changes with the human life. Thus variability or variation is something connected with human life and study is very important for mankind.

13.2 MEASURES OF DISPERSION

The word dispersion has a technical meaning in statistics. The average measures the center of the data. It is one aspect observations. Another feature of the observations is as to how the observations are spread about the center. The observation may be close to the center or they may be spread away from the center. If the observation are close to the center (usually the arithmetic mean or median), we say that dispersion, scatter or variation is small. If the observations are spread away from the center, we say dispersion is large. Suppose we have three groups of students who have obtained the following marks in a test. The arithmetic means of the three groups are also given below:

Group *A*: 46, 48, 50, 52, 54 $\overline{X}_A = 50$

Group *B*: 30, 40, 50, 60, 70 $\overline{X}_B = 50$

Group *C*: 40, 50, 60, 70, 80 $\overline{X}_C = 60$.

In a group *A* and *B* arithmetic means are equal, i.e $\overline{X}_A = \overline{X}_B = 50$. But in group *A* the observations are concentrated on the center. All students of group *A* have almost the same level of performance. We say that there is consistence in the observations in group *A*. In group *B* the mean is 50 but the observations are not closed to the center. One observation is as small as 30 and one observation is as large as 70. Thus there is greater dispersion in group *B*. In group *C* the mean is 60 but the spread of the observations with respect to the center 60 is the same as the spread of the observations in group *B* with respect to their own center which is 50. Thus in group *B* and *C* the means are different but their dispersion is the same. In group *A* and *C* the means are different and their dispersions are also different. Dispersion is an important feature of the observations and it is measured with the help of the measures of dispersion, scatter or variation. The word variability is also used for this idea of dispersion.

Definitions:

According to A.L Bowley, *"Dispersion is the measure of variation of the elements"*.

According to Simpson and Kafka, *"Dispersion is the measurement of the mass of figures in a series about an average is called measure of variation"*.

The study of dispersion is very important in statistical data. If in a certain factory there is consistence in the wages of workers, the workers will be satisfied. But if some workers have high wages and some have low wages, there will be unrest among the low paid workers and they might go on strikes and arrange demonstrations. If in a certain country some people are very poor and some are very high rich, we say there is economic disparity. It means that dispersion is large. The idea of dispersion is important in the study of wages of workers, prices of commodities, standard of living of different people, distribution of wealth, distribution of land among framers and various other fields of life. Some brief definitions of dispersion are:

1. The degree to which numerical data tend to spread about an average value is called the dispersion or variation of the data.
2. Dispersion or variation may be defined as a statistics signifying the extent of the scatteredness of items around a measure of central tendency.
3. Dispersion or variation is the measurement of the scatter of the size of the items of a series about the average.

13.2.1 Properties of Measure of Dispersion

Following are the some major and important properties of dispersion:

1. A good measure of dispersion is that which is simple to understand.
2. It should be easy to calculate
3. A good measure of dispersion should be rigidly defined.
4. Dispersion should be based on each and every item of the distribution.
5. It should be capable of further algebraic treatment.
6. A good measure should also enjoy the quality of sample at ability that means it should be little affected by fluctuation of sampling.
7. Last but not least, the measure of dispersion should not be unduly affected by extreme items.

13.2.2 Uses of Dispersion

There are number of reasons to use dispersion which is as:

1. It is helpful in determining the reliability of an average.
2. It is also helpful during the comparison of two or more series corresponding to their variability.
3. It is helpful in the use of other statistical measures such as correlation, regression etc.
4. It is useful in research.
5. Use of dispersion is beneficial in daily life.

13.3 TYPES OF MEASURES OF DISPERSION

For the study of dispersion, we need some measures which show whether the dispersion is small or large. There are two types of measure of dispersion which are:

1. Absolute Measure of Dispersion
2. Relative Measure of Dispersion

13.3.1 Absolute Measures of Dispersion

These measures give us an idea about the amount of dispersion in a set of observations. They give the answers in the same units as the units of the original observations. When the observations are in kilograms, the absolute measure is also in kilograms. If we have two sets of observations, we cannot always use the absolute measures to compare their dispersion. We shall explain later as to when the absolute measures can be used for comparison of dispersion in two or more than two sets of data. The absolute measures which are commonly used are:

1. The Range
2. Intrequartile range, Decile Range, Percentile Range
3. The Quartile Deviation
4. The Mean Deviation
5. The Standard deviation and Variance

13.3.2 Relative Measure of Dispersion

These measures are calculated for the comparison of dispersion in two or more than two sets of observations. These measures are free of the units in which the original data is measured. If the original data is in dollar or kilometers, we do not use these units with relative measure of dispersion. These measures are a sort of ratio and are called coefficients. Each absolute measure of dispersion can be converted into its relative measure. Thus the relative measures of dispersion are:

1. Coefficient of Range or Coefficient of Dispersion.
2. Coefficient of Quartile Deviation or Quartile Coefficient of Dispersion.
3. Coefficient of Mean Deviation or Mean Deviation of Dispersion.
4. Coefficient of Standard Deviation or Standard Coefficient of Dispersion.

Coefficient of Variation (a special case of Standard Coefficient of Dispersion)

13.3.3 Dispersion by the Method of Graph (Lorenz Curve)

Now we will discuss each method one by one by using suitable examples.

13.4 THE RANGE

Range is defined as the difference between the maximum and the minimum observation of the given data. If x_m denotes the maximum observation x_0 denotes the minimum observation then the range is defined as

$$\text{Range} = x_m - x_0$$

In case of grouped data, the range is the difference between the upper boundary of the highest class and the lower boundary of the lowest class. It is also calculated by using the difference between the mid points of the highest class and the lowest class. It is the simplest measure of dispersion. It gives a general idea about the total spread of the observations. It does not enjoy any prominent place in statistical theory. But it has its application and utility in quality control methods which are used to maintain the quality of the products produced in factories. The quality of products is to be kept within certain range of values.

The range is based on the two extreme observations. It gives no weight to the central values of the data. It is a poor measure of dispersion and does not give a good picture of the overall spread of the observations with respect to the center of the observations. Let us consider three groups of the data which have the same range:

Group *A*: 30, 40, 40, 40, 40, 40, 50

Group *B*: 30, 30, 30, 40, 50, 50, 50

Group *C*: 30, 35, 40, 40, 40, 45, 50

In all the three groups the range is $50 - 30 = 20$. In group A there is concentration of observations in the center. In group B the observations are friendly with the extreme corner and in group C the observations are almost equally distributed in the interval from 30 to 50. The range fails to explain these differences in the three groups of data. This defect in range cannot be removed even if we calculate the coefficient of range which is a relative measure of dispersion. If we calculate the range of a sample, we cannot draw any inferences about the range of the population.

13.4.1 Coefficient of Range

It is relative measure of dispersion and is based on the value of range. It is also called range coefficient of dispersion. It is defined as:

$$\text{Coefficient of Range} = \frac{x_m - x_0}{x_m + x_0}$$

The range $x_m - x_0$ is standardized by the total $x_m + x_0$

Let us take two sets of observations. Set A contains marks of five students in Mathematics out of 25 marks and group B contains marks of the same student in English out of 100 marks.

Set A: 10, 15, 18, 20, 20

Set B: 30, 35, 40, 45, 50

The values of range and coefficient of range are calculated as:

	Range	**Coefficient of Range**
Set A: (Mathematics)	$20 - 10 = 10$	$\frac{20-10}{20+10} = 0.33$
Set B: (English)	$50 - 30 = 20$	$\frac{50-30}{50+30} = 0.25$

In set A the range is 10 and in set B the range is 20. Apparently it seems as if there is greater dispersion in set B. But this is not true. The range of 20 in set B is for large observations and the range of 10 in set A is for small observations. Thus 20 and 10 cannot be compared directly. Their base is not the same. Marks in Mathematics are out of 25 and marks of English are out of 100. Thus, it makes no sense to compare 10 with 20. When we convert these two values into coefficient of range, we see that coefficient of range for set A is greater than that of set B. Thus there is greater dispersion or variation in set A. The marks of students in English are more stable than their marks in Mathematics.

13.4.2 Merits of Range

There are following advantages of range:

(i) Range is simple to calculate

(ii) Range is easy to understand.

(iii) Range is rigidly defined.

(iv) For calculating the values of range, only extreme values are sufficient instead of all intermediate values too.

13.4.3 Demerits of Range

There are following disadvantages of range:

(i) As we know range is not based on all the items of the series. Thus, in its computation the entire data is not used.

(ii) Range is affected by unusual items of the series. Under the circumstances, range cannot be regarded as a reliable measure of dispersion or variability

(iii) Range can't be calculated in open-end classes, because lowest class and upper class are not given.

(iv) Range is very much affected by the fluctuations of sampling.

(v) Mathematical treatment of range is not possible.

(vi) As range is based on extreme values, if we change middle values range must not Change but the reliability of observation may suffer.

SOLVED PROBLEMS

Range in case of individual observations

When the data is given as individual items the range is calculated simply as:

Problem 1: Calculate the range and coefficient of range from the following data:

X :	160	158	170	142	176	187

Solution:

The largest value is 187

The smallest value is 142

$$\text{Range} = L - S$$

$$= 187 - 142 = 45$$

$$\text{Coefficient of range} = \frac{L-S}{L+S}$$

$$= \frac{187-142}{187+142} = 0.137$$

Range in case of Discrete Series

When the data is given in case of discrete series, frequency doesn't play any role, i.e. we are mainly concerned with items only. For example,

Problem 2: From the following data calculate the range and coefficient of range.

X :	11	14	18	24	29	31
f :	100	101	91	70	61	32

Solution: As we are only concerned with items only not frequency, the largest item is 31 and smallest item is 11.

$$\text{Range} = L - S$$

$$\text{Range} = 31 - 11 = 20$$

$$\text{Coefficient of range} = \frac{L - S}{L + S}$$

$$= \frac{31 - 11}{31 + 11} = 0.476$$

Range in case of Continuous Series

In case of continuous series, range can be determined by two methods:

(i) Method of Limits : To compute the difference between the upper limit of the highest class and lowest limit of the lowest class.

(ii) Method of mid-points : To subtract the mid point of the lowest class from the mid point of the highest class.

Frequency does not play any role here also.

Problem 3: From the following data calculate the range and coefficient of range

Age :	10-20	20-30	30-40	40-50	50-60	60-70
No. of people in area :		8	12	22	38	15 5

Solution:

Age	No. of people in area	Mid-points
10-20	8	15
20-30	12	25
30-40	22	35
40-50	38	45
50-60	15	55
60-70	5	65

Method of Limits

$$\text{Range} = \text{upper limit of highest class} - \text{lower limit of lowest class}$$ 13

$$= 70 - 10 = 60 \text{ Ans.}$$

$$\text{Coefficient of range} = \frac{\text{Highest limit} - \text{lowest limit}}{\text{Highest limit} + \text{lowest limit}}$$

$$= \frac{70 - 10}{70 + 10} = \frac{60}{80} = 0.75 \text{ Ans.}$$

Method of mid-points

$$\text{Range} = \text{Highest mid point- lowest mid point}$$
$$= 65 - 15 = 50 \text{ Ans.}$$

$$\text{Coefficient of range} = \frac{\text{Highest mid point} - \text{lowest mid point}}{\text{Highest mid point} + \text{lowest mid point}}$$

$$= \frac{65-15}{65+15} = \frac{50}{80} = 0.625 \text{ Ans.}$$

Problem 4: Following are the wages of 8 workers of a factory. Find the range and the coefficient of range. Wages in (\$) 1400, 1450, 1520, 1380, 1485, 1495, 1575, 1440.

Solution:

Here Largest value $= x_m = 1575$ and Smallest Value $= x_0 = 1380$

$$\text{Range} = x_m - x_0 = 1575 - 1380 = 195.$$

$$\text{Coefficient of Range} = \frac{x_m - x_0}{x_m + x_0}$$

$$= \frac{1575-1380}{1575+1380} = \frac{195}{2955} = 0.66$$

Problem 5: The following distribution gives the numbers of houses and the number of persons per house.

No. of Persons :	1	2	3	4	5	6	7	8	9	10
No. of Houses :	26	113	120	95	60	42	21	14	5	4

Calculate the range and coefficient of range.

Solution:

Here Largest value $= x_m = 10$ and Smallest Value $= x_0 = 1$

$$\text{Range} = x_m - x_0 = 10 - 1 = 9.$$

$$\text{Coefficient of Range} = \frac{x_m - x_0}{x_m + x_0}$$

$$= \frac{10-1}{10+1} = \frac{9}{11} = 0.818$$

Problem 6: Find the range of the weight of the students of a university.

Weights (Kg) :	60-62	63-65	66-68	69-71	72-74
Number of Students :	5	18	42	27	8

Calculate the range and coefficient of range.

Solution:

Weights (Kg)	Class Boundaries	Mid Value	No. of Students
60-62	59.5 - 62.5	61	5
63-65	62.5 – 65.5	64	18
66-68	65.5 – 68.5	67	42
69-71	68.5 – 71.5	70	27
72-74	71.5 – 74.5	73	8

Method 1:

Here

x_m = Upper class boundary of the highest class = 74.5

x_0 = Lower class boundary of the lowest class = 59.5

$$\text{Range} = x_m - x_0 = 74.5 - 59.5 = 15 \text{ Kilogram}$$

$$\text{Coefficient of Range} = \frac{x_m - x_0}{x_m + x_0} = \frac{74.5 - 59.5}{74.5 + 59.5} = \frac{15}{134} = 0.1119$$

Method 2:

Here

x_m = Mid value of the highest class = 73

x_0 = Mid value of the lowest class = 61

$$\text{Range} = x_m - x_0 = 73 - 61 = 12 \text{ Kilogram}$$

$$\text{Coefficient of Range} = \frac{x_m - x_0}{x_m + x_0} = \frac{73 - 61}{73 + 61} = \frac{12}{134} = 0.0895$$

UNSOLVED PROBLEMS

1. What do you understand by measure of dispersion?
2. What are the various properties of dispersion?
3. What are the various uses of dispersion?
4. What are the various types of dispersion?
5. What are the various methods to compute dispersion?
6. What are the various positional measures?
7. What are the various deviations? Explain all?
8. What do you mean by range?
9. How can you find the range?
10. What are the various merits and demerits of range?
11. Find the range and its coefficient from the data:

Size :	5	7	9	10	11	12
Frequency:	1	3	5	7	4	3

12. Calculate coefficient of range from the following data:

Marks:	10-20	20-30	30-40	40-50	50-60
No. of Students:	8	10	12	8	4

13.5 INTERQUARTILE RANGE

In descriptive statistics, the interquartile range (IQR), also called the midspread or middle fifty, is a measure of statistical dispersion, being equal to the difference between the third and first quartiles.

Unlike the (total) range, the interquartile range is a robust statistic, having a breakdown point of 25%, and is thus often preferred to the total range.

Definition : *"Difference between the two extreme quartiles of a series"*

Divide a set of data into two groups (high and low) of equal size at the statistical median if there is an even number of data points, or two groups consisting of points on either side of the statistical median itself plus the statistical median if there is an odd number of data points. Find the statistical medians of the low and high groups, denoting these first and third quartiles by $Q_1$and Q_3. The interquartile range is then defined by

$$\text{Inter Quartile range} = Q_3 - Q_1$$

ØThe inter quartile range is based on upper quartile Q_3 and lower quartile Q_1 and equal to the difference between third and first qualities. Coefficient of Inter Quartile Range:

Inter Quartile range is an absolute measure of dispersion. Relative measure of dispersion, known as coefficient of quartile range is computed as:

$$\text{Coefficient of Quartile Deviation} = \frac{(Q_3 - Q_1)/2}{(Q_3 + Q_1)/2} = \frac{Q_3 - Q_1}{Q_3 + Q_1}$$

13.5.1 Merits of Inter Quartile Range

There are following advantages of Inter quartile range:

1. It is simple to calculate
2. It is easy to understand.
3. It is rigidly defined.
4. It is efficient in computation.
5. It gives reliable results.

13.5.2 Demerits of Inter Quartile Range

There are following disadvantages of Inter quartile range:

1. It is not based on all the items of the series.

2. It is affected by unusual items of the series.
3. It can't be computed for irregular data.
4. It is very much affected by the fluctuations of sampling.
5. Mathematical treatment is not possible.
6. It needs more time for computation.

13.6 Decile Range

In descriptive statistics, a decile is any of the nine values that divide the sorted data into ten equal parts, so that each part represents 1/10 of the sample or population. Thus:

- The 1st decile cuts off the lowest 10% of data, i. e., the 10th percentile.
- The 5th decile cuts off lowest 50% of data, i. e., the 50th percentile, 2nd quartile, or median.
- The 9th decile cuts off lowest 90% of data, i. e., the 90th percentile.

Deciles are points in a set of data which are 1/10, 2/10, etc. of the way through the data, in the same way that the median is the point which is half way through the data. It divides the series in 10 equal parts. Decile Range has the difference between first part and ninth part.

$$\text{Decile range} = D_9 - D_1$$

13.7 PERCENTILE RANGE

A percentile (or centile) is the value of a variable below which a certain percent of observations fall. So the 20th percentile is the value (or score) below which 20 percent of the observations may be found. The term percentile and the related term percentile rank are often used in descriptive statistics as well as in the reporting of scores from norm-referenced tests.

The 25th percentile is also known as the first quartile(Q_1); the 50th percentile as the median or second quartile(Q_2); the 75th percentile as the third quartile (Q_3).

A percentile is a measure that tells us what percent of the total frequency scored at or below that measure. It divides the series into 100 equal parts. Percntile Range has the difference between first part and ninty nineth part.

$$\text{Percentile range} = P_{99} - P_1$$

13.8 QUARTILE DEVIATION

It is based on the lower quartile Q_1 and the upper quartile Q_3. The difference $Q_3 - Q_1$ is called the inter quartile range. The difference $Q_3 - Q_1$ divided by 2 is called semi-inter-quartile range or the quartile deviation. Thus

$$\text{Quartile Deviation (Q.D)} = \frac{Q_3 - Q_1}{2}$$

The quartile deviation is a slightly better measure of absolute dispersion than the range. But it ignores the observation on the tails. If we take difference samples from a population and calculate their quartile deviations, their values are quite likely to be sufficiently different. This is called sampling fluctuation. It is not a popular measure of dispersion. The quartile deviation calculated from the sample data does not help us to draw any conclusion (inference) about the quartile deviation in the population.

13.8.1 Coefficient of Quartile Deviation

A relative measure of dispersion based on the quartile deviation is called the coefficient of quartile deviation. It is defined as

$$\text{Coefficient of Quartile Deviation} = \frac{Q_3 - Q_1}{2} \Big/ \frac{Q_3 + Q_1}{2} = \frac{Q_3 - Q_1}{Q_3 + Q_1}$$

It is pure number free of any units of measurement. It can be used for comparing the dispersion in two or more than two sets of data.

13.8.2 Merits of Quartile Deviation

Quartile deviation has the following advantages:

(i) Quartile deviation can also be computed, when the distribution has open end classes.

(ii) Quartile deviation is more reliable than range. The quartile values will not change much if one or more values are added or taken away from the distribution.

(iii) Quartile deviation is also easy to calculate and effective. Quartile deviation makes use of 50% of the data and in that case it is better measure than range.

(iv) Quartile deviation is not affected by extreme items.

(v) The size of quartile deviation gives an indication about the uniformity.

(vi) The calculation of Quartile deviation doesn't involve any mathematical intricacies.

(v) It is rigidly defined.

13.8.3 Demerits of Quartile Deviation

Quartile deviation has the following disadvantages:

(i) Quartile deviation is not suited to algebraic treatment.

(ii) The computation of Quartile deviation is more time consuming than Range.

(iii) It is not measure of variation from any specific average.

(iv) Quartile deviation is much affected by sampling fluctuations.

(v) Quartile deviation is not affected by the distribution of items between Q_1 and Q_2 or by the distribution outside the quartiles.

(vi) It is not based on all the observations.

(vii) It is not so simple and can not understood by layman.

13.8.4 Program to Calculate Quartile Range, Quartile Deviation, and Coefficient of Quartile Deviation for Individual Series.

```
#include<stdio.h>
#include<conio.h>
void main()
{
float x[100],a,c,d,f,q1=0,q3=0,qd=0,qr=0,cqd=0;
int i,n,j,temp=0,b,e;
clrscr();
printf(" How many items in the individual distribution ?");
scanf("%d", &n);
printf("\n Ente the data \n");
for(i=1;i<=n;i++)
{
scanf("%f",&x[i]);
}
for(i=1;i<=n-1;i++)
{
for(j=1;j<=n-i;j++)
{
if(x[j]>x[j+1])
{
temp=x[j];
x[j]=x[j+1];
x[j+1]=temp;
}
}
}
for(i=1;i<=n;i++)
{
printf("%f",x[i]);
}
a=((float)n+1)/4;
b=(int)a;
c=a-b;
```

```
if(c!=0)
q1=x[b]+(c*(x[b+1]-x[b]));
else
q1=x[b];
printf("%f",q1);
d=(3*(n+1))/4;
e=(int)d;
f=d-e;
if(f!=0)
q3=x[e]+(f*(x[e+1]-x[e]));
else
q3=x[e];
printf("%f",q3);
qd=(q3-q1)/2;
qr=(q3-q1);
cqd=(q3-q1)/(q3+q1);
printf("\nQuartile Range=%f",qr);
printf("\nQuartile Deviation=%f",qd);
printf("\nCoefficient of Quartile Deviation=%f",cqd);
getch();
}
```

SOLVED PROBLEMS

Problem 1: Calculate the inter quartile range from the following data:

Age group:	20-25	25-30	30-35	35-40	40-45	45-50	50-55	55-60
No. of persons:	50	70	100	180	150	120	70	60

Solution:

Age	No. of persons *F*	Mid points C.F
20-25	50	50
25-30	70	120
30-35	100	220
35-40	180	400
40-45	150	550
45-50	120	670
50-55	70	740
55-60	60	800

Lower Quartile:

$$Q_1 = \text{size of } \frac{N}{4} \text{ th item}$$

$$= \text{size of } \frac{800}{4}, \text{ i.e } 200^{th} \text{ item}$$

Q_1 lies in the class 30-35

Interpolation:

$$Q_1 = L1 + \frac{(N/4 - C)}{F}(L_2 - L_1)$$

$$= 30 + \frac{200 - 120}{100}(35 - 30)$$

$$= 30 + \frac{80}{100}(5)$$

$$= 30 + 4 = 34 \text{ yrs}$$

Upper Quartile:

$$Q_3 = \text{size of } \frac{3N}{4} \text{ th item}$$

$$= \text{size of } \frac{3 \times 800}{4}, \text{ i.e } 600^{th} \text{ item}$$

Q_3 lies in the class 45 – 50

Interpolation:

$$Q_3 = L1 + \frac{(3N/4 - C)(L2 - L1)}{F}(L2 - L1)$$

$$= 45 + \frac{600 - 550}{120}(50 - 45)$$

$$= 45 + \frac{50}{100}(5)$$

$$= 45 + 2.08 = 47.08 \text{ yrs}$$

Inter Quartile Range $\quad Q_3 - Q_1 = 47.08 - 34 = 13.08$ yrs.

Determination of Quartile deviation in Individual Observation:

Problem 2: The wheat production (in Kg) of 20 acres is given as: 1120, 1240, 1320, 1040, 1080, 1200, 1440, 1360, 1680, 1730, 1785, 1342, 1960, 1880, 1755, 1720, 1600, 1470, 1750, and 1885. Find the quartile deviation and coefficient of quartile deviation.

Solution:

After arranging the observations in ascending order, we get

1040, 1080, 1120, 1200, 1240, 1320, 1342, 1360, 1440, 1470,
1600, 1680, 1720, 1730, 1750, 1755, 1785, 1880, 1885, 1960.

$$Q_1 = \text{Value of } \left(\frac{n+1}{4}\right)\text{th item}$$

$$= \text{Value of } \left(\frac{20+1}{4}\right)\text{th item}$$

$$= \text{Value of (5.25)th item}$$

$$= \text{5th item} + 0.25(\text{6th item} - \text{5th item})$$

$$= 1240 + 0.25\,(1320 - 1240)$$

$$Q_1 = 1240 + 20 = 1260$$

$$Q_3 = \text{Value of } \frac{3(n+1)}{4}\text{th item}$$

$$= \text{Value of } \frac{3(20+1)}{4}\text{th item}$$

$$= \text{Value of (15.75)th item}$$

$$= \text{15th item} + 0.75(\text{16th item} - \text{15th item})$$

$$= 1750 + 0.75\,(1755 - 1750)$$

$$Q_3 = 1750 + 3.75 = 1753.75$$

$$\text{Quartile Deviation (Q.D)} = \frac{Q_3 - Q_1}{2} = \frac{1753.75 - 1260}{2} = \frac{492.75}{2} = 246.875$$

$$\text{Coefficient of Quartile Deviation } \frac{Q_3 - Q_1}{Q_3 + Q_1} = \frac{1753.75 - 1260}{1753.75 + 1260} = 0.164$$

Problem 3: Calculate the quartile deviation and coefficient of quartile deviation from the datagiven below:

Maximum Load (short-tons)	Number of Cables
9.3-9.7	2
9.8-10.2	5
10.3-10.7	12
10.8-11.2	17
11.3-11.7	14
11.8-12.2	6
12.3-12.7	3
12.8-13.2	1

Solution:

The necessary calculations are given below:

Maximum Load (short-tons)	Number of Cables f	Class Boundaries	Cumulative Frequencies
9.3-9.7	2	9.25-9.75	2
9.8-0.2	5	9.75-0.25	2 + 5 = 7
10.3-10.7	12	10.25-10.75	7 + 12 = 19
10.8-11.2	17	10.75-11.25	19 + 17 = 36
11.3-11.7	14	11.25-11.75	36 + 14 = 50
11.8-12.2	6	11.75-12.25	50 + 6 = 56
12.3-12.7	3	12.25-12.75	56 + 3 = 59
12.8-13.2	1	12.75-13.25	59 + 1 = 60

$$Q_1 = \text{Value of } \left(\frac{n}{4}\right) \text{ th item}$$

$$= \text{Value of } \left(\frac{60}{4}\right) \text{th item 15th item}$$

Q_1 lies in the class 10.25 – 10.75

$$\therefore \quad Q_1 = 1 + \frac{h}{f}\left(\frac{n}{4} - c\right)$$

Where $1 = 10.25, h = 0.5, f = 12, n/4 = 15$ and $c = 7$

$$\therefore \quad Q_1 = 10.25 + \frac{0.5}{12}(15 - 7) = 10.25 + 0.33 = 10.58$$

$$Q_3 = \text{Value of } \left(\frac{3n}{12}\right) \text{th item}$$

$$= \text{Value of } \left(\frac{3 \times 60}{4}\right) \text{th item 45th item}$$

Q_3 lies in the class 11.25 – 11.75

$$\therefore \quad Q_3 = 1 + \frac{h}{f}\left(\frac{3n}{4} - c\right)$$

Where $1 = 11.25, h = 0.5, f = 14, 3n/4 = 45$ and $c = 36$

$$\therefore \quad Q_1 = 11.25 + \frac{0.5}{14}(45 - 36) = 11.25 + 0.32 = 11.57$$

$$\text{Quartile Deviation (Q.D)} = \frac{Q_3 - Q_1}{2} = \frac{11.57 - 10.58}{2} = \frac{0.99}{2} = 0.495$$

$$\text{Coefficient of Quartile Deviation } \frac{Q_3 - Q_1}{Q_3 + Q_1} = \frac{11.57 - 10.58}{11.57 + 10.58} = \frac{0.99}{22.15} = 0.045$$

Problem 4: Calculate the quartile deviation and coefficient of quartile deviation of the following data:

Marks:	5	10	20	25	40	42	45	48	70	80	85

Solution: Determination of quartile deviation

Sr. no.	Marks	Sr. No.	Marks
1	5	7	45
2	10	8	48
3	20	9	70
4	25	10	80
5	40	11	85
6	42		

$$Q_1 = \text{size of } \frac{N+1}{4} \text{th item}$$

$$= \text{size of } \frac{11+1}{4} \text{th item}$$

$$= \text{size of } 3^{rd} \text{ item} = 20$$

$$Q_3 = \text{size of } \frac{3(N+1)}{4} \text{th item}$$

$$= \text{size of } 9^{th} \text{ item} = 70$$

$$\text{Q.D.} = ½(Q_3 - Q_1)$$

$$= ½\,(70 - 20) = 25$$

$$\text{Coeff. of Q.D.} = \frac{Q_3 - Q_1}{Q_3 + Q_1}$$

$$= \frac{70 - 0}{70 + 20} = \frac{50}{90} = 0.56$$

Problem 5: Compute quartile deviation and its coefficient from the following data:

Height (in inches):	58	59	60	61	62	63	64	65	66
No of student (f):	15	20	32	35	33	22	20	10	8

Solution: Determination of quartile deviation

Ht. (in inches)	Frequency	Cumulative Frequency
58	15	15
59	20	35
60	32	67
61	35	102
62	33	135
63	22	155
64	20	177
65	10	187
66	8	195

Here $N = \Sigma f = 195$

$$Q_1 = \text{size of } \frac{N+1}{4} \text{th item}$$

$$= \text{size of } \frac{195+1}{4} \text{th item} = \text{size of } 49^{\text{th}} \text{ item} = 60$$

$$Q_3 = \text{size of } \frac{3(N+1)}{4} \text{th item} = \text{size of } 147^{\text{th}} \text{ item} = 63$$

$$\text{Q.D.} = \tfrac{1}{2}(Q_3 - Q_1) = \tfrac{1}{2}(63 - 60) = 1.5$$

$$\text{Coeff. Of Q.D.} = \frac{Q_3 - Q_1}{Q_3 + Q_1} = \frac{63-60}{63+60} = \frac{3}{123} = 0.024$$

Problem 6: Evaluate an appropriate measure of dispersion for the following data:

Income (in Rs.) :	Less than 50	50-70	70-90	90-110	110-130	130-150	Above 150
No. of persons:	54	100	140	300	230	125	51

Solution: Since we are given the classes with open end intervals, the only measure of dispersion that we can compute is the quartile deviation.

Computation of Quartile Deviation		
Income (in Rs.)	No. of persons (f)	Less than c.f.
Less than 50	54	54
50-70	100	154
70-90	140	294
90-110	300	594
110-130	230	824
130-150	125	949
Above 150	51	1000

Here $$N = 1000, \frac{N}{4} = \frac{1000}{4} = 250, \frac{3N}{4} = 750$$

Since c.f. just greater than 250 is 294, Q_1 (first quartile) lies in the corresponding class 70-90. Similarly since c.f. just greater than 750 is 824 m the corresponding class 110-130 contains Q_3.

Hence, $$Q_1 = 70 + \frac{20}{140}(250 - 154) = 70 + \frac{96}{7}$$

$$= 70 + 13.714 = 83.714$$

$$Q_3 = 110 + \frac{20}{230}(750 - 594)$$

$$= 110 + \frac{2 \times 156}{23} = 110 + 13.565 = 123.565$$

∴ Quartile deviation is given by:

$$\text{Q.D.} = \frac{Q_3 - Q_1}{2} = \frac{123.565 - 83.714}{2} = \frac{39.851}{2} = 19.925$$

UNSOLVED PROBLEMS

1. What is Inter Quartile Range?
2. What you know about coefficient of IRQ?
3. What are the various merits and demerits of IRQ?
4. What is Decile and Percentile Range?
5. How Quartile Deviation works? Give its merits and demerits?
6. Compute coefficient of quartile deviation from the following data:

Marks:	10	20	30	40	50	80
No. of Students:	4	7	15	8	7	2

7. Compute quartile deviation and its coefficient from the following data:

Wages(in Rs.) :	20-25	25-30	30-35	35-40	40-45
No. of Students:	2	10	25	16	7

8. Age distribution of 200 employees of a firm is given below: Construct a 'less than ogive curve, and hence or otherwise calculate semi-interquartile range $\frac{Q_3 - Q_1}{2}$ of the distribution:

Age in years (less than):	25	30	35	40	45	50	55
No. of employees:	10	25	75	130	170	189	200

9. Find the mode, median, lower quartile (Q_1) and upper quartile (Q_3) and Coeff. of Q.D from the following data:

Marks obtained:	0-10	10-20	20-30	30-40	40-50
No. of students:	22	38	46	35	20

13.9 THE MEAN DEVIATION

The mean deviation or the average deviation is defined as the mean of the absolute deviations of observations from some suitable average which may be the arithmetic mean, the median or the mode. The difference (X-average) is called deviation and when we ignore the negative sign, this deviation is written as $|X\text{-average}|$ and is read as mod deviations. The mean of these mod or absolute deviations is called the mean deviation or the mean absolute deviation. Thus for sample data in which the suitable average is the $\bar{X}$, the mean deviation (M.D) is given by the relation:

$$\text{M.D.} = \frac{\Sigma|X - \bar{X}|}{n} \quad \text{or} \quad \frac{\Sigma|D|}{n} \quad \text{where} \quad |D|\ X - \bar{X}|$$

For frequency distribution, the mean deviation is given by

$$\text{M.D} = \frac{\Sigma f|X - \bar{X}|}{\Sigma f}$$

When the mean deviation is calculated about the median, the formula becomes

$$\text{M.D (about median)} = \frac{\Sigma f|X - \text{Median}|}{\Sigma f}$$

The mean deviation about the mode is

$$\text{M.D (about mode)} = \frac{\Sigma f|X - \text{Mode}|}{\Sigma f}$$

For a population data the mean deviation about the population mean *m* is

$$\text{M.D} = \frac{\Sigma f|X - \mu|}{\Sigma f}$$

The mean deviation is a better measure of absolute dispersion than the range and the quartile deviation.

A drawback in the mean deviation is that we use the absolute deviations $|X -$ average which does not seem logical. The reason for this is that $\Sigma(X - \bar{X})$ is always equal to zero. Even if we use median or mode in place of $\bar{X}$, even then the summation $\Sigma(X -$ median$)$ or $\Sigma(X -$ mode$)$ will be zero or approximately zero with the result that the mean deviation would always be either zero or close to zero. Thus the very definition of the mean deviation is possible only on the absolute deviations.

The mean deviation is based on all the observations, a property which is not possessed by the range and the quartile deviation. The formula of the mean deviation gives a mathematical impression that is a better way of measuring the variation in the data. Any suitable average among the mean, median or mode can be used in its calculation but the value of the mean deviation is minimum if the deviations are taken from the median. A series drawback of the mean deviation is that it cannot be used in statistical inference.

13.9.1 Coefficient of the Mean Deviation

A relative measure of dispersion based on the mean deviation is called the coefficient of the mean deviation or the coefficient of dispersion. It is defined as the ratio of the mean deviation to the average used in the calculation of the mean deviation. Thus

$$\text{Coefficient of M.D (about mean)} = \frac{\text{Mean Deviation from mean}}{\text{Mean}}$$

$$\text{Coefficient of M.D (about median)} = \frac{\text{Mean Deviation from median}}{\text{Median}}$$

$$\text{Coefficient of M.D (about mode)} = \frac{\text{Mean Deviation from mode}}{\text{Mode}}$$

13.9.2 Merits of Mean Deviation

Mean deviation has the following advantages:

(i) Mean deviation is clear and easy to understand.

(ii) As compared to standard deviation it is less affected by extreme values.

(iii) It is better than range and quartile deviation as it is based on all values in the distribution.

(iv) Mean deviation is based on measurement rather than estimation.

(v) Mean deviation is also rigidly defined.

(vi) Mean deviation is very much close to arithmetic mean.

(vii) It can be computed from any measure of central tendency.

13.9.3 Demerits of Mean Deviation

Mean deviation has the following disadvantages:

(i) Mean deviations can't be calculated for distribution with open end classes.

(ii) Mean deviation tends to increase as the size of the sample though not in the same ratio and not as rapidly as range.

(iii) It ignores the algebraic signs of the deviations and as such, it is not capable of further mathematical treatment.

(iv) It is not a very accurate measure of dispersion particularly when it is calculated from mode.

(v) Mean deviation is not as convenient measure of dispersion as standard deviation.

(vi) It may not give us very accurate results when we use median.

13.9.4 Program to Calculate Mean Deviation and Coefficient of Mean Deviation From Mean in Individual Series.

```
#include<stdio.h>
#include<conio.h>
#include<math.h>
void main()
{
float x[100],n,dx[100],mean,sum=0,sum1=0,md,cmd;
int i;
clrscr();
printf(" How many items in the individual distribution ?");
scanf("%f",&n);
printf("\n Enter the data \n");
for(i=1;i<=n;i++)
{
scanf("%f",&x[i]);
sum=sum+x[i];
}
mean=sum/n;
for(i=1;i<=n;i++)
{
dx[i]=fabs(x[i]-mean);
sum1=sum1+dx[i];
}
md=sum1/n;
printf("Mean Deviation=%f",md);
cmd=md/mean;
printf("Coefficient of Mean Deviation=%f",cmd);
getch();
}
```

13.9.5 Program to Calculate Mean Deviation and Coefficient of Mean Deviation From Mean in Discrete Series.

```
#include<stdio.h>
#include<conio.h>
#include<math.h>
void main()
{
float a[100],n, f[100],dx[i], fdx[100], af[100], i, md,
summ=0,sumf=0,mean,cmd,summ1=0;
clrscr();
printf(" How many items in the individual distribution ?");
scanf("%f",&n);
printf("\n Ente the X-values and frequency \n");
for(i=1;i<=n;i++)
{
scanf("%f",&a[i]);
scanf("%f",&f[i]);
af[i]=a[i]*f[i];
summ=summ+af[i];
sumf=sumf+f[i];
}
mean=summ/sumf;
for(i=1;i<=n;i++)
{
dx[i]=fabs(a[i]-mean);
fdx[i]=f[i]*dx[i];
summ1=summ1+fdx[i];
}
md=summ1/sumf;
printf("Mean Deviation=%f",md);
cmd=md/mean;
printf("Coefficient of Mean Deviation=%f",cmd);
getch();
}
```

13.9.6 Program to Calculate Mean Deviation and Coefficient of Mean Deviation From Mean in Continuous Series.

```
#include<conio.h>
#include<math.h>
void main()
{
floatn,i,x[30],y[30],z[30],sum=0,sum1=0,summ1=0,a[30],csd,mean,c[30],dx[30],fdx[30],md=0;
clrscr();
printf(" How many items in the individual distribution ?");
scanf("%f",&n);
printf("\n Ente the data \n");
for(i=1;i<=n;i++)
{
scanf("%f",&x[i]);
scanf("%f",&y[i]);
scanf("%f",&z[i]);
sum=sum+z[i];
}

for(i=1;i<=n;i++)
{
a[i]=(x[i]+y[i])/2;

c[i]=z[i]*a[i];
sum1=sum1+c[i];
}
mean=sum1/sum;
printf("mean=%f",mean);
for(i=1;i<=n;i++)
{
dx[i]=fabs(a[i]-mean);
fdx[i]=z[i]*(dx[i]);
summ1=summ1+fdx[i];
}
```

```
md=summ1/sum;
printf("Mean Deviation=%f",md);
cmd=md/mean;
printf("Coefficient of Mean Deviation=%f",cmd);
getch();
}
```

SOLVED PROBLEMS

Problem 1: Calculate the mean deviation form (1) arithmetic mean (2) median (3) mode in respect of the marks obtained by nine students gives below and show that the mean deviation from median is minimum.

Marks (out of 25):	7	4	10	9	15	12	7	9	7

Solution:

After arranging the observations in ascending order, we get

Marks: 4, 7, 7, 7, 9, 9, 10, 12, 15

$$\text{Mean} = \frac{\Sigma X}{n} = \frac{80}{9} = 8.89$$

$$\text{Median} = \text{Value of } \frac{n+1}{2} \text{ th item} = \text{Value of } \frac{9+1}{2} \text{ th item}$$

$$= \text{Value of (5)th item} = 9$$

Mode = 7 (Since 7 is repeated maximum number of times)

Marks X	$\lvert X - \bar{X} \rvert$	$\lvert X -$ Median$\rvert$	$\lvert X -$ Mode$\rvert$
4	4.89	5	3
7	1.89	2	0
7	1.89	2	0
7	1.89	2	0
9	0.11	0	2
9	0.11	0	2
10	1.11	1	3
12	3.11	3	5
15	6.11	6	8
Total	21.11	21	23

$$\text{M.D from mean} = \frac{\Sigma|X - \bar{X}|}{n} = \frac{21.11}{9} = 2.35$$

$$\text{M.D from median} = \frac{\Sigma|X - \text{Median}|}{n} = \frac{21}{9} = 2.33$$

$$\text{M.D from mean} = \frac{\Sigma|X - \text{Mean}|}{n} = \frac{23}{9} = 2.56$$

From the above calculations, it is clear that the mean deviation from the median hast the least value.

Problem 2: Calculate the mean deviation from mean and its coefficients from the following data.

Size of Items :	3-4	4-5	5-6	6-7	7-8	8-9	9-10
Frequency :	3	7	22	60	85	32	8

Solution:

The necessary calculation is given below:

Size of Items	X	f	fX	$\lvert X - \bar{X} \rvert$	$f\lvert X - \bar{X} \rvert$
3-4	3.5	3	10.5	3.59	10.77
4-5	4.5	7	31.5	2.59	18.13
5-6	5.5	22	121.0	1.59	34.38
6-7	6.5	60	390.0	0.59	35.40
7-8	7.5	85	637.5	0.41	34.85
8-9	8.5	32	272.0	1.41	45.12
9-10	9.5	8	76.0	2.41	19.28
Total		217	1538.5		198.53

$$\text{Mean} = \bar{X} = \frac{\Sigma fX}{\Sigma f} - \frac{1538.5}{217} - 7.09$$

$$\text{M.D from Mean} = \frac{\Sigma|X - \bar{X}|}{n} - \frac{198.53}{217} - 0.915$$

$$\text{Coefficient of M.D from (Mean)} = \frac{\text{M.D from Mean}}{\text{Mean}} \frac{0.915}{7.09} = 0.129$$

Problem 3: Calculate Mean deviation and coefficient of MD of the given series:

X:	6	10	26	28	30	32

Solution: MD for Mean

$$|D| = |X - \bar{x}|$$

Series(X)	(D)
6	6
10	2
26	4
28	6
30	8
32	10
$\Sigma X = 132$	$\Sigma\|D\| = 56$

$$\bar{x} = \Sigma X/N = 132/6 = 22$$

$$MD = \Sigma|D|/N = 56/6 = 9.333$$

$$\text{Coefficient of } MD = MD/\bar{x} = 9.333/22 = 0.42$$

Mean Deviation from Median

Series(X)	$(X - M) = \|D\|$
6	21
10	17
26	1
28	1
30	3
32	5
	$\Sigma\|D\| = 48$

$$M = \text{Size of } \frac{(N+1)}{2} \text{th item}$$

$$= \text{Size of } \frac{(6+1)}{2} \text{th item.} = 3.5 \text{ th item}$$

$$= \frac{3^{rd} + 4^{th}}{2} = \frac{26+28}{2} = 27$$

$$MD = \Sigma|D|/N = 48/6 = 8$$

$$\text{Coefficient of } MD = MD/M \quad \text{(As deviations are taken from median)}$$

$$= 8/27 = 0.296$$

Problem 4: Calculate Mean deviation and coefficient of MD of the given series:

X:	7	12	16	18	22

Solution: MD for Mean

$$|D| = |X - \bar{x}|$$

Series(*X*)	(*D*)
7	8
12	3
16	1
18	3
22	7
$\Sigma X = 75$	$\Sigma\|D\| = 22$

$$\bar{x} = \Sigma X/N = 75/5 = 15$$

$$MD = \Sigma|D| = 22/5 = 4.4$$

$$\text{Coefficient of } MD = MD/\bar{x} = 4.4/15 = 0.293$$

Mean Deviation from Median

Series(*X*)	$(X - M) = \|D\|$
7	9
12	4
16	0
18	2
22	6
$\Sigma X = 75$	$\Sigma\|D\| = 21$

$$M = \text{Size of } \frac{(N+1)}{2} \text{th item}$$

$$= \text{Size of } \frac{(5+1)}{2} \text{th item} = \text{Size of } 3^{rd} \text{ item} = 16$$

$$MD = \Sigma|D|/N = 21/5 = 4.2$$

$$\text{Coefficient of MD} = \text{MD}/M \quad \text{(As deviations are taken from median)}$$

$$= 4.2/16$$

$$= 0.2625$$

Calculation of Mean deviation for Discrete Series

In Discrete series mean deviation can be obtained from the following formula:

$$MD = \frac{\Sigma f|D|}{N}$$

Where $|D|$ is the deviation from median (or mean) ignoring sign.

Now Steps for calculating Mean Deviation in Discrete Series

(i) Calculate Mean or Median of the series.

(ii) Compute deviations of the items from the mean (or median).

(iii) Multiply $|D|$ by respective frequency and sum up all the products, i.e $\Sigma f|D|$

(iv) Finally compute $\frac{\Sigma f|D|}{N}$

Problem 5: Calculate MD and coefficient of MD from mean and median both for the following data.

Marks :	3	12	19	26	42	48
No. of Students :	10	12	24	8	16	11

Solution:

X	f	fX	Cf	$\|(X-M)\| = \|D\|$	$f \cdot \|D\|$	$\|X - \bar{X}\| = \|D'\|$	$f\|D'\|$
3	10	30	10	16	160	22.16	221.6
12	12	144	22	7	84	13.16	157.92
19	24	456	46	0	0	6.16	147.84
26	8	208	54	7	56	0.84	6.72
42	16	672	70	23	368	16.84	269.44
48	11	528	81	29	319	22.84	251.24
	$N = 81$	$\Sigma fX = 2038$			$\Sigma f\|D\| = 987$		$\Sigma f\|D'\| = 1054.76$

Calculation of *MD* and its coefficient from Median:

$$M = \text{size of } \frac{(N+1)}{2}$$

$$= \text{size of } \frac{(81+1)}{2} = \text{size of } 41^{\text{st}} \text{ item} = 19$$

$$MD = \frac{\Sigma f|D|}{N} = \frac{987}{81} = 12.18$$

$$\text{Coefficient of } MD = MD/M = 12.18/19 = 0.64$$

Calculation of MD and its coefficient from Mean:

$$\bar{x} = \frac{\Sigma fX}{N} = \frac{2038}{81} = 25.16$$

$$MD = \frac{\Sigma f|D'|}{N} = \frac{1054.76}{81} = 13.02$$

$$\text{Coefficient of } MD = MD/\bar{x} = 13.02/25.16 = 0.51$$

Problem 6: Calculate MD and coefficient of MD from mean and median both for the following data.

Size of item:	4	6	8	10	12	14	16
Frequency:	2	1	3	6	4	3	1

Solution:

X	f	fX	Cf	$\|(X-M)\| = \|D\|$	$f \cdot \|D\|$	$\|X - \bar{X}\| = \|D'\|$	$f\|D'\|$
4	2	8	2	6	12	6.2	12.4
6	1	6	3	4	4	4.2	4.2
8	3	24	6	2	6	2.2	6.6
10	6	60	12	0	0	0.2	1.2
12	4	48	16	2	8	1.8	7.2
14	3	42	19	4	12	3.8	11.4
16	1	16	20	6	6	5.8	5.8
	$N = 20$	$\Sigma fX = 204$			$\Sigma f\|D\| = 48$		$\Sigma f\|D'\| = 48.8$

Calculation of *MD* and its coefficient from Median:

$$M = \text{size of } \frac{(N+1)}{2}$$

$$= \text{size of } \frac{(20+1)}{2} = \text{size of } 10.5^{\text{th}} \text{ item} = 10$$

$$MD = \frac{\Sigma f|D|}{N} = \frac{48}{20} = 2.4$$

$$\text{Coefficient of } MD = MD/M = 2.4/10 = 0.24$$

Calculation of *MD* and its coefficient from Mean:

$$\bar{x} = \frac{\Sigma fX}{N} = \frac{204}{20} = 10.2$$

$$MD = \frac{\Sigma f|D'|}{N} = \frac{48.8}{20} = 2.44$$

$$\text{Coefficient of } MD = MD/\bar{x}$$

$$= 2.44/10.23 = 0.239$$

Calculation of Mean Deviation for Continuous Series:

In case of series the classes are replaced by mid values. We determine the deviation of theses mid values from the mean or median. The formula for MD or coefficient of MD is the same like discrete series.

$$MD = \frac{\Sigma f|D|}{N} \quad \text{or} \quad \frac{\Sigma f|D|}{N} \times C$$

Coefficient of $MD = MD/(\text{mean})$ or $M(\text{median})$

Problem 7: Calculate the mean deviation from mean and median for the following data:

Class Interval:	0-10	10-20	20-30	30-40	40-50	50-60	60-70	70-80
Frequency:	5	8	7	12	28	20	10	10

Solution: Calculation of Mean deviation from Mean

$A = 45, C = 10$

Class Interval(X)	Frequency(f)	Mid points	$A = 45$ $d' = (m-A)/10$	fd'	$\lvert D\rvert = \lvert m - \bar{x}\rvert$	$f\lvert D\rvert$
0-10	5	5	–4	–20	40	200
10-20	8	15	–3	–24	30	240
20-30	7	25	–2	–14	20	140
30-40	12	35	–1	–12	10	120
40-50	28	45	0	0	0	0
50-60	20	55	1	20	10	200
60-70	10	65	2	20	20	200
70-80	10	75	3	30	30	300
	$N = 100$			$\Sigma fd' = 0$		$\Sigma f\lvert D\rvert = 1400$

$$\bar{x} = A + \frac{\sum fD'}{N} \times C = 45 + \frac{0}{100} \times 10 = 45$$

$$MD = \frac{\sum f|D|}{N} = 1400/100 = 14$$

$$\text{Coefficient of } MD = MD/\bar{x} = 14/45 = 0.31$$

Calculation of Mean deviation from Median:

Class Interval(X)	m	f	cf	$\lvert m-49\rvert = \lvert D\rvert$	$f\lvert D\rvert$
0-10	5	5	5	44	220
10-20	15	8	13	34	272
20-30	25	7	20	24	168
30-40	35	12	32	14	168
40-50	45	28	60	4	112
50-60	55	20	80	6	120
60-70	65	10	90	16	160
70-80	75	10	100	26	260
		$N = 100$			$\Sigma f\lvert D\rvert = 1480$

$$M = \text{Size of } (N/2)\text{th item}$$
$$= \text{Size } (100/2)\text{th item} = \text{Size of } 50^{\text{th}} \text{ item}$$

which lies in 50-60 group

$l_1 = 50, \quad N/2 = 52, \quad cf = 32, \quad f = 20$

$$M = \frac{l_1 + N/2 - cf \times i}{f}$$

$$= 40 + \frac{50-32}{20} \times 10 = 40 + 9 = 49$$

$$MD = \frac{\Sigma f|D|}{N} = 1480/100 = 14.80$$

Problem 8: Calculate the mean deviation from mean and median for the following data:

Class Interval :	0-5	5-10	10-15	15-20	20-25	25-30	30-35	35-40
Frequency :	14	18	24	30	45	35	22	12

Solution: Calculation of Mean deviation from Mean

$A = 22.5, C = 5$

Class Interval(X)	Frequency (f)	Mid points	A = 22.5 d′ = (m–A)/5	fd′	|D| = |m – $\bar{x}$|	f|D|
0-5	14	2.5	–4	–56	18.2	254.8
5-10	18	7.5	–3	–54	13.2	237.6
10-15	24	12.5	–2	–48	8.2	196.8
15-20	30	17.5	–1	–30	3.2	96
20-25	45	22.5	0	0	1.8	81
25-30	35	27.5	1	35	6.8	238
30-35	22	32.5	2	44	11.8	259.6
35-40	12	37.5	3	36	16.8	201.6
	N = 200			Σfd′ = –73		Σf |D| = 1565.4

$$\bar{x} = A + \frac{\Sigma fD'}{N} \times C$$

$$= 22.5 + \frac{-73}{200} \times 5 = 22.5 - 18 = 20.7 \text{ yrs}$$

$$MD = \frac{\Sigma f|D|}{N} = 1565.4/200 = 7.823 \text{ yrs}$$

$$\text{Coefficient of } MD = MD - \bar{x} = 7.823/20.7 = 0.377$$

UNSOLVED PROBLEMS

1. What do you know about mean deviation?
2. How can you compute mean deviation?

3. What are the various pros and cons of mean deviations?
4. What is coeff. Of mean deviation?
5. Calculate the mean and mean deviation of the following data.

Classes:	0-5	5-10	10-15	15-20	20-25	25-30	30-35	35-40
Frequency:	14	18	24	30	45	35	22	12

13.10 STANDARD DEVIATION

The standard deviation is defined as the positive square root of the mean of the square deviations taken from arithmetic mean of the data.

For the sample data the standard deviation is denoted by Sand is defined as:

$$S = \sqrt{\frac{\Sigma(X - \bar{X})^2}{n}}$$

For a population data the standard deviation is denoted by s (sigma) and is defined as:

$$\sigma = \sqrt{\frac{\Sigma(X - \mu)^2}{N}}$$

For frequency distribution the formulas becomes

$$S = \sqrt{\frac{\Sigma f(X - \bar{X})^2}{\Sigma f}} \quad \text{or} \quad \sigma = \sqrt{\frac{\Sigma f(X - \mu)^2}{\Sigma f}}$$

The standard deviation is in the same units as the units of the original observations. If the original observations are in grams, the value of the standard deviation will also be in grams.

The standard deviation plays a dominating role for the study of variation in the data. It is a very widely used measure of dispersion. It stands like a tower among measure of dispersion. As far as the important statistical tools are concerned, the first important tool is the mean $\bar{X}$ and the second important tool is the standard deviation S. It is based on all the observations and is subject to mathematical treatment. It is of great importance for the analysis of data and for the various statistical inferences.

However some alternative methods are also available to compute standard deviation. The alternative methods simplify the computation. Moreover in discussing these methods we will confirm ourselves only to sample data because sample data rather than whole population confront mostly a statistician.

13.10.1 Actual Mean Method

In applying this method first of all we compute arithmetic mean of the given data either ungroup or grouped data. Then take the deviation from the actual mean. This method is already defined above. The following formulas are applied:

For Ungrouped Data

$$S = \sqrt{\frac{\Sigma(X-\bar{X})^2}{n}}$$

For Grouped Data

$$S = \sqrt{\frac{\Sigma f(X-\bar{X})^2}{\Sigma f}}$$

This method is also known as **direct method**

13.10.2 Assumed Mean Method

(a) We use following formulas to calculate standard deviation:

For Ungrouped Data

$$S = \sqrt{\frac{\Sigma D^2}{n} - \left(\frac{\Sigma D}{n}\right)^2}$$

For Grouped Data

$$S = \sqrt{\frac{\Sigma fD^2}{\Sigma f} - \left(\frac{\Sigma fD}{\Sigma f}\right)^2}$$

Where $D = X - \text{A}$ and A is any assumed mean other than zero. This method is also known as short-cut method.

(b) If A is considered to be zero then the above formulas are reduced to the following formulas:

For Ungrouped Data

$$S = \sqrt{\frac{\Sigma X^2}{n} - \left(\frac{\Sigma X}{n}\right)^2}$$

For Grouped Data

$$S = \sqrt{\frac{\Sigma fX^2}{\Sigma f} - \left(\frac{\Sigma fX}{\Sigma f}\right)^2}$$

(c) If we are in a position to simplify the calculation by taking some common factor or divisor from the given data the formulas for computing standard deviation are:

For Ungrouped Data

$$S = \sqrt{\frac{\Sigma U^2}{n} - \left(\frac{\Sigma U}{n}\right)^2} \times C$$

For Grouped Data

$$S = \sqrt{\frac{\Sigma fU^2}{\Sigma f} - \left(\frac{\Sigma fU}{\Sigma f}\right)^2} \times c \text{ or } h$$

Where $U = \frac{X-A}{\text{horc}} = \frac{D}{\text{horc}}$, h = Class Interval and c = Common Divisor. This method is also called method of step-deviation.

13.10.3 Merits of Standard Deviation

The various advantages of the standard deviation are as below:

(i) It is used in further statistical work

(ii) It is possible to compute combined standard deviation of two or more groups.

(iii) It is helpful while comparing variability of two or more series.

(iv) Among all measures of dispersion it is the best measure.

(v) It can be treated mathematically and algebraically.

(vi) It is rigidly defined

(vii) It is based on all observations

(viii) It is least affected by fluctuations of sampling.

(ix) It follows algebraic principles and has algebraic properties.

(x) It exhibits the scatter of dispersion.

(xi) It is highly reliable.

(xii) It is efficient in computation.

13.10.4 Demerits of Standard Deviation

The various disadvantages of the standard deviation are as below:

(i) It is not easy to compute.

(ii) It can't be computed in open end classes.

(iii) It can't be used for comparison of two phenomena.

(iv) It is not easy to understood for a layman.

(v) It is very much affected by extreme value.

(vi) Calculations are complicated.

(vii) It is time consuming method.

13.10.5 Coefficient of Standard Deviation and Variation

Coefficient of Standard Deviation

The standard deviation is the absolute measure of dispersion. Its relative measure is called standard coefficient of dispersion or coefficient of standard deviation. It is defined as:

$$\text{Coefficient of Standard Deviation} = \frac{S}{\overline{X}}$$

Coefficient of Variation

The most important of all the relative measure of dispersion is the coefficient of variation. This word is variation not variance. There is no such thing as coefficient of variance. The coefficient of variation (C.V) is defined as:

$$\text{Coefficient of Variation (C.V)} = \frac{S}{\overline{X}} \times 100$$

Thus C.V is the value of S when $\overline{X}$ is assumed equal to 100. It is a pure number and the unit of observations is not mentioned with its value. It is written in percentage form like 20% or 25%. When its value is 20%, it means that when the mean of the observations is assumed equal to 100, their standard deviation will be 20. The C.V is used to compare the dispersion in different sets of data particularly the data which differ in their means or differ in the units of measurement. The wages of workers may be in dollars and the consumption of meat in their families may be in kilograms. The standard deviation of wages in dollars cannot be compared with the standard deviation of amounts of meat in kilograms. Both the standard deviations need to be converted into coefficient of variation for comparison. Suppose the value of C.V for wages is 10% and the values of C.V for kilograms of meat are 25%. This means that the wages of workers are consistent. Their wages are close to the overall average of their wages. But the families consume meat in quite different quantities. Some families use very small quantities of meat and some others use large quantities of meat. We say that there is greater variation in their consumption of meat. The observations about the quantity of meat are more dispersed or more variant.

Calculation of Standard Deviation for Individual Series:

(1) When deviations are taken form Actual Mean (Direct method)

(i) Compute $\overline{x}$

(ii) Compute $x = (X - \overline{x})$ by taking deviations from mean

(iii) Square the deviations and obtain theier sum as Σx^2

(iv) Apply the formula

$$\sigma = \frac{\sqrt{\Sigma x^2}}{N}$$

(2) When deviations are taken from Assumed mean(Short cut method)

When we get actual mean in terms of fraction. Then we use this method by taking assumed mean. The formula for this case is

$$\sigma = \sqrt{\frac{\Sigma d^2}{N} - (\Sigma d/N)^2}$$

Where Σd^2 is the sun of squares of deviations from assumed mean.

(i) Compute N

(ii) Obtain $dx = (X - A)$ by taking devaitions from assumed mean. Sum of these devaitions is denoted by Σd.

(iii) Square these devaitions ans compute sum Σd^2

(iv) Apply the formula

$$\sigma = \frac{\sqrt{\Sigma d^2}}{N} - (\Sigma d/N)^2$$

13.10.6 Program to Calculate Standard Deviation and Coefficient of Standard Deviation in Individual Series Using Direct Method.

```
#include<stdio.h>
#include<conio.h>
```

```
#include<math.h>
void main()
{
float x[100],n,dx[100],mean,sum=0,sum1=0,sd,csd;
int i;
clrscr();
printf(" How many items in the individual distribution ?");
scanf("%f",&n);
printf("\n Enter the data \n");
for(i=1;i<=n;i++)
{
scanf("%f",&x[i]);
sum=sum+x[i];
}
mean=sum/n;
for(i=1;i<=n;i++)
{
dx[i]=(x[i]-mean)*(x[i]-mean);
sum1=sum1+dx[i];
}
sd=sqrt(sum1/n);
printf("Standard Deviation=%f",sd);
csd=sd/mean;
printf("Coefficient of Mean Deviation=%f",csd);
getch();
}
```

SOLVED PROBLEMS

Problem 1: Calculate Standard deviation by Direct and Short cut method from the following series

Score:	10	20	25	30	55	70

Solution: In case of direct method deviations are taken from Actual mean, so

Score(*X*)	x = (*X* − $\bar{x}$)	x²
10	−25	625
20	−15	225
25	−10	100
30	−5	25
55	20	400
70	35	
Σ*X* = 210	Σx = 0	Σx² = 1225

$$\bar{x} = \Sigma X/N = 210/6 = 35$$

$$\sigma = \frac{\sqrt{\Sigma X^2}}{N} = \sqrt{1225/6} = \sqrt{204.16} = 14.28$$

Coefficient of Standard Deviation $= \sigma/\bar{x} = 14.28/35 = 0.408$

Here $A = 30$

X	X − *A* = *d*	*d*²
10	−20	400
20	−10	100
25	−5	25
30	0	0
55	25	625
70	40	1600
N = 6	Σ*d* = 30	Σ*d*² = 2750

$$\bar{x} = A + \frac{\Sigma d}{N} = 30 + 30/6 = 30 + 5 = 35$$

Coefficient of Standard Deviation $= \sigma/\bar{x} = 14.28/35 = 0.408$

$$\sigma = \sqrt{\frac{\Sigma d^2}{N} - (\Sigma d/N)^2} = \sqrt{(2750/6) - (30/10)2}$$

$$= \sqrt{458.33 - 9} = \sqrt{449.33} = 21.19$$

(i) Product Moment Method

Standard deviation can also be computed by using Product moment method. The formula for this is as

$$\sigma = \sqrt{\frac{\Sigma fX^2}{N} - (\Sigma fX/N)^2}$$

Where *X* is the item.

Calculation of Stanard Deviation for Discrete Series

The following formula is used in case of Discrete series (Direct method)

$$\sigma = \frac{\sqrt{\Sigma fx^2}}{N}$$

Where $x = (X - \bar{x})$

(ii) Short cut method

$$\sigma = \sqrt{\frac{\Sigma fd^2}{N} - (\Sigma fd/N)^2}$$

Where $d = (X - A)$

Steps:

(i) Find N or Σf.

(ii) Compute d as the deviations of items from assumed mean.

(iii) Obtain squares of deviation d^2

(iv) Compute Σfd

(v) Compute Σfd^2

(iii) Step Deviation method

$$\sigma = \sqrt{\frac{\Sigma fd'^2}{N} - (\Sigma fd'/N)^2} \times c$$

Where $d' = d/c$

Coefficient of *SD*:

$$\text{Coefficient of } SD = \sigma/\bar{x}$$

Where $\bar{x} = A + \Sigma fd/N$

13.10.7 Program to Calculate Standard Deviation and Coefficient of Standard Deviation in Discrete Series Using Direct Method.

```
#include<stdio.h>
#include<conio.h>
#include<math.h>
void main()
{
float a[100],n,f[100],dx[20],fdx[100],af[100],sd,summ=0,sumf=0,mean,csd,summ1=0;
int i;
clrscr();
```

```
printf(" How many items in the individual distribution ?");
scanf("%f",&n);
printf("\n Ente the X-values and frequency \n");
for(i=1;i<=n;i++)
{
scanf("%f",&a[i]);
scanf("%f",&f[i]);
af[i]=a[i]*f[i];
summ=summ+af[i];
sumf=sumf+f[i];
}
mean=summ/sumf;
for(i=1;i<=n;i++)
{
dx[i]=(a[i]-mean)*(a[i]-mean);
fdx[i]=f[i]*(dx[i]);
summ1=summ1+fdx[i];
}
sd=sqrt(summ1/sumf);
printf("Standard Deviation=%f",sd);
csd=sd/mean;
printf("Coefficient of Mean Deviation=%f",csd);
getch();
}
```

Problem 2: Calculate the coefficient of standard deviation and coefficient of variation for the following sample data: 2, 4, 8, 6, 10, and 12.

Solution:

X	$(X-\bar{X})^2$
2	$(2-7)^2 = 25$
4	$(4-7)^2 = 9$
8	$(8-7)^2 = 1$
6	$(6-7)^2 = 1$
10	$(10-7)^2 = 9$
12	$(12-7)^2 = 25$
$\Sigma X = 42$	$\Sigma(X-\bar{X})^2 = 70$

13

$$\bar{X} = \frac{\Sigma X}{n} = \frac{42}{6} = 7$$

$$S = \sqrt{\frac{\Sigma(X - \bar{X})^2}{n}}$$

$$S = \sqrt{\frac{70}{6}} = \sqrt{\frac{35}{3}} = 3.42$$

$$\text{Coefficient of Standard Deviation} = \frac{S}{\bar{X}} = \frac{3.42}{7} = 0.49$$

$$\text{Coefficient of Variation (C.V)} = \frac{S}{\bar{X}} \times 100 = \frac{3.42}{7} \times 100 = 48.86\,\%$$

Problem 3: Calculate Standard Deviation from the following data:

X:	12	14	16	18	20	22	24
f:	6	12	18	26	16	10	8

Solution: Suppose $A = 20$

X	f	(X–A) = d	d^2	fd	fd^2
112	6	–8	64	–48	384
14	12	–6	36	–72	432
16	18	–4	16	–72	288
18	26	–2	4	–52	104
20	16	0	0	0	0
22	10	2	4	20	40
24	8	4	16	32	128
	N = 96			Σfd = –192	Σfd^2 = 1376

$$\sigma = \sqrt{\frac{\Sigma\, fd^2}{N} - (\Sigma\, fd/N)^2}$$

$$= \sqrt{(1376/96) - (-192/96)2} = \sqrt{10.33} = 3.2145$$

Calculation of Standard Deviation for Continuous Series

1. The following formula is used in case of continuous series (Direct method)

$$\sigma = \frac{\sqrt{\Sigma\, fX^2}}{N}$$

Where $x = (X - \bar{x}\ )$

Steps:

(i) Find out the mid points of different classes.

(ii) Compute aritmetic means and take deviations from the mid values.

(iii) Compute Σfx^2

(iv) Apply the formula

$$\sigma = \frac{\sqrt{\Sigma fX^2}}{N}$$

(ii) Short cut method

Steps:

(i) Find out the mid points of different classes.

(ii) Take deviations from the assumed means and compute dx.

(iii) Compute Σfd^2

(iv) Apply the formula

$$\sigma = \sqrt{\frac{\Sigma fd^2}{N} - (\Sigma fd/N)^2}$$

Where $d = (X - A)$

(iii) Step deviation method

$$\sigma = \sqrt{\frac{\Sigma fd'^2}{N} - (\Sigma fd'/N)^2} \times c$$

Steps:

(i) Find out the mid points of different classes.

(ii) Take deviations from an assumed mean and compute d.

(iii) If class intervals are equal take *a* common factor.Divide the different devaitions by the common factor and denote them as d′

i.e $$d' = (m - A)/C$$

(iv) Compute $\Sigma fd'$

(v) Compute $\Sigma fd'^2$

(iv) Apply the formula

$$\sigma = \sqrt{\frac{\Sigma fd'^2}{N} - (\Sigma fd'/N)^2} \times c$$

Problem 4: Calculate coefficient of standard deviation and coefficient of variation from the following distribution of marks:

13

Marks	No. of Students
1-3	40
3-5	30
5-7	20
7-9	10

Solution:

Marks	f	x	fx	$(X-\bar{X})^2$	$f(X-\bar{X})^2$
1-3	40	2	80	4	160
3-5	30	4	120	0	0
5-7	20	6	120	4	80
7-9	10	8	80	16	160
Total	**100**		**400**		**400**

$$X = \frac{\Sigma fX}{\Sigma f} = \frac{400}{100} = 4$$

$$S = \sqrt{\frac{\Sigma f(X-\overline{X})^2}{\Sigma f}} = \sqrt{\frac{400}{100}} = \sqrt{4} = 2 \text{ Marks}$$

$$\text{Coefficient of Standard Deviation} = \frac{S}{\overline{X}} = \frac{2}{4} = 0.5$$

$$\text{Coefficient of Variation (C.V)} = \frac{S}{\overline{X}} \times 100 = \frac{2}{4} \times 100 = 50\,\%$$

Problem 5: Calculate standard deviation from the following:

X:	31-35	36-40	41-45	46-50	51-55	56-60	61-65	66-70
f:	2	3	8	12	16	5	2	2

Solution: Take $A = 53$

X	f	Mid values(m)	$d = m - 53$	$d' = d/CC = 5$	fd'	fd'^2
31-35	2	33	−20	−4	−8	32
36-40	3	38	−15	−3	−9	27
41-45	8	43	−10	−2	−16	32
46-50	12	48	−5	−1	−12	12
51-55	16	53	0	0	0	0
56-60	5	58	5	1	5	5
61-65	2	63	10	2	4	8
66-70	2	68	15	3	6	18
	$N = 50$				$\Sigma fd' = -30$	$\Sigma fd^{2\prime} = 134$

As $N = \Sigma f = 50$, $\Sigma fd' = -30$, $\Sigma fd^2 = 134$

$$\sigma = \sqrt{\frac{\Sigma fd'^2}{N} - (\Sigma fd'/N)^2} \times c$$

$$= \sqrt{(134/50) - (-30/50)^2} \times 5 = \sqrt{2.68 - 0.36} \times 5$$

$$= \sqrt{2.32} \times 5 = 1.52 \times 5 = 7.60$$

Problem 6: Calculate standard deviation from the following:

Marks :	20	40	60	80	100
No. of Students:	8	20	50	70	80

Solution:

Marks	No. of Students	Mid values(m)	A=20, C = 20d'	fd'	fd'^2
0-20	8	10	−2	−16	32
20-40	12	30	−1	−12	12
40-60	30	50	0	0	0
60-80	20	70	1	20	20
80-100	10	90	2	20	40
	$N = 80$			$\Sigma fd' = 12$	$\Sigma fd'^2 = 104$

$$\bar{x} = A + \frac{\Sigma fd'}{N} \times c = 50 + \frac{12}{80} \times c = 53$$

$$\sigma = \sqrt{\frac{\Sigma fd'^2}{N} - (\Sigma fd'/N)^2} \times c$$

$$= \sqrt{(104/80) - (12/80)^2} \times 20$$

$$= 1.130 \times 20 = 22.605.$$

Relationship between different Measures of Dispersion

(1) The Quartile deviation is 0.6745 times the standard deviation, i.e.

$$QD = \frac{2}{3}\sigma$$

(2) The mean deviation is equal to 0.7979 times the standard deviation

$$MD = \frac{4}{5}\sigma$$

(3) The six times Quartile deviation is equal to five times the mean deviation i.e.

$$QD = \frac{5}{6}MD$$

Combining all the three relationships, we get

$$4\sigma = 5\ MD = 6\ QD$$

13

13.10.8 Program to Calculate Standard Deviation and Coefficient of Standard Deviation in Continuous Series Using Direct Method.

```
#include<stdio.h>
#include<conio.h>
#include<math.h>
void main()
{
float x[30],y[30],z[30],sum=0,sum1=0,sum2=0,a[30],csd, mean ,c[30],d[30],e[30],sd=0;
int i,n;
clrscr();
printf(" How many items in the individual distribution ?");
scanf("%d",&n);
printf("\n Ente the data \n");
for(i=1;i<=n;i++)
{
scanf("%f",&x[i]);
scanf("%f",&y[i]);
scanf("%f",&z[i]);
sum=sum+z[i];
}
for(i=1;i<=n;i++)
{
a[i]=(x[i]+y[i])/2;

c[i]=z[i]*a[i];
sum1=sum1+c[i];
}
mean=sum1/sum;
for(i=1;i<=n;i++)
{
d[i]=(a[i]-mean)*(a[i]-mean);
e[i]=z[i]*d[i];
sum2=sum2+e[i];
}
```

```
sd=sqrt((sum2/sum));
printf("Standard Deviation=%f",sd);
csd=sd/mean;
printf("coefficient of Standard Deviation=%f",csd);
getch();
}
```

Coefficient of Variation or Coefficient of Variability

It is used to compare the variability of two or more than two series. The relation between coefficient of variability and consistency is that if the coefficient of variability is higher then there will be less consistency and vice versa. It is the relative measure of dispersion.

According to Karl Pearson "Coefficient of Variation is the percentage variation in mean, standard deviation being considered as the total variation on the mean"

If we have to compare the variability of distributions, we have find coefficient of variation for that which is

$$CV = (\sigma/\bar{x}) \times 100$$

Now we will discuss *CV* for individual, discrete and continuous series one by one.

Coefficient of Variation for Individual Series

Problem 7: The following are the test scores of two batsmen *A* and B. Find which of the two batsmen is more consistent in scoring?

A:	5	7	16	27	39	53	56	61	80	101	105
B:	0	4	16	21	41	43	57	78	83	90	95

Solution: In order to find the batsman, which is more consistent we are to compare the coefficient of variation.

Batsman A			Batsman B		
X	$x = X - 50$	x^2	Y	$y = Y - 48$	y^2
5	–45	2025	0	–48	2304
7	–43	1849	4	–44	1936
16	–34	1156	16	–32	1024
27	–23	529	21	–27	729
39	–11	121	41	–7	49
53	3	9	43	–5	25
56	6	36	57	9	81
61	11	121	78	30	900
80	30	900	83	35	1225
101	51	2601	90	42	1764
105	55	3025	95	47	2209
$\Sigma X = 550$	$\Sigma x = 0$	$\Sigma x^2 = 12372$	$\Sigma Y = 528$		$\Sigma y^2 = 12246$

As $N = 11$, $\Sigma X = 550$, $\Sigma x^2 = 12372$, $\Sigma y^2 = 12246$

Batsman A,

$$\bar{x} = \Sigma X/N = 550/11 = 50$$

$$\sigma = \sqrt{\Sigma x^2/N}$$

$$= \sqrt{12372/11} = \sqrt{1124.73} = 33.54$$

$$CV = (\sigma/\bar{x}) \times 100$$

$$= (33.54/50) \times 100 = 67.08$$

Batsman *B*,

$$\bar{x}\ \bar{y} = \Sigma Y/N = 528/11 = 48$$

$$\sigma = \sqrt{\Sigma y^2/N} = \sqrt{12246/11} = \sqrt{1113.27} = 33.37$$

$$CV = (\sigma/\bar{x}\ \bar{y}) \times 100 = (33.27/48) \times 100 = 69.52$$

Batsman *A* is more consistent because *CV* for *A* is less.

Coefficient of Variation for Discrete Series

Problem 8: From the following data given below calculate, *s* and *CV*

X:	10	11	12	13	14
f:	3	12	18	12	3

Solution: Take $A = 12$

X	f	(X – A) = d	fd	fd²
10	3	-2	-6	12
11	12	-2	-12	12
12	18	0	0	0
13	12	1	12	12
14	3	2	6	12
	N = 48	Σfd = 0		Σfd² = 48

$$\bar{x} = A + \Sigma fd/N = 12 + 0/48 = 12$$

$$\sigma = \sqrt{\frac{\Sigma fd^2}{N} - (\Sigma fd/N)^2}$$

$$= \sqrt{(48/48) - (0/48)^2} = 1$$

$$CV = (\sigma/\bar{x}) \times 100 = 100/12 = 8.33\%$$

Coefficient of Variation for Continuous Series

Problem 9: Following data represents the lives of two Air conditioners *A* and *B*

	Refrigerator	
Life(in yrs)	*A*	*B*
5-10	2	9
10-15	9	11
15-20	29	18
20-25	54	32
25-30	11	27

Find which model has greater uniformity.

Solution: For determining the greater uniformity we have to compute coefficient of variation.

Life (in yrs)	*f*	Mid values (*m*)	Refrigerator *A* $d = m - 17.5$	$C = 5d' = d/5$	*fd'*	fd'^2	Refrigerator *B* *f*	*fd'*	fd'^2
5-10	2	7.5	–10	–2	–4	8	9	–18	36
10-15	9	12.5	–5	–1	–9	9	11	–1111	
15-20	29	17.5	0	0	0	0	18	0	0
20-25	54	22.5	5	1	54	54	32	32	32
25-30	11	27.5	10	2	22	44	27	54	108
30-35	5	32.5	15	3	15	45	13	39	117
	N = 110				$\Sigma fd'$ = 78	$\Sigma fd'^2$ = 160	*N* = 110	$\Sigma fd'$ = 96	$\Sigma fd'^2$ = 304

Refrigerator *A*,

$$\bar{x} = A + \frac{\Sigma fd'}{N} \times C$$

$$= 17.5 + \frac{78}{100} \times 5 = 17.5 + 3.55 = 21.05$$

$$\sigma = \sqrt{\frac{\Sigma fd'^2}{N} - (\Sigma fd'/N)^2} \times C$$

$$= \sqrt{(160/110) - (78/110)^2} \times 5$$

$$= \sqrt{1.4545 - 0.5028} \times 5 = \sqrt{0.9517} \times 5 = 4.878$$

$$CV = (\sigma/\bar{x}) \times 100$$

$$= (4.878/21.05) \times 100 = 23.17$$

Refrigerator *B*,

$$\bar{x} = A + \frac{\sum fd'}{N} \times C$$

$$= 17.5 + \frac{96}{100} \times 5 = 17.5 + 4.36 = 21.86$$

$$\sigma = \sqrt{\frac{\sum fd'^2}{N} - (\sum fd'/N)^2} \times C$$

$$= \sqrt{(304/110) - (96/110)^2} \times 5$$

$$= \sqrt{2.76 - 0.7617} \times 5 = \sqrt{1.9983} \times 5 = 7.068$$

$$CV = (\sigma/\bar{x}) \times 100 = (7.068/21.86) \times 100 = 32.33$$

Refrigerator *A* has greater unifrormity because *CV* for *A* is less.

Problem 10: Calculate the standard deviation for the following sample data using all methods: 2, 4, 8, 6, 10, and 12.

Solution:

Method-I: Actual Mean Method

X	$(X - \bar{X})^2$
2	$(2-7)^2 = 25$
4	$(4-7)^2 = 9$
8	$(8-7)^2 = 1$
6	$(6-7)^2 = 1$
10	$(10-7)^2 = 9$
12	$(12-7)^2 = 25$
$\Sigma X = 42$	$\Sigma(X - \bar{X})^2 = 70$

$$\bar{X} = \frac{\Sigma X}{n} = \frac{42}{6} = 7$$

$$S = \sqrt{\frac{\Sigma(X - \bar{X})^2}{n}}$$

$$S = \sqrt{\frac{70}{6}} = \sqrt{\frac{35}{3}} = 3.42$$

Method-II: Taking Assumed Mean as 6

X	$D = X - 6$	D^2
2	–4	16
4	–2	4
8	2	4
6	0	0
10	4	16
12	6	36
Total	$\Sigma D = 6$	$\Sigma D^2 = 76$

$$S = \sqrt{\frac{\sum D^2}{n} - \left(\frac{\sum D}{n}\right)^2}$$

$$S = \sqrt{\frac{76}{6} - \left(\frac{6}{6}\right)^2} = \sqrt{\frac{70}{6}}$$

$$S = \sqrt{\frac{35}{3}} = 3.42$$

Method-III: Taking Assume Mean as Zero

X	X^2
2	4
4	16
8	64
6	36
10	100
12	144
$\Sigma X = 42$	$\Sigma X^2 = 364$

$$S = \sqrt{\frac{\sum X^2}{n} - \left(\frac{\sum X}{n}\right)^2}$$

$$S = \sqrt{\frac{364}{6} - \left(\frac{42}{6}\right)^2}$$

$$S = \sqrt{\frac{70}{6}} = \sqrt{\frac{35}{3}} = 3.42$$

Method-IV: Taking 2as common divisor or factor

X	$U = (X - 4)/2$	U^2
2	–1	1
4	0	0
8	2	4
6	1	1
10	3	9
12	4	16
Total	$\Sigma U = 9$	$\Sigma U^2 = 31$

$$S = \sqrt{\frac{\Sigma U^2}{n} - \left(\frac{\Sigma U}{n}\right)^2} \times C$$

$$S = \sqrt{\frac{31}{6} - \left(\frac{9}{6}\right)^2} \times 2$$

$$S = \sqrt{2.92} \times 2 = 3.42$$

Problem 11: Calculate standard deviation from the following distribution of marks by using all the methods.

Marks:	1-3	3-5	5-7	7-9
No. of students:	40	30	20	10

Solution:

Method-I: Actual Mean Method

Marks	f	x	fX	$(X - \bar{X})^2$	$f(X - \bar{X})^2$
1-3	40	2	80	4	160
3-5	30	4	120	0	0
5-7	20	6	120	4	80
7-9	10	8	80	16	160
Total	100		400		400

$$X = \frac{\Sigma fX}{\Sigma f} = \frac{400}{100} = 4$$

$$S = \sqrt{\frac{\Sigma f(X - \overline{X})^2}{\Sigma f}} = \sqrt{\frac{400}{100}} = \sqrt{4} = 2 \text{ Marks}$$

Method-II: Taking assumed mean as 2

Marks	f	X	$D = (X - 2)$	fD	fD^2
1-3	40	2	0	0	0
3-5	30	4	2	60	120
5-7	20	6	4	80	320
7-9	10	8	6	60	160
Total	**100**			**200**	**800**

$$S = \sqrt{\frac{\sum fD^2}{\sum f} - \left(\frac{\sum fD}{\sum f}\right)^2} = \sqrt{\frac{800}{100} = \left(\frac{200}{100}\right)^2}$$

$$S = \sqrt{8-4} = \sqrt{4} = 2 \text{ Marks}$$

Method-III: Using assumed mean as Zero

Marks	f	x	fx	fx^2
1-3	40	2	80	160
3-5	30	4	120	480
5-7	20	6	120	720
7-9	10	8	80	640
Total	**100**		**400**	**2000**

$$S = \sqrt{\frac{\sum fX^2}{\sum f} - \left(\frac{\sum fX}{\sum f}\right)^2} = \sqrt{\frac{2000}{100} = \left(\frac{400}{100}\right)^2}$$

$$S = \sqrt{20-16} = \sqrt{4} = 2 \text{ Marks}$$

Method-IV: By taking 2 as the common divisor

Marks	f	X	$U = (X - 2)/2$	fU	fU^2
1-3	40	2	–2	–80	160
3-5	30	4	–1	–30	30
5-7	20	6	0	0	0
7-9	10	8	1	10	10
Total	**100**			**–100**	**200**

$$S = \sqrt{\frac{\sum fU^2}{\sum f} - \left(\frac{\sum fU}{\sum f}\right)^2} \times h = \sqrt{\frac{200}{100} = \left(\frac{100}{100}\right)^2} \times 2$$

$$S = \sqrt{2-1} \times 2 = \sqrt{1} \times 2 = 1 \times 2 = 2 \text{ Marks}$$

UNSOLVED PROBLEMS

1. What do you know about standard deviation?
2. How coefficient of standard deviation works?
3. What are the various advantages and disadvantages of standard deviation?
4. How can you compute standard deviation for different types of series?
5. What is the relationship between different types of measures?
6. Calculate the standard deviation from the following data:

Size of item:	6	7	8	9	10	11	12
Frequency:	3	8	9	13	8	5	4

7. find standard deviation and mean for the distribution of wages of 65 employees working in a factory

Wages(X):	55	65	75	85	95	105	115
Frequency:	8	10	16	14	10	5	2

8. Calculate standard deviation from the following data:

Marks:	1	2	3	4	5	6	7	8	9
No. of students:	32	41	57	98	123	83	46	17	3

9. Compute mean, mode and standard deviation from the following:

Year(Below):	10	20	30	40	50	60
No. of Students:	15	32	51	78	97	109

10. Calculate arithmetic mean, median and standard deviation of the following frequency distribution:

Class:	3-6	6-9	9-12	12-15	15-18	18-21	21-24
Frequency:	2	5	10	23	21	12	3

11. Find mea and Standard deviation for the following data:

Age(in yrs)	No. of teachers
25-30	30
30-35	23
35-40	20
40-45	14
45-50	10
50-55	3

12. The following table shows the daily pocket expenses of 48 students of a college in rupees.

Daily pocket expenses (inRs.):	6	7	8	9	10	11	12
No. of Students :	3	6	9	13	8	5	4

Calculate mean deviation, standard deviation and coefficient of variation of the above data.

13. Compute coefficient of variation for the following data:

Class interval:	75-80	80-85	85-90	90-95	95-100
Frequency :	14	10	12	8	6

14. Compute standard deviation and coefficient of variation from the frequency distribution given below

Class Interval	Frequency
10-20	13
20-30	3
30-40	2
40-50	49
50-60	24
60-70	2
70-80	0
80-90	1

15. Calculate the standard deviation of the following two series. Comment upon their variations.

P:	6	8	11	9	8	10	12	10	18	12
Q:	16	20	18	18	20	22	26	18	22	20

16. Calculate the coefficient of dispersion for the following age distribution.

Age:	10	20	30	40	50	60	70	80
No. of persons:	15	30	53	75	100	110	115	125

17. Calculate the mean and standard deviation of the following data.

Height:	150	153	156	159	162	165	168	171	174
Frequency:	2	0	15	29	25	12	10	4	3

13.11 THE VARIANCE

Variance is another absolute measure of dispersion. It is defined as the average of the squared difference between each of the observations in a set of data and the mean. For a sample data the variance is denoted is denoted by S^2 and the population variance is denoted by σ^2 (sigma square). The sample variance S^2 has the formula:

$$S^2 = \sqrt{\frac{\Sigma(X - \bar{X})^2}{n}}$$

Where $\bar{X}$ is sample mean and n is the number of observations in the sample.

The population variance σ^2 is defined as:

$$\sigma^2 = \sqrt{\frac{\Sigma(X - \mu)^2}{N}}$$

Where μ is the mean of the population and *N* is the number of observations in the data. It may be remembered that the population variance σ^2 is usually not calculated. The sample variance S^2 is calculated and if need be, this S^2 is used to make inference about the population variance.

The term $\Sigma(X - \bar{X})^2$ is positive, therefore S^2 is always positive. If the original observations are in centimeter, the value of the variance will be (centimeter) 2. Thus the unit of S^2 is the square of the units of the original measurement.

For a frequency distribution the sample variance S^2 is defined as:

$$S^2 = \sqrt{\frac{\Sigma f(X - \bar{X})^2}{\Sigma f}}$$

For s frequency distribution the population variance σ^2 is defined as:

$$\sigma^2 = \sqrt{\frac{\Sigma f(X - \mu)^2}{\Sigma f}}$$

In simple words we can say that variance is the square of standard deviation.

$$\text{Variance} = (\text{Standard Deviation})^2$$

SOLVED PROBLEMS

Problem 1: Calculate the variance for the following sample data: 2, 4, 8, 6, 10, and 12.

Solution:

X	$(X - \bar{X})^2$
2	$(2 - 7)^2 = 25$
4	$(4 - 7)^2 = 9$
8	$(8 - 7)^2 = 1$
6	$(6 - 7)^2 = 1$
10	$(10 - 7)^2 = 9$
12	$(12 - 7)^2 = 25$
$\Sigma X = 42$	$\Sigma(X - \bar{X})^2 = 70$

$$\bar{X} = \frac{\Sigma X}{n} = \frac{42}{6} = 7$$

$$S^2 = \frac{\Sigma(X - \bar{X})^2}{n}$$

$$S^2 = \frac{70}{6} = \frac{35}{3} = 11.67$$

$$\text{Variance} = S^2 = 11.67$$

Problem 2: Calculate variance from the following distribution of marks:

Marks	No. of Students
1-3	40
3-5	30
5-7	20
7-9	10

Solution:

Marks	f	x	fX	$(X-\bar{X})^2$	$f(X-\bar{X})^2$
1-3	40	2	80	4	160
3-5	30	4	120	0	0
5-7	20	6	120	4	80
7-9	10	8	80	16	160
Total	100		400		400

$$X = \frac{\Sigma fX}{\Sigma f} = \frac{400}{100} = 4$$

$$S^2 = \sqrt{\frac{\Sigma f(X-\overline{X})^2}{\Sigma f}} = \frac{400}{100} = 4 \text{ Marks}$$

$$\text{Variance} = S^2 = 4$$

13.12 SHEPPARD CORRECTIONS AND CORRECTED COEFFICIENT OF VARIATION

13.12.1 Sheppard Corrections

In grouped data the different observations are put into the same class. In the calculation of variation or standard deviation for grouped data, the frequency f is multiplied with X which is the mid-point of the respective class. Thus it is assumed that all the observations in a class are centered at X. But this is not true because the observations are spread in the said class. This assumption introduces some error in the calculation of S^2 and S. The value of S^2 and S can be corrected to some extent by applying Sheppard correction. Thus

$$S^2 \text{ (Corrected)} = S^2 - \frac{h^2}{12}$$

$$S^2 \text{ (Corrected)} = \sqrt{S^2 - \frac{h^2}{12}}$$

Where h is the uniform class interval.

This correction is applied in grouped data which has almost equal tails in the start and at the end of the data. If a data a longer tail on any side, this correction is not applied. If size of the class interval h is not the same in all classes, the correction is not applicable.

13.12.2 Corrected Coefficient of Variation

When the corrected standard deviation is used in the calculation of the coefficient of variation, we get what is called the corrected coefficient of variation. Thus

$$\text{Corrected Coefficient of Variation} = \frac{S(\text{corrected})}{\bar{X}} \times 100$$

SOLVED PROBLEMS

Problem 1: Calculate Sheppard correction and corrected coefficient of variation from the following distribution of marks by using all the methods.

Marks	No. of Students
1-3	40
3-5	30
5-7	20
7-9	10

Solution:

Marks	f	X	fX	$U = (X-2)/2$	fU	fU^2
1-3	40	2	80	–2	–80	160
3-5	30	4	120	–1	–30	30
5-7	20	6	120	0	0	0
7-9	10	8	80	1	10	10
Total	**100**		**400**		**–100**	**200**

$$\bar{X} = \frac{\Sigma fX}{\Sigma f} = \frac{400}{100} = 4$$

$$S^2 = \left[\frac{\Sigma fU^2}{\Sigma f} - \left(\frac{\Sigma fU}{\Sigma f}\right)^2\right] \times h$$

$$= \left[\frac{200}{100} = \left(\frac{-100}{100}\right)^2\right] \times 2$$

$$S^2 = [2-1] \times 2 = 1 \times 2 = 2$$

$$S^2\ (\text{Corrected}) = S^2 - \frac{h^2}{12} = 2 - \frac{4}{12} = \frac{5}{3} = 1.67$$

$$\text{Corrected Coefficient of Variation} = \frac{S(\text{corrected})}{\bar{X}} \times 100$$

$$= \frac{1.29}{4} \times 100 = 32.25$$

13.13 COMBINED VARIANCE

Like combined mean, the combined variance or standard deviation can be calculated for different sets of data. Suppose we have two sets of data containing n_1 and n_2 observations with means $\bar{X}_1$ and $\bar{X}_2$, and variances S_1^2 and S_2^2 and . If $\bar{X}_c$ is the combined mean and S_c^2 is the combined variance of $n_1 + n_2$ observations, then combined variance is given by

$$S_c^2 = n_1 S_1^2 + n_2 S_2^2 + n_1(\bar{X} - \bar{X}_c)^2 + n_2(\bar{X} - \bar{X}_c)^2$$

It can be written as
$$S_c^2 = \frac{n_1\left[S_1^2 + (\bar{X}_1 - \bar{X}_c)^2\right] + n_2\left[S_2^2 + (\bar{X}_2 - \bar{X}_c)^2\right]}{n_1 + n_2}$$

Where
$$X_c^2 = \frac{n_1\bar{X}_1 + n_2\bar{X}_2}{n_1 + n_2}$$

The combine standard deviation S_c can be calculated by taking the square root of S_c^2.

Problem 2: For a group of 50 male workers the mean and standard deviation of their daily wages are \$ 63 and \$ 9 respectively. For a group of 40 female workers these values are \$ 54 and \$ 6 respectively. Find the mean and variance of the combined group of 90 workers.

Solution: Here $n_1 = 50,\ \bar{X}_1 = 63,\ S_1^2 = 81$

$n_2 = 40,\ \bar{X}_2 = 54,\ S_2^2 = 36$

$$\text{Combined Arithmetic Mean} = X_c^2 = \frac{n_1\bar{X}_1 + n_2\bar{X}_2}{n_1 + n_2}$$

$$= X_c^2 = \frac{50(63) + 40(54)}{50 + 40} = \frac{5390}{90} = 59$$

$$(\text{Combined variance})\ S_c^2 = \frac{n_1\left[S_1^2 + (\bar{X}_1 - \bar{X}_c)^2\right] + n_2\left[S_2^2 + (\bar{X}_2 - \bar{X}_c)^2\right]}{n_1 + n_2}$$

$$= \frac{50 + [81\,(63 - 59)^2] + 40\,[36 + (54 - 59)^2]}{50 + 40}$$

$$= \frac{4850 + 2440}{90} = \frac{7290}{90} = 81$$

13.14 DISPERSION BY THE METHOD OF GRAPH (LORENZ CURVE)

The Lorenz curve is a graphical representation of the cumulative distribution function of a probability distribution; it is a graph showing the proportion of the distribution assumed by the bottom y % of the values. It is often used to represent income distribution, where it shows for the bottom x % of households, what percentage y % of the total income they have. The percentage of households is plotted on the x-axis, the percentage of income on the y-axis. It can also be used to show distribution of assets. In such use, many economists consider it to be a measure of social inequality. It was developed by Max O. Lorenz in 1905 for representing income distribution.

Every point on the Lorenz curve represents a statement like "the bottom 20% of all households have 10% of the total income". A perfectly equal income distribution would be one in which every person has the same income. In this case, the bottom N% of society would always have N% of the income. This can be depicted by the straight line $y = x$; called the line of perfect equality or the 45° line.

By contrast, a perfectly unequal distribution would be one in which one person has all the income and everyone else has none. In that case, the curve would be at $y = 0$ for all $x < 100\%$, and $y = 100\%$ when $x = 100\%$. This curve is called the line of perfect inequality.

The Gini coefficient is the area between the line of perfect equality and the observed Lorenz curve, as a percentage of the area between the line of perfect equality and the line of perfect inequality. This equals two times the area between the line of perfect equality and the observed Lorenz curve.

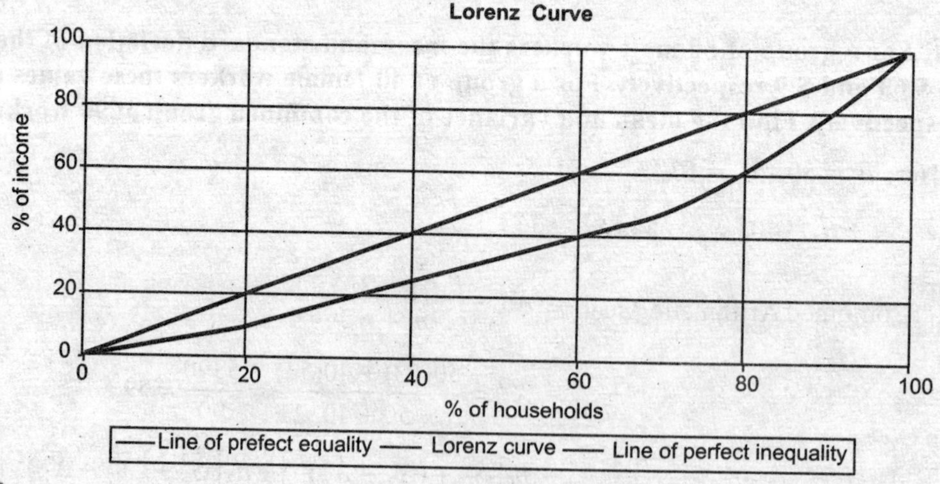

Computation

The Lorenz curve can often be represented by a function $L(F)$, where F is represented by the horizontal axis, and L is represented by the vertical axis.

For a population of size n, with a sequence of values y_i, $i = 1$ to n, that are indexed in nondecreasing order ($y_i \leq y_{i+1}$), the Lorenz curve is the continuous piecewise linear function connecting the points (F_i , L_i), $i = 0$ to n, where $F_0 = 0$, $L_0 = 0$, and for $i = 1$ to n:

$$F_i = i/n$$

$$S_i = \Sigma_{j=1}^{i} y_i$$

$$L_i = S_i/S_n$$

For a discrete probability function $f(y)$, let y_i, $i = 1$ to n, be the points with non-zero probabilities indexed in increasing order ($y_i < y_{i+1}$). The Lorenz curve is the continuous piecewise linear function connecting the points (F_i , L_i), i = 0 to n, where $F_0 = 0$, $L_0 = 0$, and for $i = 1$ to n:

$$F_i = \Sigma_{j=1}^{i} f(y_i)$$

$$S_i = \Sigma_{j=1}^{i} f(y_i)\, y_i$$

$$L_i = S_i/S_n$$

For a probability density function $f(x)$ with the cumulative distribution function $F(x)$, the Lorenz curve $L(F(x))$ is given by:

$$L(F(x)) = \frac{\int_{-\infty}^{x} t\, f(t)dt}{\int_{-\infty}^{\infty} t\, f(t)dt}$$

For a cumulative distribution function $F(x)$ with inverse $x(F)$, the Lorenz curve $L(F)$ is given by:

$$L(F(x)) = \frac{\int_{0}^{F} x(F_1)dF_1}{\int_{0}^{1} x(F_1)dF_1}$$

The inverse $x(F)$ may not exist because the cumulative density function has jump discontinuities or intervals of constant values. However, the previous formula can still apply by generalizing the definition of $x(F)$:

$$x(F_1) = \inf \{y : F(y) \geq F_1\}$$

For an example of a Lorenz curve, see the Pareto distribution.

13.14.1 Properties

A Lorenz curve always starts at (0,0) and ends at (1,1).

The Lorenz curve is not defined if the mean of the probability distribution is zero or infinite.

The Lorenz curve for a probability distribution is a continuous function. However, Lorenz curves representing discontinuous functions can be constructed as the limit of Lorenz curves of probability distributions, the line of perfect inequality being an example.

If the variable being measured cannot take negative values, the Lorenz curve:

- cannot rise above the line of perfect equality,
- cannot sink below the line of perfect inequality,
- is increasing, and
- is a convex function.

If the variable being measured can take negative values but has a positive mean, then the Lorenz curve will sink below the line of perfect inequality and is a convex function.

If the variable being measured can take negative values and has a negative mean, then the Lorenz curve will be above the line of perfect equality, except at the end points, and is a concave function.

The Lorenz curve is invariant under positive scaling. If X is a random variable, for any positive number c the random variable c X has the same Lorenz curve as X.

The Lorenz curve is flipped twice, once about $F = 0.5$ and once about $L = 0.5$, by negation. If X is a random variable with Lorenz curve $L_X(F)$, then $-X$ has the Lorenz curve:

$$L_{-X} = 1 - L_X(1-F)$$

The Lorenz curve is changed by translations so that the equality gap $F–L(F)$ changes in proportion to the ratio of the original and translated means. If X is a random variable with a Lorenz curve $L_X(F)$ and mean μ_x, then for any constant $c \neq -\mu_x$, $X + c$ has a Lorenz curve defined by:

$$F - L_{x+c}(F) = \frac{\mu_x}{\mu_x + c}(F - L_x(F))$$

For a cumulative distribution function $F(x)$ with mean μ and (generalized) inverse $x(F)$, then for any F with $0 < F < 1$:

- If the Lorenz curve is differentiable:

$$\frac{dL(F)}{dF} = \frac{x(F)}{\mu}$$

- If the Lorenz curve is twice differentiable, then the probability density function f(x) exists at that point and:

$$\frac{d^2L(F)}{dF^2} = \frac{1}{\mu\, f(x(F))}$$

- If $L(F)$ is continuously differentiable, then the tangent of $L(F)$ is parallel to the line of perfect equality at the point $F(\mu)$. This is also the point at which the equality gap $F - L(F)$, the vertical distance between the Lorenz curve and the line of perfect equality, is greatest. The size of the gap is equal to half of the relative mean deviation:

$$F(\mu) - L(F(\mu)) = \frac{\text{mean deviation}}{2\mu}$$

QUERIES

1. What do you mean by variance?
2. Why we use Sheppard corrections?
3. Write corrected coefficient of variation.
4. What is combined variance?
5. What is coefficient of variations?
6. How can you compute coefficient of variations for different series?
7. Explain Lorenz curve. What are its properties? How it is calculated?

CHAPTER 14

Moments, Skewness and Kurtosis

14.1 INTRODUCTION

In statistics, the method of moments is a method of estimation of population parameters such as mean, variance, median, etc. (which need not be moments), by equating sample moments with unobservable population moments and then solving those equations for the quantities to be estimated. The **first moment** about zero, if it exists, is the expectation of X, i.e. the mean of the probability distribution of X, designated m. In higher orders, the central moments are more interesting than the moments about zero. The **second central moment** is the variance, the positive square root of which is the standard deviation, s.

Skewness is a measure of the asymmetry of the probability distribution of a real valued random variable. Skewness is a measure of the degree of asymmetry of a distribution. If the left tail (tail at small end of the distribution) is more pronounced than the right tail (tail at the large end of the distribution), the function is said to have negative skewness. If the reverse is true, it has positive skewness. If the two are equal, it has zero skewness. The **third central moment (skewness)** is a measure of the lopsidedness of the distribution; any symmetric distribution will have a third central moment, if defined, of zero. The normalized third central moment is called the **skewness**, often g.

kurtosis is a statistical measure used to describe the distribution of observed data around the mean. The **fourth central moment (kurtosis)** is a measure of whether the distribution is tall and skinny or short and squat, compared to the normal distribution of the same variance. Since it is the expectation of a fourth power, the fourth central moment, where defined, is always non-negative; and except for a point distribution, it is always strictly positive. The fourth central moment of a normal distribution is $3\sigma^4$.The kurtosis k is defined to be the normalized fourth central moment minus 3.

14.2 MOMENTS

The concept of moment in mathematics evolved from the concept of moment in physics. The nth moment of a real-valued function $f(x)$ of a real variable about a value c is

$$\mu'_n = \int_{-\infty}^{\infty} (x-c)^n f(x)dx$$

It is possible to define moments for random variables in a more general fashion than moments for real values. See Moments in metric spaces.

The word "moment' is a mechanical term which, refers to a measure of force. But in statistics it may be defined as" The arithmetic average of certain power of the deviation of the items from their arithmetic mean".

Moments play very important role in statistics. They are used to study the nature of distribution. Moments are used to calculate the mean, variance, skewness and kurtosis. The moments about zero are usually referred to simply as the moments of a function. Usually, except in the special context of the problem of moments below, the function will be a probability density function. The nth moment (about zero) of a probability density function $f(x)$ is the expected value of Xn. The moments about its mean μ are called central moments; these describe the shape of the function, independently of translation.

If (lower-case) f is a probability density function, then the value integral above is called the nth moment of the probability distribution. More generally, if (capital) F is a cumulative probability distribution function of any probability distribution, which may not have a density function, then the nth moment of the probability distribution is given by the Riemann-Stieltjes integral

$$E(X^n) = \int_{-\infty}^{\infty} x^n dF(x)$$

where X is a random variable that has this distribution. When

$$E(|X^n|) = \int_{-\infty}^{\infty} x^n |dF(x)| = \infty$$

Then the moment is said not to exist. If the nth moment about any point exists, so does $(n-1)^{\text{th}}$ moment, and all lower-order moments, about every point.

The mean and the variance provide information on the location and variability (spread, dispersion) of a set of numbers, and by doing so, provide some information on the appearance of the distribution of the numbers.

The mean and variance are the first two *statistical moments*, and the third and fourth moments also provide information on the shape of the distribution. For comparison, the

$$\text{One moment} = \sum_{i=1}^{n} (X_i - \bar{X})^1,$$

is by definition is equal to zero. (One might think of the mean as being that value of x that makes the above statement true, and consequently indicates where the individual numbers generally lie. The second moment is recognized as the numerator of the variance

$$\text{Second moment} = \sum_{i=1}^{n}(X_i - \bar{X})^2,$$

which gives information on the spread or scale of the distribution of numbers.

The third moment, which is given as follow

$$\sum_{i=1}^{n}(X_i - \bar{X})^3,$$

is used to define the *skewness* of a distribution:

$$\text{Skewness} = \frac{\sum_{i=1}^{n}(X_i - \bar{X})^3,}{ns^3}$$

Skewness is a measure of the symmetry of the shape of a distribution. If a distribution is symmetric, the skewness will be zero. If there is a long tail in the positive direction, skewness will be positive, while if there is a long tail in the negative direction, skewness will be negative.

The fourth moment, given as following

$$\sum_{i=1}^{n}(X_i - \bar{X})^4,$$

is used to define the kurtosis of a distribution:

$$\text{Kurtosis} = \frac{\sum_{i=1}^{n}(X_i - \bar{X})^4,}{ns^4}$$

Kurtosis is a measure of the flatness or peakedness of a distribution. Flat-looking distributions are referred to as "platykurtic," while peaked distributions are referred to as "leptokurtic."

14.2.1 Uses of Moments

1. They are used to describe the central tendency of a series.
2. They are used to describe the skewness of a series.
3. They are used to describe the Kurtosis of a series.
4. They are used to describe the dispersion of a series.

14.2.2 Moment of Variable X About Origin

The *r*th moment of variable X about origin is defined as under

$$\mu'_r = \frac{1}{N}\Sigma f_i x_i^r$$

Obviously, $\mu_1' = \frac{1}{N}\Sigma f_i x_i$ = Arithmetic mean

$$\mu_2' = \frac{1}{N}\Sigma f_i x_i^2$$

$$\mu_3' = \frac{1}{N}\Sigma f_i x_i^3$$

$$\mu_4' = \frac{1}{N}\Sigma f_i X_i^4 \text{ and so on}$$

Moments of variables X about any arbitrary point A. The rth moment of variable X about any arbitrary point 'A' is defined as under

$$\mu_r' = \frac{1}{N}\Sigma f_i (X_i - A)^r \quad \text{where } r = 1,2,3,4,\ldots\ldots\ldots$$

$$\mu_1' = \frac{1}{N}\Sigma f_i (X_i - A)$$

$$= \frac{1}{N}\Sigma f_i X_i - \frac{A}{N}\Sigma f_i = \frac{1}{N}\Sigma f_i X_i - A$$

A.M. – A

$$\text{A.M.} = u_1' + \text{A}$$

$$\mu_2' = \frac{1}{N}\Sigma f_i (X_i - A)^2$$

$$u_3' = \frac{1}{N}\Sigma f_i (X_i - A)^3$$

$$u_4' = \frac{1}{N}\Sigma f_i (X_i - A)^4$$

14.2.3 Moments About A.M (or Central Moment)

The rth moment about a variable X about Arithmetic mean (called central moment) is defined as under

$$\mu_r = \frac{1}{N}\Sigma f_i (x_i - \bar{X})^r \qquad r = 1,2,3,4,\ldots\ldots..$$

$$\mu_1 = \frac{1}{N}\Sigma f_i (x_i - \bar{X})$$

$$\mu_2 = \frac{1}{N}\Sigma f_i (x_i - \bar{X})^2$$

$$\mu_3 = \frac{1}{N}\Sigma f_i(x_i - \bar{X})^3$$

$$\mu_4 = \frac{1}{N}\Sigma f_i(x_i - \bar{X})^4$$

14.2.4 Relationship Between Moment About Any Mean and About Any Point A

The central moment (about A.M) can be computed from moment about any arbitrary point A as under

$$\mu_2 = \mu_2' - \mu_1'^2$$

$$\mu_3 = \mu_3' - 3\mu_2'\mu_1' + 2\mu_1'^3$$

$$\mu_4 = \mu_4' + 4\mu_3'\mu_1' + 6\mu_2'\mu_1'^2 - 3\mu_1'^4$$

Similarly, moment about any point '*A*' can be computed from central moment.

$$\text{A.M} = \mu_1' + A$$

$$\mu_2' = \mu_2 + \mu_1'^2$$

$$\mu_3' = \mu_3 + 3\mu_2\mu_1' + \mu_1'^3$$

$$\mu_4' = \mu_4 + 4\mu_3\mu_1' + 6\mu_2\mu_1'^2 + \mu_1'^4$$

14.2.5 Moment About Origin

These moments are also another type of assumed moments which are calculated as the arithmetic mean of some power of the deviations of the items taken from the origin, i.e. zero. The formulae for the various moments of this type are as follows:

Direct Method:

First moment about the origin $\frac{\Sigma X}{N}$ or $\frac{\Sigma FX}{N}$

Second moment about the origin $\frac{\Sigma X^2}{N}$ or $\frac{\Sigma FX^2}{N}$

Third moment about the origin $\frac{\Sigma X^3}{N}$ or $\frac{\Sigma FX^3}{N}$

14.2.6 Moment About an Arbitrary Value

The moments about an arbitrary values are called an assumed moments. These are computed as arithmetic mean of certain power of deviations of items from arbitrary value. These moments are symbolically represented as follows.

First moment about an arbitrary value

$$\mu_1' = \frac{\Sigma d}{N} \text{ or } \frac{\Sigma Fd}{N}$$

Second moment about an arbitrary value or

$$\mu_2' = \frac{\Sigma d^2}{N} \text{ or } \frac{\Sigma Fd^2}{N}$$

Third moment about an arbitrary value or

$$\mu_3' = \frac{\Sigma d^3}{N} \text{ or } \frac{\Sigma Fd^3}{N}$$

Fourth moment about an arbitrary value or

$$\mu_4' = \frac{\Sigma d^4}{N} \text{ or } \frac{\Sigma Fd^4}{N}$$

14.2.7 Significance of the Moments

The first moment about zero, if it exists, is the expectation of X, i.e. the mean of the probability distribution of X, designated μ. In higher orders, the central moments are more interesting than the moments about zero. The nth central moment of the probability distribution of a random variable X is

$$\mu_n = E((X-\mu)^n)$$

The first central moment is thus 0.

14.2.8 Problem of Moments

The problem of moments seeks characterizations of sequences { $\mu' n : n = 1, 2, 3, \ldots$ } that are sequences of moments of some function f..

14.2.9 Partial Moments

Partial moments are sometimes referred to as "one-sided moments." The nth order lower and upper partial moments with respect to a reference point r may be expressed as

$$\mu_n^-(r) = \int_{-\infty}^{r} (\max\{x-r,0\})^n f(x)dx,$$

$$\mu_n^+(r) = \int_{-\infty}^{r} (-\min\{x-r,0\})^n f(x)dx.$$

Partial moments are normalized by being raised to the power $1/n$. The upside potential ratio may be expressed as a ratio of a first-order upper partial moment to a normalized second-order lower partial moment.

14.2.10 Moments in Metric Space

Let (M, d) be a metric space, and let $B(M)$ be the Borel σ-algebra on M, the σ-algebra generated by the d-open subsets of M. (For technical reasons, it is also convenient to assume that M is a separable space with respect to the metric d.) Let $1 \leq p \leq +\infty$.

The pth moment of a measure μ on the measurable space $(M, B(M))$ about a given point x_0 in M is defined to be

$$\int_m d(x, x_0)^p \, d\mu(x)$$

μ is said to have finite p^{th} moment if the pth moment of μ about x_0 is finite for some $x0$.

This terminology for measures carries over to random variables in the usual way: if (Ω, Σ, P) is a probability space and $X : \Omega \to M$ is a random variable, then the pth moment of X about $x_0 \in M$ is defined to be

$$\int_m d(x, x_0)^p \, d(X*(P))(x) \equiv \int_\Omega d(X(\omega), x_0)^p \, dP(\omega),$$

and X has finite pth moment if the pth moment of X about x_0 is finite for some $x_0 \in M$.

14.2.11 Central Moments

Central moments are defined as

$$\mu_{pq} = \int_{-\infty}^{\infty}\int_{-\infty}^{\infty} (x-\bar{x})^p \, (y-\bar{y})^q \, f(x, y) dx \, dy$$

where $\bar{x} = \dfrac{M_{10}}{M_{00}}$ and $\bar{y} = \dfrac{M_{01}}{M_{00}}$ are the components of the centroid.If $f(x, y)$ is a digital image, then the previous equation becomes

$$\mu_{pq} = \sum_x \sum_y (x-\bar{x})^p (y-\bar{y})^q f(x, y)$$

The central moments of order up to 3 are:

$$\mu_{00} = M_{00},$$

$$\mu_{01} = 0,$$

$$\mu_{10} = 0,$$

$$\mu_{11} = M_{11} - \bar{x}M_{01} = M_{11} - \bar{y}M_{10},$$

$$\mu_{20} = M_{20} - \bar{x}M_{10}$$

$$\mu_{02} = M_{02} - \bar{y}M_{01},$$

$$\mu_{21} = M_{21} - 2\bar{x}M_{11} - \bar{y}M_{20} + 2\bar{x}^2 M_{01},$$

$$\mu_{12} = M_{12} - 2\bar{y}M_{11} - \bar{x}M_{02} + 2\bar{y}^2 M_{10},$$

$$\mu_{30} = M_{30} - 3\bar{x}M_{20} - 2\bar{x}^2 M_{10},$$

$$\mu_{03} = M_{03} - 3\bar{y}M_{02} - 2\bar{y}^2 M_{01}.$$

It can be shown that:

$$\mu_{pq} = \sum_{m}^{p}\sum_{n}^{q}\binom{p}{m}\binom{q}{n}(-\bar{x})^{(p-m)}(-\bar{y})^{(q-n)}M_{mn}$$

Central moments are translational invariant.

14.2.12 Various Moment Generating Functions

1. Random Variable

Suppose that X is a random variable then the moments of X are defined as

$\mu_r(X) = E[X^r]$ $r = 1, 2, \ldots$

- The moments give a lot of useful information about the distribution of X.
- The first moment is the expected value of X.
- The variance of X can be written as

$$\mathrm{Var}(X) = \mu_2(X) \leq [\mu_1(X)]^2 .$$

Suppose that X is a random variable then the moment generating function of X is

$$M(t) = E[e^{tx}]$$

$$= \begin{cases} \sum_x e^{tx} p(x) & X \text{ discreate with mass function } p(x). \\ \int_{-\infty}^{\infty} e^{tx} f(x)dx & X \text{ is continuous with density } f(x). \end{cases}$$

This expectation is a function of the variable t.

- It is important that the expectation is finite for all t 2 $(-t_0, t_0)$ for some $t_0 > 0$.
- Clearly if $M(t)$ is the moment generating function for a random variable X then

$$M(0) = E[e^{0 \times X}] = E[1] = 1$$

Bernoulli Random Variable

Suppose that X is a Bernoulli random variable with success probability p then the moment generating function of X is 14

$$M(t) = 1 - p + pe^t$$

Binomial Random Variable

Suppose that X is a binomial random variable with parameters n and p. The moment generating function of X is

$$M(t) = (1 - pe^t)^n$$

Poisson Random Variable

Suppose that X is a Poisson random variable with rate λ. The moment generating function of X is

$$M(t) = \exp = \{\lambda (e^t - 1)\}$$

Geometric Random Variable

Suppose that X is a geometric random variable with success probability p. The moment generating function of X is

$$M(t) = \frac{pe^t}{1-(1-p)e^t}$$

Exponential Random Variable

Suppose that X is an exponential random variable with rate . The moment generating function of X is

$$M(t) = \frac{\lambda}{\lambda - t}$$

Gamma Random Variable

Suppose that X is a gamma random variable with parameters α and λ. The moment generating function of X is

$$M(t) = \left(\frac{\lambda}{\lambda - t}\right)^{\alpha}$$

2. Moment Generating Function for Exponential

The moment-generating function is so called because, if it exists on an open interval around $t = 0$, then it is the **exponential generating** function of the moments of the probability distribution:

$$E(X^n) = M_x^{(n)}(0) = \frac{d^n M_X}{dt^n}(0).$$

If the moment generating function is defined on such an interval, then it uniquely determines a probability distribution.

The moment generating function is defined as the expected value of exp (tX). For a normal distribution, the moment generating function is

$$M_x(t) = E[\exp(tX)]$$

$$= \int_{-\infty}^{\infty} \frac{1}{\sigma\sqrt{2\pi}} \exp\left(-\frac{(x-\mu)^2}{2\sigma^2}\right) \exp(tx)dx$$

$$= \exp\left(\mu t + \frac{\sigma^2 t^2}{2}\right)$$

as can be seen by completing the square in the exponent.

3. Moment Generating Function for Factorial

The **factorial moment** generating function of the probability distribution of a real-valued random variable X is defined as

$$M_x(t) = E[t^x]$$

for all complex numbers t for which this expected value exists. This is the case at least for all t on the unit circle $|t| = 1$, see characteristic function. If X is a discrete random variable taking values only in the set {0, 1, ...} of non-negative integers, then MX is also called probability-generating function of X and $MX(t)$ is well-defined at least for all t on the closed unit disk $|t| \leq 1$.

SOLVED PROBLEMS

Problem 1: Here $X = 2, 4, 6, 8, 10$ Calculate first from moment about zero.

Solution:

X	X^2	X^3	X^4
2	4	8	16
4	16	64	256
6	36	216	1296
8	64	512	4096
10	100	1000	10,000
$\Sigma x = 30$	$\Sigma x^2 = 220$	$\Sigma x^3 = 1800$	$\Sigma x^4 = 15.664$

First moment: $\mu_1 = \frac{\Sigma x}{N} = \frac{30}{5} = 6$

Second Moment: $\mu_2 = \frac{\Sigma x^2}{N} = \frac{220}{5} = 44$

Third Moment: $\mu_3 = \frac{\Sigma x^3}{N} = \frac{1800}{5} = 360$

Fourth Moment: $\mu_4 = \frac{\Sigma x^4}{N} = \frac{1564}{5} = 3123.5$

Problem 2: The first four moments about origin of a distribution about point 4 are –1.4, 16, –30 and 110. Find Central moments. Also find moment about origin

Solution: Given $A = 4,\ \mu_1' = -1.4,\ \mu_2' = 16,\ \mu_3' = -30,\ \mu_4' = 110$

Central Moment:

$$\mu_2 = \mu_2' - (\mu_1')^2 = (16) - (-1.4)^2 = 14.04$$

$$\mu_3 = \mu_3' - 3\mu_2'\mu_1' + 2(\mu_1')^3$$
$$= -30-3\times17\times(-1.4)+2(-1.4)^3$$
$$= -30 + 71.4 - 5.488=35.91$$

$$\mu_4 = \mu_4' - 4\mu_3'\mu_1' + 6\mu' 2\mu_1'^2 - 3\mu_1'^4$$
$$= 110-4(-30)(-1.4) + 6(16)(-1.4)^2-3(-1.4)^4$$
$$= 110-168 + 188.16-856.128$$
$$= 298.16-1024.128= 725.96$$

Central moment are $\mu_2 = 14.04$, $\mu_3 = 35.91$

Moments about origin

Given $$X = \mu_1' + A = -1.4 + 4 = 2.6$$
$$\mu_2 = 14.04$$
$$\mu_3 = 35.91$$
$$\mu_4 = 725.96$$

$\mu_1' = X$ bar = First moment about origin = 2.6

$$\mu_2' = \mu_2 + \mu_1'^2 = 14.04 + (2.6)^2 = 20.8$$
$$\mu_3' = \mu_3 + 3\mu_2\mu_3' + \mu_1'^3$$
$$= 35.91 + 3(14.04)(2.6) + (2.6)^3$$
$$= 35.91+109.5+17.57$$
$$= 162.9$$
$$\mu_4' = \mu_4 + 4\mu_3\mu_1' + \mu_4'$$
$$= 725.96 + 4(35.91)(2.5)+(2.6)^4$$
$$= 725.96 + 359.1+45.69$$
$$= 1130.7$$

Moments about origin are

$$\mu_1' = 2.6$$
$$\mu_2' = 20.8$$
$$\mu_3' = 162.9$$
$$\mu_4' = 1130.7$$

Problem 3:From the following series find the first four moments about the mean.

X:	5	10	15	20	25

Solution:

$(X-\bar{X})$	–10	–5	0	5	10	$\Sigma x = 0$
$(X-\bar{X})^2$	100	25	0	25	100	$\Sigma x^2 = 250$
$(X-\bar{X})^3$	–1000	–125	0	125	1000	$\Sigma x^3 = 0$
$(X-\bar{X})^4$	10000	625	0	625	10000	$\Sigma x^4 = 21250$

We have
$$N = 5, X = \frac{\Sigma \bar{X}}{N} = \frac{75}{5} = 15$$

Thus
$$\mu_1 = \frac{\Sigma X}{N} = \frac{0}{5} = 0$$

$$\mu_2 = \frac{\Sigma X^2}{N} = \frac{250}{5} = 25$$

$$\mu_3 = \frac{\Sigma X^3}{N} = \frac{0}{5} = 0$$

$$\mu_4 = \frac{\Sigma X^4}{N} = \frac{21250}{5} = 4250$$

Problem 4: From the following series find the first four moments about the Mean.

X:	5	10	15	20	25

Solution:

X	$(X-\bar{X})$	X^2	X^3	X^4
5	–10	100	–1000	10000
10	–5	25	–125	625
15	0	0	0	0
20	5	2	1	625
25	10	100	1000	1000
75	$\Sigma x = 0$	$\Sigma x^2 = 250$	$\Sigma x^3 = 0$	$\Sigma x^4 = 21250$

We have $N = 5, \quad X = \frac{\Sigma X}{N} = \frac{75}{5} = 15$

$$\mu_1 = \frac{\Sigma X}{N} = \frac{0}{5} = 0$$

$$\mu_2 = \frac{\Sigma X^2}{N} = \frac{250}{5} = 25$$

$$\mu_3 = \frac{\Sigma X^3}{N} = \frac{0}{5} = 0$$

$$\mu_4 = \frac{\Sigma X^4}{N} = \frac{21250}{5} = 4250 \text{ Ans.}$$

Problem 5:From the following series find the first four moments about an arbitrary value and convert them into the moment about the actual mean.

X:	2	3	5	8	12

Solution: Computation of first four moments about an arbitrary value, say 5

X	(X – A) d	d^2	d^3	d^4
2	–3	9	–27	81
3	–2	4	–8	16
5	0	0	0	0
8	3	9	27	81
12	7	49	343	2401
30	$\Sigma x = 5$	$\Sigma x^2 = 71$	$\Sigma x^3 = 335$	$\Sigma x^4 = 2579$

First four moments about the arbitrary value, 5

$$\mu_1' = \frac{\Sigma d}{N} = \frac{5}{5} = 1$$

$$\mu_2' = \frac{\Sigma d^2}{N} = \frac{71}{5} = 14.2$$

$$\mu_3' = \frac{\Sigma d^3}{N} = \frac{335}{5} = 67$$

$$\mu_4' = \frac{\Sigma d^4}{N} = \frac{2579}{5} = 515.8$$

Conversion into the first four moments about the mean

$$\mu_1 = \mu_1' - \mu_1' = 1 - 1 = 0$$

$$\mu_2 = \mu_2' - (\mu_1')^2$$
$$= (14.2) - (1)^2 = 13.2$$

$$\mu_3 = \mu_3' - 3\mu_2'\mu_1' + 2(\mu_1')^3$$
$$= 67 - 3 \times (14.2) \times (1) + 2(1)^3$$
$$= 67 - 42.6 + 2$$
$$= 26.4$$

$$\mu_4 = \mu_4' - 4\mu_3'\mu_1' + 6\mu_2'\mu_1'^2 - 3\mu_1'^4$$
$$= 515.8 - 4(67)(1) + 6(14.2)(1)^2 - 3(1)^4$$
$$= 515.8 - 268 + 85.2 - 3$$
$$= 601 - 271$$
$$= 330$$

Problem 6: Compute first four moments about the about the mean of the observations 4, 7, 10, 13, 16, 19, 22.

Solution:

X	$(X-\bar{X})$	X^2	X^3	X^4
4	–9	81	–729	6561
7	–6	36	–216	1296
10	–3	9	–27	81
13	0	0	0	0
16	3	9	27	81
19	6	36	216	1296
22	9	81	729	6561
91	$\Sigma x = 0$	$\Sigma x^2 = 243$	$\Sigma x^3 = 0$	$\Sigma x^4 = 15795$

$$N = 7, \quad X = \frac{\Sigma X}{N} = \frac{91}{7} = 13$$

$$\mu_1 = \frac{\Sigma X}{N} = \frac{0}{7} = 0$$

$$\mu_2 = \frac{\Sigma X^2}{N}$$

$$= \frac{243}{7} = 34.71$$

$$\mu_3 = \frac{\Sigma X^3}{N} = \frac{0}{7} = 0$$

$$\mu_4 = \frac{\Sigma X^4}{N} = \frac{15795}{5} = 2256.42 \text{ Ans.}$$

Problem 7: From the following find the first four moments about an arbitrary value and convert them into the moment about the actual mean.

X:	2	3	7	8	10

Solution:

X	(X – A) d	d^2	d^3	d^4
2	–5	25	–125	625
3	–4	16	–64	256
7	0	0	0	0
8	1	1	1	1
10	3	9	27	81
30	$\Sigma x = -5$	$\Sigma x^2 = 51$	$\Sigma x^3 = -161$	$\Sigma x^4 = 963$

First four moments about the arbitrary value, 5

$$\mu_1' = \frac{\Sigma d}{N} = \frac{-5}{5} = -1$$

$$\mu_2' = \frac{\Sigma d^2}{N} = \frac{51}{5} = 10.2$$

$$\mu_3' = \frac{\Sigma d^3}{N} = \frac{-161}{5} = -32.2$$

$$\mu_4' = \frac{\Sigma d^4}{N} = \frac{963}{5} = 192.6$$

Conversion into the first four moments about the mean

$$\mu_1 = \mu_1' - \mu_1' = -1-(-1) = 0$$

$$\mu_2 = \mu_2' - (\mu_1')^2 = 10.2-(1)^2 = 9.2$$

$$\mu_3 = \mu_3' - 3\,\mu_2'\,\mu_1' + 2(\mu_1')^3$$
$$= -32.2 - 3(10.2)(-1) + 2(-1)^3$$
$$= -32.2 + 30.6 - 2$$
$$= -3.6$$

$$\mu_4 = \mu_4' - 4\mu_3'\,\mu_1' + 6\,\mu_2'(\mu_1')^2 - 3(\mu_1')^4$$
$$= 192.6-4(-32.2)(-1)+6(10.2)(-1)^2-3(-1)^4$$
$$= 192.6-128.8+61.2-3$$
$$= 63.8+58.2$$
$$= 122.$$

Problem 8: Given following information

$N = 100$, $\Sigma fd = 100$, $\Sigma fd^2 = 2000$, $\Sigma fd^3 = 300$, $\Sigma fd^4 = 6000$. Find μ_2, μ_3 and μ_4.

Solution: With the factors given above we find the values of the first four assumed moments as follows:

$$\mu_1' = \frac{\Sigma fd}{N} = \frac{100}{100} = 1$$

$$\mu_2' = \frac{\Sigma fd^2}{N} = \frac{2000}{100} = 20$$

$$\mu_3' = \frac{\Sigma fd^3}{N} = \frac{300}{100} = 3$$

$$\mu_4' = \frac{\Sigma fd^4}{N} = \frac{6000}{100} = 60$$

From the above assumed moments we find the moments about the mean as follows:
First four moments:

$$\mu_1 = \mu_1' - \mu_1' = 0$$

$$\mu_2 = \mu_2' - (\mu_1')^2 = 20 - (1)^2 = 19$$

$$\mu_3 = \mu_3' - 3\mu_2'\mu_1' + 2(\mu_1')^3$$
$$= 3 - 3(20)(1) + 2(1)^3$$
$$= 3 - 60 + 2$$
$$= -55$$

$$\mu_4 = \mu_4' - 4\mu_3'\mu_1' + 6\mu_2'(\mu_1')^2 - 3(\mu_1')^4$$
$$= 60 - 4(3)(1) + 6(20)(1)^2 - 3(1)^4$$
$$= 60 - 12 + 120 - 3$$
$$= 180 - 1$$
$$= 165$$

UNSOLVED PROBLEMS

1. What do you understand by moment? Explain its types?
2. Why moments are used?
3. What do you mean by Moment of *X* about Origin?
4. What do you mean by moment about AM?
5. What do you mean by central moment?
6. What are various methods to find the moment? Explain uses of Moment?
7. What are the various moments generating functions?

8. From the following data calculate the first four central moments

X:	0	1	2	3	4	5	6	7	8
F:	1	8	28	56	70	56	28	8	1

9. Compute the first four moments of the following series about $A = 15$, and then ,determine the corresponding moments about mean

X:	10	11	12	13	14	15	16	17	18
F:	9	35	70	105	116	107	88	46	40

14.3 SKEWNESS

Skewness is a measure of the asymmetry of the probability distribution of a real-valued random variable.

A normal distribution is a bell-shaped distribution of data where the mean, median and mode all coincide. A frequency curve showing a normal distribution would look like this:

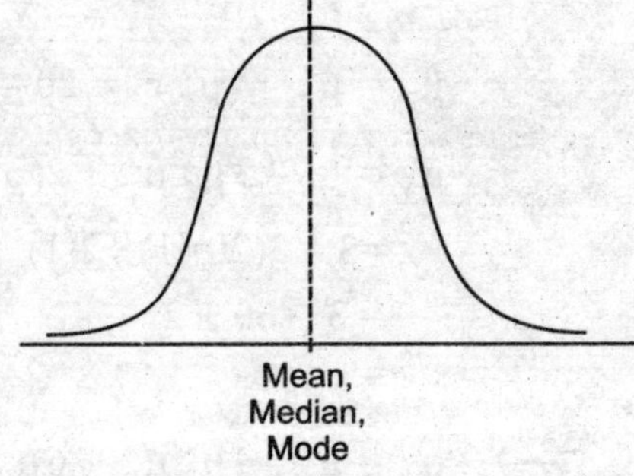

In a normal distribution, approximately 68% of the values lie within one standard deviation of the mean and approximately 95% of the data lies within two standard deviations of the mean.

A distribution is skewed if one of its tails is longer than the other. The first distribution shown has a positive skew. This means that it has a long tail in the positive direction. The distribution below it has a negative skew since it has a long tail in the negative direction. Finally, the third distribution is symmetric and has no skew. Distributions with positive skew are sometimes called "skewed to the right" whereas distributions with negative skew are called "skewed to the left."

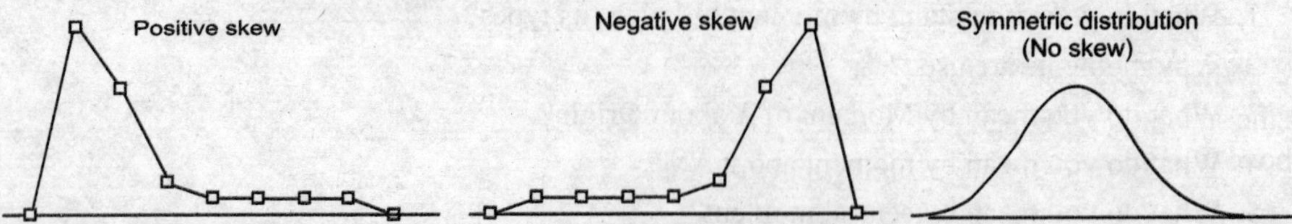

If there are extreme values towards the positive end of a distribution, the distribution is said to be **positively skewed.** In a positively skewed distribution, the mean is greater than the mode.

For example:

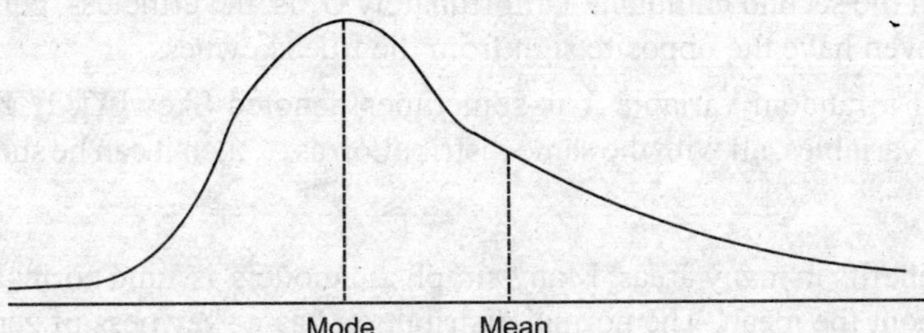

A negatively skewed distribution, on the other hand, has a mean which is less than the mode because of the presence of extreme values at the negative end of the distribution.

Definition: *"Skewness is a measure of symmetry, or more precisely, the lack of symmetry. A distribution, or data set, is symmetric if it looks the same to the left and right of the center point."*

Skewness, the third standardized moment, is written as ã1 and defined as

$$\gamma_1 = \frac{\mu_3}{\sigma^3}$$

Where μ^3 is the third moment about the mean and σ is the standard deviation. Equivalently, skewness can be defined as the ratio of the third cumulate $k3$ and the third power of the square root of the second cumulate k_2:

$$\gamma_1 = \frac{k_3}{k_2^{3/2}}.$$

This is analogous to the definition of kurtosis, which is expressed as the fourth cumulant divided by the fourth power of the square root of the second cumulant.

For a sample of n values the sample skewness is

$$g_1 = \frac{m_3}{m_2^{3/2}} = \frac{\sqrt{n}\Sigma_{i=1}^{n}(x_i - \overline{x}^3)}{\left(\Sigma_{i=1}^{n}(x_i - \overline{x})^2\right)3/2},$$

Where x_i is the ith value, $\overline{x}$ is the sample mean, m_3 is the sample third central moment, and m_2 is the sample variance. Given samples from a population, the equation for the sample skewness g1 above is a biased estimator of the population skewness. The usual estimator of skewness is

$$G_1 = \frac{k_3}{k_2^{3/2}} = \frac{\sqrt{n(n-1)}}{n-2}\, g_1,$$

Where k_3 is the unique symmetric unbiased estimator of the third cumulant and k_2 is the symmetric unbiased estimator of the second cumulant. Unfortunately G_1 is, nevertheless, generally biased. Its expected value can even have the opposite sign from the true skewness.

The skewness of a random variable X is sometimes denoted Skew[X]. If Y is the sum of n independent random variables, all with the same distribution as X, then it can be shown that Skew[Y] = Skew[X] / $\sqrt{n}$.

Skewness has benefits in many areas. Many simplistic models assume normal distribution, i.e. data is symmetric about the mean. The normal distribution has a skewness of zero. But in reality, data points are not perfectly symmetric. So, an understanding of the skewness of the dataset indicates whether deviations from the mean are going to be positive or negative.

14.4 MEASURE OF SKEWNESS

Skewness and Kurtosis

A fundamental task in many statistical analyses is to characterize the *location* and *variability* of a data set. A further characterization of the data includes skewness and kurtosis.

Skewness is a measure of symmetry, or more precisely, the lack of symmetry. A distribution, or data set, is symmetric if it looks the same to the left and right of the center point.

Kurtosis is a measure of whether the data are peaked or flat relative to a normal distribution. That is, data sets with high kurtosis tend to have a distinct peak near the mean, decline rather rapidly, and have heavy tails. Data sets with low kurtosis tend to have a flat top near the mean rather than a sharp peak. A uniform distribution would be the extreme case.

The histogram is an effective graphical technique for showing both the skewness and kurtosis of data set.

Definition of Skewness

For univariate data Y_1, Y_2, ..., Y_N, the formula for skewness is:

$$\text{skewness} = \frac{\Sigma_{i=1}^{N}(Y_i - \bar{Y})^3}{(N-1)S^3}$$

where $\bar{Y}$ is the mean, ■is the standard deviation, and N is the number of data points. The skewness for a normal distribution is zero, and any symmetric data should have a skewness near zero. Negative values for the skewness indicate data that are skewed left and positive values for the skewness indicate data that are skewed right. By skewed left, we mean that the left tail is long relative to the right tail. Similarly, skewed right means that the right tail is long relative to the left tail. Some measurements have a lower bound and are skewed right. For example, in reliability studies, failure times cannot be negative.

Definition of Kurtosis

For univariate data $Y_1, Y_2, ..., Y_N$, the formula for kurtosis is:

$$\text{kurtosis} = \frac{\Sigma_{i=1}^{N}(Y_i - \bar{Y})^4}{(N-1)S^4}$$

where $\bar{Y}$ is the mean, bsis the standard deviation, and N is the number of data points.

The kurtosis for a standard normal distribution is three. For this reason, excess kurtosis is defined as

$$\text{kurtosis} = \frac{\Sigma_{i=1}^{N}(Y_i - \bar{Y})^4}{(N-1)S^4} - 3$$

so that the standard normal distribution has a kurtosis of zero. Positive kurtosis indicates a "peaked" distribution and negative kurtosis indicates a "flat" distribution.

Examples

The following example shows histograms for 10,000 random numbers generated from a normal, a double exponential, a Cauchy, and a Weibull distribution.

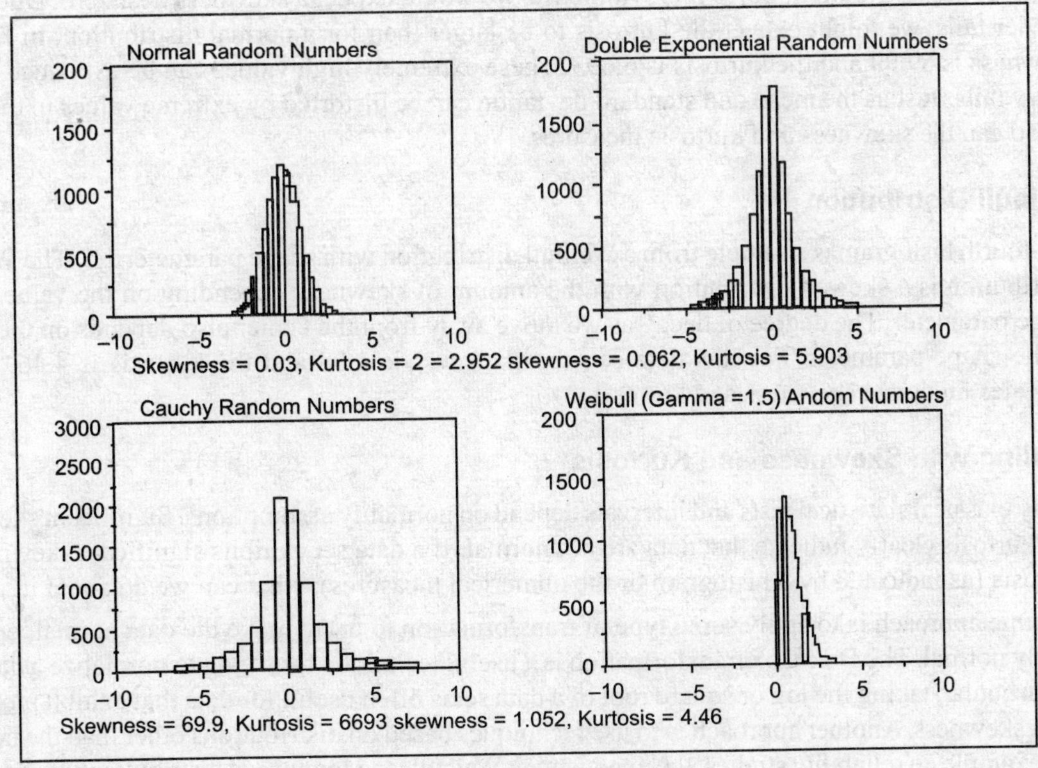

Normal Distribution

The first histogram is a sample from a normal distribution. The normal distribution is a symmetric distribution with well-behaved tails. This is indicated by the skewness of 0.03. The kurtosis of 2.96 is near the expected value of 3. The histogram verifies the symmetry.

Double Exponential Distribution

The second histogram is a sample from a double exponential distribution. The double exponential is a symmetric distribution. Compared to the normal, it has a stronger peak, more rapid decay, and heavier tails. That is, we would expect a skewness near zero and a kurtosis higher than 3. The skewness is 0.06 and the kurtosis is 5.9.

Cauchy Distribution

The third histogram is a sample from a Cauchy distribution.

For better visual comparison with the other data sets, we restricted the histogram of the Cauchy distribution to values between -10 and 10. The full data set for the Cauchy data in fact has a minimum of approximately -29,000 and a maximum of approximately 89,000.

The Cauchy distribution is a symmetric distribution with heavy tails and a single peak at the center of the distribution. Since it is symmetric, we would expect a skewness near zero. Due to the heavier tails, we might expect the kurtosis to be larger than for a normal distribution. In fact the skewness is 69.99 and the kurtosis is 6,693. These extremely high values can be explained by the heavy tails. Just as the mean and standard deviation can be distorted by extreme values in the tails, so too can the skewness and kurtosis measures.

Weibull Distribution

The fourth histogram is a sample from a Weibull distribution with shape parameter 1.5. The Weibull distribution is a skewed distribution with the amount of skewness depending on the value of the shape parameter. The degree of decay as we move away from the center also depends on the value of the shape parameter. For this data set, the skewness is 1.08 and the kurtosis is 4.46, which indicates moderate skewness and kurtosis.

Dealing with Skewness and Kurtosis

Many classical statistical tests and intervals depend on normality assumptions. Significant skewness and kurtosis clearly indicate that data are not normal. If a data set exhibits significant skewness or kurtosis (as indicated by a histogram or the numerical measures), what can we do about it?

One approach is to apply some type of transformation to try to make the data normal, or more nearly normal. The Box-Cox transformation is a useful technique for trying to normalize a data set. In particular, taking the log or square root of a data set is often useful for data that exhibit moderate right skewness. Another approach is to use techniques based on distributions other than the normal. For example, in reliability studies, the exponential, Weibull, and lognormal distributions are typically used as a basis for modeling rather than using the normal distribution. The probability plot correlation coefficient plot and the probability plot are useful tools for determining a good distributional model for the data.

Software

The skewness and kurtosis coefficients are available in most general purpose statistical software programs, including Dataplot.

SOLVED PROBLEMS

Problem 1: From the following data, calculate mean, mode and S.D. Also find Karl Pearson Coefficient of skewness.

Value	Frequency
12	8
24	14
36	18
48	36
60	30
72	20
84	10

Solution: Consider the following Table

Value(x)	F	C = x–48	D = d/c C = 12	Fd1	Fd²
12	8	–36	–3	–24	72
24	14	–24	–2	–28	56
36	18	–12	–1	–18	18
48	36	0	0	0	0
60	30	12	1	30	30
72	20	24	2	40	80
84	10	36	3	30	90
Total	**136**	**–**	**–**	**30**	**346**

Here assumed mean $A = 48$

$$C = 12$$

$N = \Sigma f = 136, \quad \Sigma fd1 = 30, \quad \Sigma fd^2 = 346$

$$\text{Mean} = A + \frac{\Sigma fd1}{N} \times C = 48 + \frac{30}{136} \times 12 = 50.64$$

As we know Standard deviation is given by

$$\text{S.D} = \sqrt{\frac{\Sigma fd1}{N} - \frac{\Sigma fd1^2}{N}} \times C$$

$$= \frac{346}{136} - \frac{(30)^2}{136} \times 12$$

$$= \sqrt{2.544 - 0.0487} \times 12 = \sqrt{2.4953} \times 12 = 18.96$$

As By inspection Mode = 48

Hence Karl Pearson Coefficient Is

$$S_k = \frac{\text{mean} - \text{mode}}{\text{S.D}} = \frac{m - z}{6}$$

$$= \frac{50.64 - 48}{18.96} = \frac{2.64}{d18.96} = 0.14$$

Problem 2: Calculate the Karl Pearson Coefficient of skewness:

Wages (More than):	5	15	25	35	45	55	65	75	85
No. of workers:	120	105	96	85	72	58	32	12	0

Solution:

Consider the following Table

Class Interval (Wages)	Frequency (f)	Mid points X	$d'x = (x - 50)/10$	$fd'x$	fd'^2x	c.f
5-15	15	10	–4	–60	240	15
15-25	9	20	–3	–27	81	24
25-35	11	30	–2	–22	44	35
35-45	13	40	–1	–13	13	48
45-55	14	50	0	0	0	62
55-65	26	60	1	26	26	88
65-75	20	70	2	40	80	108
75-85	12	80	3	36	108	120
		$N = 120$		$\Sigma fd'x = -20$	$\Sigma fd'^2x = 592$	

(a) Mean:

$$\overline{x} = A + \frac{\Sigma fd'x}{N} \times C$$

$$= 50 + \frac{-20}{120} \times 10 = 50 - 1.66$$

$$= \text{Rs } 48.34$$

(b) Standard deviation:

$$\sigma = \sqrt{\frac{\Sigma fd'^2 x}{N} - (\Sigma fd'x/N)^2} \times C$$

$$= \sqrt{(592/120) - (-20/120)^2} \times 10$$

$$= \sqrt{4.9333 - 0.0256} \times 10 = \sqrt{4.9076} = \times 10$$

= Rs 22.2

(c) Median:

M = Size of (N/2)th item

= Size (120/2)th item

= Size of 60th item

which lies in 45–55 group

$$M = l_1 + \frac{(N/2 - C)}{f} \times l_2 - l_1$$

$$= 45 + \frac{60 - 48}{14} \times 10 = 45 + 8.57$$

= Rs 53.57

(d) Karl Pearson Coefficient of skewness:

$$S_k = \frac{3(|\overline{x}| - \text{Med})}{\sigma}$$

$$= \frac{3(48.34 - 53.57)}{22.2}$$

$$S_k = \frac{-15.69}{22.2} = -0.706$$

Problem 3: Find the value of Karl Pearson and Bowley Coefficient of Skewness of following data:

Distribution.	Mean	Medium	S.D	Q3	Q1
Distribution X	160	152	40	205	72
Distribution Y	150	165	55	270	90

Solution:

As we Know Karl Pearson Coeff. Of Skew ness is given by

$$S_{k\rho} = \frac{3(\text{mean–medium})}{\text{Standard Deviation}}$$

And quartile coefficient of Skew ness is

$$S_{kp} = \frac{Q3 + Q1 - 2(\text{S.D})}{Q3 - Q1}$$

Now S_{kp} for Distribution X is

$$\frac{3(160-152)}{40} = 0.6$$

S_{kp} for Distribution is

$$\frac{3(150-165)}{55} = -0.82$$

Boyley Coeff for Distribution

$$X = \frac{Q3 + Q1 - 2(\text{S.D})}{Q3 - Q1}$$

$$= \frac{205 + 72 - 2\times 152}{205 - 72} = -0.203$$

For Distribution Y

$$\frac{270 + 90 - 2\times 165}{270 - 90} = 0.167$$

Problem 4:Find the value of Coefficient of Skewness using Bowley formula:

Income :	100-199	200-299	300-399	400-499	500-599	600-699	700-799	800-899
No. of persons:	39	25	49	62	38	37	32	18

Solution:

Calculation of Bowley's coefficient of skewness		
Class Interval (Income)	Frequency(f)	C.f
100-199	39	39
200-299	25	64
300-399	49	113
400-499	62	175
500-599	38	213
600-699	37	250
700-799	32	282
800-899	18	300

(a) Median:

$$M = \text{Size of } (N/2)\text{th item}$$
$$= \text{Size } (300/2)\text{th item}$$
$$= \text{Size of } 150^{\text{th}} \text{ item}$$

which lies in 400-499 group whose real limts are 399.5-499.5

$$M = l_1 + \frac{(N/2 - C)}{f} \times l_2 - l_1$$
$$= 399.5 + \frac{150 - 113}{62} \times 100 = 399.5 + 59.68$$
$$= \text{Rs } 459.18$$

(b) *Q*1:

$$Q1 = \text{Size of } (N/4)\text{th item}$$
$$= \text{Size } (300/4)\text{th item}$$
$$= 75^{\text{th}} \text{ item}$$

which lies in 300-399 group whose real limts are 299.5-399.5

$$M = l_1 + \frac{(N/4 - C)}{f} \times l_2 - l_1$$
$$= 299.5 + \frac{75 - 64}{49} \times 100 = 299.5 + 22.45$$
$$= \text{Rs } 321.95$$

(c) *Q*3:

$$Q3 = \text{Size of } (3N/4)\text{th item}$$
$$= \text{Size } \frac{(3 \times 300)}{4} \text{th item}$$
$$= 225^{\text{th}} \text{ item}$$

which lies in 600-699 group whose real limts are 599.5-699.5

$$Q3 = l_1 + \frac{(3N/4 - C)}{f} \times l_2 - l_1$$
$$= \frac{599.5 + 225 - 213 \times 100}{38} = 599.5 + 31.58$$
$$= \text{Rs } 631.08$$

(d) Boyley Coeff icient of skewness

$$S_k = \frac{Q3 + Q1 - 2(\text{M})}{Q3 - Q1}$$

$$S_k = \frac{631.08 + 321.95 - 2 \times 459.18}{631.08 - 321.95}$$

$$= \frac{952.03 - 918.36}{309.13} = 0.112$$

Problem 5: There is moderately Skewed Distribution of retail price for women Cosmetic; it is found that mean price is Rs 200and medium price is 17 of Coeff. Of Variation is 20%. Find Karl Pearson Coeff. Of Skewness of Distribution.

Solution:

Here

$$m = 200$$

$$\text{Median} = 17$$

$$\text{C.V} = 20\%$$

As we know that

$$\text{C.V} = \frac{6 \times 100}{X}$$

$$\rightarrow \quad 20 = \frac{6 \times 100}{200} \rightarrow 6 = 40$$

Now Karl Pearson Coeff. is

$$Sk\rho = \frac{\overline{x} - Z}{6} = \frac{\overline{x} - (3\text{Median} - 2\overline{x})}{6}$$

$$= \frac{3(\overline{x} - \text{Median})}{6} = \frac{3(200 - 17)}{40}$$

$$= \frac{3 \times 183}{40} = 13.725$$

Problem 6: In a frequency distribution the coefficient of Skewness based on quartile is 0.8. of the sum of upper and lower quartile is two and Median is 40 . Find the value of upper quartile.

Solution:

Here: Coefficient of skew ness $S_{k\rho} = 0.8$

$$Q1 + Q3 = 100 \ldots\ldots\ldots\ldots A$$

$$M = 40$$

We Know that

$$S_{k\rho} = \frac{Q3 + Q1 - 2\text{m.d}}{Q3 - Q1}$$

$$\rightarrow \qquad 0.8 = \frac{100 - 2 \times 40}{Q3 - Q1}$$

$$\rightarrow \qquad 0.8 = \frac{100 - 80}{Q3 - Q1}$$

$$\rightarrow \qquad Q3 - Q1 = \frac{20}{8}$$

$$\rightarrow \qquad Q3 - Q1 = 25 \ldots\ldots\ldots\ldots\ldots B$$

Now From A & B

$$Q1 + Q3 = 100$$
$$Q3 - Q1 = 25$$

Or

$$Q1 + Q = 100 \qquad \ldots(i)$$
$$Q3 - Q1 = 25 \qquad \ldots(ii)$$

Now add (i) and (ii)

$$2Q3 = 125$$

$$Q3 = \frac{125}{2} = 62.5$$

Put $Q3$ in (ii)

$$62.5 - Q1 = 25$$
$$62.5 - 25 = Q1$$
$$37.5 = Q$$

Hence

$$Q = 37.5$$
$$Q2 = 62.5$$

Problem 7: From the following data determine Bowley Coefficient of Skewness.

$$\varphi_3 - \varphi_1 = 9$$
$$\varphi_3 + \varphi_1 = 20$$
$$m = 10$$

Solution: As we know that Bowley Coefficient of skewness is given by

$$S_{K(B)} = \frac{\varphi_3 + \varphi_1 - 2m}{\varphi_3 - \varphi_1}$$

$$S_{K(B)} = \frac{20 - 2(10)}{9} = \frac{20 - 20}{9} = 0$$

Problem 8:Determine the value of standard deviation of a series for which mean = 50, median = 70 and Pearson Coefficient of skewness is – 0.5.

Solution: We know that Karl Person Coefficient of skewness is given by

$$Sk_{(P)} = \frac{3(\overline{x} - m)}{6}$$

Where $\overline{x}$ = Mean, M = Median, 6 = Standard deviation

$$6 = \frac{3(\overline{x} - m)}{Sk_{(P)}} = \frac{3(50 - 70)}{-0.5}$$

$$= \frac{3(-20)}{-0.5} = \frac{-60}{-0.5} = 120$$

Problem 9: Find the value of mode when Mean is 30, 6 is 8 and Pearson coefficient of skewness is 0.50.

Solution: As we know that

$$SK_{(P)} \frac{\overline{x} - z}{6} \qquad ...(i)$$

Where $\overline{x}$ = Mean , z = mode, σ = S.D

from eq (i) we get

$$Z = \overline{x} + \sigma \cdot SK_{(p)}$$

$$= 30 + 8 \times (.50)$$

$$= 30 + 4 = 34$$

$$\text{Mode} = 34$$

Problem 10: In a distribution Mean is 100 Median is 110 and coefficient of skewness is –.5. Find Mode and Coefficient of variance.

Solution: As we know that

$$\text{Mode} = 3(\text{Median}) - 2\ (\text{Mean})$$

$$Z = 3\ (110) - 2\ (100)$$

$$330 - 200 = 130$$

Also Coefficient of Skewness

$$Sk_{(P)} = \frac{\overline{x} - 2}{\sigma}$$

$$-.5 = \frac{100 - 130}{\sigma}$$

$$\sigma = \frac{-30}{-.5} = 60$$

$$\sigma = 60$$

Now Coefficient of variance

Here first for moment about 35

$$M_1^1 = \frac{\Sigma Fd}{N} = \frac{-14}{100} = -0.14$$

$$M_2^1 = \frac{\Sigma Fd^2}{N} = \frac{154}{100} = 1.54$$

$$M_3^1 = \frac{\Sigma Fd^3}{N} = \frac{-62}{100} = 0.62$$

$$M_4^1 = \frac{\Sigma Fd^4}{N} = \frac{490}{100} = -4.90$$

So First Four central moments by Conversion

$$M_1 = M_1^1 - M_1^1 = 0$$

$$\begin{aligned} M_2 &= M_2^1 - (M^1)^2 \\ &= 1.54 - (-.14)^2 \\ &= 1.54 - .0196 = 1.5204 \end{aligned}$$

$$\begin{aligned} M_3 &= M_3^1 - 3M_2^1 M_1^1 + 2(M_1^1)^3 \\ &= -.62 - 3(1.54)(0.14) + 2(0.14)^3 \\ &= .0213 \end{aligned}$$

$$\begin{aligned} M_3 &= M_4^1 - 4M_3^1 M_1^1 + 6M_2^1 (M_1^1)^2 - 3(M_1^1)^4 \\ &= 4.9 - 4(-.62)(-.14) + 6(1.54)(-.14)^2 - 3(-.14)^4 \\ &= 4.9 - .3472 + .1811 - .0012 \\ &= 4.7327 \end{aligned}$$

Now Kurtosis is

$$y \rightarrow = (\beta_2 - 3)$$

$$\beta_2 = \frac{H_4}{H_2^2} = \frac{4.7327}{2.3116} = 2.047$$

Hence

$$\begin{aligned} \gamma &= 2.047 - 3 \\ &= -0.953 \end{aligned}$$

So distribution is platykurtic

Problem 11: Find First and Second Coefficient of Skewness of following data

$\bar{x} = 80$, Median = 100, Mode = 60, S.D = 30

Solution: First Coefficient of Skewness is given by

$$SK_{(P)} = \frac{\text{Mean} - \text{Mode}}{\text{S.D}}$$

$$= \frac{80-60}{30}$$

$$= \frac{20}{30} = \frac{2}{3} = 0.666$$

Second Coefficient of Skewness

$$SK_{(P)} = \frac{3(\text{Mean} - \text{Median})}{\text{S.D}}$$

$$= \frac{3(80-100)}{30}$$

$$= \frac{3(-20)}{30} = \frac{-20}{10} = -2$$

UNSOLVED PROBLEMS

1. What is skewness? Explain?
2. What are the various types of skewness?
3. What do you mean by measure of skewness?
4. From the following distribution compute the S.D,Q.D,C.V and coefficient of skew ness

Wages in Rs :	20	40	10	13	20	45	44	56
No. of workers :	2	2	19	14	3	4	1	1

5. Determine the second moment about the mean and Pearson s coefficient of skewness of the following distribution

Values:	3.5	4.5	5.5	6.5	7.5	8.5	9.5
No. of items:	3	7	22	60	85	32	8

6. Ascertain Karl Pearson s coefficient of skewness from the following data:

Weights:	28	25	30	31	32	33	34
No. of Girls:	1	1	0	32	8	12	4

7. Find out the mean and coefficient of skewness for the following data:

Particulars	Rs.
30 men get the rate of	4.50
40 men get the rate of	5.50
48 men get the rate of	6.50
100 men get the rate of	7.50
125 men get the rate of	8.50
87 men get the rate of	9.50
43 men get the rate of	10.50
22 men get the rate of	11.50

8. From the following shown below find out the coefficient of dispersion and coefficient of skewness as propounded by Karl Pearson

Wages in Rs. :	70-80	80-90	90-100	100-110	110-120	120-130	130-140	140-150
No. of employees :	12	18	35	42	50	45	20	8

9. Compute the variance and skewness of the following two distributions:

Age group	Population	
	Town *X*	Town *Y*
0-10	18	10
10-20	16	12
20-30	15	11
30-40	12	30
40-50	10	20
50-60	5	11
60-70	2	3
Above 70	1	1

10. From the following data calculate the quartile deviation and the coefficient of skewness of Bowely:

Values:	4-8	8-12	12-16	16-20	20-24	24-28	28-32	32-36
Frequency:	6	10	18	30	15	11	10	6

11. Find Karl Pearson *s* coefficient of skewness of the following Distribution:

Marks :	10	20	30	40	50	60	70	80
No. of students :	30	70	120	160	190	350	480	500

12. Find out the mean , Mode, S.D and coefficient of skewness of the following Series:

Wages in Rs. :	Above 0	10	20	30	40	50	60	70
No. of workers :	680	500	422	389	209	70	30	0

13. From the following data find out the Karl Pearson's coefficient of Skewness

Marks above:	0	10	20	30	40	50	60	70	80
No. of candidates:	150	145	100	80	80	70	14	30	0

14. A distribution runs as under

Values :	0-20	20-40	40-60	60-80	80–100
No. of items :	13	X	20	X	16

If the Mode of the distribution be 40 Find coefficient .Sk(p)

15. Find the C.V and coefficient of skewness for the following distribution

	Digits in unit place									
Tens	0	1	2	3	4	5	6	7	8	9
70	1	2	6	-	4	2	3	-	-	2
80	8	-	5	4	6	-	9	-	5	3
90	6	5	-	4	2	5	3	4	4	5

16. Calculate an appropriate measure of skewness for the following Distribution

Size :	5-7	8-10	11-13	14-16	17-19
Frequncy :	12	20	38	20	4

17. Find pearson's coefficent of skewness of the foll. Distribution

Values of variable :	6	11	19	24	30	36	42
Frequncy :	4	10	15	18	9	7	4

18. Compute the Pearson's coefficient of skewness based upon Mean, Median and S.D from the following series:

Class :	100-	200-	400-	500-	600-	700-	800-
Frequncy :	45	88	200	70	50	30	10

19. From the following distribution find out the coefficient of skewness given by Karl Pearson:

Size of items less then :	10	20	30	40	50
No. of students:	5	12	32	44	47

20. Compute the C.V and coefficient $SK_{(p)}$ for the following. Series

Age :	10	20	30	40	50
under :	5	15	30	50	80

21. From the following distribution find Bowely's coefficient of skewness:

Profit in lakhs :	1-5	6-10	11-15	16-20	21-25	26-30	31-35
No. of firms :	20	26	29	38	40	50	70

22. Find the appropriate coefficient of skewness for the following:

Sales in Rs	Per cent of profit
Under 1000	16.2
1000-1999	11.8
2000-2999	12.1
3000-3999	24.8
4000-4999	15.9
5000-5999	11.9
6000-6999	3.7
100000 and over	2.7

14.5 KURTOSIS

In probability theory and statistics, kurtosis is a measure of the "peakedness" of the probability distribution of a real-valued random variable. Higher kurtosis means more of the variance is due to infrequent extreme deviations, as opposed to frequent modestly-sized deviations.

Kurtosis describes the concentration of data around the mean. A low value indicates that there is a high probability that a random number will be close to the mean, whilst a high one suggests that the range of values will be greater. This is illustrated by a comparison of the normal and logistic distributions. Both are symmetrical "bell shaped" curves, the coefficient of kurtosis for the normal distribution is constant with a value of 3.0, whilst for the logistic it is 4.2, for this reason, the logistic distribution is preferred for datasets where the "tails" have a greater significance. Kurtosis is based on the ratio of the fourth moment about the mean to the standard deviation:

$$\text{kurtosis} = \frac{1}{N}\sum_{i=N}^{i=1}\left(\frac{x_i - \text{mean}}{\text{stdev}}\right)^4 = \frac{M^4}{M_2^2}$$

Where M_4 is the 4th moment about the mean

M_2 is the 2nd moment about the mean (variance)

Some definitions of kurtosis reduce the moment ratio by 3 in order to provide a comparison with the normal distribution. The table below shows the "absolute" and "relative" values for distributions which have a fixed value of kurtosis:

Distribution	Absolute	Relative
Exponential	9	6
Gumbel	5.4	2.4
Laplace	6	3
Logistic	4.2	1.2
Normal	3	0
Rayleigh	3.245	0.245
Triangular	2.4	–0.6
Uniform	1.8	–1.2

Kurtosis is also referred to as the coefficient of excess.

Definition: *Kurtosis is the degree of peakedness of a distribution, defined as a normalized form of the fourth central moment* μ_4 *of a distribution. The fourth standardized moment is defined as* μ_4/σ^4*, where* μ_4 *is the fourth moment about the mean and* σ *is the standard deviation. This is sometimes used as the definition of kurtosis in older works, but is not the definition used here.*

Kurtosis is more commonly defined as the fourth cumulant divided by the square of the variance of the probability distribution,

$$\gamma_2 = \frac{k_4}{k_2^2} = \frac{\mu_4}{\sigma^4} - 3,$$

which is known as **excess kurtosis**. The "minus 3" at the end of this formula is often explained as a correction to make the kurtosis of the normal distribution equal to zero. Another reason can be seen by looking at the formula for the kurtosis of the sum of random variables. Because of the use of the cumulant, if Y is the sum of n independent random variables, all with the same distribution as X, then Kurt[Y] = Kurt[X] /n, while the formula would be more complicated if kurtosis were defined as μ_4/σ^4.

More generally, if $X_1, ..., X_n$ are independent random variables all having the **same variance**, then

$$\text{Kurt}\left(\sum_{i=1}^{n} X_i\right) = \frac{1}{n^2}\sum_{i=1}^{n} \text{Kurt}\,(X_i),$$

whereas this identity would not hold if the definition did not include the subtraction of 3.

14.5.1 Types of Kurtosis

A high kurtosis distribution has a sharper "peak" and fatter "tails", while a low kurtosis distribution has a more rounded peak with wider "shoulders".

Distributions with zero kurtosis are called mesokurtic, or mesokurtotic. The most prominent example of a mesokurtic distribution is the normal distribution family, regardless of the values of its parameters. A few other well-known distributions can be mesokurtic, depending on parameter values: for example the binomial distribution is mesokurtic for $p = 1/2 \pm \sqrt{1/12}$.

A distribution with positive kurtosis is called leptokurtic, or leptokurtotic. In terms of shape, a leptokurtic distribution has a more acute "peak" around the mean (that is, a higher probability than a normally distributed variable of values near the mean) and "fat tails" (that is, a higher probability than a normally distributed variable of extreme values). Examples of leptokurtic distributions include the Laplace distribution and the logistic distribution. Such distributions are sometimes termed "super Gaussian".

A distribution with negative kurtosis is called platykurtic, or platykurtotic. In terms of shape, a platykurtic distribution has a smaller "peak" around the mean (that is, a lower probability than a normally distributed variable of values near the mean) and "thin tails" (that is, a lower probability than a normally distributed variable of extreme values). Examples of platykurtic distributions include the continuous or discrete uniform distributions, and the raised cosine distribution. The most platykurtic distribution of all is the Bernoulli distribution with $p = ½$ (for example the number of times one obtains "heads" when flipping a coin once), for which the kurtosis is -2. Such distributions are sometimes termed "sub Gaussian".

SOLVED PROBLEMS

Problem 1: Find Kurtosis of following data

Sales :	10–20	20–30	30–40	40–50	50–60
No. of items :	18	20	30	22	10

Solution:

Class interval	Mid point	F	$(m^d - A)/10$	Fd	Fd^2	Fd^3	Fd^4
10-20	15	18	–2	–36	72	–144	288
20-30	25	20	–1	–20	20	–20	20
30-40	35	30	0	0	0	0	0
40-50	45	22	1	22	22	22	22
50-60	55	10	2	20	40	80	160
Total		$N = 100$		–14	154	–62	490

QUERIES

1. What is kurtosis? How will you measure it?
2. Explain the role of Kurtosis in analyzing frequency distribution?
3. What are the various types of Kurtosis?

UNSOLVED PROBLEMS

1. Given following information

Hourly wages	No. of workers	
10-12	3	
12-14	30	
14-16	110	
16-18	218	
18-20	275	
20-22	222	
22-24	108	
24-26	32	
26-28	2	N=1000

Compute kurtosis.

2. Compute kurtosis from the following data:

X:	2	3	4	5	6	Total
f:	1	3	7	3	1	**15**

3. Calculate kurtosis

   (i) 5, 8, 15, 12, 20

   (ii) 20, 40, 60, 80, 100

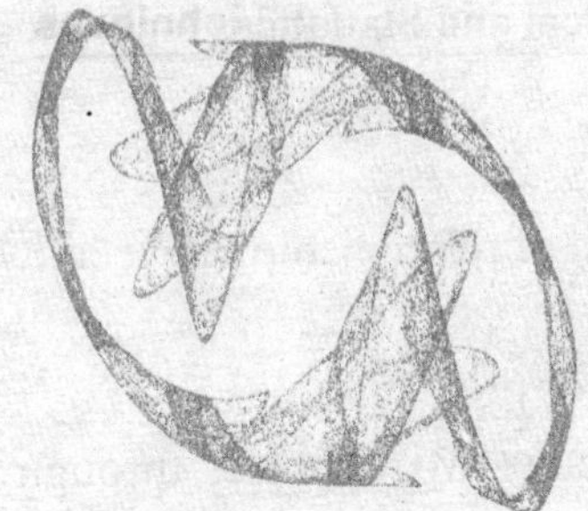

CHAPTER 15

Correlation

15.1 INTRODUCTION

So far we have discussed those statistical methods, which are concerned with the description and analysis of single variables. We have computed measures of central tendency, measures of dispersion, skewness and kurtosis, but always only one variable was involved at one time in these situations.

However, we may be interested in finding some relationship between two variables in some problems like with increase in height of a person, weight of the person also increases. In this problem height and weight are correlated with each other. Two variables are correlated with each other if change in value of one variable is accompanied by change in the value of the other. The degree of extent of relationship between two such variables is measured by a statistical tool known as correlation coefficient.

Thus correlation can be defined as a statistical tool for measuring numerically the degree or extent of co-variability between the two variables.

15.2 CORRELATION

In statistics, correlation (often measured as a correlation coefficient, ñ) indicates the strength and direction of a linear relationship between two random variables.

Some other Definations:

Croxton and Cowden: *"When the relationship is of a quantitative nature, the appropriate statistical tool for discovering and measuring the relationship and expressing it in a brief formula is known as correlation."*

A.M.Tuttle: *"Correlation is an analysis of the covariation between two or more variables."*

Two variables are said to be correlated if the change in one variable affects the result of other variable.

15.2.1 Major Uses Of Correlation

(i) Major use of correlation is to derive the degree and direction of relationship between two variables dependent on each other in some manner.

(ii) It is used in reducing the range of uncertainty in the matter of prediction.

(iii) It is used in presenting the average relationship between any two variables through a single value of coefficient of correlation.

(iv) It is very useful in regression analysis, which in turn is the basis of the theory of econometrics.

(v) It is helpful to executives to estimate profit, interest, costs, sales etc. on the basis of some other series with which these variables may be functionally related.

(vi) It compares the degree of association between two variables.

(vii) It is used in various fields like business, science and philosophy, economics etc.

15.3 TYPES OF CORRELATION

Following are the types of Correlation:

(i) Positive and negative correlation

(ii) Simple, Partial and Multiple correlations

(iii) Linear and Non-linear correlation

(i) Positive Correlation and Negative Correlation: If the values of the two variables change in the same direction (if we increase the value of one variable, it results in a corresponding increase in the value of other variable or if a decrease in the value of one variable results in a corresponding decrease in the values of other variable), correlation is said to be positive.

Some examples of positive correlation are:

(a) Heights and Weights

(b) The family income and expenditure on luxury items

On the other hand, correlation is said to be negative if the variables change in the opposite direction i.e. if the increase (decrease) in the values of one variable results in a corresponding decrease (increase) in the values of the other variable.

Some examples of negative correlation are:

(a) Price and Demand of a commodity

(b) Sale of Woolen garments and the day temperature

(ii) Simple, Partial and Multiple Correlation: Correlation is divided into three types- Simple, Partial and Multiple correlation on the basis of the number of the variables under study.

Simple correlation is the study of relation ship between two variables only. If there are more than two variables say in a multivariate distribution and we study the relationship and association between any two variables ignoring the effects of all other variables the technique used for the purpose is known as partial correlation. If there are more than two variables and we study the relationship and association between one variable and the remaining all other variables taken together, the technique used for the purpose is known as multiple correlation.

(iii) Linear and Non-linear Correlation: The correlation between two variables is said to be linear if corresponding to a unit change in one variable, there is a constant change in the other variable over the entire range of the values.

For example consider the following data:

X:	1	2	3	4	5
Y:	5	10	15	20	25

Thus for a unit change in the value of x, there is a constant change, i.e. 5 in the corresponding value of y. So a linear correlation exists between x and y. In general, two variables x and y are said to be linearly related, if there exists a relationship of the form $Y = a + bX$ between them.

The correlation between two variables is said to be non linear, when the amount of change in one variable is not in constant ratio to the change in other variables. The graph of the variables having non-linear correlation will not form a straight line, whereas the graph of the variables having linear correlation will form a straight line.

15.4 DIFFERENT METHODS OF CORRELATION

As we have discussed earlier, correlation may be linear or non-linear. In this section, we shall discuss techniques of studying only the linear relationship between two variables only, i.e. simple correlation. Non-linear correlation is beyond the scope of this book.

The techniques of studying linear correlation are:

(i) Scatter Diagram Method
(ii) Karl Pearson's Coefficient of Correlation
(iii) Spearman's Rank Correlation Coefficient
(iv) Concurrent Deviations Method

15.4.1 Scatter Diagram Method

It is the simplest way of studying correlation. In scatter diagram methods, we plot data on the graph paper with the help of dots. We represent (x, y) with the dots and thus get as many dots as the number of items. This figure is called scatter diagram. The ways in which dots lie on the scatter diagram show the type of correlation

If the paths formed by dots running between the lower left hand corner to the upper right hand corner, it means there is +ive correlation. If the paths formed by dots running between the upper left hand corner to the lower right hand corner, it means there is –ive correlation. If the path formed by dots falls in a narrow band and the dots are rising from lower left hand corner to upper right hand

15

corner, it means existence of high degree of +ive correlation. But if the path formed by dots falls in a narrow band from the upper left hand corner to lower right hand corner, it means the existence of high degree of negative correlation.

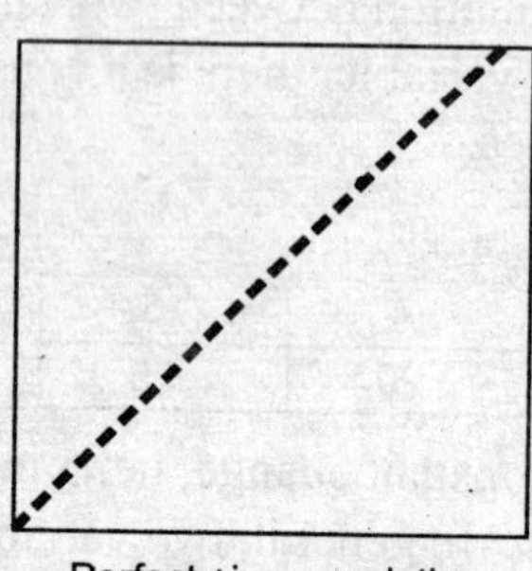

Perfect +ive correlation

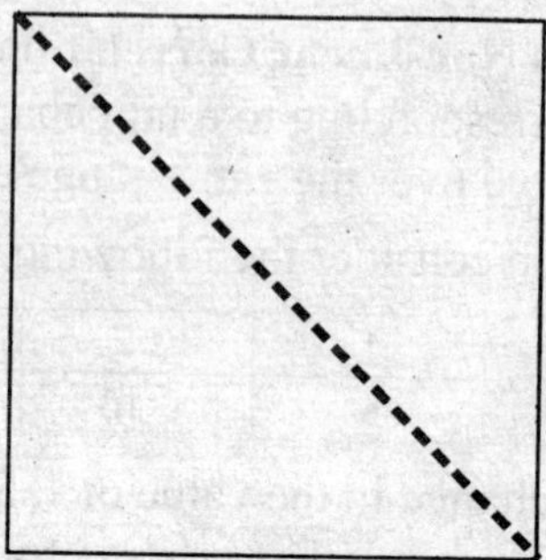

Perfect –ive correlation

On the other hand, if the dots from a wide scatter a flow ascending from the lower left corner to upper right hand corner, it means existence of low degree of positive correlation. If the dots from a wide scatter a flow falling from the upper left hand corner to the lower right hand corner, it means there is low degree of negative correlation. If the plotted dots lie scattered all over the diagram, there is no correlation between two variables.

15.4.1.1 *Merits of Scatter Diagram Method*

(i) It is the simplest method of studying correlation.

(ii) It is not affected by the presence of extreme observations.

(iii) It is suitable for rough idea about relationship between variables.

(iv) This method is a non-mathematical method.

(v) This method helps us to determine an approximate estimating line or the line of best fit by free hand method.

(vi) This method is useful in detecting abnormal points, if any, in the data.

15.4.1.2 Demerits of Scatter Diagram Method

(i) It doesn't determine the degree of correlation, it only tells about relationship between the two variables.

(ii) This method is not suitable for large number of observations.

(iii) The result produced by this method is not purely perfect.

SOLVED PROBLEMS

Problem 1: The following is the distribution of marks (out of 20) obtained by five students in Computer and Mathematics

Marks in Computer	Marks in Mathematics
10	12
15	17
11	13
18	20
13	15

Draw a scatter diagram and give an estimate of the correlation between the marks in computer and mathematics.

Solution: Take the marks of computer along X-axis and marks of mathematics along Y-axis we have

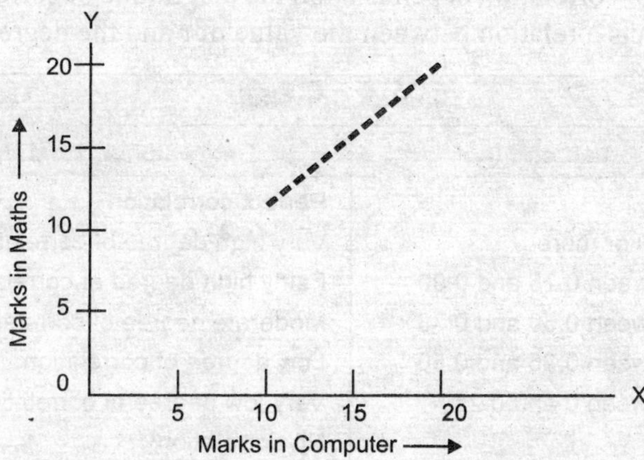

The above scatter diagram shows that given data exhibit a perfect positive correlation between marks obtained by students in computer and mathematics.

15.4.2 Karl Pearson's Coefficient of Correlation

Karl Pearson, a great British Biometrician and Statistician developed a method to measure the linear correlation between the two variables in a bivariate distribution. Karl Pearson's correlation coefficient between two variables X and Y, usually represented by r is a numerical measure of linear relationship between them and is defined as the ratio of the covariance between X and Y, written as $Cov(X, Y)$ to the product of the standard deviations of X and Y.

So symbolically, we have

$$r = \frac{Cov(X,Y)}{\sigma_x \sigma_y} \quad ...(1)$$

If $(X_1, Y_1)(X_2, Y_2)$ ....., (X_n, Y_n) are n pairs of observations of the variables X and Y in a bivariate distribution, then

$$Cov(X,Y) = \frac{1}{n}\Sigma xy, \quad \sigma_x = \sqrt{\frac{\Sigma x^2}{n}}, \quad \sigma_y = \sqrt{\frac{\Sigma y^2}{n}}$$

where $x = X - \bar{X}$, $y = Y - \bar{Y}$

Substituting these in (1), we get

$$r = \frac{\Sigma xy}{\sqrt{\Sigma x^2}\sqrt{\Sigma y^2}}$$

The coefficient of correlation is represented by letter '*r*' and value of '*r*' always lies between ±1. Positive value of *r* represents positive correlation and negative value represents negative correlation. The degree of correlation depends upon the magnitude of correlation coefficient *r*, i.e. $|r|$. Following chart depicts a relation between the value of r and the degree of correlation:

Chart of Correlation

Value of $\|r\|$	Degree of correlation
1	Perfect correlation
0.90 or more	Very high degree of correlation
between 0.75 and 0.90	Fairly high degree of correlation
between 0.50 and 0.75	Moderate degree of correlation
between 0.25 and 0.50	Low degree of correlation
between 0 and 0.25	Very low degree of correlation
0	No correlation

The different formulae for obtaining the coefficient of correlation are as under:

(1) Direct Method or Product Moment Method (Based on Values)

Product Moment method is also known as direct method based on values. In case the number of items is small, which can be easily squared and multiplied, the following formula can be applied to calculate the correlation coefficient:

$$r = \frac{N\Sigma XY - \Sigma X \cdot \Sigma Y}{\sqrt{N\Sigma X^2 - (\Sigma X)^2} \cdot \sqrt{N\Sigma Y^2 - (\Sigma Y)^2}}$$

where *r* denotes Karl Pearson's coefficient of correlation. *X* and *Y* are given values of first and second variable. *N* is the number of pairs of observations.

Note 1: *When each of the values of variable is multiplied or divided by a common factor, it is a case of change of scale, when each of the values of a variable are added or subtracted by a common factor; it is a case of change of origin. The correlation coefficient is a pure number, that is, is independent from any unit of measurement. Also it is independent of both shifting of scale and origin.*

Note 2: *If the given values are larger, reduce them to the smallest possible size by making either shifting of origin or shifting of scale or shifting of both origin and scale.*

SOLVED PROBLEMS

Problem 1: You are provided the data given below; find the coefficient of correlation between the two variables using Karl Pearson 's direct method (based on values):

Marks in Computer :	1	2	3	4	5
Marks in Maths :	6	7	8	9	10

Solution:

Marks in Computer (X)	X^2	Marks in Maths (Y)	Y^2	XY
1	1	6	36	6
2	4	7	49	14
3	9	8	64	24
4	16	9	81	36
5	25	10	100	50
$\Sigma X = 15$	$\Sigma X^2 = 55$	$\Sigma Y = 40$	$\Sigma Y^2 = 330$	$\Sigma XY = 130$

We have

$$r = \frac{N\Sigma XY - \Sigma X \cdot \Sigma Y}{\sqrt{N\Sigma X^2 - (\Sigma X)^2} \cdot \sqrt{N\Sigma Y^2 - (\Sigma Y)^2}}$$

$$r = \frac{5(130) - 15 \times 40}{\sqrt{5 \times 55 - (15)^2} \cdot \sqrt{5 \times 330 - (40)^2}}$$

$$= \frac{650 - 600}{\sqrt{275 - 225(1650 - 1600)}}$$

$$= \frac{50}{\sqrt{50 \times 50}} = \frac{50}{50} = 1$$

As the value of 'r' comes to be 1 so it indicates that correlation between the marks in Computer and marks in Maths is perfectly positive.

Problem 2: You are provided the following data, Determine coefficient of correlation using the Pearson's direct method based on the values.

U :	50	40	30	20	10
V :	120	140	160	180	200

Solution: The Problem is solved by using change of scale, as the magnitude of given values is large.

U	U/10 = X, say	X^2	V	V/20 = Y, say	Y^2	XY
50	5	25	120	6	36	30
40	4	16	140	7	49	28
30	3	9	160	8	64	24
20	2	4	180	9	81	18
10	1	1	200	10	100	10
-	$\Sigma X = 15$	$\Sigma X^2 = 55$	-	$\Sigma Y = 40$	$\Sigma Y = 330$	$\Sigma XY = 110$

Here $N = 5$

$$r = \frac{N\Sigma XY - \Sigma X \cdot \Sigma Y}{\sqrt{N\Sigma X^2 - (\Sigma X)^2} \cdot \sqrt{N\Sigma Y^2 - (\Sigma Y)^2}}$$

$$r = \frac{5 \times 110 - 15 \times 40}{\sqrt{5 \times 55 - (15)^2} \cdot \sqrt{5 \times 330 - (40)^2}}$$

$$= \frac{550 - 600}{\sqrt{275 - 225(1650 - 1600)}}$$

$$= \frac{50}{\sqrt{50 \times 50}} = \frac{-50}{50} = -1$$

As the value of '*r*' comes to be –1 so it indicates that correlation between the two variables is perfectly negative.

Problem 3: From the following pairs of data, find the Karl Pearson s coefficient of correlation.

U:	24	26	28	30	32
V:	–2	–4	–6	–8	–10

Solution: The Problem is solved by using change of origin as:

U	U-23 X, say	X^2	V	V + 15 Y, say	Y^2	XY
24	1	1	–2	13	169	13
26	3	9	–4	11	121	33
28	5	25	–6	9	81	45
30	7	49	–8	7	49	49
32	9	81	–10	5	25	45
–	$\Sigma X = 25$	$\Sigma X^2 = 165$	–	$\Sigma Y = 45$	$\Sigma Y^2 = 445$	$\Sigma XY = 185$

Here $N = 5$

$$r = \frac{N\Sigma XY - \Sigma X \cdot \Sigma Y}{\sqrt{N\Sigma X^2 - (\Sigma X)^2} \cdot \sqrt{N\Sigma Y^2 - (\Sigma Y)^2}}$$

$$r = \frac{5 \times 185 - 25 \times 45}{\sqrt{5 \times 165 - (25)^2} \cdot \sqrt{5 \times 445 - (45)^2}}$$

$$= \frac{925 - 1125}{\sqrt{825 - 625}\,\sqrt{2225 - 2025}}$$

$$= \frac{-200}{\sqrt{40000}} = \frac{-200}{200} = -1$$

As the value of '*r*' comes to be -1 so it indicates that correlation between the two variables is perfectly negative.

(2) Direct Method (Based on Deviations)

This method is used when the magnitude of the variables is large and their deviations from their respective actual means are found to be whole numbers.

Under this method, '*r*' (coefficient of correlation) is

$$r = \frac{\Sigma xy}{\sqrt{\Sigma x^2} \cdot \sqrt{\Sigma y^2}}$$

where *x* represents the deviation of the first variable from its mean and *y* represents deviations of the second variable from its mean.

Problem 4: From the given statistics, calculate Karl Pearson's coefficient of correlation by direct method based on the deviations.

X:	8	6	10	5	6
Y:	11	13	8	9	9

Solution:

X	$(X - \bar{X}) = x$	x^2	Y	$(Y - \bar{Y}) = y$	y^2	xy
8	1	1	11	1	1	1
6	–1	1	13	3	9	–3
10	3	9	8	–2	4	–6
5	–2	4	9	–1	1	2
6	–1	1	9	–1	1	1
$\Sigma X = 35$		$\Sigma x^2 = 16$	$\Sigma Y = 50$		$\Sigma y^2 = 16$	$\Sigma xy = -5$

Means of X and Y are $\bar{X} = \frac{\Sigma X}{N} = \frac{35}{5} = 7, \quad \bar{y} = \frac{\Sigma Y}{N} = \frac{50}{5} = 10$

Now Karl Pearson's coefficient is given by

$$r = \frac{\Sigma xy}{\sqrt{\Sigma x^2}.\sqrt{\Sigma y^2}} \frac{-5}{\sqrt{16 \times 16}} = -0.3125 \text{ Approx.}$$

This value of r tells us that a low degree of negative correlation exists between two variables.

Problem 5: Determine the coefficient of correlation from the following data using the Pearson's direct method based on deviations.

Y1 :	120	−100	−80	−40	120	140	−60	−100
Y2 :	300	180	−210	−360	120	240	330	−360

Solution:

Y1	Y1/20 = X	$(X - \bar{X}) = x$	x^2	Y2	Y2/30 = Y	$(Y - \bar{Y}) = y$	y^2	xy
120	6	6	36	300	10	9	81	54
−100	−5	−5	25	180	6	5	25	−25
−80	−4	−4	16	−210	−7	−8	64	32
−40	−2	−2	4	−360	−12	−13	169	26
120	6	6	36	120	4	3	9	18
140	7	7	49	240	8	7	49	49
−60	−3	−3	9	330	11	10	100	−30
−100	−5	−5	25	−360	−12	−13	169	65
	$\Sigma X = 0$		$\Sigma x^2 = 200$		$\Sigma Y = 8$		$\Sigma y^2 = 666$	$\Sigma xy = 189$

Mean of first series

$$\bar{X} = \frac{\Sigma X}{N} = \frac{0}{8} = 0$$

Mean of second series

$$\bar{Y} = \frac{\Sigma Y}{N} = \frac{8}{8} = 1$$

$$r = \frac{\Sigma xy}{\sqrt{\Sigma x^2 \Sigma y^2}}$$

Substitute the respective value in the formula, we get

$$r = \frac{189}{\sqrt{200 \times 666}} = \frac{189}{133200}$$

$$r = \frac{189}{365} = 0.52 \text{ approx.}$$

The value of r tells us that moderate degree of positive correlation exists between two variables.

(3) Short-cut Method

When the actual means are in fractions, then the calculation of correlation by direct method becomes complicated. In such cases it is better to use short cut method. Under this method, deviations are taken from assumed mean and the formula is:

$$r = \frac{N\Sigma dxdy - \Sigma dx \cdot \Sigma dy}{\sqrt{N\Sigma dx^2 - (\Sigma dx)^2}\sqrt{N\Sigma dy^2 - (\Sigma dy)^2}}$$

where dx is the deviation of X taken from its assumed mean say A and dy is the deviation of Y taken from its assumed mean say B.

Problem 6: Calculate coefficient of correlation from the following data for the variables *X* and *Y*

X:	24	27	28	29	30	32	33	33	35	35	40
Y:	18	20	22	25	22	28	28	30	27	30	22

Solution: Suppose A and B be the assumed means of series X and Y respectively.

Take $A = 32$ and $B = 25$

X	X – A = dx	dx²	Y	Y – B = dy	dy²	dxdy
24	–8	64	18	–7	49	56
27	–5	25	20	–5	25	25
28	–4	16	22	–3	9	12
28	–4	16	25	0	0	0
29	–3	9	22	–3	9	9
30	–2	4	28	3	9	–6
32	0	0	28	3	9	0
33	1	1	30	5	25	5
35	3	9	27	2	4	6
35	3	9	30	5	25	15
40	8	64	22	–3	9	–24
	$\Sigma dx = -11$	$\Sigma dx^2 = 217$		$\Sigma dy = -3$	$\Sigma dy^2 = 173$	$\Sigma dxdy = 98$

$$r = \frac{N\Sigma dxdy - \Sigma dx \cdot \Sigma dy}{\sqrt{N\Sigma dx^2 - (\Sigma dx)^2}\sqrt{N\Sigma dy^2 - (\Sigma dy)^2}}$$

$$r = \frac{11 \times 98 - (-11) \times (-3)}{\sqrt{217 \times 11 - (-11)^2}\sqrt{173 \times 11 - (-3)^2}}$$

$$r = 1045/2071.66 = 0.504$$

The value of 'r' here indicates that there is a moderate degree of positive correlation between the two variables.

Problem 7: The following observations are given to you. Compute the Karl Pearson's coefficient of correlation between age and success of candidates by short cut method.

Ages :	15	16	17	18	19	20	21
No. of candidates :	200	300	100	50	150	400	250
No. of successful candidates :	120	180	60	30	90	252	140

Solution: Out of 200 candidates number of successes = 120

So, out of 100 candidates, number of successes $= \dfrac{120 \times 100}{200} = 60$

In a similar way, the number of successful candidates per 100 for other age groups would be 60, 60, 60, 60, 60, 63, 56.

Let A and B be the assumed means of series X and Y respectively.

Take $A = 18$ and $B = 60$

X	(X – A)dx	dx²	Y	(Y – B)dy	dy²	dxdy
15	–3	9	60	0	0	0
16	–2	4	60	0	0	0
17	–1	1	60	0	0	0
18	0	0	60	0	0	0
19	1	1	60	0	0	0
20	2	4	63	3	9	6
21	3	9	56	–4	16	–12
	Σdx = 0	Σdx² = 28		Σdy = –1	Σdy² = 25	Σdxdy = –6

$$r = \frac{N\Sigma dxdy - \Sigma dx \cdot \Sigma dy}{\sqrt{N\Sigma dx^2 - (\Sigma dx)^2}\sqrt{N\Sigma dy^2 - (\Sigma dy)^2}}$$

$$r = \frac{7 \times (-6) - (0)(-1)}{\sqrt{7 \times 28}\sqrt{7 \times 25 - (-1)^2}}$$

$$= \frac{-42}{\sqrt{196}\sqrt{174}}$$

$$= \frac{-42}{\sqrt{(196 \times 174)}}$$

$$= \frac{42}{\sqrt{34104}} = \frac{-42}{184.67} = -0.22$$

The result shows that there is a very low degree of negative correlation between age and success of candidates.

Problem 8: From the following table calculate the coefficient of correlation by Karl Pearson's Method:

X:	6	2	10	4	8
y:	9	11	?	8	7

Given that the means of *X* and *Y* series are 6 and 8 respectively.

Solution: First of all find the missing value of *Y*.

Let the missing value in Y-series is a.

Then $$\bar{Y} = \frac{\Sigma Y}{N} = \frac{9+11+a+8+7}{5} = \frac{35+a}{5} = 8$$

$$35 + a = 40$$

$$a = 40 - 35 = 5$$

X	Y	$x = (X - \bar{X})$ $= x - 6$	$y = (Y - \bar{Y})$ $= y - 8$	x^2	y^2	xy
6	9	0	1	0	1	0
2	11	–4	3	16	9	–12
10	5	4	–3	16	9	–12
4	8	–2	0	4	0	0
8	7	2	–1	4	1	–2
ΣX = 30	ΣY = 40	Σx = 0	Σy = 0	$\Sigma x^2 = 40$	$\Sigma y^2 = 20$	Σxy = –26

$$\bar{X} = \frac{\Sigma X}{5} = \frac{30}{5} = 6, \qquad \bar{X} = \frac{\Sigma Y}{5} = \frac{40}{5} = 8,$$

Now $$r = \frac{\Sigma xy}{\sqrt{\Sigma x^2 \Sigma y^2}}$$

$$= \frac{-26}{\sqrt{40 \times 20}} = \frac{-26}{\sqrt{800}} = \frac{-26}{28.2843} = -0.92$$

Problem 9: Calculate Karl Pearson's coefficient of correlation from the following data:

(i) Sum of deviations of $x = 5$

(ii) Sum of deviations of $y = 4$

(iii) Sum of squares of deviations of $x = 40$

(iv) Sum of squares of deviations of $y = 50$

(v) Sum of product of deviations of x and $y = 32$

(vi) No. of pairs of observations = 10

Solution: We are given

$\Sigma dx = 5,\quad \Sigma dx^2 = 40,\quad \Sigma dxdy = 32$

$\Sigma dy = 4,\quad \Sigma dy^2 = 50,\quad N = 10$

Substituting these values in the formula:

$$r = \frac{N\Sigma dxdy - \Sigma dx \cdot \Sigma dy}{\sqrt{N\Sigma dx^2 - (\Sigma dx)^2}\sqrt{N\Sigma dy^2 - (\Sigma dy)^2}}$$

$$r = \frac{10 \times 32 - 5 \times 4}{\sqrt{10 \times 40 - 25}\sqrt{10 \times 50 - 16}}$$

$$r = \frac{320 - 20}{\sqrt{375}\sqrt{484}} = \frac{300}{\sqrt{181500}}$$

$$r = \frac{300}{426.03} = 0.7042$$

Problem 10: The coefficient of correlation between two variables X and Y is 0.48. The covariance is 36. The variance of X is 16. Find the standard deviation of Y.

Solution: We are given:

$r = 0.48$, Cov $(X,Y) = 36$, $\sigma_x^2 = 16$ Þ $\sigma_x = 4$

$$r = \frac{Cov(X,Y)}{\sigma_x \sigma_y} \Rightarrow \sigma_y = \frac{Cov.(X,Y)}{\sigma_x \cdot r}$$

$$\therefore \qquad \sigma_y = \frac{36}{4 \times 0.48} = \frac{9}{0.48} = 18.75$$

Problem 11: A computer while calculating correlation coefficient between two variables X and Y from 25 pairs of observations obtained the following results:

$N = 25, \Sigma X = 125, \Sigma X^2 = 650, \Sigma Y = 100, \Sigma Y^2 = 460, \Sigma XY = 508$

It was discovered at the time of verification that two pairs of observations were not correctly copied. They were taken as (6,14) and (8,6) while the correct values were (8,12) and (6,8). Find the correct value of the correlation coefficient.

Solution:

Corrected $\Sigma X = 125 - 6 - 8 + 8 + 6 = 125$

Corrected $\Sigma Y = 100 - 14 - 6 + 12 + 8 = 100$

Corrected $\Sigma X^2 = 650 - 6^2 - 8^2 + 8^2 + 6^2 = 650$

Corrected $\Sigma Y^2 = 460 - 14^2 - 6^2 + 12^2 + 8^2 = 436$

Corrected $\Sigma XY = 508 - 6 \times 14 - 8 \times 6 + 8 \times 12 + 6 \times 8 = 520$

$$\text{Corrected value of } r = \frac{N\Sigma XY - \Sigma X \cdot \Sigma Y}{\sqrt{N\Sigma X^2 - (\Sigma X)^2} \cdot \sqrt{N\Sigma Y^2 - (\Sigma Y)^2}}$$

$$= \frac{25 \times 520 - 125 \times 100}{\sqrt{25 \times 650 - (125)^2} \cdot \sqrt{25 \times 436 - (100)^2}}$$

$$= \frac{13000 - 12500}{\sqrt{16255 - 15625}\sqrt{10900 - 10000}}$$

$$= \frac{500}{\sqrt{625 \times 900}} = \frac{500}{25 \times 30} = \frac{2}{3} = 0.67$$

15.4.3 Correlation of Frequency Distribution

Sometimes the number of items is very large and they are classified into a two-way frequency distribution table, called bivariate frequency distribution table. Here for each variable, the values are grouped into various classes (not necessarily the same for both the variables), keeping in view the same considerations as in case of univariate distribution. The values of the *Y* variable are listed in the column headings and those of *X* are listed in the stubs at the left of the table. The frequencies for each cell of the table are determined by preparing tally bars just as in the case of individual series.

Karl Pearson's coefficient of correlation in such bivariate frequency table can be calculated by the following methods:

(i) Formation of Bivariate frequency table

(ii) Construction of correlation table

(iii) Application of frequency oriented formula, which is as:

$$r = \frac{N\Sigma fdxdy - \Sigma fdx \cdot \Sigma fdy}{\sqrt{N\Sigma fdx^2 - (\Sigma fdx)^2}\sqrt{N\Sigma fdy^2 - (\Sigma fdy)^2}} \quad \text{and} \quad N = \Sigma f$$

This formula is similar to the already discussed one for calculating '*r*' with the help of assumed mean. The only difference is that here the deviations are also multiplied by their respective frequencies. The procedure for calculating '*r*' by this formula is as:

(i) Mid values are calculated for *X* and *Y* series and deviations are taken for both the variables, which are denoted, by *dx* and dy.

(ii) The frequencies of *X* and *Y* series are multiplied by their respective deviations, i.e. *dx* and *dy*. Their sum is calculated and is denoted by Σfdx and Σfdy.

(iii) Squares of the deviations of both the variables are calculated and are multiplied by their respective frequencies to get Σfdx^2 and Σfdy^2

(iv) Multiply *dx* and *dy* with their common frequency in each cell and write the figures obtained in the right corner of the cell.

(v) Add together all the values as calculated in step (iv) and obtain their $\Sigma fdxdy$

SOLVED PROBLEMS

Problem 1: From the following frequency distribution of marks, form a bivariate frequency table with class interval of 20 in case of both the variables.

Roll No. :	1	2	3	4	5	6	7	8	9	10
Marks in Maths :	81	50	62	16	41	90	30	25	20	5
Marks in Economics :	70	75	65	35	20	22	88	74	30	18
Roll No. :	11	12	13	14	15	16	17	18	19	20
Marks in Maths :	35	18	15	18	94	81	22	30	12	8
Marks in Economics :	27	15	21	25	16	6	8	8	95	85

Solution: The Bivariate frequency table above data is

Marks in Maths / Marks in Economics	0-20	20-40	40-60	60-80	80-100	Total
0-20	| | =(2)	| | =(2)	=(0)	=(0)	| |=(2)	6
20-40	| | |=(3)	| |=(2)	| =(1)	=(0)	| = (1)	7
40-60	=(0)	=(0)	=(0)	=(0)	=(0)	0
60-80	=(0)	| = (1)	| =(1)	| =(1)	| =(1)	4
80-100	| | =(2)	| = (1)	=(0)	=(0)	(0)	3
Total	7	6	2	1	4	20

Problem 2: You are provided the following data prepare a bivariate frequency of table with class interval of 10 for computer and 20 for social science.

Roll No.	Marks in Computer	Marks in Social Science	Roll No.	Marks in Computer	Marks in Social Science
1	5	20	11	39	21
2	15	30	12	25	67
3	25	50	13	3	81
4	17	10	14	0	83
5	18	60	15	8	93
6	22	80	16	15	17
7	30	75	17	28	32
8	35	9	18	35	48
9	17	18	19	17	45
10	24	20	20	14	37

Solution:

Marks in S.Sc / Marks in Computer	0-10	10-20	20-30	30-40	Total
0-20		||| =(3)		| =(1)	4
20-40	| =(1)	|| =(2)	|| =(2)	| =(1)	6
40-60		| =(1)	| =(1)	| =(1)	3
60-80		| =(1)	| =(1)	| =(1)	3
80-100	||| =(3)		| =(1)		4
Total	**4**	**7**	**5**	**4**	**20**

Problem 3: Find the Karl Person's coefficient of Correlation for the following bivariate frequency distribution:

Marks in English → / Marks in History ↓	0-20	20-40	40-60	60-80	80-100	Total
0-20	2	2	0	0	2	6
20-40	3	2	1	0	1	7
40-60	0	0	0	0	0	0
60-80	0	1	1	1	1	4
80-100	2	1	0	0	0	3
Total	**7**	**6**	**2**	**1**	**4**	**20**

Solution:

Marks in English → / Marks in X History m ↓ dx / Y	m	dy	0-20 10 -2	20-40 30 –1	40-60 50 0	60-80 70 1	80-100 90 2	f	fdy	fdy^2	fdxdy
0–20	10	–2	**8** 2	**4** 2	**0** 0	**0** 0	**– 8** 2	6	–12	24	4
20-40	30	–1	**6** 3	**2** 2	**0** 1	**0** 0	**– 2** 1	7	–7	7	6
40-60	50	0	**0** 0	**0** 0	**0** 0	**0** 0	**0** 0	0	0	0	0
60-80	70	1	**0** 0	**– 1** 1	**0** 1	**1** 1	**2** 1	4	4	4	2

80-100	90	2	– **8** 2	– **2** 1	**0** 0	**0** 0	**0** 0	3	6	12	–10
		f	7	6	2	1	4	**Σ*f* = 20**	**Σ*fdy* = –9**	**Σ*fdy*² = 47**	**Σ*fdxdy* = 2**
		fdx	–14	–6	0	1	8	**Σ*fdx* = –11**			
		*fdx*²	28	6	0	1	16	**Σ*fdx*² = 51**			
		fdxdy	6	3	0	1	–8	**Σ*fdxdy* = 2**			

$$r = \frac{N\Sigma fdxdy - \Sigma fdx \cdot \Sigma fdy}{\sqrt{N\Sigma fdx^2 - (\Sigma fdx)^2}\sqrt{N\Sigma fdy^2 - (\Sigma fdy)^2}}$$

Substitute the respective values in the above formula

$$r = \frac{(20\times 2) - (-11\times -9)}{\sqrt{(20\times 51) - (-11)^2}\sqrt{20\times 47 - 81}}$$

$$= \frac{40 - 99}{\sqrt{(1020 - 121)(940 - 81)}}$$

$$= \frac{-59}{\sqrt{899\times 859}} = \frac{-59}{\sqrt{772241}}$$

$$= \frac{-59}{878.77} - 0.067$$

The value of '*r*' indicates that there is a very low degree of negative correlation.

15.4.4 Properties of Karl Pearson's Coefficient of Correlation (*R*)

Following are the properties of the Karl Pearson's coefficient of correlation:

(i) Correlation coefficient lies between –1 to +1.

(ii) The coefficient of correlation is independent of the change of scale and origin of the variables *X* and *Y*.

(iii) The correlation coefficient does not depend on the order of two variables, that is, *r* is symmetric with respect to variables.

(iv) It is independent of the units of measurement.

(v) It is not so rigid.

15.4.5 Merits of Karl Pearson's Coefficient of Correlation

(i) This method along with the other information helps us to know the value of one variable, given the value of the other independent variable.

(ii) This method gives direction as well as degree of relationship between the two variables at one time.

15.4.6 Demerits of Karl Pearson's Coefficient of Correlation

(i) It takes more time to reach at results as compared to other methods.

(ii) It is affected by the values of the extreme items easily.

(iii) One has to be very careful in interpreting the values of the coefficient of correlation.

(iv) It is comparatively difficult to calculate as its computation involves intricate algebraic methods of calculations.

(v) This method is based on certain assumptions.

15.4.7 Probable Error and Coefficient of Correlation

After determining the correlation coefficient, the next step is to find out the extent to which it is dependable. The probable error of the coefficient of correlation is computed for this purpose.

After adding and subtracting probable error from the coefficient of correlation, we would get two limits within which we can reasonably expect the value of coefficient of correlation. The formula for finding out the probable error of Karl Pearson's coefficient of correlation is

$$PE_{(r)} = 0.6745 \frac{1-r^2}{\sqrt{n}}$$

Where r = coefficient of correlation and n = number of pairs of the two variables.

If we omit 0.6745 the above formula can be written as

$$SE_{(r)} = \frac{1-r^2}{\sqrt{n}}$$

which is known as the standard error of the coefficient of correlation.

The formula for probable error is helpful in interpreting the significance of the coefficient of correlation as follows:

(i) Correlation is taken as not at all significant if $r < PE_{(r)}$

(ii) Correlation is taken to be significant if $r > 6$ times $PE_{(r)}$

(iii) In other situations, nothing can be concluded with certainty.

(iv) The limits of the correlation coefficient of the population are $r \pm PE_{(r)}$

SOLVED PROBLEMS

Problem 1: You are given the following data. Find out the probable error and standard error of Pearson's coefficient of correlation. Also determine the limits of the correlation coefficient of population.

15

Solution: Given that, $r = 0.5, n = 50$

$$PE_{(r)} = 0.6745 \frac{1-r^2}{\sqrt{n}}$$

$$= 0.6745 \times \frac{1-(0.5)^2}{\sqrt{50}} = 0.6745 \times \frac{1-0.25}{7.07}$$

$$= 0.6745 \times \frac{0.75}{7.07} = 0.6745 \times 0.106 = 0.07$$

$$SE_{(r)} = \frac{1-r^2}{\sqrt{n}} = \frac{1-(0.5)^2}{\sqrt{50}} = \frac{1-0.25}{7.07} = 0.106$$

Upper limit of the population correlation coefficient

$$r + PE_{(r)} = 0.5 + 0.07 = 0.57$$

Lower limit of the population correlation coefficient

$$r - PE_{(r)} = 0.5 - 0.07 = 0.4$$

15.5 SPEARMAN'S RANK CORRELATION

Sometimes we are faced with statistical series in which the variables under consideration are not capable of quantitative measurement, but can be arranged in serial order. This happens when we are dealing with qualitative features such as honesty, beauty, character, morality etc. In such circumstances Karl Pearson's coefficient of correlation can't be employed. Charles Edward Spearman, a British Psychologist, developed a formula in 1904, which is used to determine correlation coefficient between two variables, which cannot be measured quantitatively.

Spearman's rank correlation is used when data is in the form of ranks. Although spearman's coefficient is designed for use with ranked data, it can be used with interval data that have been expressed as ranks.

The formula for Spearman's Rank correlation is given by

$$r_{(s)} = 1 - \frac{6\Sigma D^2}{N^3 - N}$$

where D is difference between the ranks and N is the total number of items. This formula is used when there is no repetition of any rank.

When two or more items in a series are of equal values, they are assigned equal ranks or average ranks. In such cases, an adjustment is made in the formula.

In the formula, we add the correction factor $\frac{(m^3 - m)}{12}$ for each tie to ΣD^2

where m refers to the number of times an item is repeated in a tie. The correction factor is to be added for each and every repeated value. The formula in this case becomes

$$r_{(s)} = 1 - \frac{6\left(\Sigma D^2 + \frac{m^3 - m}{12} + \frac{m^3 - m}{12} + \ldots\right)}{N^3 - N}$$

In the rank correlation coefficient, we may have three types of problems:

(i) Rank correlation when actual ranks are given

(ii) Rank correlation when actual ranks are not given

(iii) Rank correlation in case of equal marks

(i) Rank Correlation When Actual Ranks are Given

(i) First calculate the difference between the ranks and denote it by D.

(ii) Compute ΣD^2

(iii) Now use the formula

$$r_{(s)} = 1 - \frac{6\Sigma D^2}{N^3 - N}$$

SOLVED PROBLEMS

Problem 1: The preferences of two consumers *A* and *B* for different brands of television sets are given in the following table:

T.V Brand	Consumer *A*	Consumer *B*
Weston	3	5
Konark	5	6
Binatone	8	4
Televista	12	9
Crown	10	8
Uptron	7	12
Onida	1	3
Nelco	9	11
Sears	11	7
Telerama	2	10
Sonoyna	4	1
Dyanora	6	2

Determine coefficient of rank correlation between the preference of consumers *A* and *B* and interpret your result.

Solution: Calculation of rank correlation coefficient

R_1	R_2	$D = R_1 - R_2$	D^2
3	5	−2	4
5	6	−1	1
8	4	4	16
12	9	3	9
10	8	2	4
7	12	−5	25
1	3	−2	4
9	11	−2	4
11	7	4	16
2	10	-8	64
4	1	3	9
6	2	4	16
		$\Sigma D = 0$	$\Sigma D^2 = 172$

Here $N = 12$

$$r_{(s)} = 1 - \frac{6\Sigma D^2}{N^3 - N}$$

$$r_{(s)} = 1 - \frac{6(172)}{(12)^3 - 12}$$

$$= 1 - \frac{1032}{1716}$$

$$= 0.399$$

Here, value of r shows that there is a moderate degree of positive correlation between consumer A and consumer B.

Problem 2: Three judges in the following order rank ten runners in a racing competition. Use rank correlation coefficient to discuss which pair of judges has the nearest approach to common tastes in run:

I[st] judge:	1	6	5	10	3	2	4	9	7	8
2nd judge:	3	5	8	4	7	10	2	1	6	9
3rd judge:	6	4	9	8	1	2	3	10	5	7

Solution: Calculation of Rank correlation coefficient

R_1	R_2	R_3	$D_{12}^2 = (R_1 - R_2)^2$	$D_{23}^2 = (R_2 - R_3)^2$	$D_{13}^2 = (R_3 - R_1)^2$
1	3	6	4	9	25
6	5	4	1	1	4
5	8	9	9	1	16
10	4	8	36	16	4
3	7	1	16	36	4
2	10	2	64	64	0
4	2	3	4	1	1
9	1	10	64	81	1
7	6	5	1	1	4
8	9	7	1	4	1
$N = 10$			$\Sigma D_{12}^2 = 200$	$\Sigma D_{23}^2 = 214$	$\Sigma D_{13}^2 = 60$

Rank correlation between judgment if 1st and 2nd judge

$$r_{(s)} = 1 - \frac{6\Sigma D_{12}^2}{N^3 - N}$$

After substituting the values of parameters in above formula, we get

$$r_{(s)} = -0.212$$

Rank correlation between judgment if 2nd and 3rd judge

$$r_{(s)} = 1 - \frac{6\Sigma D_{23}^2}{N^3 - N}$$

After substituting the values of parameters in above formula from table, we get

$$r_{(s)} = -0.297$$

Rank correlation between judgment if 3rd and 1st judge

$$r_{(s)} = 1 - \frac{6\Sigma D_{13}^2}{N^3 - N}$$

After substituting the values of parameters in above formula from table, we get

$$r_{(s)} = 0.636$$

Since coefficient of correlation is maximum in the judgment of 3rd and 1st judge, we conclude that they have the nearest approach.

(ii) Rank Correlation When Ranks are Not Given

In such circumstances we first convert the data into ranks. The highest (smallest) item is given the rank 1. The next highest (next lowest) item is given rank 2 and so on. But whether we start with the lowest value or the highest value we must follow the same method for both the variables.

SOLVED PROBLEMS

Problem 1: Calculate the rank correlation coefficient from the data given below:

Marks in Computer X:	20	30	40	50	60	70	80
Marks in English Y:	14	5	30	32	40	45	65

Solution: Calculation of Rank Correlation coefficient

X	R_X	Y	R_Y	$D = R_x - R_y$	D^2
20	1	14	2	–1	1
30	2	5	1	1	1
40	3	30	3	0	0
50	4	32	4	0	0
60	5	40	5	0	0
70	6	45	6	0	0
80	7	65	7	0	0
$N = 7$				$\Sigma D = 0$	$\Sigma D^2 = 2$

$$r_{(s)} = 1 - \frac{6\Sigma D^2}{N^3 - N}$$

$$r_{(s)} = 1 - \frac{6(2)}{(7)^3 - 7}$$

$$= 0.96$$

(iii) Rank Correlation Coefficient in Case of Equal Ranks

When there is more than one item with the same value, a common rank is given to such items. They are assigned the average of ranks that they occupy when there is no repetion.

In this case, following formula is used

$$r_{(s)} = 1 - \frac{6\left(\Sigma D^2 + \frac{m^3 - m}{12} + \frac{m^3 - m}{12} + \ldots\right)}{N^3 - N}$$

where $1/12\,(m^3 - m)$ occurs as many items as there are groups having equal ranks, m, is number of items having equal rank, N is total numbers of pairs.

SOLVED PROBLEMS

Problem 1: Calculate rank correlation coefficient for the following data:

X:	12	15	18	20	16	15	18	22	15	21	18	15
Y:	10	18	19	12	15	19	17	19	16	14	13	17

Solution: Calculation of rank correlation coefficient:

X	Y	R_X	R_Y	$D = R_X R_Y$	D^2
12	10	1	1	0	0
15	18	3.5	9	–5.5	30.25
18	19	8	11	–3	9
20	12	10	2	8	64
16	15	6	5	1	1
15	19	3.5	11	–7.5	56.25
18	17	8	7.5	0.5	0.25
22	19	12	11	1	1
15	16	3.5	6	–2.5	6.25
21	14	11	4	7	49
18	13	8	3	5	25
15	17	3.5	7.5	–4	16
				$\Sigma D = 0$	$\Sigma D^2 = 258.00$

$$r_{(s)} = 1 - \frac{6\left(\Sigma D^2 + \dfrac{m^3 - m}{12} + \dfrac{m^3 - m}{12} + \ldots\right)}{N^3 - N}$$

$$r_{(s)} = 1 - \frac{6\left(258 + \dfrac{(4)^3 - 4}{12} + \dfrac{(3)^3 - 3}{12} + \dfrac{(2)^3 - 2}{12} + \dfrac{(3)^3 - 3}{12}\right)}{(12)^3 - 12}$$

$$= 1 - 6\frac{(258 + 5 + 2.5 + 2)}{1716}$$

$$= 1 - 0.93$$

$$= 0.07$$

15.5.1 Features of Rank Correlation

(i) The value of coefficient of correlation lies between –1 and +1.

(ii) The sum of differences between the corresponding ranks is definitely equal to zero, i.e. $\Sigma d = 0$.

(iii) It is calculated on the basis of the ranks of individual items.

15.5.2 Merits of Rank Correlation

(i) Easy to understand and simple to calculate.

(ii) The extreme items do not affect this method.

(iii) This method is applied to qualitative type of data, where Karl Pearson's Correlation coefficient can't be applied.

(iv) It can be easily applied even when the data don't follow normal distribution.

15.5.3 Demerits of Rank Correlation

(i) This method is not suitable in case of bivariate frequency distribution, i.e. grouped data.

(ii) For $n > 30$, this formula should not be used unless the ranks are given, since in the contrary case the calculations are quite time consuming.

(iii) This method is not based on the original values of the observations.

(iv) This method is not capable of further treatment.

15.6 CONCURRENT DEVIATION METHOD

With the help of this method, we can study the direction of relationship between the two variables, i.e. whether it is positive or negative. In this method correlation is computed between the direction of deviations and not their magnitudes. As such only the direction of deviations is taken into account in the calculation of this coefficient, and their magnitude is ignored. We put a either +ve sign or –ve sign or zero for the deviation according as the value of the variable is greater than, less than or equal to the preceding value respectively. The direction of change in X and Y variables is indicated D_x and D_y respectively. The deviations in values of two variables are said to be concurrent if they have the same sign i.e. either both directions are positive or both are negative or both are equal.

The formula for the calculation of coefficient of concurrent deviations is

$$r_{(c)} = \pm\sqrt{\pm\frac{2c-n}{n}}$$

$r_{(c)}$ = coefficient of concurrent deviation

c = number of concurrent deviations

n = number of pairs of deviation.

Here if $(2c - n)$ is positive then we take positive sign in-side and out-side the square root in the above formula. On the other hand, if $(2c - n)$ is negative then we take negative sign in-side and out-side the square root in the above formula.

SOLVED PROBLEMS

Problem 1: Calculate coefficient of correlation from the following data using Concurrent deviations method.

X:	50	52	48	56	68	70	75	71
Y:	40	45	44	50	62	68	70	69

Solution: Calculation of correlation coefficient by concurrent deviation method

X	Y	D_x	D_y	D_xD_y
50	40			
52	45	+	+	+
48	44	–	–	+
56	50	+	+	+
68	62	+	+	+
70	68	+	+	+
75	70	+	+	+
71	69	–	–	+
				$\Sigma D_xD_y = c = 7$

Here $n = 8 - 1 = 7, c = 7$

$$r_{(c)} = \pm\sqrt{\pm\frac{2c-n}{n}},$$

$$r_{(c)} = \pm\sqrt{\pm\frac{2(7)-7}{7}} = 1$$

Thus value of '$r_{(c)}$' tells that there is a perfect correlation between X and Y.

15.6.1 Merits of Concurrent Deviation Method

(i) It is the simplest of all the methods of computing correlation coefficient.

(ii) This method is useful when the number of items is very large and it gives quick idea about the degree of correlation.

(iii) This method is also good for studying short-term fluctuations.

(iv) It gives the correct indication about the direction of the correlation.

15.6.2 Demerits of Concurrent Deviation Method

(i) It does not give the correct degree of correlation as given by Pearson's and Spearman's method.

(ii) It is not suitable for grouped data.

(iii) It is not suitable for further algebraic treatment.

(iv) This method gives us only the rough idea of the presence or absence of correlation.

(v) It is not suitable for studying long-term changes.

(vi) It does not differentiate between big and small changes.

(vii) Some times exact correlation cannot be determined.

QUERIES

1. What do you understand by correlation? Explain in detail?
2. What are the various uses of correlation?
3. What are the various types of correlation?
4. How do you interpret a calculated value of Karl Pearson's Coefficient of correlation? Discuss in particular the values of $r = 0$, $r = -1$ and $r = +1$.
5. What is the difference between positive and negative correlation?
6. What is the role of simple correlation?
7. Differentiate between partial and multiple correlations.
8. Give the comparison of linear and non-linear correlation.
9. What is the degree of correlation?
10. What are the various methods used to compute the correlation?
11. What do you know about Scatter diagram?
12. What are the various pros and cons of Scatter diagram method?
13. Explain Karl Pearson's coefficient of correlation?
14. What you know about product moment method?
15. How can you use direct method to compute the correlation?
16. Explain shortcut method to compute the coefficient of correlation.
17. What do you know about correlation of frequency distribution?
18. What are the various properties of Karl Pearson's coefficient of correlation?
19. What are the various advantages and disadvantages of the Karl Pearson's correlation?
20. What are the various probable errors?
21. What you know about Spearman's Rank Correlation?
22. Explain the difference between Karl Pearson's coefficient and rank correlation coefficient.
23. What are the various types of problems related with the rank correlation?
24. What are the different features of Rank correlation?
25. What are the various merits and demerits of rank correlation?
26. What do you know about concurrent deviation method?
27. What are the different pros and cons of concurrent deviation method?
28. Write a brief note on correlation table.

UNSOLVED PROBLEMS

1. From the following pairs of series indicate the nature of correlation by means of a scatter diagram:

Height (inches) :	10	20	30	45	50	60	70	80
Weight (kg) :	30	10	20	35	40	28	38	45

2. From the following data, calculate the nature of correlation and indicate if there is any sort of correlation:

X :	0	1	2	3	4	5	6	7	8
Y :	1	8	28	56	70	56	28	8	1

3. From the following data, calculate the nature of correlation and indicate if there is any sort of correlation:

X :	1	2	3	4	5	6	7	8	0
Y :	2	8	18	16	24	46	8	44	10

4. Compute the nature of correlation by means of a scatter diagram:

X :	10	11	12	13	14	15	16	17	18	19	20	21
Y :	9	35	70	105	116	107	88	46	40	30	38	5

5. Draw a scatter diagram from the given data:

X :	3.5	4.5	5.5	6.5	7.5	8.5	9.5
Y :	3	7	22	60	85	32	8

6. From the following data find out Karl Pearson's coefficient of correlation:

X :	179	170	160	168	167	166	144
Y :	171	160	176	167	155	168	140

7. Indicate through graphic method if there is any correlation between the two variables *X* and *Y* for the following data:

Year :	1989	1990	1991	1992	1993	1994	1995	1996	1997
X :	100	102	105	106	101	110	225	112	115
Y :	90	92	94	95	92	100	101	110	112

8. From the following Data calculate the Karl Pearson's coefficient of correlation:

X :	1200	–1000	–800	–400	1200	1400	–600	–1000
Y :	–3500	3300	2400	1100	2000	1000	1700	1200

9. From the following data, find out the spearman's correlation coefficient and comment upon the results:

Selling cost :	179	170	160	168	167	166	144	165
Profit :	171	160	176	167	155	168	140	159

10. From the following data, examine whether there exists any correlation between *X* and *Y*:

X :	6.9	8.5	5.8	8.6	9.6	8	9.7
Y :	2.9	3.8	6.5	2.3	5.5	3.5	3.2

11. Find the conditional value of $Y = (X-7)^5$ corresponding to marginal values of $X = 1, 2, 3, 4$ and 5. Calculate the coefficient of correlation between the variables X and Y.
12. Coefficient of correlation between X and Y for 20 items is 0.3; mean of X is 15 and that of Y is 20, standard deviations are 4 and 5 respectively. At the time of calculations one item 27 was wrongly taken as 17 in case of X series and 35 instead of 30 in case of Y series. Find the correct coefficient of correlation.
13. In order to find the correlation coefficient between two variables X and Y from 12 pairs of observations, the following calculations were made:

$$\Sigma XY = 334,\ \Sigma X = 30,\ \Sigma X^2 = 670,\ Y = 5,\ Y = 285$$

On the subsequent verification, it was found that the pair ($X = 11, Y = 4$) was copied wrongly, the correct value being ($X = 10, Y = 14$). Find the correct value of correlation coefficient.
14. Calculate Karl Pearson's correlation coefficient between the age and playing habit of the people from the following data. Also comment on the result:

Age group	No. of people	No. of players
15 and less then 20	200	100
20 and less than 25	290	160
25 and less than 30	340	170
30 and less than 35	360	180

15. Calculate the correlation coefficient between the height of fathers and height of sons from the given data:

Height of fathers :	64	65	66	67	68	69	70
Height of sons :	65	67	65	68	70	68	72

Also find its Probable Error.
16. Test the significance of the coefficient of correlation for its following Values with reference to

 (i) $N = 10, r = 0.4$ and

 (ii) $N = 50, r = 0.9$
17. A student calculates the value of r as 0.7, when the number of items n in the sample is 25. Find the limits within which r lies for another sample from the same universe.
18. Calculate Karl Pearson's Coefficient of correlation for the following series:

Price (in Rs.) :	110-111	111-112	112-113	113-114	114-115	115-116
Demand(Kg.) :	600	640	640	680	700	780
Price (in Rs.) :	116-117	117-118	118-119			
Demand(Kg.) :	830	900	1000			

Also calculate the probable error of the correlation coefficient. From your result can you conclude that the demand is correlated with price?

19. From the following data find out rank correlation coefficient to determine which pair of judges has the nearest approach to common taste in beauty

1st judge :	1	6	5	10	3	4	9	7
2nd judge :	3	5	8	47	2	1	2	6
3rd judge :	6	4	3	8	4	9	7	10

20. From the following data relating to the marks secured by a batch of 10 students in maths and computer, find out the rank correlation coefficient

Roll no. :	1	2	3	4	5	6	7	8
Marks in Maths :	30	35	40	16	17	19	14	23
Marks in computer :	24	29	19	34	23	22	27	33

21. Three judges rank ten competitors in a beauty contest as follows:

Judges	Ranks of the competitors									
X:	6	5	3	10	3	4	9	7	8	9
Y:	5	8	9	3	4	10	1	6	8	7
Z:	4	9	8	1	2	3	10	5	7	6

Indicate which pair of judges has the nearest approach to beauty?

22. Calculate the coefficient of concurrent deviation from the following data:

X:	179	170	160	168	167	166	144	165
Y:	171	160	176	167	155	168	140	159

23. From the following data determine the coefficient of correlation by the method of least square:

Supply :	80	170	160	168	167	166	144	165
Demand :	146	160	176	167	155	160	140	159

24. Find the value of Karl Pearson coefficient of correlation from the following data:

Roll no. :	1	2	3	4	5	6	7	8
Marks in Physics :	33	55	56	34	22	68	40	59
Marks in economics :	65	55	34	70	80	50	90	55

25. From the following pairs of data, calculate the Karl Pearson's correlation coefficient:

X:	79	70	60	68	67	66	44	65
Y:	120	134	150	115	110	142	100	110

26. The coefficient of correlation between two variables X and Y is 0.38. Their covariance is 10.2. The variance of X is 16. Find the standard deviation of Y series.

27. Find the number of items where $r = 0.7$, $\Sigma xy = 60$, $\sigma_y = 2.4$, $\Sigma x^2 = 90$ given that x and y represent the respective deviations of X and Y series from the means.

28. Given $r = 0.5$, $\Sigma xy = 60$, $\sigma_y = 4$ and $\Sigma x^2 = 90$, find the number of pairs of the observations where x and y denote deviations from their respective means.

CHAPTER 16

Regression Analysis

16.1 REGRESSION ANALYSIS

Regression analysis is the statistical technique that identities the relationship between two or more quantitative variables: a dependent variable whose value is to be predicted, and an independent or explanatory variable (or variables), about which knowledge is available. The technique is used to find the equation that represents the relationship between the variables. A simple regression analysis can show that the relation between an independent variable X and a dependent variable Y is linear, using the simple linear regression equation $Y = a + bX$ (where a and b are constants). Multiple regression will provide an equation that predicts one variable from two or more independent variables, $Y= a + bX_1+ cX_{2+} dX_{3.}$

Regression analysis is used to understand the statistical dependence of one variable on other variables. The technique can show what proportion of variance between variables is due to the dependent variable, and what proportion is due to the independent variables.

Regression analysis refers to techniques for modeling and analyzing several variables, when the focus is on the relationship between a dependent variable and one or more independent variables.

Regression analysis is widely used for prediction (including forecasting of time-series data). Use of regression analysis for prediction has substantial overlap with the field of machine learning. Regression analysis is also used to understand which among the independent variables are related to the dependent variable, and to explore the forms of these relationships. In restricted circumstances, regression analysis can be used to infer causal relationships between the independent and dependent variables.

A regression line is a line drawn through a scatter plot of two variables. The line is chosen so that it comes as close to the points as possible. Regression analysis, on the other hand, is more than curve fitting. It involves fitting a model with both deterministic and stochastic components. The deterministic component is called the predictor and the stochastic component is called the error term.

The simplest form of a regression model contains a dependent variable, also called the "*Y*-variable" and a single independent variable, also called the "*X*-variable".

According to M.M. Blair: *"Regression is the measure of the average relationship between two or more variables in terms of original units of the data".*

According to Ya-Lun-Chou: *"Regression analysis attempts to establish the nature of relationship between variables-that is, to study the functional relationship between the variables and thereby provide a mechanism for prediction or forecasting."*

16.2 MAIN STEPS INVOLVED

The application of a regression analysis is rooted in an initial credible explanatory model. A sound model can only be applied where the effects of the intervention are well identified and where the production process of the effects is understood.

The main steps involved in performing a regression analysis are reviewed in Box 2.

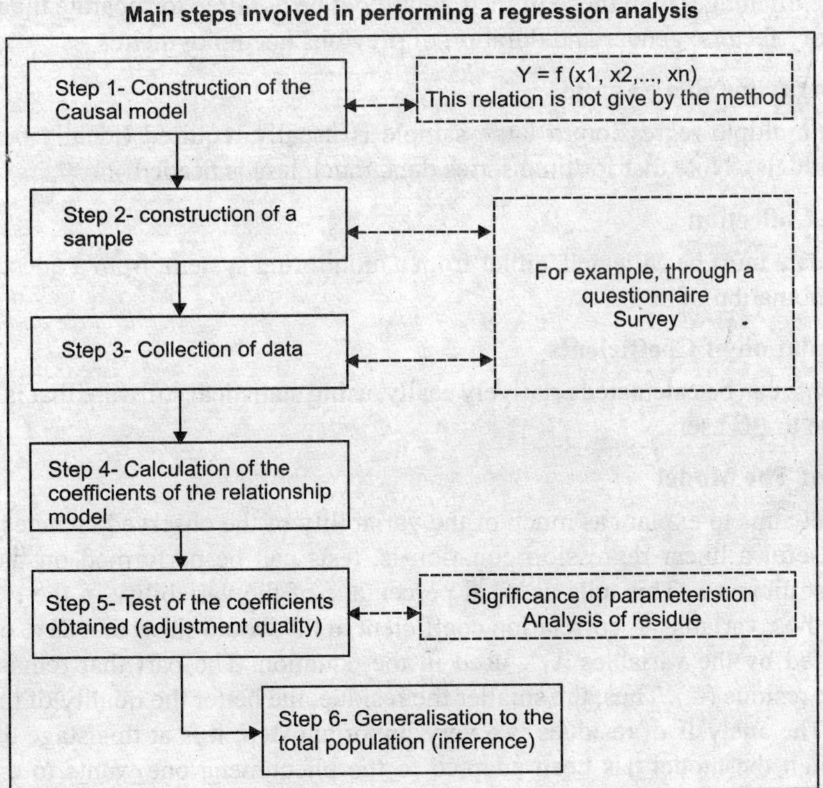

Step 1: Construction of the Causal Model

The construction of an explanatory model is a crucial step in the regression analysis. It must be defined with reference to the action theory of the intervention. It is likely that several kinds of variable exist. In some cases, they may be specially created, for example to take account of the fact that an individual has benefited from support or not (a dummy variable, taking values 0 or 1). A variable may also represent an observable characteristic (having a job or not) or an unobservable one (probability of having a job). The model may presume that a particular variable evolves in a linear, logarithmic, exponential or other way. All the explanatory models are constructed on the basis of a model, such as the following, for linear regression:

$Y = \beta_0 + \beta_1 X_1 + \beta_2 X_2 + ... + \beta_k X_k + \in$ where

Y is the change that the programme is mainly supposed to produce (e.g. employment of trainees)

$X_{1\text{-}k}$ are independent variables likely to explain the change.

$\beta_{0\text{-}k}$ are constants, and

$\in$ is the error term

Phenomena of co-linearity weaken the explanatory power. For example, when questioning women about unemployment, if they have experienced periods of previous unemployment which are systematically longer than those of men, it will not be possible to separate the influence of the two explanatory factors: gender and duration of previous unemployment.

Step 2: Construction of a Sample

To apply multiple regression, a large sample is usually required (ideally between 2,000 to 15,000 individuals). Note that for time series data, much less is needed.

Step 3: Data Collection

Reliable data must be collected, either from a monitoring system, from a questionnaire survey or from a combination of both.

Step 4: Calculation of Coefficients

Coefficients can be calculated relatively easily, using statistical software that is both affordable and accessible to PC users.

Step 5: Test of The Model

The model aims to explain as much of the variability of the observed changes as possible. To check how useful a linear regression equation is, tests can be performed on the square of the correlation coefficient r. This tells us what percentage of the variability in the y variable can be explained by the x variable. A correlation coefficient of 0.9 would show that 81% of the variability in Y is captured by the variables X_{1-k} used in the equation. The part that remains unexplained represents the residue ($\in$). Thus, the smaller the residue, the better the quality of the model and its adjustment. The analysis of residues is a very important step: it is at this stage that one sees the degree to which the model has been adapted to the phenomena one wants to explain. It is the residue analysis that also enables one to tell whether the tool has made it possible to estimate the effects in a plausible way or not. If significant anomalies are detected, the regression model should not be used to estimate effects, and the original causal model should be re-examined, to see if further predictive variables can be introduced.

16.3 STRENGTHS AND LIMITATIONS OF THE APPROACH

Strengths

1. Regression analysis provides an opportunity to specify hypotheses concerning the nature of effects (action theory), as well as explanatory factors.
2. When it is successfully executed (with a statistically valid adjustment), regression analysis can produce a quantitative estimate of net effects.

Limitations

1. The technique is demanding because it requires quantitative data relating to several thousand individuals.
2. Implementing the data collection can be time-consuming and expensive.
3. Regression analysis is likely to reach the conclusion that there is a strong link between two variables, whereas the influence of other, more important, variables may not have been estimated (this error is called "data snooping"). The tool should therefore be used with care.
4. Relations between the different explained and explanatory variables are often circular (*X* explains *Y* and *Y* explains *X*). In this case, the method is inapplicable.
5. The observations must present sufficiently contrasted evolutions to allow for adjustment. For example, if all the observations concern the 30-40 age group, it will not be possible to estimate the influence of age on employment.

16.4 TYPES OF REGRESSION ANALYSIS

The different types of regression analysis are given below:

(i) Simple And Multiple Regression Analysis
(ii) Linear And Non-linear Regression Analysis
(iii) Partial And Total Regression Analysis

(i) Simple and Multiple Regression Analysis: Regression analysis is divided into two types- Simple and Multiple based on the number of variables under study. In simple regression analysis, we study only two variables at a time, one is independent variable and the other is a dependent variable. Whereas in multiple regression analysis, we study more than two variables at a time, one variable is dependent and others are independent variables.

(ii) Linear and Non-linear Regression Analysis: In case of linear regression, the values of the dependent variable say *Y* increase by a constant absolute amount for a unit change in the value of the independent variable say X. In such cases, the simple linear relation is the equation of Y on X. In case of non-linear regression, the values of the dependent variable say Y do not increase by a constant absolute amount for a unit change in the values of the independent variable say X.

(iii) Partial and Total Regression Analysis: In case of partial regression, all variables are not considered but only few variables are under consideration. In total regression, we study the relationship among all the important variables.

16.5 LINEAR REGRESSION

Linear regression refers to any approach to modeling the relationship between one or more variables denoted y and one or more variables denoted X, such that the model depends linearly on the unknown parameters to be estimated from the data. Such a model is called a "linear model." Most commonly, linear regression refers to a model in which the conditional mean of y given the value of X is an affine function of X. Less commonly, linear regression could refer to a model in which the median, or some other quantile of the conditional distribution of y given X is expressed as a linear function of X. Like all forms of regression analysis, linear regression focuses on the conditional probability distribution of y given X, rather than on the joint probability distribution of y and X, which is the domain of multivariate analysis.

Linear regression was the first type of regression analysis to be studied rigorously, and to be used extensively in practical applications. The reason for this is that models that depend linearly on their unknown parameters are easier to fit than models that are related non-linearly to their parameters and the statistical properties of the resulting estimators are easier to determine.

Linear regression has many practical uses. Most applications of linear regression fall into one of the following two broad categories:

- If the goal is prediction, or forecasting, linear regression can be used to fit a predictive model to an observed data set of y and X values. After developing such a model, if an additional value of X is then given without its accompanying value of y, the fitted model can be used to make a prediction of the value of y.
- If we have a variable y and a number of variables $X1$, ..., Xp that may be related to y, we can use linear regression analysis to quantify the strength of the relationship between y and the Xj, to assess which Xj may have no relationship with y at all, and to identify which subsets of the Xj contain redundant information about y, so that once one of them is known, the others are no longer informative.

Linear regression models are often fit using the least squares approach, but may also be fit in other ways, such as by minimizing the "lack of fit" in some other norm, or by minimizing a penalized version of the least squares loss function as in ridge regression. Conversely, the least squares approach can be used to fit models that are not linear models. Thus, while the terms "least squares" and linear model are closely linked, they are not synonymous.

16.6 LINES OF REGRESSION

There are two methods of drawing regression lines:

(i) Scatter Diagram Method or Free Hand Curve Method

(ii) Least Square Method or Algebraic Method

(i) Scatter Diagram Method or Free Hand Curve Method: This method is very simple to use. Under this method, the points are plotted on a graph paper representing various pairs of values of the concerned variables. A straight line is drawn in such a way that it passes through the centre of scatter points either by free hand or by scale rule.

The shape of regression line depends upon the scattering of points. The estimate based on free hand curve method is not accurate.

(ii) Least Square Method or Algebraic Method: Let $P_i(x_i, y_i)$; $i = 1, 2, 3, \ldots, n$ be n points in the scatter diagram. Let $y = f(x)$ be a relation between x and y.

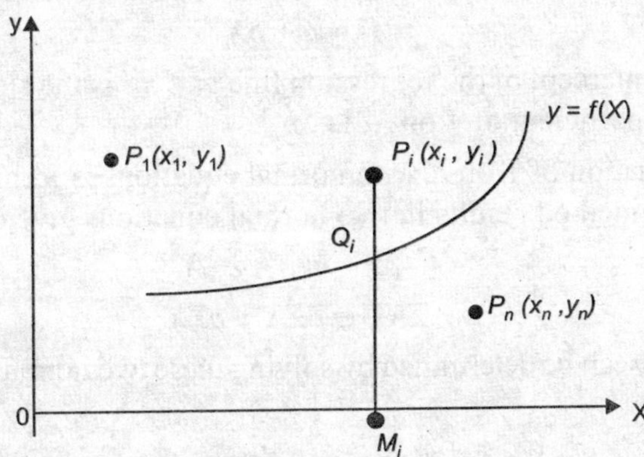

Let the ordinate at P_i meet $y = f(x)$ at Q_i and the x-axis at M_i.

Therefore

$$M_iQ_i = f(x_i) \text{ and } M_iP_i = y_i$$

and
$$Q_iP_i = M_iP_i - M_iQ_i = y_i - f(x_i)$$

Let $d_i = y_i - f(x_i)$; $i = 1, 2, 3\ldots, n$ and it is called the residual (or error) at $x = x_i$.

Now $S = \sum_{i=1}^{n} d_i^2 = \sum_{i=1}^{n} [y_i - f(x_i)]^2$ is the sum of the squares of the residuals(or errors).

Here S will be either minimum or maximum for some value of 'a' such that the partial derivative $\partial S/\partial a = 0$, where '$a$' is one of the unknown parameters in the function $y = f(x)$.

If $S = 0$, i.e. each $d_i = 0$ then all the n points P_i's will lie on the curve $y = f(x)$.

If S is not equal to 0 then we will choose $f(x)$ such that S is minimum, i.e the best-fit curve to the set of points P_i's. This principle of selecting $f(x)$ is known as the principle of least squares.

16.7 DERIVATION OF LINES OF REGRESSION

Regression equations are the algebraic expressions of regression lines. These are used for determining the value of unknown variable for a given independent variable. Assume that X is an independent variable and Y is a dependent variable. A linear relationship exists between the two variables which is of the form $Y = a + bX$. It represents a straight line called Regression line of Y on X. Similarly if Y is taken as a dependent variable and X is taken as in independent variable the regression equation representing linear relationship is given by the equation $X = a + bY$. It represents a straight line called regression line of X on Y. We can obtain any of these equations or lines by following either of the following methods:

1. Regression Equations Through Normal Equations
2. Regression Equations Through Regression Coefficients

1. **Regression Equations Through Normal Equations:** Consider a Regression equation of Y on X(X is an independent variable and Y is a dependent variable)

$$\boldsymbol{Y = a + bX}$$

where 'a' is the Y intercept of the regression line and 'b' is the slope of regression line. It is called regression coefficient of Y on X, i.e. b_{yx}.

The regression equation of Y on X with normal equations is formed through the method of least squares. This method results in two normal equations given below:

$$\Sigma Y = n\,a + b\Sigma X$$

$$\Sigma XY = a\Sigma X + b\Sigma X^2$$

The value of a and b can be determined by solving these two normal equations simultaneously which are given by:

$$a = \frac{(\Sigma Y)(\Sigma X^2) - (\Sigma X)(\Sigma XY)}{n\Sigma X^2 - (\Sigma X)^2}$$

$$b = \frac{(\Sigma XY) - (\Sigma X)(\Sigma Y)}{n\Sigma X^2 - (\Sigma X)^2}$$

Similarly, for regression equation of X on Y will be of the form

$$X = a + bY$$

where 'b' is called the regression coefficient of X on Y, i.e. b_{xy}.

We have the following normal equations for regression line of X on Y:

$$\Sigma X = n\,a + b\Sigma Y$$

$$\Sigma XY = a\Sigma Y + b\Sigma Y^2$$

To determine the value of 'a' and 'b', the two normal equations are to be solved simultaneously. On solving these equations simultaneously we get,

$$a = \frac{(\Sigma X)(\Sigma Y^2) - (\Sigma Y)(\Sigma XY)}{n\Sigma Y^2 - (\Sigma Y)^2}$$

$$b = \frac{(\Sigma XY) - (\Sigma Y)(\Sigma X)}{n\Sigma Y^2 - (\Sigma Y)^2}$$

Once value of a and b is known, we can substitute these values to get Regression equation of Y on X or Regression equation of X on Y.

Both these regression lines will intersect each other at the point ($\Sigma X/n$, $\Sigma Y/n$), i.e $(\bar{X}, \bar{Y})$.

SOLVED PROBLEMS

Problem 1: Given the data below:

X:	0.2	0.4	0.6	0.8	1.0
Y:	7.2	7.6	8.0	8.4	8.8

Prove that regression lines (y on x and x on y) intersect at the point $(\bar{X}, \bar{Y})$.

Solution:

i	X_i	Y_i	X_i^2	Y_i^2	$X_i Y_i$
1	0.2	7.2	0.04	51.84	1.44
2	0.4	7.6	0.16	57.76	3.04
3	0.6	8.0	0.36	64.00	4.8
4	0.8	8.4	0.64	70.56	6.72
5	1.0	8.8	1.0	77.44	8.8
$n = 5$	$\Sigma X = 3$	$\Sigma Y = 40$	$\Sigma X^2 = 2.2$	$\Sigma Y^2 = 321.6$	$\Sigma XY = 24.8$

As we know regression line of Y on X is

$$Y = a + bX$$

where

$$a = \frac{(\Sigma Y)(\Sigma Y^2) - (\Sigma X)(\Sigma XY)}{n\Sigma X^2 - (\Sigma X)^2}$$

$$b = \frac{(\Sigma XY) - (\Sigma X)(\Sigma Y)}{n\Sigma X^2 - (\Sigma X)^2}$$

After substitution a and b becomes

$$a = 6.8, \quad b = 2$$

$\therefore$ Regression line of Y on X becomes

$$Y = 2X + 6.8$$

Similarly regression line of X on Y will be

$$X = a + bY$$

Where

$$a = \frac{(\Sigma X)(\Sigma Y^2) - (\Sigma Y)(\Sigma XY)}{n\Sigma Y^2 - (\Sigma Y)^2}$$

$$b = \frac{(\Sigma XY) - (\Sigma Y)(\Sigma X)}{n\Sigma Y^2 - (\Sigma Y)^2}$$

After substitution, the value a and b is:

$$a = -3.4, \quad b = 0.5$$

∴ Regression line of X on Y becomes

$$X = 0.5\text{Y} - 3.4$$

If the two regression lines are satisfied by the point $(\bar{X}, \bar{Y})$, then we can say that the two lines intersect at $(\bar{X}, \bar{Y})$.

Now $$\bar{X} = \frac{\Sigma X}{n} = \frac{3}{5} = 0.6, \bar{Y} = \frac{\Sigma Y}{n} = \frac{40}{5} = 8$$

By substituting $X = 0.6$ in equation $Y = 2X + 6.8$, we get

$$Y = 2 \times 0.6 + 6.8 = 8, \text{ which matches } \bar{Y} = 8$$

Similarly for equation $X = 0.5\text{Y} - 3.4$, substitute $Y = 8$, we get

$$X = 0.5 \times 8 - 3.4 \ = 0.6, \text{ which matches } \bar{X} = 0.6$$

Hence we conclude that both the regression lines pass through the common point $(\bar{X}, \bar{Y})$.

Program 1: Write a Program to fit a linear regression line of *y* on *x* of the form *y* = *a*+*bx*.

```
#include<stdio.h>
#include<conio.h>
void main( )
{
int i,n;
float x[100],y[100],sx=0,sy=0,sx2=0,sxy=0,a,b,denom;
clrscr( );
printf("\n\t\t Program for regression line of Y on X\n\n");
printf(" How many items of data?");
scanf("%d",&n);
printf(" Enter the data as Xi Yi\n");
for(i=1;i<=n;i++)
```

```
{
scanf("%f %f",&x[i],&y[i]);
sx+=x[i];
sy+=y[i];
sx2+=x[i]*x[i];
sxy+=x[i]*y[i];
}
printf("\n Number of observations =%d",n);
printf("\n sigma x   =%.4f",sx);
printf("\n sigma y   =%.4f",sy);
printf("\n sigma x square =%.4f",sx2);
printf("\n sigma xy       =%.4f",sxy);
denom=(n*sx2-sx*sx);
if (denom= =0)
{
printf("\n Error: Invalid code, please check");
getche( );
exit(1);
}
a=(sy*sx2-sx*sxy)/denom;
b=(n*sxy-sx*sy)/denom;
printf("\n Value of a =%.4f",a);
printf("\n Value of b =%.4f",b);
printf("Regression equation of y on x is—> y = %.4f + %.4f x",a,b);
getche( );
}
```

Output:

Program for regression line of Y on X

How many items of data? 5

Enter the data as Xi Yi

0.2	7.2
0.4	7.6
0.6	8.0
0.8	8.4

16

```
1.0          8.8
Number of observations = 5
sigma x   = 3.0000
sigma y   = 40.0000
sigma x square = 2.2000
sigma xy      = 24.8000
Value of a = 6.8000
Value of b = 2.0000
Regression equation of y on x is—> y =  6.8000 + 2.0000 x
```

Program 2: Write a Program to fit a linear regression line of x on y of the form $x = a + by$.

```
#include<stdio.h>
#include<conio.h>
void main()
{
int i,n;
float x[100],y[100],sx=0,sy=0,sy2=0,sxy=0,a,b,denom;
clrscr();
printf("\n\t\t Program for regression line  Y on X\n\n");
printf(" How many items of data?");
scanf("%d",&n);
printf(" Enter the data as Xi Yi\n");
for(i=1;i<=n;i++)
{
scanf("%f %f",&x[i],&y[i]);
sx+=x[i];
sy+=y[i];
sy2+=y[i]*y[i];
sxy+=x[i]*y[i];
}
printf("\n Number of observations =%d",n);
printf("\n sigma x   =%.4f",sx);
printf("\n sigma y   =%.4f",sy);
printf("\n sigma y square =%.4f",sy2);
printf("\n sigma xy      =%.4f",sxy);
```

```
denom=(n*sy2-sy*sy);
if (denom = = 0)
{
printf("\n Error: Invalid code, please check");
getche( );
}
```

Output:

```
Program for regression line of Y on X
How many items of data? 5
Enter the data as Xi Yi
0.2            7.2
0.4            7.6
0.6            8.0
0.8            8.4
1.0            8.8
Number of observations = 5
sigma x   = 3.0000
sigma y   = 40.0000
sigma y square = 321.6000
sigma xy       = 24.8000
Value of a = -3.4000
Value of b = 0.5000
Regression equation of x on y is—> x = - 3.4000 + 0.5000 y
```

2. **Regression Equations Through Regression Coefficients:** Following are the methods to find regression equations through regression coefficients:

(i) Using Deviations from Actual Mean

(ii) Using Deviations from Assumed Mean

(iii) Using the Actual Observations

(i) Regression Equations When Deviations are Taken from the Actual Mean: Here, regression equation of Y on X of the form $Y = a + bX$, can be written as

$$Y - \bar{Y} = b_{yx}(X - \bar{X})$$

where b_{yx} is the regression coefficient of Y on X given by

$$b_{yx} = r\frac{\sigma_y}{\sigma_x} \quad \text{and} \quad r = \frac{\Sigma xy}{N\sigma_x\sigma_y}$$

with $$x = X - \bar{X},$$
$$y = Y - \bar{Y}$$

After simplification, we have

$$b_{yx} = \frac{\Sigma xy}{N\sigma_x\sigma_y} \times \frac{\sigma_y}{\sigma_x} = \frac{\Sigma xy}{N\sigma_x^2}$$

but $$\sigma_x^2 = \frac{\Sigma x^2}{N}$$

$\therefore$ $$b_{yx} = \frac{\Sigma xy}{\Sigma x^2}$$

Now, the regression equation of Y on X becomes

$$Y - \bar{Y} = \frac{\Sigma xy}{\Sigma x^2}(X - \bar{X})$$

Similarly, Regression equation of X on Y of the form $X = a + bY$, can be written as

$$X - \bar{X} = \frac{\Sigma xy}{\Sigma y^2}(Y - \bar{Y})$$

SOLVED PROBLEMTES

Problem 1: From the following data, obtain regression equations using regression coefficients.

X:	6	2	10	4	8
Y:	9	11	5	8	7

Solution:

X	*Y*	$(X - \bar{X})$	$(Y - \bar{Y})$	x^2	y^2	xy
6	9	0	1	0	1	0
2	11	–4	3	16	9	–12
10	5	4	-3	16	9	–12
4	8	–2	0	4	0	0
8	7	2	–1	4	1	–2
Σ*X* = 30	**Σ*Y* = 40**			**Σx^2 = 40**	**Σy^2 = 20**	**Σ*xy* = –26**

$$\bar{X} = \frac{\Sigma X}{N} = \frac{30}{5} = 6$$

$$\bar{Y} = \frac{\Sigma Y}{N} = \frac{40}{5} = 8$$

$$\therefore \quad b_{xy} = \frac{\Sigma xy}{\Sigma y^2}$$

$$= (-26)/20 = -1.3$$

$$\therefore \quad b_{yx} = \frac{\Sigma xy}{\Sigma x^2}$$

$$= (-26)/40 = -0.65$$

Regression equation of X on Y is

$$X - \bar{X} = b_{xy}(Y - \bar{Y})$$

$$(X - 6) = -1.3(Y - 8)$$

$$X = 16.4 - 1.3Y$$

Regression equation of Y on X is

$$Y - \bar{Y} = b_{yx}(X - \bar{X})$$

$$(Y - 8) = -0.65(X - 6)$$

$$Y = 11.9 - 0.65X$$

(ii) Regression Equations When Deviations are Taken from Assumed Mean: When deviations are taken from assumed mean, then

$$\bar{X} = A + \frac{\Sigma dx}{N},$$

$$\bar{Y} = B + \frac{\Sigma dy}{N}$$

where A is the assumed mean for X series and B is the assumed mean for Y series. The two regression coefficients are calculated by

$$b_{yx} = r\frac{\sigma_y}{\sigma_x} = \frac{N\Sigma dxdy - (\Sigma dx)(\Sigma dy)}{N\Sigma dx^2 - (\Sigma dx)^2}$$

$$b_{xy} = r\frac{\sigma_y}{\sigma_x} = \frac{N\Sigma dxdy - (\Sigma dx)(\Sigma dy)}{N\Sigma dy^2 - (\Sigma dy)^2}$$

where $dx = X - A$, $dy = Y - B$

SOLVED PROBLEMS

Problem 1: Obtain the equations of two lines of regression for the following data:

X:	43	44	46	40	44	42	45	42	38	40	52	57
Y:	29	31	19	18	19	27	27	29	41	30	26	10

Solution: Calculation of Regression equations

X	$dx = X–42$	dx^2	Y	$dy = Y–27$	dy^2	$dxdy$
43	1	1	29	2	4	2
44	2	4	31	4	16	8
46	4	16	19	–8	64	–32
40	–2	4	18	–9	81	18
44	2	4	19	–8	64	–16
42	0	0	27	0	0	0
45	3	9	27	0	0	0
42	0	0	29	2	4	0
38	–4	16	41	14	196	–56
40	–2	4	30	3	9	–6
52	10	100	26	–1	1	–10
57	15	225	10	–17	289	–255
$\Sigma X = 533$	$\Sigma dx = 29$	$\Sigma dx^2 = 383$	$\Sigma Y = 306$	$\Sigma dy = -18$	$\Sigma dy^2 = 728$	$\Sigma dxdy = -347$

After substitution of values, from table in the formula

$$b_{yx} = r\frac{\sigma_y}{\sigma_x} = \frac{N\Sigma dxdy - (\Sigma dx)(\Sigma dy)}{N\Sigma dx^2 - (\Sigma dx)^2}$$

We b_{yx} get $= -0.97$

After substitution of values, from table in the formula

$$b_{xy} = r\frac{\sigma_y}{\sigma_x} = \frac{N\Sigma dxdy - (\Sigma dx)(\Sigma dy)}{N\Sigma dy^2 - (\Sigma dy)^2}$$

we b_{xy} get $= -0.43$

$$\bar{X} = \frac{\Sigma X}{N} = \frac{533}{12} = 44.42$$

$$\bar{Y} = \frac{\Sigma Y}{N} = \frac{306}{12} = 25.5$$

Now regression equation of X on Y is

$$X - \bar{X} = b_{xy}(Y - \bar{Y})$$

i.e.

$$X - 44.42 = -0.43(Y - 35.5)$$

$$X = -0.43\text{Y} + 55.39$$

Regression equation of Y on X is

$$Y - \bar{Y} = b_{yx}(X - \bar{X})$$

i.e.

$$Y - 25.5 = -0.97(X - 44.42)$$

$$Y = -0.97X + 68.59$$

Problem 2: You are given the following data:

	x	y
Arithmetic Means	36	85
Standard deviations	11	8

Correlation coefficient between x and $y = 0.66$

(i) Find two regression equations.

(ii) Estimate value of x when $y = 75$.

Solution: We are given

$\bar{X} = 36,\ \bar{Y} = 85,\ \sigma_x = 11,\ \sigma_y = 8,\ r = 0.66$

Regression equation of Y on X is given by

$$Y - \bar{Y} = r\frac{\sigma_y}{\sigma_x}(X - \bar{X})$$

$$Y - 85 = \frac{0.66 \times 8}{11}(X - 36)$$

$$Y = 0.48X + 67.72$$

Regression equation of X on Y is given by

$$X - \bar{X} = r\frac{\sigma_x}{\sigma_y}(Y - \bar{Y})$$

$$Y - 36 = \frac{0.66 \times 11}{8}(Y - 85)$$

$$X = 0.9075Y - 41.1375$$

To estimate the value of x when $y = 75$, we use the regression equation of x on y, i.e.

$$X = 0.9075Y - 41.1375$$

$$X(\text{at } Y = 75) = 0.9075 \times 75 - 41.1375 = 26.925$$

(iii) Regression Equations by Using Actual Observations: In this case, Regression equation of Y on X is taken as

$$Y - \bar{Y} = b_{yx}(X - \bar{X})$$

where
$$b_{yx} = \frac{\Sigma XY - N\overline{XY}}{\Sigma X^2 - N(\bar{X})^2}$$

Similarly regression equation of X on Y is taken as

$$X - \bar{X} = b_{xy}(Y - \bar{Y})$$

where $$b_{xy} = \frac{\Sigma XY - N\bar{X}\bar{Y}}{\Sigma Y^2 - N(\bar{Y})^2}$$

$$\bar{X} = \frac{\Sigma X}{N}, \quad \bar{Y} = \frac{\Sigma Y}{N}$$

SOLVED PROBLEMS

Problem 1: Find the two regression equations for the following data:

X :	1	3	4	8	9	11	14
Y :	1	2	4	5	7	8	9

Solution:

X	X^2	Y	Y^2	XY
1	1	1	1	1
3	9	2	4	6
4	16	4	16	16
8	64	5	25	40
9	81	7	49	63
11	121	8	64	88
14	196	9	81	126
$\Sigma X = 50$	$\Sigma X^2 = 488$	$\Sigma Y = 36$	$\Sigma Y^2 = 240$	$\Sigma XY = 340$

Here $N = 7$, $$\bar{X} = \frac{50}{7} = 7.14,$$

$$\bar{Y} = \frac{36}{7} = 5.14$$

Substituting the values in $$b_{xy} = \frac{\Sigma XY - N\bar{X}\bar{Y}}{\Sigma Y^2 - N(\bar{Y})^2}$$

We get $$b_{yx} = \frac{340 - 7(7.14)(5.14)}{488 - 7(7.142)^2} = 0.633$$

Similarly substituting the values in $$b_{xy} = \frac{\Sigma XY - N\bar{X}\bar{Y}}{\Sigma Y^2 - N(\bar{Y})^2}$$

We get $$b_{xy} = \frac{340 - 7(7.14)(5.14)}{240 - 7(5.142)^2} = 1.509$$

Substituting the values in Regression Equation of Y on X

i.e. $$Y - \bar{Y} = b_{yx}(X - \bar{X})$$

We get $Y - 5.14 = 0.633\,(X - 7.142)$

$\Rightarrow$ $Y = 0.633\,X + 0.621$

Substituting the values in Regression Equation of X on Y

i.e. $X - \bar{X} = b_{xy}(Y - \bar{Y})$

We get $Y - 7.14 = 1.509\,(Y - 5.14)$

$\Rightarrow$ $Y = 1.509\,Y + 0.616$

Hence the two regression equations are:

$$Y = 0.633X + 0.621$$
$$X = 1.509Y - 0.616$$

16.8 PROPERTIES OF REGRESSION COEFFICIENTS

Following are the properties of regression coefficients:

(i) The geometric mean of two regression coefficients is equal to the coefficient of correlation i.e. $r = \pm\sqrt{b_{yx}\,b_{xy}}$.

(ii) The arithmetic mean of the two regression coefficients is equal to or greater than the coefficient of correlation.

(iii) Product of both the regression coefficients is always less than unity.

(iv) If the two standard deviations are equal, then the coefficient of correlation will be equal to the regression coefficients.

(v) If $r = 0$, then value of both the regression coefficients will also be zero.

(vi) Both the regression coefficients have the same algebraic signs, i.e + or -.

(vii) The algebraic sign of regression coefficients and coefficient of correlation is always the same.

(ix) Regression coefficients are mathematical measures and are capable of further algebraic treatment.

(x) Regression coefficients are independent of change of origin but not of scale.

SOLVED PROBLEMS

Problem 1: The lines of regression of a bivariate population are:

$$8x - 10y + 66 = 0$$
$$40x - 18y = 214$$

The variance of x is 9. Find

(i) The mean values of x and y.

(ii) Correlation coefficient between *x* and *y*.

(iii) Standard deviation of y.

Solution:

(i) We are given the two regression equations:

$$8x - 10y + 66 = 0 \quad ...(1)$$

$$40x - 18y = 214 \quad ...(2)$$

Since both the regression lines intersect at the point $(\bar{X}, \bar{Y})$, hence the point of intersection of these two lines will give us the values of $\bar{X}, \bar{Y}$.

On solving (1) and (2) simultaneously we get:

$$\frac{x}{(-10)(-214)+(18)(66)} = \frac{y}{(66)(40)+(8)(214)} = \frac{1}{(8)(-18)+(10)(40)}$$

$$\frac{x}{2140+1188} = \frac{y}{2640+1712} = \frac{1}{-144+400}$$

$$\therefore \quad x = 13,\ y = 17$$

Hence the mean values are given by: $\bar{X} = 13, \bar{Y} = 17$

(ii) Assume that (1) is the equation of line of regression of *y* on *x* and (2) is the equation of line of regression of *x* on y.

Rewriting (1), we get:

$$10y = 8x + 66 \Rightarrow y = \frac{8}{10}x + \frac{66}{10}$$

$$\therefore \quad b_{yx} = \text{Coefficient of regression of } y \text{ on } x = \frac{8}{10} = \frac{4}{5}$$

Similarly, rewriting (2),

we get:

$$40x = 18y + 214 \Rightarrow x = \frac{18}{40}y + \frac{214}{40}$$

$$\therefore \quad b_{xy} = \text{Coefficient of regression of } x \text{ on } y = \frac{18}{40} = \frac{9}{20}$$

Since the two regression coefficients satisfy all the properties so our assumption is true.

Hence

$$r^2 = b_{yx} \cdot b_{xy} = \frac{4}{5} \times \frac{9}{20} = \frac{9}{25}$$

$$\therefore \quad r = \pm\sqrt{\frac{9}{25}} = \pm\frac{3}{5} = \pm 0.6$$

Since both the regression coefficients are positive, *r* must be positive. Hence we take $r = 0.6$

Note: *If we had assumed that (1) is the equation of the line of regression of X on Y and (2) is the equation of the line of regression of Y on X, then rewriting (1) and (2) we get respectively:*

$$8x = 10y - 66 \qquad\qquad 18y = 40x \times 214$$

$$x = \frac{10}{8}y - \frac{66}{8} \qquad \text{And} \qquad y = \frac{40}{18}x - \frac{214}{18}$$

$$b_{xy} = \frac{10}{8} \qquad\qquad b_{yx} = \frac{40}{18}$$

Since r^2 always lies between 0 and 1 i.e.since r^2 cannot exceed 1, OUR assumption that (1) is the line of regression of x on y and (2) is the line of regression of y on x is wrong. So we have to take (1) as the line of regression of y on x.

(iii) We are given that $\sigma_x^2 = 9 \Rightarrow \sigma_x = 3$

Now $$b_{yx} = r\frac{\sigma_y}{\sigma_x} \Rightarrow \frac{4}{5} = \frac{3}{5}\cdot\frac{\sigma_y}{3} \Rightarrow \sigma_y = 4$$

16.9 REGRESSION EQUATIONS FOR A BIVARIATE FREQUENCY

In a correlation table, the equations of the two lines of regression, i.e. line of regression of y on x and x on y are respectively:

$$Y - \bar{Y} = b_{yx}(X - \bar{X}) = r \cdot \frac{\sigma_y}{\sigma_x}(X - \bar{X})$$

$$X - \bar{X} = b_{xy}(Y - \bar{Y}) = r \cdot \frac{\sigma_y}{\sigma_x}(Y - \bar{Y})$$

$$b_{yx} = \frac{k}{h}\left[\frac{N\Sigma fdxdy - (\Sigma fdx)(\Sigma fdy)}{N\Sigma fdx^2 - (\Sigma fdx)^2}\right]$$

where

$$b_{xy} = \frac{h}{k}\left[\frac{N\Sigma fdxdy - (\Sigma fdx)(\Sigma fdy)}{N\Sigma fdy^2 - (\Sigma fdy)^2}\right]$$

$$dx = \frac{X - A}{h} \quad \text{and} \quad dy = \frac{Y - B}{k}$$

Here, A and B are the assumed means for X and Y series respectively; h and k are any constants.

SOLVED PROBLEMS

Problem 1: Following table gives the marks in Computer and marks in Maths of 20 students. Find the two regression lines. Also estimate the marks in Computer when the marks in Maths = 75.

Marks in Computer → Marks in Maths ↓	0-20	20-40	40-60	60-80	80-100	Total
0-20	2	2	0	0	2	6
20-40	3	2	1	0	1	7
40-60	0	0	0	0	0	0
60-80	0	1	1	1	1	4
80-100	2	1	0	0	0	3
Total	**7**	**6**	**2**	**1**	**4**	**20**

Solution:

Marks in Computer →			0-20	20-40	40-60	60-80	80-100				
			10	30	50	70	90				
Marks in *X* Maths *m* ↓ *dx* *Y* *m* *dy*			-2	–1	0	1	2	f	*fdy*	*fdy*^{2}	*fdxdy*
0–20	10	–2	**8** 2	**4** 2	**0** 0	**0** 0	– **8** 2	6	–12	24	4
20-40	30	–1	**6** 3	**2** 2	**0** 1	**0** 0	– **2** 1	7	–7	7	6
40-60	50	0	**0** 0	**0** 0	**0** 0	**0** 0	**0** 0	0	0	0	0
60-80	70	1	**0** 0	– **1** 1	**0** 1	**1** 1	**2** 1	4	4	4	2
80-100	90	2	– **8** 2	– **2** 1	**0** 0	**0** 0	**0** 0	3	6	12	–10
		f	7	6	2	1	4	**Σ*f* = 20**	**Σ*fdy* = –9**	**Σ*fdy*^{2} = 47**	**Σ*fdxdy* = 2**
		fdx	–14	–6	0	1	8	**Σ*fdx* = –11**			
		fdx^2	28	6	0	1	16	**Σ*fdx*^{2} = 51**			
		fdxdy	6	3	0	1	–8	**Σ*fdxdy* = 2**			

Here $A = 50, B = 50, h = 20, k = 20, N = 20$

$$\bar{X} = A + \frac{\Sigma fdx}{N} \times h$$

$$= 50 + \frac{-11}{20} \times 20 = 39$$

$$\bar{Y} = B + \frac{\Sigma fdy}{N} \times k$$

$$= 50 + \frac{-9}{20} \times 20 = 41$$

$$b_{yx} = \frac{k}{h}\left[\frac{N\Sigma fdxdy - (\Sigma fdx)(\Sigma fdy)}{N\Sigma fdx^2 - (\Sigma fdx)^2}\right]$$

$$= \frac{20}{20}\left[\frac{20 \times 2 - (-11)(-9)}{20 \times 51 - (-11)^2}\right] = -0.065$$

$$b_{xy} = \frac{k}{h}\left[\frac{N\Sigma fdxdy - (\Sigma fdx)(\Sigma fdy)}{N\Sigma fdy^2 - (\Sigma fdy)^2}\right]$$

$$= \frac{20}{20}\left[\frac{20 \times 2 - (-11)(-9)}{20 \times 47 - (-9)^2}\right] = -0.068$$

Line of regression of Y on X:

$$Y - \bar{Y} = b_{yx}(X - \bar{X})$$

$$Y - 41 = -0.065\,(X - 39)$$

$$\Rightarrow \quad y = -0.065X + 43.535$$

Line of regression of Y on X:

$$X - \bar{X} = b_{xy}(Y - \bar{Y})$$

$$X - 39 = -0.068\,(Y - 41)$$

$$\Rightarrow \quad X = -0.068Y + 41.788$$

Estimation of the marks in Computer when the marks in Maths = 75 is obtained by substituting $Y = 75$ in regression equation of X on Y.

We get, $X = 36.688$ when $Y = 75$.

16.10 DIFFERNECE BETWEEN CORRELATION AND REGRESSION

Following are the differences between correlation and regression:

Correlation	Regressoion
1. Correlation means the relationship between two or more variables under study.	1. It is a mathematical measure expressing the average relationship between two or more variables.
2. There is a functional relationship between two variables under study.	2. There is a mutual dependence of variables under study.
3. Regression is not nonsense.	3. There may be nonsense correlation between two variables.
4. Correlation has limited applicability because of only linear relationship.	4. Regression has wider applicability because it considers linear as well as non-linear relationship.
5. Regression coefficients are absolute measure representing the change in the value of the variable $Y(X)$ for a unit change in $X(Y)$.	5. Correlation coefficient is a relative measure of linear relationship between X and Y and is independent of the measurement.
6. It also reflects upon the nature of the variable, i.e. which is dependent variable and which is independent variable.	6. In correlation, it is immaterial which of x and y is dependent variable and which is independent variable.

16.11 NONLINEAR REGRESSION

Nonlinear regression is a form of regression analysis in which observational data are modeled by a function which is a nonlinear combination of the model parameters and depends on one or more independent variables. the data are fitted by a method of successive approximations.

Nonlinear regression is a regression in which the dependent or criterion variables are modeled as a non-linear function of model parameters and one or more independent variables. There are several common models for nonlinear regression such as Asymptotic Regression/Growth Model, which is given by $b1 + b2 * \exp(b3 * x)$, Logistic Population Growth Model, which is given by $b1/(1 + \exp(b2 + b3 * x))$, Asymptotic Regression/Decay Model, which is given by $b1 - (b2 * (b3 ** x))$ etc. The reason that these models are called nonlinear regression is because the relationships between the dependent and independent parameters are nonlinear.

16.11.1 Non-linear Least Squares

Non-linear least squares is the form of least squares analysis which is used to fit a set of m observations with a model that is non-linear in n unknown parameters ($m > n$). It is used in some forms of non-linear regression. The basis of the method is to approximate the model by a linear one and to refine the parameters by successive iterations. There are many similarities to linear least squares, but also some significant differences.

16.11.2 Polynomial Fit: Non-linear Regression

We use polynomial models to estimate and predict the shape of response values over a range of input parameter values. Polynomial models are a great tool for determining which input factors drive responses and in what direction. These are also the most common models used for analysis of designed experiments.

Let (x_i, y_i) be a point where $i = 1, 2, 3........, n$ be the n sets of observations

Suppose we want to fit a polynomial of degree m as

$$y = f(x) = a_0 + a_1x + a_2x^2 + ...+ a_mx^m \qquad ...(1)$$

Where $a_0, a_1, a_2, ..., a_m$ are the constants that are to be computed for giving the best polynomial fit.

The sum of squares of the residuals is given by

$$E = \sum_{i=1}^{n} d_i^2$$

$$= \sum_{i=1}^{n} [(y_i - f(x_i)]^2$$

$$= \sum_{i=1}^{n} [y_i - (a_0 + a_1x_i + a_2x_i^2 + \cdots + a_mx_i^m)]^2$$

By principles of least squares, E is minimum.

$\therefore \quad \partial E/\partial a_0 = 0, \partial E/\partial a_1 = 0, ..., \partial E/\partial a_m = 0$

$$\partial E/\partial a_0 = \sum_{i=1}^{n} [y_i - (a_0 + a_1x_i + a_2x_i^2 + \cdots + a_mx_i^m)](-1) = 0$$

$$\partial E/\partial a_1 = \sum_{i=1}^{n} [y_i - (a_0 + a_1x_i + a_2x_i^2 + \cdots + a_mx_i^m)](-x_i) = 0$$

$$\partial E/\partial a_m = \sum_{i=1}^{n} [y_i - (a_0 + a_1x_i + a_2x_i^2 + \cdots + a_mx_i^m)](-x_i^m) = 0$$

After eliminating suffices and simplification of the above equations we get the normal equations as:

$$na_0 + a_1 \Sigma x + a_2 \Sigma x^2 + ...+ a_m \Sigma x^m = \Sigma y$$

$$a_0 \Sigma x + a_1 \Sigma x^2 + a_2 \Sigma x^3 ...+ a_m \Sigma x^{m+1} = \Sigma xy$$

....

....

$$a_0 \Sigma x^m + a_1 \Sigma x^{m+1} + a_2 \Sigma x^{m+3} ....+ a_m \Sigma x^{m+m} = \Sigma x^m y$$

These equations can be written in matrix form as

$$\begin{bmatrix} n & \Sigma x & \Sigma x^2 & \dots\dots & \Sigma x^m \\ \Sigma x & \Sigma x^2 & \Sigma x^3 & \dots\dots & \Sigma x^{m+1} \\ \dots & \dots & \dots & \dots & \dots \\ \dots & \dots & \dots & \dots & \dots \\ \Sigma x^m & \Sigma x^{m+1} & \Sigma x^{m+3} & \dots\dots & \Sigma x^{2m} \end{bmatrix} \begin{bmatrix} a_0 \\ a_1 \\ \dots \\ \dots \\ a_m \end{bmatrix} = \begin{bmatrix} \Sigma y \\ \Sigma xy \\ \dots \\ \dots \\ \Sigma x^m y \end{bmatrix}$$

$$AX = B$$

The elements of X and B matrices can be calculated by the following formula

$$X_{ij} = \sum_{k=1}^{n} X_k^{i+j-2}$$

Where $i = 1, 2, 3, \dots\dots, m$ and $j = 1, 2, 3, \dots\dots, m$

$$B_i = \sum_{k=1}^{n} X_k^{i-1} y_k$$

Where $i = 1, 2, 3, \dots\dots, m$.

SOLVED PROBLEMS

Problem 1: (a) Fit a second degree parabola to the following data:

x :	0.0	1.0	2.0
y :	1.0	6.0	17.0

(b) Fit a second degree curve of regression of *y* on *x* to the following data:

x :	1.0	2.0	3.0	4.0
y :	6.0	11.0	18.0	27

(c) Fit a second degree parabola in the following data:

x :	0.0	1.0	2.0	3.0	4.0
y :	1.0	4.0	10.0	17.0	30.0

Solution: The equation of second degree parabola is given by

$$y = a + bx + cx^2 \qquad \dots(1)$$

Normal equations are

$$\Sigma y = ma + b\Sigma x + c\Sigma x^2 \qquad \dots(2)$$

$$\Sigma xy = a\Sigma x + b\Sigma x^2 + c\Sigma x^3 \qquad \dots(3)$$

and

$$\Sigma x^2 y = a\Sigma x^2 + b\Sigma x^3 + c\Sigma x^4 \qquad \dots(4)$$

(a) Here $m = 3$

The table is as follows:

x	y	x^2	x^3	x^4	xy	x^2y
0	1	0	0	0	0	0
1	6	1	1	1	6	6
2	17	4	8	16	34	68
Total	24	5	9	17	40	74

Substituting in eqns. (2), (3) and (4), we get

$$24 = 3a + 3b + 5c \quad ...(5)$$

$$40 = 3a + 5b + 9c \quad ...(6)$$

$$74 = 5a + 9b + 17c \quad ...(7)$$

Solving eqns. (5), (6) and (7), we get

$$a = 1, \ b = 2, \ c = 3$$

Hence the required second degree parabola is

$$y = 1 + 2x + 3x^2$$

(b) Here $m = 4$

The table is as follows:

x	y	x^2	x^3	x^4	xy	x^2y
1	6	1	4	1	6	6
2	11	4	8	16	22	44
3	18	9	27	81	54	162
4	27	16	64	256	108	432
$\Sigma x = 10$	$\Sigma y = 62$	$\Sigma x^2 = 30$	$\Sigma x^3 = 100$	$\Sigma x^4 = 354$	$\Sigma xy = 190$	$\Sigma x^2y = 644$

Substituting values in eqns. (2), (3) and (4), we get

$$62 = 4a + 10b + 30c \quad ...(8)$$

$$190 = 10a + 30b + 100c \quad ...(9)$$

$$644 = 30a + 100b + 354c \quad ...(10)$$

Solving equations (8), (9) and (10), we get

$$a = 3, \ b = 2, \ c = 1$$

Hence the required second degree parabola is

$$y = 3 + 2x + x^2$$

(c) Here $m = 5$

The table is as follows:

x	y	x^2	x^3	x^4	xy	x^2y
0.0	1.0	0	0	0	0	0
1.0	4.0	1	1	1	4	4
2.0	10.0	4	8	16	20	40
3.0	17.0	9	27	81	51	153
4.0	30.0	16	64	256	120	480
$\Sigma x = 10$	$\Sigma y = 62$	$\Sigma x^2 = 30$	$\Sigma x^3 = 100$	$\Sigma x^4 = 354$	$\Sigma xy = 195$	$\Sigma x^2y = 677$

Substituting values in eqns. (2), (3) and (4), we get

$$2 = 5a + 10b + 30c \qquad ...(11)$$

$$195 = 10a + 30b + 100c \qquad ...(12)$$

$$677 = 30a + 100b + 354c \qquad ...(13)$$

Solving eqns. (11), (12) and (13), we get

$$a = 1.2, \ b = 1.1 \text{ and } c = 1.5$$

Hence the required second degree parabola is

$$y = 1.2 + 1.1x + 1.5x^2$$

Problem 2: Employ the method of least squares to fit a parabola $y = a + bx + cx^2$ in the following data:

(x, y): (– 1, 2), (0, 0), (0, 1), (1, 2)

Solution: Normal equations to the parabola

$$y = a + bx + cx^2 \text{ are}$$

$$\Sigma y = ma + b\Sigma x + c\Sigma x^2 \qquad ...(1)$$

$$\Sigma xy = a\Sigma x + b\Sigma x^2 + c\Sigma x^3 \qquad ...(2)$$

and

$$\Sigma x^2y = a\Sigma x^2 + b\Sigma x^3 + c\Sigma x^4 \qquad ...(3)$$

Here $m = 4$

The table is as follows:

x	y	x^2	x^3	x^4	xy	x^2y
–1	2	1	–1	1	–2	2
0	0	0	0	0	0	0
0	1	0	0	0	0	0
1	2	1	1	1	2	2
$\Sigma x = 0$	$\Sigma y = 5$	$\Sigma x^2 = 2$	$\Sigma x^3 = 0$	$\Sigma x^4 = 2$	$\Sigma xy = 0$	$\Sigma x^2y = 4$

Substituting these values in equations (1), (2) and (3); we get

$$5 = 4a + 2c \qquad ...(4)$$

$$0 = 2b \qquad ...(5)$$

and $$4 = 2a + 2c \qquad ...(6)$$

Solving (4), (5) and (6), we get

$$a = 0.5,\ b = 0 \text{ and } c = 1.5$$

Hence the required second degree parabola is

$$y = 0.5 + 1.5x^2$$

16.12 MULTIPLE LINEAR REGRESSION

In most research problems where regression analysis is applied, more than one independent variable is needed in the regression model. The complexity of most scientific mechanisms is such that in order to be able to predict an important response, a **multiple regression model** is needed. When this model is linear in the coefficients, it is called a **multiple linear regression model.**Now, the case where the dependent variable is a function of two or more linear or non-linear independent variables. Consider such a linear function as

$$y = a + bx + cz$$

The sum of the squares of residual is

$$U = \sum_{i=1}^{n}(y_i - a - bx_i - cz_i)^2$$

Differentiating U partially with respect to a, b, c; we get

$$\frac{\partial U}{\partial a} = 0 \Rightarrow 2\sum_{i=1}^{n}(y_i - a - bx_i - cz_i)(-1) = 0$$

$$\frac{\partial U}{\partial b} = 0 \Rightarrow 2\sum_{i=1}^{n}(y_i - a - bx_i - cz_i)(-x_i) = 0$$

$$\frac{\partial U}{\partial c} = 0 \Rightarrow 2\sum_{i=1}^{n}(y_i - a - bx_i - cz_i)(-z_i) = 0$$

which on simplification and omitting the suffix i, yields.

$$\Sigma y = ma + b\Sigma x + c\Sigma z$$

$$\Sigma xy = a\Sigma x + b\Sigma x^2 + c\Sigma xz$$

$$\Sigma yz = a\Sigma z + b\Sigma xz + c\Sigma z^2$$

Solving the above three equations, we get values of a, b and c. Consequently, we get the linear function $y = a + bx + cz$ called **regression plane.**

SOLVED PROBLEMS

Problem 1: Obtain a regression plane by using multiple linear regression to fit the data given below:

x:	1	2	3	4
z:	0	1	2	3
y:	12	18	24	30

Solution: Let $y = a + bx + cz$ be the required regression plane where a, b, c are the constants to be determined by following equations:

$$\Sigma y = ma + b\Sigma x + c\Sigma z$$

$$\Sigma yx = a\Sigma x + b\Sigma x^2 + c\Sigma zx$$

and $$\Sigma yz = a\Sigma z + b\Sigma zx + c\Sigma z^2$$

Here $m = 4$

x	z	y	x^2	z^2	yx	zx	yz
1	0	12	1	0	12	0	0
2	1	18	4	1	36	2	18
3	2	24	9	4	72	6	48
4	3	30	16	9	120	12	90
$\Sigma x = 10$	$\Sigma z = 6$	$\Sigma y = 84$	$\Sigma x^2 = 30$	$\Sigma z^2 = 14$	$\Sigma yx = 240$	$\Sigma zx = 20$	$\Sigma yz = 156$

Substitution yields, $84 = 4a + 10b + 6c$

$$240 = 10a + 30b + 20c$$

and $$156 = 6a + 20b + 14c$$

Solving, we get $a = 10,\ b = 2,\ c = 4$

Hence the required regression plane is

$$y = 10 + 2x + 4z.$$

QUERIES

1. What is regression analysis? How does it differ from correlation? Why there are in general two regression equations for bivariate data?
2. Write a note on the different types of regression.
3. What are regression lines? When do they coincide? Explain their uses.
4. Distinguish between correlation and regression.
5. Why regression line is known as the line of best fit? Explain the properties of regression lines.
6. What are regression lines? Is it possible to draw coefficient of correlation from them?
7. What are the various methods to find the regression line?
8. What is the difference between correlation and regression coefficients? Can correlation coefficient be computed out of regression coefficients? If yes, how?

9. A student obtained the two regression lines as:

$$Y \text{ on } X: 2x - 5y - 7 = 0$$

$$X \text{ on } Y: 3x + 2y - 8 = 0$$

Do you agree with him?

10. Comment on the following statements:

The correlation coefficient between x and y is 0.90 and the regression coefficient is b_{xy} is –1.

If the two coefficients of regression are negative then their correlation coefficient is positive.

$r = 0.9$, $b_{xy} = 2.04$, $b_{yx} = -3.2$

UNSOLVED PROBLEMS

1. From the following data, obtain the two regression coefficients:

Income:	91	97	108	121	67	124	51	73	111	57
Expenditure:	71	75	69	97	70	91	39	61	80	47

2. Fit a least square line to the following data:

   (a) Using X as independent variable

   (b) Using X as dependent variable

X:	0	100	200	300	400
Y:	50	40	35	15	0

3. Obtain two equations of the lines of regression for the following data and calculate the value of Y when X is equal to 10.

X:	4	5	6	7	8
Y:	18	20	22	21	23

4. Calculate two regression equations from the following data and calculate the value of X when Y is equal to 100.

	x	y
Arithmetic Means	25	30
Standard deviations	5	7

Coefficient of correlation = 0.85

5. From the data given below, find:

   (i) Coefficient of correlation between the age of husbands and the age of wives

   (ii) The regression equations

Age of husbands	Age of wives	Age of husbands	Age of wives
22	18	27	22
23	20	27	23
23	21	28	24
24	20	30	25
26	21	30	26

6. Given the following information, find the regression of X on
   $N = 6,\ \bar{X} = 5,\ \bar{Y} = 6.5,\ \Sigma Y^2 = 271,\ \Sigma XY = 166$
7. The weights of a calf taken at weekly intervals are given below. Fit a straight line using the method of least squares and calculate the average rate of growth per week.

Age:	1	2	3	4	5	6	7	8	9
Weight:	52.5	58.3	60	70.3	81.1	87.2	95.5	10.2	107.3

8. From the following data, find the two regression coefficients:

X:	6	2	10	4	5	6
Y:	5	8	60	7	11	9

9. From the given data, find the two regression coefficients

X:	4	1	3	4	7	6
Y:	4	7	6	2	11	19

10. You are provided the following data , find out the regression coefficients.

X:	6	2	10	4	5	7
Y:	5	8	16	7	12	19

11. You are given the data below, from this:
    (i) Fit regression line of Y on X and hence predict Y when X=12
    (ii) Fit a regression line of X and Y then predict X when Y=25
    (iii) Calculate the correlation coefficient between the two variables.

X:	1	5	3	4	5	6
Y:	5	8	0	0	11	9

12. From the given data, find out Regression line Y on X and X on Y.

X:	1	5	3	4	5	6
Y:	4	7	3	0	10	9

13. Given below is the data relating to the age and weight of some men:

Age:	56	42	36	47	27	33
Weight:	80	70	40	90	75	45

From this data calculate:

(i) Correlation coefficient

(ii) Estimate the weight of man whose age is 44 years

14. You are provided the data below relating to profit and loss of *XYZ* Company in the year 2000:

Profit:	50	42	47	123	143	156
Loss:	140	125	100	50	110	9

From the above data find:

(i) The regression coefficient of profit on loss

(ii) The correlation coefficient between profit and loss

15. Using the following data:

(i) Plot the given observations on a graph paper

(ii) Fit a line of Regression of *X* on Y, using least square method

X:	1	5	3	4	5	6
Y:	5	8	0	0	11	9

16. You are provided the data given below:

(i) Fit a line of regression of *N* on *M*

(ii) Plot the observations on a graph paper

N:	1	2	3	4	5	6
M:	2	3	4	4	8	9

17. Using the following data:

(i) Fit a line of regression of Expenditure on Household size and

(ii) State what is likely to be the expenditure for a household of size 7?

(iii)

Expenditure:	200	300	400	450	500	600
Household size:	1	8	3	2	11	9

18. Given the bivariate Data:

Monthly Income ('000):	8	15	13	14	15	16
Percent allocation for investment:	5	38	30	20	31	29

From the above data:

(a) Develop a regression equations that would describe the data in a best way

(b) Estimate the per cent allocation for investment of a family earning Rs.2, 000 per annum.

19. You are provided the bivariate data given below:

Profit:	11	15	13	14	5	6
Loss:	9	8	0	0	11	9

From the above data develop a regression equation that would describe the data in a best way.

20. From the data given below:
   (a) Draw a scatter diagram
   (b) Find the least square regression of the maintenance cost on age
   (c) Predict the maintenance cost for a machine of this type, which is 40 years old

Machine:	1	2	3	4	5	6
Age:	5	10	15	20	30	40
Maintenance cost (in Rs.):	190	200	250	300	310	350

21. You are given the following data:

	x	y
Arithmetic Means	36	85
Standard deviations	11	8

Correlation coefficient between x and y is 0.66
   (i) Find two regression coefficients
   (ii) Estimate value of x when $y = 75$.

22. In the estimation of regression equations of two variables X and Y, the following results were obtained:

$\bar{X} = 90$, $\bar{Y} = 70$, $n = 10$

$\Sigma x^2 = 6360$, $\Sigma Y^2 = 2860$, $\Sigma xy = 3900$,

where x and y are deviations from respective means. Obtain the two regression equations.

23. Find the mean values of the variables X and Y and correlation coefficient between them from the following regression equations:

$$2Y - X - 50 = 0$$
$$3Y - 2X - 10 = 0$$

24. If the two lines of regression are:

$$4x - 5y + 30 = 0$$

and
$$20x - 9y - 107 = 0,$$

Which of these is the line of regression of x on y? Find r_{xy} and σ_y whwb $\sigma_x = 3$.

25. The two regression equations between x and y are:

$$10x - 20y - 14 = 0$$
$$5x - 6y = 47$$

The standard deviation of x is 9. Find
   (i) The mean values of x and y

(ii) The variance of y

(iii) The coefficient of correlation between x and y

26. Regression equations of two variables X and Y are as follows:

$$3x - 2y - 26 = 0$$

$$6x - y - 31 = 0$$

Find:

(i) The mean of X and Y

(ii) The regression coefficient of X on Y

(iii) The coefficient of correlation between X and Y

(iv) s_y if $\sigma_x^2 = 25$

27. From the following bivariate frequency distribution table

(i) Find the two coefficients of regression

(ii) Obtain the two regression equations

	Weights in lbs			
Height in inches	**90-100**	**100-110**	**110-120**	**120-130**
50-55	4	5	6	5
55-60	6	11	12	10
60-65	6	12	10	7
65-70	3	8	6	3

28. Obtain the lines of regression for the following bivariate frequency distribution:

	Expenditure ('000 Rs.)				
Sales ('000Rs.)	**5-15**	**15-25**	**25-35**	**35-45**	**Total**
75-125	4	1	-	-	5
125-175	7	6	2	1	16
175-225	1	3	4	2	10
225-275	1	1	3	4	9
Total	**13**	**11**	**9**	**7**	**40**

29. In a study of a relationship between yield of wheat and rainfall the following results were obtained:

	Yield	Rainfall
Means	900	10"
Standard deviations	70	2"

Correlation coefficient = 0.5

Calculate the most likely yield when rainfall is 14".

30. You are provided the data relating to marks secured by 200 students in economics and English. Draw the two lines of regression on a graph paper.

Subject	Mean	S.D	r
Economics	18	3	0.8
English	50	5	

31. Fit a second degree parabola to the following data taking x as the independent variable:

x:	0	1	2	3	4
y:	1	5	10	22	38

32. Fit a second degree parabola to the following data by Least Squares Method:

x:	0	1	2	3	4
y:	1	1.8	1.3	2.5	6.3

33. Determine the constants a, b, and c by the Method of Least Squares such that $y = ax^2 + bx + c$ fits the following data:

x:	2	4	6	8	10
y:	4.01	11.08	30.12	81.89	222.62

Index

H

I

J

K

L

M

N

T

U

V

W